BALANCED MICROWAVE FILTERS

WILEY SERIES IN MICROWAVE AND OPTICAL ENGINEERING

Professor Kai Chang, Editor
Texas A&M University

A complete list of the titles in this series appears at the end of this volume.

BALANCED MICROWAVE FILTERS

Edited by
FERRAN MARTÍN
LEI ZHU
JIASHENG HONG
FRANCISCO MEDINA

WILEY

This edition first published 2018

Registered Offices
John Wiley & Sons, Inc., 111 River Street, Hoboken, NJ 07030, USA

Editorial Office
111 River Street, Hoboken, NJ 07030, USA

For details of our global editorial offices, customer services, and more information about Wiley products visit us at www.wiley.com.

Wiley also publishes its books in a variety of electronic formats and by print-on-demand. Some content that appears in standard print versions of this book may not be available in other formats.

Library of Congress Cataloging-in-Publication Data

Names: Martin, Ferran, 1965– editor. | Zhu, Lei, 1963– editor. | Hong, Jiasheng, editor. | Medina, Francisco, 1960– editor.
Title: Balanced microwave filters / edited by Ferran Martin, Lei Zhu, Jiasheng Hong, Francisco Medina.
Description: Hoboken, NJ : John Wiley & Sons, 2018. | Series: Wiley series in microwave and optical engineering | Includes bibliographical references and index. | Description based on print version record and CIP data provided by publisher; resource not viewed.
Identifiers: LCCN 2017036439 (print) | LCCN 2017046155 (ebook) | ISBN 9781119237624 (pdf) | ISBN 9781119238232 (epub) | ISBN 9781119237617 (cloth)
Subjects: LCSH: Microwave filters.
Classification: LCC TK7872.F5 (ebook) | LCC TK7872.F5 B35 2018 (print) | DDC 621.381/3224–dc23
LC record available at https://lccn.loc.gov/2017036439

Cover design by Wiley
Cover image: © Vectorig/Gettyimages

Set in 11/13pt TimesTen by SPi Global, Pondicherry, India

Printed in the United States of America

10 9 8 7 6 5 4 3 2 1

To our families
Anna, Alba, and Arnau
Kai and Haide
Huizheng and Yi Hang
Carmen, Marta, Santos, Juan, and Lola

The editors would like to acknowledge the effort of many people directly or indirectly involved in the preparation and writing of this book, not only the chapter contributors but also the members of their respective groups, without whom this book had never been written.

CONTENTS

LIST OF CONTRIBUTORS

JORDI BONACHE, CIMITEC, Departament d'Enginyeria Electrònica, Universitat Autònoma de Barcelona, Bellaterra, Spain

VICENTE E. BORIA, Departamento de Comunicaciones-iTEAM, Universitat Politècnica de València, Valencia, Spain

WENQUAN CHE, Department of Communication Engineering, Nanjing University of Science and Technology, Nanjing, China

FU-CHANG CHEN, School of Electronic and Information Engineering, South China University of Technology, Guangzhou, China

JIAN-XIN CHEN, School of Electronics and Information, Nantong University, Nantong, China

CHIU-CHIH CHOU, Graduate Institute of Communication Engineering, National Taiwan University, Taipei, Taiwan

QING-XIN CHU, School of Electronic and Information Engineering, South China University of Technology, Guangzhou, China

WENJIE FENG, Department of Communication Engineering, Nanjing University of Science and Technology, Nanjing, China

ARMANDO FERNÁNDEZ-PRIETO, Departamento de Electrónica y Electromagnetismo, Universidad de Sevilla, Sevilla, Spain

Christophe Fumeaux, School of Electrical and Electronic Engineering, The University of Adelaide, Adelaide, SA, Australia

Xin Guo, Department of Electrical and Computer Engineering, Faculty of Science and Technology, University of Macau, Macau SAR; Ministerial Key Laboratory, JGMT, Nanjing University of Science and Technology, Nanjing, China

Jiasheng Hong, Institute of Sensors, Signals and Systems, School of Engineering and Physical Sciences, Heriot-Watt University, Edinburgh, UK

Ali Karami-Horestani, School of Electrical and Electronic Engineering, The University of Adelaide, Adelaide, SA, Australia

Teck Beng Lim, School of Engineering, Nanyang Polytechnic, Ang Mo Kio, Singapore

Jun-Fa Mao, Key Laboratory of Ministry of Education of Design and Electromagnetic Compatibility of High-Speed Electronic Systems, Shanghai Jiao Tong University, Shanghai, PR China

Jesús Martel, Departamento de Física Aplicada II, Universidad de Sevilla, Sevilla, Spain

Ferran Martín, CIMITEC, Departament d'Enginyeria Electrònica, Universitat Autònoma de Barcelona, Bellaterra, Spain

Francisco Medina, Departamento de Electrónica y Electromagnetismo, Universidad de Sevilla, Sevilla, Spain

Jordi Naqui, CIMITEC, Departament d'Enginyeria Electrònica, Universitat Autònoma de Barcelona, Bellaterra, Spain

Wei Qin, School of Electronics and Information, Nantong University, Nantong, China

Ana Rodríguez, Departamento de Comunicaciones-iTEAM, Universitat Politècnica de València, Valencia, Spain

Marc Sans, CIMITEC, Departament d'Enginyeria Electrònica, Universitat Autònoma de Barcelona, Bellaterra, Spain

Jordi Selga, CIMITEC, Departament d'Enginyeria Electrònica, Universitat Autònoma de Barcelona, Bellaterra, Spain

Jin Shi, School of Electronics and Information, Nantong University, Nantong, China

PARIS VÉLEZ, CIMITEC, Departament d'Enginyeria Electrònica, Universitat Autònoma de Barcelona, Bellaterra, Spain

JIANPENG WANG, Ministerial Key Laboratory, JGMT, Nanjing University of Science and Technology, Nanjing, China

LIN-SHENG WU, Key Laboratory of Ministry of Education of Design and Electromagnetic Compatibility of High-Speed Electronic Systems, Shanghai Jiao Tong University, Shanghai, PR China

TZONG-LIN WU, Graduate Institute of Communication Engineering, National Taiwan University, Taipei, Taiwan

WEN WU, Ministerial Key Laboratory, JGMT, Nanjing University of Science and Technology, Nanjing, China

BIN XIA, Key Laboratory of Ministry of Education of Design and Electromagnetic Compatibility of High-Speed Electronic Systems, Shanghai Jiao Tong University, Shanghai, PR China

QUAN XUE, School of Electronic and Information Engineering, South China University of Technology, Guangzhou, China

SHI-XUAN ZHANG, School of Electronic and Information Engineering, South China University of Technology, Guangzhou, China

CHUNXIA ZHOU, Ministerial Key Laboratory, JGMT, Nanjing University of Science and Technology, Nanjing, China

LI-HENG ZHOU, School of Electronics and Information, Nantong University, Nantong, China

LEI ZHU, Department of Electrical and Computer Engineering, Faculty of Science and Technology, University of Macau, Macau SAR, China

PREFACE

Differential, or balanced, transmission lines and circuits have been traditionally applied to low-frequency analog systems and to high-speed digital systems. As compared with single-ended signals, differential-mode signals exhibit lower electromagnetic interference (EMI) and higher immunity to electromagnetic noise and crosstalk. Consequently, a better signal integrity and a higher signal-to-noise ratio (SNR) can be achieved in differential systems. These aspects are especially critical in modern digital systems, where logic signal swing and noise margin have dramatically decreased and hence are less immune to the effects of noise and EMI. However, differential systems are implemented through balanced circuits and transmission lines (interconnects), representing further design and fabrication complexity as compared with single-ended systems. For this main reason, in radiofrequency (RF) and microwave applications, unbalanced structures have dominated the designs for decades, still being more common than differential circuits. Nevertheless, recent technological advances are pushing differential circuits into the RF and microwave frequency domain, and balanced lines and devices are becoming increasingly common not only in high-speed digital circuits but also in modern communication systems.

Despite the inherent advantages of differential signals over their single-ended counterparts, in a real scenario, perfect circuit symmetry cannot be guaranteed, and the applied signals may exhibit certain level of time skew. Therefore, the presence of common-mode noise due to cross-mode coupling (from the differential signals of interest) is almost unavoidable. This common-mode noise is the source of most of the radiation and EMI problems in differential systems and may degrade the desired differential signals. Therefore, the design of differential lines and circuits should be preferably focused on suppressing the common mode and, at the same time, preserving the integrity of the differential mode within the frequency range of interest.

The increasing research activity devoted to the design of common-mode suppressed balanced transmission lines and microwave circuits (especially filters) in the last decade has motivated this book proposal. Filters are key components in any communication system, and the fact that balanced systems are increasingly penetrating into the high-frequency domain has focused the attention of many microwave researchers working on planar passive components on the design of balanced microwave filters, the main topic of this book. Efficient common-mode suppression preserving the integrity of the differential signals, compact dimensions, wideband and ultra-wideband differential-mode filter responses, multiband functionality and the implementation of more complex devices (e.g., balanced diplexers, power dividers, etc.) are some of the challenging aspects covered by this book. The subject is so wide and the research activity is so intense that this book has been conceived from contributions by the main relevant researchers and groups worldwide working on the topics covered by the different book chapters.

After an introductory chapter (Part I of the book), devoted to the fundamentals of balanced transmission lines, circuits, and networks, the book has been structured by grouping the chapters in further four parts. In Part II, the main strategies for common-mode suppression in balanced transmission lines are reviewed (Chapter 2). It is also shown in this part that these common-mode rejection filters can be applied to enhance the common-mode rejection level of balanced filters with limited common-mode suppression efficiency (Chapter 3).

Part III of the book is focused on the design of balanced filters exhibiting wideband and ultra-wideband differential-mode responses with inherent common-mode rejection. Several strategies to achieve this challenging objective (i.e., the intrinsic and efficient suppression of the common mode over the wide or ultrawide differential-mode transmission bands) are reviewed. The general idea behind the different considered approaches is the conception of filter topologies able to provide the required wide or ultra-wideband differential-mode responses (subjected to certain specifications) and, at the same time, efficient common-mode suppression in the region of interest. This selective mode transmission/suppression is typically achieved by using symmetry properties and topologies providing opposite behavior for the differential (passband) and common (stopband) modes. Typically, circuit elements insensitive to the differential mode, but providing controllable transmission zeros for the common mode, are used. The different studied approaches/structures include branch-line topologies (Chapter 4), coupled line sections (Chapter 5), T-shaped structures (Chapter 6), multilayer and defect ground structures (Chapter 7), signal interference techniques (Chapter 8), multi-section mirrored stepped-impedance resonators (Chapter 9), metamaterial-inspired resonators (Chapter 10), and slotline resonators (Chapter 11).

In Part IV, several strategies to achieve narrowband and dual-band differential-mode filter responses with inherent common-mode suppression are reviewed. The challenge here is to achieve the maximum possible common-mode suppression covering the differential-mode band, or bands. Strategies based on coupled resonators implemented in planar technology are reviewed in Chapters 12 and 13, whereas in Chapter 14, dual-band balanced filters based on substrate integrated waveguide (SIW) technology are introduced.

Finally, in Part V of the book, different common-mode suppressed balanced circuits are studied for completeness. This includes power dividers/combiners (Chapter 15) and equalizers (Chapter 16).

To end this preface, the book editors would like to mention that the contents of this book have been determined in order to provide a wide and balanced overview of the international activity and state of the art in the field of balanced microwave filters and related topics. Nevertheless, the designated contributors for the different chapters have been given full freedom to conceive and structure their respective assigned chapters at their convenience. For this reason, and because the different book chapters are self-sustaining for easy reading, some (but few) overlapping between different parts of the book has been accepted by the book editors. Some aspects related to the terminology may also vary from chapter to chapter due to the same reason. It is the editors' hope that the present manuscript constitutes a reference book in the topic of balanced microwave filters and some other passive devices and that the book can be of practical use to students, researchers, and engineers involved in the design/optimization of RF/microwave components and filters.

Barcelona, Spain FERRAN MARTÍN
Macao, China LEI ZHU
Edinburg, UK JIASHENG HONG
Seville, Spain FRANCISCO MEDINA
May, 2017

PART 1

Introduction

CHAPTER 1

INTRODUCTION TO BALANCED TRANSMISSION LINES, CIRCUITS, AND NETWORKS

Ferran Martín,[1] Jordi Naqui,[1] Francisco Medina,[2] Lei Zhu,[3] and Jiasheng Hong[4]

[1]CIMITEC, Departament d'Enginyeria Electrònica, Universitat Autònoma de Barcelona, Bellaterra, Spain
[2]Departamento de Electrónica y Electromagnetismo, Universidad de Sevilla, Sevilla, Spain
[3]Department of Electrical and Computer Engineering, Faculty of Science and Technology, University of Macau, Macau SAR, China
[4]Institute of Sensors, Signals and Systems, School of Engineering and Physical Sciences, Heriot-Watt University, Edinburgh, UK

1.1 INTRODUCTION

This chapter is an introduction to the topic of balanced lines, circuits, and networks. The main objectives are (i) to point out the advantages and limitations of balanced versus unbalanced systems; (ii) to analyze the origin and effects of the main source of noise in differential systems, that is, the common-mode noise; (iii) to provide the fundamentals of

Balanced Microwave Filters, First Edition. Edited by Ferran Martín, Lei Zhu, Jiasheng Hong, and Francisco Medina.

differential transmission lines, with special emphasis on microstrip lines (the most common), including the main topologies and fundamental propagating modes; and (iv) to present the mixed-mode scattering parameters, suitable for microwave differential circuit characterization. We will also point out the two main approaches for the implementation of balanced microwave filters with common-mode noise suppression (end of Section 1.3), which will be further discussed along this book.

1.2 BALANCED VERSUS SINGLE-ENDED TRANSMISSION LINES AND CIRCUITS

Unbalanced systems transmit single-ended signals. In such systems, one of the two conductors is connected to the common ground, being the signal referenced to ground. Alternatively, signal propagation can be made on the basis of balanced or differential systems, where each wire (or conductor) has the same impedance to the circuit common, which is typically grounded. Differential signals are transmitted as complementary pairs, driving a positive voltage on one wire and an equal but opposite voltage on the other wire. The signal of interest is the potential difference between the two conductors, called differential-mode signal, which is no longer referenced to ground [1].

The main advantages of differential over single-ended signals are lower electromagnetic interference (EMI) and higher immunity to electromagnetic noise and crosstalk. Due to the previous advantages, a better signal integrity and a higher signal-to-noise ratio (SNR) can be achieved in differential systems [2, 3]. The cancelation of the fields, resulting from opposite current flowing, is the reason for the low EMI in differential systems [4]. The high noise immunity is related to the fact that voltages and currents induced from interfering sources (noise) tend to be identical on both conductors, and hence this noise couples to the differential line as a common-mode signal (to be discussed later in this chapter) [1, 3, 5]. The main drawback of differential systems is the need for balanced circuits and interconnects (transmission lines),[1] representing further complexity in terms of layout and number of elements [6].

Traditionally, differential circuits have been used in low-frequency analog and digital systems. In radio-frequency (RF) and microwave applications, unbalanced structures have dominated the designs for decades and are still more common than differential circuits. Nevertheless,

[1] Note that a complete differential system involves a differential transmitter, a differential interconnect, and a differential receiver.

recent technological advances are pushing differential circuits into the RF and microwave frequency domain [7]. Thus, balanced lines and devices are becoming increasingly common in high-speed digital circuits, as well as in modern balanced communication systems [8, 9].

1.3 COMMON-MODE NOISE

In differential transmission lines and circuits, the main contribution to noise is the so-called common-mode noise [1]. Common-mode noise is originated from electromagnetic radiation (through crosstalk or through an external source) and from the ground terminal [3, 10, 11]. Moreover, common-mode signals (also viewed as noise for the differential signals) can also be generated as consequence of time skew, amplitude unbalance, and/or different rising/falling times of the differential signals. These latter effects are ultimately caused by imperfect balance, resulting in conversion from the differential mode to the common mode. Similarly, in practice, conversion from the common mode to the differential mode always exists. Therefore, a perfect balance of the two signal conductors with respect to the reference conductor is necessary to avoid (or minimize) the conversion from common-mode noise to differential-mode noise (always representing a degradation in signal integrity).

Although, ideally, the differential mode is fully independent of the common mode, in actual differential systems, the circuits are sensitive to the common mode (e.g., in differential-mode receivers, the common-mode noise is rejected up to a certain limit that defines the ability of the receiver to work properly up to a defined amount of common-mode noise) [3]. The presence of common-mode signals in differential lines and circuits may also cause radiated emission [3]. The reason is that common-mode currents flow in the same direction (contrary to differential-mode currents, which flow in opposite directions, thus preventing far-field radiation, provided the two conductors are closely spaced). A method to reduce dramatically common-mode radiated emission is to place a metallic plane beneath and parallel to the differential line pair [3]. Such metallic plane produces image currents flowing in opposite direction to the original common-mode currents, generating fields that tend to cancel the fields resulting from the original wires. Nevertheless, due to the limited dimensions of the ground plane, a perfect image is not achieved, causing the ground plane to radiate.

Due to the negative effects of common-mode noise in differential systems (mainly signal integrity degradation and common-mode radiation), it must be reduced as much as possible. Traditionally, solutions that use common-mode chokes with high permeability ferrite cores have been proposed [12–14], but chokes represent a penalty in terms of size and frequency operating range, not being useful for high-speed, high-density, and microwave systems. Recently, many approaches fully compatible with planar fabrication technology for the design of differential lines able to suppress the common mode in the range of interest, and simultaneously preserving the integrity of the differential signals, have been reported. These common-mode filters are exhaustively reviewed in Chapter 2. Such filters may be used not only for differential-mode interconnects but also to improve the common-mode rejection of balanced bandpass filters by cascading both components, as will be pointed out in Chapter 3. Nevertheless, the design of balanced filters with inherent (and efficient) common-mode suppression without the need to cascade common-mode filters is by far the optimum solution for common-mode suppressed microwave filters, the main objective of this book (Parts III and IV of this book are dedicated to this topic).

1.4 FUNDAMENTALS OF DIFFERENTIAL TRANSMISSION LINES

Since most of the balanced filters and circuits studied in this book are implemented in microstrip technology, the present analysis is entirely focused on microstrip differential lines. Such lines are able to propagate both differential- and common-mode signals. Therefore, a comprehensive analysis of both modes is carried out in this section. The first part of the section is devoted to the topology of these lines.

1.4.1 Topology

Transmission lines may be classified according to the currents flowing on it. For comparative purposes let us first consider a two-wire unbalanced transmission line (see Figure 1.1a). In such lines, the conductors have different impedance to ground, and they are fed by single-ended ports

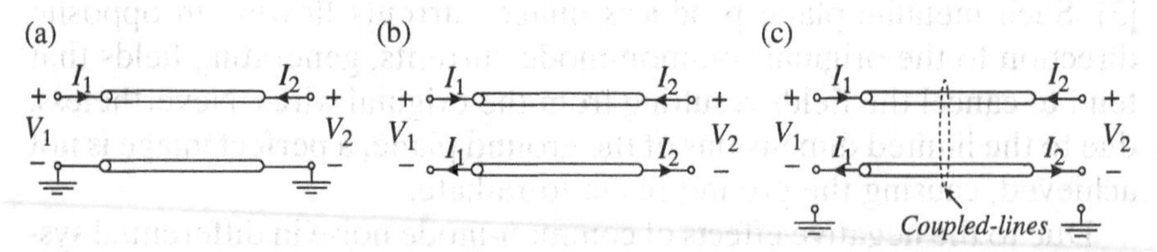

Figure 1.1 Schematic of two-port transmission lines. (a) Two-wire unbalanced line; (b) two-wire balanced line; (c) three-wire balanced line.

in which there are an active terminal and a ground terminal (or, equivalently, one of the conductors is fed, whereas the other one is tied to ground potential) [15]. One conductor is used for transporting signal current and the other one is the return current path. By contrast, in a two-wire balanced line (Figure 1.1b), the conductors have equal potential respect to ground with 180° phase shift [16]. The signal on one line is referenced to the other, which means that each conductor provides the signal return path for the other and the currents flowing on the conductors have the same magnitude but opposite direction. Such lines are fed by differential ports consisting of two terminals, neither of which is explicitly tied to ground. In a balanced line, also called differential line, the conductors have the same impedance to ground, if it exists. It is important to highlight that the nature (balanced or unbalanced) of a transmission line is determined by the currents, not only by the physical structure. Essentially, a balanced line carries balanced currents. Microstrip lines (Figure 1.2a), coplanar waveguides (CPW), and coaxial lines are well-known examples of two-wire unbalanced lines. Conversely, coplanar strips (CPS), such as those depicted in Figure 1.2(b), or slotlines are balanced structures by nature. Nevertheless, these balanced lines can be regarded as either balanced or unbalanced depending on whether the feeding is either balanced or unbalanced, respectively [15].

Most practical implementations of balanced lines incorporate a third conductor acting as a ground plane. Such three-wire line is composed of

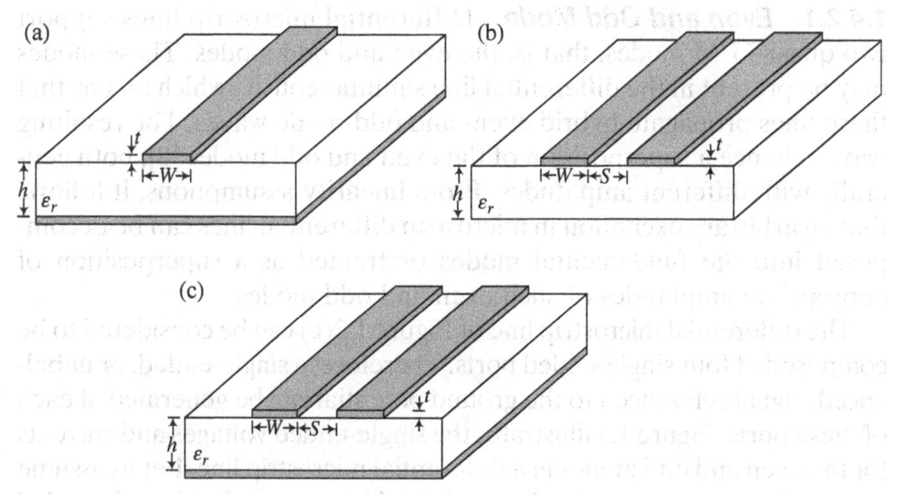

Figure 1.2 Examples of two-port transmission lines. (a) Microstrip line as an example of a two-wire unbalanced line; (b) coplanar strips (CPS) as an example of a two-wire balanced line; (c) symmetric microstrip coupled lines as an example of a three-wire balanced line.

a pair of coupled lines over a ground plane (Figure 1.1c). If this is not perfectly balanced due to the presence of the ground plane, currents flowing on it can unbalance the currents on the wires. Conversely, if the three-wire structure is perfectly balanced, the active wires carry equal and opposite currents because the impedances of each line to ground are identical. Figure 1.2(c) is an example of a three-wire balanced line, implemented by means of a pair of coupled microstrip lines. It can also be seen as a CPS transmission line with a ground plane.

1.4.2 Propagating Modes

Three-wire balanced transmission lines support two fundamental propagation modes: the balanced mode and the unbalanced mode.[2] The balanced or differential mode is the fundamental mode, equivalent to the so-called odd mode, in which the line is driven differentially. The unbalanced mode, also called common mode, is equivalent to the so-called even mode. In the common mode, equal signals (in magnitude and phase) propagate at both individual lines. For the balanced structure of Figure 1.2(c), all these modes are quasi-TEM modes [16], provided the separation between the ground plane and the lines (substrate thickness) is very small as compared with the wavelength. Let us now discuss the subtle differences between the even/odd modes and the common/differential modes, which are indeed related to signal definitions.

1.4.2.1 Even and Odd Mode Differential microstrip lines support two quasi-TEM modes, that is, the even and odd modes. These modes may be present in the differential line simultaneously, which means that these lines propagate hybrid even- and odd-mode waves. The resulting wave is hence a superposition of the even and odd modes [8], both generally with different amplitudes. From linearity assumptions, it follows that an arbitrary excitation in microstrip differential lines can be decomposed into the fundamental modes or treated as a superposition of appropriate amplitudes of such even and odd modes.

The differential microstrip line of Figure 1.2(c) can be considered to be composed of four single-ended ports. Therefore, a single-ended, or unbalanced, signal referenced to the ground potential can be generated at each of these ports. Figure 1.3 illustrates the single-ended voltages and currents for the even and odd modes in a differential microstrip line. Let us assume that the lines are appropriately terminated by a source/load single-ended reference (matched) impedance, Z_0, such that no reflections exist. Under

[2] Contrarily, two-wire balanced transmission lines support only the balanced mode.

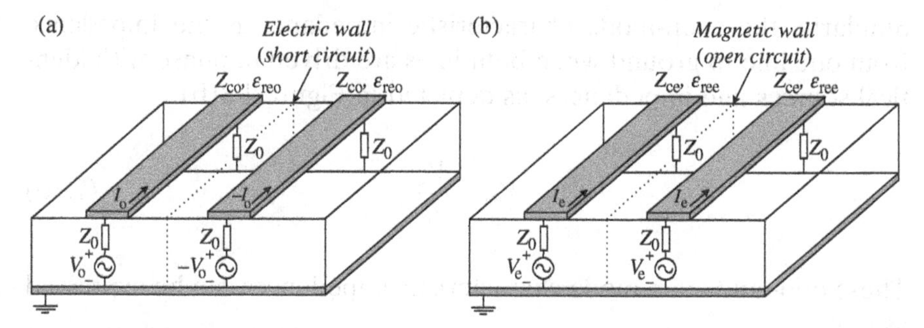

Figure 1.3 Voltages and currents in a differential microstrip line and generation of the odd (a) and even (b) modes.

these conditions, we can consider only the propagation of forward waves. The odd and even voltages are defined, respectively, as [17]

$$V_o = \frac{1}{2}(V_1 - V_2) \tag{1.1a}$$

$$V_e = \frac{1}{2}(V_1 + V_2) \tag{1.1b}$$

whereas the odd and even currents are given by

$$I_o = \frac{1}{2}(I_1 - I_2) \tag{1.2a}$$

$$I_e = \frac{1}{2}(I_1 + I_2) \tag{1.2b}$$

In the previous expressions, V_1 (V_2) and I_1 (I_2) are the voltage and current, respectively, in line 1 (line 2) of the differential pair. Note that for the pure fundamental modes, it follows that $V_o = V_1 = -V_2$ and $V_e = 0$ for the odd mode and $V_o = 0$ and $V_e = V_1 = V_2$ for the even mode. Similar results are obtained for the currents.

The characteristic impedance of each mode can be computed as the ratio between voltage and the current on each line. The odd-mode characteristic impedance is defined as the impedance from one line to ground when both lines are driven out of phase from identical sources of equal impedances and voltages (or currents), that is, odd-mode excitation, as shown in Figure 1.3(a):

$$Z_{co} = \frac{V_o}{I_o} \tag{1.3a}$$

Similarly, the even-mode characteristic impedance is the impedance from one line to ground when both lines are driven in phase with identical sources and impedances, as depicted in Figure 1.3(b):

$$Z_{\mathrm{ce}} = \frac{V_{\mathrm{e}}}{I_{\mathrm{e}}} \tag{1.3b}$$

These odd- and even-mode characteristic impedances can be expressed as [16]

$$Z_{\mathrm{co}} = \frac{1}{v_{\mathrm{po}} C_{\mathrm{o}}} = \frac{1}{c\sqrt{C_{\mathrm{o}} C_{0\mathrm{o}}}} \tag{1.4a}$$

$$Z_{\mathrm{ce}} = \frac{1}{v_{\mathrm{pe}} C_{\mathrm{e}}} = \frac{1}{c\sqrt{C_{\mathrm{e}} C_{0\mathrm{e}}}} \tag{1.4b}$$

where c is the speed of light in vacuum; C_{o} and C_{e} are the per-unit-length odd- and even-mode capacitances, respectively; $C_{0\mathrm{o}}$ and $C_{0\mathrm{e}}$ denote the per-unit-length odd- and even-mode capacitance, respectively, of each line by replacing the relative dielectric constant of the substrate by unity; and v_{po} and v_{pe} are the odd- and even-mode phase velocities, given by

$$v_{\mathrm{po}} = \frac{c}{\sqrt{\varepsilon_{\mathrm{re,o}}}} \tag{1.5a}$$

$$v_{\mathrm{pe}} = \frac{c}{\sqrt{\varepsilon_{\mathrm{re,e}}}} \tag{1.5b}$$

where the effective dielectric constants for the odd and even mode are, respectively [16],

$$\varepsilon_{\mathrm{re,o}} = \frac{C_{\mathrm{o}}}{C_{0\mathrm{o}}} \tag{1.6a}$$

$$\varepsilon_{\mathrm{re,e}} = \frac{C_{\mathrm{e}}}{C_{0\mathrm{e}}} \tag{1.6b}$$

In general, the even and odd modes exhibit different characteristic impedances and effective dielectric constants. However, the values are identical if the lines are uncoupled. In this case (uncoupled lines), the characteristic impedance is the same than the one of the individual (isolated) line.[3]

1.4.2.2 Common and Differential Mode The differential and common modes are quasi-TEM modes equivalent to the odd and even mode, respectively. The difference between such modes comes just from signal definitions. In the differential and common modes, two single-ended ports are driven as a pair that is called composite port [8]. Any single-ended signal pair applied to a composite port can be decomposed into its differential- and common-mode portions [7], which are equivalent (but not equal) to the odd and even portions, respectively.

The decomposition of hybrid differential- and common-mode voltages and currents into its differential and common-mode portions is depicted in Figure 1.4. Let us assume that the lines are appropriately terminated (so that only forward waves are present in the lines) by a differential source/load reference impedance Z_{0d} (for the differential mode) and by a common source/load reference impedance Z_{0c} (for the common mode). The differential voltage is defined as the difference between the voltages in the pair of lines [7]:

$$V_d = V_1 - V_2 \tag{1.7}$$

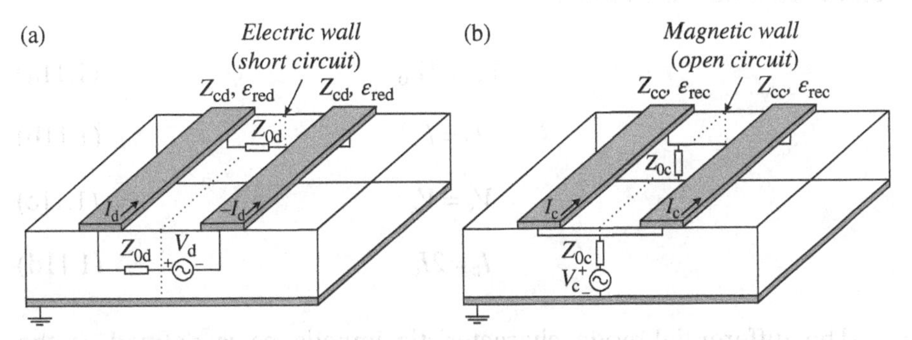

Figure 1.4 Voltages and currents in a differential microstrip line and generation of the differential (a) and common (b) modes.

[3] Note that the definition of Z_0 in Figure 1.3 corresponds to Z_{co} (Figure 1.3a) and Z_{ce} (Figure 1.3b), not to the characteristic impedance of the isolated line. Z_{co} and Z_{ce} are indeed the characteristic impedances of the isolated line with the presence of an electric wall and magnetic wall, respectively, in the symmetry plane of the lines. Unless the lines are uncoupled, there is not a single-ended impedance, Z_0, that guarantees lack of reflections at the output ports (i.e., matching), regardless of the single-ended signals injected at the input ports. To this end, a π-network is necessary, and such network is composed by the impedance Z_{ce} in the shunt branches and by the impedance $2Z_{ce}Z_{co}/(Z_{ce} - Z_{co})$ in the series branch.

With this definition, the signal is no longer referenced to ground potential. Rather than this, the signal on one line is referenced to the other. The magnitude of the current entering a single-ended port is expected to be the same than the one leaving the other single-ended port. Hence, the differential-mode current is defined as one half the difference between currents entering the single-ended ports [7]:

$$I_d = \frac{1}{2}(I_1 - I_2) \tag{1.8}$$

The common-mode voltage is defined as the average between the voltages at each line, whereas the common-mode current is given by the total current flowing on the lines,[4] that is,

$$V_c = \frac{1}{2}(V_1 + V_2) \tag{1.9}$$

$$I_c = I_1 + I_2 \tag{1.10}$$

According to the previous definitions, the voltages and currents for the differential and common mode are related to those for the odd and even mode as follows:

$$V_d = 2V_o \tag{1.11a}$$

$$I_d = I_o \tag{1.11b}$$

$$V_c = V_e \tag{1.11c}$$

$$I_c = 2I_e \tag{1.11d}$$

The differential-mode characteristic impedance is defined as the impedance between the pair of lines when both lines are driven out of phase from equal sources of equal impedances and voltages (or currents). The common-mode characteristic impedance is defined as the impedance seen from both lines to ground. The characteristic impedances for the differential and common mode can be expressed in terms

[4] Note that the return current for the common-mode signal flows through the ground plane. Ideally, in a pure differential signal, $V_1 = -V_2$ and $I_1 = -I_2$, and the common mode is canceled.

of the characteristic impedances for the even and odd modes (ground referenced) according to [7]

$$Z_{cd} = 2Z_{co} \tag{1.12a}$$

$$Z_{cc} = \frac{Z_{ce}}{2} \tag{1.12b}$$

Note that the differential- and common-mode reference impedances of Figure 1.4 are indeed those given by 1.12(a) and 1.12(b), respectively.

1.5 SCATTERING PARAMETERS

The scattering parameters (*S*-parameters), or scattering matrix (*S*-matrix), are typically used for the characterization of microwave networks. These parameters give relative information of the amplitude and phase of the transmitted and reflected wave with reference to incident wave, at least in the small-signal limit (linear regime). Let us first present the single-ended *S*-parameters, applicable to any arbitrary network, and then the mixed-mode *S*-parameters, specific of differential networks. The relation between these sets of parameters will be given at the end of this section.

1.5.1 Single-Ended *S*-Parameters

Single-ended *S*-parameters provide characterization for networks driven by single-ended or unbalanced signals. A conceptual diagram of single-ended *S*-parameters, providing the so-called single-ended *S*-matrix, $\mathbf{S_{se}}$, for a four-port structure is depicted in Figure 1.5. Such parameters satisfy

$$\mathbf{b_{se}} = \mathbf{S_{se}} \cdot \mathbf{a_{se}} \rightarrow \begin{pmatrix} b_1 \\ b_2 \\ b_3 \\ b_4 \end{pmatrix} = \begin{pmatrix} S_{11} & S_{12} & S_{13} & S_{14} \\ S_{21} & S_{22} & S_{23} & S_{24} \\ S_{31} & S_{32} & S_{33} & S_{34} \\ S_{41} & S_{42} & S_{43} & S_{44} \end{pmatrix} \begin{pmatrix} a_1 \\ a_2 \\ a_3 \\ a_4 \end{pmatrix} \tag{1.13}$$

where a_i and b_i are the normalized voltages corresponding to waves entering (V_i^+) or being reflected (V_i^-) from the different ports, respectively, that is,

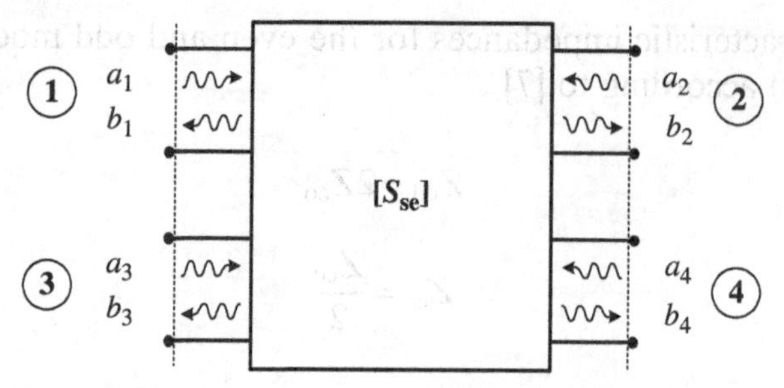

Figure 1.5 Single-ended four-port circuit described by the single-ended S-parameter matrix and indication of the normalized voltages (a_i, b_i) at device ports.

$$a_i = \frac{V_i^+}{\sqrt{Z_0}};\ b_i = \frac{V_i^-}{\sqrt{Z_0}} \tag{1.14}$$

For the calculation of the S-parameters, all ports except the stimulus port must be terminated with the port reference impedance Z_0 (matched port).[5]

If the considered network is symmetric, it can be bisected into two identical halves with respect to the symmetry plane. Using the symmetry properties [16, 18], the analysis of the N-port network is reduced to the analysis of two $N/2$-port networks. Particularly, the analysis of symmetric differential lines driven by single-ended four ports is reduced to the analysis of two single-ended two-port networks. As indicated in Figure 1.6(a), when an odd excitation is applied to the network, the symmetry plane is an electric wall (short circuit), and the two halves become the same two-port network, namely, the odd-mode network, with S-matrix defined as

$$\mathbf{S_o} = \begin{pmatrix} S_{11}^{o} & S_{21}^{o} \\ S_{12}^{o} & S_{22}^{o} \end{pmatrix} \tag{1.15}$$

Similarly, under an even excitation (Figure 1.6b), the symmetry plane is a magnetic wall (open circuit), and the two identical halves constitute the even-mode network, with scattering matrix given by

$$\mathbf{S_e} = \begin{pmatrix} S_{11}^{e} & S_{21}^{e} \\ S_{12}^{e} & S_{22}^{e} \end{pmatrix} \tag{1.16}$$

[5] We assume that the reference impedance is identical in all ports.

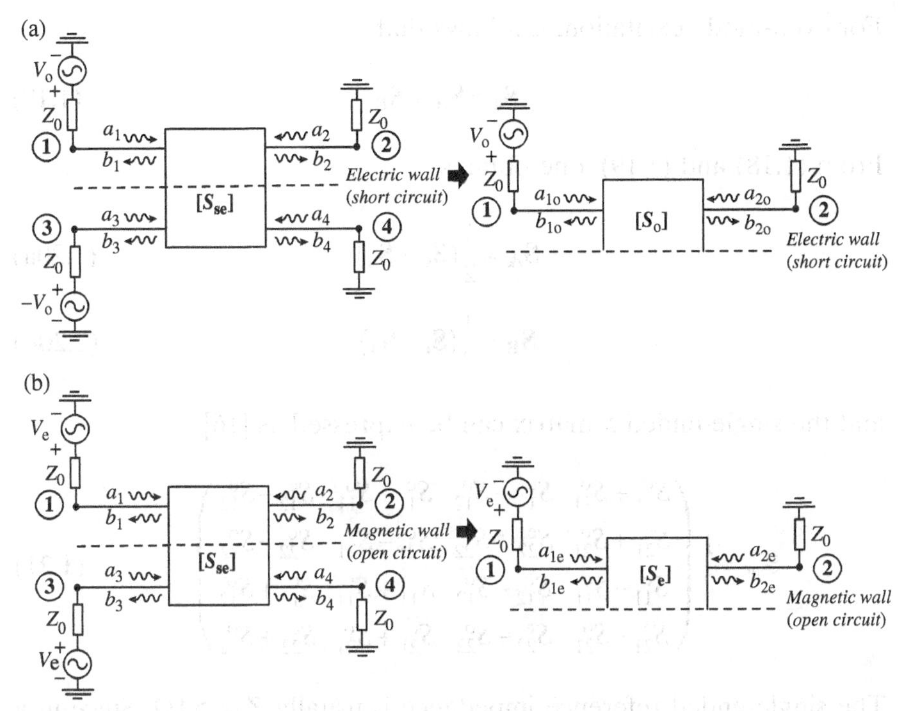

Figure 1.6 Single-ended S-parameters in a symmetric single-ended four-port network under (a) odd- and (b) even-mode excitations. The four-port circuit is reduced to two two-port circuits.

On the other hand, from symmetry considerations, it follows that the four-port S-matrix of the symmetric differential lines can be expressed as [16]

$$\mathbf{S_{se}} = \begin{pmatrix} \mathbf{S_A} & \mathbf{S_B} \\ \mathbf{S_B} & \mathbf{S_A} \end{pmatrix} \tag{1.17}$$

where $\mathbf{S_A}$ and $\mathbf{S_B}$ are order-2 matrices. For odd-mode excitation, the normalized voltages at both sides of the symmetry plane are identical in magnitude and have different sign. The resulting order-4 matrix equation can be reduced to an order-2 matrix equation. By comparing $\mathbf{S_o}$ with the such matrix, it follows that

$$\mathbf{S_o} = \mathbf{S_A} - \mathbf{S_B} \tag{1.18}$$

For even-mode excitation, it follows that

$$\mathbf{S_e} = \mathbf{S_A} + \mathbf{S_B} \tag{1.19}$$

From (1.18) and (1.19), one obtains

$$\mathbf{S_A} = \frac{1}{2}(\mathbf{S_e} + \mathbf{S_o}) \tag{1.20a}$$

$$\mathbf{S_B} = \frac{1}{2}(\mathbf{S_e} - \mathbf{S_o}) \tag{1.20b}$$

and the single-ended S-matrix can be expressed as [16]

$$\mathbf{S_{se}} = \frac{1}{2}\begin{pmatrix} S_{11}^{e} + S_{11}^{o} & S_{12}^{e} + S_{12}^{o} & S_{11}^{e} - S_{11}^{o} & S_{12}^{e} - S_{12}^{o} \\ S_{21}^{e} + S_{21}^{o} & S_{22}^{e} + S_{22}^{o} & S_{21}^{e} - S_{21}^{o} & S_{22}^{e} - S_{22}^{o} \\ S_{11}^{e} - S_{11}^{o} & S_{12}^{e} - S_{12}^{o} & S_{11}^{e} + S_{11}^{o} & S_{12}^{e} + S_{12}^{o} \\ S_{21}^{e} - S_{21}^{o} & S_{22}^{e} - S_{22}^{o} & S_{21}^{e} + S_{21}^{o} & S_{22}^{e} + S_{22}^{o} \end{pmatrix} \tag{1.21}$$

The single-ended reference impedance is usually $Z_0 = 50\,\Omega$. Since in a differential microstrip line the characteristic impedances of the even and odd modes are in general different, it follows that both modes cannot be, in general, matched simultaneously to that reference impedance.

1.5.2 Mixed-Mode *S*-Parameters

A differential microwave network consists of and can be analyzed by an even number of N single-ended ports or $N/2$ composite ports (i.e., two single-ended ports driven as a pair [19, 20]). While standard S-parameters (1.13) or (1.21) for a symmetric network can be used to characterize the network, they do not provide information about the propagation properties for the differential and common modes. The simultaneous propagation of differential and common mode is referred to as mixed-mode propagation [7]. Mixed-mode S-parameters are convenient for microwave differential circuit characterization. In this chapter, the mixed-mode S-parameters are limited to the two-port case, but the generalized theory for N-port circuits can be found in Ref. [19].

A conceptual diagram of mixed-mode S-parameters in a differential two-port circuit is shown in Figure 1.7. These mixed-mode S-parameters can be seen as corresponding to a traditional four-port network where two single-ended ports are driven as a pair. However, these four ports

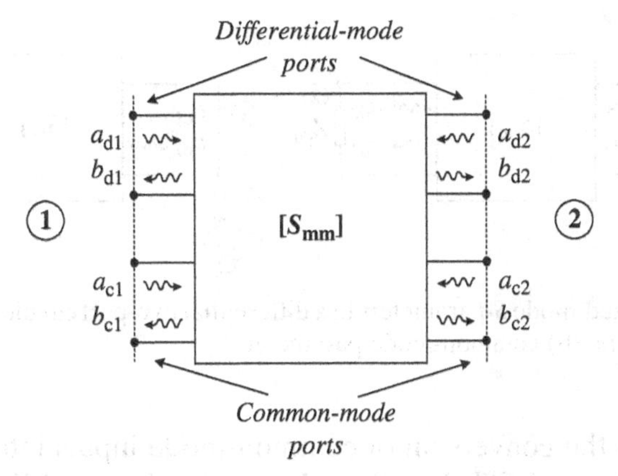

Figure 1.7 Conceptual diagram of mixed-mode S-parameters in a differential two-port network.

are conceptual tools only, rather than physically separated ports. Mixed-mode S-parameters can be arranged in matrix form as [7, 8, 20]

$$\mathbf{b_{mm}} = \mathbf{S_{mm}} \cdot \mathbf{a_{mm}} \rightarrow \begin{pmatrix} b_{d1} \\ b_{d2} \\ b_{c1} \\ b_{c2} \end{pmatrix} = \begin{pmatrix} S_{11}^{dd} & S_{12}^{dd} & S_{11}^{dc} & S_{12}^{dc} \\ S_{21}^{dd} & S_{22}^{dd} & S_{21}^{dc} & S_{22}^{dc} \\ S_{11}^{cd} & S_{12}^{cd} & S_{11}^{cc} & S_{12}^{cc} \\ S_{21}^{cd} & S_{22}^{cd} & S_{21}^{cc} & S_{22}^{cc} \end{pmatrix} \begin{pmatrix} a_{d1} \\ a_{d2} \\ a_{c1} \\ a_{c2} \end{pmatrix} \tag{1.22}$$

where b_{di} and b_{ci} are the normalized differential- and common-mode voltages corresponding to waves reflected from the two-port differential network ($i = 1,2$) and a_{di} and a_{ci} are the same variables but for waves impinging the network. In order to calculate the S-parameters, all ports except the stimulus port must be terminated with the port reference impedance, Z_{0d} (for the differential mode), or Z_{0c} (for the common mode). The mixed-mode S-matrix can be expressed as [7, 8]

$$\mathbf{S_{mm}} = \begin{pmatrix} \mathbf{S^{dd}} & \mathbf{S^{dc}} \\ \mathbf{S^{cd}} & \mathbf{S^{cc}} \end{pmatrix} \tag{1.23}$$

where $\mathbf{S^{dd}}$ is the differential-mode S-parameter matrix (Figure 1.8a), $\mathbf{S^{cc}}$ is the common-mode S-matrix (Figure 1.8b), and $\mathbf{S^{dc}}$ and $\mathbf{S^{cd}}$ are the mode-conversion or cross-mode S-matrices. The interpretation of the previous matrices is very clear: $\mathbf{S^{dd}}/\mathbf{S^{cc}}$ determine the differential-/common-mode responses to differential-/common-mode inputs,

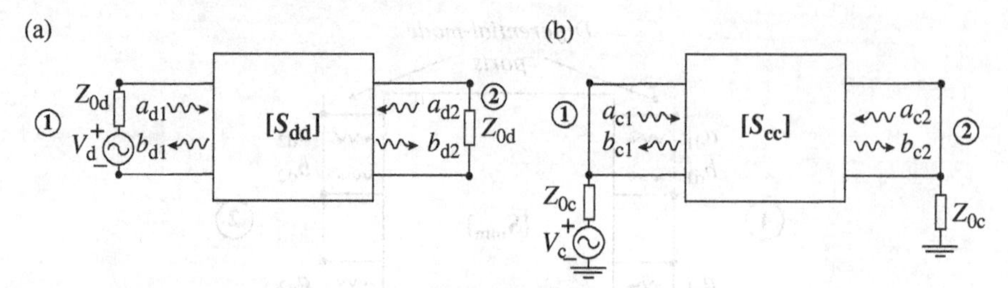

Figure 1.8 Mixed-mode S-parameters in a differential two-port circuit. (a) Differential-mode parameters; (b) common-mode parameters.

$\mathbf{S^{dc}}$ describes the conversion of common-mode inputs into differential-mode outputs, and $\mathbf{S^{cd}}$ describes the conversion of differential-mode inputs into common-mode outputs. Mode conversion occurs as consequence of imbalances in the circuit. An ideal balanced device is characterized by perfect (ideal) symmetry. Actual devices are not perfectly balanced, in part due to manufacturing imperfections, and energy transfer between the differential and common modes is unavoidable.

The relationship between the single-ended S-parameters and the mixed-mode S-parameters is given by [8][6]

$$\mathbf{S^{dd}} = \frac{1}{2}\begin{pmatrix} S_{11}-S_{13}-S_{31}+S_{33} & S_{12}-S_{14}-S_{32}+S_{34} \\ S_{21}-S_{23}-S_{41}+S_{43} & S_{22}-S_{24}-S_{42}+S_{44} \end{pmatrix} \quad (1.24a)$$

$$\mathbf{S^{cc}} = \frac{1}{2}\begin{pmatrix} S_{11}+S_{13}+S_{31}+S_{33} & S_{12}+S_{14}+S_{32}+S_{34} \\ S_{21}+S_{23}+S_{41}+S_{43} & S_{22}+S_{24}+S_{42}+S_{44} \end{pmatrix} \quad (1.24b)$$

$$\mathbf{S^{dc}} = \frac{1}{2}\begin{pmatrix} S_{11}+S_{13}-S_{31}-S_{33} & S_{12}+S_{14}-S_{32}-S_{34} \\ S_{21}+S_{23}-S_{41}-S_{43} & S_{22}+S_{24}-S_{42}-S_{44} \end{pmatrix} \quad (1.24c)$$

$$\mathbf{S^{cd}} = \frac{1}{2}\begin{pmatrix} S_{11}-S_{13}+S_{31}-S_{33} & S_{12}-S_{14}+S_{32}-S_{34} \\ S_{21}-S_{23}+S_{41}-S_{43} & S_{22}-S_{24}+S_{42}-S_{44} \end{pmatrix} \quad (1.24d)$$

It is important to mention that if a differential two-port network is symmetric, the single-ended S-parameters for the odd and even networks are identical to the S-parameters for the differential and common mode, that is,

[6] The derivation of the mixed-mode S-parameters can be found in Chapter 6 as well. However, note that the ports' designation in that chapter is different than in Figure 1.6.

$$\mathbf{S^{dd}} = \mathbf{S_o} \tag{1.25a}$$

$$\mathbf{S^{cc}} = \mathbf{S_e} \tag{1.25b}$$

Thus, from (1.24) and (1.25), the *S*-parameters for the even and odd modes, of special interest along this book, can be easily obtained by measuring the single-ended *S*-parameters of the complete network.

1.6 SUMMARY

In this introductory chapter, the advantages of differential circuits over their single-ended counterparts have been reviewed, and the need to suppress the common mode as much as possible has been justified. Moreover, this chapter has dealt with the fundamentals of differential-mode transmission lines, including the main topologies and propagating modes. The chapter ends with the mixed-mode *S*-parameters, useful for the analysis and characterization of differential circuits, and their relation to the single-mode *S*-parameters.

REFERENCES

1. J. F. White, *High Frequency Techniques: An Introduction to RF and Microwave Engineering*, John Wiley & Sons, Inc., Hoboken, 2004.
2. C. H. Tsai, T. L. Wu, "A broadband and miniaturized common-mode filter for gigahertz differential signals based on negative permittivity metamaterials," *IEEE Trans. Microw. Theory Technol.*, vol. **58**, pp. 195–202, Jan. 2010.
3. S. H. Hall, G. W. Hall, J. A. McCall, *High-Speed Digital System Design: A Handbook of Interconnect Theory and Design Practices*. John Wiley & Sons, Inc., New York, 2000.
4. E. Fronbarg, M. Kanda, C. Lasek, M. Piket-May, S. H. Hall, "The impact of a non-ideal return path on differential signal integrity," *IEEE Trans. Electromagn. Compat.*, vol. **44**, pp. 11–15, Feb. 2002.
5. R. Schmitt, *Electromagnetic Explained: A Handbook for Wireless/RF, EMC, and High-Speed Electronics*, Newnes, Amsterdam, 2002.
6. S. Caniggia, F. Maradei, *Signal Integrity and Radiated Emission of High-Speed Digital Systems*, John Wiley & Sons, Ltd, Chichester, 2008.
7. D. E. Bockelman, W. R. Eisenstadt, "Combined differential and common-mode scattering parameters: theory and simulation," *IEEE Trans. Microw. Theory Technol.*, vol. **43**, pp. 1530–1539, Jul. 1995.

8. G. D. Vendelin, A. M. Pavio, U. L. Rodhe, *Microwave Circuit Design Using Linear and Nonlinear Techniques*, John Wiley & Sons, Inc., Hoboken, 2nd Ed., 2005.
9. K. C. Gupta, R. Carg, I. Bahl, P. Barthia, *Microstrip Lines and Slotlines*, Artech House, Dedham, 2nd Ed., 1996.
10. R. J. Collier, A. D. Skinner, *Microwave Measurements*, Institution of Engineering and Technology (IET), Stevenage, 3rd Ed., 2007.
11. C. H. Wu, C. H. Wang, C. H. Chen, "Novel balanced coupled line bandpass filters with common-mode noise suppression," *IEEE Trans. Microw. Theory Technol.*, vol. **55**, pp. 287–295, Feb. 2007.
12. J. D. Gavenda, "Measured effectiveness of a toroid choke in reducing common-mode current," *Proceedings of the IEEE International Electromagnetic Compatibility Symposium*, 23–25 May 1989, Denver, CO, pp. 208–210.
13. K. Yanagisawa, F. Zhang, T. Sato, and Y. Miura, "A new wideband common-mode noise filter consisting of Mn–Zn ferrite core and copper/polyimide tape wound coil," *IEEE Trans. Magn.*, vol. **41**, no.10, pp. 3571–3573, Oct. 2005.
14. J. Deng and K. Y. See, "In-circuit characterization of common-mode chokes," *IEEE Trans. Electromagn. Compat.*, vol. **49**, no. 5, pp. 451–454, May 2007.
15. T. C. Edwards, M. B. Steer, *Foundations of Interconnect and Microstrip Design*, John Wiley & Sons, Ltd, Chichester, 3rd Ed., 2000.
16. R. Mongia, I. Bahl, P. Bhartia, *RF and Microwave Coupled Line Circuits*, Artech House, Boston, 1999.
17. J. C. Freeman, *Fundamentals of Microwave Transmission Lines*, John Wiley & Sons, Inc., New York, 1996.
18. D. M. Pozar, *Microwave Engineering*, John Wiley & Sons, Inc., Hoboken, 2012.
19. A. Ferrero, M. Pirola, "Generalized mixed-mode S parameters," *IEEE Trans. Microw. Theory Technol.*, vol. **54**, pp. 458–463, Jan. 2006.
20. W. R. Eisenstadt, B. Stengel, B. M. Thompson, *Microwave Differential Circuit Design Using Mixed-mode S-Parameters*, Artech House, Norwood, 2006.

PART 2

Balanced Transmission Lines with Common-Mode Noise Suppression

CHAPTER 2

STRATEGIES FOR COMMON-MODE SUPPRESSION IN BALANCED LINES

Ferran Martín,[1] Paris Vélez,[1] Armando Fernández-Prieto,[2] Jordi Naqui,[1] Francisco Medina,[2] and Jiasheng Hong[3]

[1]CIMITEC, Departament d'Enginyeria Electrònica, Universitat Autònoma de Barcelona, Bellaterra, Spain
[2]Departamento de Electrónica y Electromagnetismo, Universidad de Sevilla, Sevilla, Spain
[3]Institute of Sensors, Signals and Systems, School of Engineering and Physical Sciences, Heriot-Watt University, Edinburgh, UK

2.1 INTRODUCTION

Within the framework of this book, common-mode suppression in balanced transmission lines, achieved through common-mode stopband filters, may be of interest to enhance the rejection of that mode in balanced bandpass filters. Most balanced microwave filters are designed in such a way that common-mode noise is inherently suppressed in the region of interest, that is, the differential-mode passband (see parts III and IV of this book). However, in certain designs, the common-mode rejection ratio (CMRR), defined as the ratio between the transmission

Balanced Microwave Filters, First Edition. Edited by Ferran Martín, Lei Zhu, Jiasheng Hong, and Francisco Medina.

coefficients for the differential and common modes expressed in dB, is limited within the differential-mode passband. To improve the CMRR, one solution is to cascade common-mode stopband filters. These filters are the subject of this chapter, whereas some examples of common-mode suppressed balanced bandpass filters that use common-mode stopband filters are reported in Chapter 3.

Enhancement of common-mode suppression in balanced filters is a potential application of common-mode stopband filters (as indicated in the previous paragraph). In this application, the filters deal with differential signals in radio-frequency/microwave circuitry. However, common-mode stopband filters are also of high interest for high-speed digital circuit applications, where differential signals are widely employed due to the high immunity to noise, electromagnetic interference (EMI), and crosstalk of differential interconnects. In practice, undesired common-mode noise in balanced lines is inevitable due to timing skew or amplitude imbalance between the differential lines. This may degrade the signal quality and cause EMI emission [1], and for that reason broadband common-mode stopband filters are required. However, in order to transmit differential-mode signals with high quality, it is necessary that the common-mode filters exhibit an all-pass response for the differential mode, from DC up to frequencies beyond those required in high-speed data transmission (multi-Gb/s), with low insertion loss and preservation of signal integrity for that mode.

In summary, regardless of the application, common-mode stopband filters are required to exhibit a broad common-mode rejection bandwidth, keeping at the same time the differential signals unaltered over a wide frequency band. To achieve that purpose, in most approaches, lumped, semi-lumped, or distributed elements insensitive to differential signals, but able to prevent the propagation of the common mode, are added to the differential lines. Other approaches are based on periodic structures, acting as effective electromagnetic bandgaps (EBGs)[1] [2, 3] for the common mode, but being transparent for the differential signals. Some of these strategies for common-mode suppression in balanced

[1] Electromagnetic bandgaps (EBGs), referred to as photonic bandgaps (PBGs) or photonic crystals (PCs) in the optical domain, are periodic structures able to inhibit signal propagation at certain frequencies as a consequence of the Bragg effect, derived from periodicity. In transmission lines, periodicity may be introduced, for example, by modulating the transverse dimensions of the line or by etching patterns in the ground plane. As consequence of this, such periodic EBG-based transmission lines act as reflectors (or stopband filters), efficiently preventing the propagation of signals in the vicinity of the so-called Bragg frequency (corresponding to a wavelength equal to twice the period of the EBG-based line). An exhaustive analysis of these periodic transmission lines and applications can be found in Ref. [4] (Chapter 2) and references therein.

transmission lines are reviewed in this chapter, with special emphasis on those based on patterned ground structures (PGSs) [5–13] or EBGs [14–18] that make use of two metal levels. Solutions that use common-mode chokes with high permeability ferrite cores [19–21] are not considered, and other approaches based on multilayer or LTCC technology [22–30] are succinctly reviewed in the last part of the chapter. Chokes represent a penalty in terms of size and frequency operating range, not being useful for high-speed and high-density systems. Multilayer technology may provide compact solutions, but at the expense of higher fabrication costs and complexity (especially if vias are used [23–25]). For these reasons, the main focus of the chapter is on fully planar bilayer structures.[2]

2.2 SELECTIVE MODE SUPPRESSION IN DIFFERENTIAL TRANSMISSION LINES

As will be shown in Section 2.3, common-mode suppression in differential lines can be achieved by loading the line with symmetric resonant elements. Let us now review the general principle to achieve selective mode suppression (common or differential) in differential lines through symmetric resonators. It is well known that a transmission line (single-ended or differential) loaded with a resonator coupled to it exhibits a stopband (transmission zero, or notch) at the fundamental resonance frequency (and eventually at higher-order harmonics). Signal propagation is inhibited at these frequencies because the injected power is reflected back to the source. However, depending on the topology of the resonator and the relative orientation between the resonator and the line, it is possible to prevent resonance (and hence the notch in the transmission coefficient) [4, 31].

Let us consider that the line is symmetric and it is loaded with a symmetric resonator. Moreover, let us assume that the symmetry plane of the resonator is perfectly aligned with the symmetry plane of the transmission line, so that the structure is perfectly symmetric with respect to the bisection midplane. If the symmetry planes of the line and resonator are of the same electromagnetic nature (either electric or magnetic walls), the resonator is excited, the structure is resonant, and signal propagation is inhibited. Conversely, if the symmetry plane of the line is an electric wall and the one of the resonator is a magnetic wall, or vice

[2] Nevertheless, it will be shown that certain multilayer common-mode filters are described by circuit models identical to those that describe some of the considered bilayer structures.

versa, the resonator cannot be excited. Due to perfect symmetry, the net electric and magnetic field components illuminating the resonator exactly cancel (i.e., the electric and magnetic fluxes over the resonator area are zero) if the symmetry planes of the line and resonator are of different (or complementary) nature. Under these conditions, the structure is all-pass, at least in the vicinity of the resonance frequency under consideration.[3] Obviously, if the symmetry planes of the line and resonator are misaligned and both elements are close enough, line-to-resonator coupling activates, and the resulting notch depth in the transmission coefficient is determined by the level of misalignment.[4]

Let us now consider the particular case of microstrip differential lines. According to the previous paragraph, either the common or the differential mode can be selectively suppressed while preserving the other mode unaltered (or almost unaltered) by aligning a resonator exhibiting a magnetic wall or an electric wall, respectively. This filtering functionality, referred to as selective mode suppression [38–40], is illustrated in Figure 2.1, where a microstrip differential line is loaded with a

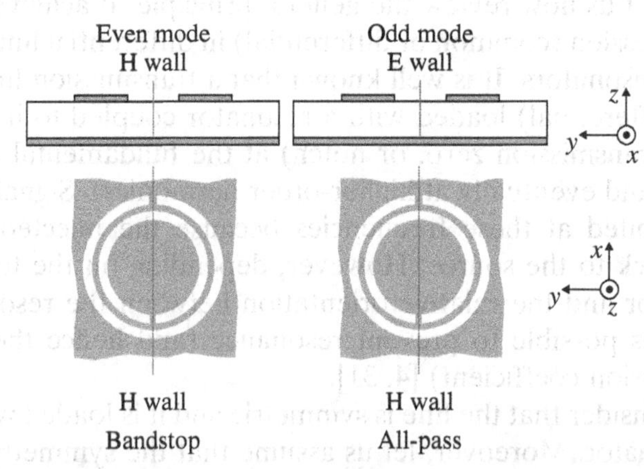

Figure 2.1 Illustration of selective mode suppression in microstrip differential lines by using CSRRs (etched in the ground plane). Note that bandstop and all-pass are in reference to a region in the vicinity of the fundamental CSRR resonance (further stopbands may appear for both modes at higher frequencies).

[3] Note that the electromagnetic nature (electric or magnetic wall) of the symmetry plane of a symmetric resonator depends on the resonance order.

[4] This dependence of the notch depth on the relative orientation/position between the resonator and the line has been exhaustively exploited for the implementation of sensors based on symmetry properties (symmetry disruption) [32–37].

complementary split ring resonator (CSRR) [41]. Such resonant element exhibits a magnetic wall at the fundamental resonance, and hence it inhibits signal propagation at that frequency for the common mode, whereas the line is transparent for the differential mode. Thus, CSRRs are useful for the implementation of common-mode suppression filters, as will be demonstrated in Section 2.3.2. Mode suppression can be reversed by loading the differential line with split ring resonators (SRRs) symmetrically placed at the upper metal level, between the pair of lines [42], because SRRs exhibit an electric wall in their symmetry plane at the fundamental resonance frequency [43]. The suppression of the differential mode may be of interest in certain applications, but it is out of the scope of this chapter, focusing on common-mode suppression filters. Many other resonant elements can be used for selectively suppressing the common mode in differential lines, including magnetic and electric LC resonators [10, 40], dumbbell-shaped resonators [5], C-shaped resonators [11], folded slotted stepped-impedance resonators, and so on. The application of some of these resonant elements to the implementation of common-mode rejection filters is reviewed in the next section.

2.3 COMMON-MODE SUPPRESSION FILTERS BASED ON PATTERNED GROUND PLANES

Four examples of common-mode suppression filters for differential microstrip lines, based on patterned ground planes (or defected ground structures), are presented in this section. The ground planes of the different implementations exhibit specific patterns that efficiently suppress the common mode, leaving the differential mode unaltered (or almost unaltered). These common-mode filters, all of them exhibiting significant common-mode rejection bandwidths, are studied in the next subsections and presented chronologically (special emphasis is dedicated to CSRR-based common-mode filters, developed by the authors of the present chapter).

2.3.1 Common-Mode Filter Based on Dumbbell-Shaped Patterned Ground Plane

Slotted dumbbell resonators symmetrically etched in the ground plane of a microstrip line can be modeled, to a first-order approximation, as series-connected parallel resonators [44]. These resonators are thus able to inhibit signal propagation in the vicinity of resonance. The symmetry planes of the line and resonator are magnetic walls, and this gives another interpretation to the stopband functionality of these lines, according to symmetry considerations (see Section 2.2).

By symmetrically etching slotted dumbbell-shaped resonators in a differential microstrip line, it is expected that the common mode is efficiently suppressed, while signal propagation for the differential mode is preserved. Besides the previous explanation for selective mode suppression, based on symmetry considerations, it can also be argued that the resonant elements open the return current path through the ground plane for the common mode, whereas the presence of resonators has negligible effect on the differential signals since relatively small current density returns through the ground plane for such signals [5]. Regardless of the specific interpretation for selective common-mode suppression in differential lines loaded with slotted dumbbell resonators, such structures behave as efficient common-mode filters. Figure 2.2 depicts a specific common-mode filter topology and the frequency response. The structure is able to efficiently suppress the common mode with broad stopband in the GHz range (Figure 2.2b), yet keeping good signal integrity performance for the differential signals. In addition, mode conversion from differential mode to common mode, also shown in the figure, is very small.

An interesting time-domain experiment carried out in Ref. [5] to verify the common-mode noise suppression capability consists of adding a delay line in one of the differential lines (see inset of Figure 2.3). This delay line generates a signal skew that in turn excites the common mode in the differential line. By fabricating an identical structure with solid ground plane (reference board), the time-domain responses to a differential signal can be compared. In Ref. [5], two pulse trains of 800 mV peak-to-peak voltage and 8 Gb/s are differentially launched into port 1 and port 2. The output waveforms are measured at port 3 and port 4 using a digital oscilloscope. The common-mode noise, V_{common}, defined as half of the sum of the measured voltage at these two output ports, reveals that the peak-to-peak output common-mode voltage for the reference board is 412 mV, whereas it is reduced to 64 mV by employing the dumbbell-shaped resonators (i.e., over 80% improvement is achieved); see Figure 2.3. Thus, the common-mode filter based on these PGS structures is revealed to be efficient in rejecting the common mode while preserving the integrity of the differential signals. Measured differential eye diagrams not shown here, but depicted in Ref. [5], indicate that the eye diagram parameters are roughly the same in the structure

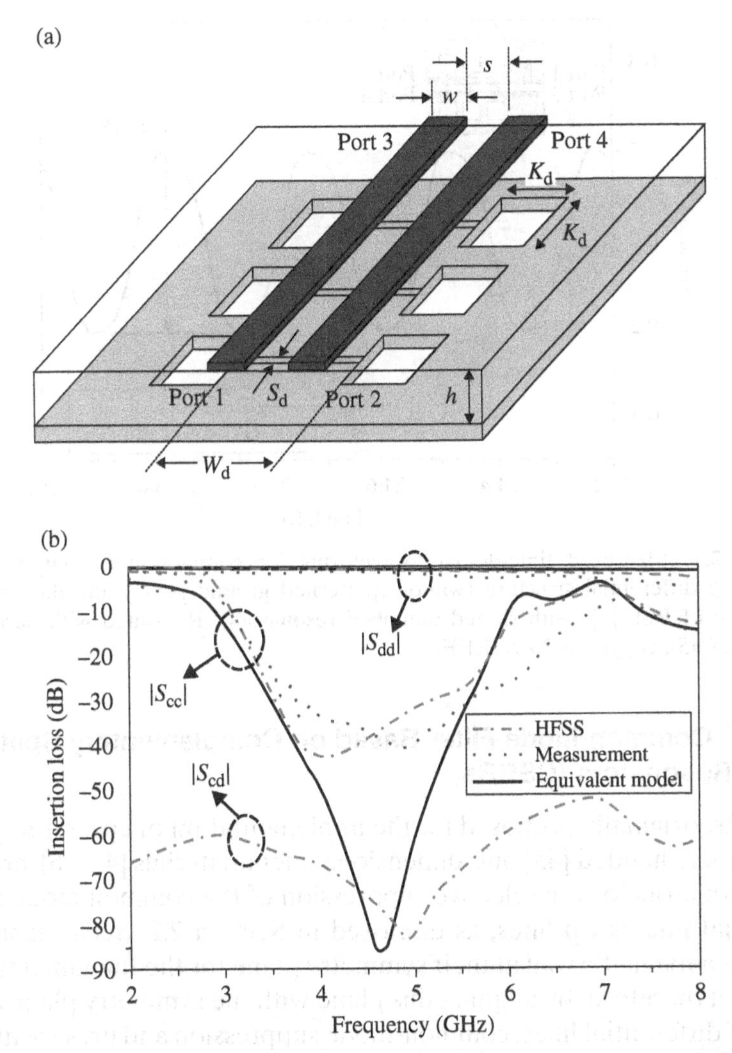

Figure 2.2 Common-mode stopband filter implemented by slotted dumbbell-shaped resonators (a) and frequency response for the differential (S_{dd}) and common modes (S_{cc}) (b). The simulated transmission of mode conversion from differential mode to common mode, S_{cd}, is also shown in (b). Reprinted with permission from Ref. [5]; copyright 2008 IEEE.

of Figure 2.2 and in the same structure with solid ground plane (indicating that the differential signal quality is not degraded when the common-mode filter is present).[5]

[5] In the common-mode filter of Section 2.3.2, the differential eye diagrams are provided and discussed in detail.

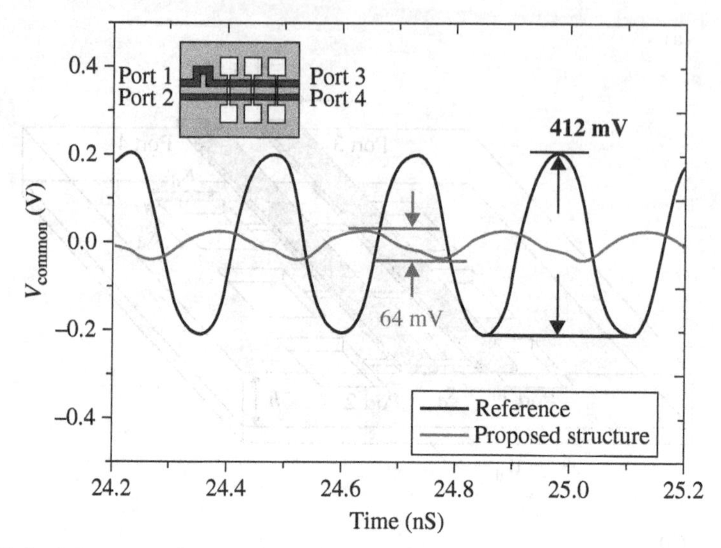

Figure 2.3 Measured time-domain waveforms for common-mode output of the reference differential structure (without patterned ground plane) and the proposed structure of Ref. [5] (with slotted dumbbell resonators). Reprinted with permission from Ref. [5]; copyright 2008 IEEE.

2.3.2 Common-Mode Filter Based on Complementary Split Ring Resonators (CSRRs)

CSRRs, originally proposed for the implementation of epsilon-negative [41] or left-handed [45] one-dimensional metamaterials [43, 46], are useful resonators for the selective suppression of the common mode in differential microstrip lines, as discussed in Section 2.2. Such resonators exhibit a magnetic wall at their symmetry plane for the fundamental resonance; therefore, by aligning this plane with the symmetry plane of the pair of differential lines, common-mode suppression and preservation of signal integrity for the differential mode are expected. This was confirmed for the first time in Ref. [7], where it was argued that for the differential mode there is not a net axial electric field in the inner metallic region of the CSRR able to excite the resonators, whereas for the common mode the electric field below the transmission lines is codirectional, and the CSRRs are excited.

In order to obtain a significant common-mode rejection bandwidth, several CSRRs with slightly different dimensions (resulting in different, but close, resonance frequencies) can be etched in the ground plane of the differential lines. This approach, first reported in coplanar waveguides (CPWs) loaded with slightly different SRRs [47] and then in CSRR-loaded microstrip lines [48], is effective, but many resonant elements are required to achieve a wide stopband. Alternatively, common-mode suppression bandwidth can be enhanced by tightly coupling identical resonant elements, and further broadening is achieved in CSRR-loaded differential lines by increasing the coupling capacitance between the pair of lines and the CSRR and by decreasing the inductance and capacitance of the CSRR. This enhances the stopband bandwidth of an individual resonator, providing a wide stopband when several tightly coupled CSRRs are considered. As reported in Ref. [7], the requirements to achieve small CSRR inductance and capacitance are wide ring widths and separation. To enhance the coupling capacitance between the pair of lines and the CSRR, the lines must be as wide as possible and hence as much uncoupled as possible. Finally, by closely spacing the CSRRs and by using square-shaped or rectangular CSRR geometries, inter-resonator coupling is favored. Figure 2.4(a) depicts a fabricated differential line loaded with symmetrically etched,

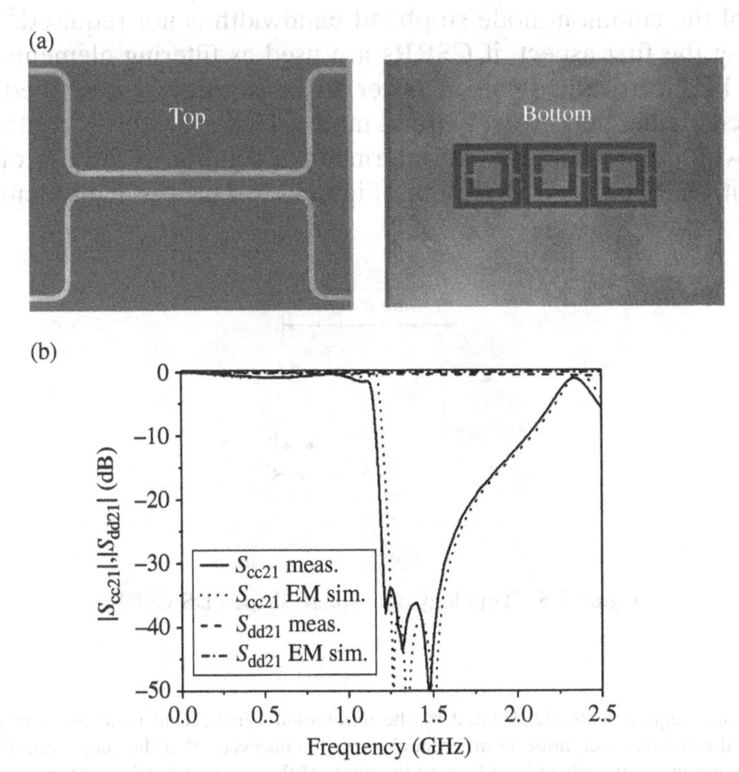

Figure 2.4 Photograph (a) and common-mode ($|S_{cc21}|$) and differential-mode ($|S_{dd21}|$) insertion loss (b) of a differential line with wideband common-mode rejection based on CSRRs. Reprinted with permission from Ref. [7]; copyright 2011 IEEE.

square-shaped, and tightly coupled CSRRs. The common-mode and differential-mode insertion loss is depicted in Figure 2.4(b) and reveals that CSRRs are efficient elements to suppress the common mode over a wide band, while the differential-mode is transmitted in that band.

To design CSRR-based common-mode filters following a systematic design approach, a simple circuit model that describes the CSRR-loaded differential line, including inter-resonator coupling, is necessary. In this regard, two limitations arise. First of all, the convenient CSRRs to achieve a wide common-mode stopband (with wide slots and inter-slot distance) cannot be considered to be electrically small, and hence they cannot be described by an accurate lumped element circuit model over a wide band. Secondly, the CSRRs must be oriented with their symmetry plane aligned with the symmetry plane of the differential line. Under these conditions, mixed coupling (electric and magnetic) may arise, as discussed in Ref. [49], and the circuit model (provided the CSRRs are electrically small) is even more complex. This second aspect can be ignored to a first-order approximation, since a very accurate determination of the common-mode stopband bandwidth is not required.[6] Concerning the first aspect, if CSRRs are used as filtering elements, they must be electrically small in order to be accurately described by a lumped element equivalent circuit model. However, this is contrary to bandwidth enhancement. An intermediate solution is the use of double-slit CSRRs (DS-CSRRs); see Figure 2.5. This resonant element is

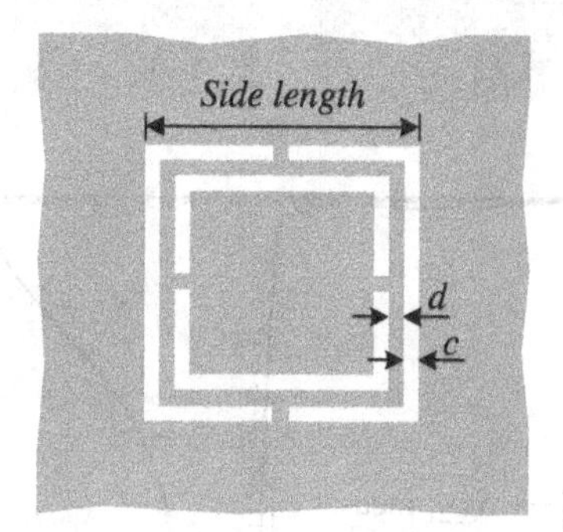

Figure 2.5 Topology of a square-shaped DS-CSRR.

[6] Bandwidth requirements are dictated by the differential signals. That is, in order to prevent unwanted common-mode noise in differential lines, it is necessary that the suppressed band for the common mode extends at least beyond the limits of the required band for differential signal transmission. Therefore, rather than pursuing a specific rejection bandwidth, common-mode filters must be typically designed to satisfy a predefined common-mode rejection level within a certain frequency band.

derived by application of duality to the DS-SRR [50].[7] By driving the slot width and separation to the minimum value allowed by the available technology, DS-CSRRs can be considered to be electrically small, and the circuit model provides a reasonable description of the loaded differential lines. Such circuit model (unit cell), valid for electrically small CSRRs or DS-CSRRs, neglecting magnetic coupling between the line and the resonators, is depicted in Figure 2.6(a).[8] In this model, L and C are the inductance and capacitance of the lines, respectively, the resonant element (CSRR or DS-CSRR) is described by the resonant tank L_c–C_c, the coupling capacitance between resonant elements is given by C_R, and L_m and C_m account for the magnetic and electric coupling, respectively, between the pair of lines.

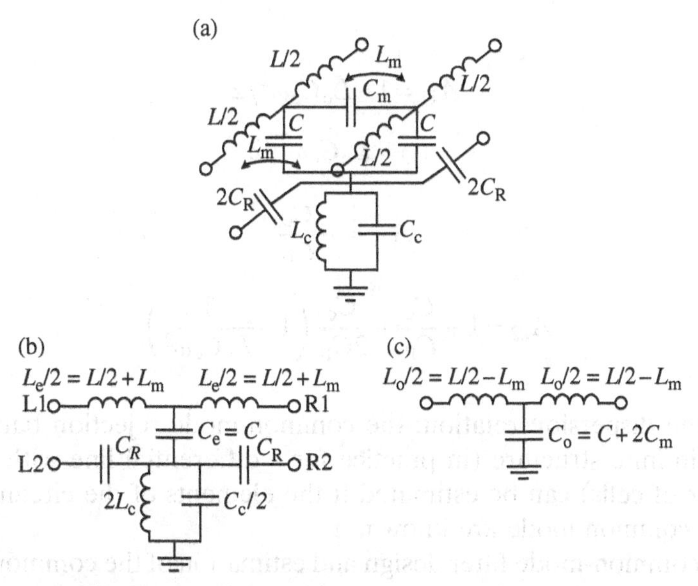

Figure 2.6 Lumped element equivalent circuit model (unit cell) of a differential microstrip line loaded with CSRRs or DS-CSRRs (a), equivalent circuit model for the common mode (b), and equivalent circuit model for the differential mode (c).

[7] As compared with a CSRR with identical dimensions and etched on the same substrate, the DS-CSRR exhibits identical capacitance, but four times smaller inductance [4, 43]. Hence the DS-CSRR is electrically larger than the CSRR by a factor of 2 (i.e., for identical dimensions, the resonance frequency of the DS-CSRR is twice the one of the CSRR).

[8] Indeed, this approximation is not needed for differential lines loaded with DS-CSRR, since this resonant particle is non-bianisotropic and magnetic coupling between the line and the DS-CSRR is not possible for symmetric orientations.

The lumped element equivalent circuit models for the common and differential modes are depicted in Figure 2.6(b) and (c), respectively. For the differential mode, the resulting circuit model is simply the circuit model of an ordinary line with modified parameters. The circuit model for the common mode is formally identical to that of CSRR-loaded single-ended microstrip lines with inter-resonator coupling [51]. Therefore, the dispersion relation, derived in Ref. [51] for CSRR-loaded single-ended lines, is also valid for the common mode in CSRR- or DS-CSRR-loaded differential lines. It is given by Refs. [4, 31, 51] (see Appendix 2.A):

$$\cosh(\gamma l) = \frac{1}{2}\left(A_{11} + A_{22} \pm \sqrt{(A_{11} - A_{22})^2 + 4A_{12}A_{21}}\right) \tag{2.1}$$

with

$$A_{11} = 1 - L_e C_e \omega^2/2 \tag{2.2a}$$

$$A_{12} = L_e C_e \omega^2/2 \tag{2.2b}$$

$$A_{21} = -\frac{C_e}{C_R} \tag{2.2c}$$

$$A_{22} = 1 + \frac{C_e}{C_R} + \frac{C_c}{2C_R}\left(1 - \frac{1}{L_c C_c \omega^2}\right) \tag{2.2d}$$

From the dispersion relation, the common-mode rejection bandwidth for an infinite structure (in practice for a differential line with a large number of cells) can be estimated if the elements of the circuit model for the common mode are known.

For common-mode filter design and estimation of the common-mode rejection bandwidth, the procedure is as follows [8]. For the reasons explained before, relative to the validity of the models of Figure 2.6 and fabrication tolerances, the slot width and separation are set to a minimum implementable value with the technology in use (e.g., $c = d = 200$ μm [8]). Notice that this reduces the degrees of freedom and eases the common-mode filter design. Square-shaped resonators (rather than circular) are considered in order to enhance the electric coupling between the differential line and the resonators and between adjacent resonators as well. To further enhance inter-resonator coupling, the separation between adjacent resonators is also set to a limiting value (200 μm). In order to achieve a strong electric coupling between the pair of lines and the resonator, the lines must be fitted inside the CSRR (or DS-CSRR) region and must be as wide as possible, and hence as uncoupled as possible, as mentioned earlier (line dimensions can easily be inferred from a transmission-line calculator). The side length of the resonator is determined from the model of the CSRR reported in Ref. [46] (which gives an estimation of L_c and C_c) and the per unit length capacitance of the coupled lines for the even mode (which gives C_e). The transmission zero frequency in the absence of inter-resonator coupling, given by[9]

$$f_z = \frac{1}{2\pi\sqrt{2L_c(C_e + C_c/2)}} \tag{2.3}$$

is adjusted to the required central filter frequency,[10] and this provides the CSRR side length (for a DS-CSRR the side length can also be determined by taking into account that the inductance is four times smaller than the inductance of the CSRR). Obviously, optimization of the resonator side length in order to fit the required central frequency is necessary.

With the previous procedure the common-mode filter dimensions are perfectly determined. To predict the maximum achievable bandwidth, that is, the bandwidth obtained by considering an infinite number of cells, the dispersion relation given by (2.1) is used. However, since optimization at the layout level is required, it is necessary to extract the parameters of the circuit model (e.g., following the procedure reported in Ref. [52] and detailed in the Appendix G of Ref. [4]). Thus, L_e, C_e, L_c, and C_c are first extracted by considering a single cell structure, and then C_R is adjusted to fit the EM simulation of a second-order common-mode filter. Once the circuit parameters are known, expression (2.1) can be evaluated and the common-mode stopband can be estimated.

Following the previous approach, the maximum achievable rejection bandwidth for different CSRR and DS-CSRR common-mode filters can be obtained. The rejection bandwidth can also be obtained from full-wave EM simulations (the results are depicted in Figure 2.7). The extracted circuit parameters and estimated fractional bandwidths are shown in Table 2.1.

[9] Using CSRRs, the transmission zero frequency also depends on the line-to-resonator magnetic coupling, but its contribution is neglected.

[10] The transmission zero frequency for an isolated unit cell provides a reasonable estimate of the filter's central frequency.

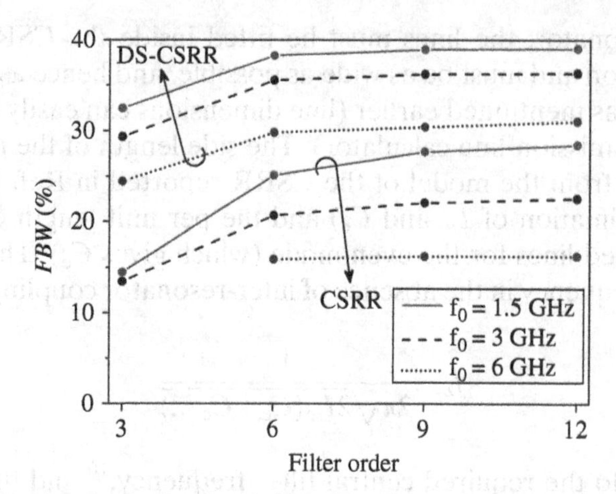

Figure 2.7 Fractional rejection bandwidth (*FBW*) at −20 dB for the common mode given by EM simulation for CSRR- and DS-CSRR-loaded differential lines. Dimensions are as follows: for the CSRRs and DS-CSRRs, $c = d = 0.2$ mm and *inter-resonator distance* = 0.2 mm; for the CSRRs, *side length* = 7.3 mm ($f_0 = 1.5$ GHz), 4.3 mm ($f_0 = 3$ GHz), and 2.6 mm ($f_0 = 6$ GHz); for the DS-CSRRs, *side length* = 13.8 mm ($f_0 = 1.5$ GHz), 7.5 mm ($f_0 = 3$ GHz), and 4.3 mm ($f_0 = 6$ GHz); and for the differential line, $2W + S = side\ length - 2(2c - d) + 0.4$ mm, exhibiting a 50 Ω characteristic impedance (odd mode). The considered substrate is the *Rogers RO3010* with thickness $h = 1.27$ mm and dielectric constant $\varepsilon_r = 10.2$. Reprinted with permission from Ref. [8]; copyright 2012 IEEE.

Table 2.1 Extracted Parameters and Maximum Fractional Bandwidth Inferred from the Circuit Model

f_0 (GHz)	L_e (nH)	C_e (pF)	L_c (nH)	C_c (pF)	C_R (pF)	FBW (%)
CSRR-loaded differential lines						
1.5	6.3	1.1	2.1	3.2	0.1	30.7
3	4.2	0.5	1.0	2.0	0.06	26.4
6	3.0	0.2	0.4	1.4	0.08	20.7
DS-CSRR-loaded differential lines						
1.5	15.9	1.6	1.0	8.6	0.64	32.6
3	9.0	0.8	0.45	4.8	0.18	29.4
6	6.9	0.3	0.15	3.6	0.1	23.6

By comparing the fractional bandwidths predicted by the reported approach with the saturation values of Figure 2.7, it can be concluded that the reported approach is more accurate for CSRR-loaded lines, as expected. Figure 2.8 compares the circuit and EM simulation (common mode) of the first-, second-, and third-order CSRR-based common-mode filters designed to exhibit a central frequency of 1.5 GHz (in the circuit simulation the inter-resonator capacitance at input and

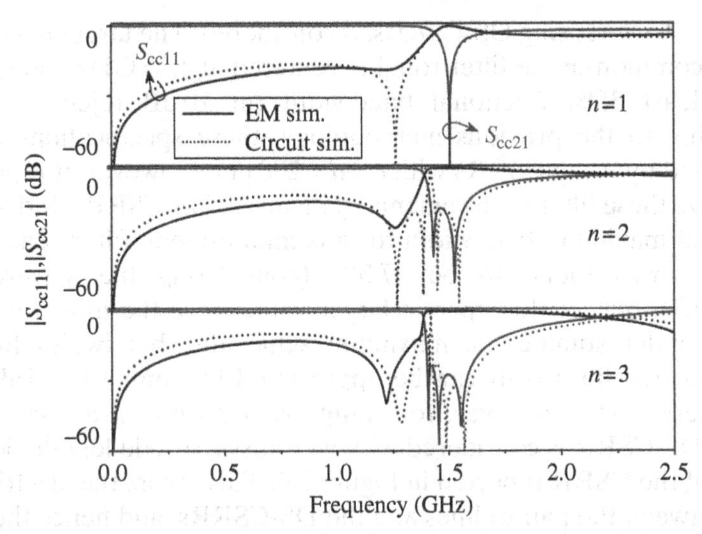

Figure 2.8 Common-mode return loss $|S_{cc11}|$ and insertion loss $|S_{cc21}|$ given by the EM and circuit simulation for the first-, second-, and third-order common-mode filters (1.5 GHz central frequency) based on CSRRs. Dimensions and substrate are indicated in the caption of Figure 2.7. Circuit parameters are given in Table 2.1. Reprinted with permission from Ref. [8]; copyright 2012 IEEE.

output ports has been left open since the CSRRs of the input and output cells are not externally fed, resulting in a two-port circuit). There is good agreement between the circuit and EM simulation, pointing out the accuracy of the model for CSRR-loaded lines with narrow slots, *c*, and inter-slot distance, *d*.

As it can be seen in Figure 2.7, six resonators are enough to nearly achieve the maximum rejection bandwidth. Obviously, common-mode filter size can be reduced by decreasing the number of resonators, but at the expense of a reduced common-mode rejection bandwidth. In summary, following a systematic approach based on the circuit model of the common mode, it can be inferred whether a specified rejection bandwidth and central frequency can be roughly fulfilled or not.[11] If the required bandwidth is wider, we are forced to consider resonators with wider slots (*c*) and inter-slot distance (*d*), or, alternatively, multiple tuned resonators. In these cases, however, filter design and maximum bandwidth estimation are not so straightforward.

In order to illustrate the design methodology and the prediction of the maximum bandwidth, a common-mode filter similar to that reported

[11] The reported approach allows us to infer the maximum achievable common-mode rejection bandwidth, rather than setting the bandwidth to a particular value.

in Figure 2.4, but using DS-CSRRs, is considered. The target is to implement a common-mode filter roughly centered at 1.35 GHz and exhibiting at least 35% fractional bandwidth (at 20 dB rejection level). According to the previous methodology, these specifications cannot be fulfilled by using CSRRs with $c = d = 200\ \mu\text{m}$. However, it is possible to achieve these filter requirements by means of DS-CSRRs. Indeed, the estimated maximum bandwidth for a common-mode filter centered at 1.35 GHz was found to be 37.3% (considering the substrate of Figure 2.7), but we do expect a larger value since the model tends to slightly underestimate the maximum achievable bandwidth for DS-CSRR-loaded lines (this can be appreciated by comparing Table 2.1 and Figure 2.7). Moreover, for comparison purposes, a rectangular-shaped DS-CSRR is considered with its transverse side length identical to that of the CSRR reported in Figure 2.4. This favors the electric coupling between the pair of lines and the DS-CSRRs, and hence the common-mode stopband expansion (because the DS-CSRR longitudinal side is longer than the transverse one, and this increases the coupling capacitance C_e as compared with that of a square-shaped DS-CSRR with identical transmission zero frequency). The longitudinal side length is thus the single design parameter, and this has been determined following the same approach applied to square-shaped particles (the geometrical parameters of the structure are given in the caption of Figure 2.9a).

The photograph and frequency responses of the device, a third-order common-mode filter, are depicted in Figure 2.9(a)–(c) (this filter order was found to be enough to satisfy the bandwidth requirements). As it can be seen, the differential signal is almost unaltered, while the common mode is rejected within a fractional bandwidth (41%) comparable with that achieved in Figure 2.4 by using CSRRs with wide and widely spaced rings. The DS-CSRR-based structure is a bit larger than that reported in Figure 2.4, but the design follows a systematic procedure.

Figure 2.10 shows the measured differential eye diagrams[12] with the excitation of 0.2 V amplitude in 2.5 Gb/s for the differential line of

[12] An eye diagram is an indicator of the quality of signals in high-speed digital transmissions. It is constructed from a digital waveform by folding the parts of the waveform corresponding to each individual bit into a single graph with signal amplitude on the vertical axis and time on horizontal axis. By repeating this construction over many samples of the waveform, the resultant graph will represent the average statistics of the signal and will resemble an eye. The parameters indicated in Table 2.2 are defined in many sources focused on signal integrity. In brief, eye height is a measure of the vertical opening of an eye diagram, and it is determined by noise, which "closes" the eye. Eye width is a measure of the horizontal opening of an eye diagram, and it is calculated by measuring the difference between the statistical mean of the crossing points of the eye. Jitter is the time deviation from the ideal timing of a data-bit event. To compute jitter, the time deviations of the transitions of the rising and falling edges of an eye diagram at the crossing point are measured.

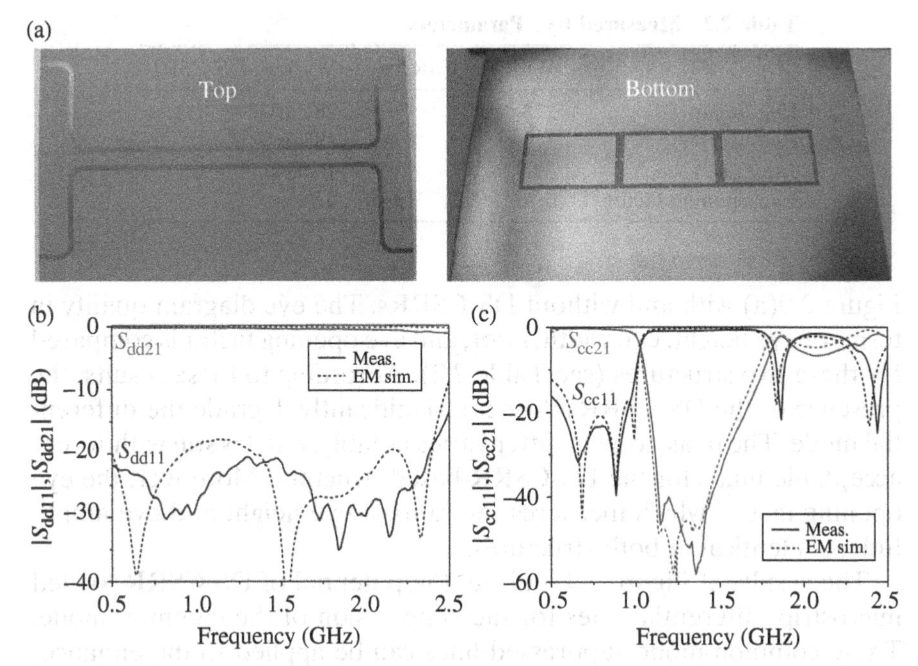

Figure 2.9 Photograph (a), differential-mode return loss $|S_{dd11}|$ and insertion loss $|S_{dd21}|$ (b), and common-mode return loss $|S_{cc11}|$ and insertion loss $|S_{cc21}|$ (c) of the designed differential line with common-mode suppression based on DS-CSRRs. Dimensions are as follows: for the DS-CSRRs, $c = d = 0.2$ mm, *longitudinal side length* = 17.6 mm, *transverse side length* = 10.8 mm, and *inter-resonator distance* = 0.2 mm and for the differential line, $W = 1$ mm, and $S = 2.5$ mm. The considered substrate is the *Rogers RO3010* with thickness $h = 1.27$ mm, dielectric constant $\varepsilon_r = 10.2$, and loss tangent $\tan\delta = 0.0023$. Reprinted with permission from Ref. [8]; copyright 2012 IEEE.

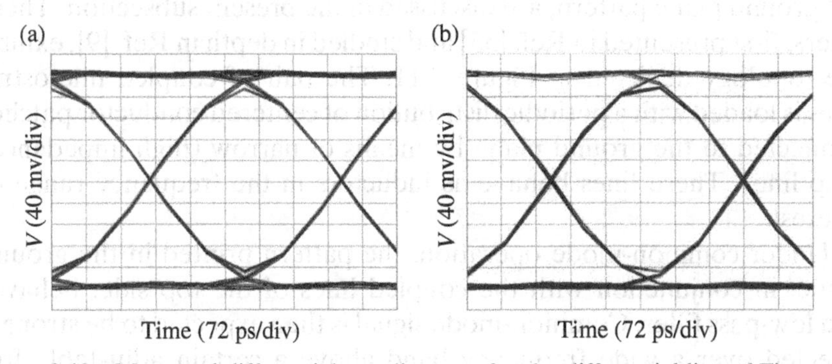

Figure 2.10 Measured differential eye diagrams for the differential line of Figure 2.9 with (a) and without (b) DS-CSRRs. Reprinted with permission from Ref. [8]; copyright 2012 IEEE.

Table 2.2 Measured Eye Parameters

	With DS-CSRRs	Without DS-CSRRs
Eye height	278 mV	281 mV
Eye width	371 ps	383 ps
Jitter (PP)	29.3 ps	16.9 ps
Eye opening factor	0.76	0.76

Figure 2.9(a) with and without DS-CSRRs. The eye diagram quality in terms of eye height, eye width, jitter, and eye opening factor is compared for these two structures (see Table 2.2). According to these results, the presence of the DS-CSRRs does not significantly degrade the differential mode. The peak-to-peak jitter varies notably, but it is still within very acceptable limits for the DS-CSRR-based structure. Moreover, the eye opening factor, which measures the ratio of eye height and eye amplitude, is identical in both structures.

The results of Figure 2.9 indicate the potential of DS-CSRR-loaded microstrip differential lines for the suppression of the common mode. These common-mode suppressed lines can be applied to the enhancement of common-mode suppression in differential-mode bandpass filters, as will be reported in Chapter 3.

2.3.3 Common-Mode Filter Based on Defected Ground Plane Artificial Line

Common-mode filters conceptually similar to those presented in the previous subsection, but implemented by means of a completely different ground plane pattern, are discussed in the present subsection. These filters, first presented in Ref. [53] and studied in depth in Ref. [9], exhibit the topology depicted in Figure 2.11. The pair of coupled microstrip lines is loaded with a periodic distribution of centered conductor patches connected to the ground plane by means of narrow (high impedance) strip lines. These lines behave as inductors in the frequency range of interest.

Under common-mode operation, the pattern printed in the ground plane, in conjunction with the coupled lines of the top side, behaves as a low-pass filter. Common-mode signal is then expected to be strongly rejected over a wide frequency band above a certain adjustable frequency. To gain insight of this, let us consider the equivalent circuit of the structure (unit cell), depicted in Figure 2.12(a). In this circuit, L_s is the inductance of the lines, C_p is the capacitance between each

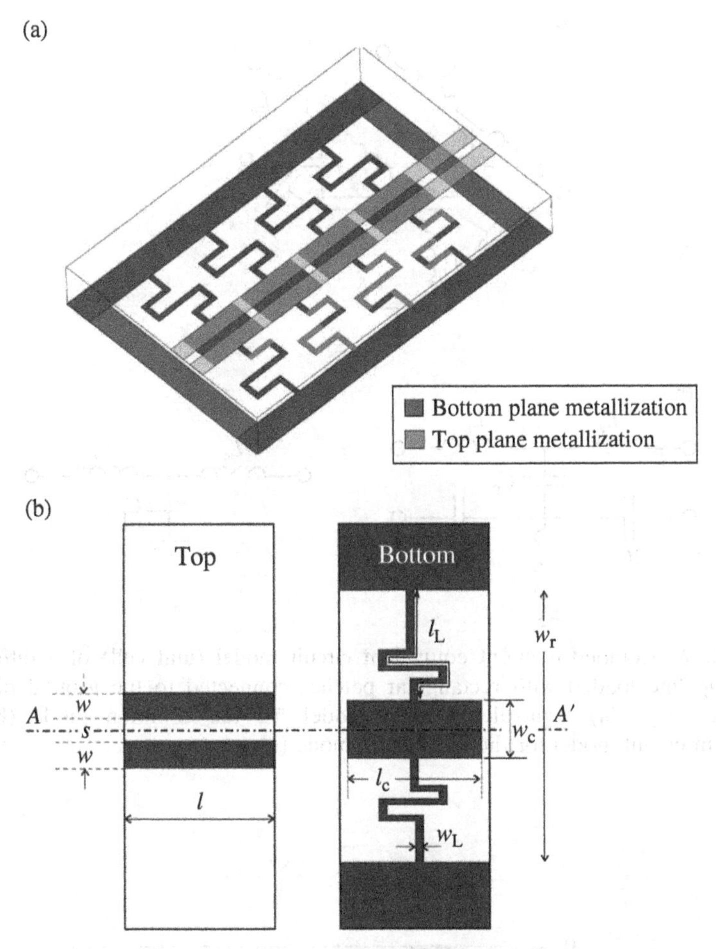

Figure 2.11 3D view (a) and layout (unit cell) (b) of the common-mode suppression filter based on rectangular patches connected to the ground plane through meandered inductive strips. The metallic regions in (b) are indicated in gray.

individual line and the central patches, L_p is the inductance of the narrow strips, and C_s is the capacitance between adjacent patches. The equivalent circuits for the common and differential modes are depicted in Figure 2.13(b) and (c), respectively. For the differential mode, the lumped element equivalent circuit is the one of an all-pass structure (similar to the circuit model of the CSRR- (or DS-CSRR)-loaded differential lines for the differential mode). Conversely, for the common mode, the equivalent circuit indicates that the signal is inhibited in certain frequency bands. Indeed, if the distance between adjacent patches is significant, the coupling capacitance C_s can be neglected, and

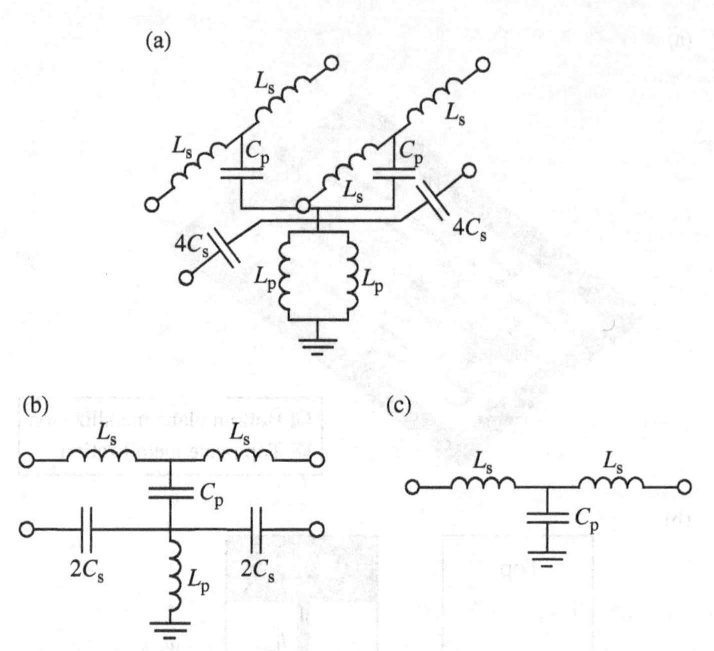

Figure 2.12 Lumped element equivalent circuit model (unit cell) of a differential microstrip line loaded with rectangular patches connected to the ground plane by inductive strips (a), equivalent circuit model for the common mode (b), and equivalent circuit model for the differential mode (c).

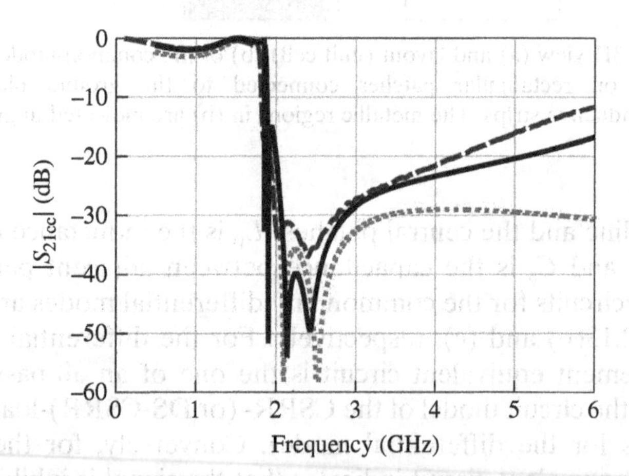

Figure 2.13 Common mode insertion loss of the four-cell common mode filter with dimensions indicated in the text. Full wave simulated (solid line), measured (dashed line), and circuit simulated (dotted line).

transmission zeros at the frequencies of the resulting shunt-connected series resonators (C_p–L_p) are expected.[13] The analysis is not so simple if such capacitances cannot be neglected. However, the resulting multi-port network (common mode) can be studied by using a Bloch–Floquet analysis similar to the one considered in Section 2.3.2 (and detailed in Appendix 2.A). Note that the equivalent circuit of Figure 2.12(b) resembles the circuit of Figure 2.6(b) that describes the CSRR- (or DS-CSRR)-loaded differential lines for the common mode. However, the inductance L_p in the circuit of Figure 2.12(b) replaces the resonant tank L_c–C_c in the circuit of Figure 2.6(b). This fact is useful for achieving broader rejection bandwidths for the common mode.

In Ref. [9], three prototypes (with different number of cells) of these common-mode rejection filters, fabricated on a substrate with thickness $h = 0.49$ mm and dielectric constant $\varepsilon_r = 2.43$, are reported. The geometrical dimensions (see Figure 2.11) are $w = 1.4$ mm, $s = 2.4$ mm, $l = 5.4$ mm, $w_c = 5.2$ mm, $l_c = 5$ mm, $w_L = 0.2$ mm, $l_L = 24.2$ mm, $w_r = 14.6$ mm, and $s_g = 0.2$ mm. The common-mode insertion loss for the filter with four cells is depicted in Figure 2.13, where the results predicted by the circuit model with extracted parameters are also represented. Such parameters are extracted in Ref. [9] according to the following procedure: C_p is roughly approximated as the capacitance of a parallel plate capacitor of plate dimensions $l \times w$ with the dielectric substrate between the plates. C_s is approximated as the series gap capacitance of the equivalent π-circuit of a microstrip gap discontinuity between two strips of width w_c separated at a distance s_g. The value of L_p is extracted from the location of the transmission zero close to 2 GHz. Note that the location of this zero can be tuned by controlling the length l_L of the meandered line inductor. Finally, L_s is chosen to fit the transmission level in the common-mode passband. The values of the lumped parameters for this case example are found to be $L_s = 4.20$ nH, $C_p = 0.35$ pF, $C_s = 0.044$ pF, and $L_p = 16$ nH.

The common-mode insertion loss is above 20 dB in the frequency range 2–4 GHz. In terms of fractional bandwidth, this result is better than those corresponding to the common-mode filters of the previous subsections. However the maximum CMRR is worse in the filter with rectangular patches and inductive strips. It is worth mentioning that the presence of patches and meandered lines in the ground plane does not significantly affect the differential-mode operation, as revealed by the differential-mode measurements reported in Ref. [9], including

[13] Note that if C_s is neglected, the resulting two-port of the unit cell is the building block of elliptic low-pass filters.

the differential-mode insertion loss and eye diagram, which are compared with those of the filter with solid ground plane (not shown in this chapter).

As mentioned in reference to the differential lines loaded with slotted split rings of the previous subsection, the dispersion relation is a useful tool to predict the common-mode stopband bandwidth. For the common-mode filters considered in this subsection, the dispersion relation is also given by expression (2.1). However, the elements of the order-two **A** matrix are in this case

$$A_{11} = 1 - L_s C_p \omega^2 \tag{2.4a}$$

$$A_{12} = L_s C_p \omega^2 \tag{2.4b}$$

$$A_{21} = -\frac{C_p}{2C_s} \tag{2.4c}$$

$$A_{22} = 1 + \frac{C_p}{2C_s} - \frac{1}{2\omega^2 L_p C_s} \tag{2.4d}$$

Note that, except (2.4d), these expressions are identical to (2.2), as expected on account of the similarity of the common-mode circuit models of the CSRR-loaded and patch-loaded differential lines. However, note that the last term in (2.2d) is resonant, whereas the one in (2.4d) is not. Due to this fact, the structure of Figure 2.11 exhibits a low-pass filtering behavior for the common mode, whereas a stopband is generated for that mode in CSRR- (or DS-CSRR)-loaded differential lines (see in Appendix 2.B the dispersion relation for the considered case example).

2.3.4 Common-Mode Filter Based on C-Shaped Patterned Ground Structures

The common-mode filters presented in this subsection (reported in Ref. [11]) are also based on patterned ground planes, but they are described by a combination of distributed and lumped components. The configuration, shown in Figure 2.14, consists of a C-shaped PGS and two meandered signal lines. The C-shaped slot comprises two main components, that is, the patch and bridge (see Figure 2.14a). The patch can be considered to be a floating ground connected to the ground plane by means of the bridge. The differential pair is composed of two meandered

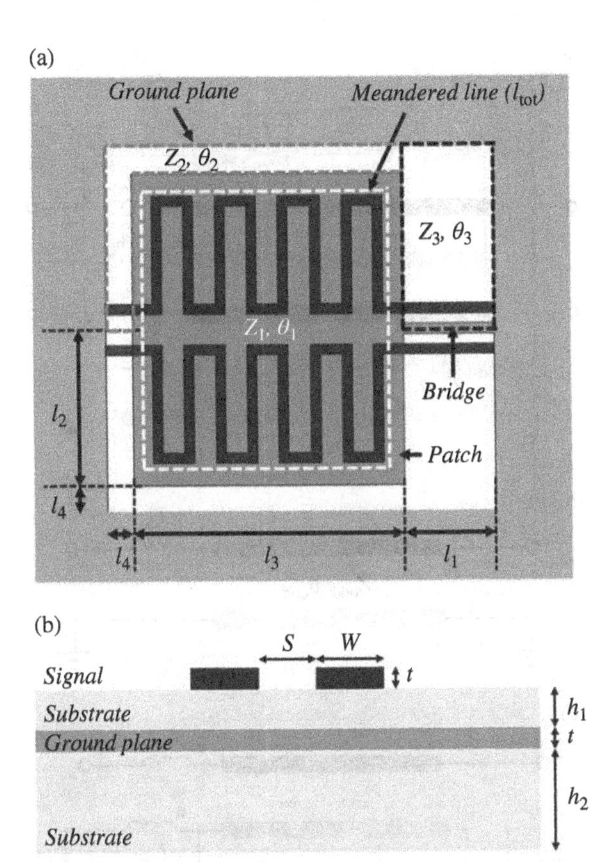

Figure 2.14 Topology (a) and cross section (b) of the C-shaped patterned ground structure common-mode stopband filter. Reprinted with permission from Ref. [11]; copyright 2014 IEEE.

microstrip lines located on top of the patch, and their coupling is very weak, provided they are distant enough.

The circuit schematic of this structure is depicted in Figure 2.15(a). Distributed components are needed since the C-shaped slot is not electrically small. The patch and the meandered lines on top of it are modeled as a coupled transmission line with single-ended characteristic impedance Z_1 and corresponding electrical length θ_1. The C-shaped slot can be considered as four slotlines: two of them with impedance and electrical length Z_2 and θ_2, respectively, and the other two with impedance and electrical length Z_3 and θ_3, respectively. Note that the two different slotlines are in series connection and shunt between the patch and ground plane, as shown in Figure 2.15(a). The bridge is described by a partial inductance L_b. Similarly, the signal line sections on top of the

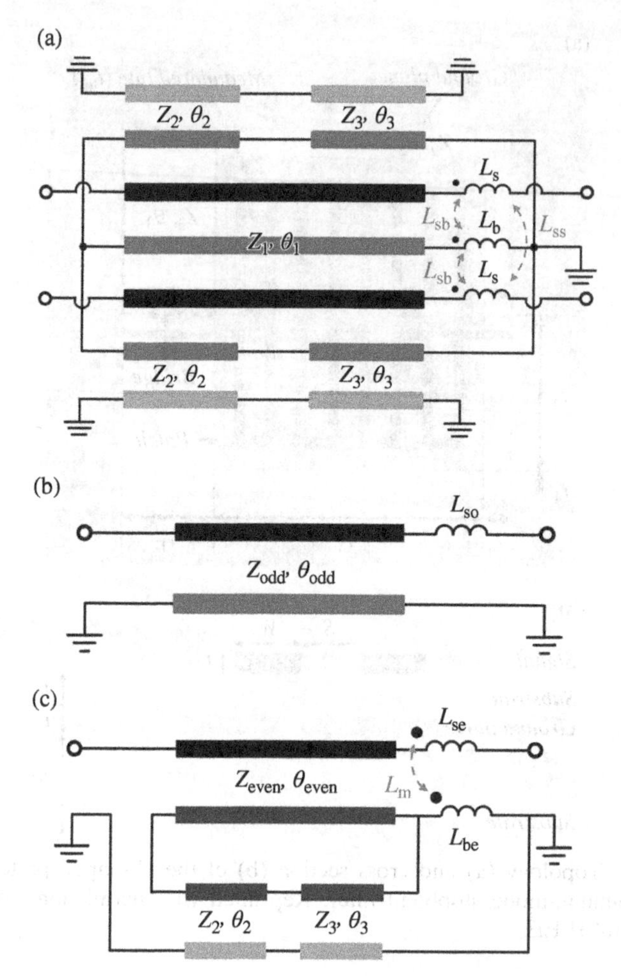

Figure 2.15 (a) Equivalent circuit model of the common-mode filter of Figure 2.14. (b) Differential-mode circuit. (c) Common-mode circuit. Reprinted with permission from Ref. [11]; copyright 2014 IEEE.

bridge are also modeled as partial inductances L_s due to the lack of ground metallization below them. It is worth noting that mutual inductances between the signal lines and between the signal line and bridge are also considered in Ref. [11]. Such mutual inductances are denoted as L_{ss} and L_{sb}.

The circuit models under differential- and common-mode excitation are depicted in Figures 2.15(b) and (c), respectively, where

$$L_{so} = L_s - L_{ss} \tag{2.5a}$$

$$L_{se} = L_s + L_{ss} - L_{sb} \tag{2.5b}$$

$$L_{be} = 2L_b - L_{sb} \tag{2.5c}$$

$$L_m = L_{sb} \tag{2.5d}$$

For the differential mode, the structure is all-pass, according to the circuit model. Conversely, the common mode is rejected in a certain frequency band due to the interaction of the signal lines, the slotlines, and the bridge. Note that this rejection mechanism is different than those of the previous common-mode stopband filters, where the common mode is rejected by a resonance-associated phenomenon.

The common-mode frequency response can be inferred from the circuit model of Figure 2.15(c). Transmission zeros appear when the transfer impedance of the impedance matrix equals zero, that is, $Z_{21} = 0$. Consequently, an analytical formula for common-mode transmission zeros is obtained by solving the network shown in Figure 2.15(c) [11]:

$$\frac{Z_{even}}{\sin\theta_{even}} = \frac{\omega(L_{be} - L_m)}{A - \omega L_{be} B} \tag{2.6}$$

with

$$A = \cos\theta_2 \cos\theta_3 - \frac{Z_3}{Z_2}\sin\theta_2 \sin\theta_3 \tag{2.7a}$$

$$B = \frac{\sin\theta_2 \cos\theta_3}{Z_2} - \frac{\cos\theta_2 \sin\theta_3}{Z_3} \tag{2.7b}$$

Derivation of (2.6) is cumbersome and the details, provided in Ref. [11], are not reproduced here. A systematic synthesis method for common-mode stopband filters based on these C-shaped slot structures, with transmission zeros for that mode at desired frequencies, is detailed in Ref. [11]. The method makes use of expression (2.6) and provides the geometry of the structure. The validity of the model of Figure 2.15(c) is demonstrated by comparison of the circuit response to the electromagnetic simulation and measured response of a designed common-mode stopband filter reported in Ref. [11]. Specifically, the transmission zeros of that filter were set to 2.6 GHz and 6.6 GHz.

The photograph of the designed filter (fabricated on an *FR4* substrate with dielectric constant $\varepsilon_r = 4.3$, loss tangent $\tan\delta = 0.02$, and thicknesses $h_1 = 0.1$ mm and $h_2 = 0.9$ mm; see Figure 2.14) and the frequency response are depicted in Figure 2.16, where it can be appreciated the

(a)

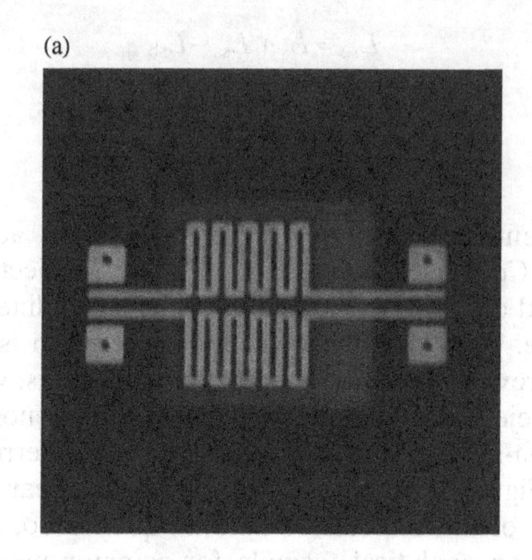

(b)

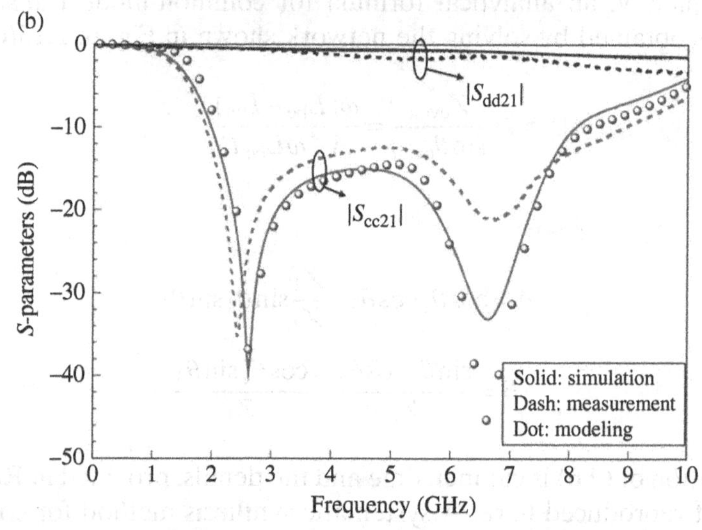

Figure 2.16 Photograph (a) and frequency response (b) of the C-shaped patterned ground plane common-mode filter. Geometrical parameters are $W = 0.18$ mm, $S = 0.36$ mm, $t = 0.035$ mm, $b = 0.1$ mm, $l_1 = 1.5$ mm, $l_2 = 2.3$ mm, $l_3 = 3.67$ mm, $l_4 = 0.5$ mm, and $l_{tot} = 26$ mm. Circuit parameters are $Z_{even} = 48.48\ \Omega$, $Z_2 = 97.7\ \Omega$, $Z_3 = 130.8\ \Omega$, $L_{be} = 1.95$ nH, $L_m = 0.51$ nH, $L_{se} = 0.7$ nH, $t_{deven} = 0.108$ ns, $t_{d2} = 0.033$ ns, and $t_{d3} = 0.012$ ns, where t_{deven}, t_{d2}, and t_{d3} denote time delays of transmission lines (i.e., $\theta_{even} = \beta_{even} \cdot l_{even} = \omega \cdot t_{deven}$, $\theta_2 = \beta_2 \cdot l_2 = \omega \cdot t_{d2}$, and $\theta_3 = \beta_3 \cdot l_3 = \omega \cdot t_{d3}$). Reprinted with permission from Ref. [11]; copyright 2014 IEEE.

good agreement between the circuit simulation (model), the electromagnetic simulation, and the measurement. It is remarkable that the filter is transparent for the differential mode, with differential-mode insertion loss lower than 3 dB from DC up to 9 GHz. However, for the common mode, a −10 dB fractional bandwidth of 130% is obtained, thanks to the two transmission zeros for that mode. It is also worth mentioning that dimensions are as small as $5.6 \times 5.65\ \text{mm}^2$ (i.e., $0.04\,\lambda^2$, where λ is the wavelength of the stopband central frequency).

2.4 COMMON-MODE SUPPRESSION FILTERS BASED ON ELECTROMAGNETIC BANDGAPS (EBGs)

EBGs are periodic structures able to inhibit signal propagation at certain frequencies as consequence of the Bragg effect, derived from periodicity [2–4]. The Bragg frequency, that is, the frequency of maximum attenuation within the first rejection band, is given by

$$f_{\max} = \frac{v_{\mathrm{p}}}{2l} \tag{2.8}$$

where l is the period and v_{p} is the phase velocity. Expression (2.8), known as the Bragg condition, indicates that the first stopband appears at the frequency satisfying that the period is half the wavelength (further bandgaps may appear at harmonic frequencies, depending on the harmonic content of the periodic function describing the periodicity).

In planar transmission lines, EBG-based reflectors can be implemented by periodically modulating the transverse dimensions (or characteristic impedance) of the line. In this case, the Bragg condition can be rewritten as

$$f_{\max} = \frac{c}{2l\sqrt{\varepsilon_{\mathrm{eff}}}} \tag{2.9}$$

where c is the speed of light in vacuum and $\varepsilon_{\mathrm{eff}}$ is an averaged effective dielectric constant. According to the coupled-mode theory [4, 54, 55], the reflection characteristics of the nonuniform periodic transmission line are actually dictated by the coupling coefficient between the forward and backward traveling waves associated to the operation mode. This coupling coefficient is also a periodic function given by

$$K(z) = \frac{1}{2Z_{\mathrm{o}}(z)}\frac{dZ_{\mathrm{o}}(z)}{dz} \tag{2.10}$$

where $Z_{\mathrm{o}}(z)$ is the point-to-point characteristic impedance of the line and z indicates the position along the line. Specifically, the coupled-mode theory determines that the rejection bands are given by the harmonic content of the coupling coefficient. Particularly, if the coupling coefficient is a sinusoidal function, only a bandgap centered at the Bragg frequency is expected.

Let us denote by K_n the weighting coefficients of the Fourier expansion of $K(z)$, each one giving a rejection band that can be characterized by the order nth. The maximum attenuation and bandwidth (delimited by the first reflection zeros around the maximum reflectivity) for each stopband are approximately given by [4, 55]

$$|S_{21}|_{\min,n} = \operatorname{sech}(|K_n| \cdot L) \tag{2.11}$$

$$BW_n = \frac{c|K_n|}{\pi\sqrt{\varepsilon_{\text{eff}}}}\sqrt{1+\left(\frac{\pi}{|K_n|\cdot L}\right)^2} \tag{2.12}$$

where L is the length of the structure (i.e., $L = l \cdot m$, m being the number of cells), and the averaged effective dielectric constant is calculated according to [4, 55]

$$\varepsilon_{\text{eff}} = \left(\frac{1}{l}\int_0^l \sqrt{\varepsilon_{\text{eff}}(z)} \cdot dz\right)^2 \tag{2.13}$$

where $\varepsilon_{\text{eff}}(z)$ is the point-to-point effective dielectric constant. Using expressions (2.9)–(2.13), it is possible to design periodic transmission lines with specific value of central stopband frequency (or frequencies, if $K(z)$ has harmonic content), rejection level, and bandwidth [4, 56].

2.4.1 Common-Mode Filter Based on Nonuniform Coupled Lines

In this subsection, it is shown that it is possible to suppress the common mode in balanced lines by means of EBGs, keeping the ground plane unaltered. To this end, the common-mode characteristic impedance of the line is periodically modulated, and, at the same time, the differential-mode impedance is kept uniform along the line (and equal to the reference impedance of the differential ports) [17]. Thus, the differential line acts as a reflector for the common mode, opening a bandgap in the vicinity of the Bragg frequency for that mode. However, as long as the differential-mode impedance is uniform along the line and equal to the reference impedance of the differential ports, the line is transparent for the differential mode.

Through (2.9)–(2.13), it is possible to generate an even-mode characteristic impedance profile useful to obtain a bandgap (or bandgaps) for the common mode in differential lines. The degrees of freedom in differential microstrip lines are enough to simultaneously achieve line impedance modulation for the common mode and uniform characteristic impedance for the differential mode. This is essential to achieve common-mode noise suppression and keep the differential signals unaltered. However, since the differential-mode impedance is always smaller than the common-mode impedance, and the line must be matched to the ports for the differential mode (i.e., the differential-mode impedance must typically be $Z_{oo}(z) = 50\,\Omega$), it follows that the common-mode impedance must be periodically modulated, satisfying $Z_{oe}(z) \geq Z_{oo}(z) = 50\,\Omega$. If the coupling coefficient is sinusoidal, and the common-mode impedance is forced to be $50\,\Omega$ at the extremes of the EBG structure (i.e., $Z_{oe}(0) = Z_{oe}(L) = 50\,\Omega$), integration of (2.10) with these boundary conditions gives

$$Z_{oe}(z) = Z_{oe}(0) \cdot e^{-\frac{2|K_1| \cdot l}{\pi}\left[\cos\left(\frac{2\pi}{l}z\right) - 1\right]} \tag{2.14}$$

Let us consider the implementation of a common-mode filter with maximum rejection level of 19 dB in the vicinity of $f_{max} = 2.4$ GHz, by considering the *Rogers RO3010* substrate with dielectric constant $\varepsilon_r = 10.2$ and thickness $h = 1.27$ mm. This frequency gives a period (using 2.9) of $l = 2.38$ cm. Considering four cells ($m = 4$), $L = 9.53$ cm and, using (2.11), the weighting factor is found to be $K_1 = 0.305\ \text{cm}^{-1}$. Applying (2.12), the common-mode rejection bandwidth is 1.63 GHz (i.e., 68%). From (2.14), it follows that the common-mode characteristic impedance varies between 50 and $131.5\,\Omega$, which are implementable values. The transverse geometry of the differential line must give a common-mode impedance given by (2.14) and, simultaneously, a differential-mode impedance of $Z_{oo}(z) = 50\,\Omega$ along the line. Such geometry can be easily inferred by means of any commercial transmission-line calculator. Since the common-mode impedance is a continuously varying function, in practice the transverse geometry is calculated at a given number of discrete points along the period (40 points in the reported example), and then the corresponding extremes of the strips are connected by straight lines, resulting actually in a linear piecewise function.

The photograph of the fabricated line, as well as the differential- and common-mode insertion and return loss, is depicted in Figure 2.17. The maximum simulated rejection level for the common mode, obtained at the design frequency f_{max}, that is, 19 dB, is in agreement with the design

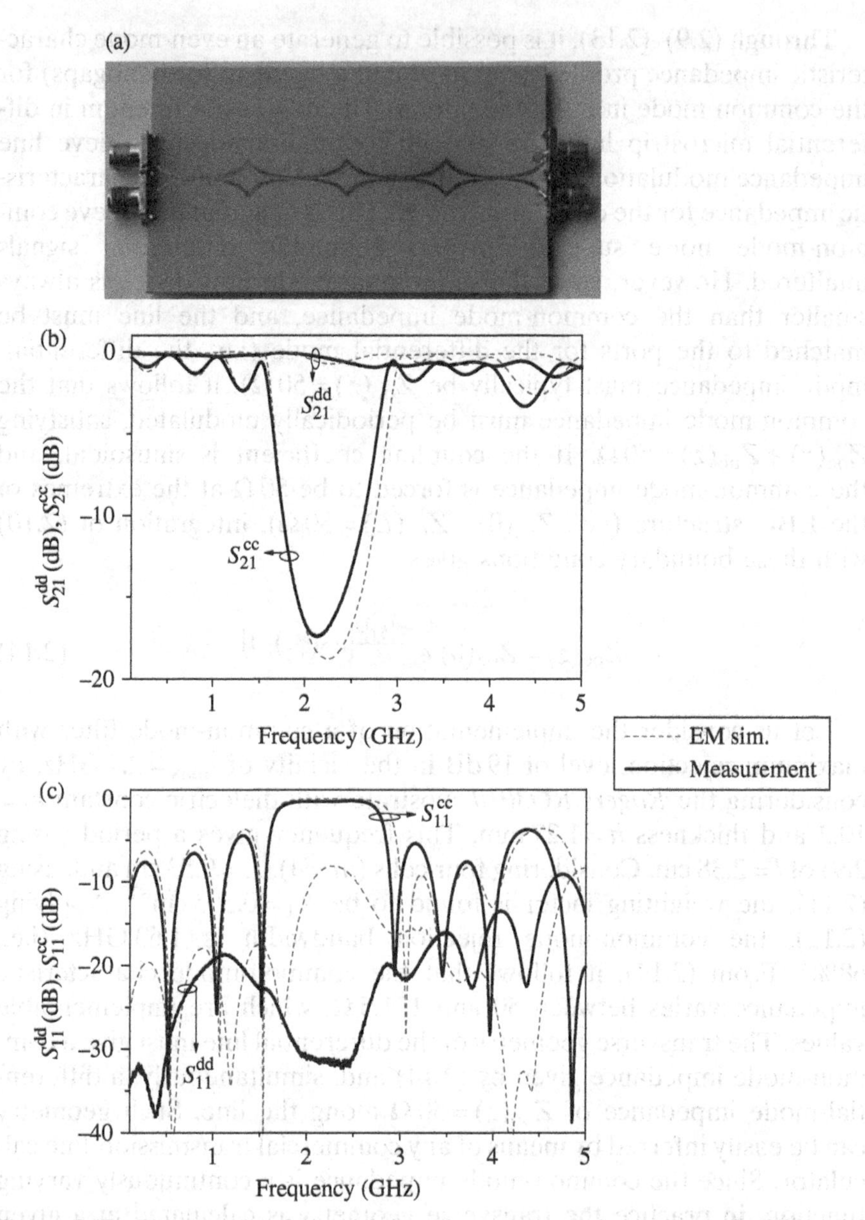

Figure 2.17 Photograph of the fabricated common-mode suppressed differential transmission line (a) and differential- and common-mode insertion (b) and return (c) loss. Line width W and separation S at the planes of maximum (131.5 Ω) and minimum (50 Ω) value of Z_{oe} are $W = 0.16$ mm, $S = 0.32$ mm, $W = 1.06$ mm, and $S = 6.4$ mm, respectively. Reprinted with permission from Ref. [17]; copyright 2015 IEEE.

value given previously, and the fractional bandwidth (62.5%) is reasonably predicted by expression (2.12). The designed line is roughly transparent to the differential mode, as desired (the differential-mode insertion loss is better than 0.7 dB in the whole common-mode rejection band). Some level of common-mode rejection at the first harmonic frequency can also be appreciated. This can be attributed to the fact that a perfect sinusoidal coupling coefficient is never obtained in practice. Nevertheless, this does not significantly affect the rejection level at the design frequency.

According to (2.11) and (2.12), once the maximum rejection level is set to a certain value, bandwidth can be enhanced by increasing K_n, at the expense of reducing L in order to preserve the product $|K_n|\cdot L$. Reduction of L is in favor of miniaturization; however, the increase of $|K_n|$ is limited by the maximum implementable characteristic impedance for the common mode. This means that bandwidth enhancement is limited by using these simple EBG structures. Bandwidth can be improved by cascading more than one EBG structure with different periods, by implementing impedance profiles (for the common mode) corresponding to the superposition of two or more coupling coefficients (multiple tuned structures [57]), or by using chirping functions [58]. To illustrate these possibilities, a common-mode suppressed differential line with two cascaded EBG structures (with $f_{max,1} = 2$ GHz and $f_{max,2} = 2.8$ GHz) is reported [17]. Both EBGs have been designed to exhibit a maximum attenuation of 13 dB with bandwidths of 66% ($f_{max,1}$) and 93.6% ($f_{max,2}$) considering three cells. Using the previous equations, the parameters of both EBG structures are determined, giving $l_1 = 2.85$ cm, $m_1 = 3$, $K_1 = 0.254$ cm^{-1}, $l_2 = 2.04$ cm, $m_2 = 3$, and $K_2 = 0.356$ cm^{-1}, where the sub-index denotes the EBG, rather than the index of the weighting coefficient of the series expansion of the coupling coefficient. The photograph of the EBG-based differential lines is depicted in Figure 2.18(a), whereas the electromagnetic simulation and measurement are compared in Figure 2.18(b) and (c). In this case, the differential insertion loss is better than 0.9 dB in the common-mode rejection band. The common-mode insertion loss (S_{21}^{cc}) of a designed single-tuned EBG-based line with $m = m_1 + m_2 = 6$ and comparable rejection level (30 dB) at 2.4 GHz is also included for comparison purposes. It can be appreciated that bandwidth can be notably improved by using two cascaded EBGs, without a penalty in device size (both are comparable). Common-mode suppression filters based on chirped EBGs can be found in Ref. [18].

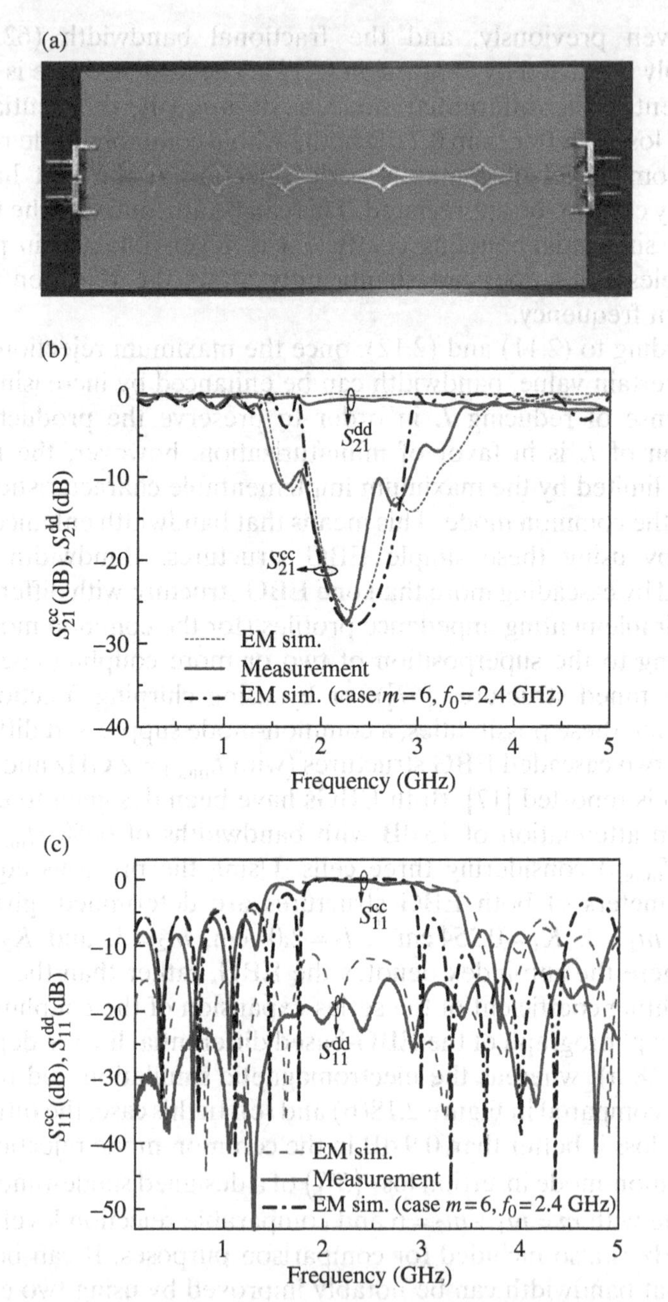

Figure 2.18 Photograph (a), differential- and common-mode insertion (b) and return (c) loss of the 6-cell nonuniform transmission line based on EBGs with different periods. The considered substrate is the *Rogers RO3010* with dielectric constant $\varepsilon_r = 10.2$ and thickness $h = 1.27$ mm. Reprinted with permission from Ref. [17]; copyright 2015 IEEE.

2.4.2 Common-Mode Filter Based on Uniplanar Compact Photonic Bandgap (UC-PBG) Structure

In this section, EBG-based common-mode suppressed differential lines with ground plane etching are presented. Specifically, the structures (first presented in Ref. [16]) are based on the uniplanar compact photonic bandgap (UC-PBG) ground plane concept [59]. As will be shown, these common-mode filters exhibit wide stopband bandwidths for the common mode, with good signal integrity for the differential mode, but at the expense of etching the ground plane.

The UC-PBG is a two-dimensional array of complementary Jerusalem crosses that behaves as a parallel LC tank, thereby presenting a high impedance path for the currents traveling through the structure. The capacitance and inductance are provided by the thin gap between the adjacent metallic patches and the narrow branches connecting each unit cell, respectively. This structure exhibits a significant slow-wave effect [59] that substantially increases the phase constant in the propagation bands. The result is that compact EBG stopband structures can be achieved with this approach. Note that the Bragg condition (2.8) in these UC-PBG structures must be expressed with the reduced phase velocity that results as consequence of the slow-wave effect, not with the phase velocity corresponding to the line with solid ground plane.

The structure reported in Ref. [16], where 10 unit cells are etched in the ground plane, is depicted in Figure 2.19(a), whereas the frequency response is shown in Figure 2.19(b). The design procedure is based on the dispersion analysis, using eigenmode solvers, rather than on the coupled-mode theory (difficult to apply due to the lack of knowledge of the common-mode impedance variation with position). It can be appreciated that differential-mode signals are insensitive to the presence of the UC-PBG structure up to roughly 8.5 GHz, whereas the common mode is rejected with a suppression level below –20 dB between 3.3 GHz and 7 GHz, representing a fractional bandwidth of 70%.

Other EBG-based common-mode stopband filters that use three metal levels have been presented in Refs. [14, 15], but these filters are able to suppress specific frequencies, as compared with the realizations presented so far.

2.5 OTHER APPROACHES FOR COMMON-MODE SUPPRESSION

In the previous approaches, common-mode suppression is achieved by considering structures with only two metal levels. In this section, other

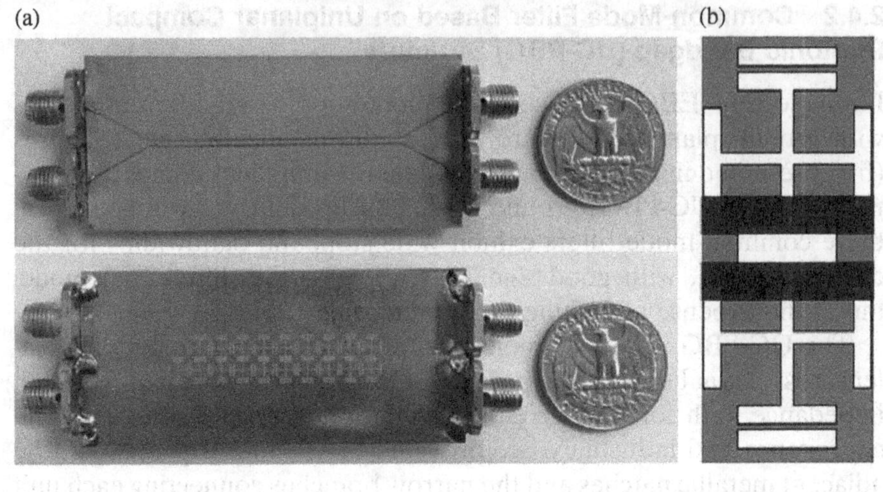

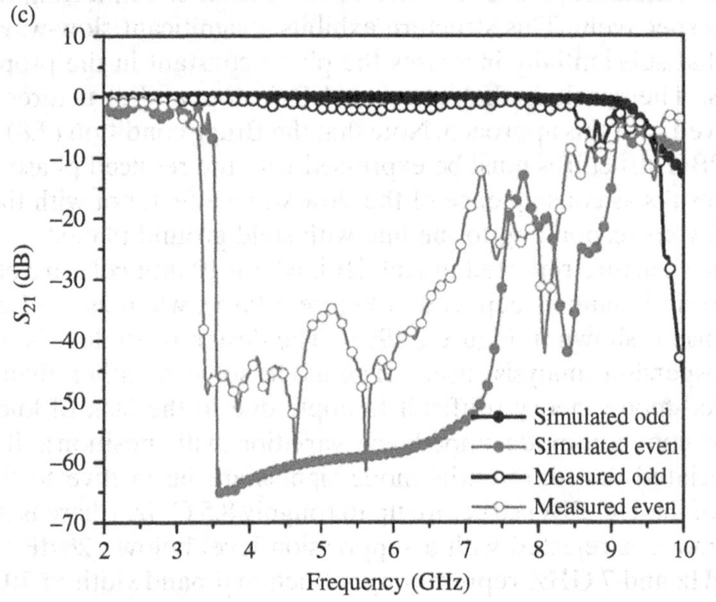

Figure 2.19 Photograph of the UC-PBG common-mode stopband filter (a), zoom view of the unit cell (b), and frequency response (c). The structure was fabricated on the *Rogers RT/duroid 6010.2* substrate with thickness $h = 635\ \mu\text{m}$, dielectric constant $\varepsilon_r = 10.2$, and loss tangent $\tan\delta = 0.0023$. Strip width and spacing of the lines is $W = 0.381$ mm and $S = 0.635$ mm, respectively, giving differential- and common-mode impedances of 51 Ω and 70 Ω, respectively. Reprinted with permission from Ref. [16]; copyright 2014 IEEE.

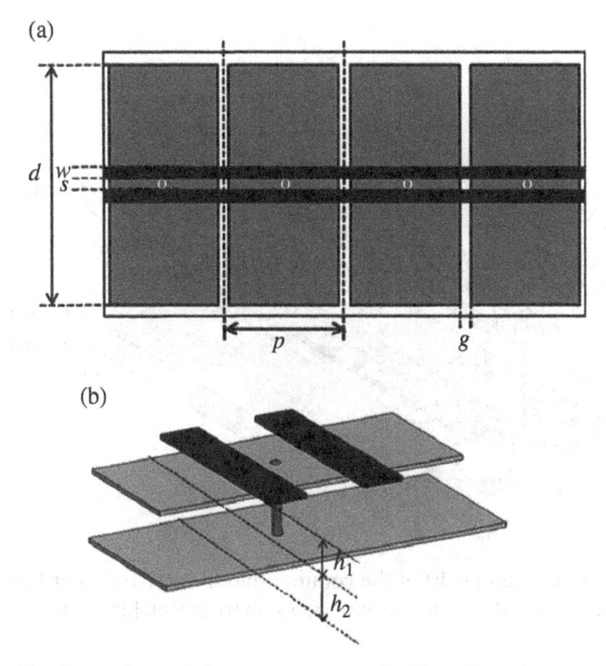

Figure 2.20 Configuration of the common-mode filter based on metallic patches ground connected through vias: (a) top view and (b) 3D view. Reprinted with permission from Ref. [23]; copyright 2010 IEEE.

common-mode stopband filters based on multilayer structures are succinctly reviewed, with particular emphasis on the common-mode filters reported in Refs. [23] and [25]. In Ref. [23], the structure consists of a pair of differential lines on top of metallic patches connected to the ground through metallic vias (Figure 2.20). The circuit model of the unit cell is identical to the one depicted in Figure 2.6(a), describing the unit cell of the CSRR- or DS-CSRR-based common-mode stopband filters discussed in Section 2.3.2. Thus, the structure of Figure 2.20 can be considered to be a negative permittivity artificial line for the common mode, similarly to CSRR-loaded differential lines. In reference to Figure 2.6 (a), C is the capacitance between the line strip and the patch of the central metal layer, whereas the inductance L_c and the capacitance C_c account for the via and the broadside capacitance between the patch and the ground plane, respectively. The common-mode suppression principle of the structure of Figure 2.20 is identical to the one detailed in Section 2.3.2 in reference to CSRR and DS-CSRR common-mode filters. Ground plane etching is not required, but at least three metal levels are needed.

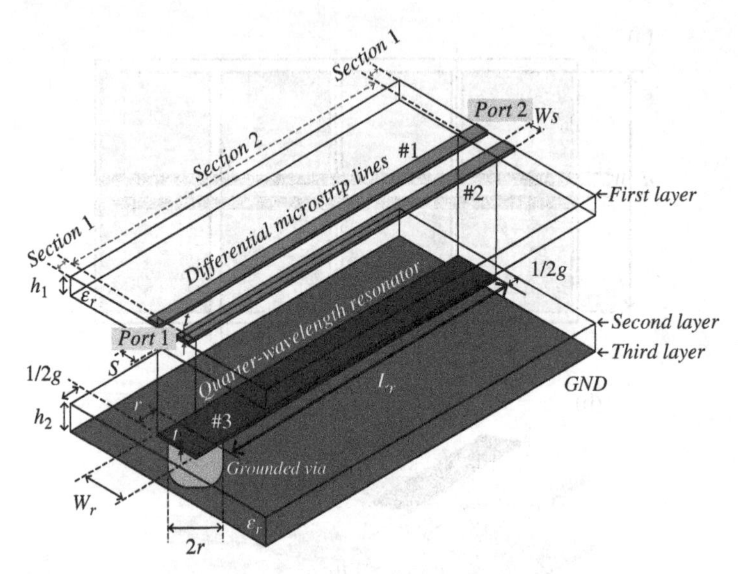

Figure 2.21 3D view (unit cell) of the common-mode stopband filter based on quarter wavelength open stubs. Reprinted with permission from Ref. [25]; copyright 2014 IEEE.

In Ref. [25], the authors present a common-mode stopband filter that uses three metal layers. The differential pair strips are etched in the top layer, whereas in the central layer a quarter-wavelength open stub, connected to the ground plane (third layer) through a metallic via, is defined and located beneath the differential lines (see the unit cell configuration in Figure 2.21). The structure is described by the distributed circuit depicted in Figure 2.22(a). For differential-mode excitation, the symmetry plane is an electric wall, and the effects of the open stub are canceled. Thus, the structure is transparent for that mode (see the equivalent circuit in Figure 2.22b). However, for the common mode (Figure 2.22c), the pair of lines is efficiently coupled to the open stub, resulting in a stopband for that mode in the vicinity of the design frequency (the one giving a quarter wavelength stub). By cascading three unit cells, it is demonstrated in Ref. [25] that a stopband bandwidth from 3.7 GHz to 7.3 GHz with common-mode rejection better than 20 dB can be achieved, preserving good signal integrity for the differential signals.

The solution proposed in Ref. [24] is conceptually similar to the one proposed in Ref. [11] and discussed in Section 2.3.4. In Ref. [26], common-mode filters based on CSRRs similar to those discussed in Section 2.3.2 (and reported in Ref. [8]), are presented. The common-mode filters of Ref. [12] are indeed those reported in Ref. [5], based on dumbbell-shaped slots (presented in Section 2.3.1). However, to reduce the size and improve the quality of the differential signals, a

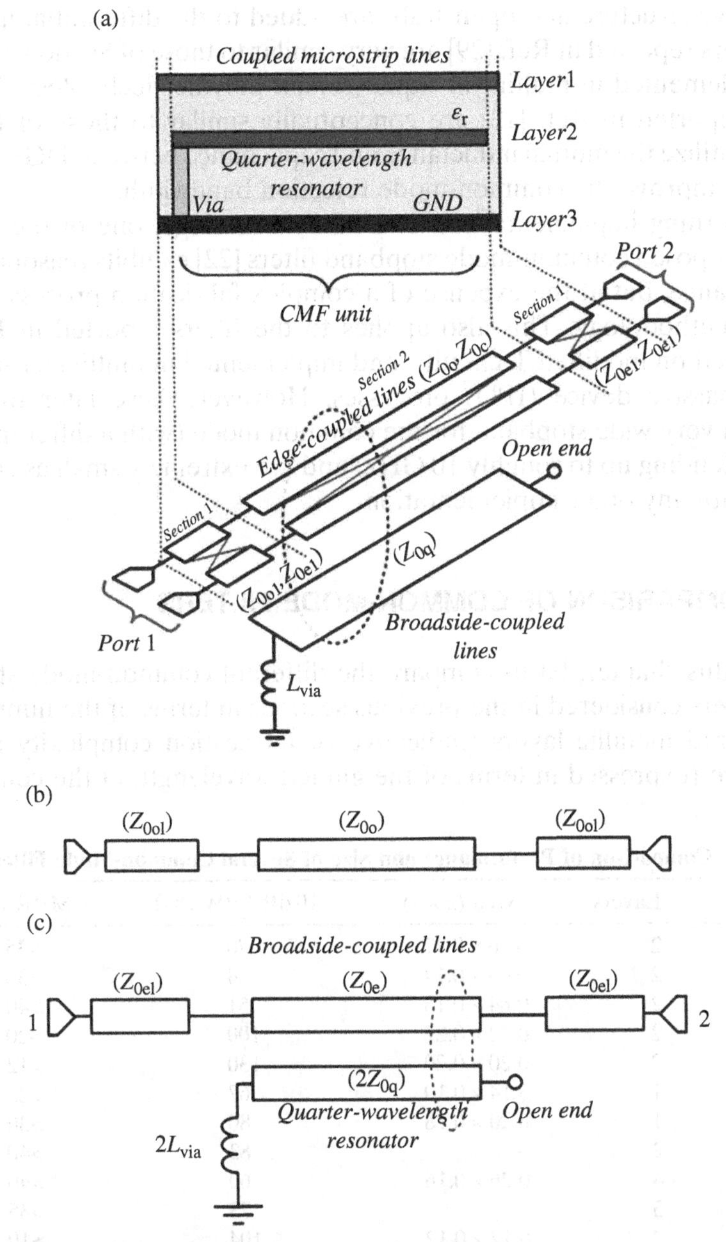

Figure 2.22 (a) Equivalent circuit model (unit cell) of the common-mode filter of Figure 2.21. Equivalent circuit models of (b) differential mode and (c) common mode. Reprinted with permission from Ref. [25]; copyright 2014 IEEE.

slow-wave structure and open stubs are added to the differential lines. The filters reported in Ref. [29] are very similar to those of Section 2.3.3, but implemented in multilayer liquid crystal polymer technology. The filters reported in Ref. [13] are conceptually similar to those of Ref. [5], but utilize the mutual inductance and capacitance between DGS patterns to improve the common-mode rejection bandwidth.

Concerning implementations in LTCC technology, one of the formerly proposed common-mode stopband filters [22] exhibits reasonable performance, but at the expense of a complex fabrication process and design methodology. This also applies to the filters reported in Ref. [28], based on modified T-circuits, and implemented in multilayer integrated passive device (IPD) processes. However, these later filters exhibit a very wide stopband for the common mode (with a differential band extending up to roughly 10 GHz) and are extremely small as compared with any other implementation.

2.6 COMPARISON OF COMMON-MODE FILTERS

To end this chapter, let us compare the different common-mode stopband filters considered in the previous sections in terms of the number of required metallic layers (indicative of fabrication complexity and cost), size (expressed in terms of the guided wavelength at the central

Table 2.3 Comparison of Performance and Size of Several Common-Mode Filters

Reference	Layers	Area ($\lambda \times \lambda$)	−10 dB FBW (%)	CMRR (dB)
[5]	2	0.76 × 0.47	70	>35
[7]	2	0.43 × 0.14	54	>30
[8]	2	0.64 × 0.13	51	>40
[9]	2	0.32 × 0.22	100	>20
[11]	2	0.20 × 0.20	130	>12
[17]	1	3.14 × 0.19	67	>25
[18]	1	6.20 × 0.28	80	>30
[16]	2	—	83	>40
[23]	4	0.26 × 0.16	60	>40
[25]	3	—	73	>35
[24]	4	0.12 × 0.12	104	>10
[12]	2	—	95	>25
[22]	LTCC	—	150	>25
[6]	2	0.44 × 0.44	87	>15
[28]	3	0.017 × 0.016	100	>12
[29]	3	0.43 × 0.19	27	>25
[13]	2	—	140	>25

frequency, λ), common-mode rejection fractional bandwidth, and maximum CMRR. These data are summarized in Table 2.3 (further details are given in Ref. [60]).

2.7 SUMMARY

In this chapter, the main strategies for common-mode suppression in differential-mode transmission lines have been reviewed. The main focus has been on defected ground structures, where it has been shown that, through an appropriate design, the differential signals remain unaltered (i.e., signal integrity is preserved for the differential mode), whereas the common mode can be efficiently rejected, at least in the frequency region of interest. Alternatively, it has also been shown that periodic structures insensitive to the differential mode can act as efficient EBGs for the common mode, providing strong attenuation for that mode. The chapter ends with a succinct review of other approaches for common-mode suppression, mainly based on multilayer structures, and with a comparison between the different common-mode stopband filters considered.

APPENDIX 2.A: DISPERSION RELATION FOR COMMON-MODE REJECTION FILTERS WITH COUPLED CSRRs OR DS-CSRRs

The equivalent circuit model of the CSRR- (or DS-CSRR)-loaded differential microstrip line for the common mode (Figure 2.6b) is a multiport network that can be generalized as indicated in Figure 2.23. To obtain the dispersion relation for this generalized circuit, we can appeal to the multiconductor line theory [61] in order to obtain the eigenmodes of the structure and hence the propagation constants. Let us denote V_{L1}, V_{L2}, I_{L1}, and I_{L2} as the voltages and currents at the ports (1 and 2) of the left-hand side (subscript L) of the unit cell and V_{R1}, V_{R2}, I_{R1} and I_{R2} as the variables at the right-hand side ports. The variables at both sides of

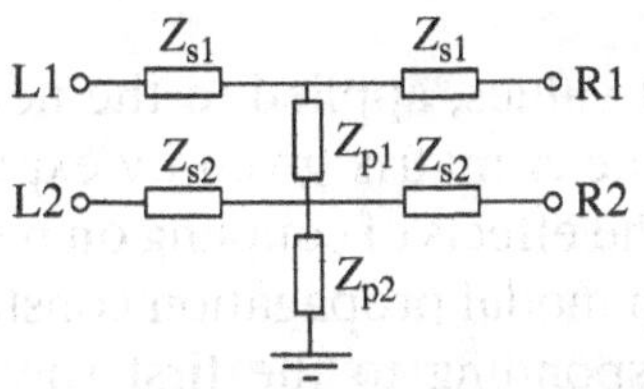

L1 Z_{s1} Z_{s1} R1
L2 Z_{s2} Z_{p1} Z_{s2} R2
Z_{p2}

Figure 2.23 Generalized impedance model of the multiport network of Figure 2.6(b).

the network are linked through a generalized order-4 transfer matrix according to

$$\begin{pmatrix} \boldsymbol{V}_{\mathbf{L}} \\ \boldsymbol{I}_{\mathbf{L}} \end{pmatrix} = \begin{pmatrix} \boldsymbol{A} & \boldsymbol{B} \\ \boldsymbol{C} & \boldsymbol{D} \end{pmatrix} \begin{pmatrix} \boldsymbol{V}_{\mathbf{R}} \\ \boldsymbol{I}_{\mathbf{R}} \end{pmatrix} \tag{2.15}$$

where $\boldsymbol{V}_{\mathbf{L}}$, $\boldsymbol{I}_{\mathbf{L}}$, $\boldsymbol{V}_{\mathbf{R}}$, and $\boldsymbol{I}_{\mathbf{R}}$ are column vectors composed of the pair of port variables and $\boldsymbol{A}$, $\boldsymbol{B}$, $\boldsymbol{C}$, and $\boldsymbol{D}$ are order-2 matrices. The dispersion relation is obtained from the eigenmodes of the system (2.15), that is,

$$\det \begin{pmatrix} \boldsymbol{A} - e^{\gamma l} \cdot \boldsymbol{I} & \boldsymbol{B} \\ \boldsymbol{C} & \boldsymbol{D} - e^{\gamma l} \cdot \boldsymbol{I} \end{pmatrix} = 0 \tag{2.16}$$

where $\boldsymbol{I}$ is the identity matrix, the phase-shift factor $e^{\gamma l}$ is the eigenvalue, $\gamma = \alpha + j\beta$ is the complex propagation constant, and l is the unit cell length. For reciprocal, lossless, and symmetric networks, the eigenvalues can be simplified to the solutions of [62, 63]

$$\det(\boldsymbol{A} - \cosh(\gamma l) \cdot \boldsymbol{I}) = 0 \tag{2.17}$$

which gives (2.1), that is,

$$\cosh(\gamma l) = \frac{1}{2}\left(A_{11} + A_{22} \pm \sqrt{(A_{11} - A_{22})^2 + 4A_{12}A_{21}}\right) \tag{2.18}$$

The elements of the $\boldsymbol{A}$ matrix (inferred from the network of Figure 2.23) are

$$\boldsymbol{A} = \boldsymbol{D}^t = \begin{pmatrix} 1 + \dfrac{Z_{s1}}{Z_{p1}} & -\dfrac{Z_{s1}}{Z_{p1}} \\ -\dfrac{Z_{s2}}{Z_{p1}} & 1 + \dfrac{Z_{s2}}{Z_{p1}} + \dfrac{Z_{s2}}{Z_{p2}} \end{pmatrix} \tag{2.19}$$

The previous element values, applied to the network of Figure 2.6(b), give the elements of the $\boldsymbol{A}$ matrix given by expressions (2.2).

To gain insight on the effects of coupling on bandwidth enhancement, let us evaluate the two modal propagation constants by considering the element values corresponding to the first row of Table 2.1 [8]. The results, shown in Figure 2.24, indicate that, for each mode, there are

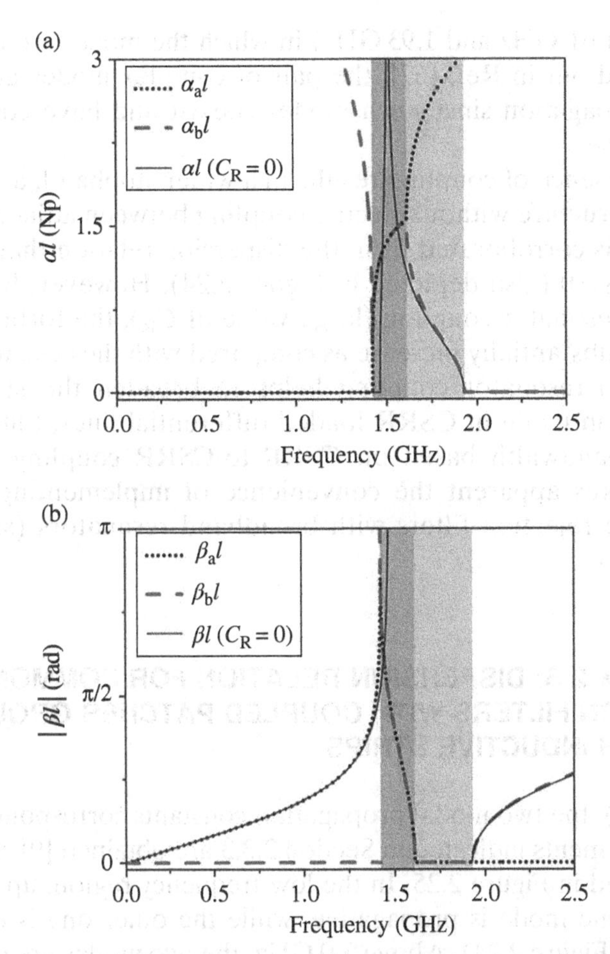

Figure 2.24 (a) Attenuation constant and (b) magnitude of the phase constant of γl corresponding to the two modes of the network of Figure 2.6(b). If inter-resonator coupling is canceled ($C_R = 0$), the two modes degenerate in a single mode, also depicted. The evanescent and complex mode regions are highlighted in light and dark gray, respectively. Reprinted with permission from Ref. [8]; copyright 2011 IEEE.

regions where γl is purely real (evanescent mode), purely imaginary (propagating mode), or complex (complex mode). Complex modes appear as conjugate pairs [64–67], and this is exactly the case in the region comprised between 1.42 GHz and 1.61 GHz, that is, $\alpha_a = \alpha_b$ and $\beta_a = -\beta_b$, where $\gamma = \alpha + j\beta$ and the subscripts a and b are used to differentiate the two modes. The forbidden band includes the region where complex modes are present (these modes do not carry net power [64–67]), plus an additional region where $\beta_a = \beta_b = 0$, $\alpha_a \neq 0$, and $\alpha_b \neq 0$

(between 1.61 GHz and 1.93 GHz) in which the modes are evanescent. As pointed out in Ref. [67], the pair of complex modes also inhibits signal propagation since such modes coexist and have contra directional phase.

The presence of coupling results in a wider stopband, as compared with the structure without electric coupling between adjacent resonators. This is corroborated from the dispersion relation that results by forcing $C_R = 0$ (also depicted in Figure 2.24). However, by magnifying inter-resonator coupling (large value of C_R), the forbidden band does not substantially increase as compared with the case with 0.1 pF. Thus, inter-resonator coupling helps to broaden the stopband of the common mode in CSRR-loaded differential lines, but the sensitivity in bandwidth based on CSRR-to-CSRR coupling is limited, which makes apparent the convenience of implementing the common-mode rejection filters with broadband resonators (such as the DS-CSRR).

APPENDIX 2.B: DISPERSION RELATION FOR COMMON-MODE REJECTION FILTERS WITH COUPLED PATCHES GROUNDED THROUGH INDUCTIVE STRIPS

Using (2.4), the two modal propagation constants corresponding to the lumped elements indicated in Section 2.3.3 are obtained [9]. The results are depicted in Figure 2.25. In the low-frequency region, up to roughly 1.8 GHz, one mode is propagative, while the other one is evanescent (similar to Figure 2.24). Above 3.0 GHz, the two modes are purely evanescent. Finally, in the middle interval (between 1.8 and 3 GHz), the four-port equivalent circuit predicts the existence of complex modes. Note the wide frequency band corresponding to complex modes, as compared with the one of Figure 2.24. All the frequencies above 1.8 GHz can be considered as part of a forbidden region, since the model does not predict the presence of propagative modes above that frequency. Therefore, the low-pass filtering functionality of these structures for the common mode is verified by the dispersion relation.[14] Note also that the cutoff frequency of Figure 2.13 is roughly predicted by the dispersion relation.

[14] Nevertheless, distributed effects, not included in this simplified circuit model, will modify this prediction at frequencies above the range of interest (S and C bands).

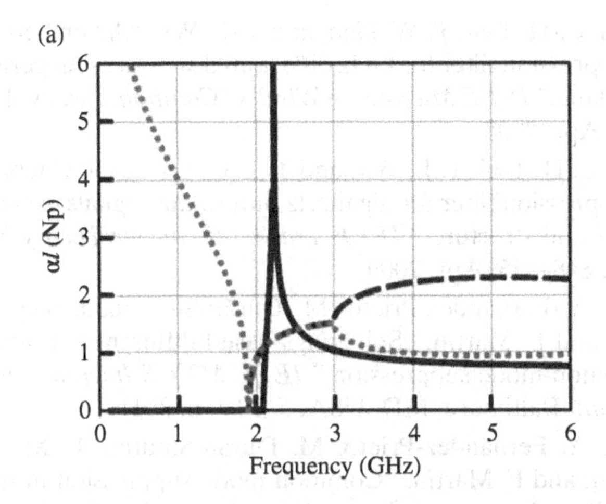

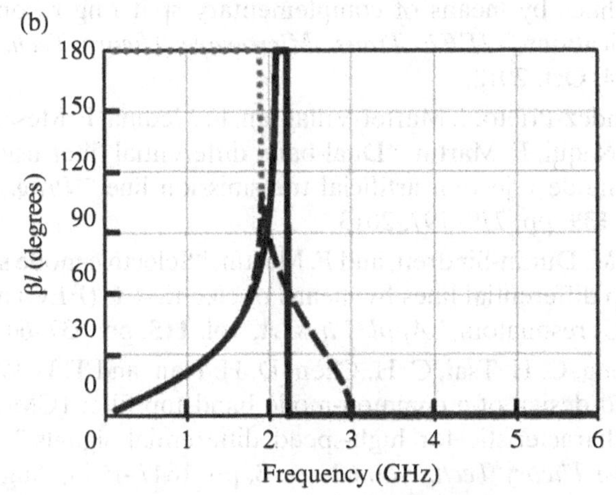

Figure 2.25 (a) Attenuation constant and (b) magnitude of the phase constant of γl corresponding to the two modes of the four-cell structure with unit cell given by Figure 2.11(b). Reprinted with permission from Ref. [9]; copyright 2013 PIER.

REFERENCES

1. C. R. Paul, "A comparison of the contributions of common-mode and differential-mode currents in radiated emissions," *IEEE Trans. Electromagn. Compat.*, vol. **31**, no. 2, pp. 189–193, May 1989.
2. E. Yablonovitch, "Photonic band gap structures," *J. Opt. Soc. Am. B*, **10**, pp. 283–295, 1993.
3. J. D. Joannopoulos, R. D. Meade, and J. N. Winn, *Photonic Crystals: Molding the Flow of Light*, Princeton University Press, Princeton, NJ, 1995.
4. F. Martín, *Artificial Transmission Lines for RF and Microwave Applications*, John Wiley & Sons, Inc., Hoboken, NJ, 2015.

5. W. T. Liu, C.-H. Tsai, T.-W. Han, and T.-L. Wu, "An embedded common-mode suppression filter for GHz differential signals using periodic defected ground plane," *IEEE Microwave Wireless Compon. Lett.*, vol. **18**, no. 4, pp. 248–250, Apr. 2008.
6. S. J. Wu, C. H. Tsai, T. L. Wu, and T. Itoh, "A novel wideband common-mode suppression filter for gigahertz differential signals using coupled patterned ground structure," *IEEE Trans. Microwave Theory Tech.*, vol. **57**, no. 4, pp. 848–855, Apr. 2009.
7. J. Naqui, A. Fernández-Prieto, M. Durán-Sindreu, J. Selga, F. Medina, F. Mesa, and F. Martín, "Split rings-based differential transmission lines with common-mode suppression," *IEEE MTT-S International Microwave Symposium*, Baltimore, MD, USA, 5–10 Jun. 2011.
8. J. Naqui, A. Fernández-Prieto, M. Durán-Sindreu, F. Mesa, J. Martel, F. Medina, and F. Martín, "Common mode suppression in microstrip differential lines by means of complementary split ring resonators: theory and applications," *IEEE Trans. Microwave Theory Tech.*, vol. **60**, pp. 3023–3034, Oct. 2012.
9. A. Fernandez-Prieto, J. Martel-Villagrán, F. Medina, F. Mesa, S. Qian, J.-S. Hong, J. Naqui, F. Martin, "Dual-band differential filter using broadband common-mode rejection artificial transmission line," *Prog. Electromagn. Res.*, vol. **139**, pp. 779–797, 2013.
10. J. Naqui, M. Durán-Sindreu, and F. Martín, "Selective mode suppression in microstrip differential lines by means of electric-LC (ELC) and magnetic-LC (MLC) resonators," *Appl. Phys. A*, vol. **115**, pp. 637–643, 2014.
11. T. W. Weng, C. H. Tsai, C. H. Chen, D. H. Han, and T. L. Wu, "Synthesis model and design of a common-mode bandstop filter (CM-BSF) with an all-pass characteristic for high-speed differential signals," *IEEE Trans. Microwave Theory Tech.*, vol. **62**, no. 8, pp. 1647–1656, Aug. 2014.
12. F. X. Yang, M. Tang, L. S. Wu, and J. F. Mao, "A novel wideband common-mode suppression filter for differential signal transmission," *2014 IEEE Electrical Design of Advanced Packaging & Systems Symposium* (EDAPS), Bangalore, Dec. 16, 2014, pp. 129–132.
13. Z. Zeng, Y. Yao, and Y. Zhuang, "A wideband common-mode suppression filter with compact-defected ground structure pattern," *IEEE Trans. Electromagn. Compt.*, vol. **57**, pp. 1277–1280, Oct. 2015.
14. F. de Paulis, L. Raimondo, Sam Connor, B. Archambeault, and A. Orlandi, "Design of a common mode filter by using planar electromagnetic bandgap structures," *IEEE Trans. Adv. Packag.*, vol. **33**, no. 4, pp. 994–1002, Nov. 2010.
15. F. de Paulis, L. Raimondo, S. Connor, B. Archambeault, and A. Orlandi, "Compact configuration for common mode filter design based on planar electromagnetic bandgap structures," *IEEE Trans. Electromagn. Compat.*, vol. **54**, no. 3, pp. 646–654, Jun. 2012.

16. J. H. Choi, P. W. C. Hon, and T. Itoh, "Dispersion analysis and design of planar electromagnetic bandgap ground plane for broadband common-mode suppression," *IEEE Microwave Wireless Compon. Lett.*, vol. **24**, no. 11, pp. 772–774, Nov. 2014.
17. P. Vélez, J. Bonache, and F. Martín, "Differential microstrip lines with common-mode suppression based on electromagnetic bandgaps (EBGs)," *IEEE Antennas Wireless Propag. Lett.*, vol. **14**, pp. 40–43, 2015.
18. P. Vélez, M. Valero, L. Su, J. Naqui, J. Mata-Contreras, J. Bonache, and F. Martín, "Enhancing common-mode suppression in microstrip differential lines by means of chirped electromagnetic bandgaps (EBGs)," *Microw. Opt. Technol. Lett.*, vol. **58**, pp. 328–332, Feb. 2016.
19. J. D. Gavenda, "Measured effectiveness of a toroid choke in reducing common-mode current," in *Proceedings of the IEEE International Electromagnetic Compatibility Symposium*, Denver, CO, May 23–25, 1989, pp. 208–210.
20. K. Yanagisawa, F. Zhang, T. Sato, and Y. Miura, "A new wideband common-mode noise filter consisting of Mn–Zn ferrite core and copper/polyimide tape wound coil," *IEEE Trans. Magn.*, vol. **41**, no. 10, pp. 3571–3573, Oct. 2005.
21. J. Deng and K. Y. See, "In-circuit characterization of common-mode chokes," *IEEE Trans. Electromagn. Compat.*, vol. **49**, no. 5, pp. 451–454, May 2007.
22. B. C. Tseng and L. K. Wu, "Design of miniaturized common-mode filter by multilayer low-temperature co-fired ceramic," *IEEE Trans. Electromagn. Compat.*, vol. **46**, no. 4, pp. 571–579, Nov. 2004.
23. C.-H. Tsai and T.-L. Wu, "A broadband and miniaturized common mode filter for gigahertz differential signals based on negative-permittivity metamaterials," *IEEE Trans. Microwave Theory Tech.*, vol. **58**, no. 1, pp. 195–202, Jan. 2010.
24. C. Y. Hsiao, C. H. Tsai, C. N. Chiu, and T. L. Wu, "Radiation suppression for cable-attached packages utilizing a compact embedded common-mode filter," *IEEE Trans. Compon. Packag. Manuf. Technol.*, vol. **2**, no. 10, pp. 1696–1703, Oct. 2012.
25. G. H. Shiue, C. M. Hsu, C. L. Yeh, and C. F. Hsu, "A comprehensive investigation of a common-mode filter for gigahertz differential signals using quarter-wavelength resonators," *IEEE Trans. Compon. Packag. Manuf. Technol.*, vol. **4**, no. 1, pp. 134–144, Jan. 2014.
26. S. G. Kang, G. Shaffer, C. Kodama, C. O'Daniel, and E. Wheeler, "CSRR common-mode filtering structures in multilayer printed circuit boards," *2015 IEEE International Symposium on Electromagnetic Compatibility* (EMC 2015), Dresden, Germany, Aug. 2015, pp. 1300–1303.
27. C. H. Cheng and T. L. Wu, "A compact dual-band common-mode filtering component for EMC in wireless communication," *2015 Asia-Pacific International EMC Symposium* (APEMC 2015), Taipei, Taiwan, 26–29 May 2015.

28. C. Y. Hsiao, Y. C. Huang, and T. L. Wu, "An ultra-compact common-mode bandstop filter with modified-T circuits in integrated passive device (IPD) process," *IEEE Trans. Microwave Theory Tech.*, vol. **63**, no 11, pp. 3624–3631, Nov. 2015.
29. A. Fernández-Prieto, S. Qian, J. Hong, J. Martel, F. Medina, F. Mesa, J. Naqui, and F. Martín "Common-mode suppression for balanced bandpass filters in multilayer liquid crystal polymer technology," *IET Microwave Antennas Propag.*, vol. **9**, pp. 1249–1253, Sep. 2015.
30. B.-F. Su and T.-G. Ma, "Miniaturized common-mode filter using coupled synthesized lines and mushroom resonators for high-speed differential signals," *IEEE Microwave Wireless Compon. Lett.*, vol. **25**, pp. 112–114, Feb. 2015.
31. J. Naqui, *Symmetry properties in Transmission Lines Loaded with Electrically Small Resonators: Circuit Modeling and Applications*, Springer, Heidelberg, Germany, 2016.
32. J. Naqui, M. Durán-Sindreu, and F. Martín, "Novel sensors based on the symmetry properties of split ring resonators (SRRs)," *Sensors*, vol. **11**, pp. 7545–7553, 2011.
33. J. Naqui, M. Durán-Sindreu, and F. Martín, "Alignment and position sensors based on split ring resonators," *Sensors*, vol. **12**, pp. 11790–11797, 2012.
34. J. Naqui and F. Martín, "Transmission lines loaded with bisymmetric resonators and their application to angular displacement and velocity sensors," *IEEE Trans. Microwave Theory Tech.*, vol. **61**(12), pp. 4700–4713, Dec. 2013.
35. A. Karami-Horestani, C. Fumeaux, S. F. Al-Sarawi, and D. Abbott, "Displacement sensor based on diamond-shaped tapered split ring resonator," *IEEE Sensors J.*, vol. **13**, pp. 1153–1160, 2013.
36. J. Naqui and F. Martín, "Angular displacement and velocity sensors based on electric-LC (ELC) loaded microstrip lines," *IEEE Sensors J.*, vol. **14**(4), pp. 939–940, Apr. 2014.
37. A. K. Horestani, J. Naqui, D. Abbott, C. Fumeaux, and F. Martín, "Two-dimensional displacement and alignment sensor based on reflection coefficients of open microstrip lines loaded with split ring resonators," *Electron. Lett.*, vol. **50**, pp. 620–622, Apr. 2014.
38. J. Naqui, M. Durán-Sindreu, and F. Martín, "Selective mode suppression in coplanar waveguides using metamaterial resonators," *Appl. Phys. A Mater. Sci. Process.*, vol. **109**, pp. 1053–1058, Dec. 2012.
39. J. Naqui, M. Durán-Sindreu, and F. Martín, "Differential and single-ended microstrip lines loaded with slotted magnetic-LC (MLC) resonators," *Int. J. Antennas Propag.*, article ID 640514, 8 p., 2013.
40. J. Naqui, M. Durán-Sindreu, and F. Martín, "Selective mode suppression in microstrip differential lines by means of electric-LC (ELC) and magnetic-LC (MLC) resonators," META'13, 4th International Conference on

Metamaterials, Photonic Crystals and Plasmonics, Sharjah, UAE, 18–22 Mar. 2013.

41. F. Falcone, T. Lopetegi, J. D. Baena, R. Marqués, F. Martín, and M. Sorolla, "Effective negative-epsilon stop-band microstrip lines based on complementary split ring resonators," *IEEE Microwave Wireless Compon. Lett.*, vol. **14**, pp. 280–282, Jun. 2004.
42. J. Naqui, M. Durán-Sindreu, A. Fernández-Prieto, F. Mesa, F. Medina, and F. Martín, "Differential transmission lines loaded with split ring resonators (SRRs) and complementary split ring resonators (CSRRs)," *Fifth International Congress on Advanced Electromagnetic Materials in Microwaves and Optics* (Metamaterials 2011), Barcelona, Spain, 10–15 Oct. 2011.
43. R. Marqués, F. Martín, and M. Sorolla, *Metamaterials with Negative Parameters: Theory, Design and Microwave Applications*, John Wiley & Sons, Inc., Hoboken, NJ, 2008.
44. D. Ahn, J.-S. Park, C.-S. Kim, J. Kim, Y. Qian, and T. Itoh, "A design of the low-pass filter using the novel microstrip defected ground structure," *IEEE Trans. Microwave Theory Tech.*, vol. **49**, pp. 86–93, 2001.
45. F. Falcone, T. Lopetegi, M. A. G. Laso, J. D. Baena, J. Bonache, R. Marqués, F. Martín, and M. Sorolla, "Babinet principle applied to the design of metasurfaces and metamaterials," *Phys. Rev. Lett.*, vol. **93**, paper 1 D, Nov. 2004.
46. J. D. Baena, J. Bonache, F. Martín, R. Marqués, F. Falcone, T. Lopetegi, M. A. G. Laso, J. García, I. Gil, M. Flores-Portillo, and M. Sorolla, "Equivalent circuit models for split ring resonators and complementary split rings resonators coupled to planar transmission lines," *IEEE Trans. Microwave Theory Tech.*, vol. **53**, pp. 1451–1461, Apr. 2005.
47. F. Martín, F. Falcone, J. Bonache, R. Marqués, and M. Sorolla, "Miniaturized CPW stop band filters based on multiple tuned split ring resonators," *IEEE Microwave Wireless Compon. Lett.*, vol. **13**, pp. 511–513, Dec. 2003.
48. J. García-García, F. Martín, F. Falcone, J. Bonache, J.D. Baena, I. Gil, E. Amat, T. Lopetegi, M.A.G. Laso, J. A. Marcotegui-Iturmendi, M. Sorolla, and R. Marqués, "Microwave filters with improved stop band based on subwavelength resonators," *IEEE Trans. Microwave Theory Tech.*, vol. **53**, pp. 1997–2006, Jun. 2005.
49. J. Naqui, M. Durán-Sindreu, and F. Martín, "Modeling split ring resonator (SRR) and complementary split ring resonator (CSRR) loaded transmission lines exhibiting cross polarization effects," *IEEE Antennas Wireless Propag. Lett.*, vol. **12**, pp. 178–181, 2013.
50. R. Marqués, J.D. Baena, J. Martel, F. Medina, F. Falcone, M. Sorolla, and F. Martín, "Novel small resonant electromagnetic particles for metamaterial and filter design," *Proceedings of the International Conference on Electromagnetic Advanced Applications.*, ICEAA, 2003, Torino, Italy, pp. 439–443, Sep. 8–12, 2003.

51. J. Naqui, M. Durán-Sindreu, A. Fernández-Prieto, F. Mesa, F. Medina, and F. Martín, "Multimode propagation and complex waves in CSRR-based transmission line metamaterials," *IEEE Antennas Wireless Propag. Lett.*, vol. **11**, pp. 1024–1027, 2012.
52. J. Bonache, M. Gil, I. Gil, J. Garcia-García, and F. Martín, "On the electrical characteristics of complementary metamaterial resonators," *IEEE Microwave Wireless Compon. Lett.*, vol. **16**, pp. 543–545, Oct. 2006.
53. A. Fernández-Prieto, J. Martel, J. S. Hong, F. Medina, S. Qian, and F. Mesa, "Differential transmission line for common-mode suppression using double side MIC technology," *Proceedings of the 41st European Microwave Conference* (EuMC), Manchester, England, pp. 631–634, Oct. 10–13, 2011.
54. B. Z. Katsenelenbaum, L. Mercader, M. Pereyaslavets, M. Sorolla, and M. Thumm, *Theory of Nonuniform Waveguides: The Cross-Section Method*, IEE Electromagnetic Waves Series, vol. **44**, The Institution of Electrical Engineers, London, UK, 1998.
55. T. Lopetegi, M. A. G. Laso, M. J. Erro, M. Sorolla, and M. Thumm, "Analysis and design of periodic structures for microstrip lines by using the coupled mode theory," *IEEE Microwave Wireless Compon. Lett.*, vol. **12**, no. 11, pp. 441–443, Nov. 2002.
56. I. Arnedo, M. Chudzik, J. Schwartz, I. Arregui, A. Lujambio, F. Teberio, D. Benito, M.A.G. Laso, D. Plant, J. Azana, and T. Lopetegi, "Analytical solution for the design of planar EBG structures with spurious-free frequency response," *Microw. Opt. Technol. Lett.*, vol. **54** no. 4, pp. 956–960, Apr. 2012.
57. M. A. G. Laso, T. Lopetegi, M. J. Erro, D. Benito, M. J. Garde, and M. Sorolla, "Multiple-frequency-tuned photonic bandgap microstrip structures," *IEEE Microwave Guided Wave Lett.*, vol. **10**, no. 6, pp. 220–222, Jun. 2000.
58. M. A. G. Laso, T. Lopetegi, M. J. Erro, D. Benito, M. J. Garde, M. A. Muriel, M. Sorolla, and M. Guglielmi, "Real-time spectrum analysis in microstrip technology," *IEEE Trans. Microwave Theory Tech.*, vol. **51**, pp. 705–717, 2003.
59. F. R. Yang, K. P. Ma, Y. Qian, and T. Itoh, "A uniplanar compact photonic-bandgap (UC-PBG) structure and its applications for microwave circuits," *IEEE Trans. Microwave Theory Tech.*, vol. **47**, pp. 1509–1514, Aug. 1999.
60. F. Martín, J. Naqui, A. Fernández-Prieto, P. Vélez, J. Bonache, J. Martel, and F. Medina, "The beauty of symmetry: common-mode rejection filters for high-speed interconnects and band microwave circuits," *IEEE Microw. Mag.*, vol. **18**, no. 1, pp. 42–55, Jan. 2017.
61. R. Mongia, I. Bahl and P. Barthia, *RF and Microwave Coupled Line Circuits*, Artech House, Norwood, MA, 1999.
62. J. Shekel "Matrix analysis of multi-terminal transducers," *Proc. IRE*, vol. **42**, pp. 840–847, 1954.

63. R. Islam, M. Zedler, and G. V. Eleftheriades, "Modal analysis and wave propagation in finite 2D transmission-line metamaterials," *IEEE Trans. Antennas Propag.*, vol. **59**, pp. 1562–70, 2011.
64. F. A. Fernández, Y. Lu, J. B. Davies, and S. Zhu, "Finite element analysis of complex modes in inhomogeneous waveguides," *IEEE Trans. Magn.*, vol. **29**, pp. 1601–1604, Mar. 1993.
65. U. Crombach, "Complex waves on shielded lossless rectangular dielectric image guide," *Electron. Lett.*, vol. **19**, pp. 557–558, Jul. 1983.
66. M. Mrozowski and J. Mazur, "Matrix theory approach to complex waves," *IEEE Trans. Microwave Theory Tech.*, vol. **40**, no. 4, pp. 781–785, Apr. 1992.
67. F. Elek and G. V. Eleftheriades, "Dispersion analysis of the shielded Sievenpiper structure using multiconductor transmission line theory," *IEEE Microwave Wireless Compon. Lett.*, vol. **14**, no. 9, pp. 434–436, Sep. 2004.

CHAPTER 3

COUPLED-RESONATOR BALANCED BANDPASS FILTERS WITH COMMON-MODE SUPPRESSION DIFFERENTIAL LINES

Armando Fernández-Prieto,[1] Jordi Naqui,[2] Jesús Martel,[3] Ferran Martín,[2] and Francisco Medina[1]

[1]Departamento de Electrónica y Electromagnetismo, Universidad de Sevilla, Sevilla, Spain
[2]CIMITEC, Departament d'Enginyeria Electrònica, Universitat Autònoma de Barcelona, Bellaterra, Spain
[3]Departamento de Física Aplicada II, Universidad de Sevilla, Sevilla, Spain

3.1 INTRODUCTION

The periodic or "artificial" differential lines studied in the previous chapter were designed with the purpose of achieving strong common-mode (CM) rejection over a desired frequency range. Simultaneously, these structures are expected to behave as conventional matched transmission lines for the differential-mode (DM) signal. In other words, they should exhibit, ideally, an all-pass response under DM operation and a

Balanced Microwave Filters, First Edition. Edited by Ferran Martín, Lei Zhu, Jiasheng Hong, and Francisco Medina.

stopband response for the CM signal. Thus no frequency selectivity (filtering response) was required for the DM. However, to obtain the stopband behavior for the CM signal, the host transmission line (commonly a pair of microstrip lines) is supplemented with elements that might also perturb the path of the DM signal. It is then a requirement for this kind of CM rejection filters to keep the differential signal integrity unaltered while passing through the differential pair, at least within the frequency range of interest. However, as it will be demonstrated in this chapter, the CM stopband differential pairs analyzed in Chapter 2 can be used to achieve proper filtering functions (for DM operation) when combined with balanced filters. These balanced filters can be designed without paying attention to their CM response, since the CM rejection is provided by a few sections of the periodic structures studied in Chapter 2. Thus, it will be shown in this chapter that the use of CM rejection stages of the differential lines studied in Chapter 2 is a straightforward strategy to improve the common-mode rejection ratio (CMRR) of balanced filters. Two or more sections of such structures can be used as input/output stages in combination with coupled-resonator balanced filters, thus allowing for the design of balanced filters with good DM response and strong CM rejection.

We will start this chapter by designing a balanced single-band bandpass filter based on conventional coupled resonators. The CM response of this filter is not particularly good. Next, several types of CM rejection stages (which are the basic unit cells of artificial differential lines with CM rejection) are cascaded with the basic filter to verify how effectively CM rejection is significantly improved without altering the DM response. Finally, the same idea will be applied to reject the CM in a balanced dual-band bandpass filter based on coupled resonators as well.

3.2 BALANCED COUPLED-RESONATOR FILTERS

Coupled-resonators are widely used for implementing narrowband bandpass filters, which are essential components for many RF/microwave applications. A detailed discussion of narrowband balanced filters based on the use of coupled resonators is provided in Chapter 12 in this book. The design procedure for this type of filters in its single-ended form is carefully described in Refs. [1, 2]. However, the method reported in those references is quite general and can be applied in the same manner for single-ended and balanced filters. We will not go deep into the physical details of such procedure, since this is not the aim of this chapter and the topic is clearly treated in Refs. [1, 2]. We will simply illustrate the application of the method using a couple of balanced bandpass filter examples. Then, we will focus our efforts on how the differential lines presented in Chapter 2 can be used to improve the CM rejection without having deleterious effects on the DM response.

3.2.1 Single-Band Balanced Bandpass Filter Based on Folded Stepped-Impedance Resonators

Coupled resonators having the shape of open square loops have been widely used to design bandpass filters (see Ref. [1] and references therein). If, instead of using square loops, folded stepped-impedance resonators (FSIRs) are chosen, more compact structures with a better out-of-band performance can be achieved. Folded resonators of this kind have been used, for instance, in Ref. [3] to implement single-band and dual-band bandpass filters in a single-ended architecture. An example of the typical layout based on FSIRs is shown in Figure 3.1(a). The layout of the balanced counterpart of this filter only requires to introduce symmetry in the excitation and the structure. A layout composed of two classical electrically coupled FSIRs intended to operate as a differential balanced device is depicted in Figure 3.1(b). This structure is the one chosen in Ref. [4] to be combined with CM bandstop cells to design good single-band balanced bandpass filters. A similar geometry using magnetic coupling (instead of electric coupling) has been used in Ref. [5] with similar purposes, but the working principle is different and its study is carried out in Chapter 12.

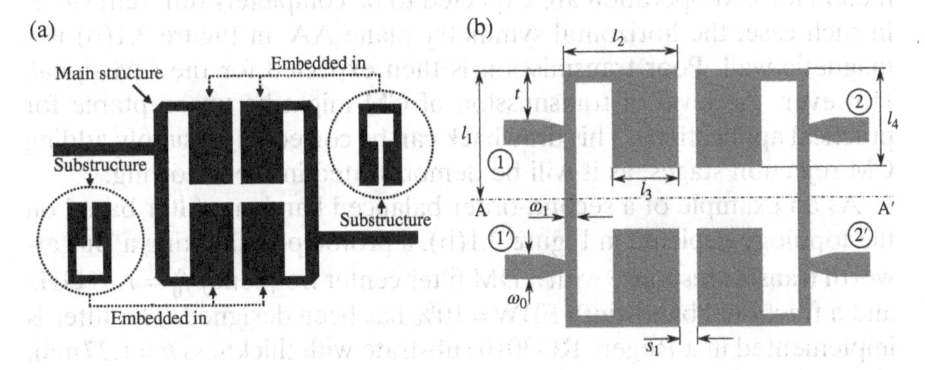

Figure 3.1 (a) Layout proposed in Ref. [3] for the design of single-band and dual-band single-ended bandpass filters. Reprinted with permission from Ref. [3]; copyright 2006 IEEE. (b) Layout of the proposed balanced bandpass filter composed of two symmetrically excited electrically coupled FSIRs in Ref. [4]. Reprinted with permission from Ref. [4]; copyright 2012 IEEE.

As it has been mentioned before, the methodology described in Ref. [1] to design coupled-resonator filters will be used here. Such a technique is based on the appropriate selection of two main parameters: (i) the external quality factor, Q_e, and (ii) the inter-resonator coupling coefficient, M. For a given set of specifications for the DM response, it is easy to determine the values of those parameters following the steps reported in Ref. [1]. In general, for simple geometries like the one shown in Figure 3.1(b), both the DM external quality factor, Q_e^{dd}, and the coupling coefficient, M^{dd}, can be expressed in terms of the filter fractional bandwidth (FBW) and the low-pass prototype element values, g_i $(i=1,\ldots,n-1)$ in the following manner:

$$M_{i,i+1}^{dd} = \frac{\text{FBW}}{\sqrt{g_i g_{i+1}}}, \quad \text{for } i=1,\ldots,n-1 \tag{3.1}$$

$$Q_{e1}^{dd} = \frac{g_0 g_1}{\text{FBW}} \quad Q_{en}^{dd} = \frac{g_n g_{n+1}}{\text{FBW}} \tag{3.2}$$

n being the order of the filter. The superscripts "dd" stand for DM operation. Under DM excitation the symmetry plane AA′ in Figure 3.1(b) is an electric wall. This means that the folded SIRs have a virtual short circuit in their center points. The obtaining of the values of the coupling coefficients and external quality factors as a function of the geometry of the layout and the dielectric permittivity and thickness of the chosen substrate should be carried out enforcing such a boundary condition for the DM operation. Obviously the values of those parameters for CM operation are expected to be completely different since, in such case, the horizontal symmetry plane AA′ in Figure 3.1(b) is a magnetic wall. Poor transmission is then expected for the CM signal. However, the level of transmission of CM might be unacceptable for practical applications. This drawback can be corrected by simply adding CM rejection stages, as it will be demonstrated in the following.

As an example of a second-order balanced bandpass filter based on the topology depicted in Figure 3.1(b), a prototype exhibiting a Butterworth transfer response with a DM filter center frequency $f_0^d = 1.37\,\text{GHz}$ and a fractional bandwidth FBW = 10% has been designed. The filter is implemented in a Rogers RO3010 substrate with thickness $h = 1.27\,\text{mm}$, dielectric constant $\varepsilon_r = 10.2$, and loss tangent $\tan\delta = 0.0023$. For the aforementioned specifications, the required theoretical values for Q_e^{dd} and M^{dd} derived from (3.1) and (3.2) turn out to be $M_{1,2}^{dd} = 0.071$ and $Q_{e1}^{dd} = Q_{e2}^{dd} = 14.14$. Note that, obviously, the center frequency, f_0^d, is mainly controlled by the resonator features (mainly its electrical size and the ratio of the characteristic impedances of the two different sections conforming the FSIR). Therefore, the first step in the design process consists in determining all the physical dimensions of the resonators that fix the DM resonance frequency at the desired value $f_0^{\mathrm{d}} = 1.37\,\mathrm{GHz}$. This task can be easily accomplished with the help of a commercial full-wave simulator refining an initial guess that can be obtained by using a transmission-line model of the FSIR [6]. This transmission-line model might include the edge effects (edge capacitances) if desired. After applying this procedure, the obtained dimensions for the two identical resonators in Figure 3.1(b) result to be $w_1 = 0.7\,\mathrm{mm}$, $l_1 = 6.4\,\mathrm{mm}$, $l_2 = 7.5\,\mathrm{mm}$, $l_3 = 3.7\,\mathrm{mm}$, and $l_4 = 5.4\,\mathrm{mm}$. Once the dimensions of the resonators have been obtained, the next step consists in determining the tap position (location of the input and output microstrip transmission lines, i.e., the value of t in Figure 3.1(b)) and the separation between resonators (s_1 in Figure 3.1(b)) required to obtain the appropriate values of external quality factor and coupling coefficient, respectively. Let us give more details about this step:

- **DM external quality factor** $Q_{\mathrm{e}}^{\mathrm{dd}}$: The external quality factor is a function of the feeding point, that is, the variable t in Figure 3.1 (b), as mentioned earlier [1]. This parameter can be calculated as the following ratio:

$$Q_{\mathrm{e}}^{\mathrm{dd}} = \frac{f_0^{\mathrm{d}}}{f_{+90}^{\mathrm{d}} - f_{-90}^{\mathrm{d}}} \tag{3.3}$$

  where f_{+90}^{d} and f_{-90}^{d} stand for the frequencies at which S_{11}^{dd} has a phase shift of $+90°$ and $-90°$ (respectively) with respect to the phase of such parameter at the filter center frequency. The expression (3.3) can be used to calculate $Q_{\mathrm{e}}^{\mathrm{dd}}$ for several values of t. These values can be then used to plot a curve like the one shown in Figure 3.2(a). This curve allows to find the proper feeding point to achieve the desired DM external quality factor imposed by the design specifications. From this figure it is clear that a value of, approximately, $t = 3\,\mathrm{mm}$ fulfills the requirement $Q_{\mathrm{e}} = 14.14$.
- **Coupling coefficient** $M_{1,2}^{\mathrm{dd}}$: The coupling coefficient for the differential signal, $M_{1,2}^{\mathrm{dd}}$, is controlled by the gap distance between resonators, s_1, in Figure 3.1(b). When two synchronous resonators are coupled, the resonance frequency f_0^{d} splits into two different peaks,

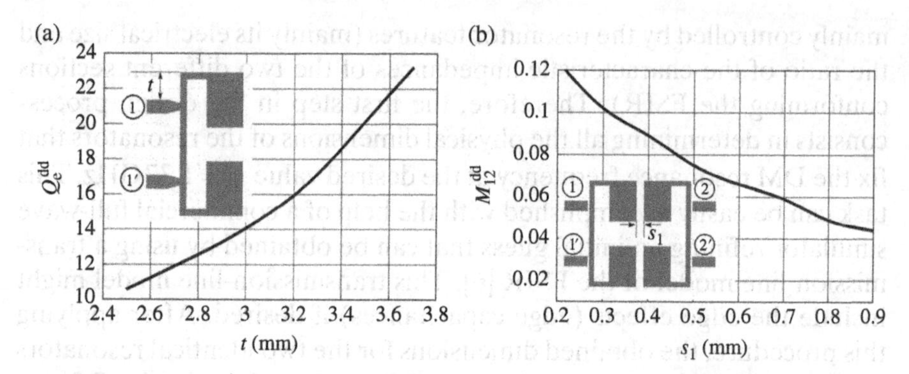

Figure 3.2 (a) Differential-mode external quality factor, Q_e^{dd}, versus the tap position, t, for the chosen FSIR resonator; (b) differential-mode coupling coefficient, M_{12}^{dd}, as a function of the separation, s_1, between the chosen resonators.

f_{p1} and f_{p2}, below and above f_0^d, respectively [1]. The separation between such shifted peaks depends on the coupling level. More precisely, the larger the separation between resonators, the lower the coupling strength, and the lower the separation between split peaks. It is straightforward to calculate $M_{1,2}^{dd}$ as a function of f_{p1} and f_{p2} by means of the following formula extracted from Hong's textbook [1]:

$$M_{1,2}^{dd} = \frac{f_{p2}^2 - f_{p1}^2}{f_{p2}^2 + f_{p1}^2} \tag{3.4}$$

Using (3.4) and the resonance frequencies obtained with a full-wave simulator for several values of the gap distance, s_1, a curve such as the one plotted in Figure 3.2(b) can be drawn for the coupling coefficient as a function of its control parameter, the gap distance s_1. The curve shows that the required value for our particular design, $M_{1,2}^{dd} = 0.071$, is attained if $s_1 = 0.5$ mm. This step ends the design process. The layout of the filter is completely determined for the desired specs for the DM response.

The electrical response of the designed filter has been simulated under DM and CM excitations. The simulated results are shown in Figure 3.3. It is observed that the CM rejection level is barely about 18 dB within the differential passband. The CMRR required in many applications is much more stringent. Unfortunately, the CM response is determined once the DM one has been specified; thus it is not possible to improve CMRR using the proposed layout. However, an obvious

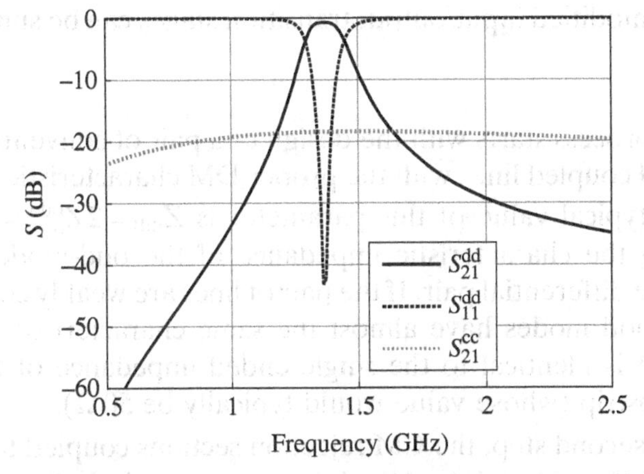

Figure 3.3 Simulated results for the designed basic (solid ground plane) balanced filter in Figure 3.1(b) (amplitude in dB of the scattering parameters). Differential-mode reflection (S_{11}^{dd}) and transmission (S_{21}^{dd}) coefficients as well as the common-mode transmission coefficient (S_{21}^{cc}) are shown.

manner of increasing the CMRR is to add CM bandstop filters to the proposed structure. These additional stages should be typically designed to provide strong CM rejection in the differential passband of the filter and frequencies around. Thus, in order to reduce CM transmission, the strategy explored in this chapter will consist in replacing the conventional input/output differential pair of uniform transmission lines by two (or more) stages of the periodic differential structures with CM rejection studied in depth in Chapter 2.

3.2.2 Balanced Filter Loaded with Common-Mode Rejection Sections

In this subsection some of the examples of CM rejection stages presented in Chapter 2 will be combined with the differential filter designed in Subsection 3.2.1 with the aim of providing extra CM noise rejection. Of course, the improvement of the CM rejection level has to be achieved without significantly perturbing the DM response. Without loss of generality, we will test the CM rejection stages proposed in Refs. [4, 7, 8]. These are just a few illustrative examples, and other geometries found in the literature can be used instead. Considerations about the enhancement of the rejection level (i.e. of the CMRR), bandwidth, size, additional losses, and influence on the DM response have to be handled before making a particular choice. The procedure to design the filter with modified input/output transition stages can be summarized as follows:

1. The process starts with the design of a pair of conventional differential coupled lines with the proper DM characteristic impedance. The typical value of this parameter is $Z_{\text{diff}} = 2Z_0^{\text{odd}} = 100\,\Omega$, Z_{odd} being the characteristic impedance of the odd-mode supported by the differential pair. If the pair of lines are weakly coupled, even and odd modes have almost the same characteristic impedance, which is identical to the single-ended impedance of the isolated microstrip (whose value would typically be 50 Ω).
2. As a second step, the CM rejection sections coupled to the differential lines are introduced in a symmetrical disposition. The dimensions of the used slotted ground pattern are chosen in such a way that the CM rejection band is within the frequency range of interest, usually covering, at least, the passband of the differential filter. Due to the symmetrical location of the ground plane perturbations, the introduced CM rejection stages will not meaningfully affect the DM signal path, as it was explained in Chapter 2.
3. Finally, once the modified differential lines have been designed, the conventional input/output microstrip lines of the filter studied in the previous subsection should be replaced by the new ones. This should improve the filter CM rejection barely affecting the DM performance, as we will see in the following.

The structures in Refs. [4, 7, 8] have been designed and optimized to reject the CM signal at frequencies around 1.37 GHz. The layout and electromagnetic response of the structures are shown in Figures 3.4 and 3.5, respectively. For comparison purposes, Figure 3.5(d) contains all the CM responses together in a single graph. Figure 3.5(d) reveals that the dumbbell-shaped DGS provides the larger rejection bandwidth when it is compared with the other two structures. Anyway, if a wider CM rejection bandwidth is required, it can be achieved in all cases by simply adding and coupling more sections, as discussed in Chapter 2 and also in Refs. [4, 7, 8].

The CM rejection filters sketched in Figure 3.4(a)–(c) have been combined with the filter shown in Figure 3.1(b). This is schematically represented in Figure 3.6(a). The electromagnetic response of each case is depicted in Figure 3.6(b)–(d). For comparison purposes, the response of the conventional filter for both DM and CM excitations is also shown. It is clear that, in all the analyzed scenarios, the CM rejection level has

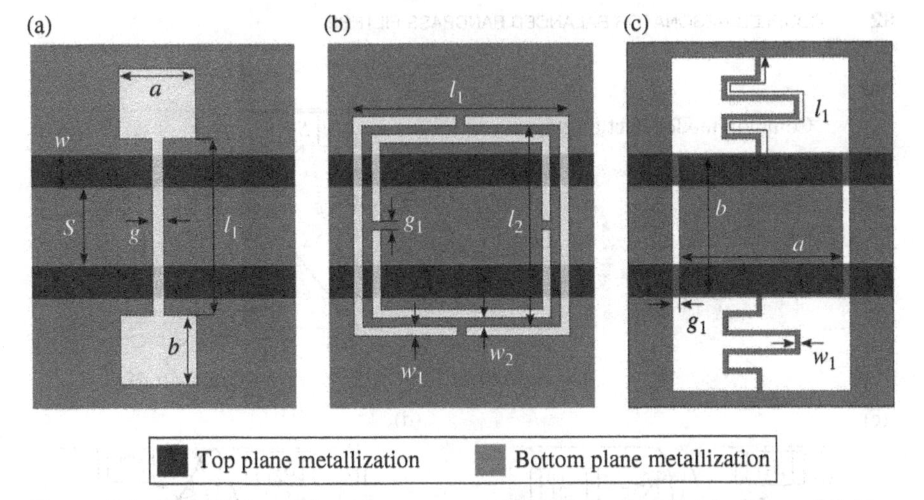

Figure 3.4 Unit cells of three common-mode rejection artificial differential lines chosen to be cascaded with the balanced standard filter proposed in Figure 3.1(b). All dimensions are given in millimeter; (a) dumbbell-shaped DGS [7]: $a = 12.6$, $b = 9.5$, $g = 0.4$, and $l_1 = 14.8$. (b) Double-slit complementary split ring resonator DGS (DS-CSRRs) [4]: $l_1 = 17.6$, $l_2 = 10.4$, $w_1 = w_2 = 0.2$, and $g_1 = 0.4$. (c) Capacitive patch with grounded meander line [8]: $a = 6.6$, $b = 6.7$, $l_1 = 20.5$, and $w_1 = g_1 = 0.3$. Top layer dimensions are the same for all the structures: $w = 1.2$ and $S = 4.6$.

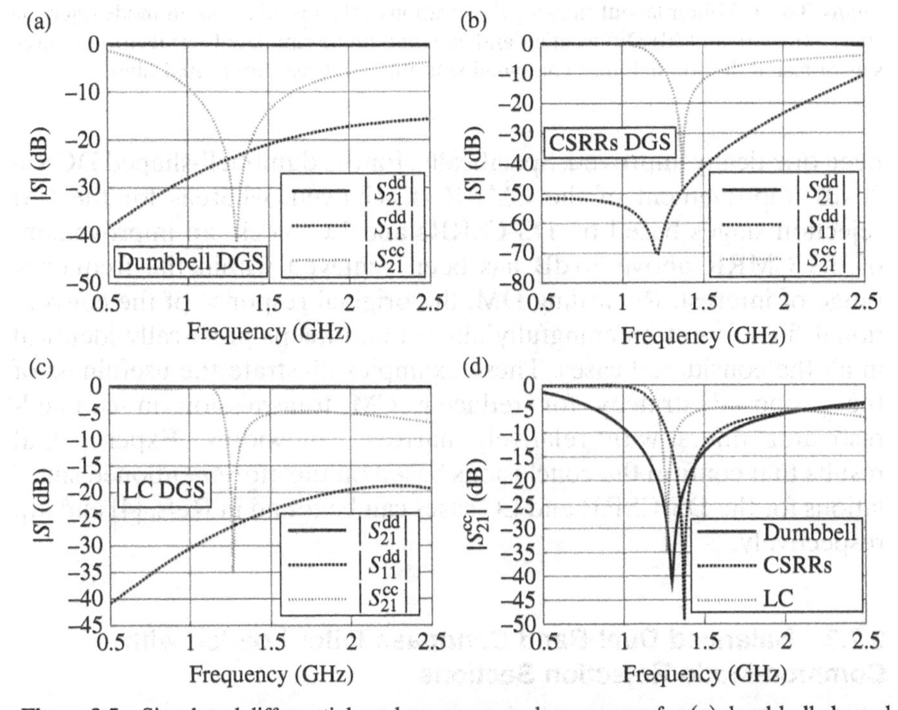

Figure 3.5 Simulated differential- and common-mode responses for (a) dumbbell-shaped input/output unit cells; (b) DS-CSRR-loaded differential lines; and (c) LC-loaded differential lines. (d) Comparison between the full-wave simulated common-mode responses of the three analyzed solutions.

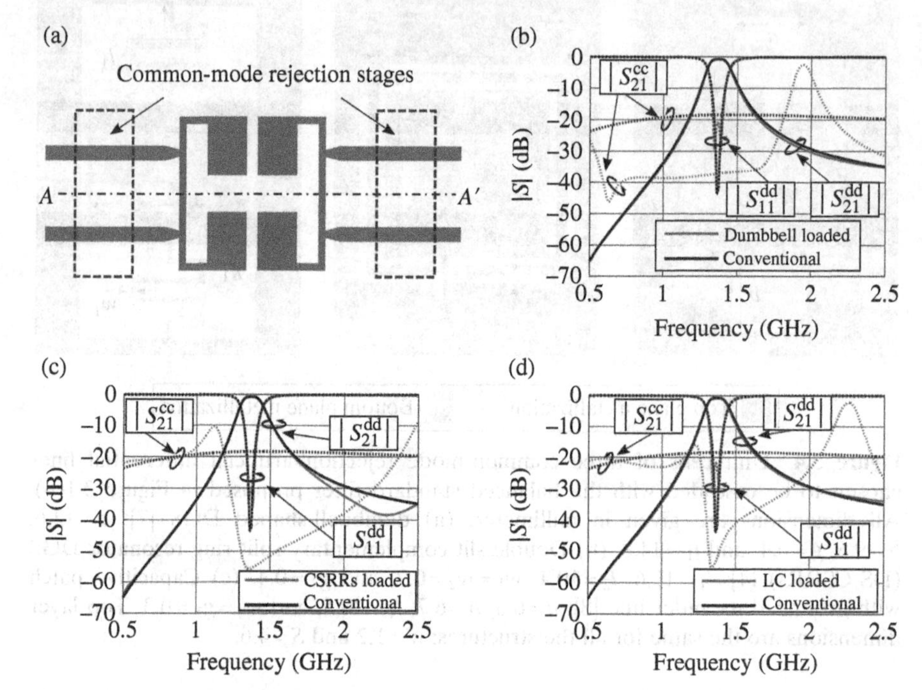

Figure 3.6 (a) Filter layout showing the location of the tested common-mode rejection stages. (b), (c), and (d) Differential- and common-mode simulated results for the three sets of tested differential lines combined with the bandpass differential filter.

been drastically improved. Specifically, for the dumbbell-shaped DGS, a 20 dB improvement of the CMRR is achieved, whereas for the CM rejection stages based on DS-CSRRs and LC block, an improvement of the CMRR above 30 dB has been achieved within the frequency range of interest. Regarding DM, the original response of the conventional filter is not meaningfully altered and keeps practically identical in all the considered cases. These examples illustrate the usefulness of the proposed strategy for reducing CM transmission in coupled-resonator filters with relatively narrow bandwidths. Experimental results that confirm the conclusions based on the aforementioned simulations for the DS-CSRR and LC cases can be found in Refs. [4] and [8], respectively.

3.2.3 Balanced Dual-Band Bandpass Filter Loaded with Common-Mode Rejection Sections

Let us now extend the idea explored in the previous section to a balanced filter with two differential passbands. This is a very interesting application of the DGS-loaded differential lines, since multiband wireless communication systems requiring the use of multiband devices are of common use nowadays. In order to reject the CM signal within two passbands, it is necessary to use a structure that cancels the CM transmission over a relatively wide frequency range. We can notice that the topologies in Figure 3.4(a) and (b) furnish a wider CM rejection band when compared with the structure in Figure 3.4(c) around resonance. However, the performance of the dumbbell-shaped and CSRR-loaded lines correspond to a bandstop filter for CM [4, 7], while the electric response of the LC-loaded lines is the typical one of an elliptic low-pass filter [8] for the CM signal. This is illustrated in Figure 3.7, which depicts the CM electromagnetic response of the differential lines in Figure 3.4(a)–(c) in a frequency range wider than the one shown in Figure 3.5(a)–(c).

This figure reveals that the performance of the LC-loaded lines (when compared with both dumbbell-shaped and DS-CSRR loaded lines) is the best when considering frequencies well above the resonance frequency. Due to this reason, it seems that, for relatively wideband applications, the topology in Figure 3.4(c) is more suitable than those sketched in Figure 3.4(a) and (b).

As it has already been mentioned, in the context of this section, the LC-loaded lines are intended to be applied to reject the CM transmission in a balanced dual-band bandpass filter. The filter itself will be implemented using two pairs of coupled resonators. Thus, to this aim, the filter that has been chosen is the one shown in Figure 3.8. Basically,

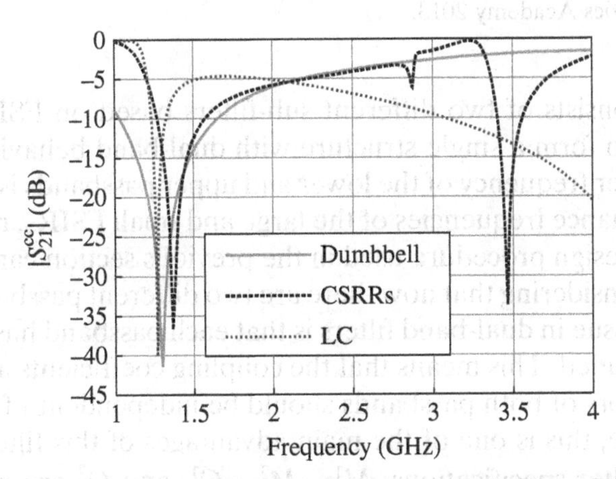

Figure 3.7 Common-mode insertion loss for the structures in Figure 3.4(a)–(c). These are simulated full-wave results over a relatively wide frequency region.

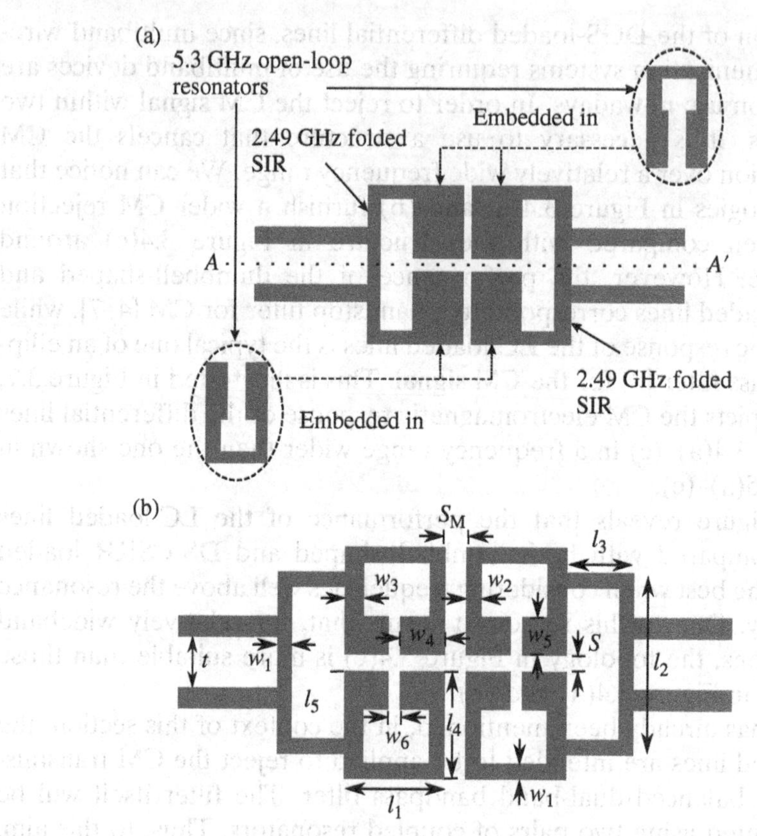

Figure 3.8 Layout of the balanced dual-band bandpass filter based on coupled resonators considered in this chapter. Dimensions (in mm) are $s=1.0$, $S_M=0.2$, $t=3.2$, $w_1=0.3$, $w_2=0.3$, $w_3=0.6$, $w_4=2.0$, $w_5=2.5$, $w_6=0.2$, $l_1=4.2$, $l_2=10.4$, $l_3=1.3$, $l_4=5.3$, and $l_5=4.7$. Reprinted with permission from Ref. [9]; courtesy of the Electromagnetics Academy 2013.

the filter consists of two different sub-filters based on FSIRs that are combined to form a single structure with dual-band behavior [3, 9].

The center frequency of the lower and upper passbands is established by the resonance frequencies of the large and small FSIRs, respectively. The same design procedure used in the previous section can be applied here, but considering that now there are two different passbands. A very important issue in dual-band filters is that each passband has to be independently tuned. This means that the coupling coefficients and external quality factors of both passbands should be independent of each other. In this sense, this is one of the main advantages of this filter topology; given the filter specifications, $M_{1,2}^1$, $M_{1,2}^2$, Q_e^1, and Q_e^2 are independent and can be tuned one by one (the superscripts (1) and (2) stand for the lower and upper bands, respectively). Let us then start by designing a balanced dual-band bandpass filter with conventional input/output lines, as it was done in the previous section.

As an example, a second-order filter ($n=2$) of Butterworth type with center frequencies located at $f_{01}^{\mathrm{d}}=2.5\mathrm{GHz}$ and $f_{02}^{\mathrm{d}}=5.3\mathrm{GHz}$ and fractional bandwidths $\mathrm{FBW}_1=10\%$ and $\mathrm{FBW}_2=4\%$ is designed using a substrate with dielectric constant $\varepsilon_r=2.43$, thickness $h=0.49\mathrm{mm}$, and loss tangent $\tan\delta=0.0025$. According to filter specifications, the required external quality factors and coupling coefficients result to be $M_{1,2}^1=0.071$, $M_{1,2}^2=0.028$, $Q_e^1=14.2$, $Q_e^2=35.4$. First, the dimensions of the resonators are determined in order to fix the desired resonance frequencies at $f_{01}^{\mathrm{d}}=2.5\mathrm{GHz}$ and $f_{02}^{\mathrm{d}}=5.3\mathrm{GHz}$ (the values of the dimensions can be found in the caption of Figure 3.8). Next, the required external quality factors and coupling coefficients must be adjusted to their respective theoretical values. For the first differential passband, Q_e^1 and $M_{1,2}^1$ depend on the tapping position, t, and the gap distance S_{M}, respectively, as occurred in the example studied in the previous section. For the second passband, it can be demonstrated that the Q_e^2 and $M_{1,2}^2$ can be controlled independently by means of l_5 and the ratio w_3/w_2, respectively, without significantly affecting Q_e^1 and $M_{1,2}^1$ [3, 9]. These facts are illustrated in Figure 3.9(a) and (b). Using the curves in Figure 3.9(a) and (b), it can be found that the theoretical values of Q_e^1, Q_e^2, $M_{1,2}^1$, and $M_{1,2}^2$ are achieved if $S_{\mathrm{M}}=0.2\mathrm{mm}$, $t=3.2\mathrm{mm}$, $l_5=4.7\mathrm{mm}$, and $w_3=0.6\mathrm{mm}$ (for $w_2=0.3\mathrm{mm}$).

Once the conventional filter (conventional in the sense that no extra elements have been added to reject the CM) has been designed, the input/output lines are replaced by the differential pair with CM rejection stages redesigned for this new substrate. Referring to Figure 3.4(c), the new dimensions of the LC-loaded lines are as follows:

1. Top layer: $w=1.4\mathrm{mm}$, $s=3.2\mathrm{mm}$
2. Bottom layer: $a=5.0\mathrm{mm}$, $b=5.2\mathrm{mm}$, $g_1=0.2\mathrm{mm}$, $w_1=0.2\mathrm{mm}$, and $l_1=24.2\mathrm{mm}$

The final layout of the cascaded LC-loaded lines and the balanced filter is depicted in Figure 3.10 [9].

The simulated and measured results for the filter in Figure 3.8 are shown in Figure 3.11(a) and (b) (DM) and Figure 3.12 (CM) [9]. Figure 3.11 shows that the introduction of the CM rejection stages within the feeding lines do not alter the DM response when compared with the

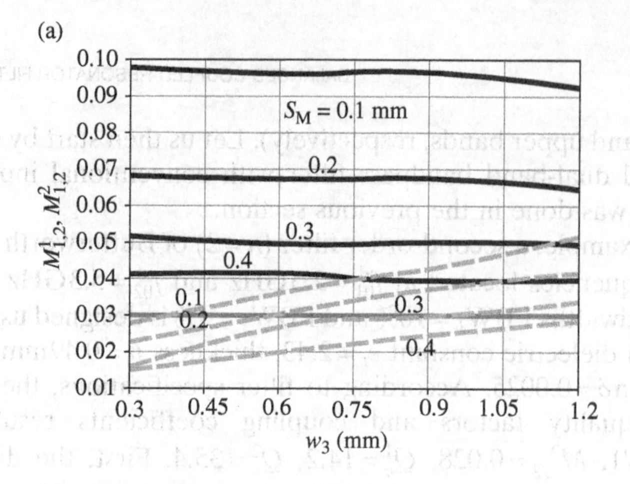

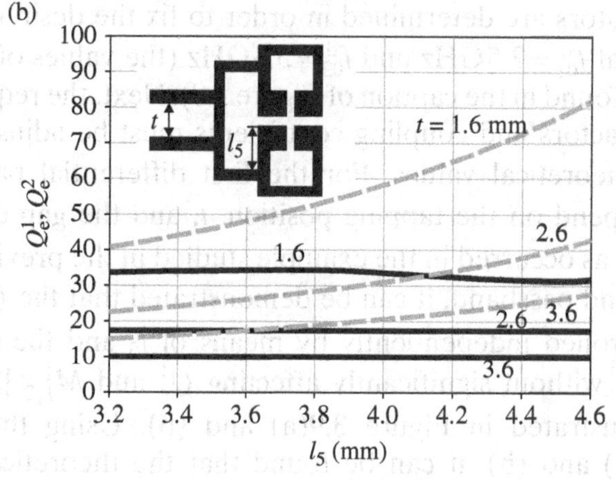

Figure 3.9 (a) Coupling coefficients $M^1_{1,2}$ (solid lines) and $M^2_{1,2}$ (dashed lines) versus w_3/w_2 keeping w_2 constant and equal to 0.3 mm using S_M as parameter. (b) External quality factors Q^1_e (solid lines) and Q^2_e (dashed lines) versus l_5 using t as parameter. Reprinted with permission from Ref. [9]; courtesy of the Electromagnetics Academy 2013.

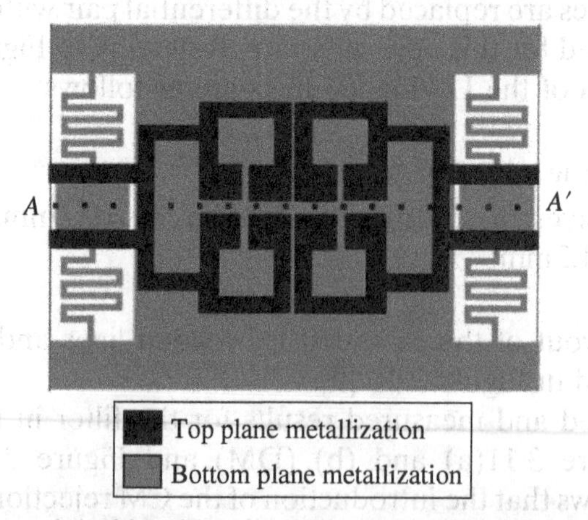

Figure 3.10 Balanced dual-band bandpass filter with modified input/output lines for common-mode rejection. Reprinted with permission from Ref. [9]; courtesy of the Electromagnetics Academy 2013.

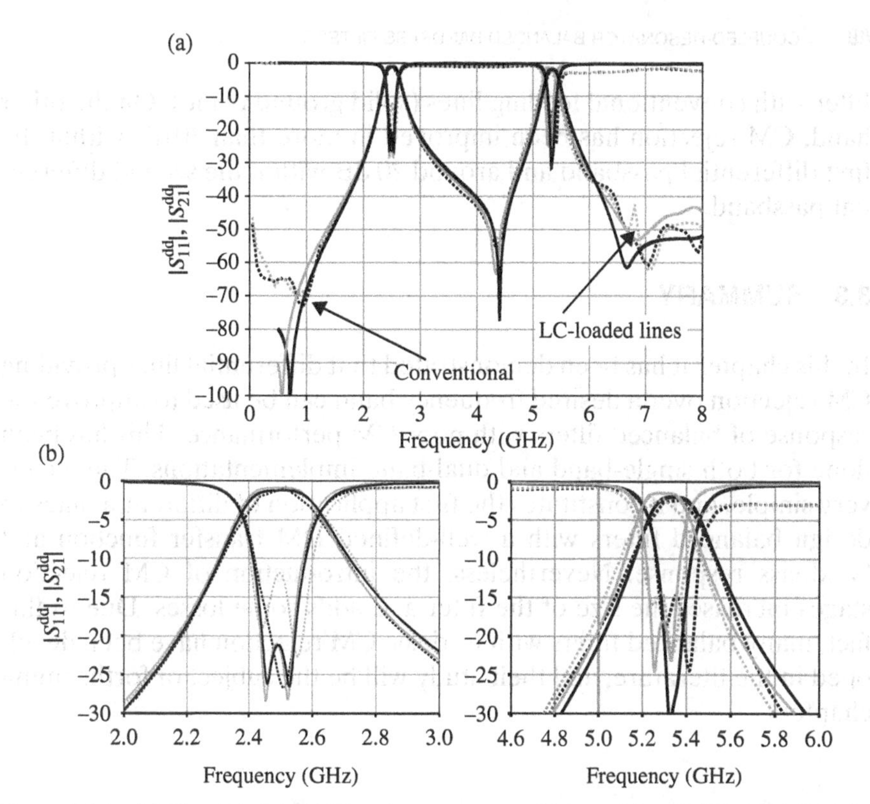

Figure 3.11 (a) Simulated (solid lines) and measured (dotted lines) differential-mode response of both the conventional filter (black color lines) and the filter with modified input/output lines (light gray lines). (b) Detail of the two differential passbands. Reprinted with permission from Ref. [9]; courtesy of the Electromagnetics Academy 2013.

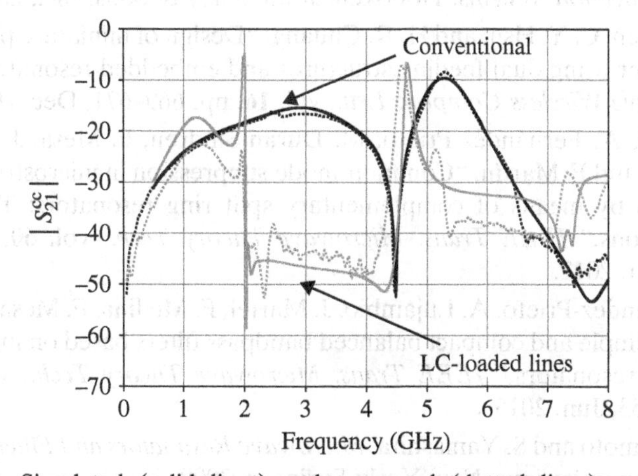

Figure 3.12 Simulated (solid lines) and measured (dotted lines) common-mode response of both the conventional filter (black color lines) and the filter with modified input/output lines (light gray lines). Reprinted with permission from Ref. [9]; courtesy of the Electromagnetics Academy 2013.

filter with conventional feeding lines (solid ground plane). On the other hand, CM rejection has been improved in more than 30 dB within the first differential passband and around 20 dB within the second differential passband.

3.3 SUMMARY

In this chapter it has been demonstrated that differential lines providing CM rejection over a desired frequency band can be used to improve the response of balanced filters with poor CM performance. This has been done for both single-band and dual-band implementations. The idea is very simple, and it constitutes the first application of differential lines to design balanced filters with a well-defined DM transfer function and bandpass response. Nevertheless, the introduction of CM rejection stages increases the size of the filter and adds some losses. Due to this fact, many balanced filters with intrinsic CM rejection have been developed in the literature, and their study will be the subject of forthcoming chapters.

REFERENCES

1. J.-S. Hong, *Microstrip Filters for RF/Microwave Applications*, 2nd edition. Hoboken: John Wiley & Sons, Inc., 2011.
2. R. J. Cameron, C. M. Kudsia, and R. R. Mansour, *Microwave Filters for Communication Systems*. Hoboken: John Wiley & Sons, Inc., 2007.
3. C. Y. Chen, C. Y. Hsu, and H. R. Chuang, "Design of miniature planar dual-band filter using dual-feeding structures and embedded resonators," *IEEE Microwave Wireless Compon. Lett.*, vol. **16**, pp. 669–671, Dec. 2006.
4. J. Naqui, A. Fernández-Prieto, M. Durán-Sindreu, F. Mesa, J. Martel, F. Medina, and F. Martín, "Common-mode suppression in microstrip differential lines by means of complementary split ring resonators: Theory and applications," *IEEE Trans. Microwave Theory Tech.*, vol. **60**, pp. 3023–3034, Oct. 2012.
5. A. Fernández-Prieto, A. Lujambio, J. Martel, F. Medina, F. Mesa, and R. R. Boix, "Simple and compact balanced bandpass filters based on magnetically coupled resonators," *IEEE Trans. Microwave Theory Tech.*, vol. **63**, pp. 1843–1853, Jun. 2015.
6. M. Makimoto and S. Yamashita, *Microwave Resonators and Filters for Wireless Communications*. New York: Springer, 2000.
7. W. T. Liu, C. H. Tsai, T. W. Han, and T. L. Wu, "An embedded common-mode suppression filter for GHz differential signals using periodic defected

ground plane," *IEEE Microwave Wireless Compon. Lett.*, vol. **18**, pp. 248–250, Apr. 2008.

8. A. Fernández-Prieto, J. Martel, J.-S. Hong, F. Medina, S. Qian, and F. Mesa, "Differential transmission line for common-mode suppression using double side MIC technology," in *Proceedings of the 41st European Microwave Conference (EuMC)*, Manchester, UK, pp. 631–634, Oct. 10–13, 2011.
9. A. Fernández-Prieto, J. Martel-Villagrán, F. Medina, F. Mesa, S. Qian, J.-S. Hong, J. Naqui, and F. Martín, "Dual-band differential filter using broadband common-mode rejection artificial transmission line," *Prog. Electromag. Res.*, vol. **139**, pp. 779–797, May 2013.

PART 3

Wideband and Ultra-wideband (UWB) Balanced Band Pass Filters with Intrinsic Common-Mode Suppression

CHAPTER 4

WIDEBAND AND UWB BALANCED BANDPASS FILTERS BASED ON BRANCH-LINE TOPOLOGY

Teck Beng Lim[1] and Lei Zhu[2]

[1]School of Engineering, Nanyang Polytechnic, Ang Mo Kio, Singapore
[2]Department of Electrical and Computer Engineering, Faculty of Science and Technology, University of Macau, Macau SAR, China

4.1 INTRODUCTION

In many existing wireless communication systems, microwave integrated devices and circuits are usually developed under the operation of single-ended mode. Figure 4.1(a) depicts a two-port network of a single-ended mode or unbalanced device, and it in general has one input port and one output port with all the signals to be referenced with reference to its ground plane as have been well known nowadays. As a signal is launched between two terminals at the same port, such a microwave circuit is considered to operate under differential mode, and it is referred as to the balanced or differential-mode circuit or network. Figure 4.1(b) depicts the schematic of a two-port balanced circuit under differential-mode operation, where two pairs of grouped ports

Balanced Microwave Filters, First Edition. Edited by Ferran Martín, Lei Zhu, Jiasheng Hong, and Francisco Medina.

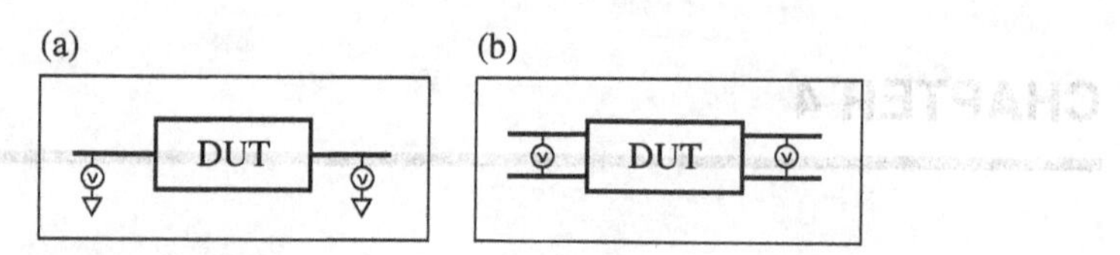

Figure 4.1 Microwave two-port network or device under test (DUT) under the operation: (a) single-ended mode and (b) differential mode.

at two sides make up the input and output ports. These balanced circuits are normally composed of two identical halves or bisections. When the two terminals at a port are excited by differential-mode signal, the signals at these two terminals of each port are equal in amplitude and 180° different in phase. Under the common-mode excitation, the signals at two terminals of each port are equal in both the amplitude and phase.

As has been well addressed, electrical characteristics of all the balanced devices or circuits under differential-mode operation are primarily dependent on the imbalanced property in amplitude and phase between two terminals of each differential port rather than absolute amplitude or level of the signals at each port. Therefore, these balanced circuits are basically insensitive to a few intrinsic problems existed in the single-ended mode circuits, such as ground bounce, noise signals, electromagnetic interference (EMI), and crosstalk coupling [1–3]. For instance, if the paired ports are routed closely together, any externally coupled noise will be coupled into two terminals of each port equally, and thus the coupled noise becomes the "common-mode" noise to which the circuit can be ideally immune due to its cancelation at two differential terminals of a port.

As intensively studied in Refs. [4, 5], the balanced or differential devices can provide higher system performance, lower noise, and better power consumption than their single-ended counterparts. Till now, they have been mostly used in low-frequency analog and digital devices. However, they have not been widely applied for design and exploration of microwave devices and circuits due to many challenging factors for these balanced devices to be operated in high frequency.

With the rapid advancement of integrated circuit (IC) technology in recent years, much demand has been aroused in exploration of differential devices with a lot of improved functionalities. In many IC applications, the unintended coupling or crosstalk within the whole circuit or module has become a critical factor or limitation in achieving the performance as desired. Differential circuit topology as adopted in ICs has its increased crosstalk immunity and increased dynamic range rather than single-ended circuits, so it is now highly demanded to develop various differential-mode or balanced radio-frequency (RF) and microwave devices and circuits for exploration of the whole module under differential-mode operation.

Among various RF and microwave circuits, low-noise amplifiers (LNA), power amplifiers (PA), mixers, and voltage-controlled oscillators (VCO) have successfully been developed to operate under differential mode, but very few reported works have been carried out to develop the balanced or differential-mode filters. With a single-ended mode filter to be connected to a dipole antenna or balanced amplifier, a balun is definitely required to convert a balanced signal into an unbalanced one or vice versa as illustrated in Figure 4.2. To establish a fully balanced system, there is a prompt need in development of a balanced or differential-mode filter so as to eliminate the use of any balun as depicted in Figure 4.3. This balanced filter has gained a few attractive features such as size reduction, low insertion loss, and flat group delay.

As the first approach, a single-ended mode wideband or ultra-wideband (UWB) bandpass filter can be straightforwardly implemented by directly cascading the low- and high-pass filters together. The works in Refs. [6–8] are typical examples for this type of wideband filters. In Ref. [6], a low-pass filter is realized by forming up a conventional stepped-impedance structure, while a high-pass filter is constructed by attaching a few short-ended stubs in shunt along a connecting transmission line. As a result, the lower and higher cutoff frequencies of the resultant bandpass filter are fully decided by those of the high- and low-pass filters, respectively.

In this context, a UWB bandpass filter with good out-of-band performance [8] is designed and explored on a basis of this proposed approach. However, this suffers from periodic frequency response due to its frequency-distributive nature. To suppress the harmonic passbands,

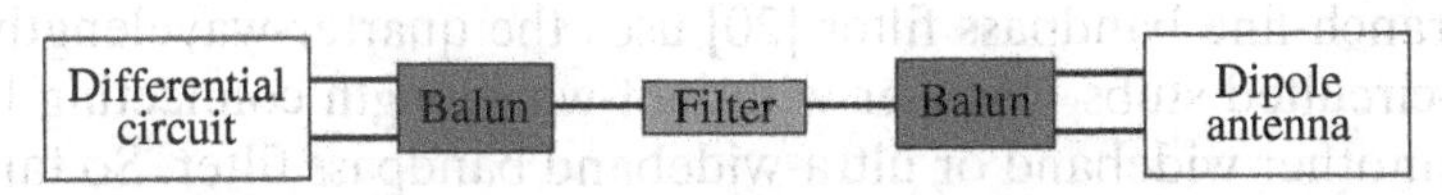

Figure 4.2 Block diagram of an RF receiver with a single-ended mode filter and two baluns.

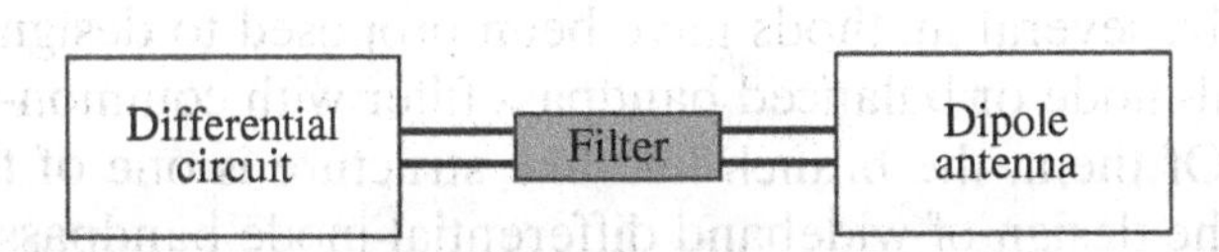

Figure 4.3 Block diagram of an RF receiver with a balanced filter.

electromagnetic bandgaps (EBGs) are formed by the capacitively loaded transmission lines, and they are appropriately placed between the two adjacent short-circuited stubs. As such, this proposed filter can achieve a desired wide passband with widened upper stopband.

As the second approach, a UWB bandpass filter can be realized by making use of a multiple-mode resonator (MMR). As was firstly proposed in Ref. [9], a wide or ultrawide passband of this filter is achieved under simultaneous excitation of a few resonant modes in a single MMR. Unlike the traditional parallel-coupled resonator filters [10], this proposed filter utilizes the first few resonant modes of an MMR to construct the passband. By properly adjusting the impedance ratio of the MMR, the resonant frequencies are able to be located within the targeted passband. Together with the input/output parallel-coupled lines, a wide or ultrawide passband with five transmission poles [11–16] can be satisfactorily achieved. In Ref. [11], the MMR is composed of a half-wavelength low characteristic impedance section in center and two quarter-wavelength high characteristic impedance sections in two sides. The first three resonant modes of MMR can be appropriately excited and used to realize the UWB passband. In this aspect, its first and third resonant modes can be used to determine the upper and lower cutoff frequencies of the desired UWB passband, thus constructing the first UWB bandpass filter under the concept of multiple-mode resonance in the MMR. Using the same concept in design, the backside aperture is installed underneath the quarter-wavelength coupled microstrip-line section to highly enhance the desired coupling degree [12] with relaxed requirement in fabrication tolerance. This MMR filter on a thin substrate has the reduced overall size. Later on, a few UWB filters are developed on the hybrid microstrip/CPW structure [13] and short-circuited CPW [14] and also exhibit good UWB passband performance.

In Refs. [17–19], a high-pass filter with short-circuited stubs in shunt is extended to form an alternative class of wideband filters. Another similar branch-line bandpass filter [20] uses the quarter-wavelength shunt short-circuited stubs together with half-wavelength connecting lines to form another wideband or ultra-wideband bandpass filter. So far, these three advanced approaches have been successfully employed to develop a variety of wideband or ultra-wideband bandpass filters with enhanced in-band and out-of-band performances.

Recently, several methods have been proposed to design wideband differential-mode or balanced bandpass filter with common-mode suppression. Of them, the branch-line-like structure is one of the earliest works in the design of wideband differential-mode bandpass filter with common-mode suppression. This method has been still widely used because of its simple geometry, easy design rule, and compact size. In another method, a wideband bandpass differential-mode filter with enhanced common-mode rejection can be constituted by using the transversal signal interference technique [21]. This design concept stems from the distributed active filter theory [22].

Meanwhile, the 180° phase shifter or parallel-coupled line with short circuit ended together with slotline resonators could be used in design of another class of wideband balanced bandpass filters. Moreover, an open/short T-shaped structure [23] is used as a resonator to design a highly selective wideband balanced bandpass filter with good common-mode suppression. In addition, the half-wavelength ring resonator with four open stubs is applied in Ref. [24] to design another wideband balanced bandpass filter.

4.2 BRANCH-LINE BALANCED WIDEBAND BANDPASS FILTER

It is desired to design a wideband bandpass filter for upper band of UWB from 3 to 5 GHz. Figure 4.4 depicts the equivalent transmission-line schematic of the proposed wideband balanced bandpass filter that is based on a modified three-stub branch-line structure [25] with a pair of open-circuited stubs, marked by L_A. All the transmission lines in Figure 4.4 are set as about one quarter-wavelength with respect to the central frequency at $f_0 = 4.0$ GHz, and the characteristic impedances of these lines or stubs are set as follows: $Z_1 = 30.1\,\Omega$, $Z_2 = 24.1\,\Omega$, $Z_{12} = 49.0\,\Omega$, and $Z_A = 100.0\,\Omega$. As shown in Figure 4.4, this four-port filter is ideally symmetric with respect to the central plane in horizon (dotted line). Therefore, this horizontal symmetrical line can be considered as a perfect electric or magnetic wall if one of the paired ports, ports 1 and 1′

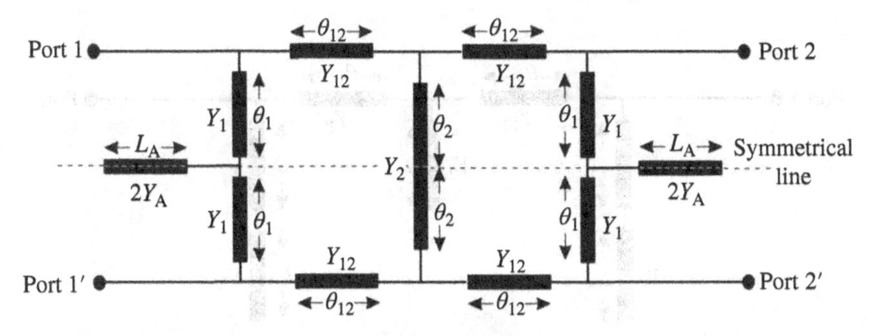

Figure 4.4 Schematic of the proposed 4-port balanced wideband bandpass filter. Reprinted with permission from Ref. [25]; copyright 2009 IEEE.

or ports 2 and 2′, are driven or excited by differential- or common-mode signal. Figures 4.5 and 4.6 show the two symmetrical bisections of this filter under differential and common-mode operation, respectively. Due to existence of perfect electric wall in the former case [26, 27], the introduced stubs at the first and third vertical arms are electrically short-circuited, thus having no influence on the differential-mode frequency response. However, in the latter case, these two stubs actually lead to extension and reconstruction of the first and third vertical arms, thereby providing us with an expected capacity in reshaping the frequency response under common-mode operation.

Looking at Figure 4.5 again, this two-port bisection with shunt short-circuited stubs are exactly the same as the transmission-line topology of the bandpass filters [28–30] or ultra-wideband bandpass filters [6]. Thus, its wide passband performance can be easily designed using an efficient synthesis approach. In Figure 4.5, the three shunt short-circuited stubs and connecting lines are $\lambda_g/4$ long, where λ_g is the guided wavelength of the relevant microstrip line. For a given filter degree (n), the frequency response of this filter is primarily dependent on characteristic admittances, Y_i ($i=1$ to n) and $Y_{i,i+1}$ ($i=1$ to $n-1$), of the three short-circuited stubs and two connecting lines. If a wide bandwidth is specified for this filter, all of these characteristic admittances can be

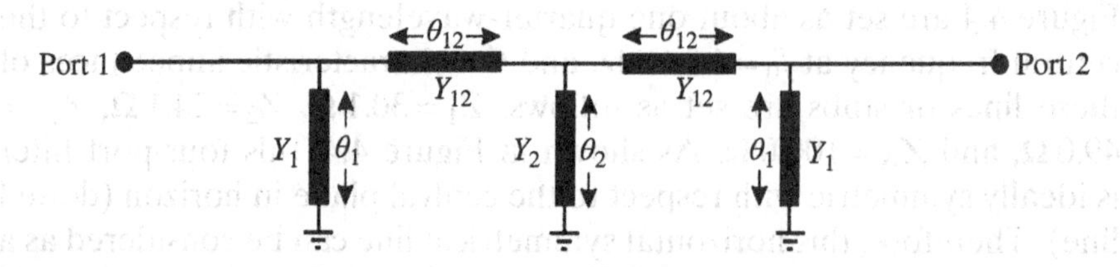

Figure 4.5 Schematic of the 2-port differential-mode bisection of the 4-port filter in Figure 4.4. Reprinted with permission from Ref. [25]; copyright 2009 IEEE.

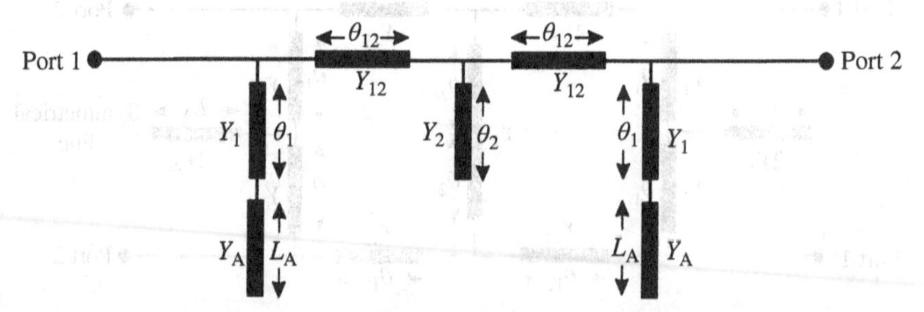

Figure 4.6 Schematic of the 2-port common-mode bisection of the 4-port filter in Figure 4.4. Reprinted with permission from Ref. [25]; copyright 2009 IEEE.

explicitly determined using the closed-form design equations in Refs. [28–30], which are reproduced here for completeness:

$$\theta = \frac{\pi}{2}\left(1 - \frac{\text{FBW}}{2}\right) \tag{4.1a}$$

$$\frac{J_{1,2}}{Y_0} = g_0\sqrt{\frac{hg_1}{g_2}} \quad \frac{J_{n-1,n}}{Y_0} = g_0\sqrt{\frac{hg_1 g_{n+1}}{g_0 g_{n-1}}} \tag{4.1b}$$

$$\frac{J_{i,i+1}}{Y_0} = \frac{hg_0 g_1}{\sqrt{g_i g_{i+1}}} \quad \text{for } i = 2 \text{ to } n-2 \tag{4.1c}$$

$$N_{i,i+1} = \sqrt{\left(\frac{J_{i,i+1}}{Y_0}\right)^2 + \left(\frac{hg_0 g_1 \tan\theta}{2}\right)^2} \quad \text{for } i = 1 \text{ to } n-1 \tag{4.1d}$$

$$Y_1 = g_0 Y_0\left(1 - \frac{h}{2}\right) g_1 \tan\theta + Y_0\left(N_{1,2} - \frac{J_{1,2}}{Y_0}\right) \tag{4.1e}$$

$$Y_n = Y_0\left(g_n g_{n+1} - g_0 g_1 \frac{h}{2}\right)\tan\theta + Y_0\left(N_{n-1,n} - \frac{J_{n-1,n}}{Y_0}\right) \tag{4.1f}$$

$$Y_i = Y_0\left(N_{i-1,i} + N_{i,i+1} - \frac{J_{i-1,i}}{Y_0} - \frac{J_{i,i+1}}{Y_0}\right) \quad \text{for } i = 2 \text{ to } n-1 \tag{4.1g}$$

$$Y_{i,i+1} = Y_0\left(\frac{J_{i,i+1}}{Y_0}\right) \tag{4.1h}$$

where g_i is the ith element value of a ladder-type low-pass prototype filter, such as a Chebyshev-type filter, given for a normalized cutoff $\Omega_c = 1.0$ and h is a dimensionless constant, which may be assigned to another value so as to give a convenient admittance level in the interior of the filter.

In Figure 4.6, the central open-circuited stub, marked by θ_2, is 90° long at the center frequency, thus creating a transmission zero at the central frequency. Meanwhile, the first and third stubs in Figure 4.6 are lengthened due to shunt attachment of the horizontal stubs as shown in Figure 4.4, thus making up two identical stepped-impedance open-circuited stubs [25]. By properly choosing the length (L_A) and adjusting the ratio of impedances (Z_A/Z_1), the first and second zeros of these two stubs can be moved downward and further symmetrically located at the two sides of the first zero. As a result, a wide common-mode stopband can be achieved by using these transmission zeros.

Figures 4.7 and 4.8 illustrate the frequency responses of the two transmission coefficients, $|S_{21}^{dd}|$ and $|S_{21}^{cc}|$, under differential- and common-mode operation. First of all, we can see from Figures 4.7 and 4.8 that $|S_{21}^{dd}|$ is practically unchanged, regardless of a large variation in the length (L_A) or impedance (Z_A). As such, it is confirmed that the attached stubs hardly affect the differential-mode passband response of this filter. Now, let us closely look at Figures 4.7 and 4.8 to understand how these attached stubs can reshape the common-mode stopband performances. As shown in Figure 4.7, there exists only a single transmission zero at the mid-band frequency of $f_0 = 4.0$ GHz of the constituted differential-mode passband for zero length of two attached stubs, that is, $L_A = 0$ mm. As L_A is increased to 4.0 mm and then to 6.5 mm, the second zero appears at low frequencies and moves downward. Meanwhile, the third zero moves downward to be reallocated at the opposite side of the second zero. As such, a wide common-mode stopband with sharp rejection skirts is realized.

Figure 4.8 illustrates that these lower and higher zeros can be moved outward or inward with each other by varying the impedance Z_A so as to achieve the specified bandwidth of the realized common-mode stopband. As shown in Figure 4.8, as Z_A is decreased from 150 to 50 Ω, the first and third transmission zeros can be further moved far away from each other. As a result, this common-mode stopband can

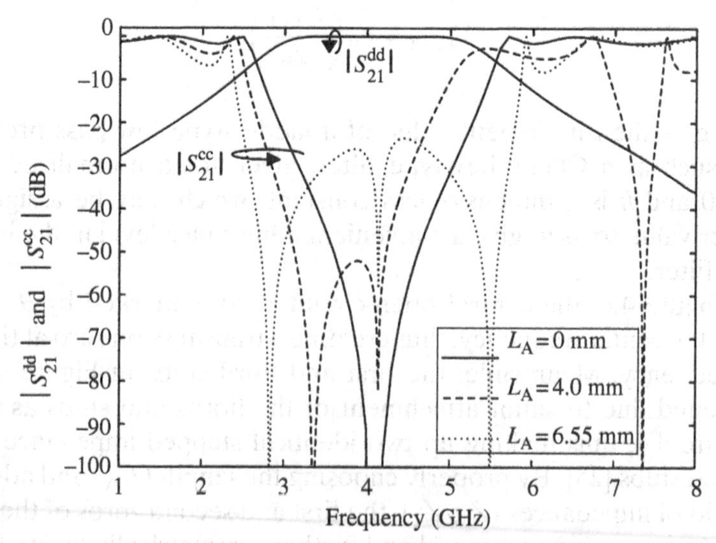

Figure 4.7 Simulated frequency-dependent transmission coefficient magnitude, $|S_{21}|$, for differential- and common-mode cases for the stub length of $L_A = 0$, 4 and 6.55 mm. Reprinted with permission from Ref. [25]; copyright 2009 IEEE.

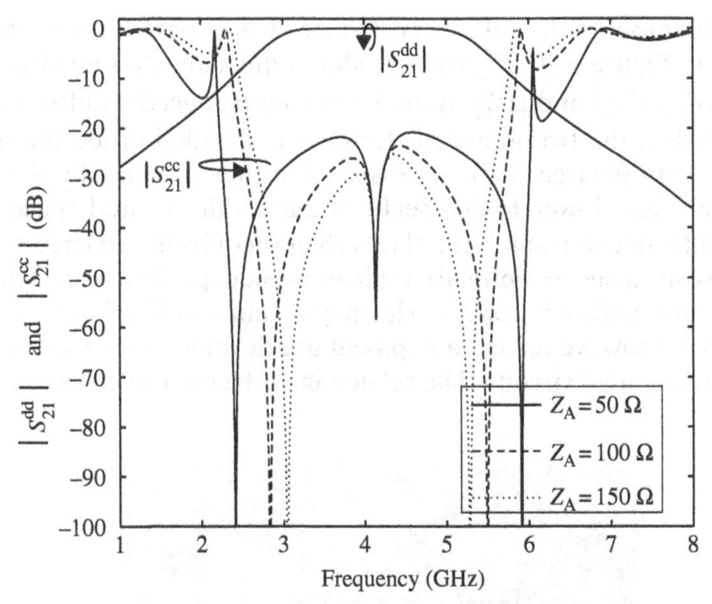

Figure 4.8 Simulated frequency-dependent transmission coefficient magnitude, $|S_{21}|$, for differential- and common-mode cases for the stub impedance of $Z_A = 50$, 100 and 150 Ω. Reprinted with permission from Ref. [25]; copyright 2009 IEEE.

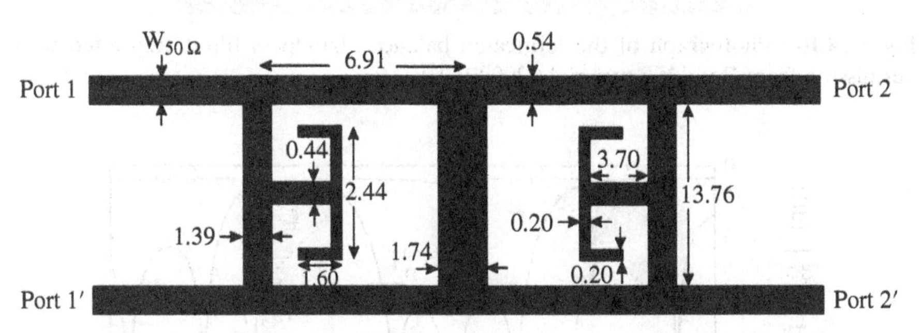

Figure 4.9 Layout of the optimized balanced bandpass filter. Reprinted with permission from Ref. [25]; copyright 2009 IEEE.

be easily widened and narrowed in order to cover the entire differential-mode passband.

The microstrip-line filter is designed and fabricated on the RT/Duroid 6010 with a substrate thickness of 0.635 mm and dielectric constant of 10.8. After its initial dimensions are determined via simple transmission-line model as illustrated in Figure 4.4, its final layout is optimally designed using the full-wave ADS simulator. Figure 4.9 indicates the physical layout with all the dimensions provided in millimeter.

A photograph of the fabricated balanced bandpass filter is also presented in Figure 4.10. Figure 4.11 shows the simulated results, that is, $|S_{11}^{dd}|$, $|S_{21}^{dd}|$, $|S_{11}^{cc}|$ and $|S_{21}^{cc}|$ of the fabricated balanced bandpass filter.

In design, the two attached stubs are embedded inside the filter in order not to enlarge the overall size, and they are equally divided to the upper and lower folded sections due to the limited space within the branch-line portion. Next, the evaluation of balanced circuits is necessary so as to ensure optimal circuit and system performance, but it presents many challenging issues. The major stumbling block is that most of RF and microwave testing equipment are developed for two- or multiport single-ended circuits. The related infrastructure such as calibration

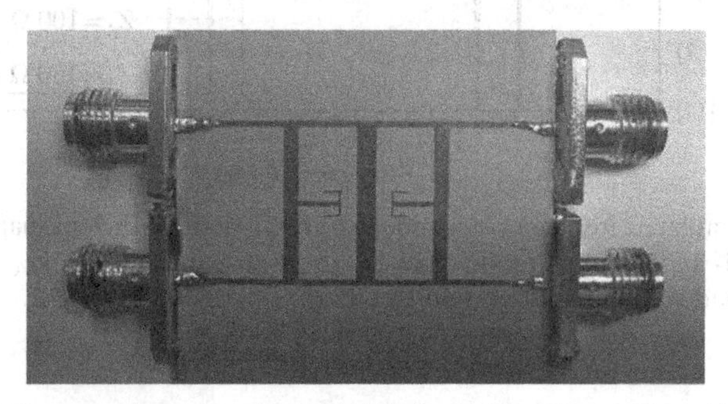

Figure 4.10 Photograph of the fabricated balanced bandpass filter. Reprinted with permission from Ref. [25]; copyright 2009 IEEE.

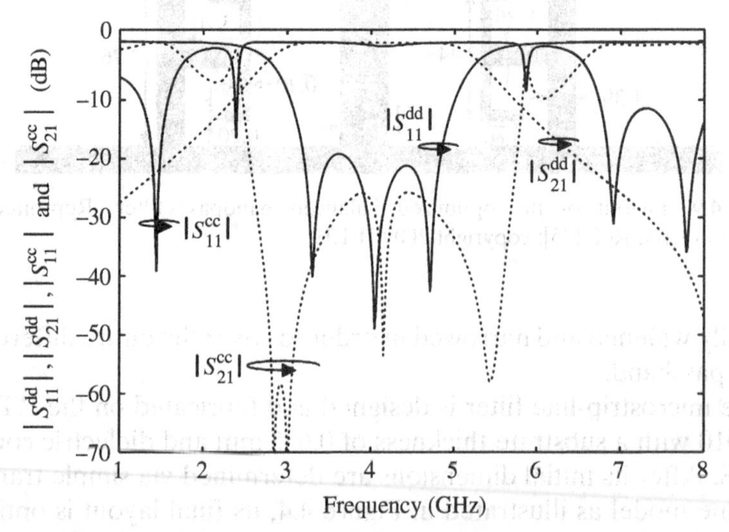

Figure 4.11 Simulation results of $|S_{11}^{dd}|$, $|S_{21}^{dd}|$, $|S_{11}^{cc}|$, and $|S_{21}^{cc}|$. Reprinted with permission from Ref. [25]; copyright 2009 IEEE.

standards, transmission lines, connectors, and even industry-standard reference impedances are set in the unbalanced format.

To our best knowledge, the two well-developed approaches are usually applied to measure the balanced circuits using the standard vector network analyzer (VNA). One is to convert each balanced port to a single-ended port using a balun and to measure this network on a single-ended VNA. As one disadvantage, this approach is inaccurate because the reference plane for calibration is located at the single-ended test port of the VNA, while the desired measurement reference plane is located at the balanced port of the device under test (DUT). The balun in between usually does not exhibit the ideal electrical performance in a wide operating band, thus probably degrading the accuracy in measurement. As the other disadvantage, this approach is not comprehensive in general since it can only portray the pure differential mode of operation, but it is not valid for the other three modes. The preferred method needs to be developed by experimentally characterizing the four-port DUT using the mixed-mode *S*-parameters, S^{mm}. The four-port *S*-parameters should be at first obtained so that the two sets of two-port differential- and common-mode *S*-parameters can then be extracted using the symmetrical properties of the considered four-port network as extensively discussed in Ref. [31].

In a balanced circuit, two terminals constitute a single port that can support both differential- and common-mode signals if no amplitude and phase in them are not ideally enforced. In this aspect, the overall four-port network performance can be in general described using the four-port mixed-mode *S*-parameters. Thereafter, the four-port device or circuit is at first measured by the VNA to derive its four-port parameters. The two-port differential- and common-mode parameters (S^{dd} and S^{cc}) can then be extracted from the standard four-port scattering matrix, $[\mathbf{S}^{std}]$, as given by Equations (4.2a)–(4.2f):

$$[\mathbf{S}^{std}] = \begin{bmatrix} S_{11} & S_{12} & S_{13} & S_{14} \\ S_{21} & S_{22} & S_{23} & S_{24} \\ S_{31} & S_{32} & S_{33} & S_{34} \\ S_{41} & S_{42} & S_{43} & S_{44} \end{bmatrix} \tag{4.2a}$$

$$[S^{mm}] = \begin{bmatrix} S_{11}^{dd} & S_{12}^{dd} & S_{11}^{dc} & S_{12}^{dc} \\ S_{21}^{dd} & S_{22}^{dd} & S_{21}^{dc} & S_{22}^{dc} \\ S_{11}^{cd} & S_{12}^{cd} & S_{11}^{cc} & S_{12}^{cc} \\ S_{21}^{cd} & S_{22}^{cd} & S_{21}^{cc} & S_{22}^{cc} \end{bmatrix} \tag{4.2b}$$

$$[S^{mm}] = [M]\,[\mathbf{S}^{std}]\,[M]^{-1} \tag{4.2c}$$

$$[M] = \frac{1}{\sqrt{2}}\begin{bmatrix} 1 & -1 & 0 & 0 \\ 0 & 0 & 1 & -1 \\ 1 & 1 & 0 & 0 \\ 0 & 0 & 1 & 1 \end{bmatrix} \tag{4.2d}$$

$$[S^{dd}] = \begin{bmatrix} S_{11}^{dd} & S_{12}^{dd} \\ S_{21}^{dd} & S_{22}^{dd} \end{bmatrix} \tag{4.2e}$$

$$[S^{cc}] = \begin{bmatrix} S_{11}^{cc} & S_{12}^{cc} \\ S_{21}^{cc} & S_{22}^{cc} \end{bmatrix} \tag{4.2f}$$

From Equation (4.2c), we can simply extract the two 2×2 S-matrix, $[S^{dd}]$ and $[S^{cc}]$, in correspondence to the differential- and common-mode frequency responses, respectively, from the standard 4-port measurement. Figure 4.12 gives the comparison between the simulated and measured transmission coefficients for differential- and common-mode operation. Good agreement between them is obtained. As demonstrated in Figure 4.2, a wide differential-mode passband is achieved in the frequency range of 2.7–5.3 GHz with a fractional bandwidth of about 65% at center frequency of 4.0 GHz. Over this frequency range, the

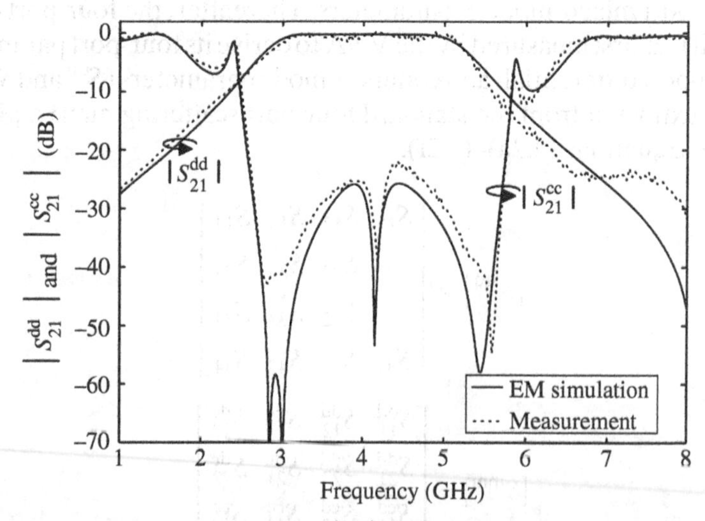

Figure 4.12 Simulated and measured $|S_{21}|$ under differential- and common-mode operations. Reprinted with permission from Ref. [25]; copyright 2009 IEEE.

common-mode attenuation with three transmission zeros is higher than 20.0 dB as can be seen in both simulation and experiment. In design, the principle of this proposed filter is quantitatively explained via transmission-line theory, and its final layout is optimized via a full-wave simulator. In final, predicted results are well validated by measuring the fabricated filter over a wide frequency range.

4.3 BALANCED BANDPASS FILTER FOR UWB APPLICATION

Figure 4.13 depicts the schematic of another balanced bandpass filter that is targeted to achieve 110% fractional bandwidth [32]. This proposed balanced filter is constituted based on a branch-line structure with two pairs of open-circuited stubs, which are attached along the symmetrical plane. Again, as shown in Figure 4.13, this four-port filter is ideally symmetric with respect to the horizontal central plane (dotted line). Thus, we can well figure out that this horizontal symmetric plane becomes a perfect electric or magnetic wall if either of differential- or common-mode signals is launched. Figures 4.14 and 4.15 show the two symmetrical bisections of this filter under differential- and common-mode operation, respectively. Owing to the existence of perfect electric wall in the former case as discussed earlier on, the introduced stubs at the first, second, fifth, and sixth vertical arms are electrically short-circuited, thus having no influence on the differential-mode

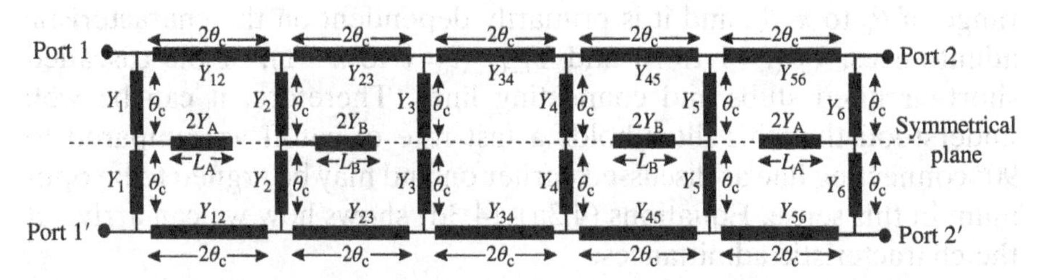

Figure 4.13 Schematic of the proposed balanced UWB bandpass filter. Reprinted with permission from Ref. [32]; copyright 2009 IEEE.

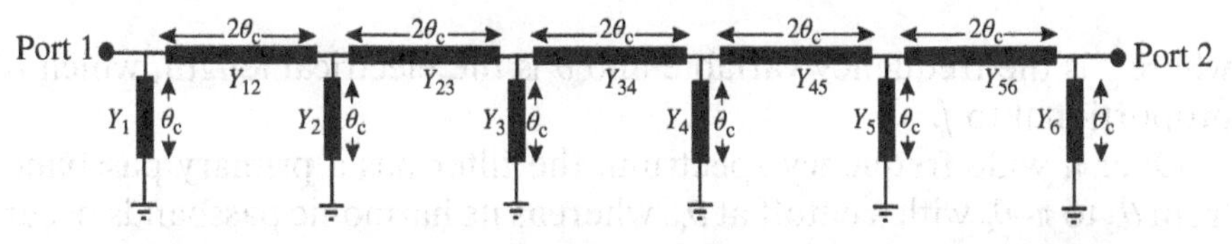

Figure 4.14 Schematic of two-port differential-mode bisection of proposed filter in Figure 4.13. Reprinted with permission from Ref. [32]; copyright 2009 IEEE.

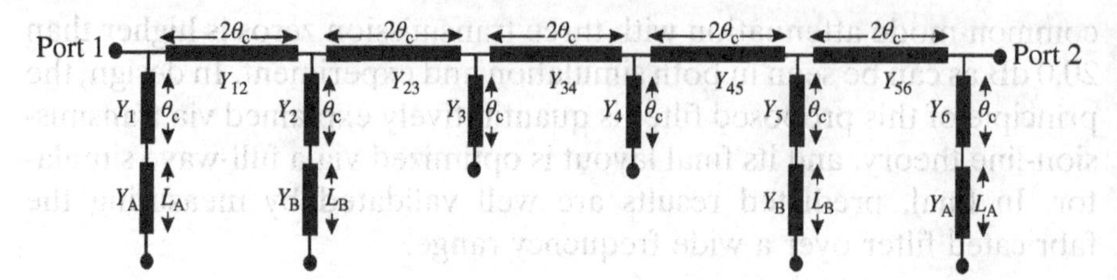

Figure 4.15 Schematic of two-port common-mode bisection of proposed filter in Figure 4.13. Reprinted with permission from Ref. [32]; copyright 2009 IEEE.

frequency response. However, under the common-mode excitation, these two pairs of stubs actually give us an extra degree of freedom to redesign this balanced filter as a stopband filter under the common-mode excitation.

In our design, frequency response of this proposed balanced UWB bandpass filter is synthesized based on an optimum distributed high-pass filter. Under the differential-mode operation, the bisection of this four-port filter is exactly the same as the two-port filter topology in Refs. [17–19], and its filtering performance can be synthesized with resorting to the shunt short-circuited stubs of electrical length θ_c and connecting lines of electrical length $2\theta_c$ at the cutoff frequency f_c of the prototype high-pass filter as discussed in Ref. [29]. Although the filter consists of only n stubs, it has an insertion loss function of degree $2n-1$ in frequency so that its high-pass filtering response has the $2n-1$ transmission poles. The frequency response of this filter has a desired passband in a range of θ_c to $\pi-\theta_c$, and it is primarily dependent on the characteristic admittances, Y_i ($i=1$ to n) and $Y_{i,i+1}$ ($i=1$ to $n-1$), of the cascaded short-circuited stubs and connecting lines. Therefore, it can be well understood that this filter holds a fast rate of cutoff as compared to 90° connecting line as discussed earlier on and may be argued to be optimum in this sense. Equations (4.3a)–(4.3d) shows how we can arrive at the characteristic admittances:

$$\theta = \theta_c \frac{f}{f_c} \tag{4.3a}$$

where f is the frequency variable and θ is the electrical length, which is proportional to f.

Over a wide frequency spectrum, the filter has a primary passband from θ_c to $\pi-\theta_c$ with a cutoff at θ_c, whereas its harmonic passbands occur periodically, centered at $\theta = 3\pi/2, 5\pi/2,\ldots$ and separated by attenuation poles located at $\theta = \pi, 2\pi,\ldots$:

$$|S_{21}(\theta)|^2 = \frac{1}{1+\varepsilon^2 F_N^2(\theta)} \tag{4.3b}$$

where ε is the passband ripple constant, θ is the electrical length as defined as in Equation (4.3a), and F_N is the polynomial function given by

$$F_N = \frac{(1+\sqrt{1-x_c^2})T_{2n-1}(x/x_c) - (1-\sqrt{1-x_c^2})T_{2n-3}(x/x_c)}{2\cos((\pi/2)-\theta)} \tag{4.3c}$$

where n is the number of short-circuit stubs

$$x = \sin\left(\frac{\pi}{2}-\theta\right) \quad x_c = \sin\left(\frac{\pi}{2}-\theta_c\right) \tag{4.3d}$$

and $T_n(x) = \cos(n\cos^{-1} x)$ is the Chebyshev function of the first kind of degree n.

For the differential-mode bisection with six short-circuited stubs in Figure 4.14, the lower cutoff frequency f_c is set to 3.1 GHz, while the upper cutoff frequency $(\pi/\theta_c - 1)\, f_c$ is designated at 10.6 GHz in order to meet the UWB spectrum mask. Based on the design formulas, Equation (4.3a) that is discussed earlier, θ_c is derived as 40.73° and characteristic impedances of all the stubs and connecting lines are determined from Equations (4.3b)–(4.3d) as $Z_1 = Z_6 = 79.73\ \Omega$, $Z_2 = Z_5 = 52.88\ \Omega$, $Z_3 = Z_4 = 47.12\ \Omega$, $Z_{12} = Z_{56} = 51.36\ \Omega$, $Z_{23} = Z_{45} = 53.71\ \Omega$, and $Z_{34} = 54.24\ \Omega$, respectively [29].

As illustrated in Figures 4.16 and 4.17, a wide common-mode stopband can be constructed by properly selecting the length (L_A and L_B) and adjusting the ratio of impedances (Z_A/Z_1 and Z_B/Z_2). As also demonstrated in Figures 4.16 and 4.17, no matter what is the value of L_A and L_B, Z_A and Z_B, $|S_{21}^{dd}|$ remains unchanged. Figure 4.16 presents three cases. Under $L_A = 0$ and $L_B = 0$ (case A), there only exists a single transmission zero at the central frequency. As L_A is stretched to 90° with $Z_A = 100\ \Omega$ (case B), two additional transmission zeros are excited, and they are symmetrically located at the two sides of the central frequency. For the case C with $L_A = L_B = 90°$ and $Z_A = Z_B = 100\ \Omega$, two additional zeros are excited. Figure 4.17 shows the effect of characteristic impedances of these stubs on the locations of these two transmission zeros. By changing Z_B from 100 Ω (case A) to 50 Ω (case B) with the unchanged $L_A = L_B = 90°$ and $Z_A = 100\ \Omega$, the second and fourth transmission zeros move outward as expected. By changing Z_A from 100 Ω (case A) to 50 Ω (case C), the first and fifth transmission zeros are forced

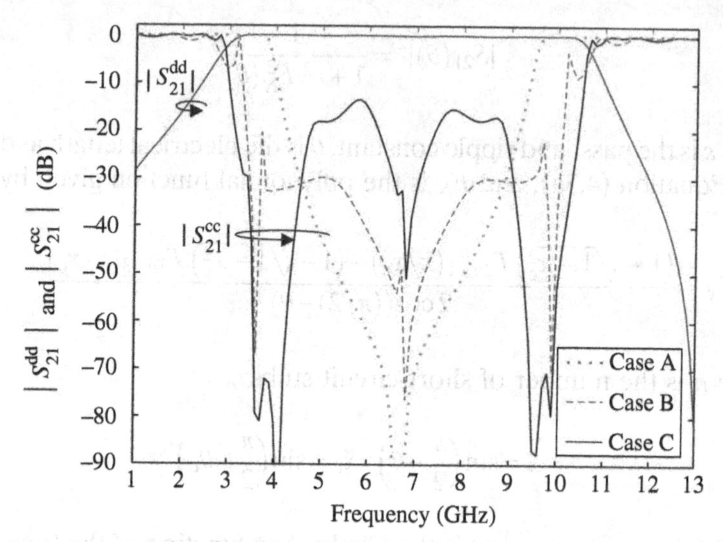

Figure 4.16 Transmission coefficients, $|S_{21}|$, for differential- and common-mode operations. Case A ($L_A = 0$ and $L_B = 0$), Case B ($L_A = 90°$, $L_B = 0$ and $Z_A = 100\,\Omega$), and Case C ($L_A = L_B = 90°$ and $Z_A = Z_B = 100\,\Omega$). Reprinted with permission from Ref. [32]; copyright 2009 IEEE.

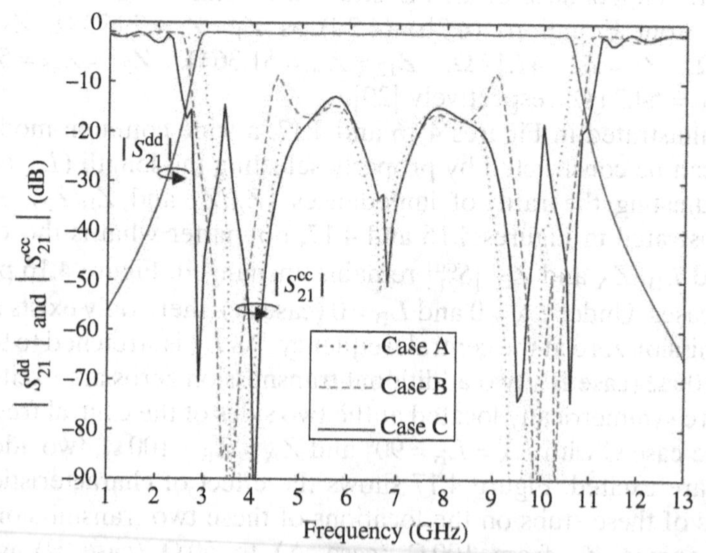

Figure 4.17 Transmission coefficients, $|S_{21}|$, for differential- and common-mode operations. Case A ($L_A = L_B = 90°$ and $Z_A = Z_B = 100\,\Omega$), Case B ($L_A = L_B = 90°$, $Z_A = 100\,\Omega$, and $Z_B = 50\,\Omega$), and Case C ($L_A = L_B = 90°$, $Z_A = 50\,\Omega$, and $Z_B = 100\,\Omega$). Reprinted with permission from Ref. [32]; copyright 2009 IEEE.

to move outward. These results provide us with a qualitative hint and design guideline on how the common-mode stopband can be created in the entire UWB band with good in-band return loss.

In this section, the proposed balanced UWB bandpass filter is optimally designed via the full-wave *keysight ADS* simulator with the target of good differential-mode transmission and good common-mode attenuation in the desired UWB range. Layout of the designed UWB filter and its photograph are depicted in Figures 4.18 and 4.19, respectively. The filter is implemented on the *RT/Duroid 6010* with a substrate thickness of 0.635 mm and relative dielectric constant of 10.8.

Figure 4.20 shows the comparison between the simulated and measured transmission coefficients for differential- and common-mode operation. Good agreement between them is obtained. Desired insertion loss configurations of the ideal UWB filters for the indoor and handheld UWB system defined by FCC are also plotted in Figure 4.20 for better comparison. As demonstrated in Figure 4.20, a differential-mode ultra-wide passband is achieved in the frequency range of 2.9–10.7 GHz with a fractional bandwidth of about 117%. Over this frequency range, the common-mode attenuation is better than 10.0 dB with resorting to 5 transmission zeros. On the other hand, some unexpected discrepancies

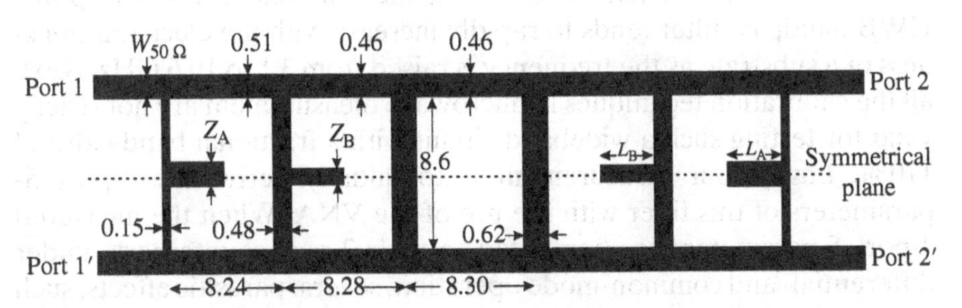

Figure 4.18 Layout of proposed balanced UWB bandpass filter (all dimensions are in millimeter). Reprinted with permission from Ref. [32]; copyright 2009 IEEE.

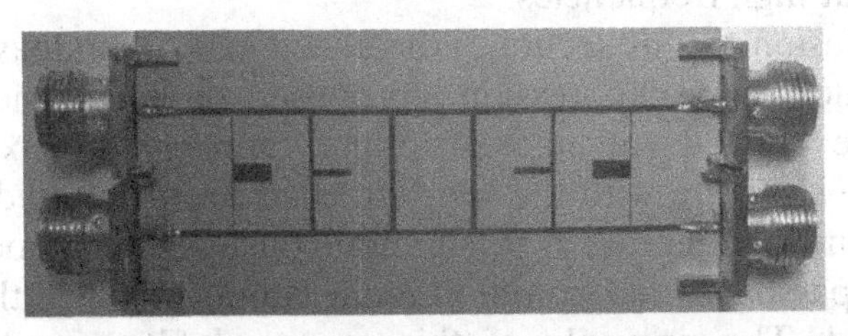

Figure 4.19 Photograph of the fabricated balanced UWB bandpass filter (top view). Reprinted with permission from Ref. [32]; copyright 2009 IEEE.

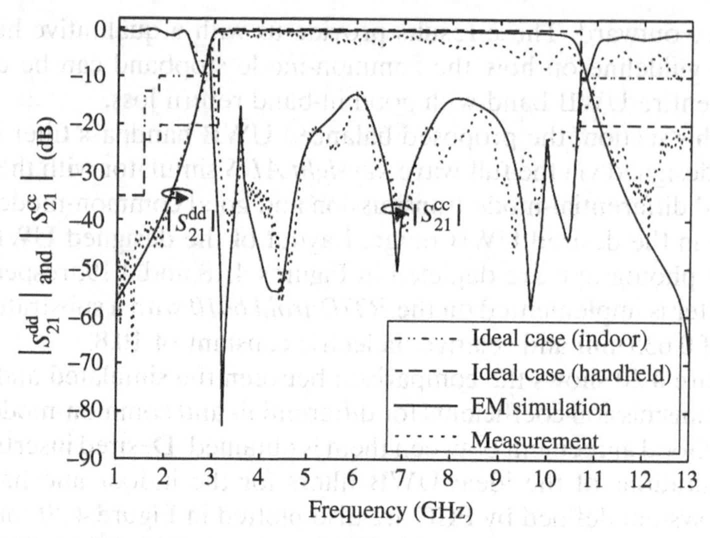

Figure 4.20 Simulated and measured $|S_{21}|$ under differential- and common-mode operations. Reprinted with permission from Ref. [32]; copyright 2009 IEEE.

between simulation and experiment are observed especially at high frequencies as will be investigated in the following text.

First of all, the radiation loss from the unbounded microstrip-line UWB bandpass filter tends to rapidly increase with the electrical thickness of a substrate as the frequency is raised from 3.1 to 10.6 GHz. Next, all the calibration techniques in microwave measurement are not strictly valid for testing such a wideband circuit with a fractional bandwidth of 110%. Lastly, our measurement is to initially derive the 4-port *S*-parameters of this filter with the use of the VNA. When the measured 4-port *S*-parameters are converted to their 2-port counterparts under differential- and common-mode operation, several parasitic effects, such as coaxial-to-microstrip discontinuities at SMA connectors and slight variation in lengths/widths of four feeding lines due to fabrication tolerance, have unfortunately affected the resultant 2-port *S*-parameters especially at high frequencies.

In conclusion to this section, a balanced UWB bandpass filter with good common-mode suppression is proposed, designed, and implemented on the single-layer microstrip-line topology. A six shunt-stub branch-line structure with the four attached open-circuited stubs at its central plane is constructed to simultaneously achieve good differential-mode passband and common-mode stopband over the specified UWB band. The principle of this proposed filter is quantitatively explained based on the equivalent model, and its final layout is optimally designed via full-wave simulator. Predicted results are well validated by measuring a fabricated filter over a wide frequency range.

4.4 BALANCED WIDEBAND BANDPASS FILTER WITH GOOD COMMON-MODE SUPPRESSION

Figure 4.21 depicts the transmission-line schematic of another proposed balanced wideband bandpass filter that is also based on a branch-line structure with a connecting line between the two vertical arms [32]. A pair of open-circuited stubs is connected to the central position of the connecting line so as to widen the spacing between the two transmission zeros. The transmission line sections in Figure 4.21, denoted by θ_1, θ_A, and θ_B, are all equal to quarter-wavelength in length, and θ_2 is around half-wavelength with respect to central frequency of $f_0 = 6.85$ GHz. The final characteristic impedances of the line sections are derived [29] as $Z_1 = 30\ \Omega$, $Z_2 = 70\ \Omega$, $Z_A = 80\ \Omega$, and $Z_B = 50\ \Omega$ from Equations (4.3b)–(4.3d). As shown in Figure 4.21, this four-port filter is again ideally symmetric with respect to the central plane in horizon (dotted line). Therefore, this horizontal symmetrical plane can be considered as a perfect electric or magnetic wall if one of the paired ports, ports 1 and 1′ or ports 2 and 2′, are driven or excited by differential- or common-mode signals.

Figures 4.22 and 4.23 show the two symmetrical bisections of this filter under differential- and common-mode operation, respectively. Due to existence of electric wall in the former case, the introduced connecting line and open-circuited stubs are electrically short-circuited with the ground plane, thus giving no influence on the differential-mode

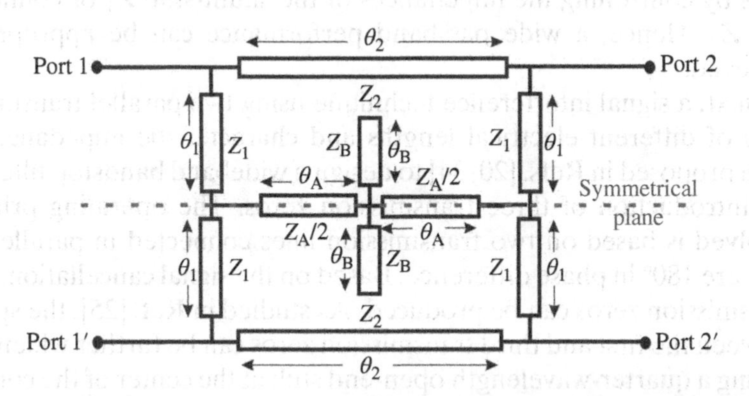

Figure 4.21 Transmission-line schematic of another proposed balanced wideband bandpass filter. Reprinted with permission from Ref. [33]; copyright 2009 IEEE.

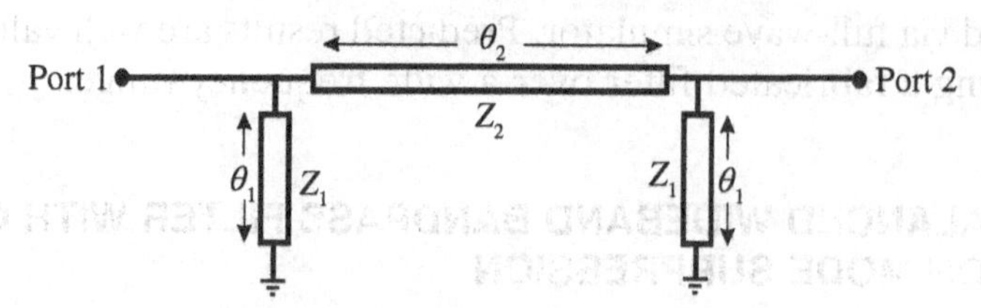

Figure 4.22 Transmission-line schematic of the differential-mode half bisection of the filter in Figure 4.21.

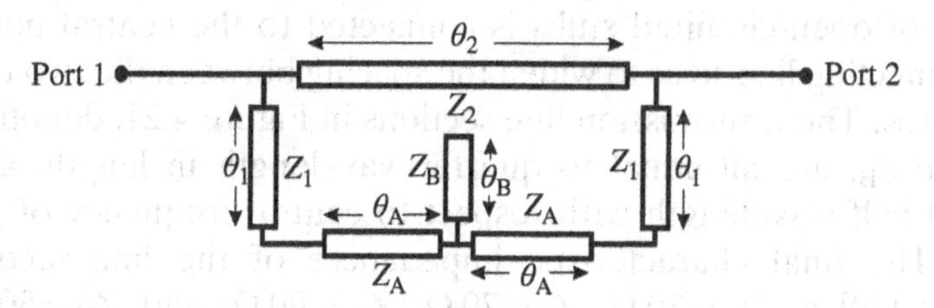

Figure 4.23 Transmission-line schematic of the common-mode half bisection of the filter in Figure 4.21. Reprinted with permission from Ref. [33]; copyright 2009 IEEE.

frequency response. However, in the latter case, this connecting line and open-circuited stubs actually provide us with an expected capacity in reshaping the frequency response to a bandstop under common-mode operation.

Figure 4.22 shows that the two-port bisection with shunt short-circuited stubs are almost the same as the high-pass filter or UWB band-pass filter in Ref. [33]. The work in Ref. [33] exhibits that the locations of first and third transmission poles can be relocated at the desired positions by controlling the impedances of the shunt-stub Z_1 or connecting line Z_2. Hence, a wide passband performance can be appropriately achieved.

Next, a signal interference technique using two parallel transmission lines of different electrical lengths and characteristic impedances has been proposed in Refs. [20, 34] to design a wideband bandstop filter with the introduction of three transmission zeros. The operating principle involved is based on two transmission lines connected in parallel, and they are 180° in phase difference. Based on the signal cancellation, three transmission zeros can be produced. As studied in Ref. [25], the spacing between the first and third transmission zeros can be further widened by adding a quarter-wavelength open-end stub at the center of the connecting line. Looking back at Figure 4.23, the three transmission zeros in common mode can be created by introducing a connecting line, denoted by θ_A, between two vertical arms. Then, the first and third transmission zeros can be moved further away from each other as illustrated in Ref. [25] by adding a pair of open-circuited stubs at the center of this connecting line as marked by θ_B.

Figures 4.24 and 4.25 illustrate the simulated frequency responses of two distinctive transmission coefficients, $|S_{21}^{dd}|$ and $|S_{21}^{cc}|$ under differential- and common-mode operation. First of all, we can see from Figures 4.24 and 4.25 that the differential-mode $|S_{21}^{dd}|$ is unchanged regardless of a large variation in the lengths of θ_A or θ_B and the impedances of Z_A and Z_B. As such, it is confirmed that the introduced connecting line and open-circuited stub can hardly affect the differential-mode passband response of this filter regardless of these varied line and stub length. Now, let us look at Figures 4.24 and 4.25 to further investigate how these lines and stub lengths can quantitatively reshape the common-mode stopband performance.

Figure 4.24 shows three sets of simulated frequency responses of the filter under the conditions denoted by the embedded schematic diagram in Figure 4.24. When $\theta_A = 0°$ and $Z_A = 0\ \Omega$ (case A), only a single transmission zero emerges at the mid-band frequency of 6.85 GHz. When $\theta_A = 180°$ and $Z_A = 30\ \Omega$ (case B), there is still only a single transmission zero at this frequency. However, when the impedance Z_A is reduced to 15 Ω, three transmission zeros can be generated as shown in case

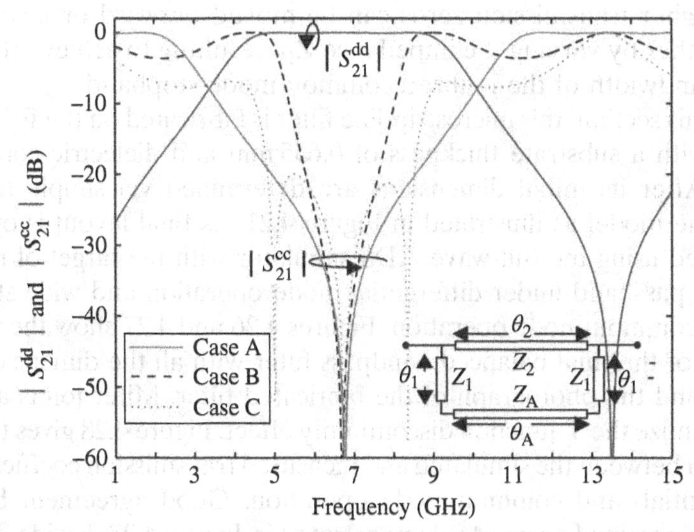

Figure 4.24 Frequency-dependent transmission coefficient magnitude, $|S_{21}|$, for differential- and common-mode cases with an additional transmission line of θ_A. Reprinted with permission from Ref. [26]; copyright 2013 IEEE.

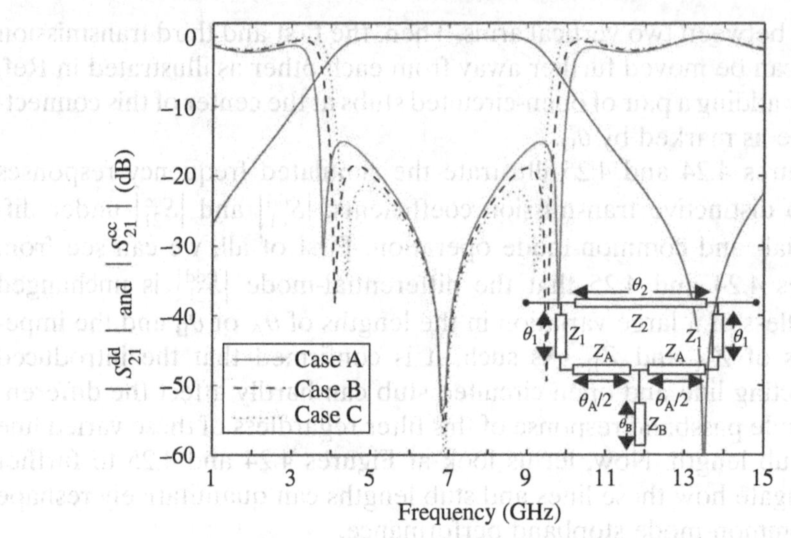

Figure 4.25 Frequency-dependent transmission coefficient magnitude, $|S_{21}|$, for differential- and common-mode cases with additional transmission line of θ_A and open-ended stubs of θ_B. Reprinted with permission from Ref. [33]; copyright 2009 IEEE.

C ($\theta_A = 180°$ and $Z_A = 15\ \Omega$). The spacing between two adjacent zeros can be further improved by adding a quarter-wavelength open-circuited stub as shown in Figure 4.25. When all the parameters are kept constant ($\theta_A = \theta_B = 180°$ and $Z_A = 80\ \Omega$) and Z_B is reduced from $60\ \Omega$ (case C), $50\ \Omega$ (case B) and $40\ \Omega$ (case A), the results illustrate that these lower and higher transmission zeros can be moved outward or inward with each other by varying the impedance Z_B, resulting to achieve the specified bandwidth of the realized common-mode stopband.

In this section, this microstrip-line filter is fabricated on the RT/Duroid 6010 with a substrate thickness of 0.635 mm and dielectric constant of 10.8. After its initial dimensions are determined via simple transmission-line model as illustrated in Figure 4.21, its final layout is optimally designed using the full-wave ADS simulator with the target of realizing a wide passband under differential-mode operation and wide stopband under common-mode operation. Figures 4.26 and 4.27 show the physical layout of the final balanced bandpass filter with all the dimensions provided and the photograph of the fabricated filter. Miter joints are used to minimize the T-junction discontinuity effect. Figure 4.28 gives the comparison between the simulated and measured transmission coefficients for differential- and common-mode operation. Good agreement between them is obtained again. As demonstrated in Figure 4.28, a wide differential-mode passband is achieved in the frequency range of 4.2–9.2 GHz with a fractional bandwidth of about 80%. Over this frequency range, the common-mode attenuation due to the three transmission zeros in

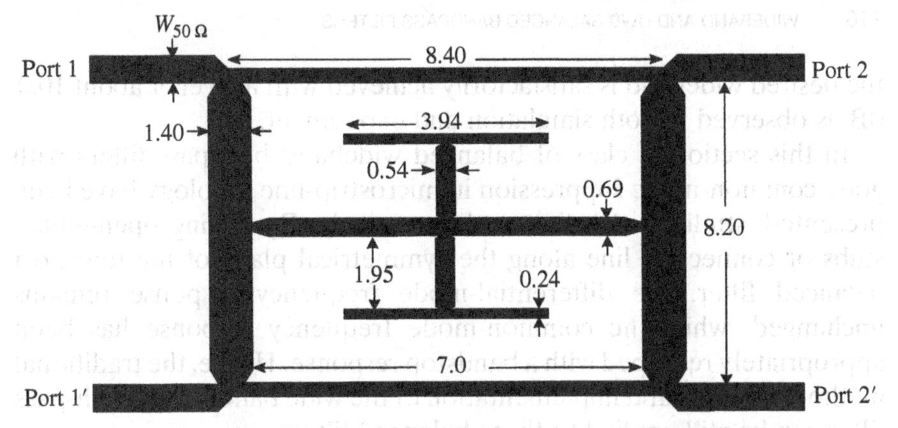

Figure 4.26 Layout of the optimized balanced wideband bandpass filter. Reprinted with permission from Ref. [33]; copyright 2009 IEEE.

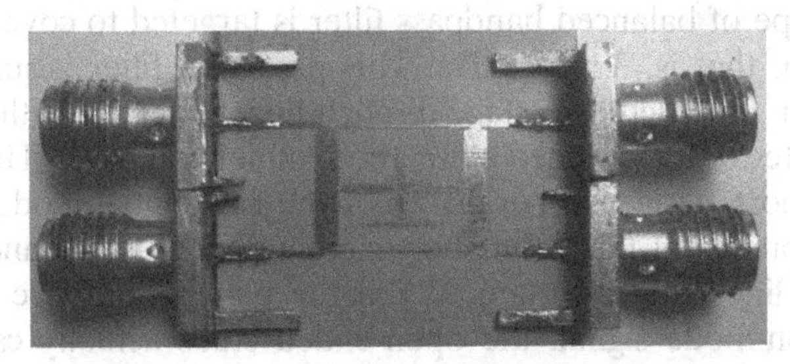

Figure 4.27 Photograph of the fabricated wideband balanced bandpass filter. Reprinted with permission from Ref. [33]; copyright 2009 IEEE.

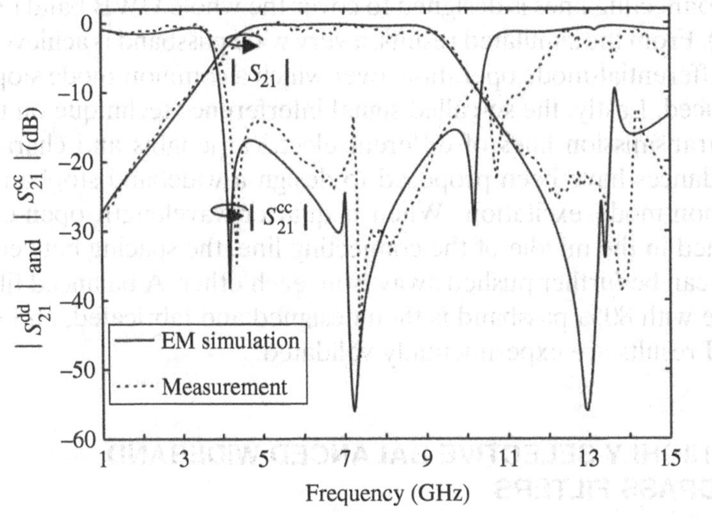

Figure 4.28 Comparison between simulated and measured transmission coefficients, $|S_{21}|$, in the differential- and common-mode cases. Reprinted with permission from Ref. [33]; copyright 2009 IEEE.

the desired wideband is satisfactorily achieved with a level of about 10.0 dB as observed in both simulation and experiment.

In this section, a class of balanced wideband bandpass filters with good common-mode suppression in microstrip-line topology have been presented, studied, and discussed extensively. By adding open-ended stubs or connecting line along the symmetrical plane of the four-port balanced filter, the differential-mode frequency response remains unchanged, while the common-mode frequency response has been appropriately reshaped with a bandstop response. Hence, the traditional method in design and implementation of the wide bandpass or high-pass filter can be still applied to these balanced filters.

More specifically, the three types of balanced filters have been successfully proposed, designed, and implemented in Sections 4.2, 4.3, and 4.4. The first type of balanced bandpass filter is targeted to cover the lower UWB band, that is, 3.1–5.1 GHz. This proposed filter is implemented based on a modified two-stage branch-line structure with a pair of open-circuited stubs attached along the symmetrical plane. Hence, when the paired ports are excited or driven by the differential-mode signal, the structure consists of three 90° short-circuited shunt stubs and two 180° connecting line between them. When the paired ports are excited by the common-mode signal, the open-ended stubs actually can reshape the common-mode frequency response as a bandstop response. The second type of balanced filter is formed up based on a branch-like structure. Instead of a bandpass topology, a high-pass filter with six shunt stubs and five connecting lines is designed to cover the whole UWB band (3.1–10.6 GHz). From the simulated results, a very wide passband is achieved under the differential-mode operation, over which a common-mode stopband is produced. Lastly, the so-called signal interference technique on two parallel transmission lines of different electrical lengths and characteristic impedances have been proposed to design a wideband stopband under common-mode excitation. When a quarter-wavelength open-end stub is joined in the middle of the connecting line, the spacing between these zeros can be further pushed away from each other. A balanced filter prototype with 80% passband is then designed and fabricated, and the predicted results are experimentally validated.

4.5 HIGHLY SELECTIVE BALANCED WIDEBAND BANDPASS FILTERS

The bandpass filter is one of the key circuit blocks in all the transmitters and receivers in order to select or pass certain frequency or group of frequencies from a source to a load while attenuating all other frequencies outside this passband. Selectivity is one of the major specifications of any receiver, and it determines whether the receiver can pick up the signals and receive them at a sufficient strength. There are two main areas of interest for a filter, that is, the passband where it accepts signals and allows them through and the stopband where it rejects them. In an ideal world, a bandpass filter would have a frequency response as shown in Figure 4.29. Here, it can be seen that there is an immediate transition between the passband and the stopband. The passband of the filter also does not introduce any loss. In the stopband, no signal transmission is allowed through and all signals in the stopband are completely attenuated. In reality, it is not possible to realize such a bandpass filter with these ideal characteristics. A typical filter frequency response looks more like what is depicted in Figure 4.30.

As illustrated in Figures 4.29 and 4.30, there exists significant difference between frequency responses of the ideal and typical filters. Looking at Figure 4.30, we can firstly figure out that there is certain loss in the passband, namely, insertion loss. Secondly, the frequency response does

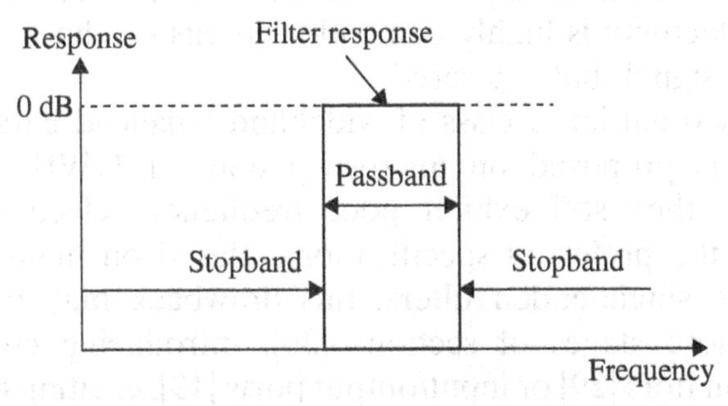

Figure 4.29 Ideal frequency response of a perfect bandpass filter.

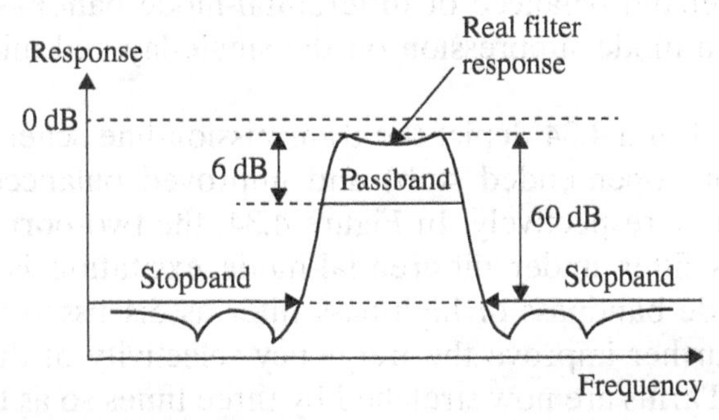

Figure 4.30 Typical frequency response of a practical bandpass filter.

not fall down infinitely fast. Thirdly, the stopband attenuation is not infinite, even though it is very large. Finally, there are some unexpected ripples in both the passband and stopband. Coming to the first point, the attenuation in the passband is normally relatively small. For a well-designed planar filter, about 1–3 dB, insertion loss is fairly typical. Fortunately it is quite easy to counteract this loss by installing an extra amplifier in connection with this filter block. Next, it can be seen that the frequency response does not fall away infinitely fast, and it is necessary to define the cutoff frequency points between which the passband lies.

As an alternative parameter, the shape factor is referred to as a ratio of the bandwidths of the passband and the stopband. Hence, a filter with a passband of 3 kHz at –6 dB and a figure of 6 kHz at –60 dB for the stopband would have a shape factor of 6/3 = 2. Shape factor is a degree of measure of the steepness of the skirts. Note that the perfect bandpass filter has its frequency response with the shape factor of 1, which is the ultimate and impossible to be achieved in practical case. As well recognized, frequency selectivity is particularly important on the today's crowded bands, and it is necessary to ensure that any receiver is able to select the wanted signal as possible as it can. The filter with realizable frequency selectivity is highly demanded to ensure the best chance of receiving the signal that is desired.

As discussed earlier, a class of wideband balanced bandpass filters [25, 32, 33] is proposed on microstrip line for UWB applications. Nevertheless, they still exhibit poor frequency selectivity and can hardly meet the preferred specifications. Based on many pioneering works on the single-ended filters, this drawback may be overcome by setting more stages of sections [30], introducing cross-coupling between resonators [29] or input/output ports [19], exciting transmission zeros via shunt stubs [35], or utilizing the signal interference technique [36]. But, so far, no literature has been reported to develop such a highly selective wideband balanced or differential-mode bandpass filter with good common-mode suppression on the single-layered microstrip-line structure.

Figures 4.31 and 4.34 depict the transmission-line schematics of the initial (without open-ended stub) and improved balanced wideband bandpass filters, respectively. In Figure 4.34, the two-port bisection of the bandpass filter under differential-mode excitation is exactly the same as a wide bandpass or high-pass filter as discussed in Refs. [25, 32, 33]. To further improve the frequency selectivity of this filter, the first and third arms are now stretched by three times so as to constitute the improved filter structure as shown in Figure 4.31.

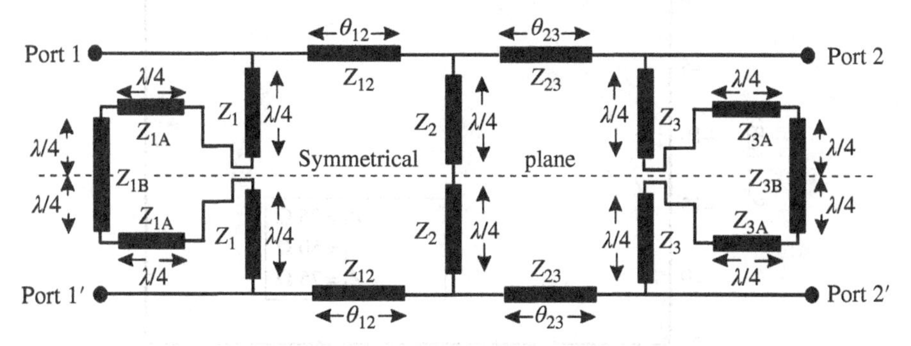

Figure 4.31 Schematic of the improved three-stage balanced wideband bandpass filters. Reprinted with permission from Ref. [37]; copyright 2011 IEEE.

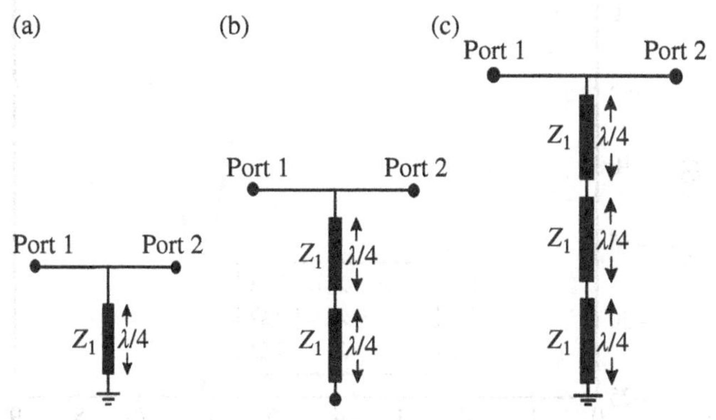

Figure 4.32 Schematics of three distinctive shunt stubs. (a) 90° short-circuited stub, (b) two-section 90° open-circuited stub, and (c) three-section 90° short-circuited stub.

To explain their filtering performances in comparison, let us firstly study the frequency responses of the three-stub structures with varied sections and terminations as illustrated in Figure 4.32. All the three structures with different loads or lengths are designed with the same central frequency at 4.1 GHz that is the central frequency of the lower UWB band, that is, 3.1–5.1 GHz. Meanwhile, each section involved in them is set as one quarter-wavelength, that is, 90°, in electrical length. Figure 4.33 plots three sets of simulated frequency responses. For the short-circuited stub in Figure 4.32(a) with 90° in length, a bandpass response centered at $f_0 = 4.1$ GHz is observed in Figure 4.33(a) with transmission zeros at 0 and 2 f_0. As Z_1 increases, the $|S_{21}|$-curve rises up in the lower and higher frequency ranges so as to achieve an almost flat frequency response close to 0 dB.

For the open-circuited stub with two sections of 90° shown in Figure 4.32(b), two transmission zeros appear at $f_0/2$ and $3f_0/2$ as shown

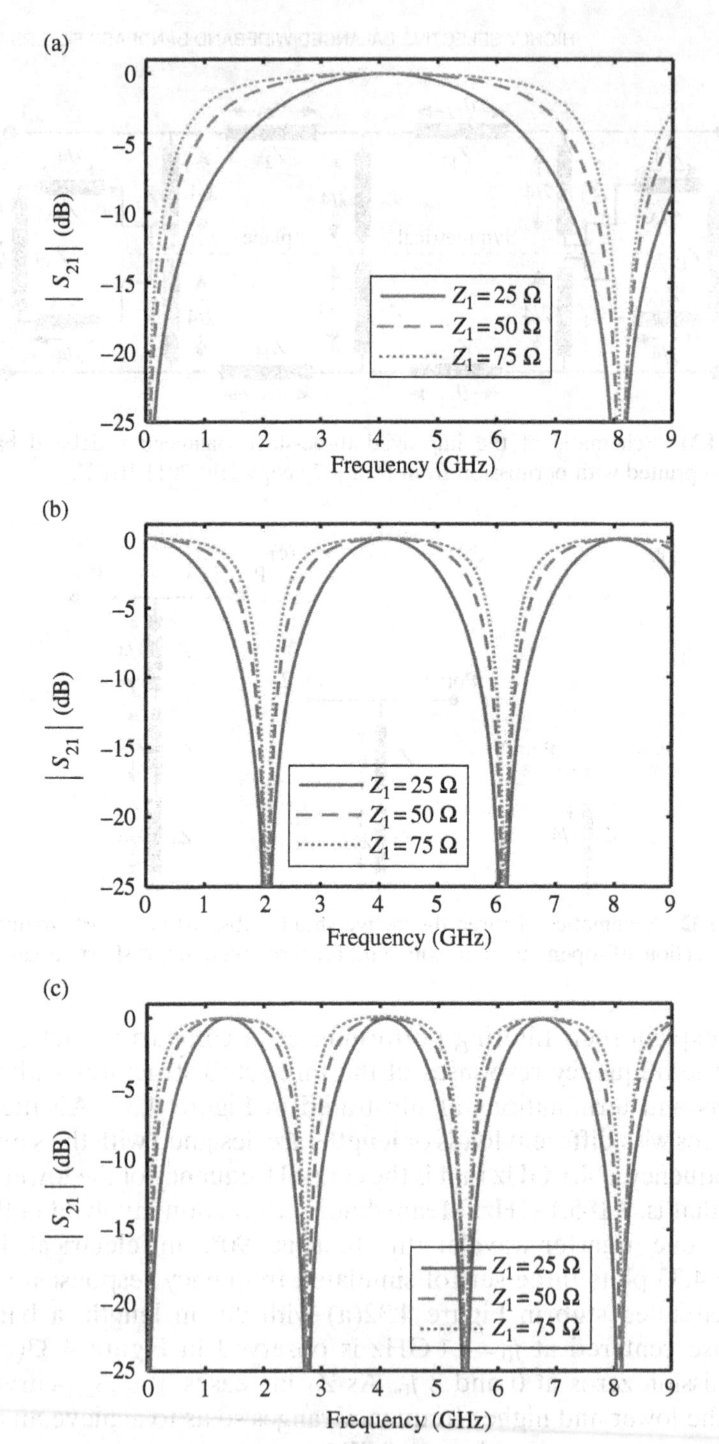

Figure 4.33 Frequency responses of the three shunt-stub circuit networks in Figure 4.32 with fixed central frequency. (a) 90° stub, (b) two-section 90° stub, and (c) three-section 90° stub.

in Figure 4.33(b). To meet the need in design of a balanced bandpass filter, the third stub structure in Figure 4.32(c) is carefully studied and its frequency response is illustrated in Figure 4.33(c). Instead of two transmission zeros, we can now see that the four transmission zeros are generated at 0, $2f_0/3$, $4f_0/3$, and $2f_0$, respectively. Again, the second passband becomes more flat in frequency response as Z_1 increases. Furthermore, its bandwidth can be appropriately widened by moving its second and third transmission zeros far away from each other via proper selection of unequal characteristic impedances of the three sections as will be discussed later on (Figure 4.34).

Under the differential-mode operation, the improved balanced filter in Figure 4.31 can be decomposed into the two bisections with a perfect electric wall at the symmetrical plane (dotted line in Figure 4.31). Figure 4.35(a) indicates the relevant upper two-port bisection under differential-mode excitation. It is composed of three shunt short-circuited stubs of 90° shown in Figure 4.35(a) and two connecting lines of 90° between them.

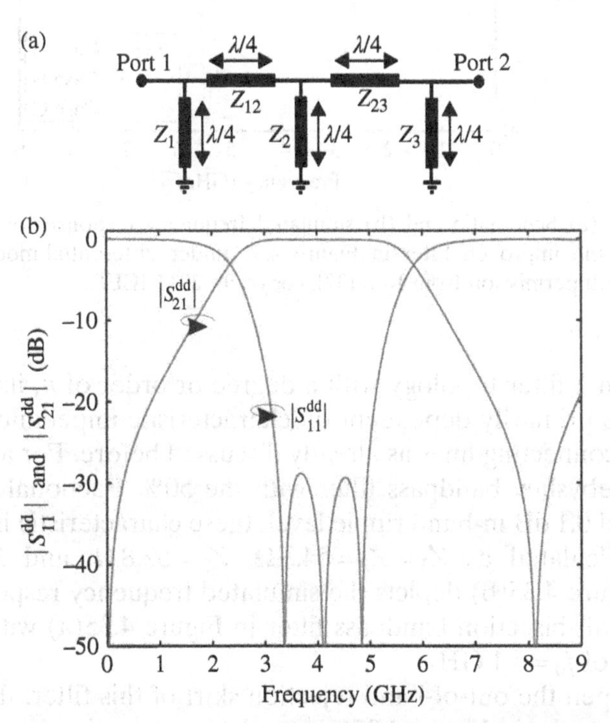

Figure 4.34 (a) Schematic and (b) simulated frequency response of the two-port bisection of initial filter in Figure 4.4 under differential-mode excitation.

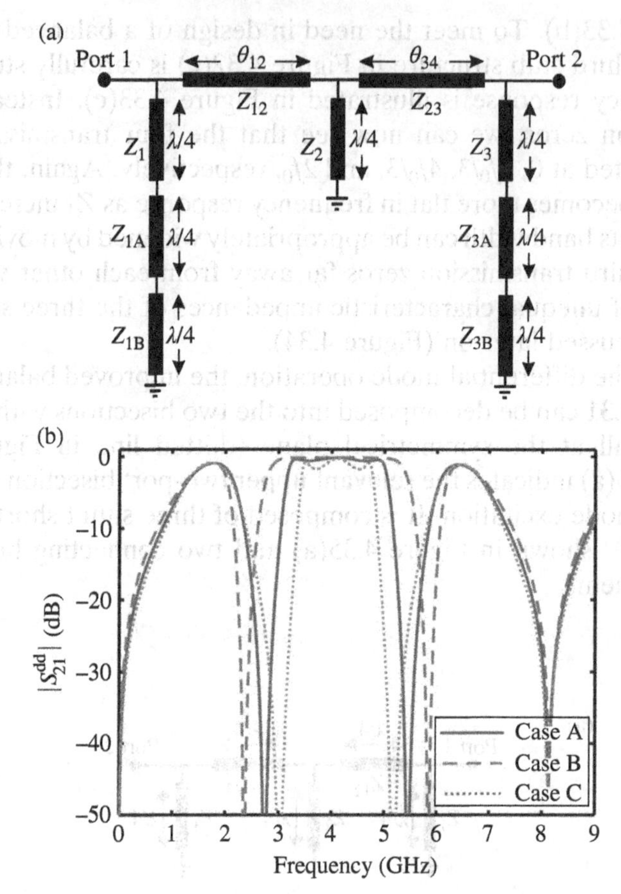

Figure 4.35 (a) Schematic and (b) simulated frequency responses of the two-port bisection of the improved filter in Figure 4.31 under differential-mode excitation. Reprinted with permission from Ref. [37]; copyright 2011 IEEE.

For such a filter topology with a degree or order of n, its frequency response is primarily dependent on characteristic impedances of these stubs and connecting lines as already discussed before. For a three-pole ($n = 3$) Chebyshev bandpass filter with the 50% fractional bandwidth (FBW) and 0.1 dB in-band ripple level, these characteristic impedances can be calculated as $Z_1 = Z_3 = 54.3\,\Omega$, $Z_2 = 52.8\,\Omega$, and $Z_{12} = Z_{23} = 52.4\,\Omega$. Figure 4.35(b) depicts the simulated frequency response of this two-port half-bisection bandpass filter in Figure 4.35(a) with a central frequency of $f_0 = 4.1$ GHz.

To sharpen the out-of-band rejection skirt of this filter, the first and third shunt stubs in Figure 4.35(a) are then appropriately stretched by three times, that is, from 90° to 270° so as to make up the two-port bisection of an alternative differential-mode bandpass filter as shown in Figure 4.35(a). Under the case A where $Z_1 = Z_{1A} = Z_{1B} = Z_3 = Z_{3A} = Z_{3B} = 54.3\ \Omega$ and all the other parameters remain unchanged, two additional transmission zeros can be produced.

The second zero occurs at around 2.8 GHz while the third zero is about two times the second one. By changing Z_1, Z_{1B}, Z_3 and Z_{3B} from 54.3 to 90 Ω (case B) and then 35 Ω (case C), these two transmission zeros are found to be able to move outward and inward as illustrated in Figure 4.35(b). As for the first designed wideband balanced filter, its desired passband is constructed to cover the only lower UWB band (3.1–5.1 GHz). As a consequence, case B with $Z_1 = Z_{1B} = Z_3 = Z_{3B} = 90\ \Omega$ is selected.

Obviously, there exist two undesired passbands appearing in the lower and upper sides of the desired passband centered at 4.1 GHz. Our investigation shows that transmission level in these two passbands can be dramatically pushed down with the reduction in impedance Z_2. Figure 4.36 depicts the three sets of graphs in correspondence to $Z_2 = 52.8\ \Omega$ (case A), $Z_2 = 35\ \Omega$ (case B), and $Z_2 = 17\ \Omega$ (case C). As Z_2 is reduced, the level in these two parasitic passbands can be decreased from 0.5 to 4.0 dB, while their bandwidths become narrower and narrower. But it seems very difficult in practice to further reduce their transmission level to a specified extent, such as −10 dB.

For this reason, an alternative bandpass filter topology with the connecting lines of 180° instead of 90° is constituted based on the works in

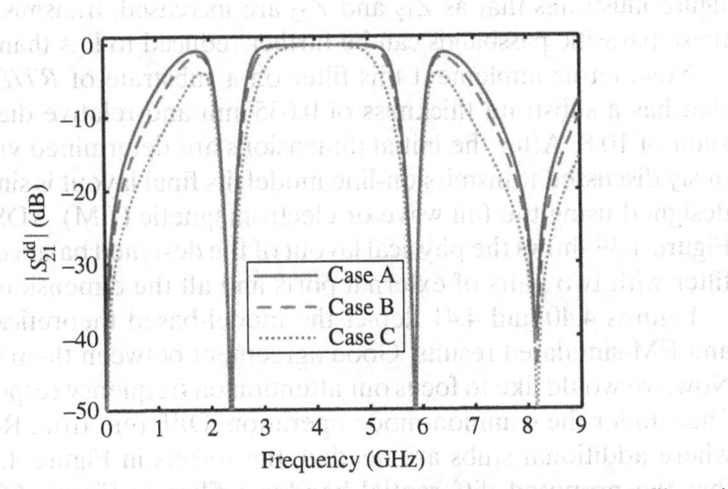

Figure 4.36 Frequency responses of the proposed balanced bandpass filters in Figure 4.35(a) with 90° connecting line. Case A: $Z_2 = 52.8\ \Omega$; Case B: $Z_2 = 35\ \Omega$; Case C: $Z_2 = 17\ \Omega$. Reprinted with permission from Ref. [37]; copyright 2011 IEEE.

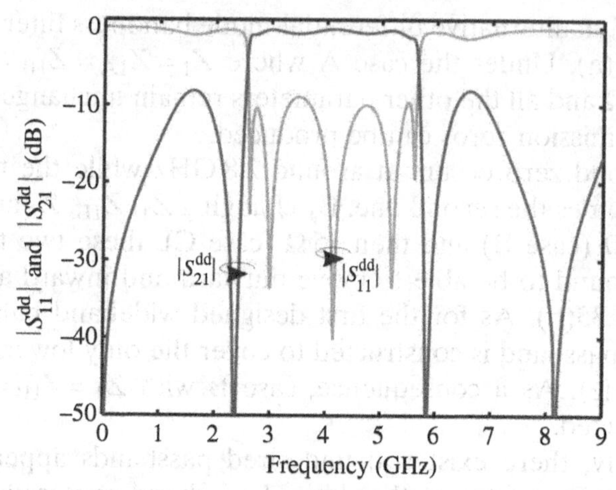

Figure 4.37 Frequency response of the balanced filters in Figure 4.33 with 180° connecting line. Reprinted with permission from Ref. [37]; copyright 2011 IEEE.

Refs. [20, 33]. For an order of $n = 3$ with all the other parameters to be unchanged, frequency response of this filter is plotted in Figure 4.37, showing emergence of five transmission poles in the dominant passband instead of three poles in the previous filter. In the meantime, the transmission level in the two parasitic passbands is now reduced to −8.0 dB. To further investigate this filter, the three cases with $Z_{12} = Z_{23} = 52.4\ \Omega$ (case A), $Z_{12} = Z_{23} = 65\ \Omega$ (case B), and $Z_{12} = Z_{23} = 90\ \Omega$ (case C) are taken into account and their simulated frequency responses are plotted in Figure 4.38. This figure illustrates that as Z_{12} and Z_{23} are increased, transmission level in these parasitic passbands can be further reduced to less than −15 dB.

Now, let us implement this filter on a substrate of *RT/Duroid 6010* that has a substrate thickness of 0.635 mm and relative dielectric constant of 10.8. After the initial dimensions are determined via the previously discussed transmission-line model, its final layout is simulated and designed using the full wave or electromagnetic (EM) ADS simulator. Figure 4.39 shows the physical layout of the designed balanced bandpass filter with two pairs of external ports and all the dimensions given.

Figures 4.40 and 4.41 depict the model-based theoretical-predicted and EM-simulated results. Good agreement between them is obtained. Now, we would like to focus our attention on frequency responses of this filter under the common-mode operation. Different from Refs. [25, 32] where additional stubs are needed, the results in Figure 4.41 show us that the proposed differential bandpass filter in Figure 4.39 can self-contain a third-order wide stopband in the common-mode case. Of three transmission zeros, the first and third ones correspond to the two

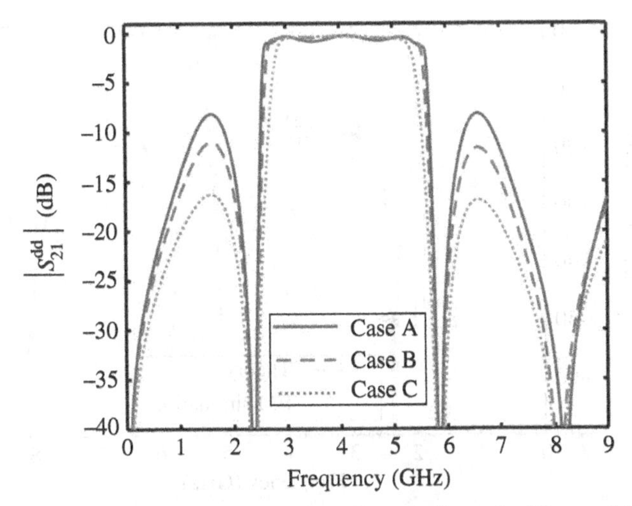

Figure 4.38 Frequency responses of the balanced filters in Figure 4.35 with 180° connecting lines. Case A: $Z_{12} = Z_{23} = 52.4\ \Omega$; Case B: $Z_{12} = Z_{23} = 65\ \Omega$; Case C: $Z_{12} = Z_{23} = 90\ \Omega$. Reprinted with permission from Ref. [37]; copyright 2011 IEEE.

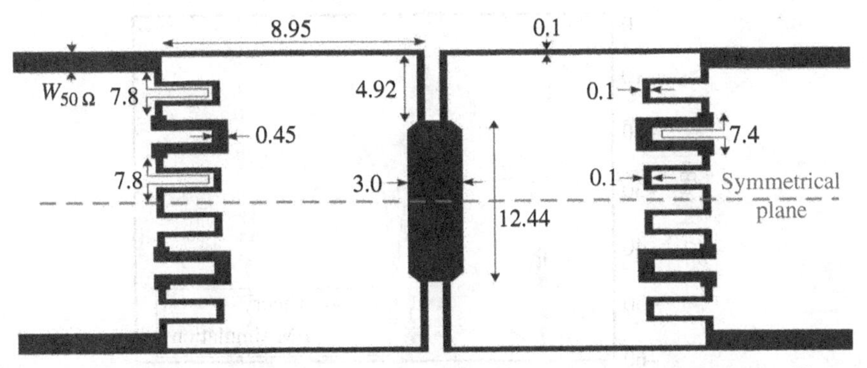

Figure 4.39 Layout of the optimized first-type balanced wideband bandpass filter on microstrip line. (All the dimensions in millimeter). Reprinted with permission from Ref. [37]; copyright 2011 IEEE.

frequencies where the longer open-circuited stubs in two sides become 90° and 270° in length, whereas the second one occurs at a frequency where the short stub in middle is 90° long. To further apply this filter topology to cover the entire UWB band, that is 3.1–10.6 GHz with an FBW of 110%, this filter needs to be reconstructed with a raised order of n and an enlarged ratio of impedances of stepped-impedance arms. Figure 4.42 shows an improved filter structure that has two 90° arms in middle and two 270° arms on both the input and output sides, while two identical open-circuited stubs are attached to the two long arms.

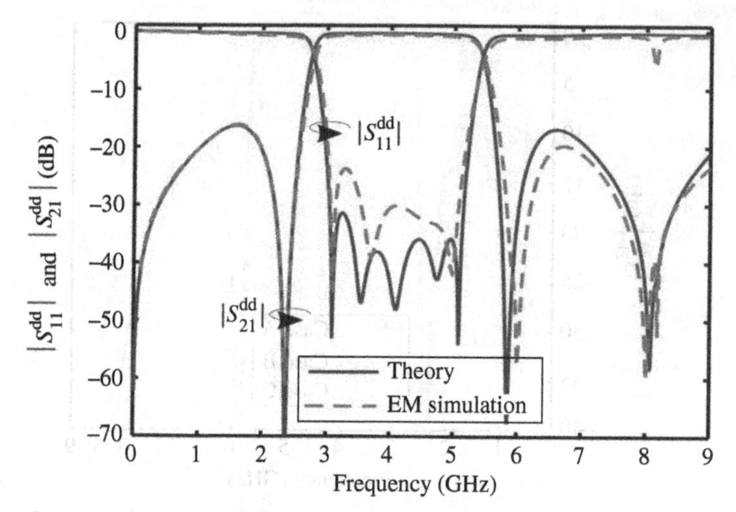

Figure 4.40 Theoretical-predicted and EM-simulated results of the designed first-type balanced wideband filter in Figure 4.39 under differential-mode operation. Reprinted with permission from Ref. [37]; copyright 2011 IEEE.

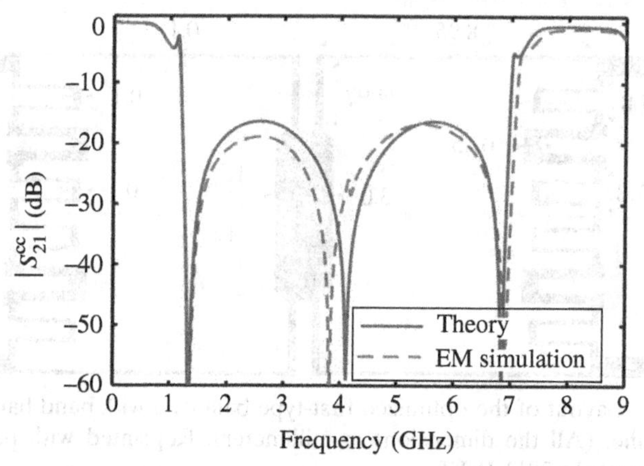

Figure 4.41 Theoretical-predicted and EM-simulated results of the designed first-type balanced wideband filter in Figure 4.39 under common-mode operation. Reprinted with permission from Ref. [37]; copyright 2011 IEEE.

This filter can be considered as a modified filter with $n = 4$ and two additional 90° stubs are introduced herein in order to control the common-mode stopband [25, 32]. Due to existence of a perfect electric wall in the symmetrical plane under the differential-mode excitation, these two stubs hardly affect the differential-mode frequency response, but their length extension in the common-mode case provides us with an expected capacity in reshaping the common-mode frequency response.

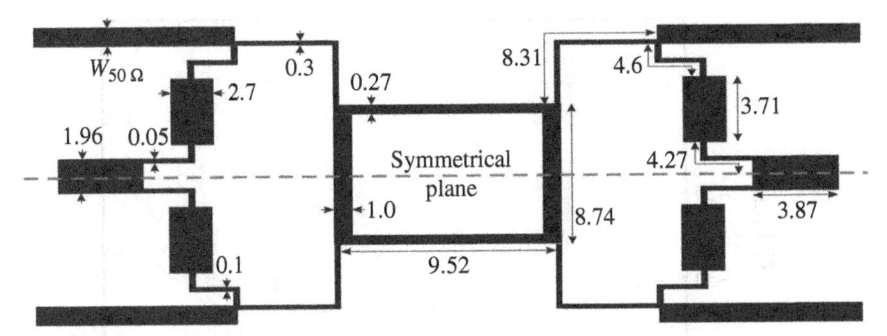

Figure 4.42 Layout of the optimized second-type balanced wideband bandpass filter on microstrip line. (All dimensions in millimeter). Reprinted with permission from Ref. [37]; copyright 2011 IEEE.

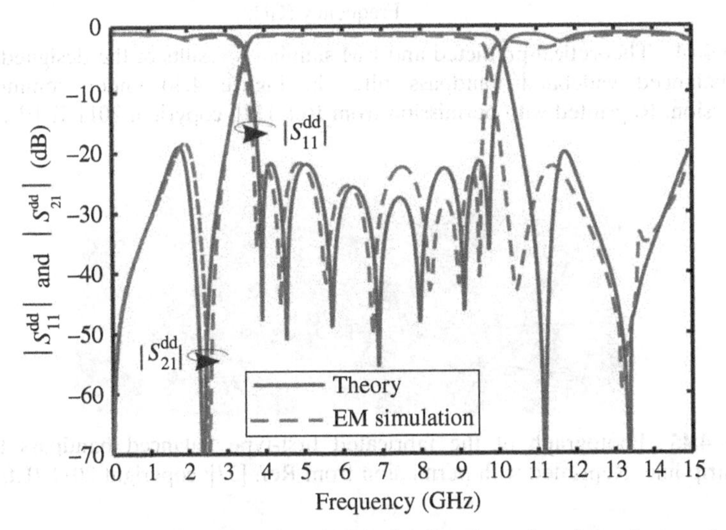

Figure 4.43 Theoretical-predicted and EM-simulated results of the designed second-type balanced wideband bandpass filter in Figure 4.40 under differential-mode operation. Reprinted with permission from Ref. [37]; copyright 2011 IEEE.

Figures 4.43 and 4.44 depict the two sets of simulated results under differential- and common-mode excitations, which are derived from the simple transmission theory and ADS EM simulation. These results demonstrate good selectivity differential-mode bandpass behavior with good common-mode suppression. From Figure 4.43, an ultrawide passband with the expected order of $2n - 1 = 7$ is derived in the specified 3.1–10.6 GHz UWB band, the two parasitic passbands in two sides are suppressed with the transmission level less than −19.0 dB and two transmission zeros in the lower and higher cutoff frequencies sharpen

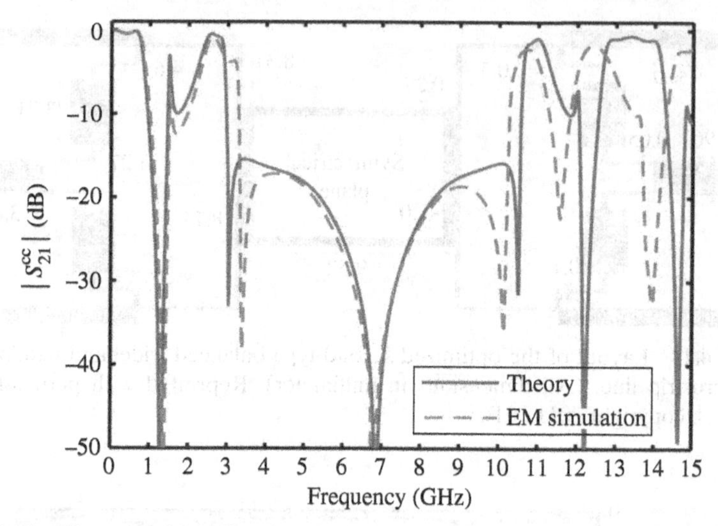

Figure 4.44 Theoretical-predicted and EM-simulated results of the designed second-type balanced wideband bandpass filter in Figure 4.40 under common-mode suppression. Reprinted with permission from Ref. [37]; copyright 2011 IEEE.

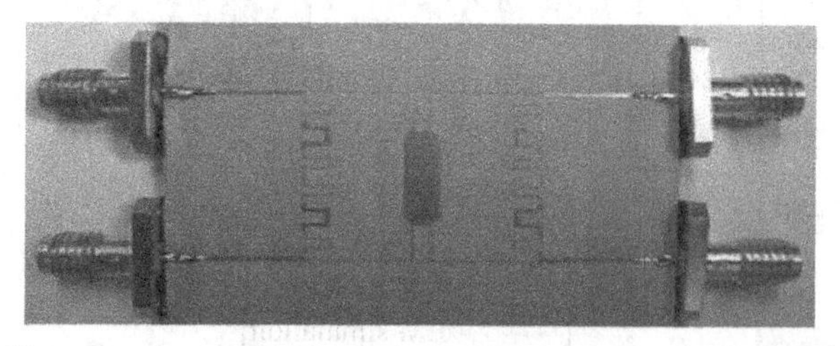

Figure 4.45 Photograph of the fabricated first-type balanced bandpass filter on microstrip line. Reprinted with permission from Ref. [37]; copyright 2011 IEEE.

the out-of-band rejection skirts. From Figure 4.44, an ultra-wide common-mode stopband is well-produced with the transmission level less than −15 dB. Based on the previous analysis and design, the two balanced bandpass filters in Figures 4.39 and 4.42 are constituted, and the photograph of the two fabricated filter prototypes are given in Figures 4.45 and 4.46, respectively. In measurement with a two-port VNA, we execute our test on these four-port networks by connecting the two of four ports with two VNA coaxial cables while terminating the two remaining ports with the 50 Ω matched loads. After similar tests are carried out three times, the 4 × 4 standard S-matrix, that is, $[\mathbf{S}^{std}]$ of these two four-port filter circuits can be obtained in experiment. Based on the well-known theory discussed earlier on, the corresponding

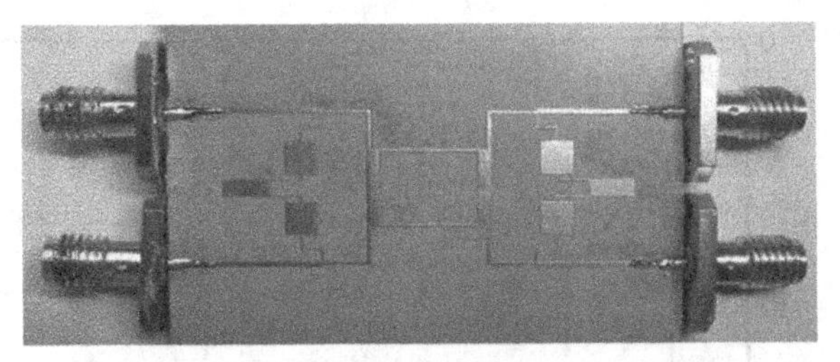

Figure 4.46 Photograph of the fabricated second-type balanced bandpass filter on microstrip line. Reprinted with permission from Ref. [37]; copyright 2011 IEEE.

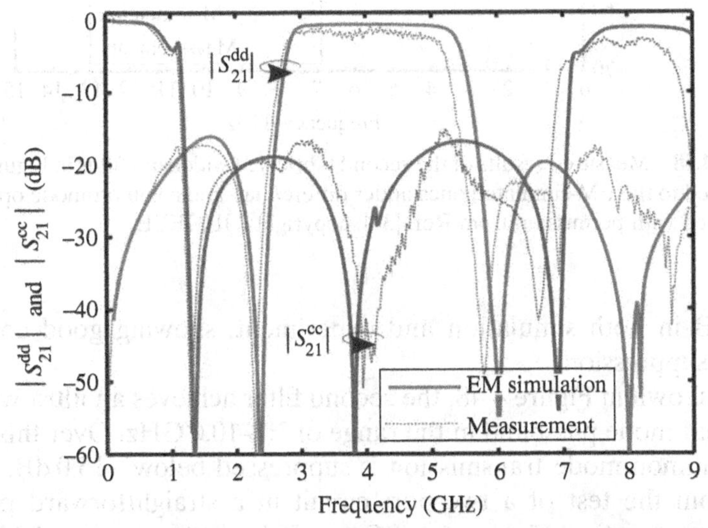

Figure 4.47 Measured results of the first fabricated wideband filter in Figure 4.45 as compared to the EM-simulated ones under differential- and common-mode operations. Reprinted with permission from Ref. [37]; copyright 2011 IEEE.

4×4 mixed-mode S-matrix, $[S^{mm}]$, can be accordingly deduced using the formulae in the previous section. As such, we can eventually extract the two 2×2 S-matrix, $[S^{dd}]$ and $[S^{cc}]$, in correspondence to the differential- and common-mode frequency responses, respectively.

Figures 4.47 and 4.48 plot the two measured differential- and common-mode transmission coefficients as compared to the EM-simulated ones. Reasonable agreement between them is obtained. As shown in Figure 4.47, the first filter achieves the differential-mode passband in the range of 2.8–5.3 GHz with a maximum insertion loss of about 1.1 dB in experiment against 0.2 dB in theory. Over this range, the common-mode attenuation reaches to a maximum value of about

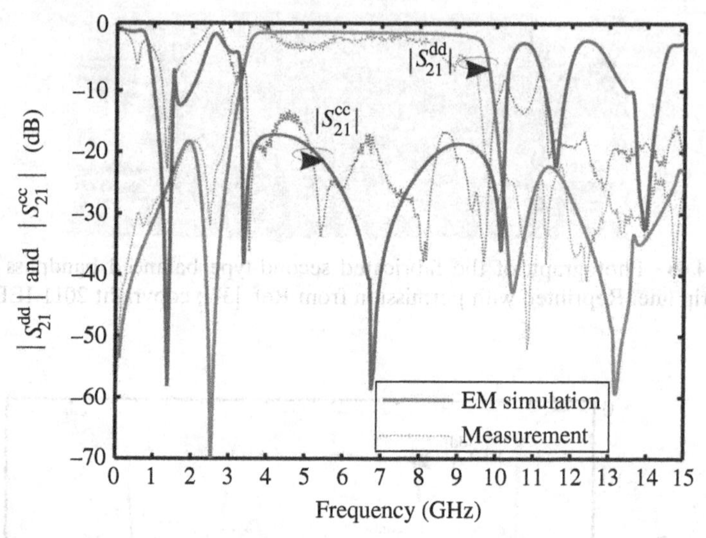

Figure 4.48 Measured results of the second fabricated wideband filter in Figure 4.46 as compared to the EM-simulated ones under differential- and common-mode operations. Reprinted with permission from Ref. [37]; copyright 2011 IEEE.

20.0 dB in both simulation and experiment, showing good common-mode suppression.

As shown in Figure 4.48, the second filter achieves an ultra-wide differential-mode passband in the range of 3.1–10.0 GHz. Over this range, the common-mode transmission is suppressed below −15.0 dB. Different from the test of a two-port circuit in a straightforward manner, explicit extraction of a two-port *S*-matrix from the measured four-port *S*-matrix always brings out some unexpected dissimilarity against the predicted results due to many uncertain factors in practical test. First, all the four ports cannot achieve perfect impedance matching over a wide frequency range due to non-50 Ω actual impedance of loads and cables as well as discontinuities at SMA connectors and microstrip-to-cable transitions. Second, the microstrip feeding lines are impossibly constructed with the exact 50 Ω characteristic impedance due to frequency dispersion of this inhomogeneous transmission line and fabrication tolerance in etching process. Third, all the four ports cannot be identically constructed due to unrepeatedly soldering process and inconsistent offset effect between the microstrip conductor and inner conductor of an SMA connector. All these three aspects of uncertainties are cumulated and thus bring out some unexpectedly small but visible discrepancies between the predicted and measured results as can be observed in Figures 4.47 and 4.48.

4.6 SUMMARY

In this chapter, a class of ultra-wideband balanced bandpass filters on microstrip line with good common-mode suppression is designed and implemented. The proposed four-port balanced bandpass filter is ideally symmetric with respect to the central plane in horizon. Therefore, this symmetrical plane can be considered as a perfect electric or magnetic wall if one of the paired ports are driven or excited by differential- or common-mode signals, respectively. By adding a pair of open-ended stubs along this symmetrical plane, the introduced stubs are electrically short-circuited in differential-mode and thus giving no influence on the differential-mode frequency response. Under the common-mode excitation, these introduced stubs actually give us the extra degree of freedom to redesign and reshape the frequency response. Instead of introducing an open-ended stub along the symmetrical plane, a connecting line is used to join the shunt stubs together. With this connecting line, three transmission zeros are produced in the common-mode frequency response because the phase of the two lines are designed with the out-of-phase property. In this aspect, they are canceled with each other, thus producing three transmission zeros for this case. This chapter also presents a novel method to improve the filter selectivity performance of a wideband filter by stretching a traditional 90° branch-line stub to 270° in two sides. With these additional lengthened stubs, the out-of-band rejection skirt can be satisfactorily sharpened with resorting to the two additional transmission zeros that are then relocated to the lower and upper sides of the core passband by virtue of the impedance ratio of the stubs, resulting in the improvement of the frequency selectivity of the proposed balanced wideband bandpass filters under differential-mode operation.

REFERENCES

1. W. R. Eisenstant, B. Stengel, and B. M. Thompson, *Microwave Differential Circuit Design Using Mixed-Mode S-Parameters*. Boston, MA: Artech House, 2006.
2. A. Ziroff, M. Nalezinski, and W. Menzel, "A 40 GHz LTCC receiver module using a novel submerged balancing filter structure," in *Proceedings of the Radio Wireless Conference*, Boston, pp. 151–154, Aug. 13, 2003.
3. Wu, S.-J., Tsai, C.-H., Wu, T.-L., and Itoh, T., "A novel wideband common-mode suppression filter for gigahertz differential signals using coupled patterned ground structure," *IEEE Trans. Microw. Theory Tech.*, vol. **57**, no. 4, pp. 848–855, Mar. 2009.

4. Razavi, B., *Design of analog CMOS integrated circuits*, McGraw Hill, New York, 2001.
5. W.R. Eisenstadt, R. Stengel, and B.M. Thompson, *Microwave Differential Circuit Design Using Mixed-Mode S Parameter*, Artech House, Boston, 2006.
6. C.-L. Hsu, F.-C. Hsu, and J.-T. Kuo, "Microstrip bandpass filters for ultra-wideband (UWB) wireless communications," in *IEEE MTT-S International Microwave Symposium Digest*, Institute of Electrical and Electronic Engineers, New York, pp. 679–682, Jun. 2005.
7. W. Menzel, M. S. R. Tito, and L. Zhu, "Low-loss ultra-wideband (UWB) filters using suspended stripline," in *Microwave Conference Proceedings, 2005. APMC 2005. Asia-Pacific Conference Proceedings*, Suzhou, China, vol. **4**, pp. 2148–2151, 4–7 Dec. 2005.
8. J. Garcia-Garcia, J. Bonache, and F. Martin," Application of electromagnetic bandgaps to the design of ultra-wide bandpass filters with good out-of-band performance," *IEEE Trans. Microw. Theory Tech.*, vol. **54**, no. 12, pp. 4136–4140, Dec. 2006.
9. L. Zhu, H. Bu, and K. Wu, "Aperture compensation technique for innovative design of ultra-broadband microstrip bandpass filter," in *Microwave Symposium Digest. 2000 IEEE MTT-S International*, Boston, MA, vol. **1**, pp. 315–318, 11–16 Jun. 2000.
10. M. Makimoto and S. Yamashita, "Bandpass filters using parallel coupled stripline stepped impedance resonators," *IEEE Trans. Microw. Theory Tech.*, vol. MTT-28, no. 12, pp. 1413–1417, Dec. 1980.
11. L. Zhu, S. Sun, and W. Menzel, "Ultra-wideband (UWB) bandpass filters using multiple-mode resonator," *IEEE Microw. Wireless Compon. Lett.*, vol. **15**, no. 11, pp. 796–798, Nov. 2005.
12. L. Zhu and H. Wang, "Ultra-wideband bandpass filter on aperture-backed microstrip line," *Electron. Lett.*, vol. **41**, no. 18, pp. 1015–1016, Sep. 2005.
13. H. Wang, L. Zhu, and W. Menzel, "Ultra-wideband (UWB) bandpass filters with hybrid microstrip/CPW structure," *IEEE Microw. Wireless Compon. Lett.*, vol. **15**, no. 12, pp. 844–846, Dec. 2005.
14. J. Gao, L. Zhu, W. Menzel, and F. Bogelsack, "Short-circuited CPW multiple-mode resonator for ultra-wideband (UWB) bandpass filter," *IEEE Microw. Wireless Compon. Lett.*, vol. **16**, no. 3, pp. 104–106, Mar. 2006.
15. S. Sun and L. Zhu, "Capacitive-ended interdigital coupled lines for UWB bandpass filters with improved out-of-band performances," *IEEE Microw. Wireless Compon. Lett.*, vol. **16**, no. 8, pp. 440–442, Aug. 2006.
16. R. Li and L. Zhu, "Compact UWB bandpass filter using stub-loaded multiple-mode resonator," *IEEE Microw. Wireless Compon. Lett.*, vol. **17**, no. 1, pp. 40–42, Jan. 2007.
17. W.-T. Wong, Y.-S. Lin, C.-H. Wang, and C. H. Chen, "Highly selective microstrip bandpass filters for ultra-wideband (UWB) applications," in

Microwave Conference Proceedings, 2005. APMC 2005. Asia-Pacific Conference Proceedings, Suzhou, China, vol. **5**, pp. 2850–2853, 4–7 Dec. 2005.

18. J.-S. Hong and H. Shaman, "An optimum ultra-wideband microstrip filter," *Microw. Opt. Technol. Lett.*, vol. **47**, no. 3, pp. 230–233, Nov. 2005.
19. H. Shaman and J.-S. Hong, "A novel ultra-wideband (UWB) bandpass filter (BPF) with pairs of transmission zeroes," *IEEE Microw. Wireless Compon. Lett.*, vol. **17**, no. 2, pp. 121–123, Feb. 2007.
20. C.-W. Tang and M.-G. Chen, "A microstrip ultra-wideband bandpass filter with cascaded broadband bandpass and bandstop filters," *IEEE Trans. Microw. Theory Tech.*, vol. **55**, no. 11, pp. 2412–2418, Nov. 2007.
21. W. J. Feng, W. Q. Che, and Q. Xue, "Transversal signal interaction: Overview of high-performance wideband bandpass filters," *IEEE Microw. Mag.*, vol. **15**, no. 2, pp. 84–96, Mar. 2014.
22. W. J. Feng, W. Q. Che, and Q. Xue, "The Proper Balance," *IEEE Microw. Mag.*, vol. **16**, no. 5, pp. 1587–1594, Jun. 2015.
23. W. J. Feng and W. Q. Che, "Novel wideband differential bandpass filters based on T-shaped structure," *IEEE Trans. Microw. Theory Tech.*, vol. **60**, no. 6, pp. 1560–1568, Jun. 2012.
24. W. J. Feng, W. Q. Che, Y. L. Ma, and Q. Xue, "Compact wideband differential bandpass filter using half-wavelength ring resonator," *IEEE Microw. Wireless Compon. Lett.*, vol. **23**, no. 2, pp. 81–83, Feb. 2013.
25. T. B. Lim and L. Zhu, "A differential-mode wideband bandpass filter on microstrip line for UWB application," *IEEE Microw. Wireless Compon. Lett.*, vol **19**, no. 10, pp. 632–634, Oct. 2009.
26. L. Zhu, H. Bu, K. Wu, and M. S. Leong, "Miniaturized multi-pole broad-band microstrip bandpass filter: Concept and Verification," in *Proceedings of 30th European Microwave Conference*, vol. **3**, Paris, pp. 334–337, Oct. 2000.
27. X. Y. Zhang, J.-X. Chen, Q. Xue, and S.-M. Li, "Dual-band bandpass filter using stub-loaded resonators," *IEEE Microw. Wireless Compon. Lett.*, vol. **17**, no. 8, pp. 583–585, Aug. 2007.
28. D. M. Pozar, *Microwave Engineering*, 2nd edition, John Wiley & Sons, Inc., New York, 1998.
29. J. S. Hong and M. J. Lancaster, *Microstrip Filters for RF/Microwave Applications*. New York: John Wiley & Sons, Inc., 2001.
30. G. L. Matthaei, L. Young, and E.M.T. Jones, *Microwave Filters, Impedance Matching Networks and Coupling Structures*. New York: McGraw-Hill, 1964.
31. D. E. Bockelamn and W. R. Eisenstant, "Combined differential and common-mode scattering parameters: Theory and simulation," *IEEE Trans. Microw. Theory Tech.*, vol. **43**, no. 7, pp. 1530–1539, Jul. 1995.
32. T. B. Lim and L. Zhu, "Differential-mode ultra-wideband bandpass filter on microstrip line," *Electron. Lett.*, vol. **45**, no. 22, pp. 1124–1125, Oct. 22, 2009.

33. T. B. Lim and L. Zhu, "Differential-mode wideband bandpass filter with three transmission zeros under common-mode operation," in *Microwave Conference, 2009. APMC 2009. Asia Pacific*, Singapore, pp. 159–162, 7–10 Dec. 2009.
34. K. Divyabramham, M. K. Mandal, and S. Sanyal, "Sharp-rejection wideband bandstop filters," *IEEE Microw. Wireless Compon. Lett*, vol. **18**, no. 10, pp. 662–664, Oct. 2008.
35. L. H. Hsieh and K. Chang, "Compact, low insertion-loss, sharp-rejection, and wideband microstrip bandpass filters," *IEEE Trans. Microw. Theory Tech.*, vol. **51**, no. 4, pp. 1241–1246, Apr. 2003.
36. R. Gómez-García and J. I. Alonso, "Design of sharp-rejection and low-loss wide-band planar filters using signal-interference techniques," *IEEE Microw. Wireless Compon. Lett.*, vol. **15**, no. 8, pp. 530–532, Aug. 2005.
37. T. B. Lim and L. Zhu, "Highly selective differential-mode wideband bandpass filter for UWB application," *IEEE Microw. Wireless Compon. Lett.*, vol. **21**, no. 3, pp. 133–135, Feb. 2011.

CHAPTER 5

WIDEBAND AND UWB COMMON-MODE SUPPRESSED DIFFERENTIAL-MODE FILTERS BASED ON COUPLED LINE SECTIONS

Qing-Xin Chu, Shi-Xuan Zhang, and Fu-Chang Chen

School of Electronic and Information Engineering, South China University of Technology, Guangzhou, China

A balanced filter presents two different filtering responses in the same structure. Therefore, it could be considered as a combination of a differential-mode bandpass filter (BPF) and a common-mode bandstop filter (BSF) to some extent. In this chapter, several balanced filters based on coupled lines are described.

5.1 BALANCED UWB FILTER BY COMBINING UWB BPF WITH UWB BSF

Figure 5.1(a) shows the schematic of a differential UWB BPF [1]. The four-port circuit is ideally symmetric with respect to the central plane AA′. Here, all the transmission lines including the parallel-coupled lines are $\lambda_g/4$ with respect to the central frequency f_0 of the differential

Balanced Microwave Filters, First Edition. Edited by Ferran Martín, Lei Zhu, Jiasheng Hong, and Francisco Medina.

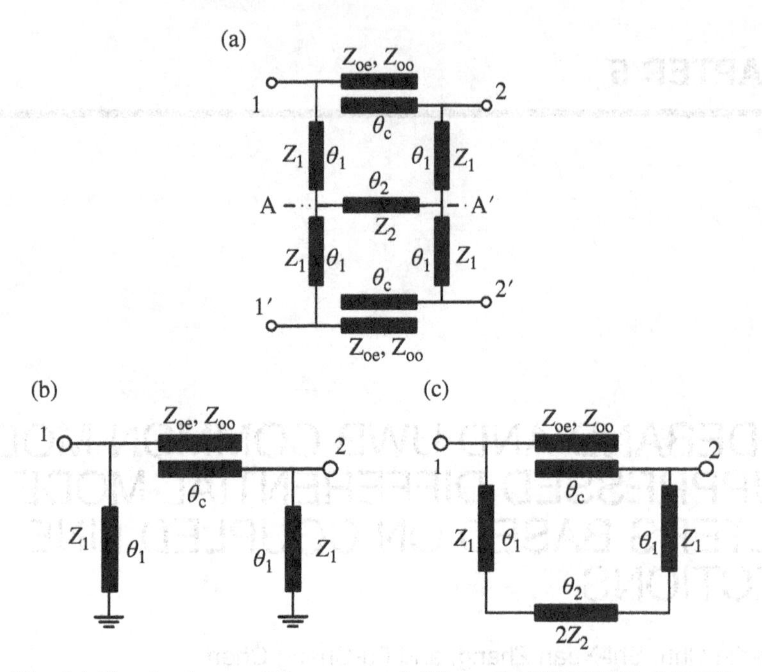

Figure 5.1 (a) Circuit scheme of the proposed differential UWB BPF. (b) Differential-mode circuit. (c) Common-mode circuit. Reprinted with permission from Ref. [1]; copyright 2012 IEEE.

passband, that is, $\theta_1 = \theta_2 = \theta_c = 90°$. For differential-mode operation, the central plane becomes a perfect electric wall (EW), and the differential-mode circuit is as shown in Figure 5.1(b). It is a two-stage branch line structure; however, the connecting structure is an open-ended parallel-coupled line pair [2]. The common-mode circuit, as shown in Figure 5.1(c), is a UWB BSF, which has been presented in Ref. [3]. Figure 5.2 is the differential- and common-mode circuit responses, where the spurious response is located at $3f_0$. The differential-mode circuit has a UWB passband response, and four reflection zeros (f_{r1}^{dd}, f_{r2}^{dd}, f_{r3}^{dd}, and f_{r4}^{dd}) can be seen in the passband, supplying a good passband transmission and reflection performance. Two transmission zeros f_{z1}^{dd} (0 GHz) and f_{z2}^{dd} ($2f_0$ GHz) due to the $\lambda_g/4$ short-ended stubs are generated in the lower and upper stopbands, respectively. Besides, a transmission zero caused by the $\lambda_g/4$ parallel-coupled line is also superposed at f_{z2}^{dd}. For the common-mode circuit [3], four transmission zeros (f_{z1}^{cc}, f_{z2}^{cc}, f_{z3}^{cc}, and f_{z4}^{cc}) can be obtained in the stopband and good common-mode suppression could be realized.

Figure 5.3 depicts the differential- and common-mode responses with varied even-mode impedance Z_{oe} of the parallel-coupled lines. As Z_{oe}

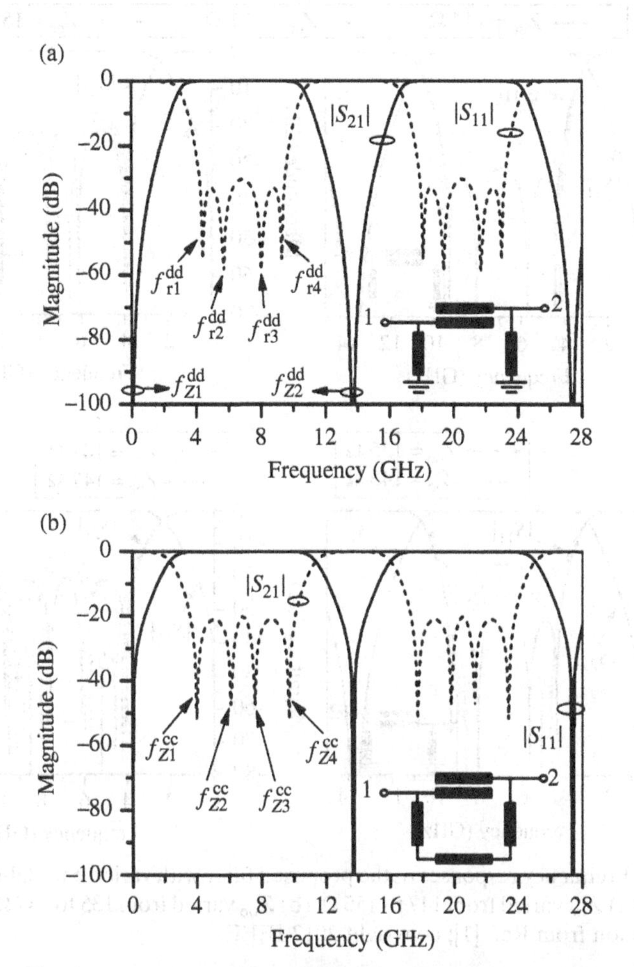

Figure 5.2 Circuit responses ($Z_1 = 61\,\Omega$, $Z_2 = 73\,\Omega$, $Z_{oo} = 50\,\Omega$, $Z_{oe} = 147\,\Omega$). (a) Differential-mode circuit responses. (b) Common-mode circuit responses. Reprinted with permission from Ref. [1]; copyright 2012 IEEE.

increases from 147 Ω to 155 Ω, f_{r2}^{dd} moves closer to f_{r3}^{dd}, leading to the worse return loss of differential passband, while f_{z2}^{cc} moves closer to f_{z3}^{cc}, which results in a worse attenuation level of common-mode stopband. However, as Z_{oe} decreases from 147 Ω to 135 Ω, f_{r1}^{dd} and f_{r4}^{dd} move close to f_{r2}^{dd} and f_{r3}^{dd}, respectively, and the return loss of differential passband decreases alternatively. Also the common-mode attenuation at the center frequency decreases.

Figure 5.4 depicts the frequency responses of the proposed filter with varied impedance Z_1. For the given coupled line parameters Z_{oo} and Z_{oe}, the return loss of the differential mode at the center frequency

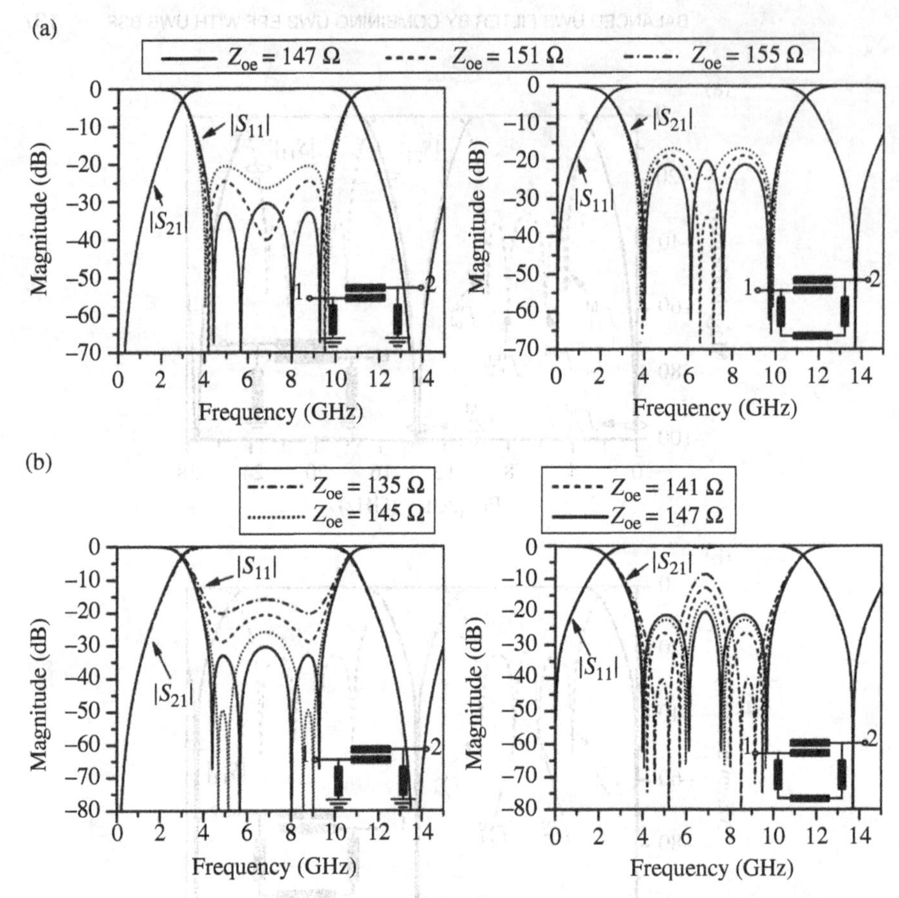

Figure 5.3 Frequency responses of the proposed filter with varied parallel-coupled line parameter. (a) Z_{oe} varied from 147 to 155 Ω. (b) Z_{oe} varied from 135 to 147 Ω. Reprinted with permission from Ref. [1]; copyright 2012 IEEE.

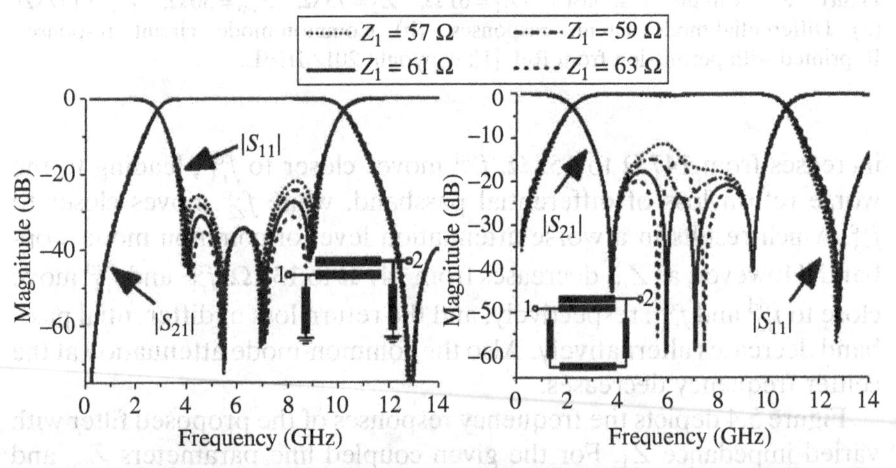

Figure 5.4 Frequency responses of the proposed filter with varied impedance Z_1. Reprinted with permission from Ref. [1]; copyright 2012 IEEE.

6.85 GHz remains the same as Z_1 decreases, while the return loss at other frequency decreases, yielding responses with worse return loss. For the common mode, as Z_1 increases from 57 Ω to 63 Ω, f_{z2}^{cc} moves closed to f_{z3}^{cc} and the attenuation of the common-mode stopband decreases. For the center connecting line, since it is a perfect EW under differential-mode operation, the center connecting line has no effect on differential operation and can be applied to determine the common-mode responses independently. As shown in Figure 5.5, when the center connecting line impedance Z_2 increases, the differential-mode responses remain the same. However, the attenuation at the common-mode stopband varies.

Consequently, the design procedures can be summarized as follows. Firstly, all the transmission lines are set to be $\lambda_g/4$. Then, by fine-tuning the coupled line parameter (Z_{oo} or Z_{oe}) and Z_1, good return loss performance of differential mode and attenuation level of common mode could be obtained. At last, the common-mode responses can be further optimized independently by properly choosing the parameter Z_2.

To validate the design concept described in the previous section, a microstrip balanced UWB BPF, based on the circuit schematic of Figure 5.1, is demonstrated through experimentally measured results, which are described as follows. For achieving the strong coupling in Figure 5.1, the three interdigital-coupled line structure, instead of the parallel-coupled two-line structure, is employed in this design [3]. Figure 5.6(a) shows the physical layout of the developed balanced UWB filter, which is fabricated on the substrate with a relative dielectric constant of 2.55, a thickness of 0.8 mm, and a loss tangent of 0.0029.

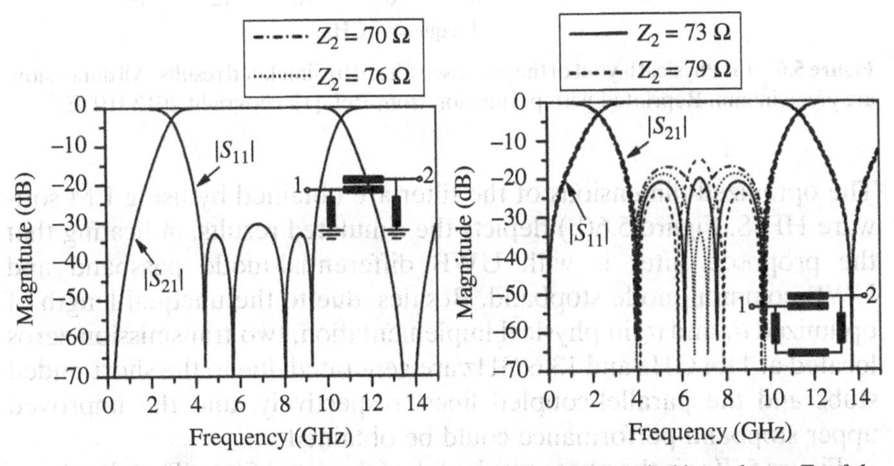

Figure 5.5 Frequency responses of the proposed filter with varied impedance Z_2 of the connecting line. Reprinted with permission from Ref. [1]; copyright 2012 IEEE.

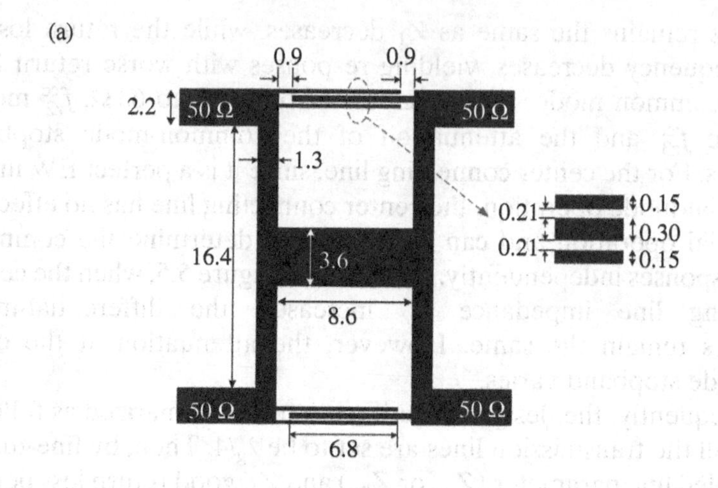

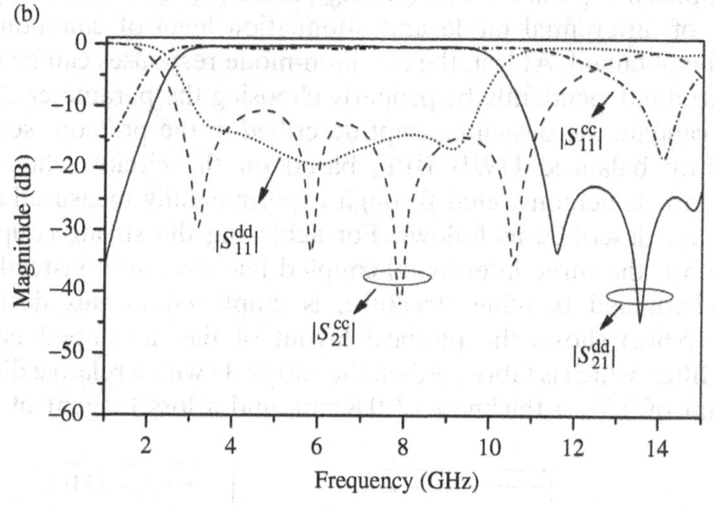

Figure 5.6 (a) Physical layout of the proposed filter. (b) Simulated results. All dimensions are given in mm. Reprinted with permission from Ref. [1]; copyright 2012 IEEE.

The optimized dimensions of the filter are obtained by using EM software HFSS. Figure 5.6(b) depicts the simulated results, indicating that the proposed filter is with UWB differential-mode passband and UWB common-mode stopband. Besides, due to the unequal length of optimized θ_1 and θ_c in physical implementation, two transmission zeros located at 11.4 GHz and 13.6 GHz are generated due to the short-ended stubs and the parallel-coupled lines, respectively, and the improved upper stopband performance could be obtained.

Figure 5.7(a) is the photograph of the fabricated filter. The filter has a compact size of 10.4 mm × 20.4 mm, corresponding to 0.35 λ_g × 0.7 λ_g,

(a)

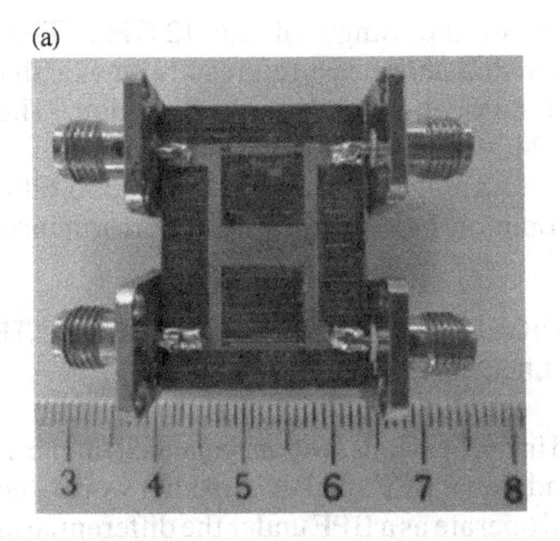

(b)

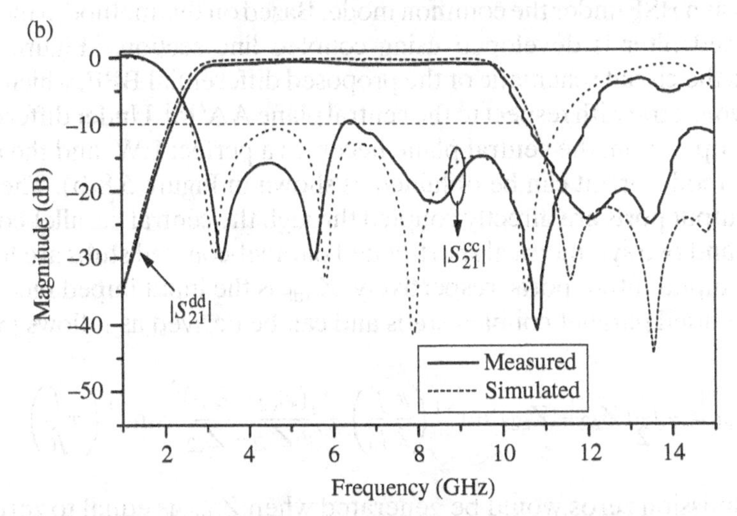

Figure 5.7 (a) Photograph of the fabricated filter. (b) Simulated and measured transmission coefficient $|S_{21}|$ under differential-mode and common-mode cases. Reprinted with permission from Ref. [1]; copyright 2012 IEEE.

where λ_g is the guided wavelength at the center frequency of the differential passband. Figure 5.7(b) shows the comparison between the simulated and measured transmission coefficients under differential- and common-mode operations, that is, $|S_{21}^{dd}|$ and $|S_{21}^{cc}|$, respectively. The measured differential passband is with minimum insertion loss of 0.8 dB and the 3 dB fractional bandwidth measured is 119%, covering the frequency range of 2.6–10.3 GHz. In the common mode, it performs as a UWB BSF and the common-mode signal is suppressed to be lower than −9.6 dB in the range of 2.6–12 GHz. The slight discrepancy between simulated and measured results may be due to the unexpected tolerance of fabrication and implementation. The ratio of Δf_{3dB} to Δf_{20dB} is 0.774, which shows good passband selectivity. The proposed BPF has the advantages of compact size, good differential UWB passband, and common-mode suppression performances.

5.2 BALANCED WIDEBAND BANDPASS FILTER USING COUPLED LINE STUBS

As discussed in the previous section, balanced filter can be realized by combining the traditional BPF and BSF. In other words, the proposed balanced circuit should operate as a BPF under the differential mode, whereas it performs as a BSF under the common mode. Based on this method, a modified balanced filter is developed using coupled line sections. Figure 5.8(a) shows the circuit schematic of the proposed differential BPF, which is ideally symmetric with respect to the central plane AA′ [4]. Under differential-mode operation, the central plane becomes a perfect EW, and the differential-mode circuit can be obtained as shown in Figure 5.8(b). The input and output ports are directly coupled through the central parallel-coupled lines, and two symmetrical short-ended parallel-coupled stubs are located in the input/output ports, respectively. Z_{stub} is the input impedance of the short-ended parallel-coupled stubs and can be derived as follows [5]:

$$Z_{stub} = -j\frac{1}{2}(Z_{o2}+Z_{e2})\tan^{-1}\left(\frac{\pi}{2}\frac{f}{f_0}\right) + j\frac{(Z_{e2}-Z_{o2})^2}{Z_{o2}+Z_{e2}}\sin^{-1}\left(\pi\frac{f}{f_0}\right) \quad (5.1)$$

Transmission zeros would be generated when Z_{stub} is equal to zero, and the transmission zero frequencies can be obtained as

$$f_{za} = \frac{2}{\pi} \times \arcsin\left(\sqrt{1-k_2^2}\right) \times f_0 < f_0; \quad (5.2a)$$

$$f_{zb} = \left(2 - \frac{2}{\pi} \times \arcsin\sqrt{1-k_2^2}\right) \times f_0 > f_0 \quad (5.2b)$$

where $k_2 = (Z_{e2}-Z_{o2})/(Z_{e2}+Z_{o2})$.

For common-mode operation, the central plane becomes a perfect magnetic wall (MW), and the common-mode circuit is shown in Figure 5.8(c). Figure 5.9 displays the differential- and common-mode circuit responses, where the spurious responses are located at $3f_0$. It

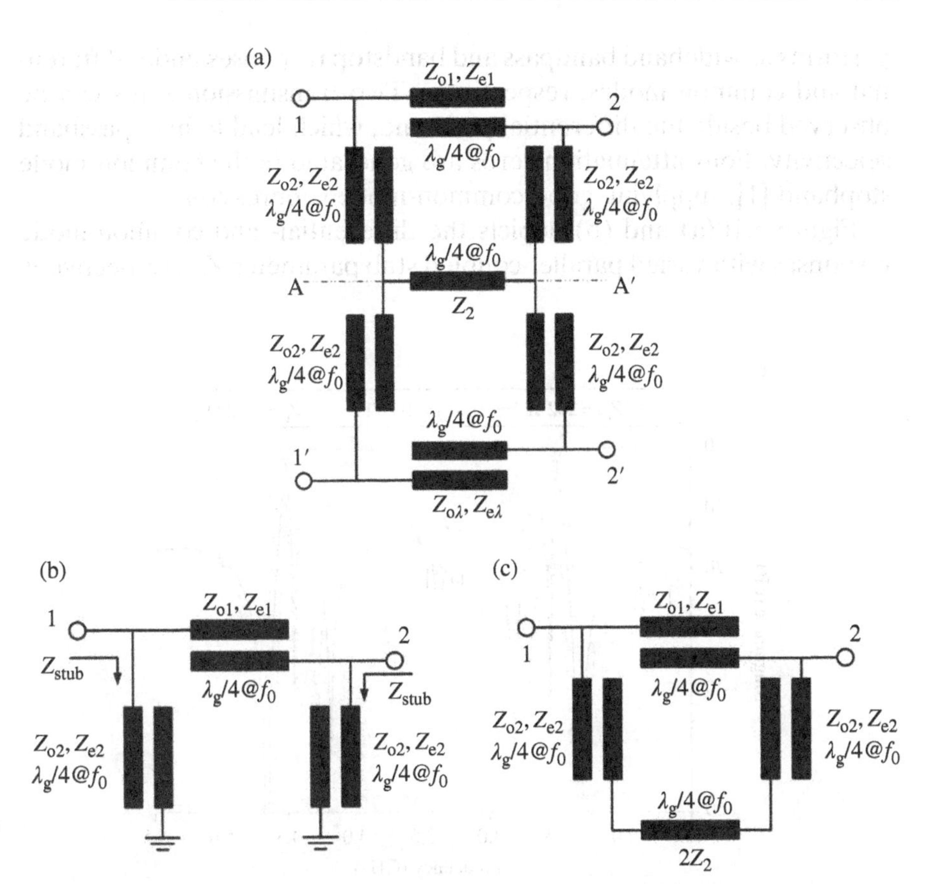

Figure 5.8 (a) The proposed differential BPF circuit. (b) Differential-mode circuit. (c) Common-mode circuit. Reprinted with permission from Ref. [4]; copyright 2013 IEEE.

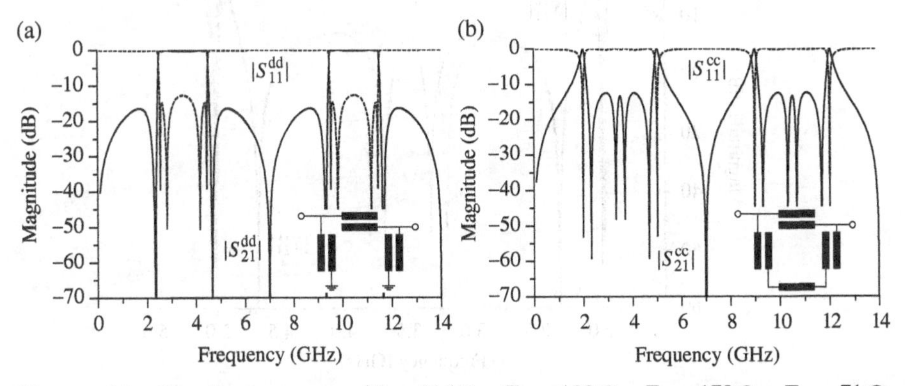

Figure 5.9 Circuit responses ($Z_2 = 115\,\Omega$, $Z_{o1} = 100\,\Omega$, $Z_{e1} = 179\,\Omega$, $Z_{o2} = 71\,\Omega$, $Z_{e2} = 212\,\Omega$). (a) Differential-mode circuit responses. (b) Common-mode circuit responses. Reprinted with permission from Ref. [4]; copyright 2013 IEEE.

performs as wideband bandpass and bandstop responses under differential and common modes, respectively. Two transmission zeros can be observed beside the differential passband, which lead to high passband selectivity. Four attenuation zeros are generated in the common-mode stopband [1], supplying good common-mode suppression.

Figure 5.10(a) and (b) depicts the differential- and common-mode responses with varied parallel-coupled stub parameter Z_{e2}, respectively.

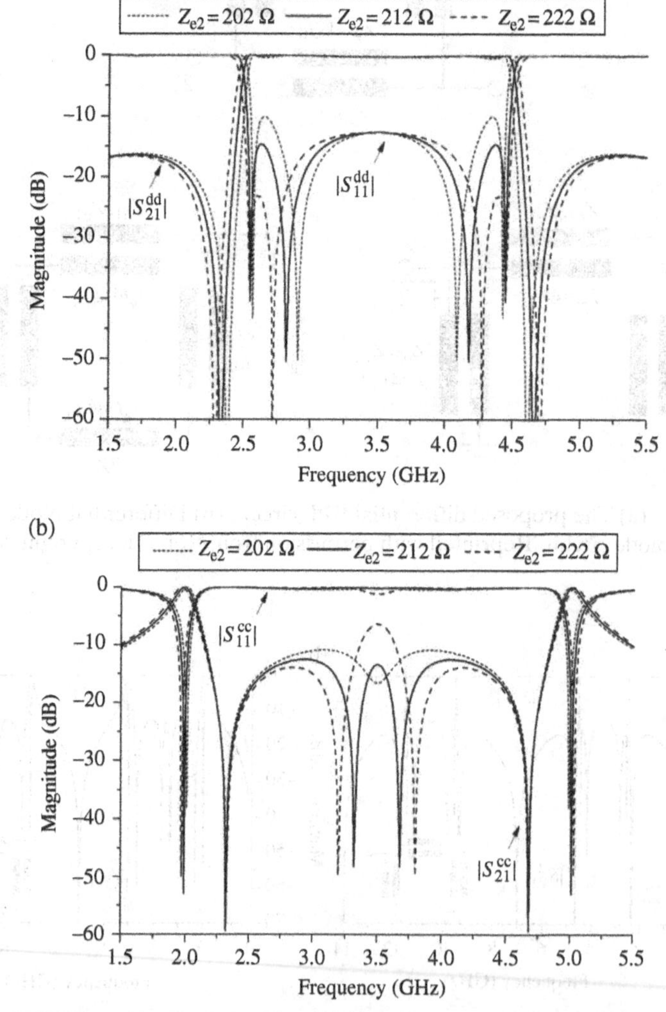

Figure 5.10 Frequency responses of the circuit with varied parallel-coupled stub parameter Z_{e2}. (a) Differential-mode circuit responses. (b) Common-mode circuit responses. Reprinted with permission from Ref. [4]; copyright 2013 IEEE.

As Z_{e2} increases, absolute increment of the differential passband bandwidth can be observed, and the two transmission zeros (f_{za} and f_{zb}) change simultaneously. Larger coupling of the parallel-coupled stub would result in a larger differential passband bandwidth. For common-mode responses, as shown in Figure 5.10(b), Z_{e2} has an absolute impact on the attenuation at f_0. Note that flat suppression level can be obtained when Z_{e2} is 212 Ω.

Figure 5.11 depicts the frequency responses of the developed circuit with varied impedance parameter Z_2. It can be observed that the common-mode performance varies when Z_2 changes, while the differential-mode performance keeps unchanged when Z_2 varies. Since the symmetry plane acts as a perfect EW under differential-mode operation, Z_2 can be employed to tune the common-mode performance independently without affecting the differential-mode performance.

For an experimental demonstration, the presented circuit in Figure 5.8(a) is implemented using microstrip-line technology. The substrate is with a relative dielectric constant of 2.55, a thickness of 0.8 mm, and loss tangent of 0.0029. Besides, to achieve strong coupling of the parallel-coupled line, the parallel-coupled line with defected ground is applied here [6]. Figure 5.12(a) displays the physical layout of the developed balanced wideband filter, which is optimally designed by using HFSS. The gray shielding is etched out in the ground. Figure 5.12(b) depicts the simulated results, indicating that the proposed filter has

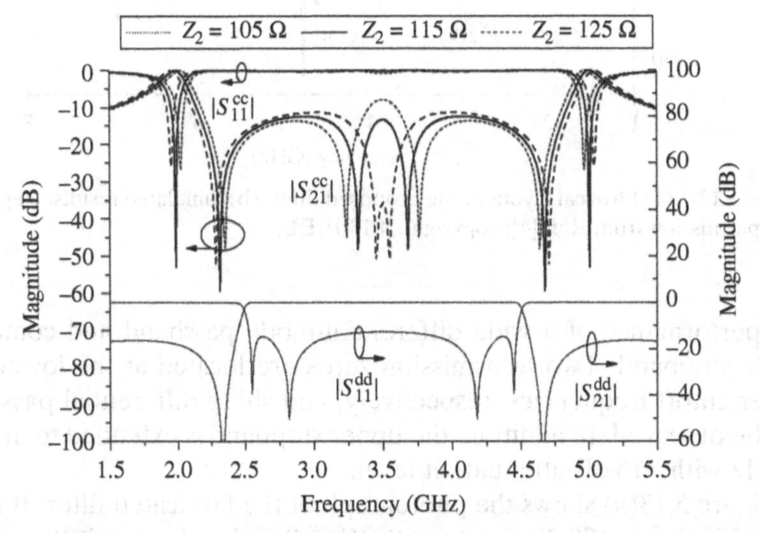

Figure 5.11 Frequency responses of the circuit with varied impedance Z_2. Reprinted with permission from Ref. [4]; copyright 2013 IEEE.

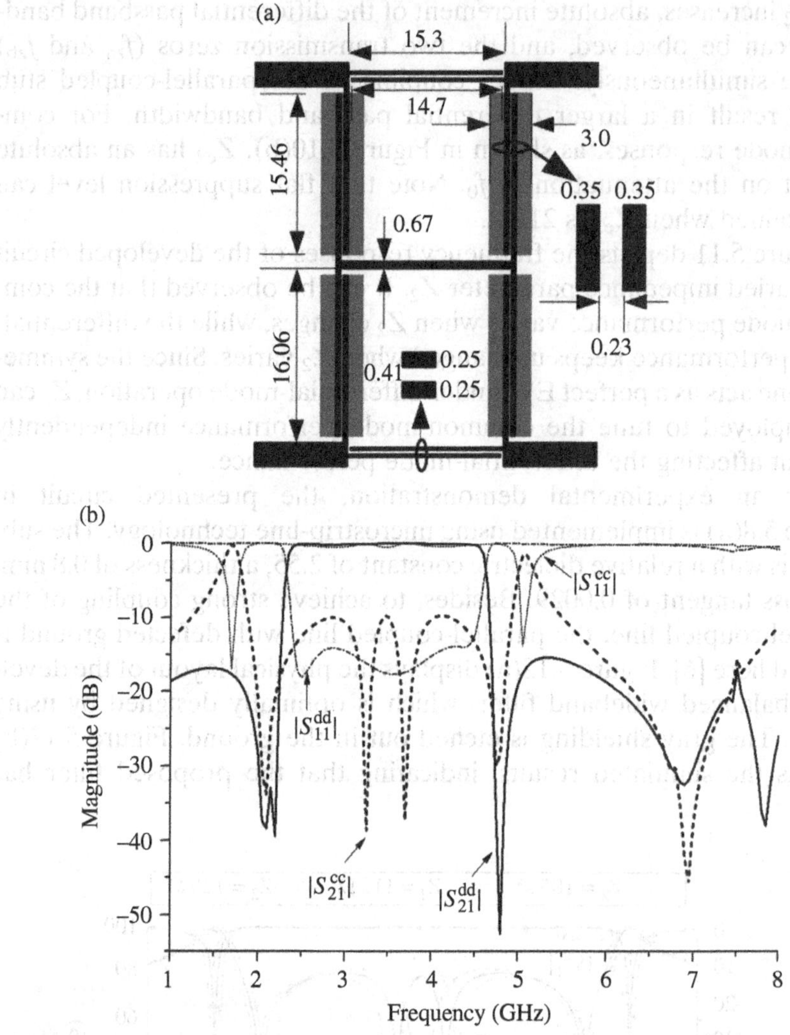

Figure 5.12 (a) Physical layout of the proposed filter. (b) Simulated results. Reprinted with permission from Ref. [4]; copyright 2013 IEEE.

the performance of a wide differential-mode passband and common-mode stopband. Two transmission zeros are located at the lower and upper cutoff frequencies, respectively, and sharp differential passband can be observed. In addition, the upper stopband is extended to at least 8 GHz with −15 dB attenuation level.

Figure 5.13(a) shows the photograph of the fabricated filter. It has a size of 37.2 mm × 20.3 mm, about $0.60\lambda_g \times 0.33\,\lambda_g$, where λ_g is the guided wavelength at the center frequency of the differential passband. The

(a)

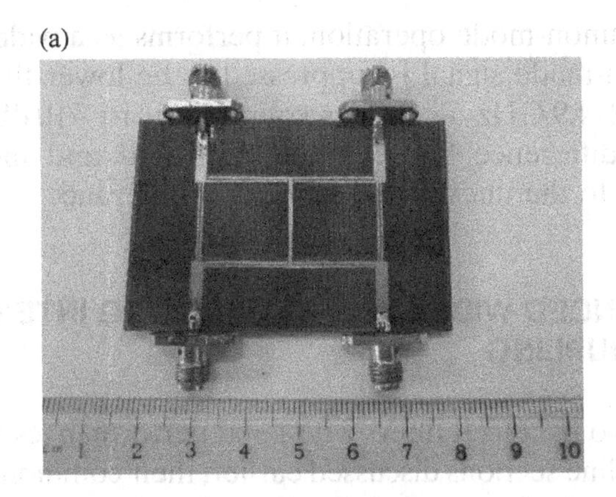

(b)

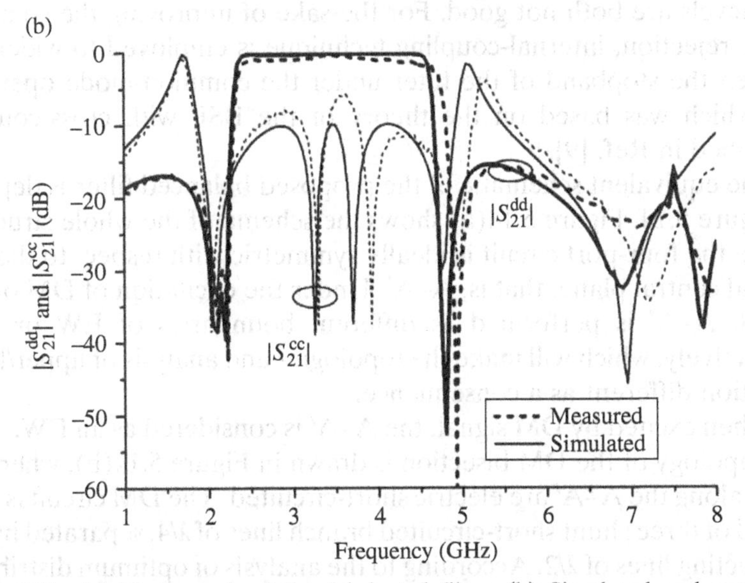

Figure 5.13 (a) Photograph of the fabricated filter. (b) Simulated and measured transmission coefficient $|S_{21}|$ under differential-mode and common-mode cases. Reprinted with permission from Ref. [4]; copyright 2013 IEEE.

two-port differential-/common-mode *S*-parameters can be extracted from the four-port *S*-parameters [7]. The simulated and measured transmission coefficients under differential- and common-mode operations are given in Figure 5.13(b). For the differential-mode operation, the measured 3 dB fractional bandwidth is 66%, covering the frequency range of 2.34–4.65 GHz with the measured minimum insertion across the passband of 0.36 dB. Two transmission zeros can be observed at 2.2 GHz and 4.75 GHz, which supply a sharp differential passband.

For the common-mode operation, it performs as a wideband BSF, and the common-mode signal is suppressed to be lower than −6 dB in the range of 1.8–4.9 GHz against its counterpart to −10 dB in simulation. The small difference between the simulated and measured results may be due to the unexpected fabrication tolerance.

5.3 BALANCED WIDEBAND FILTER USING INTERNAL CROSS-COUPLING

Although two balanced filters with good performances are achieved by the coupled line sections discussed earlier, their common-mode suppression levels are both not good. For the sake of improving the common-mode rejection, internal-coupling technique is employed to widen and deepen the stopband of the filter under the common-mode operation [8], which was based on the theory of the BSF with cross-coupling described in Ref. [9].

The equivalent schematic of the proposed balanced filter is depicted in Figure 5.14. Figure 5.14(a) shows the scheme of the whole structure, where the four-port circuit is ideally symmetric with respect to the horizontal central plane, that is, A–A′. Under the excitation of DM or CM signal, A–A′ is performed as different boundaries of EW or MW, respectively, which will make the topologies and analysis of upper/lower bisection different as a consequence.

When excited by DM signal, the A–A′ is considered as an EW. Thus, the topology of the DM bisection is drawn in Figure 5.14(b), where the stubs along the A–A′ are electric short-circuited. The DM circuit is comprised of three shunt short-circuited branch lines of $\lambda/4$, separated by two connecting lines of $\lambda/2$. According to the analysis of optimum distributed highpass filter in Ref. [10], n stubs will lead to $2n-1$ transmission poles (TPs), forming a wideband passband with this prototype. Hence, the DM circuit is a wideband BPF with 5 DM TPs herein as shown in Figure 5.15. Besides, a pair of DM TZs will be generated using $\lambda/4$ coupled line stubs with the even-mode/odd-mode impedance of Z_{1e}/Z_{1o}, whose mechanism is the same as the filter shown in Figure 5.8.

Corresponding to the circumstance in this design, the computed frequency ratios using (5.2a) and (5.2b) are $f_{za}/f_0 = 0.656$ and $f_{zb}/f_0 = 1.344$. As shown in Figure 5.15, for the differential mode, two transmission zeros are located at 4.16 and 8.54 GHz. So simulated frequency ratios with respect to the central frequency f_0 of 6.35 GHz are 0.655 and 1.345, which equate with the computed ones and indicate the controllability of transmission zeros for differential mode. Consequently, with the

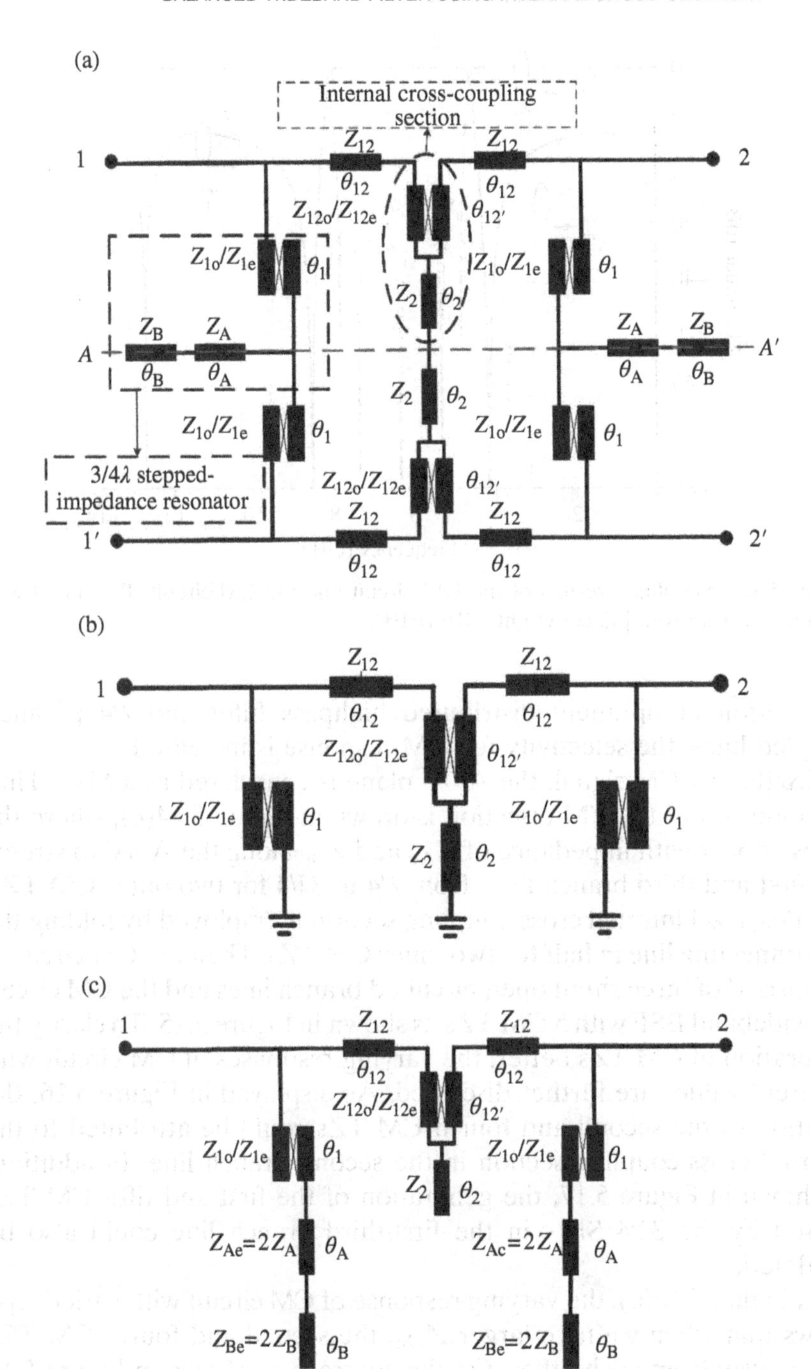

Figure 5.14 Equivalent schematic of the proposed balanced filter. (a) The whole structure, (b) the DM circuit, and (c) the CM circuit ($Z_{12} = 96\,\Omega$, $Z_{12e}/Z_{12o} = 116\,\Omega/76\,\Omega$; $Z_{1e}/Z_{1o} = 206\,\Omega/66\,\Omega$, $Z_2 = 16\,\Omega$, $Z_A = 126\,\Omega$, $Z_B = 70\,\Omega$, $\theta_{12} = \theta_{12'} = \theta_1 = \theta_2 = \theta_A = \theta_B = \pi/2$). Reprinted with permission from Ref. [8]; copyright 2016 IEEE.

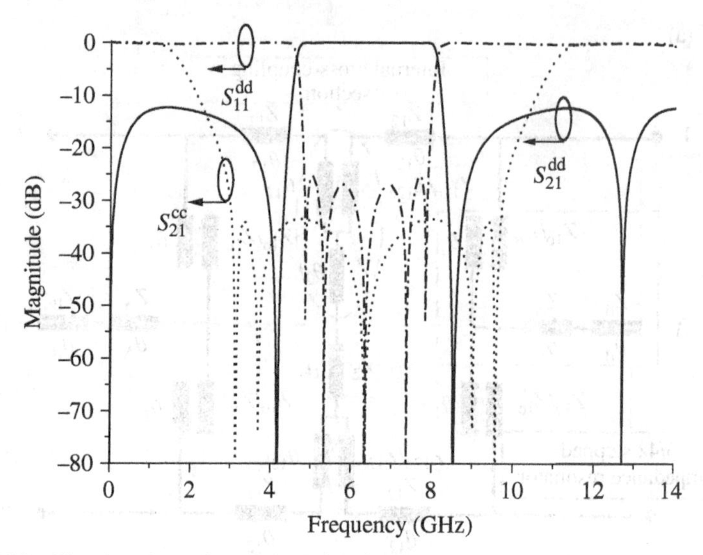

Figure 5.15 Simulated results of the DM circuit and the CM circuit. Reprinted with permission from Ref. [8]; copyright 2016 IEEE.

application of optimum distributed highpass filter and $\lambda/4$ parallel-coupled lines, the selectivity for DM response is increased.

Excited by CM signal, the A–A′ plane is considered as a MW. Thus the topology of the CM bisection is drawn in Figure 5.14(c), where the stubs appear with impedance of Z_{Ae} and Z_{Be} along the A–A′ to stretch the first and third branch lines from $\lambda/4$ to $3\lambda/4$ for two outer CM TZs. Besides, a $\lambda/4$ internal cross-coupling section is employed by folding the $\lambda/2$ connecting line in half for two inner CM TZs. Then the CM circuit is comprised of three shunt open-circuited branch lines and the CM circuit is a wideband BSF with 5 CM TZs as shown in Figure 5.15. To clarify the generation of CM TZs better, the varying responses of CM circuit with different values are further discussed. As displayed in Figure 5.16, the creation of the second and fourth CM TZs could be attributed to the internal cross-coupling section in the second branch line. In addition, as shown in Figure 5.17, the generation of the first and fifth CM TZs caused by the $3\lambda/4$ SIRs in the first/third branch line could also be validated.

In Figure 5.16(a), the varying response of CM circuit with varied Z_{12e} shows that when we take larger Z_{12e}, the second and fourth CM TZs move away from each other. On the contrary, as shown in Figure 5.16 (b) or (c), when Z_{12o} or Z_2 is larger, these CM TZs move toward each other. While in Figure 5.17(a), the response of CM circuit against variable R_{z2} indicates that the larger R_{z2} we utilize, the farther the first and

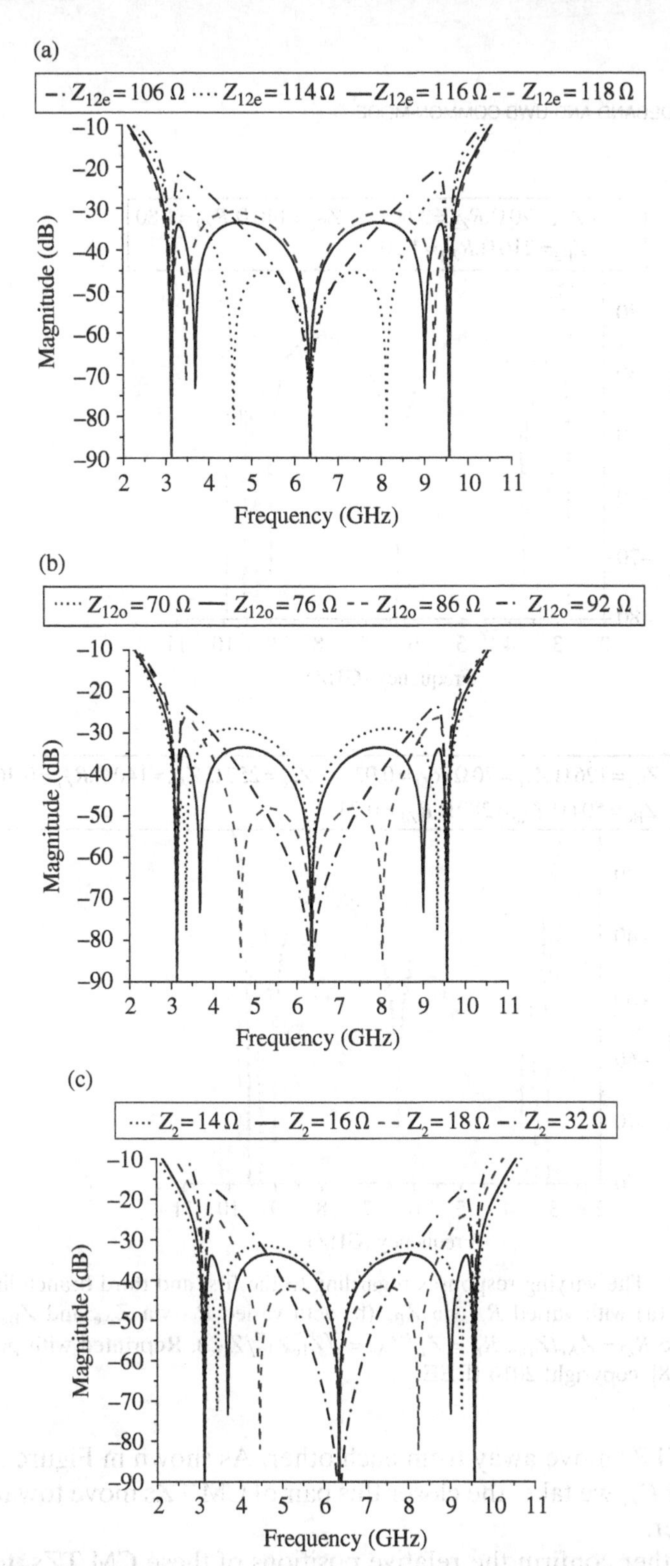

Figure 5.16 The varying responses regarding to the second branch line of the CM circuit, (a) with variable Z_{12e}, (b) with variable Z_{12o}, and (c) with variable Z_2. Reprinted with permission from Ref. [8]; copyright 2016 IEEE.

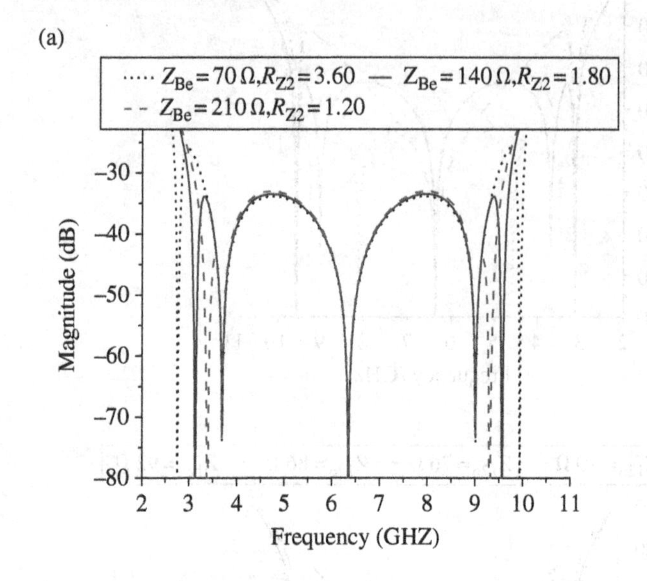

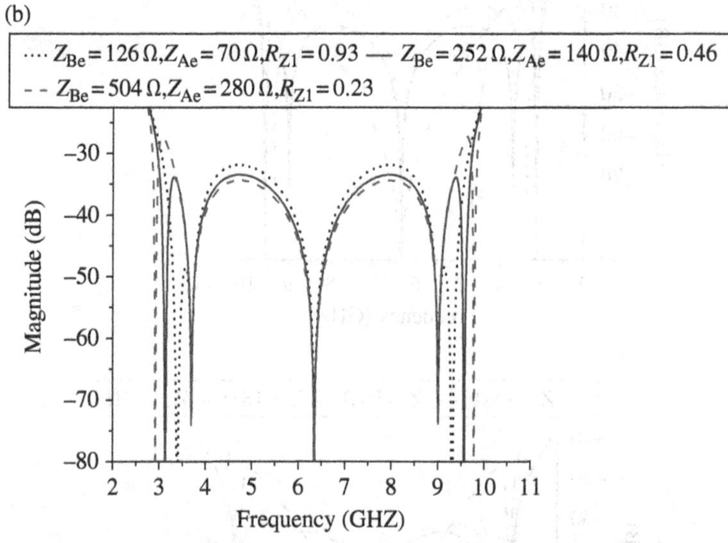

Figure 5.17 The varying responses regarding to the first and third branch line of the CM circuit (a) with varied R_{z2} via Z_{Be} (b) with varied R_{z1} via Z_{Ae} and Z_{Be} in same ratio (where $R_{z2} = Z_{Ae}/Z_{Be}$, $R_{z1} = Z_1/Z_{Ae} = \sqrt{Z_{1e}Z_{1o}}/Z_{Ae}$). Reprinted with permission from Ref. [8]; copyright 2016 IEEE.

fifth CM TZs move away from each other. As shown in Figure 5.17(b), the larger R_{z1} we take, the closer this pair of CM TZs move toward from each other.

To further confirm the relative positions of these CM TZs, four frequency ratios can be simplified through the input impedance of the cross-coupling section ($i = 2, j = 4$, where i and j indicate the numbering of the CM TZs) and $3\lambda/4$ open-circuited SIR ($i = 1, j = 5$) as

$$\frac{f_{ci}}{f_0} = \frac{\theta_{ci}}{\theta_0} = \frac{\theta_{ci}}{\pi/2} = \frac{2}{\pi}\theta_{ci} \tag{5.3}$$

$$\frac{f_{cj}}{f_0} = \frac{\theta_{cj}}{\theta_0} = \frac{\pi - \theta_{ci}}{\pi/2} = 2 - \frac{2}{\pi}\theta_{ci} \tag{5.4}$$

where $i = 1$ and 2, $j = 4$ and 5, and θ_{c1} and θ_{c2} are given as follows:

$$\theta_{c1} = \arcsin\sqrt{\frac{2(Z_{1e} + Z_{1o})Z_{Ae}Z_{Be} + 4Z_{1e}Z_{1o}(Z_{Ae} + Z_{Be})}{2(Z_{1e} + Z_{1o})Z_{Ae}Z_{Be} + 2(Z_{1e} + Z_{1o})\, Z_{Ae}^2 + (Z_{1e} + Z_{1o})^2(Z_{Ae} + Z_{Be})}} \tag{5.5}$$

$$\theta_{c2} = \arctan\sqrt{\frac{2Z_{12e}Z_2}{Z_{12e}^2 - Z_{12e}Z_{12o} - 2Z_{12o}Z_2}} \tag{5.6}$$

Corresponding to the circumstance in this design, when $Z_{Ae} = 2Z_A = 252\ \Omega$ and $Z_{Be} = 2Z_B = 140\ \Omega$ are set, the computed frequency ratios using (5.3) and (5.4) are $f_{c2}/f_0 = 0.582$, $f_{c4}/f_0 = 1.418$, $f_{c1}/f_0 = 0.493$, and $f_{c5}/f_0 = 1.507$, respectively. From Figure 5.15, it can be observed that these four CM TZs (f_{c2}, f_{c4}, f_{c1}, f_{c5}) are located at 3.69/9.01 GHz and 3.13/9.57 GHz, respectively. So simulated frequency ratios with respect to 6.35 GHz are 0.581/1.419 and 0.493/1.507, which are equal to the computed ones and indicate the controllability of CM TZs. Therefore, with appropriate impedance in internal cross-coupling and $3\lambda/4$ SIRs, enhanced CM rejection with 5 TZs is finally gained.

For demonstration, this balanced filter is optimized and fabricated on a substrate with a relative dielectric constant of 2.55, a thickness of 0.8 mm, and a loss tangent of 0.0029. Figure 5.18 shows the physical layout with all dimensions provided and photograph of the fabricated filter. The comparisons between the results of EM simulation and the measurement are shown in Figure 5.19, where solid line and dot line represent the EM simulated and measured results, respectively. In EM simulated (measured) result, a wide DM passband with 3 dB fractional bandwidth of 58% (59%) and a wide CM stopband with fractional bandwidth of 123% (118%) more than 15.5 dB (18.2 dB) are simultaneously attained. The visible discrepancies between simulated and measured results may be due to the increase of radiation loss, as well as the unexpected tolerance of fabrication.

(a)

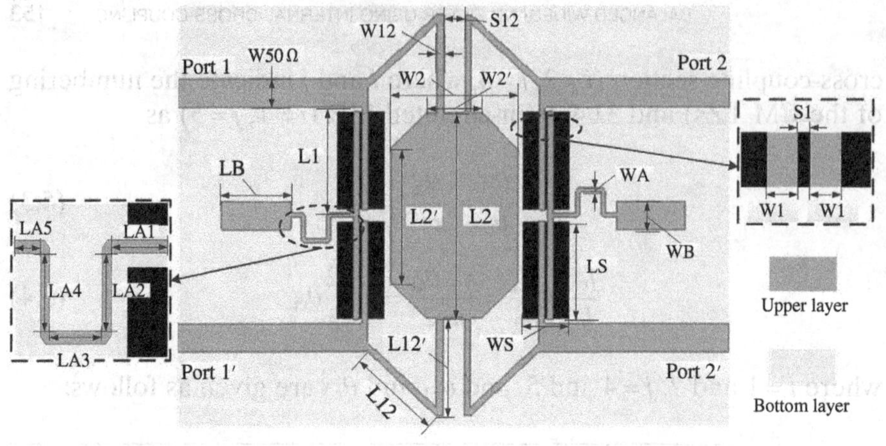

(b)

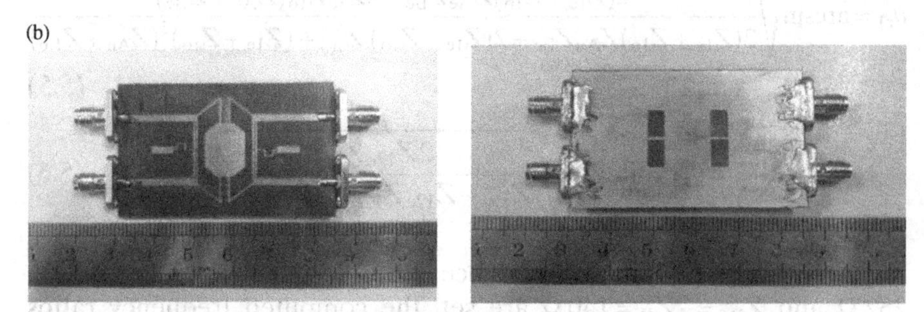

Figure 5.18 (a) Physical layout. (b) Top/bottom-view photographs of the proposed filter. (W_{12} = 0.5 mm, L_{12} = 7.44 mm, $L_{12'}$ = 7.76 mm, S_{12} = 1.68 mm, W_1 = 0.45 mm, S_1 = 0.23 mm, L_1 = 8.4 mm, W_S = 3.75 mm, L_S = 7.8 mm, W_2 = 10 mm, $W_{2'}$ = 4.35 mm, L_2 = 16.3 mm, $L_{2'}$ = 10.59 mm, W_A = 0.3 mm, L_{A1} = 1.85 mm, L_{A2} = L_{A3} = L_{A4} = 1.7 mm, L_{A5} = 0.85 mm, W_B = 2.25 mm, L_B = 5.46 mm). Reprinted with permission from Ref. [8]; copyright 2016 IEEE.

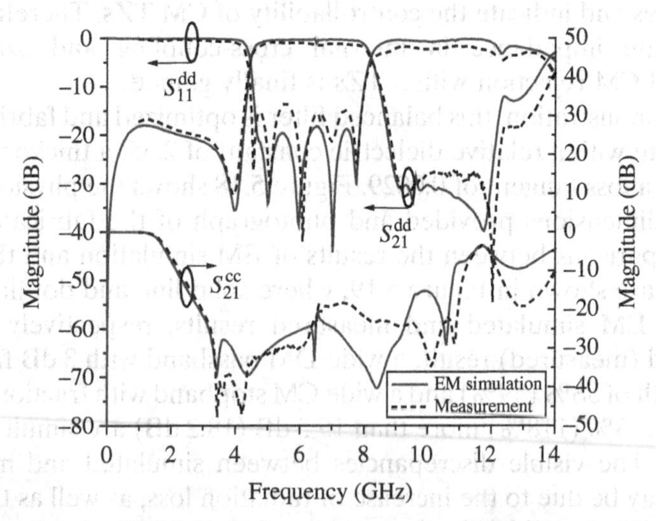

Figure 5.19 Results of EM simulation and measurement. Reprinted with permission from Ref. [8]; copyright 2016 IEEE.

5.4 BALANCED WIDEBAND FILTER USING STUB-LOADED RING RESONATOR

To achieve high selectivity in differential mode and enhanced rejection in common mode, more transmission zeros should be introduced. Another simple and innovative approach is proposed to design balanced wideband filter with high selectivity and common-mode suppression.

Figure 5.20 exhibits the circuit model and layout of the proposed balanced BPF, which contains two stub-loaded ring resonators (SLRRs) [11]. The proposed SLRRs with centrally loaded with stub (Z_2, θ_2) are coupled (Z_{oe}/Z_{oo}, θ_0) with the coupled feed lines (CFLs). Moreover, the CFLs are also loaded with stubs (Z_1, θ_1). All input/output (I/O) lines are with the characteristic impedance 50 Ω and cascaded with $\lambda/4$ impedance transformers (Z_3, θ_3). The entire structure is symmetric with respect to the center line AA′.

Under differential-mode operation, AA′ plane is equivalent to a virtual EW, and the equivalent circuit of the filter is shown in Figure 5.21(a).

(a)

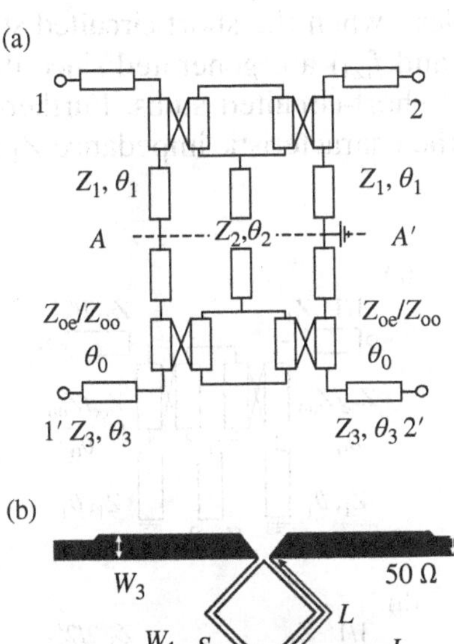

(b)

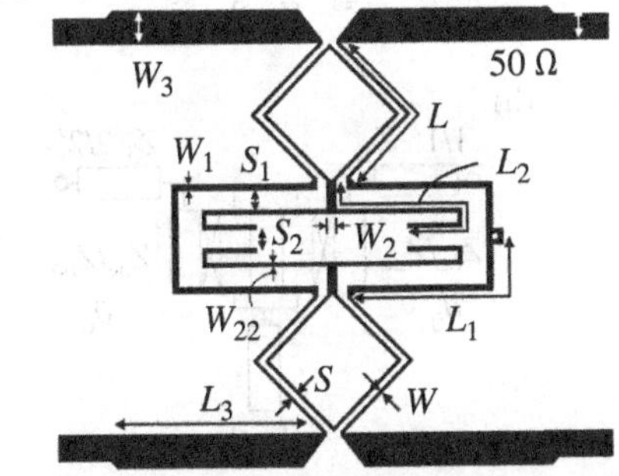

Figure 5.20 (a) Equivalent circuit and (b) layout of the proposed balanced filter ($L = 24$, $W = 0.25$, $S = 0.25$, $L_1 = 22.6$, $W_1 = 0.5$, $L_2 = 25.3$, $W_2 = 0.35$, $L_3 = 24.8$, $W_3 = 2.95$, all in millimeters). Reprinted with permission from Ref. [11]; copyright 2015 IEEE.

The proposed filter is composed of two SLRRs and short-circuited stub-loaded CFLs. For simplicity, we consider that $\theta_0 = \theta_1 = \theta_2 = \theta_3 = \theta$. When $\theta = \lambda/4$, the short-circuited loaded stubs are equivalent to an open point at the center frequency, as shown in Figure 5.21(b). Therefore, the passband is determined by the SLRR. In fact, the SLRR is a single-mode resonator and the resonant mode is calculated as $\cot\theta = 0$ [12]. However, three TPs can be achieved due to the cross-coupling of the CFLs [13]. Figure 5.22(a) displays the response curves of the circuit in Figure 5.21(a) with variable even-mode impedances Z_{oe}. It can be seen that the coupling coefficient ($K = (Z_{oe} - Z_{oo})/(Z_{oe} + Z_{oo})$) of the CFLs becomes stronger, as Z_{oe} increase. That is to say that three TPs can be tuned by Z_{oe}, helping to achieve controllable bandwidth. Moreover, $\lambda/4$ impedance transformers are cascaded to the input/output ports, which can further improve the passband ripple level. The short-circuited stubs loaded to the CFLs achieve a balanced equivalent in DM and also improve the characteristics of the filter [12]. As shown in Figure 5.22 (b), it can be observed that, without loading stubs (Z_1, θ_1), there are only two TZs (f_{tz2} and f_{tz3}) out of the passband due to the cross-coupling of the CFLs. Nevertheless, when the short-circuited stubs are loaded, two additional TZs (f_{tz1} and f_{tz4}) are generated since the loaded CFLs can be considered as $\lambda/4$ short-circuited stubs. Furthermore, the passband becomes steeper as the characteristic impedance Z_1 decreases.

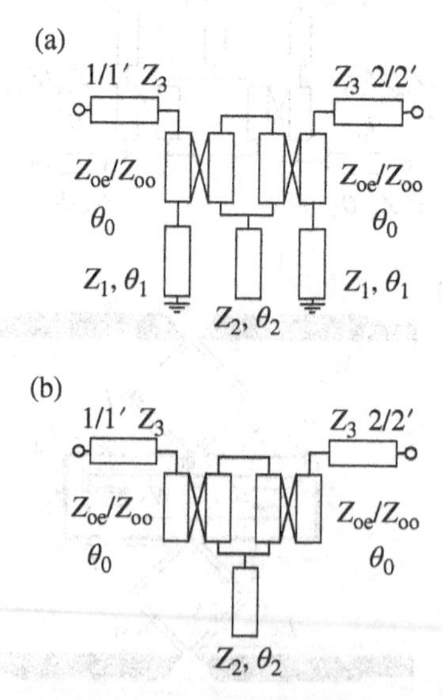

Figure 5.21 (a) Equivalent circuit in DM and (b) equivalent circuit in DM at f_0 without stubs (Z_1, θ_1). Reprinted with permission from Ref. [11]; copyright 2015 IEEE.

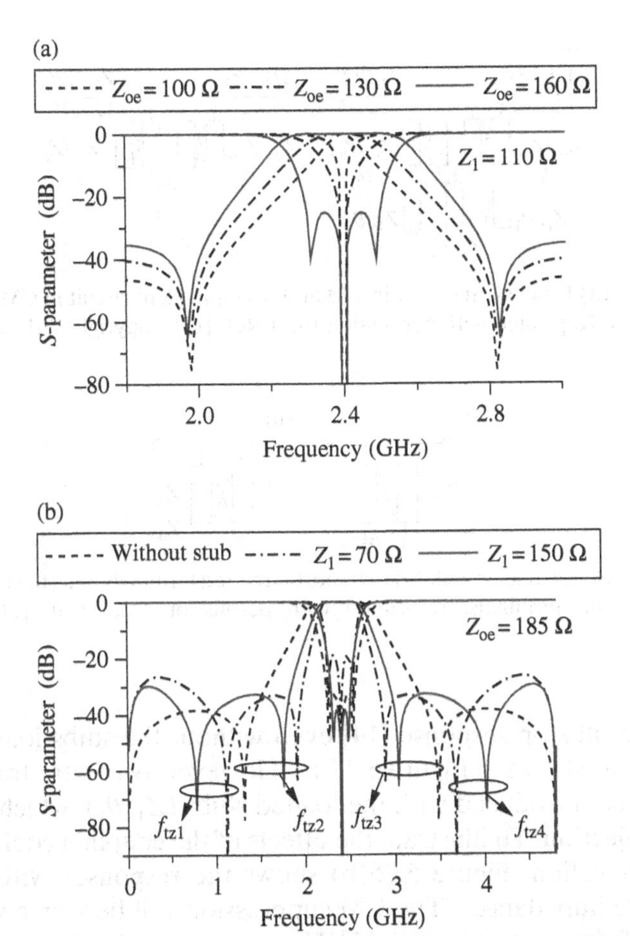

Figure 5.22 (a) Simulated DM responses of the proposed filter with varying even-mode impedances. (b) Variation of frequency responses with/without stubs (Z_1, θ_1). ($Z_2 = 92\,\Omega$, $Z_3 = 44\,\Omega$, $Z_{oo} = 85\,\Omega$). Reprinted with permission from Ref. [11]; copyright 2015 IEEE.

Alternatively, under common-mode operation, AA′ plane is equivalent to a virtual MW. The equivalent circuit is shown in Figure 5.23(a) and the input/output feed lines are loaded with open-/short-circuited stubs, respectively. Since the electrical length θ_1 is of $\lambda/4$, the circuit in Figure 5.23(a) is equivalent to the circuit in Figure 5.23(b) at the center frequency f_0. Similarly, when the loaded stub (Z_2, θ_2) is quarter-wavelength, the circuit in Figure 5.23(b) is equivalent to two cascaded coupled lines shown in Figure 5.24. According to its ABCD matrix [10], the condition for the generation of transmission zeros can be obtained as $\theta_0 = 90°$. Thus, the common-mode equivalent circuit can

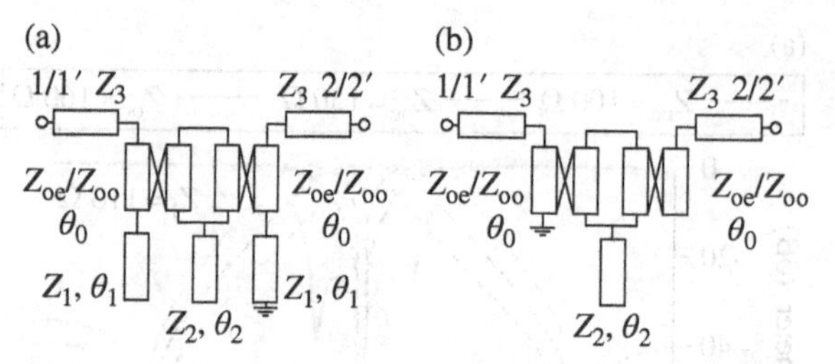

Figure 5.23 (a) Equivalent circuit in CM and (b) equivalent circuit in CM at f_0 without stubs (Z_1, θ_1). Reprinted with permission from Ref. [11]; copyright 2015 IEEE.

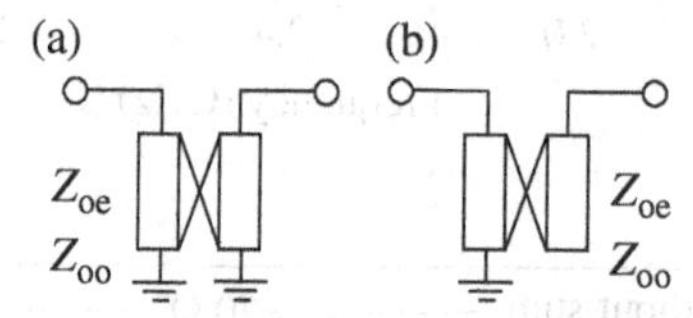

Figure 5.24 Coupled lines with (a) two short-circuited terminals and (b) short-circuited and open-circuit terminals. Reprinted with permission from Ref. [11]; copyright 2015 IEEE.

achieve an all-stop response [14] even without the stubs loaded to the feed lines, as shown in Figure 5.25(a). However, two extra transmission zeros can be introduced with the loaded stubs (Z_1, θ_1), which improves the CM rejection. To illustrate the effects of the coupling coefficients on the CM rejection, Figure 5.25(b) shows the responses with different even-mode impedances. The CM suppression will be worse when coupling coefficient gets stronger [15]. However, since the differential mode is designed for narrowband or appropriate wideband application, high CM rejection can be achieved with suitable coupling coefficients.

To verify the proposed circuit model in Figure 5.20(a), an experimental balanced wideband BPF is implemented based on the substrate with a dielectric constant of 2.55, a thickness of 0.8 mm, and a loss tangent of 0.0029. The proposed balanced BPF centered at 2.4 GHz with 3 dB fractional bandwidth of 19% is designed with the parameters of $Z_1 = 110\ \Omega$, $Z_2 = 92\ \Omega$, $Z_3 = 44\ \Omega$, $Z_{oo} = 85\ \Omega$, $Z_{oe} = 185\ \Omega$, and $\theta_0 = \theta_1 = \theta_2 = \theta_3 = 90°$. As shown in Figure 5.20(b), the loaded stubs of the resonator construct as T-shape to introduce appropriate coupling between the loaded stubs of the CFLs in differential-mode and common-mode operation. Taking differential-mode operation as an example, extra TZ f_{tz5} is generated due to the additional coupling path and the coupling strength is mainly decided by coupling gap S_1. Although the TZs f_{tz2} and f_{tz3} can also be changed since the source/load cross-coupling is altered, well passband

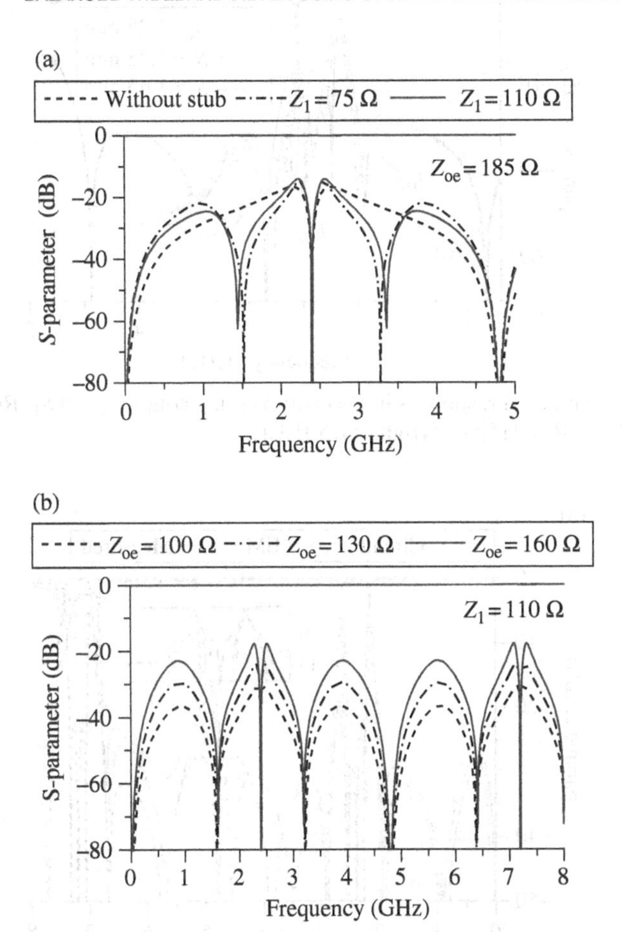

Figure 5.25 (a) Variation of frequency responses with/without stubs (Z_1, θ_1). (b) Simulated CM responses of the proposed filter with varying even-mode impedances. ($Z_2 = 92\ \Omega$, $Z_3 = 44\ \Omega$, $Z_{oo} = 85\ \Omega$). Reprinted with permission from Ref. [11]; copyright 2015 IEEE.

performance can be also achieved by appropriate value of S_1, as plotted in Figure 5.26. Thus, additional transmission zeros in DM (CM) can improve the filter performance apparently.

The comparisons between the simulated (circuit/EM) and measured results are given in Figure 5.27, which are in good agreement. In DM operation, as shown in Figure 5.27(a), the measured (EM simulated) center frequency is 2.41 (2.39) GHz with 3 dB fractional bandwidth of 17% (18.7%), while the minimum insertion loss is 0.92 (0.3) dB. Nine transmission zeros are introduced to achieve 16 dB (18 dB) rejection out of the passband.

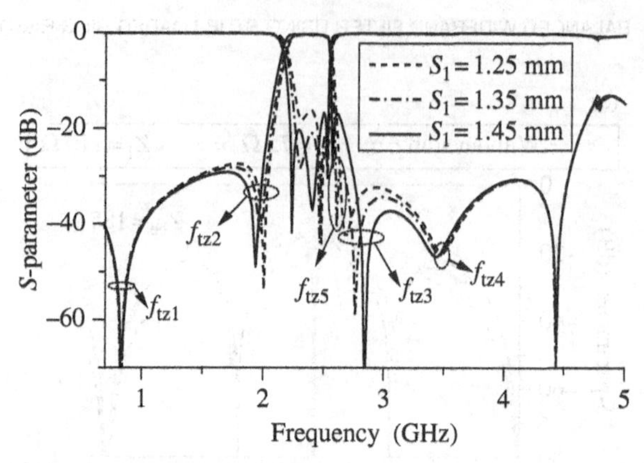

Figure 5.26 Simulated responses in DM with varying coupling gap S_1. Reprinted with permission from Ref. [11]; copyright 2015 IEEE.

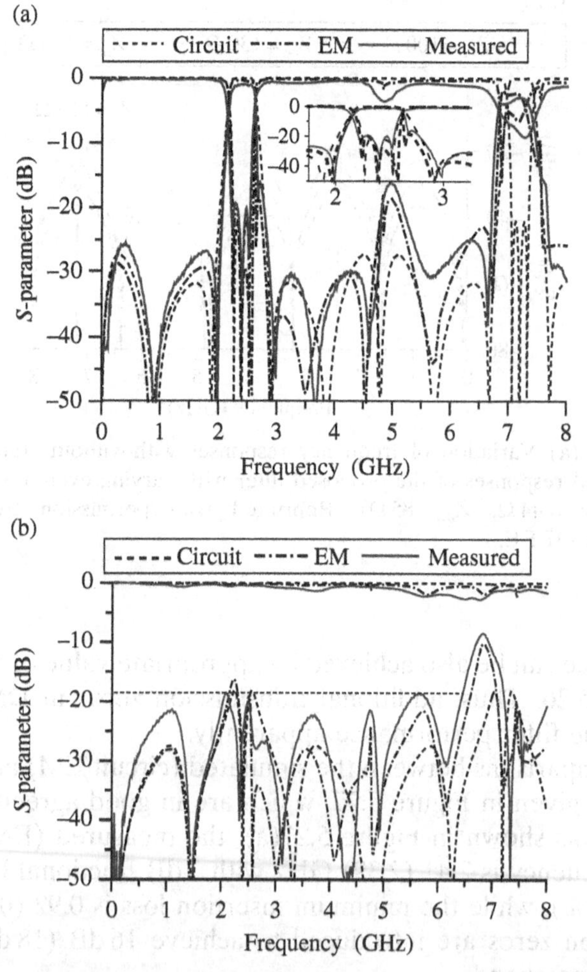

Figure 5.27 Simulated and measured results of the proposed filter. (a) DM response and (b) CM response. Reprinted with permission from Ref. [11]; copyright 2015 IEEE.

In CM operation, as shown in Figure 5.27(b), nine TZs are measured. CM suppression can be extended from 0 to 6.5 GHz with a rejection level of 18.8 dB. It should be emphasized that due to the additional coupling path introduced by the T-shaped stubs of the resonator, an extra transmission zero at 2.67/2.6 GHz in DM/CM is measured, which not only improves the selectivity of DM passband but also enhances the CM rejection. The proposed filter shows the advantages of high selectivity, low insertion loss, and high CM suppression.

5.5 BALANCED WIDEBAND FILTER USING MODIFIED COUPLED FEED LINES AND COUPLED LINE STUBS

Figure 5.28(a) and (b) shows the traditional parallel-coupled feed lines, indicated as type A and type B with the response of bandpass and bandstop, respectively [16, 17]. As shown in Figure 5.28(c), modified coupled feed line (MCFL) type C can be realized by loading a short-circuited stub with electrical length θ_1 to the open-circuited end of type A. Similar to the method used in Ref. [18], two-port impedance matrix is derived to calculate the ABCD matrix M_{d1} in (5.7), where K_1 is the coupling coefficient defined as $K_1 = (Z_{oe1} - Z_{oo1})/(Z_{oe1} + Z_{oo1})$. The resonant frequency can be obtained as $\theta_1 = \pi/2$. Therefore, type A and type C have equivalent bandpass response at the center frequency f_0. Similarly, the MCFL type D is formed if the short-circuited end of type B is replaced by an open-circuited stub with suitable electrical length θ_s, as shown in Figure 5.28(d). From its ABCD matrix M_{c1} in (5.8),

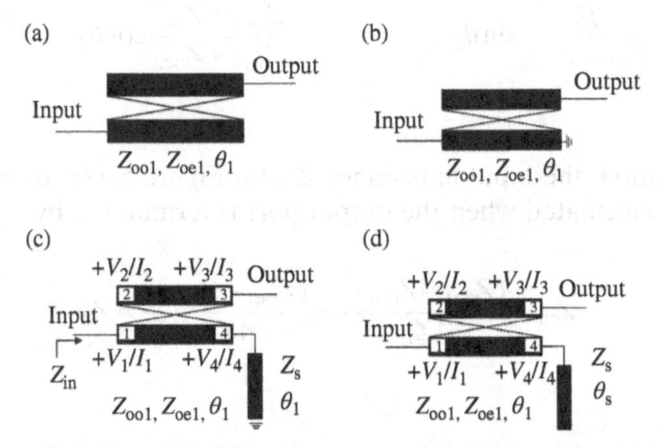

Figure 5.28 Traditional parallel-coupled lines model (a) type A and (b) type B and the MCFLs (c) type C and (d) type D. Reprinted with permission from Ref. [16]; copyright 2015 IEEE.

the transmission zero is obtained as $\cos\theta_s = 0$. By controlling the electrical length θ_s of the loaded stub, a short-circuited point at the specific frequency is provided to suppress the unwanted spurious.

$$M_{d1} = \begin{bmatrix} A_{d1} & B_{d1} \\ C_{d1} & D_{d1} \end{bmatrix} = \begin{bmatrix} \dfrac{\cos\theta_1(Z_{oo1} + Z_s - K_1 Z_s)}{K_1 Z_s(1-K_1)} & \dfrac{Z_{oo1}\left(Z_s K_1{}^2 - (Z_{oo1}K_1 + Z_{oo1} + Z_s)\cos^2\theta_1\right)}{K_1 Z_s \sin\theta_1 (1-K_1)} i \\ \dfrac{Z_s \sin^2\theta_1 + Z_{oo1}\sin^2\theta_1 - Z_{oo1}}{K_1 Z_{oo1} Z_s \sin\theta_1} i - \dfrac{\sin\theta_1}{Z_{oo1}} i & \dfrac{\cos\theta_1}{K_1} - \dfrac{Z_{oo1}\cos^3\theta_1(1+K_1)}{K_1 Z_s \sin^2\theta_1} \end{bmatrix} \tag{5.7}$$

$$M_{c1} = \begin{bmatrix} A_{c1} & B_{c1} \\ C_{c1} & D_{c1} \end{bmatrix} = \begin{bmatrix} \dfrac{\cos\theta_1}{K_1} - \dfrac{Z_{oo1}\sin\theta_1 \tan\theta_s}{K_1 Z_s(1-K_1)} & \dfrac{Z_{oo1}\left(Z_s K_1{}^2 \cos\theta_s + 0.5 Z_{oo1}(K_1+1)\sin(2\theta_1)\sin\theta_s - Z_s\cos^2\theta_1\cos\theta_s\right)}{K_1 Z_s \sin\theta_1 \cos\theta_s (1-K_1)} i \\ \dfrac{(1-K_1)\sin\theta_1}{K_1 Z_{oo1}} i + \dfrac{\cos\theta_1 \tan\theta_s}{K_1 Z_s} i & \dfrac{\cos\theta_1}{K_1} + \dfrac{Z_{oo1}\cos^2\theta_1 \sin\theta_s (1+K_1)}{K_1 Z_s \sin\theta_1 \cos\theta_s} \end{bmatrix} \tag{5.8}$$

$$M_1 = \begin{bmatrix} A_1 & B_1 \\ C_1 & D_1 \end{bmatrix} = \begin{bmatrix} \dfrac{Z_{oe2} + Z_{oo2}}{Z_{oe2} - Z_{oo2}}\cos\theta_2 & \dfrac{j}{2}\dfrac{(Z_{oe2} + Z_{oo2})^2}{Z_{oe2} - Z_{oo2}}\sin\theta_2 - \dfrac{j2Z_{oe2}Z_{oo2}}{(Z_{oe2} - Z_{oo2})\sin\theta_2} \\ \dfrac{j2}{Z_{oe2} - Z_{oo2}}\sin\theta_2 & \dfrac{Z_{oe2} + Z_{oo2}}{Z_{oe2} - Z_{oo2}}\cos\theta_2 \end{bmatrix} \tag{5.9}$$

In addition, the input impedance Z_{in} (in Figure 5.28c) of the MCFL type C is calculated when the output port is terminated by Z_A:

$$Z_{in} = \frac{(Z_{oe1} - Z_{oo1})^2}{4Z_L} = \frac{(Z_{oe1} - Z_{oo1})^2}{4Z_A} = Z_0 \tag{5.10}$$

with

$$Z_A = \frac{(Z_{oe1} - Z_{oo1})^2}{4Z_0} \tag{5.11}$$

where Z_A is the impedance of the MCFL type C when the output port is terminated by Z_0. Based on (5.10) and (5.11) [19], the matching condition is always satisfied at the center frequency. Thus, wideband can be achieved if the TPs are properly allocated within the passband.

Coupled line stubs (CLSs) are also proposed for designing balanced BPF. For the short-circuited CLS as shown in Figure 5.29(a), its impedance Z_{stubs} can be deduced from a two-port coupled line (ABCD matrix M_1 as (5.9), [5]) terminated with a short-circuited end, which is shown as follows:

$$Z_{stubs} = \frac{B_1}{D_1} = \frac{j}{2}(Z_{oe2} + Z_{oo2})\tan\theta_2 - \frac{j2Z_{oe2}Z_{oo2}}{(Z_{oe2} + Z_{oo2})\sin\theta_2\cos\theta_2} \tag{5.12}$$

Then, when $Z_{stubs} = 0$, the transmission zeros (f_{za}, f_{zb}) are obtained as

$$f_{za} = \frac{2}{\pi} \times \arcsin\left(\sqrt{1-K_2^2}\right) \times f_0 < f_0;\ f_{zb} = 2f_0 - f_{za} > f_0 \tag{5.13}$$

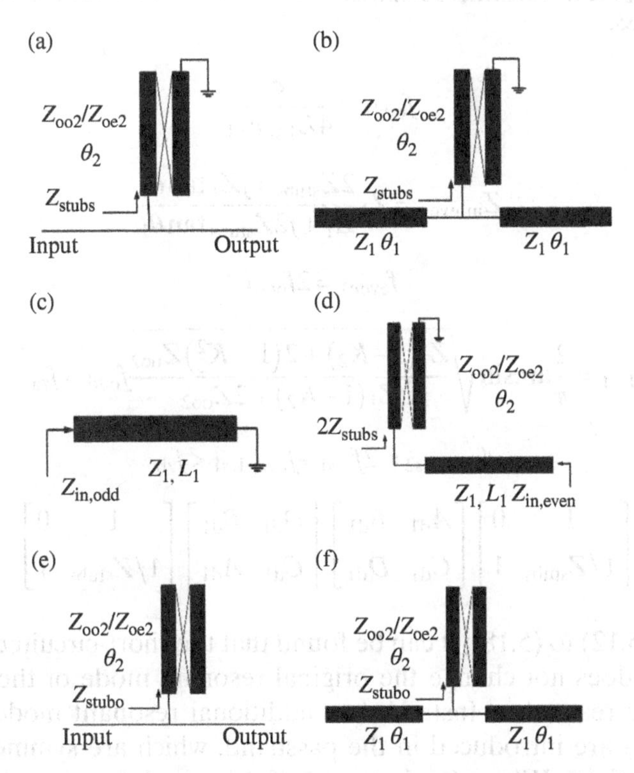

Figure 5.29 (a) Short-circuited CLS. (b) Short-circuited CLSLR and its (c) odd-mode/(d) even-mode equivalent circuit model. (e) Open-circuited CLS and (f) open-circuited CLSLR. Reprinted with permission from Ref. [16]; copyright 2015 IEEE.

where K_2 is the coupling coefficient defined as $K_2 = (Z_{oe2} - Z_{oo2})/(Z_{oe2} + Z_{oo2})$. Compared with the traditional quarter-wavelength stub, the short-circuited CLS can produce two symmetric transmission zeros to improve the passband selectivity.

Furthermore, the short-circuited CLS can be also combined with uniform impedance resonator to design multimode stub-loaded resonator with invariant transmission zeros. As shown in Figure 5.29(b), the coupled lines stub-loaded resonator (CLSLR) consists of a uniform impedance resonator with characteristic impedance Z_1 and electrical length $2\theta_1$, which is centrally loaded with a short-circuited CLS. The odd-/even-mode method is employed to analyze its characteristics, and the corresponding equivalent circuits are shown in Figure 5.29(c) and (d). In odd-mode, the central symmetric plane is short-circuited, and the fundamental frequency is f_{odd1} as (5.14). In contrast, the central symmetric plane is open in even-mode, and the input impedance circuit is $Z_{in,even}$ as (5.15). For simplicity, $\theta_1 = \theta_2 = \theta$ are chosen. Resonant frequencies (f_{even1}, $f_{even1,s1}$, $f_{even1,s2}$) are obtained as (5.16)–(5.18), when $Z_{in,even} = \infty$.

$$f_{odd1} = \frac{c}{4L_1\sqrt{\varepsilon_{eff}}} \tag{5.14}$$

$$Z_{in,even} = Z_1 \frac{2Z_{stubs} + jZ_1 \tan\theta_1}{Z_1 + j2Z_{stubs} \tan\theta_1} \tag{5.15}$$

$$f_{even1} = 2f_{odd1} \tag{5.16}$$

$$f_{even1,s1} = \frac{2}{\pi} \arcsin\sqrt{\frac{Z_1(1-K_2) + 2(1-K_2^2)Z_{oo2}}{Z_1(1-K_2) + 2Z_{oo2}}} f_{odd} > f_{za} \tag{5.17}$$

$$f_{even1,s2} = 2f_{odd} - f_{even1,s1} < f_{zb} \tag{5.18}$$

$$M_d = \begin{bmatrix} 1 & 0 \\ 1/Z_{stubs} & 1 \end{bmatrix} \begin{bmatrix} A_{d1} & B_{d1} \\ C_{d1} & D_{d1} \end{bmatrix} \begin{bmatrix} D_{d1} & B_{d1} \\ C_{d1} & A_{d1} \end{bmatrix} \begin{bmatrix} 1 & 0 \\ 1/Z_{stubs} & 1 \end{bmatrix} \tag{5.19}$$

From (5.12) to (5.18), it can be found that the short-circuited coupled lines stub does not change the original resonant mode of the uniform impedance resonator. Instead, two additional resonant modes f_{even_s1} and f_{even_s2} are introduced in the passband, which are symmetric with respect to f_{odd1}. When K_2 decreases, f_{even_s1} and f_{even_s2} are close to the transmission zeros f_{za} and f_{zb} to form a sharper roll-off rejection of the passband.

Similarly, for the open-circuited CLS and CLS-loaded resonator, as shown in Figure 5.29(e) and (f), the input impedance of the stub is $Z_{stubo} = 0.5j(Z_{oe2} + Z_{oo2})\cot(\theta_2)$. CLS has the transmission zero f_{odd1}, while the CLSLR resonates at nf_{odd1} with transmission zeros $(2n-1)f_{odd1}$, where $n = 1,2,3,\ldots$.

It is clear that traditionally coupled lines can transmit not only the differential-mode signals but also the common-mode signals. However, modified coupled lines with open- and short-circuited stub-loaded resonator have an attractive capacity in achieving common-mode rejection, helping to design balanced BPF with high common-mode suppression. According to the bandpass and bandstop characteristic of modified coupled lines type C and D, the MCFLs are very suitable for balanced filter design. Furthermore, two additional TPs and impedance matching can be achieved in wide DM passband due to the short-circuited stubs [18].

Based on the MCFLs and CLSs/CLSLRs, two wideband balanced filters with high selectivity and CM suppression are designed for demonstration. Figure 5.30(a) and (b) shows the configuration and equivalent circuit of the developed filter I, which is composed of two MCFLs (block-1) and two cascaded CLSs (block-2). The four-port circuit is ideally symmetric with respect to AA′. Under differential-mode or common-mode excitation, the central plane becomes a perfect electric or MW, respectively. So the differential-mode and common-mode

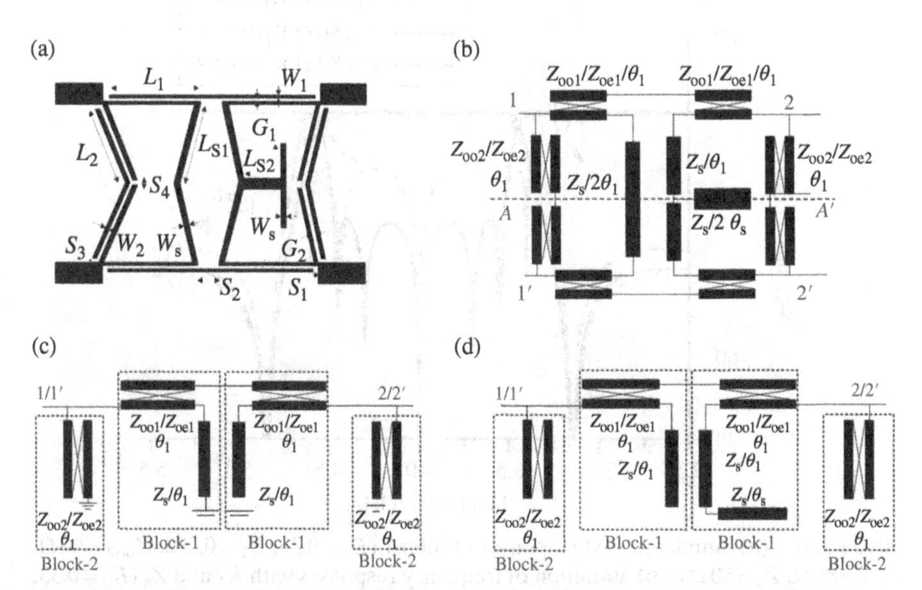

Figure 5.30 The proposed balanced filter I (a) configuration, (b) circuit, (c) DM circuit, and (d) CM circuit. Reprinted with permission from Ref. [16]; copyright 2015 IEEE.

equivalent circuits of the proposed filter can be derived in Figure 5.30(c) and (d), respectively.

For DM operation, as shown in Figure 5.31(a), block-1 without loaded stubs are actually the traditional parallel-coupled lines (type A),

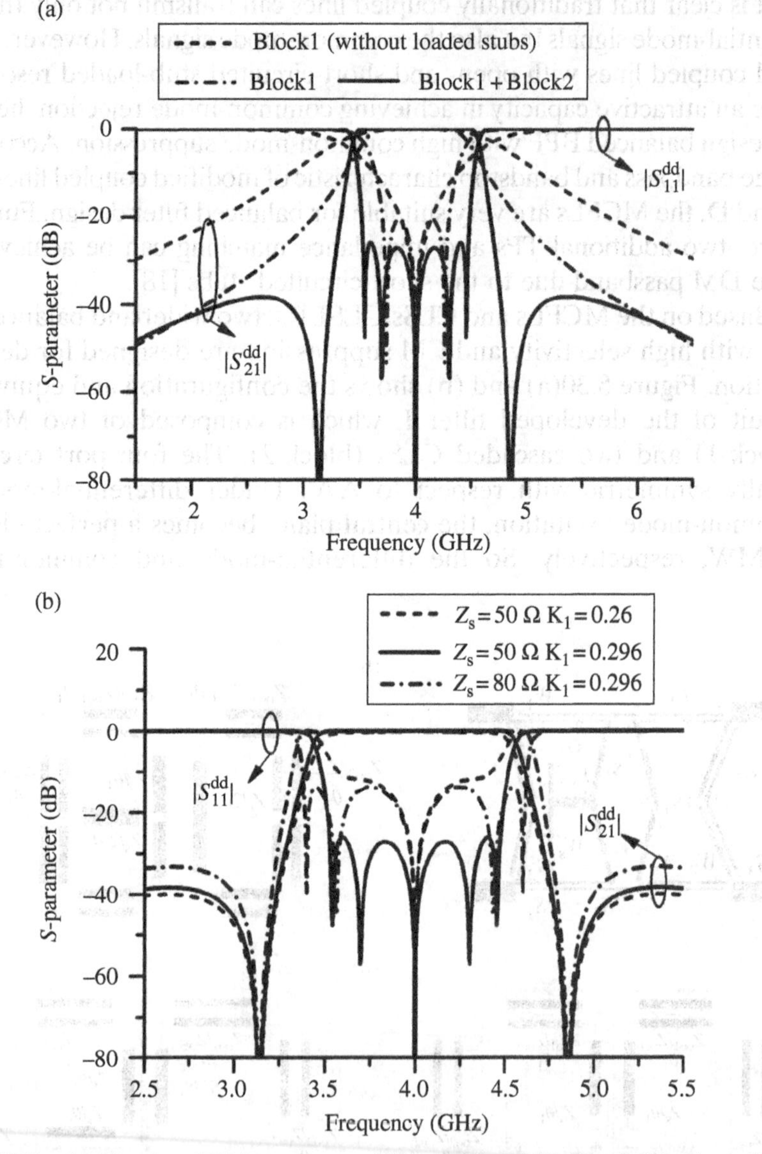

Figure 5.31 (a) Simulated DM responses of filter I ($K_2 = 0.33$, $K_1 = 0.296$, $Z_{oo2} = 90\ \Omega$, $Z_{oo1} = 80\ \Omega$, $Z_s = 50\ \Omega$). (b) Variation of frequency responses with K_1 and Z_s ($K_2 = 0.33$, $Z_{oo2} = 90\ \Omega$, $Z_{oo1} = 80\ \Omega$). Reprinted with permission from Ref. [16]; copyright 2015 IEEE.

which have a single pole. When the short-circuited stubs are loaded, a passband with three TPs are formed. The center TP comes from the parallel resonator, while the other two are generated by the loaded short-circuited stubs [18]. Block-2 is loaded to the input/output port, leading to two TPs/zeros allocated in/out of the passband. Thus, the passband ripple and selectivity are both improved. Deduced from the ABCD matrix (as (5.19)) of the equivalent circuit in DM, the transmission zeros f_{za} and f_{zb} are obtained in (5.13) and are only related to the coupling coefficient K_2. The decrease of K_2 will lead to a sharper passband skirt. Figure 5.31 (b) depicts the simulated DM response with variable coupling coefficient K_1 and characteristic impedance Z_s of the loaded stub of the MCFLs. It can be observed that the bandwidth will increase with better selectivity, as Z_s varies from 50 Ω to 80 Ω, while other parameters keep unchanged. Meanwhile, when the coupling coefficient K_1 increases from 0.26 to 0.296, the return loss changes from 10 dB to 30 dB, leading to a flatter passband.

Under CM operation, as shown in Figure 5.32, block-1 loaded with $\lambda_g/4$ stubs will achieve 20 dB suppression. Also, 20 dB rejection can be obtained even when block-1 is cascaded with block-2. Furthermore, in order to enhance the CM suppression, MCFLs loaded with different electrical length of stubs are utilized in the common-mode equivalent circuit. Note that when $\theta_s = \pi/2$, additional transmission zeros are allocated at $f_0/2$ and $3f_0/2$, which can improve the CM suppression up to

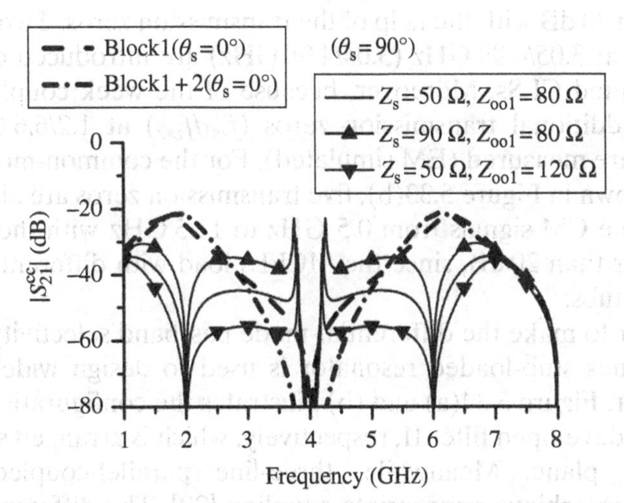

Figure 5.32 Simulated CM responses of the filter I with $\theta_s = 0°$ ($Z_{oo1} = 80\ \Omega$, $Z_s = 50\ \Omega$) and with $\theta_s = 90°$. ($K_2 = 0.33$, $K_1 = 0.296$, $Z_{oo2} = 90\ \Omega$). Reprinted with permission from Ref. [16]; copyright 2015 IEEE.

Table 5.1 Parameters for the Balanced Wideband Structure Shown in Figure 5.29

Circuit parameters		Physical parameters (mm)
FBW = 35% f_0 = 4 GHz	Z_{oo1} = 89 Ω, Z_{oo2} = 86 Ω, Z_s = 117 Ω, K_1 = 0.308, K_2 = 0.37, θ_1 = 90°, θ_s = 90°	L_1 = 12.9, L_2 = 14.2, L_{s1} = 14.8, L_{s2} = 10.9, G_1 = 0.36, G_2 = 0.25, S_1 = 0.45, S_2 = 2, S_3 = 0.3, S_4 = 0.74, W_1 = 0.25, W_2 = 0.25, Ws = 0.4

the level of 30 dB. In addition, Figure 5.32 also displays the simulated CM response with variable odd-impedance Z_{oo1} and the characteristic impedance of the loaded stub Z_s, respectively. It can be seen that larger Z_{oo1} and lower Z_s will help to achieve better CM suppression without affecting the location of transmission zeros. Consequently, the response of differential mode and common mode can be independently controlled by using MCFLs.

The developed filter I is fabricated on the substrate with a relative dielectric constant of 2.55, a thickness of 0.8 mm, and a loss tangent of 0.0029. The optimized parameters for the proposed filter I are given in Table 5.1.

Figure 5.33 shows the comparison of the simulated and measured responses of developed balanced BPF-I. For the differential mode, as shown in Figure 5.33(a), the measured (EM simulated) center frequency is 4 GHz (4.01 GHz) with 3 dB fractional bandwidth of 35% (35.5%) and minimum insertion loss of 0.7 dB (0.3 dB). It can be seen that the stopband of the DM is extended from 4.8 GHz to 11 GHz with a rejection better than 20 dB with the help of the transmission zeros. Two transmission zeros at 3.05/4.99 GHz (3.04/4.98 GHz) are introduced due to the short-circuited CLSs. Moreover, because of the week coupling of the MCFLs, additional transmission zeros (f_{zs1}/f_{zs2}) at 1.2/6.6 GHz (1.2/6.3 GHz) are measured (EM simulated). For the common-mode operation, as shown in Figure 5.33(b), five transmission zeros are allocated to suppress the CM signals from 0.5 GHz to 11.5 GHz with the rejection level larger than 20 dB, since the MCFLs load with different electrical length of stubs.

In order to make the differential-mode passband selectivity sharper, coupled lines stub-loaded resonator is used to design wideband balanced filter. Figure 5.34(a) and (b) illustrates the configuration and circuit of the developed filter II, respectively, which is arranged symmetric with AA′ plane. Meanwhile, three-line parallel-coupled line is employed to achieve appropriate coupling [20]. The differential-mode and common-mode equivalent circuits of the developed filter II are shown in Figure 5.34(c) and (d), separately.

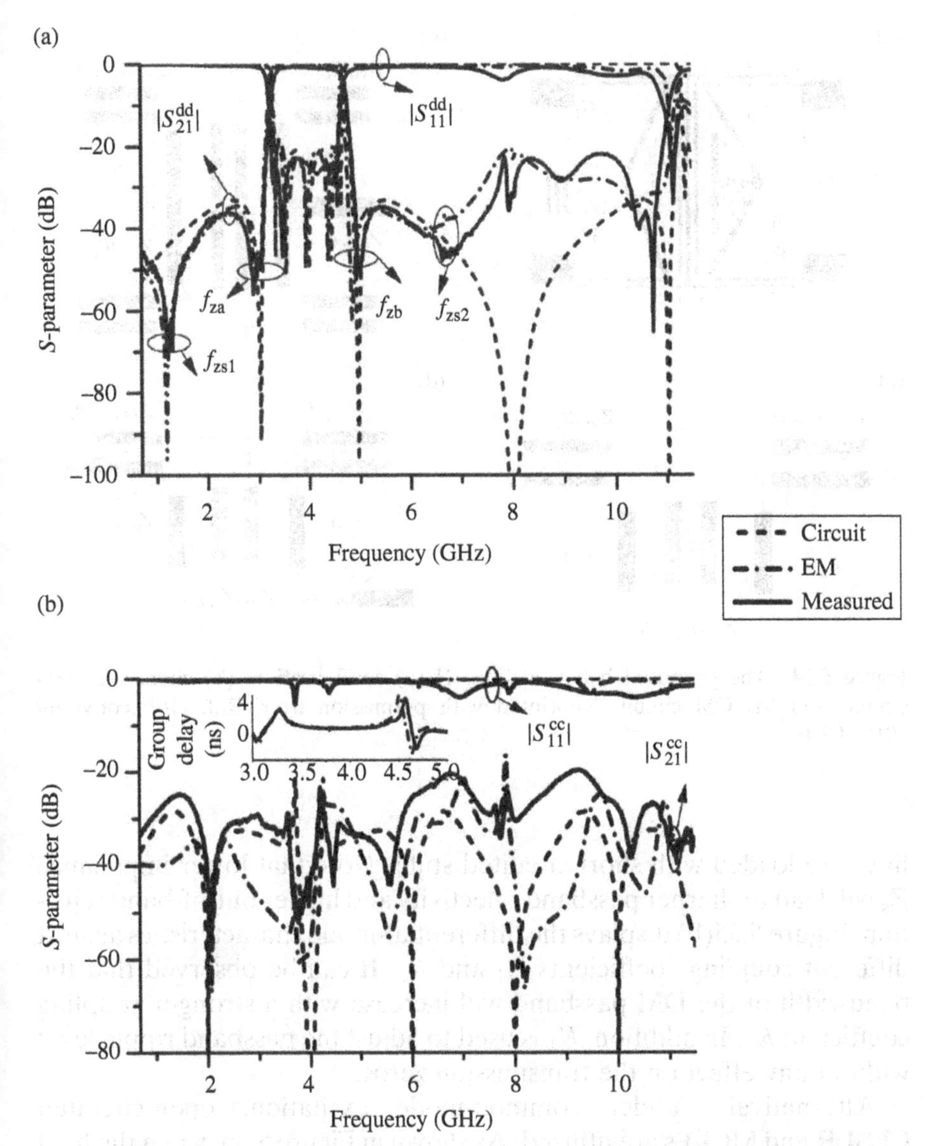

Figure 5.33 Simulated and measured results of the proposed filter I. (a) DM response. (b) CM response. Reprinted with permission from Ref. [16]; copyright 2015 IEEE.

Under differential-mode excitation, the developed filter II consists of short-circuited CLSLR and MCFLs. According to the analysis, the CLSLR $\left(Z_1 = \sqrt{Z_{oo1}Z_{oe1}}\right)$ will form a passband with three TPs and two transmission zeros, as shown in Figure 5.35(a). Furthermore, it also can be seen that two additional TPs will be introduced when the feed

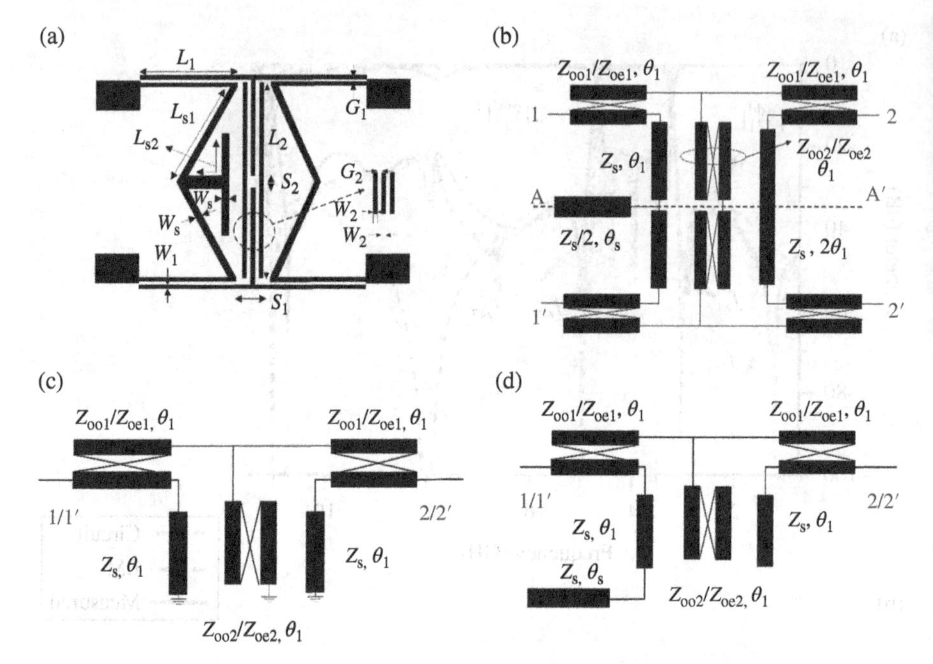

Figure 5.34 The proposed balanced filter II (a) configuration, (b) circuit, (c) DM circuit, and (d) CM circuit. Reprinted with permission from Ref. [16]; copyright 2015 IEEE.

lines are loaded with short-circuited stubs. Note that lower impedance Z_s will lead to sharper passband selectivity and higher out-of-band rejection. Figure 5.35(b) displays the differential-mode characteristics against different coupling coefficients K_1 and K_2. It can be observed that the bandwidth of the DM passband will increase with a stronger coupling coefficient K_2. In addition, K_1 is used to adjust the passband ripple level without any effect on the transmission zeros.

Alternatively, under common-mode excitation, open-circuited CLSLR and MCFLs are utilized. As shown in Figure 5.36, when the feed lines are loaded with $\lambda_g/4$ open stubs, single transmission zero at the frequency f_{ood1} is introduced to achieve CM suppression with rejection level better than 20 dB. To further enhance the CM rejection, MCFLs with $\lambda_g/4$ and $\lambda_g/2$ open stubs are applied, since additional transmission zeros are distributed at $f_{ood1}/2$ and $3f_{ood1}/2$. In addition, the simulated responses with variable odd-mode impedance Z_{oo1} and characteristic impedance of the loaded stub Z_s are also given in Figure 5.36. It can be observed that larger Z_{oo1} and lower Z_s will help to achieve better CM suppression without affecting the location of transmission zeros.

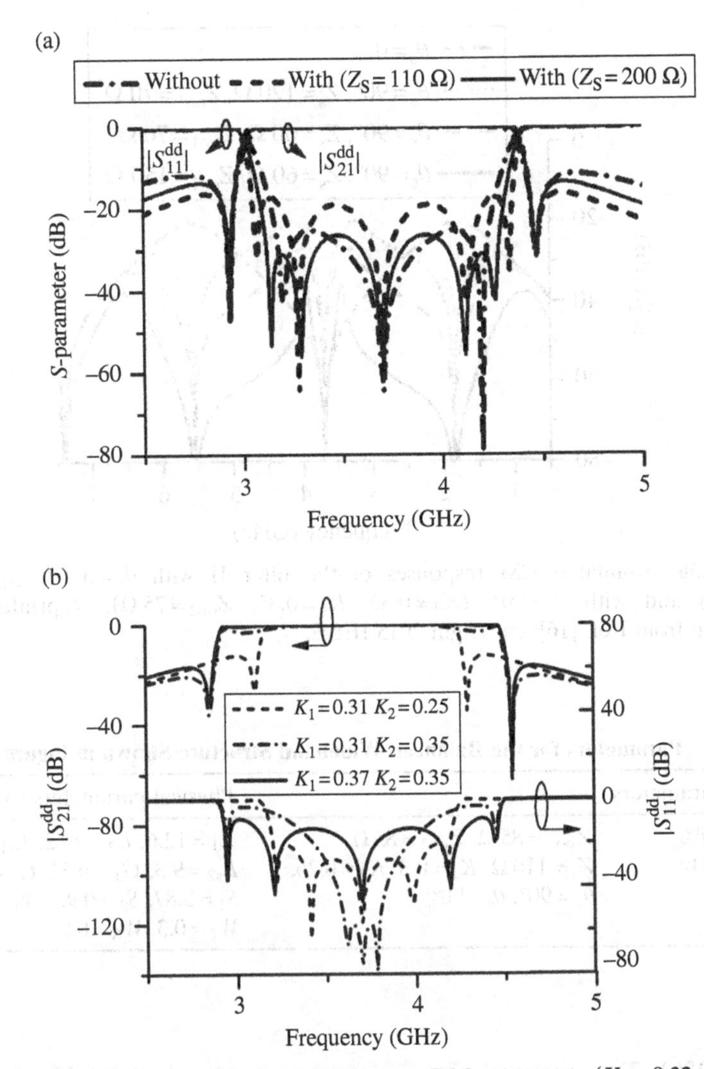

Figure 5.35 (a) Effects of the MCFLs on DM response ($K_2 = 0.32$, $K_1 = 0.35$, $Z_{oo2} = 75\ \Omega$, $Z_{oo1} = 70\ \Omega$). (b) Variation of frequency responses with K_1 and K_2 ($Z_{oo2} = 107\ \Omega$, $Z_{oo1} = 75\ \Omega$, $Z_s = 110\ \Omega$). Reprinted with permission from Ref. [16]; copyright 2015 IEEE.

Circuit and full-wave EM simulations are used for determining the geometry parameters of the developed balanced wideband filter shown in Figure 5.34, which are listed in Table 5.2. The simulated and measured results of filter are plotted in Figure 5.37. The measured (EM simulated) center frequency in differential mode is 3.65 GHz (3.7 GHz) with minimum insertion loss of 0.7 dB (0.3 dB) and 3 dB relative bandwidth of

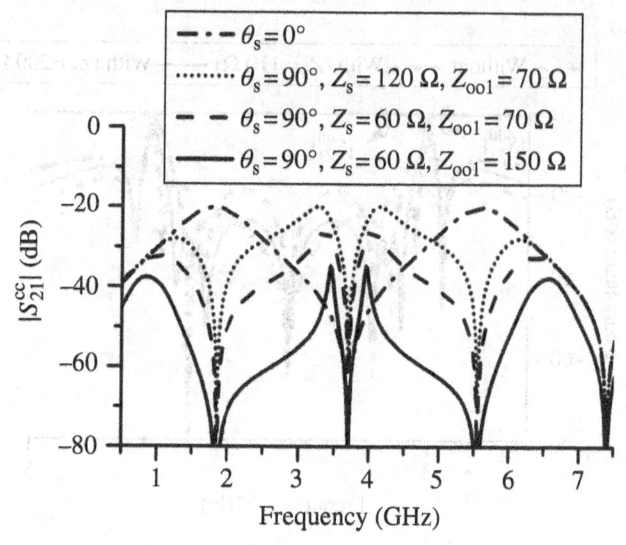

Figure 5.36 Simulated CM responses of the filter II with $\theta_s = 0°$. ($Z_{oo1} = 70\ \Omega$, $Z_s = 60\ \Omega$) and with $\theta_s = 90°$. ($K_2 = 0.32$, $K_1 = 0.35$, $Z_{oo2} = 75\ \Omega$). Reprinted with permission from Ref. [16]; copyright 2015 IEEE.

Table 5.2 Parameters for the Balanced Wideband Structure Shown in Figure 5.33

Circuit parameters		Physical parameters (mm)
FBW = 30% $f_0 = 3.7$ GHz	$Z_{oo1} = 85\ \Omega$, $Z_{oo2} = 85\ \Omega$, $Z_s = 110\ \Omega$, $K_1 = 0.3$, $K_2 = 0.26$, $\theta_1 = 90°$, $\theta_s = 90°$	$L_1 = 12.6$, $L_2 = 29.2$, $L_{s1} = 16.3$, $L_{s2} = 9.3$, $G_1 = 0.32$, $G_2 = 0.3$, $S_1 = 3.87$, $S_2 = 0.92$, $W_1 = 0.3$, $W_2 = 0.3$, $W_S = 0.5$

31% (30%). Two transmission zeros are measured at 3 GHz and 4.3 GHz, which causes a high selectivity of the passband edge ($\Delta f_{3dB}/\Delta f_{20dB}$ = 0.97). It can be seen that from 4.7 GHz to 8 GHz, the rejection level is higher than 20 dB. In general, the differential mode of the developed filter II exhibits good selectivity at both the lower and upper passband edges and high stopband rejection level in the stopband. For the common-mode operation, four transmission zeros located at 2.27/3.73/6.3/7.22 GHz (2.29/3.7/6.03/7.3 GHz) are measured (EM simulated), which help to achieve the CM suppression greater than 20 dB from 1 GHz to 6.9 GHz. The slightly discrepancy between measurement and simulation in CM is mainly due to the coupling between the resonator and loaded stubs of the MCFLs.

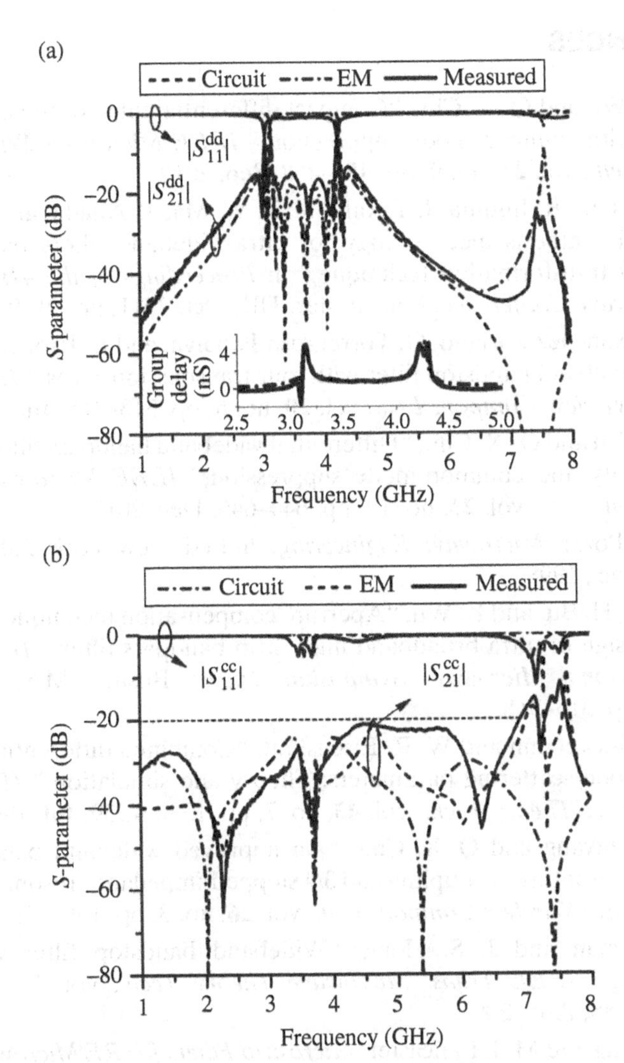

Figure 5.37 Simulated and measured results of the proposed filter II. (a) DM response. (b) CM response. Reprinted with permission from Ref. [16]; copyright 2015 IEEE.

5.6 SUMMARY

In this chapter, different kinds of coupled lines to design balanced filters with UWB or wideband differential-mode bandpass responses and high common-mode suppression have been introduced. The novel filters discussed include UWB balanced filters combining UWB BPF and UWB BSF, wideband balanced filters using coupled line stubs, wideband balanced filters with internal cross-couplings, wideband balanced filters using SLRRs, and wideband balanced filters using MCFLs and CLSs.

REFERENCES

1. X.-H. Wu and Q.-X. Chu, "Compact differential ultra-wideband bandpass filter with common-mode suppression," *IEEE Microwave Wireless Compon. Lett.*, vol. **22**, no. 9, pp. 456–458, Sep. 2012.
2. C. P. Chen, R. Iinuma, J. Takahashi, Z. W. Ma, T. Anada, and S. Takeda, "Novel synthesis methodology for ultra-wideband filters based on frequency transformation technique," in *Proceedings of the 41st European Microwave Conference*, Manchester, UK, Oct. 2011, pp. 91–94.
3. M. A. Sanchez-Soriano, G. Torregrosa-Penalva, and E. Bronchalo, "Compact wideband bandstop filter with four transmission zeros," *IEEE Microwave Wireless Compon. Lett.*, vol. **20**, no. 6, pp. 313–315, Jun. 2010.
4. X.-H. Wu and Q.-X. Chu, "Differential wideband bandpass filter with high-selectivity and common-mode suppression," *IEEE Microwave Wireless Compon. Lett.*, vol. **23**, no. 12, pp. 644–646, Dec. 2013.
5. D. M. Pozar, *Microwave Engineering*, 2nd ed. New York: John Wiley & Sons, Inc., 1998.
6. L. Zhu, H. Bu, and K. Wu, "Aperture compensation technique for innovative design of ultra-broadband microstrip bandpass filter," *IEEE MTT-S International Microwave Symposium Digest*, Boston, MA, 11–16 Jun. 2000, pp. 315–318.
7. D. E. Bockelamn and W. R. Eisenstant, "Combined differential and common-mode scattering parameters: Theory and simulation," *IEEE Trans. Microwave Theory Tech.*, vol. **43**, no. 7, pp. 1530–1539, Jul. 1995.
8. Z.-A. Ouyang and Q.-X. Chu, "An improved wideband balanced filter using internal cross-coupling and 3/4 stepped-impedance resonator," *IEEE Microwave Wireless Compon. Lett.*, vol. **26**, no. 3, pp. 156–158, Mar. 2016.
9. H. Shaman and J. S. Hong, "Wideband bandstop filter with cross-coupling," *IEEE Trans. Microwave Theory Tech.*, vol. **55**, no. 8, pp. 1780–1785, Aug. 2007.
10. J. S. Hong and M. J. Lancaster, *Microstrip Filters for RF/Microwave Applications*. New York: John Wiley & Sons, Inc., 2001.
11. L.-L. Qiu and Q.-X. Chu, "Balanced bandpass filter using stub-loaded ring resonator and loaded coupled feed-line," *IEEE Microwave Wireless Compon. Lett.*, vol. **25**, no. 10, pp. 654–656, Oct. 2015.
12. J.-L. Olvera-Cervantes, A. Corona-Chavez, "Microstrip balanced bandpass filter with compact size, extended-stopband and common-mode noise suppression," *IEEE Microwave Wireless Compon. Lett.*, vol. **23**, no. 10, pp. 530–532, Oct. 2013.
13. H. Wang, L.-M. Gao, K.-W. Tam, W. Kang, W. Wu, "A wideband differential BPF with multiple differential- and common-mode transmission zeros using cross-shaped resonator," *IEEE Microwave Wireless Compon. Lett.*, vol. **24**, no. 12, pp. 854–856, Dec. 2014.

14. M.-Y. Hsieh and S.-M. Wang, "Compact and wideband microstrip bandstop filter," *IEEE Microwave Wireless Compon. Lett.*, vol. **15**, no. 7, pp. 472–474, Jul. 2005.
15. M. K. Mandal, K. Divyabramham, and S. Sanyal, "Compact, wideband bandstop filters with sharp rejection characteristic," *IEEE Microwave Wireless Compon. Lett.*, vol. **18**, no. 10, pp. 665–667, Oct. 2008.
16. Q.-X. Chu and L.-L. Qiu, "Wideband balanced filters with high selectivity and common-mode suppression," *IEEE Trans. Microwave Theory Tech.*, vol. **63**, no. 10, pp. 3462–3468, Oct. 2015.
17. C.-H. Wu, C.-H. Wang, and C.-H. Chen, "Novel balanced coupled-line bandpass filters with common mode noise suppression," *IEEE Trans. Microwave Theory Tech.*, vol. **55**, no. 2, pp. 287–295, Feb. 2007.
18. M. Moradian and H. Oraizi, "Optimum design of microstrip parallel coupled-line band-pass filters for multi-spurious pass-band suppression," *IET Microwave Antennas Propag.*, vol. **1**, no. 2, pp. 488–495, Apr. 2007.
19. T. H. Duong and I. S. Kim, "New elliptic function type UWB BPF based on capacitively coupled open T resonator," *IEEE Trans. Microwave Theory Tech.*, vol. **57**, no. 12, pp. 3089–3098, Dec. 2009.
20. J. T. Kuo and E. Shih, "Wideband bandpass filter design with three-line microstrip structures," *IEE Proc. Microwave Antennas Propag.*, vol. **149**, no. 516, pp. 243–247, Oct./Dec. 2002.

CHAPTER 6

WIDEBAND DIFFERENTIAL CIRCUITS USING T-SHAPED STRUCTURES AND RING RESONATORS

Wenquan Che and Wenjie Feng

Department of Communication Engineering, Nanjing University of Science and Technology, Nanjing, China

6.1 INTRODUCTION

With the development of wireless communication technology, radio-frequency (RF) circuits, and even millimeter-wave circuits, systematic integrations are becoming quite complicated, while more functionality and signals are required to integrate into a limited space, and there exist high-level electromagnetic interactions among circuit nodes and interference/crosstalk from substrate coupling and free space [1]. When compared with the single-ended technology, balanced/differential circuit technology takes the advantages of good common-mode rejection and relatively high immunity to environmental noise [1–3], which has thus become more important in modern communication systems. Figure 6.1(a) and (b) shows the simplified architecture of one balanced

Balanced Microwave Filters, First Edition. Edited by Ferran Martín, Lei Zhu, Jiasheng Hong, and Francisco Medina.

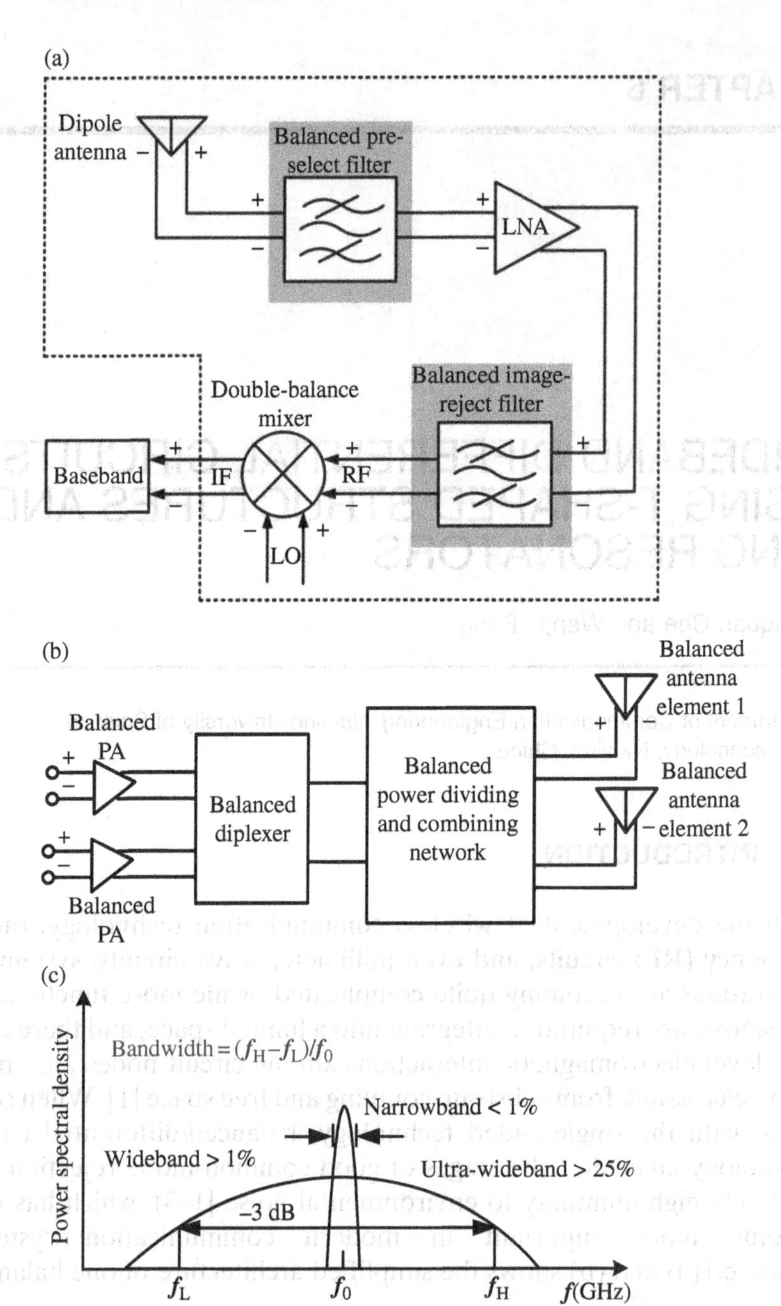

Figure 6.1 (a) Simplified architecture of the balanced transceiver with balanced filters [1]. (b) Fully balanced RF front end with balanced network [1]. (c) The definition of the bandwidth of bandpass filter [4]. Reprinted with permission from Ref. [3]; copyright 2015 IEEE.

transceiver with balanced filters and a fully balanced RF front end with balanced network, respectively. For RF and microwave applications, when sinusoidal signals are transmitted in balanced/differential circuits, DC and low-frequency terms are not allowed to pass through for differential mode, and an all-stop or bandstop response should be realized for the common mode.

As shown in Figure 6.1(c), the definition of the fractional bandwidth, that is, $(f_H - f_L)/f_0$ (f_H, high frequency of the 3-dB passband; f_L, low frequency of the 3-dB passband; f_0, center frequency of the passband), for the narrowband bandpass filter is less than 1%, and the fractional bandwidths for wideband/ultra-wideband (UWB) bandpass filter are greater than 1 and 25%, respectively [4, 5]. The main desired features of the high-performance wideband bandpass filters include simple design, compact size, low loss and good linearity, enhanced out-of-band rejection, and easy integration with other circuits/antennas. During the past years, many microstrip wideband filters employing multimode resonators, multilayer aperture-coupled patches, and transversal signal interference concept have been analyzed and designed [6–8]. The bandwidth for former wideband filters and networks are usually less than 25%, and only the in-passband common-mode suppression is realized. High-performance balanced circuits with simple structures, high selectivity, and wideband harmonic suppression for differential-mode and wideband common-mode suppression are highly desired for modern wideband communication systems. In this chapter, several wideband differential filters and networks using T-shaped structures and ring resonators will be introduced, and the design strategies and procedures are described in detail. The in-band and out-of-band performance of all the wideband circuits will be investigated. The objective of this approach is to provide an alternatively new design methodology of these wideband filters and networks based on different novel resonators. These balanced circuits have many attractive features, such as simple design, compact size, enhanced out-of-band rejection, and easy integration with other circuits/antennas.

6.2 WIDEBAND DIFFERENTIAL BANDPASS FILTERS USING T-SHAPED RESONATORS

6.2.1 Mixed-Mode *S*-Parameters for Four-Port Balanced Circuits

An *S*-parameter is defined as the ratio of two normalized power waves, saying, the response divided by the stimulus. A full *S*-matrix describes every possible combination of input/output relationships. The matrix is arranged in such a way that each column represents a particular stimulus condition and each row represents a particular response condition. The general single-ended S-parameter matrix relation is defined as hereafter in terms of power waves for the four-port circuit shown in Figure 6.2(a):

$$\begin{bmatrix} b_1 \\ b_2 \\ b_3 \\ b_4 \end{bmatrix} = \begin{bmatrix} S_{11} & S_{12} & S_{13} & S_{14} \\ S_{21} & S_{22} & S_{23} & S_{24} \\ S_{31} & S_{32} & S_{33} & S_{34} \\ S_{41} & S_{42} & S_{43} & S_{44} \end{bmatrix} \begin{bmatrix} a_1 \\ a_2 \\ a_3 \\ a_4 \end{bmatrix} \tag{6.1}$$

$$\bar{b} = [\mathbf{S}_{\text{std}}]\bar{a} \tag{6.2}$$

In (6.1) and (6.2), $\mathbf{S}_{\text{std}}$ is the four-port S-parameter matrix and $\bar{b}$ and $\bar{a}$ are the reflected and input power wave vectors at the four ports. Expanding the single-ended $\mathbf{S}_{\text{std}}$ matrix into standard input/output algebraic relations results in

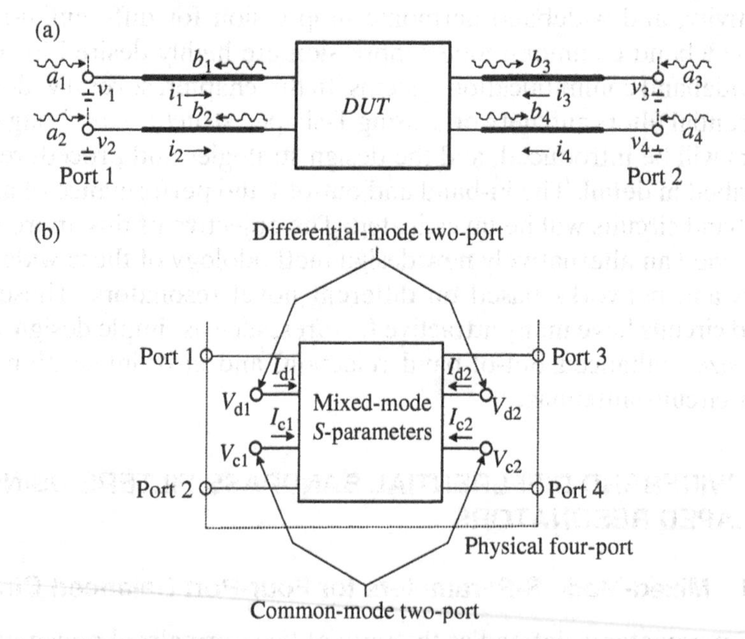

Figure 6.2 (a) RF differential two-port or single-ended four-port circuit. (b) Circuit used to obtain mixed-mode S-parameters of balanced device. Reprinted with permission from Ref. [3]; copyright 2015 IEEE.

$$
\begin{aligned}
b_1 &= S_{11}a_1 + S_{12}a_2 + S_{13}a_3 + S_{14}a_4 \\
b_2 &= S_{21}a_1 + S_{22}a_2 + S_{23}a_3 + S_{24}a_4 \\
b_3 &= S_{31}a_1 + S_{32}a_2 + S_{33}a_3 + S_{34}a_4 \\
b_4 &= S_{41}a_1 + S_{42}a_2 + S_{43}a_3 + S_{44}a_4
\end{aligned}
\tag{6.3}
$$

In (6.3), the individual S-parameter matrix elements are calculated numerically in a simulator or measured by test equipment under the following conditions:

$$
S_{xy} = \left.\frac{b_x}{a_y}\right|_{a(x \neq y) = 0} \tag{6.4}
$$

In (6.4), all the a power wave sources are turned off except a_y and then the b_x power wave is characterized. In S_{11} measurement, the a_1 power wave source is applied and the b_1 reflection from the circuit is measured as shown in (6.5):

$$
S_{11} = \left.\frac{b_1}{a_1}\right|_{a_2, a_3, \ldots = 0} \tag{6.5}
$$

As shown in Figure 6.2(b), for a balanced device, differential- and common-mode voltages and currents can be defined at each balanced port. Differential- and common-mode impedances can be defined as well. A mixed-mode S-matrix in (6.6) can be organized in a way similar to the single-ended S-matrix, where each column (row) represents a different stimulus (response) condition. The mode information as well as port information must be included in the mixed-mode S-matrix:

$$
\begin{bmatrix} b_{\mathrm{d1}} \\ b_{\mathrm{d2}} \\ b_{\mathrm{c1}} \\ b_{\mathrm{c2}} \end{bmatrix}
=
\begin{bmatrix}
S_{\mathrm{dd11}} & S_{\mathrm{dd12}} & S_{\mathrm{dc11}} & S_{\mathrm{dc12}} \\
S_{\mathrm{dd21}} & S_{\mathrm{dd22}} & S_{\mathrm{dc21}} & S_{\mathrm{dc22}} \\
S_{\mathrm{cd11}} & S_{\mathrm{cd12}} & S_{\mathrm{cc11}} & S_{\mathrm{cc12}} \\
S_{\mathrm{cd21}} & S_{\mathrm{cd22}} & S_{\mathrm{cc21}} & S_{\mathrm{cc22}}
\end{bmatrix}
\begin{bmatrix} a_{\mathrm{d1}} \\ a_{\mathrm{d2}} \\ a_{\mathrm{c1}} \\ a_{\mathrm{c2}} \end{bmatrix}
\tag{6.6}
$$

$S_{\mathrm{dd}ij}$ and $S_{\mathrm{cc}ij}$ (i, $j = 1$, 2) are the differential- and common-mode S-parameters, respectively. $S_{\mathrm{dc}ij}$ and $S_{\mathrm{cd}ij}$ (i, $j = 1$, 2) are the mode-conversion or cross-mode S-parameters. The parameters $S_{\mathrm{dd}ij}$ ($i, j = 1, 2$) in the upper left corner of the mixed-mode S-matrix (6.6) describe the performance with a differential stimulus and differential response. $S_{\mathrm{dc}ij}$ ($S_{\mathrm{cd}ij}$)

$(i, j = 1, 2)$ describes the conversion of common-mode (differential-mode) wave into differential-mode (common-mode) waves.

The mixed-mode S-parameters in (6.6) can be directly related to the standard four S-parameters in (6.3). If nodes 1 and 2 in Figure 6.2(a) are paired as a single differential port, and nodes 3 and 4 are also paired as another differential port [3], the relations between the response and stimulus of standard mode and mixed mode are shown in (6.7) and (6.8), where a_i and b_i $(i = 1–4)$ are the waves measured at ports 1–4 in Figure 6.2(a):

$$a_{\mathrm{d1}} = \frac{1}{\sqrt{2}}(a_1 - a_3), \quad a_{\mathrm{c1}} = \frac{1}{\sqrt{2}}(a_1 + a_3)$$
$$b_{\mathrm{d1}} = \frac{1}{\sqrt{2}}(b_1 - b_3), \quad b_{\mathrm{c1}} = \frac{1}{\sqrt{2}}(b_1 + b_3) \tag{6.7}$$

$$a_{\mathrm{d2}} = \frac{1}{\sqrt{2}}(a_2 - a_4), \quad a_{\mathrm{c2}} = \frac{1}{\sqrt{2}}(a_2 + a_4)$$
$$b_{\mathrm{d2}} = \frac{1}{\sqrt{2}}(b_2 - b_4), \quad b_{\mathrm{c2}} = \frac{1}{\sqrt{2}}(b_2 + b_4) \tag{6.8}$$

The transformation matrix between standard S-parameters and mixed-mode S-parameters (6.9) can be derived from the relationship of the following equations: mixed-mode incident waves A_{mm} (6.10), mixed-mode response wave B_{mm} in (6.10), mixed-mode S-parameter matrix $\mathbf{S}_{\mathrm{mm}}$ in (6.11), standard four-port S-parameter matrix $\mathbf{S}_{\mathrm{std}}$ in (6.1), and the conversion matrix $\mathbf{M}$ in (6.11) and M^{-1} in (6.11):

$$\begin{bmatrix} S_{11} & S_{12} & S_{13} & S_{14} \\ S_{21} & S_{22} & S_{23} & S_{24} \\ S_{31} & S_{32} & S_{33} & S_{34} \\ S_{41} & S_{42} & S_{43} & S_{44} \end{bmatrix} \Leftrightarrow \text{Transformation}$$
$$\Leftrightarrow \begin{bmatrix} \begin{bmatrix} S_{\mathrm{dd11}} & S_{\mathrm{dd12}} \\ S_{\mathrm{dd21}} & S_{\mathrm{dd22}} \end{bmatrix} & \begin{bmatrix} S_{\mathrm{dc11}} & S_{\mathrm{dc12}} \\ S_{\mathrm{dc21}} & S_{\mathrm{dc22}} \end{bmatrix} \\ \begin{bmatrix} S_{\mathrm{cd11}} & S_{\mathrm{cd12}} \\ S_{\mathrm{cd21}} & S_{\mathrm{cd22}} \end{bmatrix} & \begin{bmatrix} S_{\mathrm{cc11}} & S_{\mathrm{cc12}} \\ S_{\mathrm{cc21}} & S_{\mathrm{cc22}} \end{bmatrix} \end{bmatrix} \tag{6.9}$$

$$A_{\text{mm}} = M\bar{a} = \begin{bmatrix} a_{\text{d1}} \\ a_{\text{d2}} \\ a_{\text{c1}} \\ a_{\text{c2}} \end{bmatrix} = \frac{1}{\sqrt{2}} \begin{bmatrix} 1 & 0 & -1 & 0 \\ 0 & 1 & 0 & -1 \\ 1 & 0 & 1 & 0 \\ 0 & 1 & 0 & 1 \end{bmatrix} \begin{bmatrix} a_1 \\ a_2 \\ a_3 \\ a_4 \end{bmatrix},$$
$$B_{\text{mm}} = M\bar{b} = \begin{bmatrix} b_{\text{d1}} \\ b_{\text{d2}} \\ b_{\text{c1}} \\ b_{\text{c2}} \end{bmatrix} = \frac{1}{\sqrt{2}} \begin{bmatrix} 1 & 0 & -1 & 0 \\ 0 & 1 & 0 & -1 \\ 1 & 0 & 1 & 0 \\ 0 & 1 & 0 & 1 \end{bmatrix} \begin{bmatrix} b_1 \\ b_2 \\ b_3 \\ b_4 \end{bmatrix} \quad (6.10)$$

$$\mathbf{S}_{\text{mm}} = \mathbf{MS}_{\text{std}} M^{-1}, \quad \mathbf{M} = \frac{1}{\sqrt{2}} \begin{bmatrix} 1 & 0 & -1 & 0 \\ 0 & 1 & 0 & -1 \\ 1 & 0 & 1 & 0 \\ 0 & 1 & 0 & 1 \end{bmatrix}, \quad (6.11)$$
$$M^{-1} = \frac{M^*}{|M|} = \frac{1}{\sqrt{2}} \begin{bmatrix} 1 & 0 & 1 & 0 \\ 0 & 1 & 0 & -1 \\ -1 & 0 & 1 & 0 \\ 0 & -1 & 0 & 1 \end{bmatrix}$$

$$S_{\text{dd}} = \frac{1}{2} \begin{bmatrix} S_{\text{dd11}} = (S_{11} - S_{12} - S_{21} + S_{22}) & S_{\text{dd12}} = (S_{13} - S_{14} - S_{23} + S_{24}) \\ S_{\text{dd21}} = (S_{31} - S_{32} - S_{41} + S_{42}) & S_{\text{dd22}} = (S_{33} - S_{34} - S_{43} + S_{44}) \end{bmatrix} \quad (6.12)$$

$$S_{\text{dc}} = \frac{1}{2} \begin{bmatrix} S_{\text{dc11}} = (S_{11} + S_{12} - S_{21} - S_{22}) & S_{\text{dc12}} = (S_{13} + S_{14} - S_{23} - S_{24}) \\ S_{\text{dc21}} = (S_{31} + S_{32} - S_{41} - S_{42}) & S_{\text{dc22}} = (S_{33} + S_{34} - S_{43} - S_{44}) \end{bmatrix} \quad (6.13)$$

$$S_{\text{cd}} = \frac{1}{2} \begin{bmatrix} S_{\text{cd11}} = (S_{11} - S_{12} + S_{21} - S_{22}) & S_{\text{cd12}} = (S_{13} - S_{14} + S_{23} - S_{24}) \\ S_{\text{cd21}} = (S_{31} - S_{32} + S_{41} - S_{42}) & S_{\text{cd22}} = (S_{33} - S_{34} + S_{43} - S_{44}) \end{bmatrix} \quad (6.14)$$

$$S_{\text{cc}} = \frac{1}{2} \begin{bmatrix} S_{\text{cc11}} = (S_{11} + S_{12} + S_{21} + S_{22}) & S_{\text{cc12}} = (S_{13} + S_{14} + S_{23} + S_{24}) \\ S_{\text{cc21}} = (S_{31} + S_{32} + S_{41} + S_{42}) & S_{\text{cc22}} = (S_{33} + S_{34} + S_{43} + S_{44}) \end{bmatrix} \quad (6.15)$$

For a measurement system, when the balanced filters and networks are connected to the four-port vector network analyzer (VNA), the differential- and common-mode can be excited directly, and the four 2×2 mixed-mode S-parameters can be read directly from the VNA [2, 3]. For a two-port VNA, two balanced circuit ports can be measured at the same time, while the other ports are terminated with 50-Ω broadband loads. The standard four-port S-parameter matrix $\mathbf{S}_{\text{std}}$ can be read step by step, and then the mixed-mode S-parameter matrix $\mathbf{S}_{\text{mm}}$ can be converted using equations (6.9)–(6.15). The analysis of a series of microstrip wideband balanced filters and networks using different techniques will be presented.

6.2.2 T-Shaped Structures with Open/Shorted Stubs

The T-shaped structure with controllable resonance frequencies has been widely used in dual-/tri-band filters, couplers, and power dividers [9–11]. However, little research has described the application of the T-shaped structure in the wideband differential filters. In this section, two novel wideband differential bandpass filters based on T-shaped structure are proposed and investigated [12]. In the following, the characteristics of the T-shaped structures with open/shorted stubs will be analyzed firstly, in which the control of resonance frequencies will be addressed, and then followed by the design of the wideband bandpass filters. When the proposed T-shaped structures with the shorted/open stubs are used to design a wideband differential bandpass filter, the three resonant frequencies of the T-shaped structure with the shorted stub are used to realize a wide passband for the differential mode. In addition, because the three resonant frequencies of the T-shaped structure with the open stub do not change with the characteristic impedance of the stubs, wide stopband can be easily achieved for the common mode. To realize more feasibility to control the resonance characteristics of the T-shaped structure and introduce transmission zeros, open and/or shorted stubs are loaded, respectively, which will be addressed in the following text.

6.2.2.1 *T-Shaped Structure with Shorted Stubs* Figure 6.3(a) shows the T-shaped structure with a shorted stub (θ_2, Z_2) shunt connected in the center of two transmission lines (θ_1, Z_1), whose input admittance Y_{ins} is expressed as

$$Y_{\text{ins}} = \frac{j(2\tan\theta_1 - Z_1/(Z_2\tan\theta_2))}{Z_1 + Z_1^2\tan\theta 1/(Z_2\tan\theta_2) - Z_1^2\tan^2\theta_1} \tag{6.16}$$

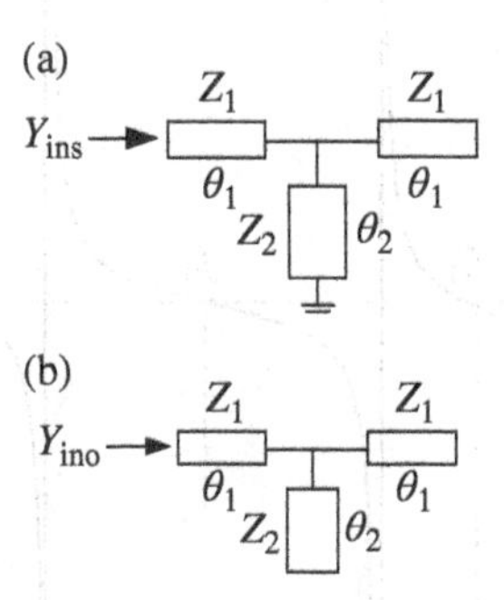

Figure 6.3 (a) T-shaped structure with a shorted stub. (b) T-shaped structure with an open stub. Reprinted with permission from Ref. [12]; copyright 2012 IEEE.

For simplicity, $\theta_1 = \theta_2 = \theta$ are chosen, and

$$Y_{ins} = \frac{j(2\tan\theta - Z_1/(Z_2\tan\theta))}{Z_1 + Z_1^2/Z_2 - Z_1^2\tan^2\theta} \tag{6.17}$$

When $\theta = 90°$, we have

$$Y_{ins} = \lim_{\theta \to 90°} \frac{j(2\tan\theta - Z_1/(Z_2\tan\theta))}{Z_1 + Z_1^2/Z_2 - Z_1^2\tan^2\theta = 0} \tag{6.18}$$

In this way, one resonance will occur at the center frequency f_0 of T-shaped structure. In addition, when $Y_{ins} = 0$, there exist another two resonant frequencies, and the corresponding electric lengths are expressed as follows:

$$f_{r1}(\theta) = \arctan\left(\sqrt{Z_1/2Z_2}\right),\ f_{r2}(\theta) = \pi - \arctan\left(\sqrt{Z_1/2Z_2}\right) \tag{6.19}$$

To demonstrate the control scheme of the resonant frequencies, Figure 6.4(a) and (b) plots the resonant frequencies versus θ and different impedance ratios Z_1/Z_2 for f_{r1}/f_0 and f_{r2}/f_0. When the electrical length for the T-shaped structure is fixed, the two resonant frequencies, f_{r1} and f_{r2}, can be adjusted by changing the characteristic impedances Z_1 and Z_2 within a wide range.

6.2.2.2 *T-Shaped Structure with Open Stubs* In addition, the T-shaped structure with an open stub is shown in Figure 6.3(b). The input admittance Y_{ino} of the T-shaped structure with an open stub is

$$Y_{ino} = \frac{j(2\tan(\theta_1) + Z_1\tan\theta_2/Z_2)}{Z_1(1 - \tan^2\theta_1 - Z_1^2\tan\theta_1\tan\theta_2/Z_2)} \tag{6.20}$$

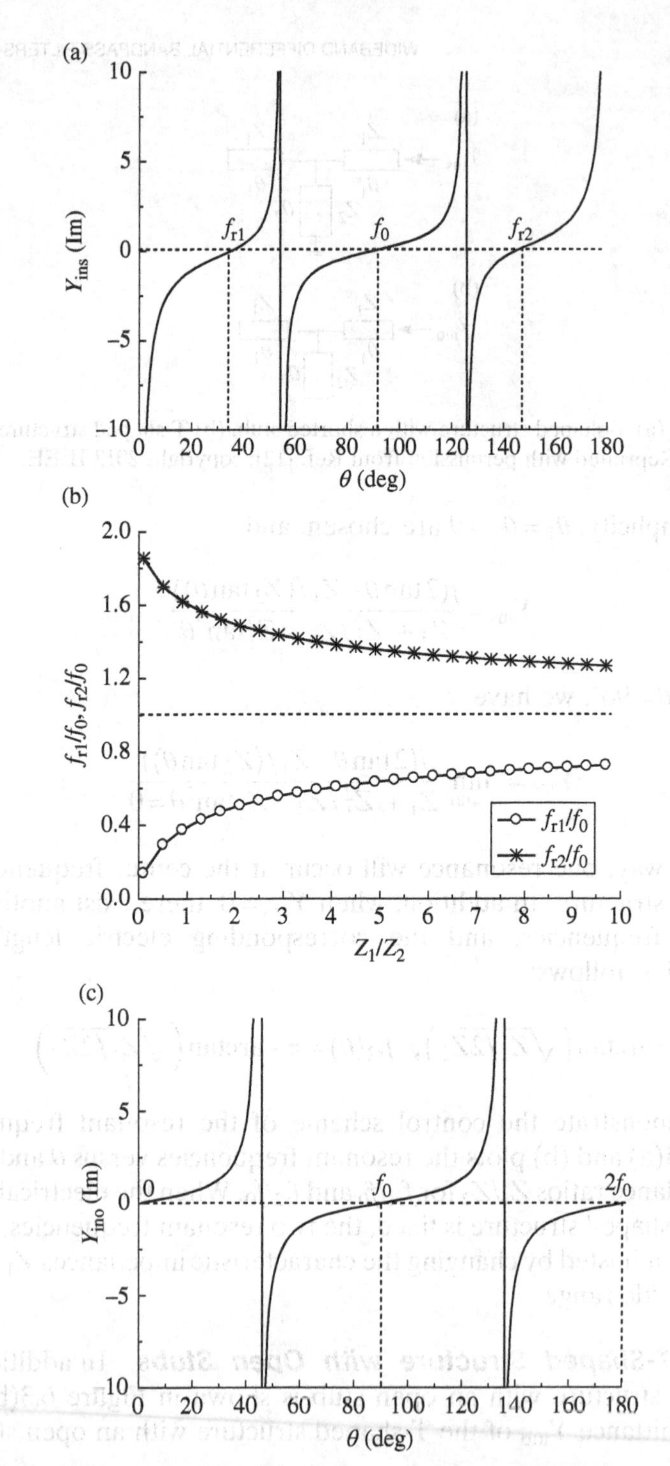

Figure 6.4 (a) Analysis of resonator frequencies versus θ ($Z_1/Z_2 = 1$, shorted stub). (b) Normalized resonant frequencies versus Z_1/Z_2 (shorted stubs). (c) Analysis of resonator frequencies versus θ ($Z_1/Z_2 = 1$, open stubs). Reprinted with permission from Ref. [12]; copyright 2012 IEEE.

When $\theta_1 = \theta_2 = \theta$,

$$Y_{ino} = \frac{j(2\tan\theta + Z_1 \tan\theta / Z_2)}{Z_1(1 - \tan^2\theta - Z_1^2 \tan^2\theta / Z_2)} \tag{6.21}$$

When $\theta = 90°$,

$$Y_{ino} = \lim_{\theta \to 90°} \frac{j(2\tan\theta + Z_1 \tan\theta / Z_2)}{Z_1(1 - \tan^2\theta - Z_1^2 \tan^2\theta / Z_2)} = 0 \tag{6.22}$$

Similarly, a resonance will occur at the center frequency f_0 of T-shaped structure with the open stub. Moreover, considering the numerator of equation (6.21), when $\theta = 0°$ and $180°$, $|2\tan\theta + Z_1 \tan\theta / Z_2| \geqq 0$, another two resonant frequencies will be realized at DC and $2f_0$ for the T-shaped structure. Figure 6.4(c) plots the resonant frequencies versus θ. Obviously, three resonant frequencies are found to locate at DC, f_0 and $2f_0$, respectively, for the T-shaped structure with the open stub.

Based on aforementioned analyses, when the T-shaped structure with shorted/open stubs are used to design wideband balanced filters, the tri-mode resonator mode for the T-shaped structure with shorted stub can be used to realize a wide passband for the differential mode, and the three transmission zeros of the T-shaped structure with open stub can be used to suppress the common mode over three octaves. For further demonstration, detailed theoretical design for the two wideband balanced filters based on T-shaped structures with open/shorted stubs will be given next.

6.2.3 Wideband Bandpass Filters without Cross Coupling

The top view and the ideal circuit of the wideband differential bandpass filter without cross coupling are shown in Figure 6.5(a). From port 1 to port 2 (from port 1′ to port 2′ also), two coupling transmission lines with electrical lengths θ_1 and characteristic impedance Z_1 are located symmetrically. A load Z_2 with electrical length $2\theta_2$ is attached to the center of the T-shaped structure. Four microstrip lines with characteristic impedance $Z_0 = 50\ \Omega$ are connected to ports 1, 1′ and ports 2, 2′. Due to the symmetry of the circuit in Figure 6.5(b), the equivalent half circuits of the differential/common mode can be used for theoretical analysis conveniently [1], as shown in Figure 6.5(c) and (d). Next we will discuss the two excitation cases with differential mode and common mode, respectively.

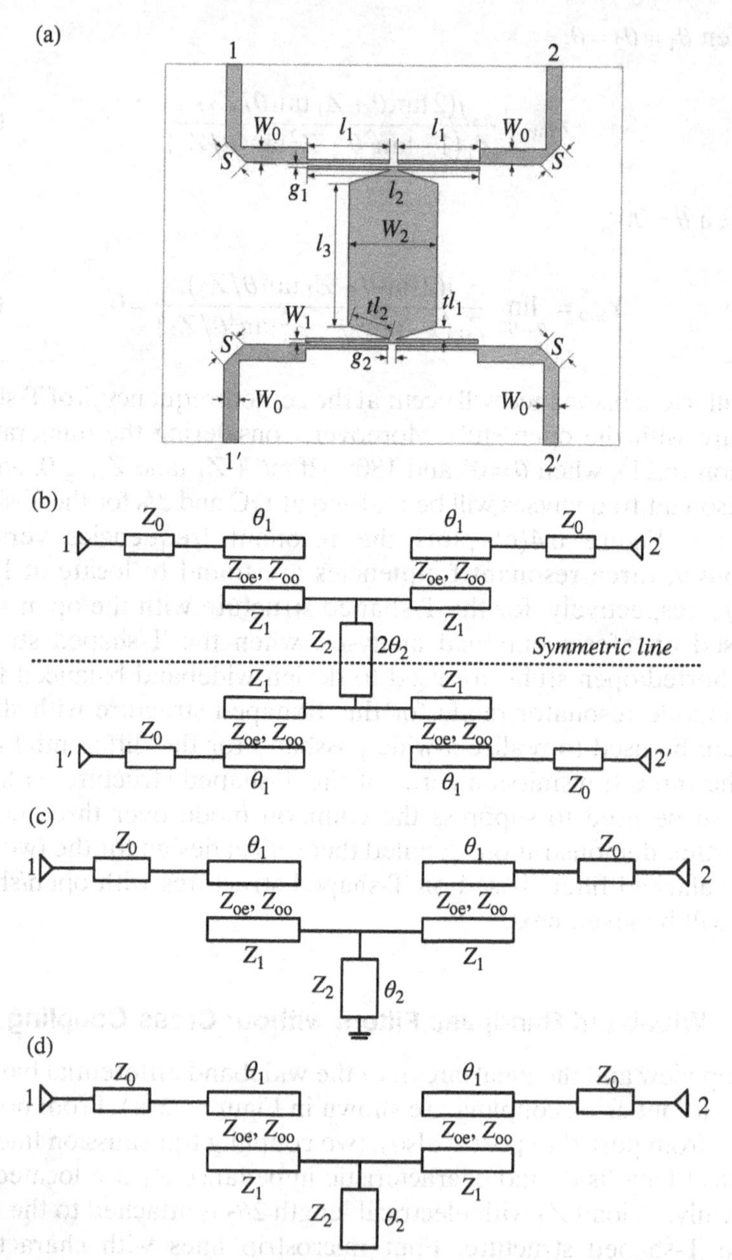

Figure 6.5 (a) Top view of the wideband differential bandpass filter without cross coupling. (b) The ideal circuit of the wideband differential filter. (c) Equivalent circuit for the differential mode. (d) Equivalent circuit for the common mode. Reprinted with permission from Ref. [12]; copyright 2012 IEEE.

6.2.3.1 *Differential-Mode Excitation* When the differential-mode signals are excited from ports 1 and 1′ in Figure 6.5(b), a virtual short appears along the symmetric line, and the center coupling structure for the differential filter is a T-shaped structure with a shorted stub, as shown in Figure 6.5(c).

As discussed in Refs. [13–15], in order to realize a wide passband for differential mode, tight couplings of the coupled lines (even-/odd-mode characteristic impedance Z_{oe} and Z_{oo}) between the two sides of the T-shaped structure are required. High impedance of the two parallel-coupled lines could be applied to realize tight coupling. The external quality factor Q_{ex} of the parallel-coupled line can be given by [14, 15]

$$Q_{ex} = \frac{2\pi Z_0^2}{(Z_{oe} - Z_{oo})^2} \tag{6.23}$$

Once Z_{oe} and Z_{oo} of the parallel-coupled line are known, the line width W_1 and gap g_1 can be simulated from software of Ansoft Designer or Advanced Design System (ADS).

In addition, the loaded quality factor Q_L and the 3-dB bandwidth Δf of the parallel-coupled line is related by

$$Q_L = \frac{f_0}{\Delta f} \tag{6.24}$$

Figure 6.6 plots the simulated Q_L versus the gap g_1 for different line widths. For the lossless case, when $Q_L = Q_{ex}$ and f_0 and Δf are specified, Z_{oe} and Z_{oo} of the parallel-coupled line can be determined, and the parameters of the parallel-coupled line can be obtained accordingly. For example, when $f_0 = 6.85$ GHz and $\Delta f = 4.1$ GHz, the gap $g_1 = 0.2$ mm can be obtained for the case of $Q_L = Q_{ex} \approx 1.67$. From (6.23), we can obtain Z_{oe} and Z_{oo}, and then the line width W_1 can be determined.

The simulated frequency responses of Figure 6.5(c) are shown in Figure 6.7(a) and (b), where we choose $f_0 = 6.85$ GHz. k is the coupling coefficient of the two coupled lines, where $k = (Z_{oe} - Z_{oo})/(Z_{oe} + Z_{oo})$. Obviously, the lower and upper cutoff frequencies of the wide passband for the differential mode are mainly determined by the two resonant frequencies, f_{r1} and f_{r2}, and the transmission zero appearing at $2f_0$ is created by the shorted stub of the T-shaped structure. Actually, when the characteristic impedance Z_2 of the shorted stub becomes smaller, the T-shaped structure can be viewed as an improved multimode resonator [13].

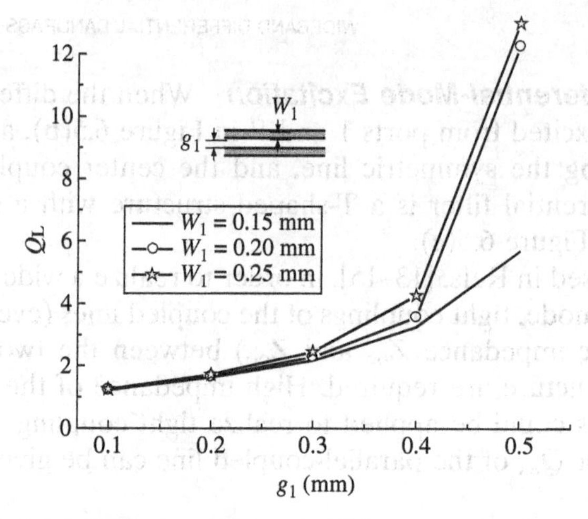

Figure 6.6 Simulated Q_L for parallel-coupled line versus gap size g_1. Reprinted with permission from Ref. [12]; copyright 2012 IEEE.

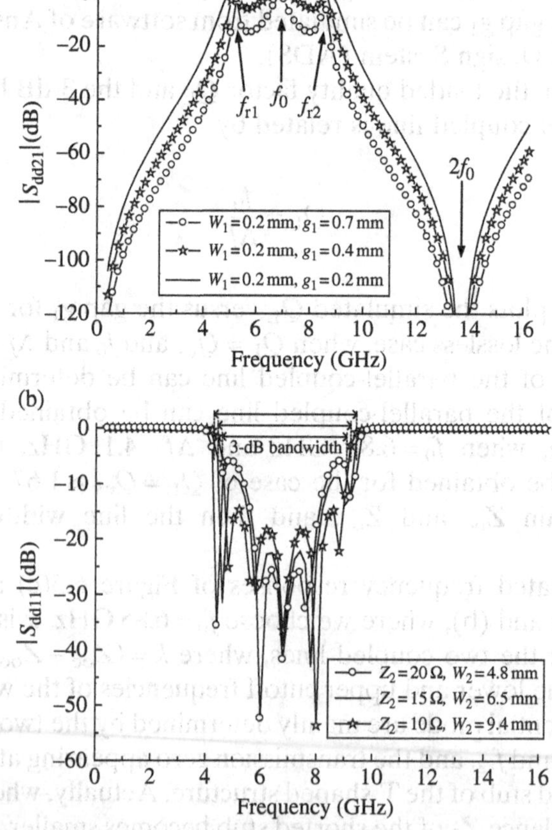

Figure 6.7 Simulated frequency responses of the differential mode. (a) $|S_{dd21}|$ ($Z_1 = 124\,\Omega$, $Z_2 = 10\,\Omega$) (b) $|S_{dd11}|$ ($Z_{oe} = 166.6\,\Omega$, $Z_{oo} = 85.6\,\Omega$, $Z_1 = 124\,\Omega$). Reprinted with permission from Ref. [12]; copyright 2012 IEEE.

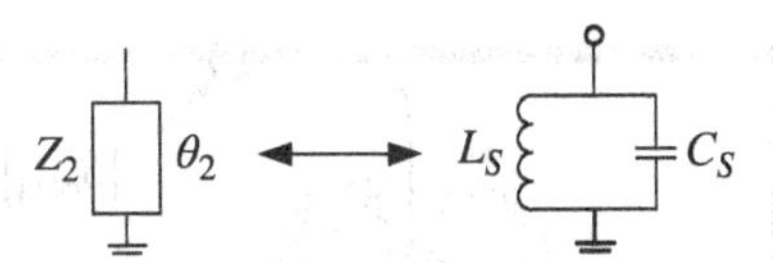

Figure 6.8 The ideal lossless equivalent circuit of the shorted stub in the center frequency of f_0. Reprinted with permission from Ref. [12]; copyright 2012 IEEE.

Because the shorted stub of the T-shaped structure is quarter wavelength at the center frequency f_0, half size reduction can be achieved for the T-shaped resonator structure when compared with the multimode resonators [13].

In addition, the bandwidth for the differential mode increases with the increase of the characteristic impedance Z_2 (as shown in Figure 6.7b). As a matter of fact, when the characteristic impedance Z_2 of the shorted stub for the T-shaped structure becomes smaller, the capacitance of the shorted stub increases, while the inductance decreases, and the shorted stub can be seen as an ideal lossless parallel resonance circuit in the center frequency f_0 (as shown in Figure 6.8). Because the Q factor of the passband for the shorted stub is proportional to the susceptance slope parameter $\sqrt{C_S/L_S}$, when the Q factor increases, the bandwidth for the passband will then become narrower [16].

6.2.3.2 *Common-Mode Excitation* When the common-mode signals are excited from ports 1 and 1′, a virtual open appears along the symmetric line in Figure 6.5(b), and the center coupling structure for the differential filter is a T-shaped structure with an open stub [1], as shown in Figure 6.5(d). Figure 6.9 shows the simulated frequency responses of the Figure 6.5(d), indicating that broadband good common-mode suppression can be easily achieved with the decrease of characteristic impedance Z_2. In addition, the three resonant frequencies at DC, f_0, and $2f_0$ for the T-shaped structure with the open stub do not change with the characteristic impedance Z_1 and Z_2.

Figure 6.10 shows the open stub and its ideal lossless equivalent circuit at the center frequency f_0 of the common mode [16], and the open stub can be seen as an ideal lossless series resonance circuit in the center frequency f_0. When the characteristic impedance Z_2 of the open stub becomes smaller, the capacitance of the open stub increases and the inductance decreases. Because the Q factor of the stopband for the stub is proportional to the reactance slope parameter $\sqrt{L_O/C_O}$, the Q factor thus decreases and results in a wider stopband bandwidth [16].

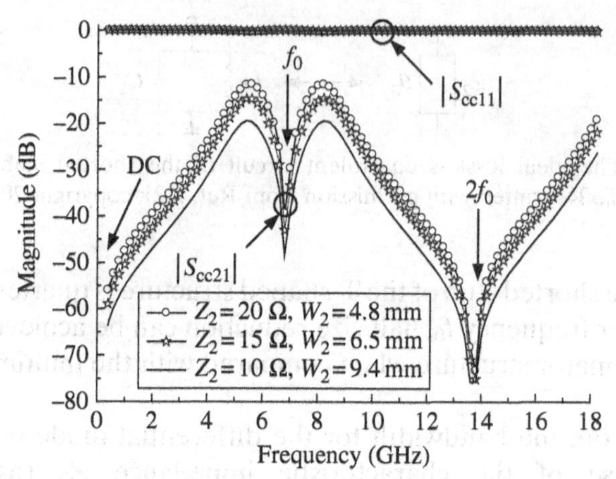

Figure 6.9 Simulated frequency responses of the common mode (Z_{oe} = 166.6 Ω, Z_{oo} = 85.6 Ω, Z_1 = 124 Ω). Reprinted with permission from Ref. [12]; copyright 2012 IEEE.

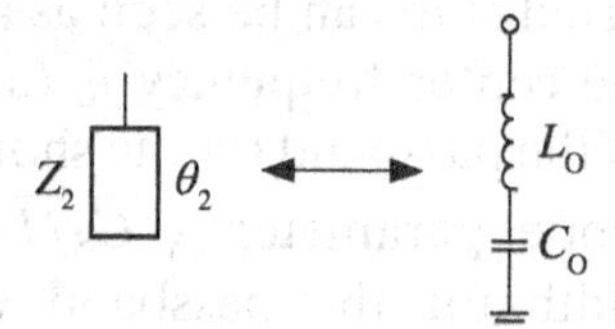

Figure 6.10 The ideal lossless equivalent circuit of the open stub in the center frequency of f_0. Reprinted with permission from Ref. [12]; copyright 2012 IEEE.

Therefore, broadband common-mode suppression can be easily achieved with the decrease of characteristic impedance Z_2 due to the three transmission zeros at DC, f_0, and $2f_0$ for the T-shaped structure with the open stub.

To clarify the proposed filter design, the design procedures of wideband differential bandpass filters are summarized as follows:

1. Based on (6.23) and (6.24), choose the desired center frequency f_0 of the differential bandpass filter and the 3-dB bandwidth Δf of the parallel-coupled line, and determine the width W_1 of the parallel-coupled line and the gap g_1.
2. Choose the characteristic impedance Z_2 of the T-shaped structure to determine specified resonant frequencies f_{r1} and f_{r2} in (6.19), and obtain the desired 3-dB bandwidth for the differential mode.
3. Adjust the characteristic impedance Z_2 to maximize the common-mode suppression and further optimize the transmission characteristic for the differential mode.

Referring to the aforementioned discussions and the simulated results, the three resonant frequencies for the common mode do not change with the characteristic impedance Z_1 and Z_2, so the differential-mode response and common-mode suppression can be tuned independently to a certain degree, which implies design feasibility of the differential bandpass filter with quite good common-mode suppression. Based on the previous theoretical analysis, the final parameters for the filter circuit of Figure 6.5(b) are $Z_0 = 50\,\Omega$, $Z_1 = 124\,\Omega$, $Z_2 = 12.5\,\Omega$, $Z_{oe} = 168.9\,\Omega$, $Z_{oo} = 82.7\,\Omega$, and $f_0 = 6.85$ GHz. The structure parameters for the differential filter (35 mm × 27 mm, $1.17\lambda_0 \times 0.9\lambda_0$) in Figure 6.5(a) are $W_0 = 1.37$ mm, $W_1 = 0.2$ mm, $W_2 = 8.1$ mm, $l_1 = 7.5$ mm, $l_2 = 15.8$ mm, $l_3 = 12.8$ mm, $g_1 = 0.2$ mm, $g_2 = 0.8$ mm, $tl_1 = 1.0$ mm, $tl_2 = 3.88$ mm, $S =$ 2.54 mm, $\varepsilon_r = 2.65$, $h = 0.5$ mm, and $\tan\delta = 0.002$. The simulated results with Ansoft HFSS v.10 and Ansoft Designer v3.0 are shown in Figure 6.11. The two-port differential-/common-mode S-parameters for the differential filter are deduced from the simulated four-port S-parameters [1]. As shown in Figure 6.11(a), for the differential mode, four transmission poles are achieved in the passband (3-dB fractional bandwidth is approximately 71%, 4.45–9.3 GHz). The insertion loss is less than 0.5 dB, while the return loss is over 15 dB from 4.7 to 9 GHz. Furthermore, over 35-dB upper stopband suppression is obtained from 10.5 to 15.5 GHz ($2.41f_0$). For the common mode, the insertion loss is greater than 15 dB from 0 to 19 GHz ($2.77f_0$), indicating very good wideband rejection, as shown in Figure 6.11(b).

Figure 6.12 shows the prototype photograph of the wideband differential filter. The measured results for the differential/common mode of the wideband differential bandpass filters are shown in Figure 6.12. As shown in Figure 6.12(a), for the differential mode of the wideband bandpass filter without cross coupling, the measured insertion loss is less than 1.65 dB, while the return loss is over 15 dB from 4.9 to 8.9 GHz (3-dB fractional bandwidth is approximately 70%). Furthermore, over 30-dB upper stopband suppression is achieved from 10 to 17.9 GHz ($2.63f_0$). The measured group delay is less than 0.38 ns from 5 to 9 GHz. For the common mode shown in Figure 6.12(b), over 14.5 dB stopband suppression is obtained from 0 to 19 GHz ($2.77f_0$), indicating very good wideband common-mode rejection.

6.2.4 Wideband Bandpass Filter with Cross Coupling

In order to further improve the selectivity and reduce the size of the differential bandpass filter without cross coupling, another compact wideband differential bandpass filter using cross coupling is introduced.

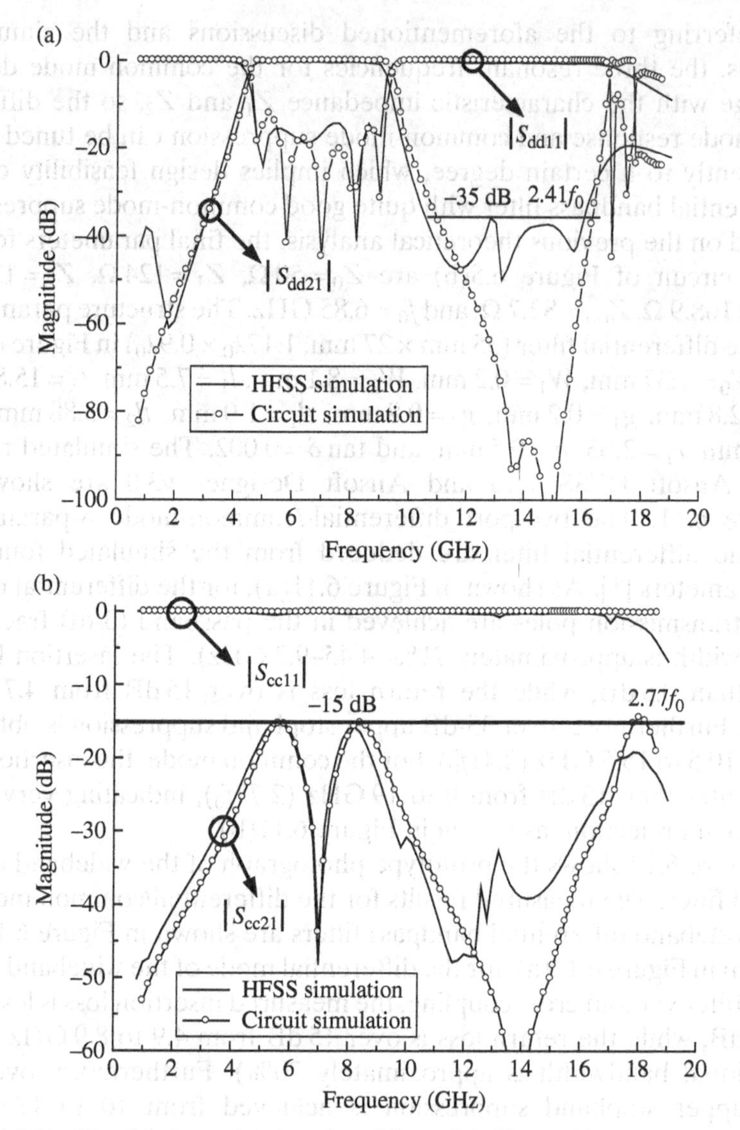

Figure 6.11 Simulated results for the differential filter without cross coupling. (a) Differential mode. (b) Common mode. Reprinted with permission from Ref. [12]; copyright 2012 IEEE.

The top view and the circuit of the differential filter with cross coupling are shown in Figure 6.13(a) and (b), with cross coupling realized by the input/output side-coupled lines of $\lambda_0/4$. In addition, the half equivalent circuits for the differential-/common-mode cases are shown in Figure 6.12(c) and (d) [1]. The simulated results for the differential/

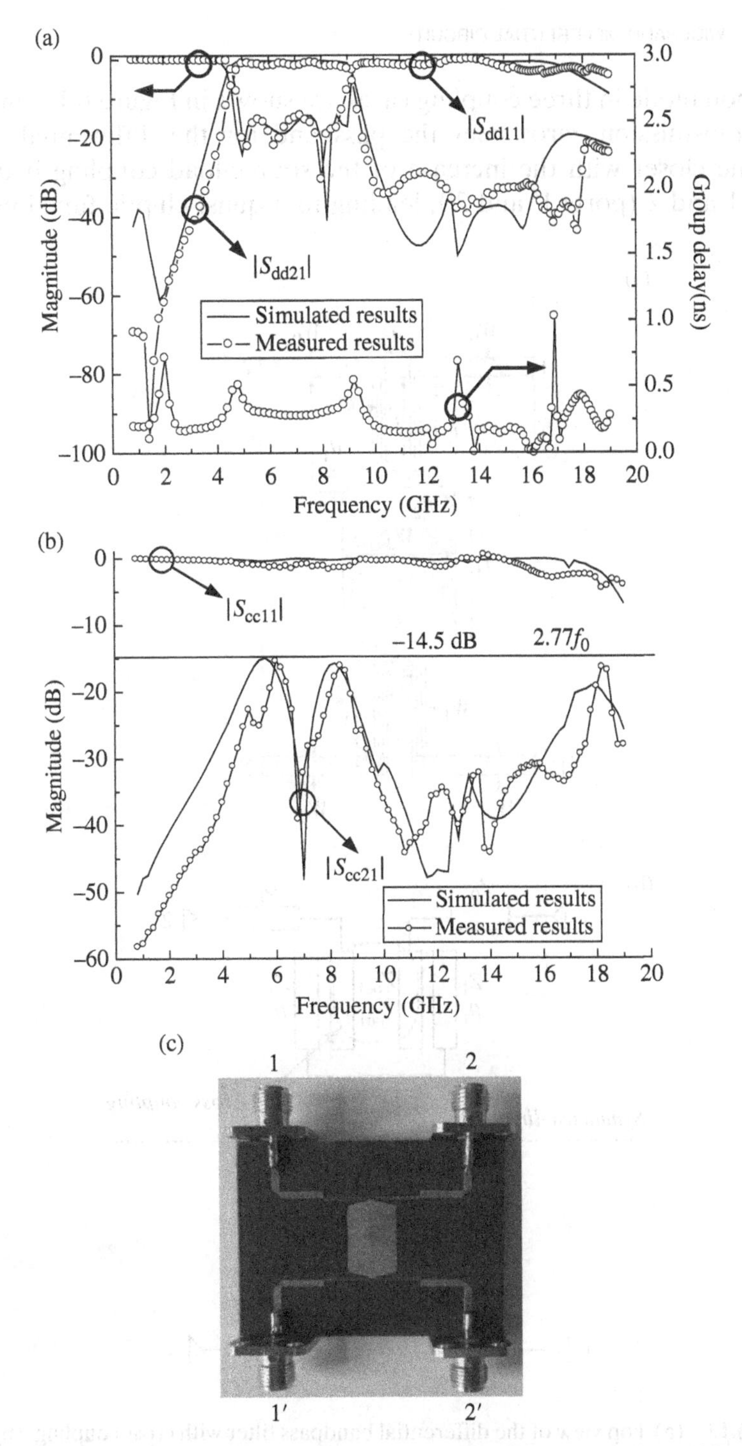

Figure 6.12 Measured and simulated results for the differential filter without cross coupling. (a) Differential mode (b) Common mode. (c) Paragraph. Reprinted with permission from Ref. [12]; copyright 2012 IEEE.

common mode in three coupling cases are shown in Figure 6.14, and the two transmission zeros near the passband for the differential mode become closer with the increase of the source/load coupling between ports 1 and 2 (ports 1′ and 2′), leading to a quasi-elliptic function that

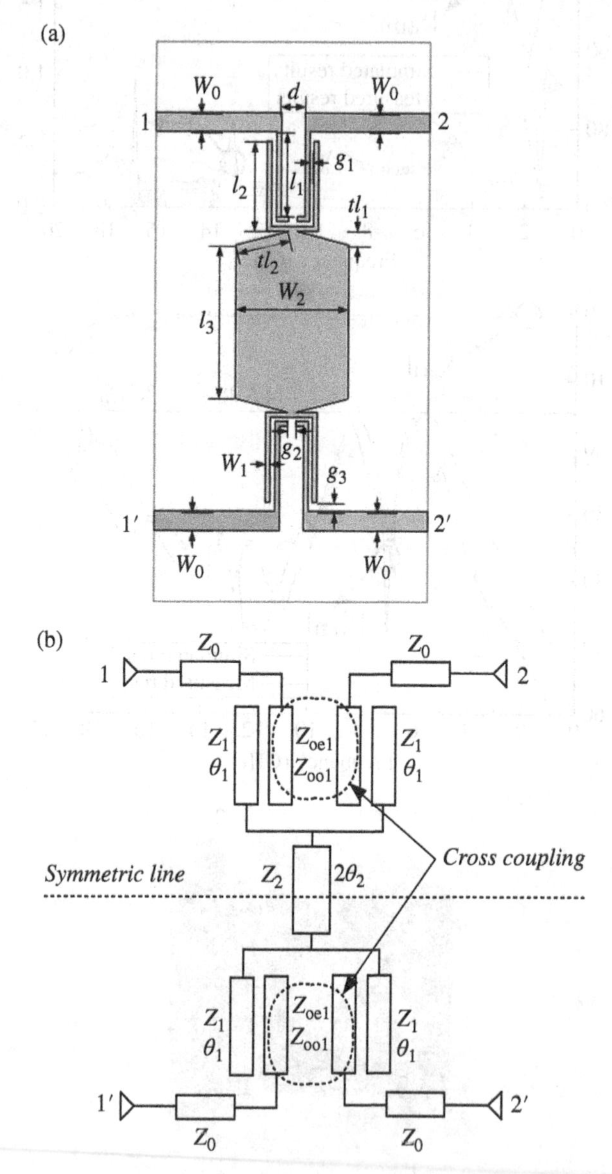

Figure 6.13 (a) Top view of the differential bandpass filter with cross coupling. (b) The ideal circuit of the wideband differential filter. (c) Equivalent circuit for the differential mode. (d) Equivalent circuit for the common mode. Reprinted with permission from Ref. [12]; copyright 2012 IEEE.

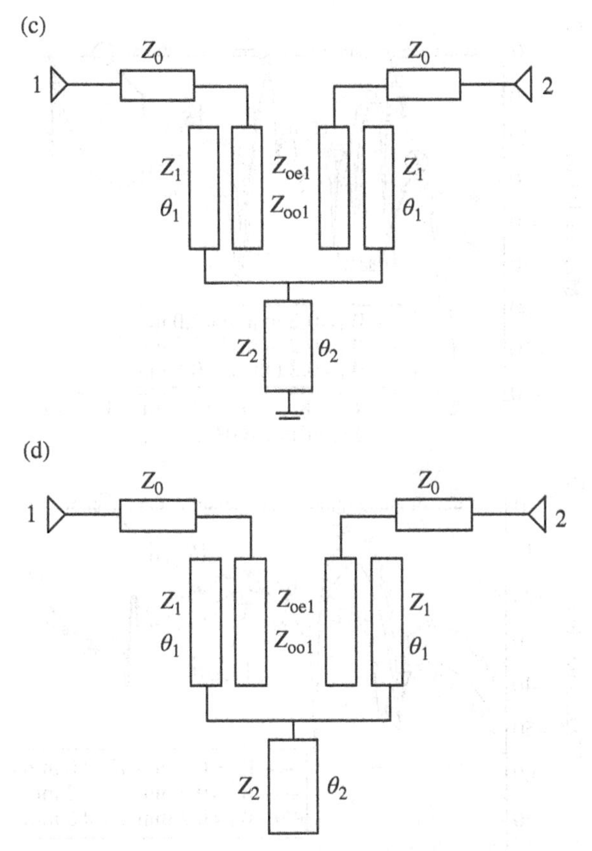

Figure 6.13 (Continued)

improves the passband and out-of-band performances for the differential mode. However, in order to realize better broadband common-mode suppression, the coupling strength between ports 1 and 2 (ports 1′ and 2′) should be controlled to a certain degree.

Based on the previous discussion and the theoretical analysis, the final parameters for the differential filter with cross coupling are $Z_0 = 50\ \Omega$, $Z_1 = 124\ \Omega$, $Z_2 = 12.2\ \Omega$, $Z_{oe} = 169.4\ \Omega$, $Z_{oo} = 82.1\ \Omega$, $Z_{oe1} = 130.8\ \Omega$, $Z_{oo1} = 124.1\ \Omega$, $f_0 = 6.85$ GHz. The final structure parameters for the differential filter (36 mm × 22 mm, $1.2\lambda_0 \times 0.73\lambda_0$, $\varepsilon_r = 2.65$, $h = 0.5$ mm, and $\tan\delta = 0.002$.) in Figure 6.13(a) are $W_0 = 1.37$ mm, $W_1 = 0.2$ mm, $W_2 = 8.3$ mm, $l_1 = 6.8$ mm, $l_2 = 6.8$ mm, $l_3 = 12.2$ mm, $g_1 = 0.17$ mm, $g_2 = 1.0$ mm, $g_3 = 0.4$ mm, $tl_1 = 1.0$ mm, $tl_2 = 3.98$ mm, $d = 1.83$ mm. The simulated results for the structure and circuits of Figure 6.13 are shown in Figure 6.15. Good agreement can be observed between the HFSS and circuit simulation results. For the

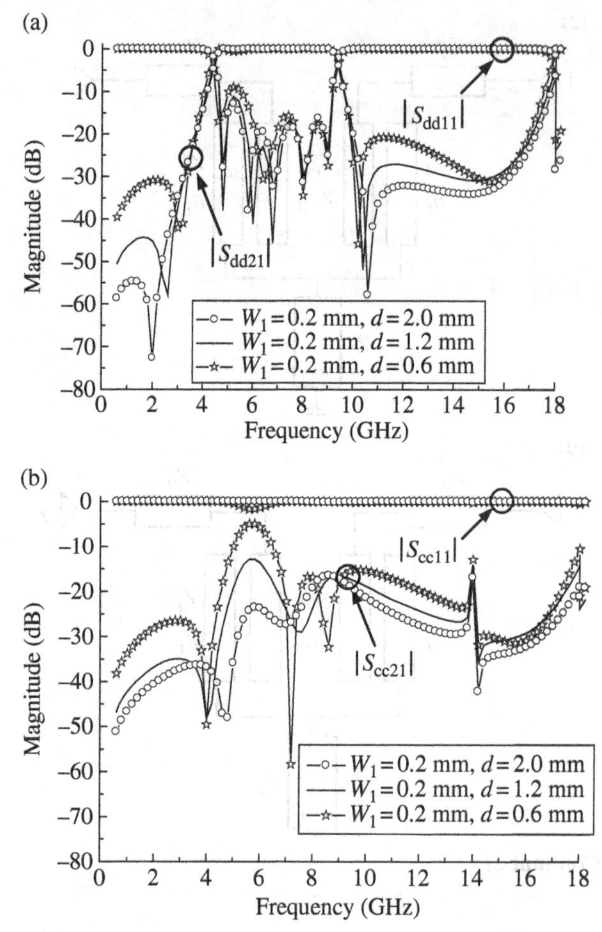

Figure 6.14 Simulated frequency responses for the differential filter with cross coupling. (a) Differential mode. (b) Common mode ($Z_1 = 124\,\Omega$, $Z_2 = 12.5\,\Omega$). Reprinted with permission from Ref. [12]; copyright 2012 IEEE.

differential-mode circuit shown in Figure 6.15(a), two transmission zeros are located at 2.5 and 10.3 GHz, respectively, while four transmission poles are also realized in the passband (3-dB fractional bandwidth is approximate 72%, 4.3–9.25 GHz). The simulated insertion loss is less than 0.8 dB, while the return loss is over 12.5 dB from 4.4 to 8.9 GHz; over 25-dB upper stopband suppression is achieved from 10 to 17.6 GHz ($2.62f_0$). For the common-mode circuit shown in Figure 6.15(b), the simulated insertion loss is greater than 13.5 dB from 0 to 19 GHz ($2.77f_0$). Compared with the proposed wideband differential bandpass filter without cross coupling, the selectivity for the differential mode has been improved, while the circuit size reduction has also been achieved.

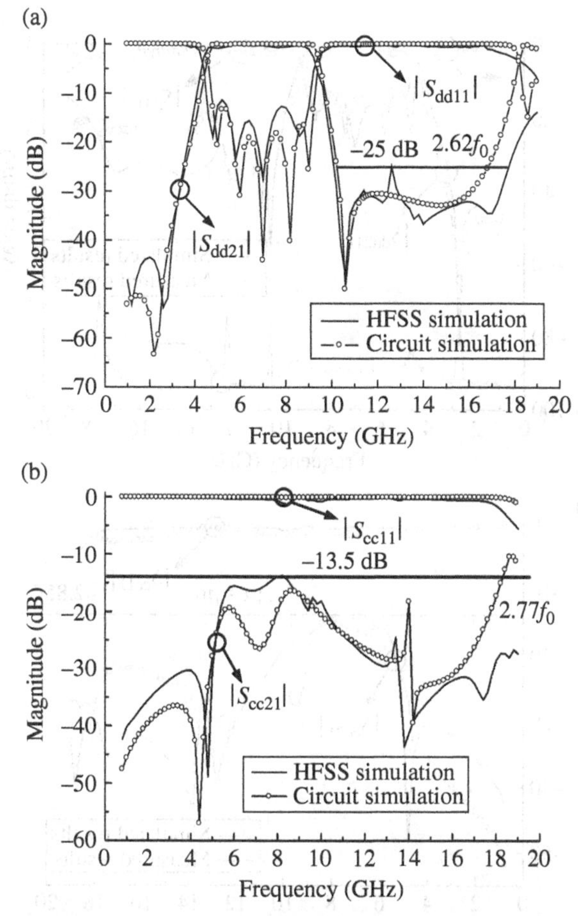

Figure 6.15 Simulated results for the differential filter with cross coupling. (a) Differential mode. (b) Common mode. Reprinted with permission from Ref. [12]; copyright 2012 IEEE.

Figure 6.16 shows the prototype photograph and measured results of the wideband differential filter with cross coupling. From Figure 6.16(a), for the differential mode of the wideband bandpass filter with cross coupling, two transmission zeros are located at 2.5 and 10.9 GHz (3-dB fractional bandwidth is approximately 70.7%), respectively. The measured insertion loss is less than 1.8 dB, while the return loss is over 12 dB from 4.6 to 8.7 GHz. Furthermore, over 20-dB upper stopband suppression is achieved from 11 to 19 GHz ($2.77f_0$). The measured group delay is less than 0.4 ns from 4.8 to 9.1 GHz. For the common-mode circuit shown in Figure 6.16(b), over 13-dB stopband suppression is obtained from 0 to 19.5 GHz ($2.85f_0$). The slight

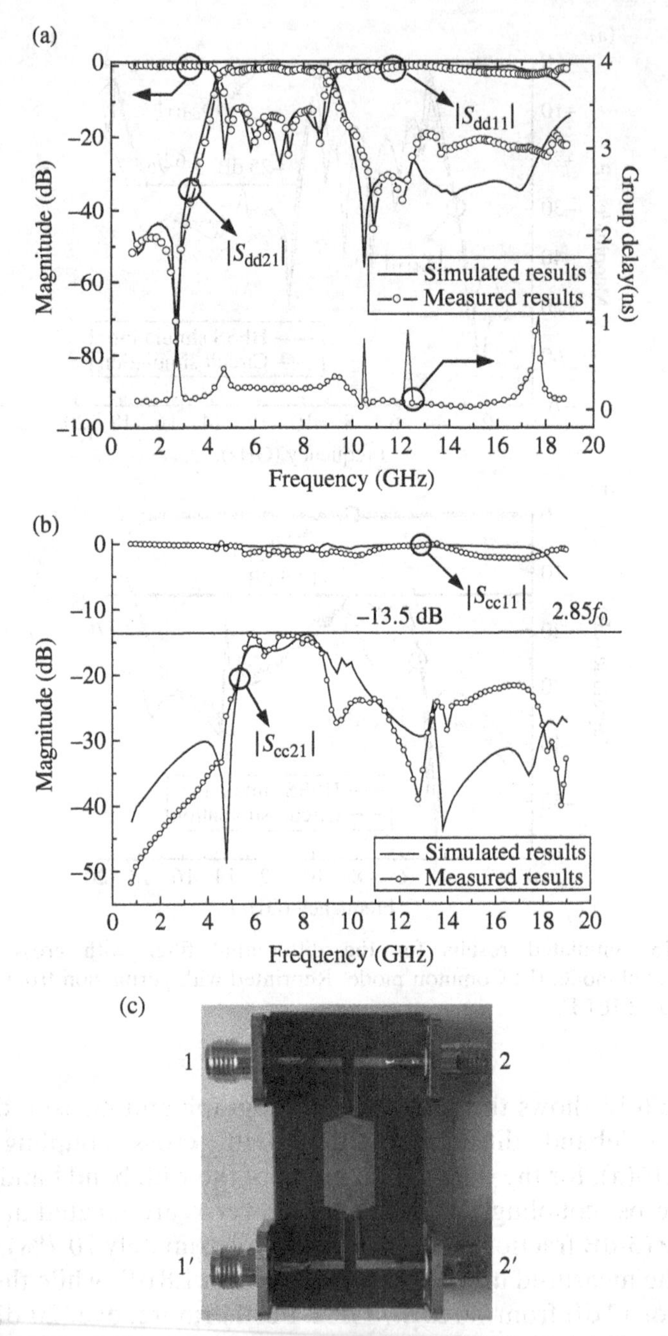

Figure 6.16 Measured and simulated results for the differential filter with cross coupling in Figure 6.11. (a) Differential mode. (b) Common mode. (c) Paragraph. Reprinted with permission from Ref. [12]; copyright 2012 IEEE.

frequency discrepancies between the measured and simulated results are mainly caused by the limited fabrication precision and measurement errors. In addition, to further improve the out-of-band performance for the differential mode and common mode, cascaded T-shaped multimode resonators proposed in Refs. [17, 18] can be also used to realize wideband differential filters.

6.3 WIDEBAND DIFFERENTIAL BANDPASS FILTERS USING HALF-/FULL-WAVELENGTH RING RESONATORS

The dual-mode resonator was firstly introduced in Ref. [19], and two degenerated modes can be excited by introducing a perturbation element along an orthogonal plane of the resonator. Different ring resonators with open stubs or notches were used to design bandpass filters with high selectivity [20–23]. In this section, three wideband differential bandpass filters using half-/full-wavelength ring resonators are proposed [24, 25], and multiple transmission zeros for the differential/common mode can be easily realized for the wideband differential filters. The half-wavelength ring resonators can be used to realize compact balanced filter circuits, and more transmission zeros realized by the full-wavelength ring resonators can be used to further suppress the common mode over five octaves for the balanced circuits. Next four wideband balanced filters using half-/full-wavelength ring resonators with high performance will be demonstrated, respectively.

6.3.1 Differential Filter Using Half-Wavelength Ring Resonators

The wideband differential bandpass filter with four open stubs is shown in Figure 6.17(a). Four open stubs with electrical length 2θ and characteristic impedance Z_1 are attached to the side-coupled quarter-wavelength lines (electrical length θ, even-/odd-mode characteristic impedance Z_{oe}, Z_{oo}). The equivalent half circuits of the differential-/common-mode can be used for theoretical analysis conveniently [1], as shown in Figure 6.17(b) and (c).

When the differential-mode signals are excited from ports 1 and 1′ in Figure 6.17(a), a virtual short appears along the half-wavelength ring resonator (as shown in Figure 6.17b), and a passband performance can be realized by the open/shorted coupled lines [26]. The **ABCD** matrix of the differential circuit for Figure 6.17(b) is $M_{stub} \times M_c \times M_c \times M_{stub}$ (M_{stub}, open stub; M_c, side-coupled line). After **ABCD**-, Y-,

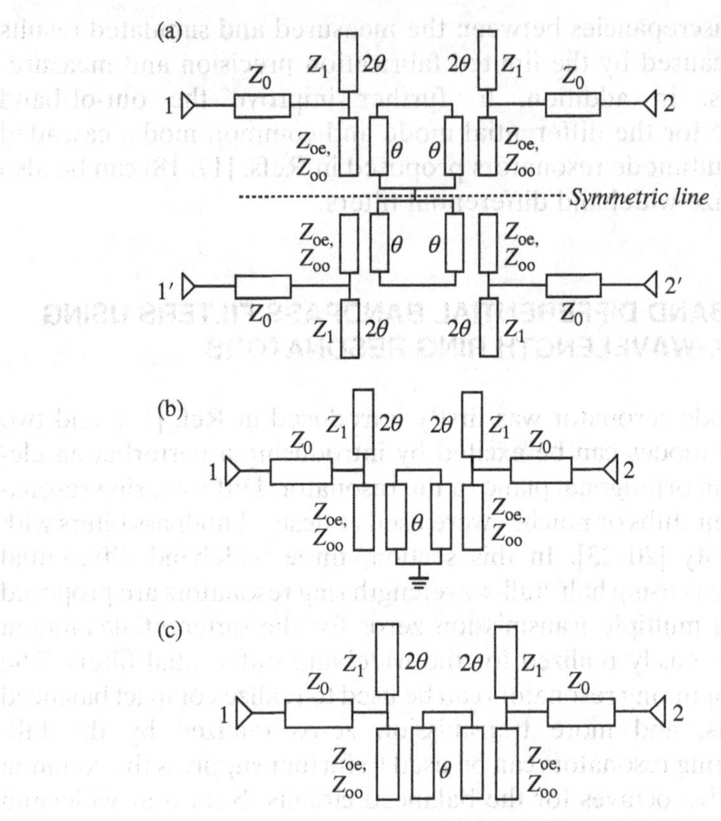

Figure 6.17 (a) The ideal circuit of the wideband differential filter. (b) Equivalent circuit for the differential mode. (c) Equivalent circuit for the common mode. Reprinted with permission from Ref. [24]; copyright 2013 IEEE.

and S-parameter conversions, when $S_{dd21} = 0$, two transmission zeros can be obtained:

$$\theta_{tz1} = \frac{\pi}{4}, \quad \theta_{tz2} = \frac{3\pi}{4} \tag{6.25}$$

In addition, when $S_{dd11} = 0$, a fifth-order equation for θ versus Z_1, Z_{oe}, and Z_{oo} can be also obtained. When Z_0 is fixed, five roots for $S_{dd11} = 0$ can be acquired by properly choosing Z_1, Z_{oe}, and Z_{oo}, and then five transmission poles in the passband can be achieved.

The simulated frequency responses of Figure 6.17(b) are shown in Figure 6.18(a) and (b) (Ansoft Designer v3.0). Obviously, the bandwidth of differential mode increases as k increases ($k = (Z_{oe} - Z_{oo})/(Z_{oe} + Z_{oo})$), and the five in-band transmission poles (f_0, f_{tp1}, f_{tp2}, f_{tp3}, f_{tp4}) reflect the fact that $S_{dd11} = 0$ has five real solutions when Z_1, Z_{oe}, and

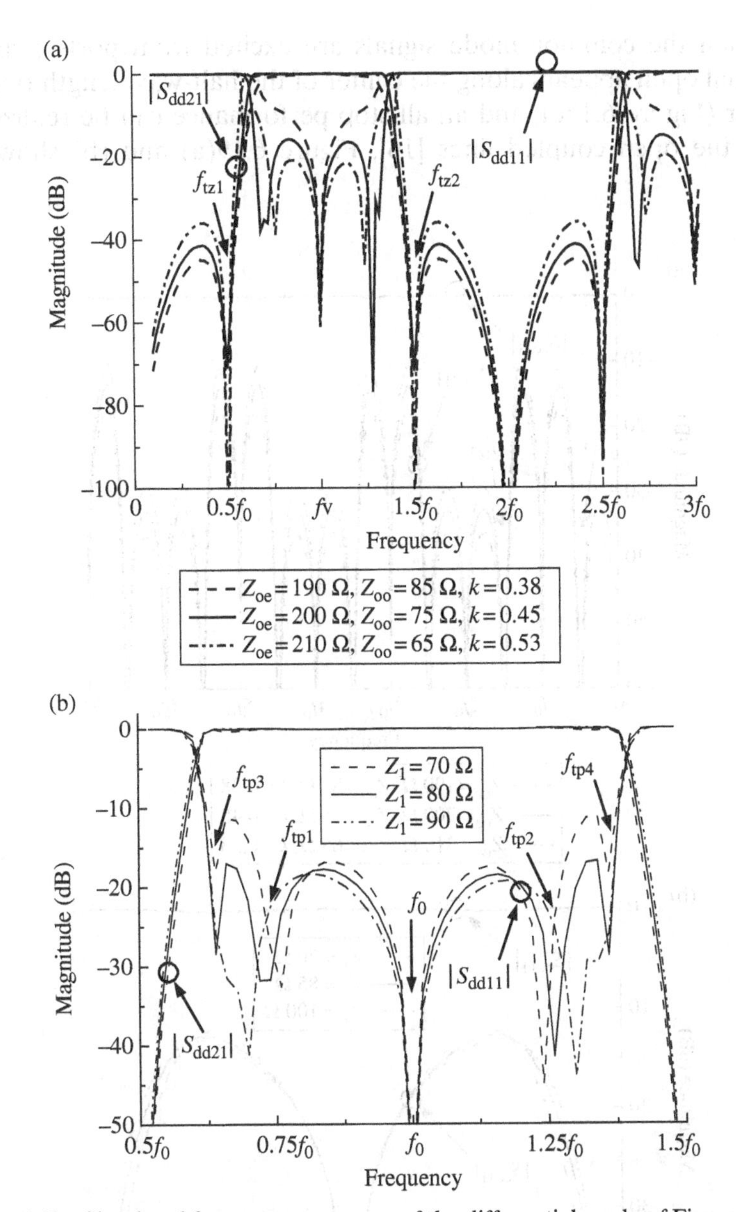

Figure 6.18 Simulated frequency responses of the differential mode of Figure 6.16(b). (a) Versus Z_{oe} and Z_{oo}, $Z_1 = 75\ \Omega$. (b) Versus Z_1, $Z_{oe} = 202.5\ \Omega$, $Z_{oo} = 70.1\ \Omega$ ($Z_o = 50\ \Omega$). Reprinted with permission from Ref. [24]; copyright 2013 IEEE.

Z_{oo} are properly selected. In addition, two transmission zeros ($f_{tz1} = 0.5f_0$, $f_{tz2} = 1.5f_0$) are realized to improve the selectivity and harmonic suppression. Moreover, the in-band return loss can be also adjusted by changing the characteristic impedance Z_1.

When the common-mode signals are excited from ports 1 and 1′, a virtual open appears along the center of the half-wavelength ring resonator (Figure 6.17c), and an all-stop performance can be realized by using the open coupled lines [16]. Figure 6.19(a) and (b) shows the

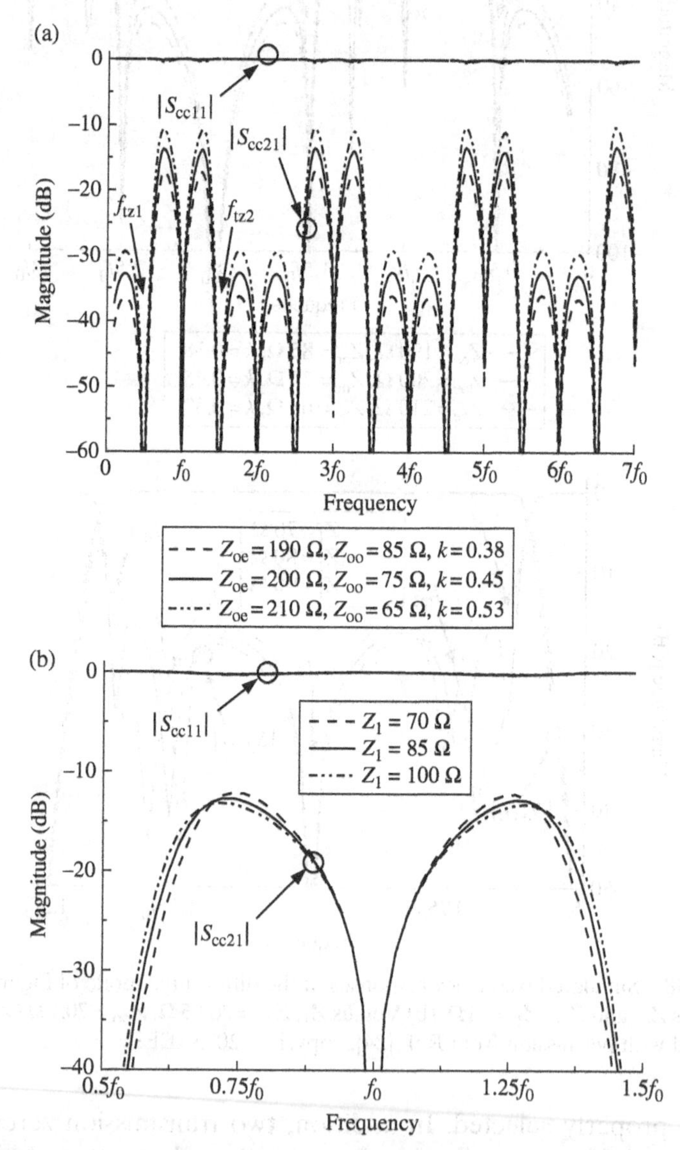

Figure 6.19 Simulated frequency responses of the common mode of Figure 6.16(c). (a) Versus Z_{oe} and Z_{oo}, $Z_1 = 75\,\Omega$. (b) Versus Z_1, $Z_{oe} = 202.5\,\Omega$, $Z_{oo} = 70.1\,\Omega$ ($Z_o = 50\,\Omega$). Reprinted with permission from Ref. [24]; copyright 2013 IEEE.

simulated frequency responses of Figure 6.16(c) (Ansoft Designer v3.0). As we can see, from 0 to ∞ (GHz), besides the numbers of transmission zeros ($f_{tzn} = nf_0$, $n = 0, 1, 2, \ldots$), some other transmission zeros ($f_{tzn} = (n - 0.5)f_0$, $n = 1, 2, \ldots$) generated by the half-wavelength open stubs can further improve the common-mode suppression. In addition, the suppression of the common mode becomes better as Z_1 increases. In this way, a wide differential filter with high selectivity and wideband common-mode suppression can be easily realized.

To clarify the proposed filter design, the design procedures of wideband differential bandpass filter are summarized as follows:

1. Choose the desired center frequency f_0 and 3-dB bandwidth of the differential mode, and determine the width W_1 of the parallel-coupled line and gap g_1.
2. Adjust the characteristic impedance Z_1 to maximize the common-mode suppression and further optimize the transmission characteristic for the differential mode.

Based on the previous theoretical analysis, the final parameters for the filter circuit of Figure 6.17(a) are $Z_0 = 50\ \Omega$, $Z_1 = 75\ \Omega$, $Z_{oe} = 202.2\ \Omega$, $Z_{oo} = 72.3\ \Omega$, and $f_0 = 3.1$ GHz. Figure 6.20(a) and (b) shows the top view and photograph of the proposed differential filter. The simulated results are shown in Figure 6.21 (Ansoft HFSS v.10). For the differential mode, a five-order passband with 3-dB bandwidth 85.4% (1.77–4.35 GHz) is realized. The insertion loss ($|S_{dd21}|$) is less than 0.7 dB, while the return loss ($|S_{dd11}|$) is over 10.5 dB (1.8–4.3 GHz). Two transmission zeros are found to locate at 1.5 and 4.55 GHz. Furthermore, over 20-dB upper stopband is obtained (4.5–7.6 GHz, $2.45f_0$). For the common mode, the insertion loss ($|S_{cc21}|$) is greater than 10 dB (0–12 GHz, $4f_0$).

Figure 6.21 shows the measured results of the wideband differential filter. For the differential mode of Figure 6.21(a), five transmission zeros are located at 0, 0.5, 1.6, 4.57, and 5.9 GHz (3-dB fractional bandwidth 79%, 1.85–4.23 GHz), respectively. The measured insertion loss is less than 1.35 dB, while the return loss is over 10 dB from 2.0 to 4.05 GHz. Furthermore, over 25-dB upper stopband is achieved from 4.4 to 7.8 GHz ($2.6f_0$). For the common mode shown in Figure 6.21(b), over 13-dB stopband is obtained from 0 to 8.0 GHz ($2.67f_0$) and 10-dB from 0 to 15 GHz ($5f_0$). The slight frequency discrepancies (<0.15 GHz) between the measured and simulated results are mainly caused by the limited fabrication precision and measurement errors. To further design balanced filters with multiple independently adjustable transmission

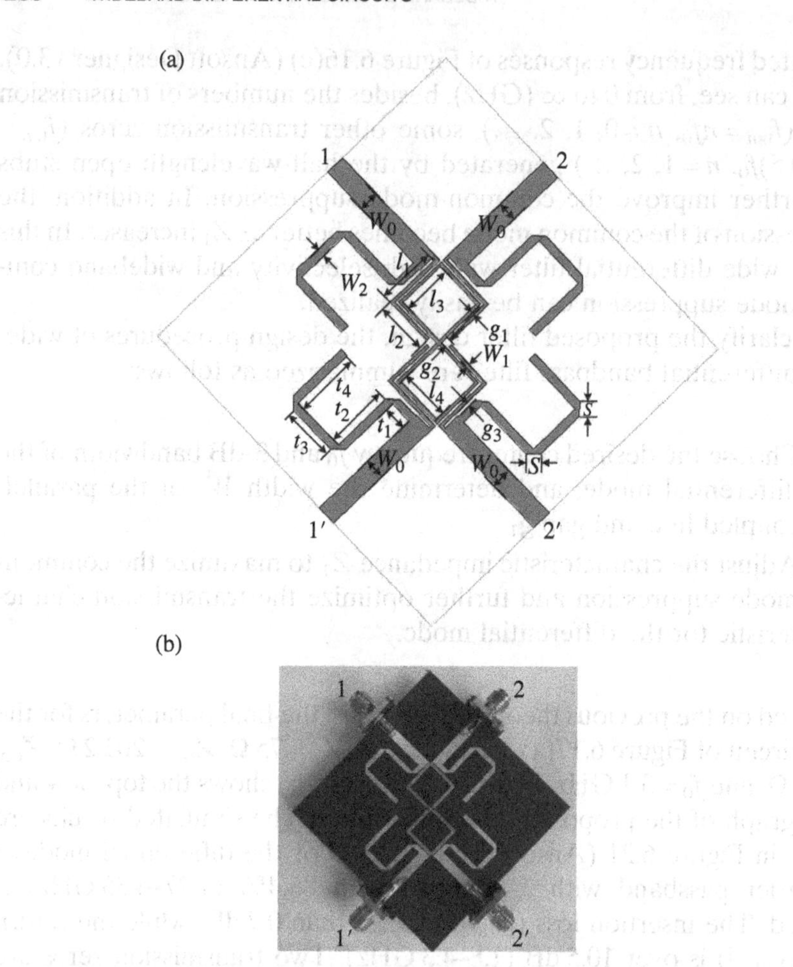

Figure 6.20 Top view and photograph of the wideband differential bandpass filter. (a) Top view, (b) photograph. ($W_0 = 2.7$ mm, $W_1 = 0.30$ mm, $W_2 = 1.3$ mm, $l_1 = 9.3$ mm, $l_2 = 8.5$ mm, $l_3 = 8.55$ mm, $l_4 = 8.55$ mm, $t_1 = 2.5$ mm, $t_2 = 9.2$ mm, $t_3 = 6.2$ mm, $t_4 = 12.7$ mm, $g_1 = 0.15$ mm, $g_2 = 0.8$ mm, $g_3 = 0.3 =$ mm, $S = 2.54$ mm, 54 mm × 54 mm, $0.82\lambda_0 \times 0.82\lambda_0$, $\varepsilon_r = 2.6$, $h = 1.0$ mm, and $\tan\delta = 0.002$.) Reprinted with permission from Ref. [24]; copyright 2013 IEEE.

zeros close to the differential-mode passband and wideband common-mode suppression, another two high-performance balanced filters using full-wavelength ring resonators will be given next.

6.3.2 Differential Filter Using Full-Wavelength Ring Resonators

Figure 6.22(a) and (b) shows two conventional two-stage open coupled line filter circuit and dual-mode filter in Refs. [22, 27], and the proposed

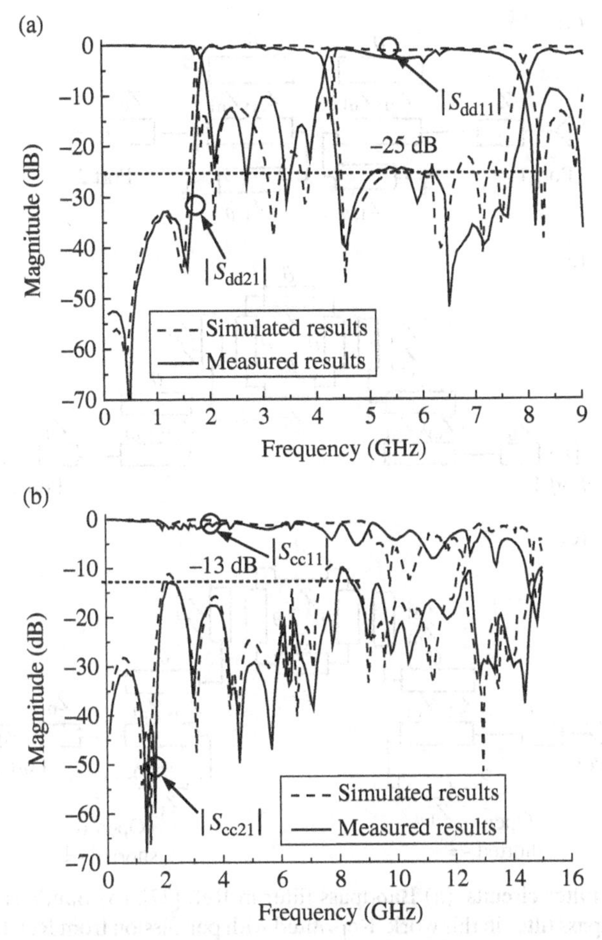

Figure 6.21 Measured and simulated results for the differential filter with four open stubs. (a) Differential mode. (b) Common mode. Reprinted with permission from Ref. [24]; copyright 2013IEEE.

bandpass filter based on dual-mode resonator with open/shorted loaded stubs is shown in Figure 6.22(c). The simulated results of Figure 6.22 (a)–(c) are shown in Figure 6.23(a) and (b). Due to the two shorted loaded stubs (Z_1), the passband order can be increased from third to sixth, and the locations of two transmission zeros realized by the dual-mode resonator do not change. In addition, Figure 6.23(b) shows the simulated results of Figure 6.22(c) with one/two open stubs. One transmission zero is found to locate at f_0 [26] due to the unsymmetrical/symmetrical stubs (Z_1 shorted/open, open/open); however, for the symmetrical stubs (Z_1, open/open), four resonator modes are produced near the center frequency f_0. Therefore, when the filter circuit of Figure 6.22(c) is used

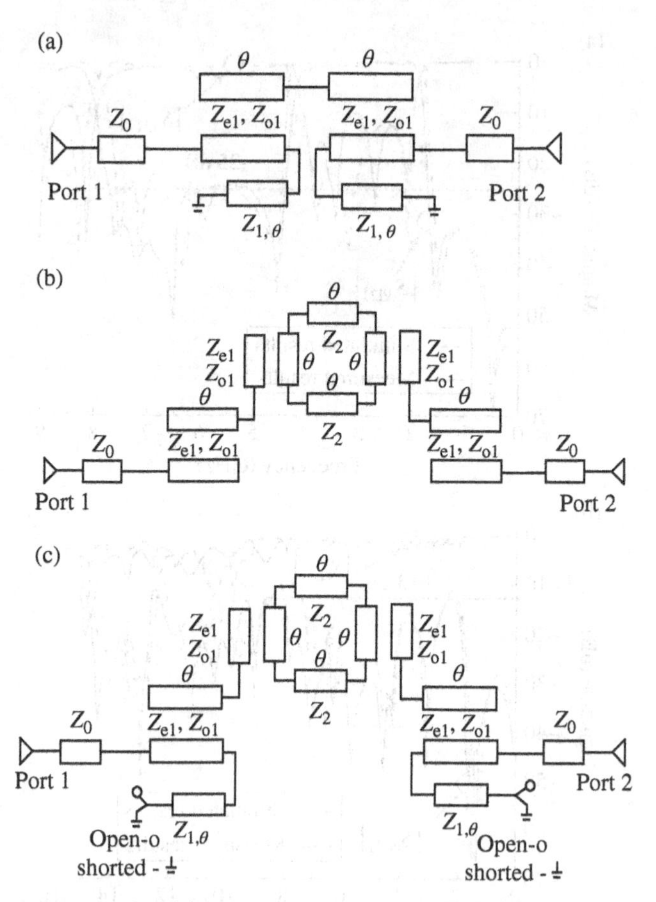

Figure 6.22 Filter circuits. (a) Bandpass filter in Ref. [27]. (b) Bandpass filter in Ref. [22]. (c) Bandpass filter in this work. Reprinted with permission from Ref. [25]; copyright 2015 IEEE.

to design balanced filters, the dual-mode resonator with symmetrical shorted stubs can be used to design a high selectivity passband for the differential mode, and the dual-mode ring resonator with asymmetrical open/shorted stubs can be used to suppress common mode over a wide frequency band. Next, detailed theoretical design for the two differential filters will be given.

The ideal circuit of the balanced filter with two transmission zeros close to the passband is shown in Figure 6.24(a). Two dual-mode ring resonators are attached to two quarter-wavelength side-coupled lines (electrical length θ, even-/odd-mode characteristic impedance Z_{e1}, Z_{o1}). Two open coupled lines (Z_{e1}, Z_{o1}, θ) with two loaded shorted stubs (Z_1, θ) are located in the middle of the balanced filter circuit, and four

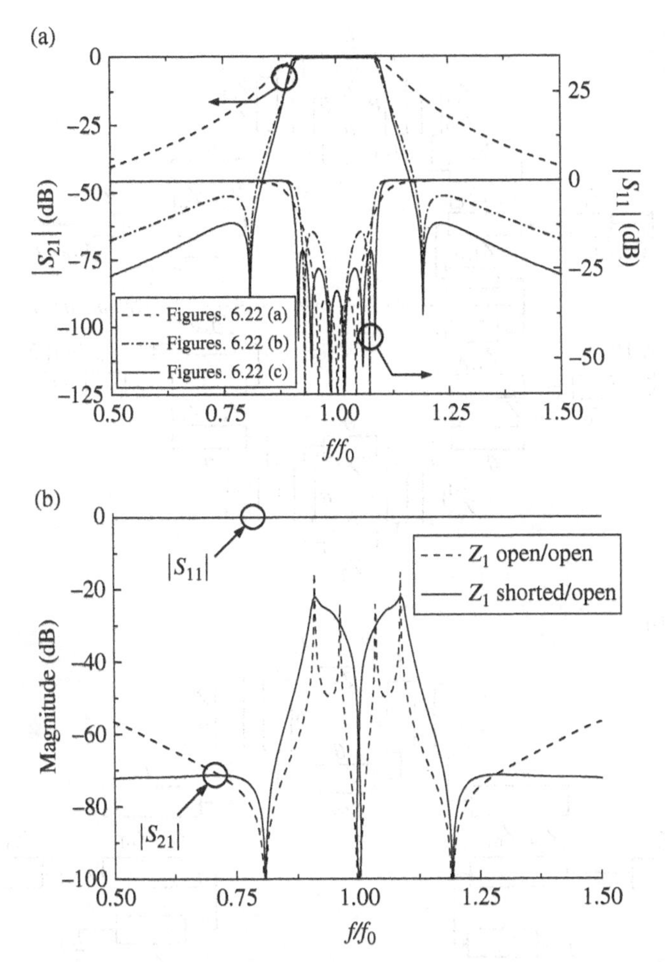

Figure 6.23 Simulated frequency responses of Figure 6.22. (a) Z_1-shorted/shorted. (b) Figure 6.22(c) with Z_1 shorted/open, ($Z_0 = 50\,\Omega$, $Z_1 = 118\,\Omega$, $Z_2 = 120\,\Omega$, $Z_{e1} = 180\,\Omega$, $Z_{o1} = 110\,\Omega$). Reprinted with permission from Ref. [25]; copyright 2015 IEEE.

microstrip lines with characteristic impedance $Z_0 = 50\,\Omega$ are connected to ports 1, 1′ and ports 2, 2′. In addition, the equivalent half circuits of the differential/common mode are shown in Figure 6.24(b) and (c), which can be used for theoretical analysis conveniently [1].

As shown in Figure 6.24(b), a virtual short appears along the symmetric line when the differential mode are excited from ports 1 and 1′, and a shorted stub with characteristic impedance Z_1 and electrical length θ is loaded in the end of the input open coupled lines. As discussed in Ref. [22], the **ABCD** matrices of the differential-mode circuit can be defined as $M_{cs1} \times M_s \times M_t \times M_s \times M_{cs2}$ (M_{cs}, open coupled lines loaded with

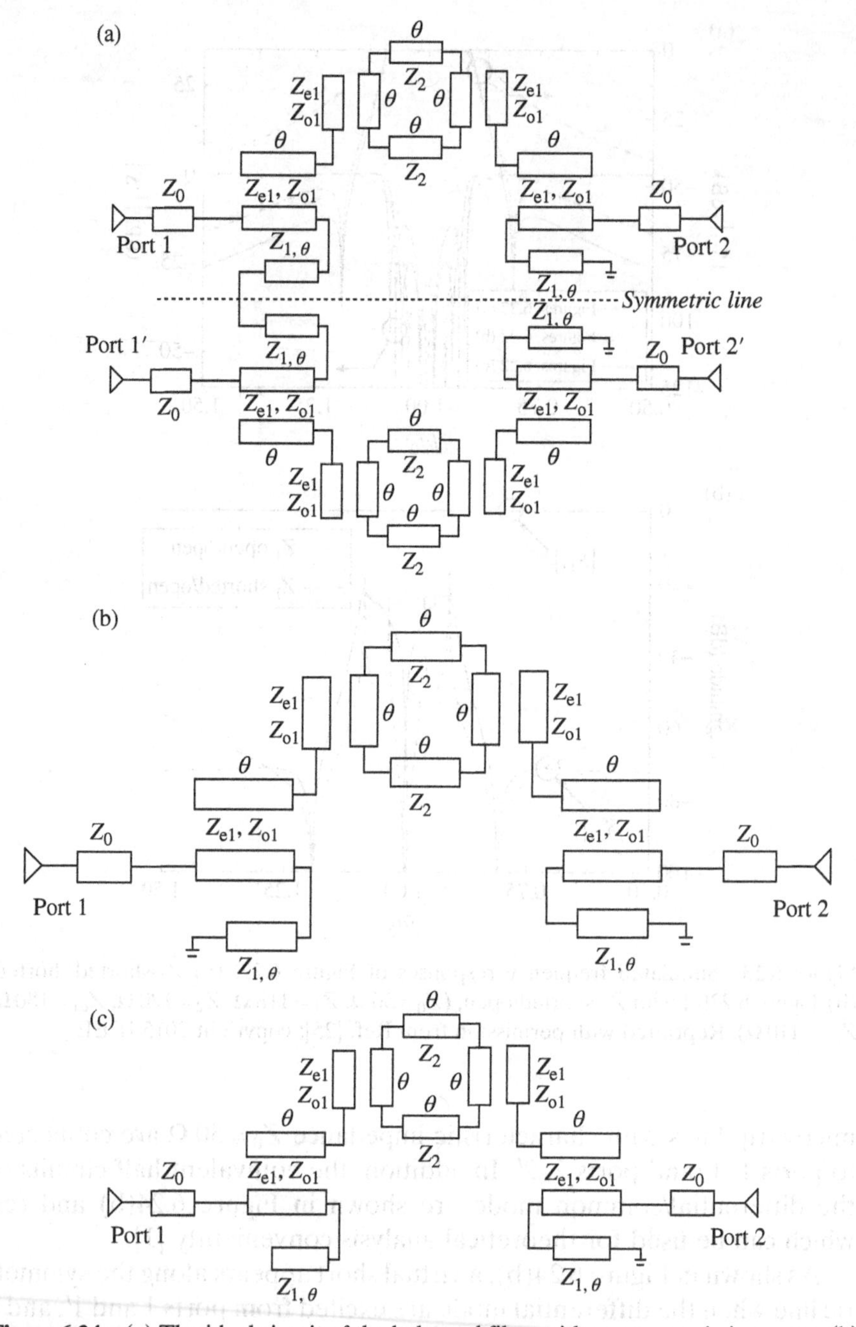

Figure 6.24 (a) The ideal circuit of the balanced filter with two transmission zeros. (b) Equivalent circuit for the differential mode. (c) Equivalent circuit for the common mode. Reprinted with permission from Ref. [25]; copyright 2015 IEEE.

shorted stub Z_1; M_s, side-coupled lines; M_r, center two transmission lines Z_2), and $M_{cs1/2}$, M_s, and M_t can be directly obtained from Ref. [27]. After **ABCD**- and Y-parameter conversions, when $S_{dd21}=0$, we can get

$$\frac{4}{Z_{e1}+Z_{o1}}Z_2\cos^2\theta+\frac{4}{(Z_{e1}+Z_{o1})^2}Z_2^2\cos^2\theta+\frac{4}{(Z_{e1}+Z_{o1})^2}Z_2^2+\cos^2\theta-1=0 \tag{6.26}$$

and the transmission zeros created by the dual-mode ring resonator for the differential mode are expressed as

$$\theta_{tz1}=\arccos\sqrt{\frac{Z_{e1}+Z_{o1}-2Z_2}{Z_{e1}+Z_{o1}+2Z_2}},\quad \theta_{tz2}=\pi-\theta_{tz1} \tag{6.27}$$

Moreover, when $S_{dd11}=0$, a sixth-order equation for θ versus Z_1, Z_2, Z_{e1}, and Z_{o1} can be obtained, and six roots for $S_{dd11}=0$ can be found by properly choosing the relationships of Z_1, Z_2, Z_{e1}, and Z_{o1}, and six transmission poles in the passband can be achieved. The two transmission poles near the passband edge are realized by the dual-mode ring resonator, and the two poles near the center frequency (f_0) are produced by the open coupled lines [27].

The simulated frequency responses of Figure 6.24(b) are shown in Figure 6.25(a) (simulated with ANSYS Designer v3.0). Two transmission zeros (f_{tz1}, f_{tz2}) close to the passband can be produced by the dual-mode ring resonator [22]. In addition, for a high selectivity balanced filter, the 3-dB bandwidth (Δf, %), maximal $|S_{dd21}|$ (T_{stop}, dB) in the stopband and the maximal in-band $|S_{dd11}|$ (T_{pass}, dB) referring to the responses in Figure 6.25(a) are the mainly concerned filter characteristics. The corresponding levels of Δf, T_{stop}, and T_{pass} versus Z_1, Z_2, Z_{e1}, and Z_{o1} are shown and listed in Figure 6.25(b) and Table 6.1, and the 3-dB bandwidth of Δf decreases as Z_2 increases, and T_{stop} increases as Z_2 increases. The two transmission zeros (f_{tz1}, f_{tz2}) move toward f_0 with the decrease of the sum for Z_{e1}, Z_{o1}. Due to the limitation of PCB fabrication precisions, the width for the transmission lines and coupled lines is always greater than 0.20 mm, so the maximum coupling coefficient k ($k=(Z_{e1}-Z_{o1})/(Z_{e1}+Z_{o1})$) of the coupling lines in this work is nearly 0.33 ($\varepsilon_r=2.65$, $h=1.0$ mm), and the characteristic impedance of the transmission lines is always less than 130 Ω. For the 3-dB bandwidth of the differential mode, here we can choose the return loss in the passband is greater than 10 dB, the maximum 3-dB bandwidth of

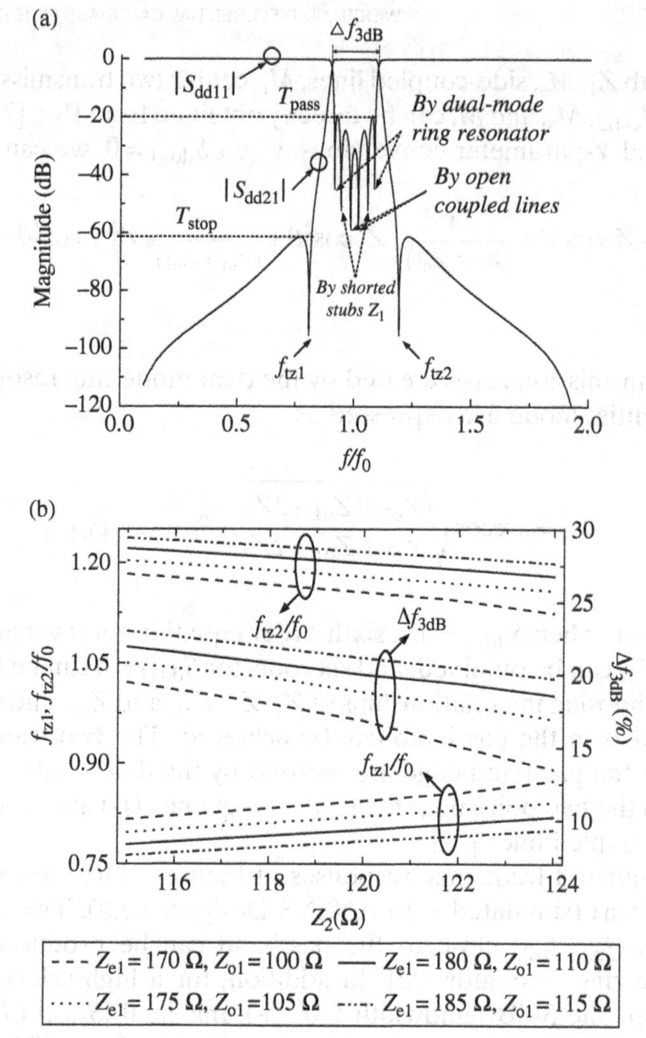

Figure 6.25 Simulated frequency responses of Figure 6.24(b). (a) $|S_{dd21}|$ and $|S_{dd11}|$, $Z_0 = 50\ \Omega$, $Z_1 = 118\ \Omega$, $Z_2 = 120\ \Omega$, $Z_{e1} = 180\ \Omega$, $Z_{o1} = 110\ \Omega$. (b) f_{tz1}, f_{tz2}/f_0, Δf versus Z_{e1}, Z_{o1}, Z_2, $Z_0 = 50\ \Omega$, $Z_1 = 118\ \Omega$. Reprinted with permission from Ref. [25]; copyright 2015 IEEE.

Table 6.1 **Δf, T_{pass}, and T_{stop} of Figure 6.24(b) versus Z_1, Z_2**

$Z_{e1} = 180\ \Omega$ and $Z_{o1} = 110\ \Omega$		Z_1: 110 Ω			120 Ω			130 Ω		
		Δf (%)	T_{pass}	T_{stop}	Δf (%)	T_{pass}	T_{stop}	Δf (%)	T_{pass}	T_{stop}
Z_2	115 Ω	22	−9.5	−68.5	20.4	−11.6	−63.2	21.9	−10.8	−66.9
	120 Ω	21	−17.4	−64.2	19.2	−20.4	−60.6	20.1	−11.3	−63.2
	125 Ω	18	−11.2	−60.1	16.4	−13.9	−57.7	18.2	−14.5	−58.6

Reprinted with permission from Ref. [25]; copyright 2015 IEEE.

the differential-mode passband is nearly 26.5%, or the minimum 3-dB bandwidth of the differential mode is nearly 16%.

For the common-mode excitation from ports 1 and 1′, a virtual open appears along the symmetric line in Figure 6.24(c), and the balanced filter for the common mode is an unsymmetrical circuit with a loaded open stub attached in the end of the left coupled lines. Using the same analysis method for the differential-mode circuit, the matrices of the common-mode circuit can be also defined as $M_{cs1} \times M_s \times M_t \times M_s \times M_{cs2}$, and for the **ABCD** matrix M_{cs1} with one loaded open stub, $Z = -jZ_1 \cot\theta$. After **ABCD**- and Y-parameter conversions, when $S_{cc21} = 0$, three transmission zeros can be also obtained as f_{tz1}, f_0, f_{tz2}; the transmission zero located at f_0 (center frequency f_0 for the balanced filter) is introduced by the open loaded stub (Z_1); and the expressions of the transmission zeros f_{tz1}, f_{tz2} realized by the dual-mode ring resonator are similar as (6.12). The simulated frequency responses of the common-mode circuit for Figure 6.24(c) are shown in Figure 6.26(a) and (b) (simulated with ANSYS Designer v3.0). Obviously, wideband common-mode suppression can be easily achieved due to the introduced transmission zeros nf_{tz1}, nf_0, nf_{tz2}, $(n \geq 1)$, which do not change with the characteristic impedance Z_1, and the common-mode suppression becomes better with the increase of the sum for Z_{e1}, Z_{o1}. It should be noted that the characteristic impedance Z_1 of the loaded stubs is an independent parameter for adjusting the passband-order and out-of-band harmonic suppression of the differential mode, and also the wideband common-mode suppression nearly do not change. Moreover, the bandwidth of the differential mode increases and common-mode suppression becomes better with the increase of the sum for Z_{e1}, Z_{o1}; this transmission characteristic can be viewed as an advantage for this kind of balanced filter with two transmission zeros, unlike the former balanced filters [23, 24].

Referring to the aforementioned discussions and the simulated results, the 3-dB bandwidth of the balanced filters is chosen as 23.5%, and the final parameters for the filters of Figure 6.24 are listed as follows: $Z_0 = 50\,\Omega$, $Z_1 = 118\,\Omega$, $Z_2 = 121\,\Omega$, $Z_{e1} = 181\,\Omega$, $Z_{o1} = 109\,\Omega$. The structure parameters for the balanced filter (80 mm × 70 mm, $\varepsilon_r = 2.65$, $h =$ 1.0 mm, and $\tan\delta = 0.002$.) shown in Figure 6.27(a) and (b) are $l_1 = 17.2$ mm, $l_2 = 17.0$ mm, $l_3 = 17.1$ mm, $m_1 = 17.2$ mm, $m_2 = 5.5$ mm, $m_3 = 7.1$ mm, $m_4 = 10$ mm, $m_5 = 12.4$ mm, $w_0 = 2.7$ mm, $w_1 = 0.2$ mm, $w_2 = 0.42$ mm, $w_3 = 0.47$ mm, $g_1 = 0.55$ mm, $d_1 = 0.7$ mm. The measured results of the balanced filter are illustrated in Figure 6.28. Good agreements can be observed between the simulation and the experiments. As shown in Figure 6.28(a), for the differential mode, the 3-dB bandwidth is 22.9% (2.74–3.45 GHz) with return loss greater than 10.5 dB (2.78–3.37 GHz);

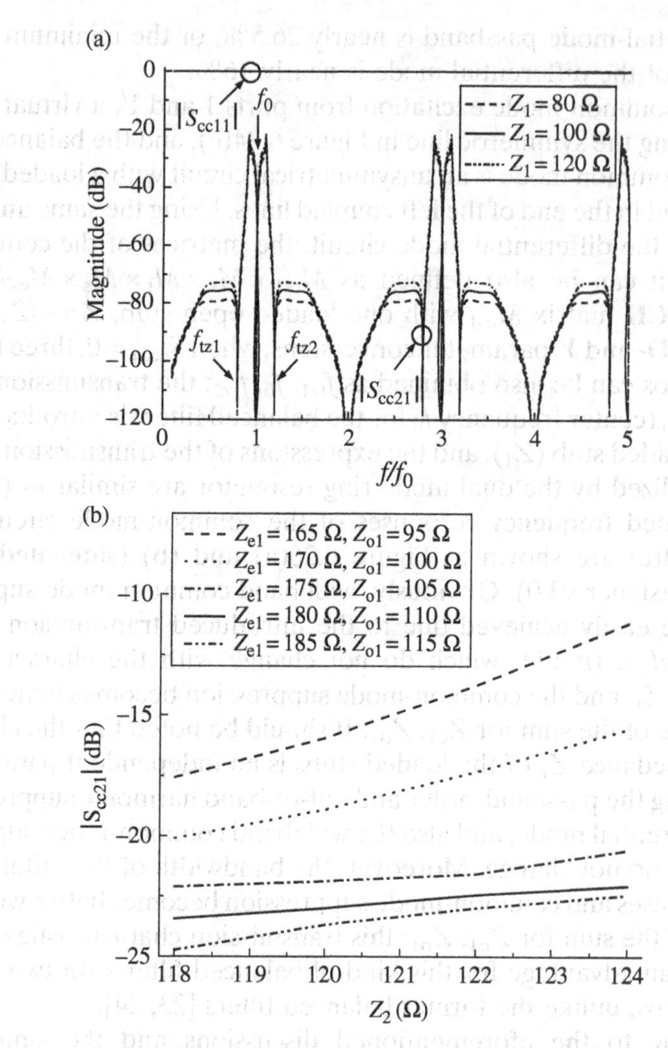

Figure 6.26 Simulated frequency responses of Figure 6.24(c). (a) Versus Z_1, $Z_0 = 50\ \Omega$, $Z_2 = 120\ \Omega$, $Z_{e1} = 180\ \Omega$, $Z_{o1} = 110\ \Omega$. (b) $|S_{cc21}|$ versus Z_{e1}, Z_{o1}, $Z_0 = 50\ \Omega$, $Z_1 = 120\ \Omega$, $Z_2 = 120\ \Omega$. Reprinted with permission from Ref. [25]; copyright 2015 IEEE.

three measured transmission zeros are located at 2.48, 3.68, and 4.3 GHz; over 20-dB upper stopband is realized from 3.56 to 9.06 GHz ($2.92f_0$); for the common mode of Figure 6.28(b), over 20-dB common-mode suppression is achieved from 0 to 15.9 GHz ($5.13f_0$). Next, to further improve the selectivity of the balanced filter with two transmission zeros, another high selectivity balanced filter structure using open/shorted coupled lines with four transmission zeros close to the differential-mode passband and wideband common-mode suppression will be presented.

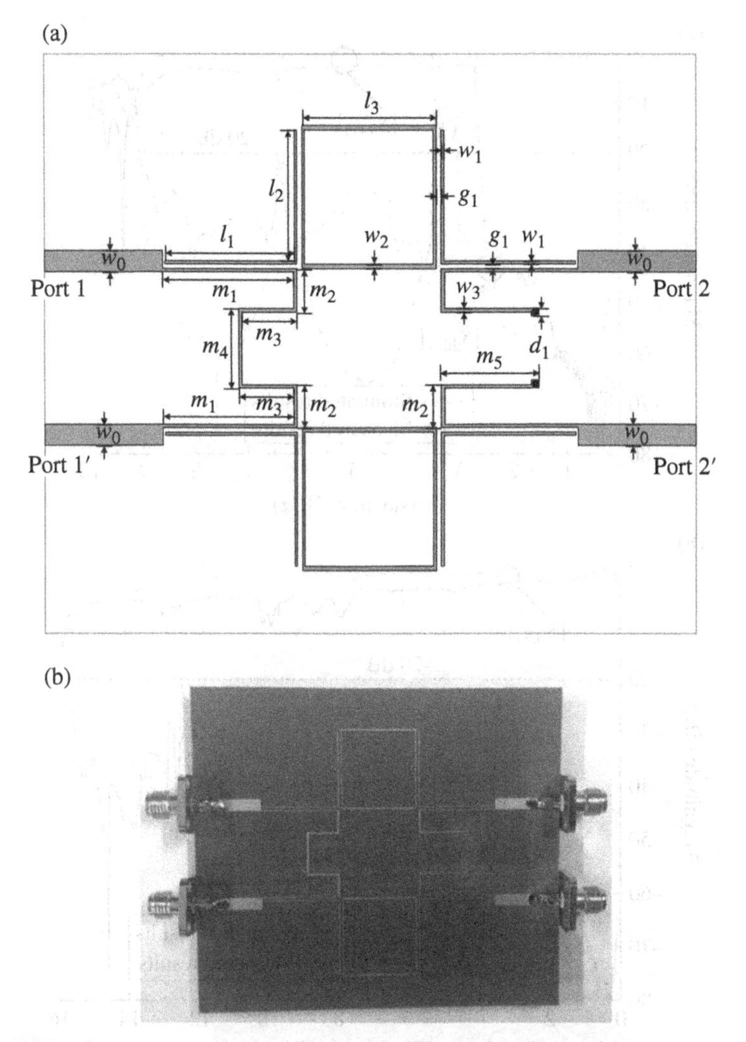

Figure 6.27 Geometry and photograph of differential filter with two transmission zeros. (a) Geometry. (b) Photograph. Reprinted with permission from Ref. [25]; copyright 2015 IEEE.

6.3.3 Differential Filter Using Open/Shorted Coupled Lines

Figure 6.29(a) shows ideal transmission circuit of the balanced filter with four transmission zeros, and four open/shorted coupled lines (Z_{e2}, Z_{o2}, θ) are shunted connected in the input/output ports 1, 1′, 2, 2′, and the other part is the same as the balanced filter in Figure 6.24(a).

When the differential-mode signals and common-mode signals are excited from ports 1 and 1′ in Figure 6.29(a), a virtual short/open

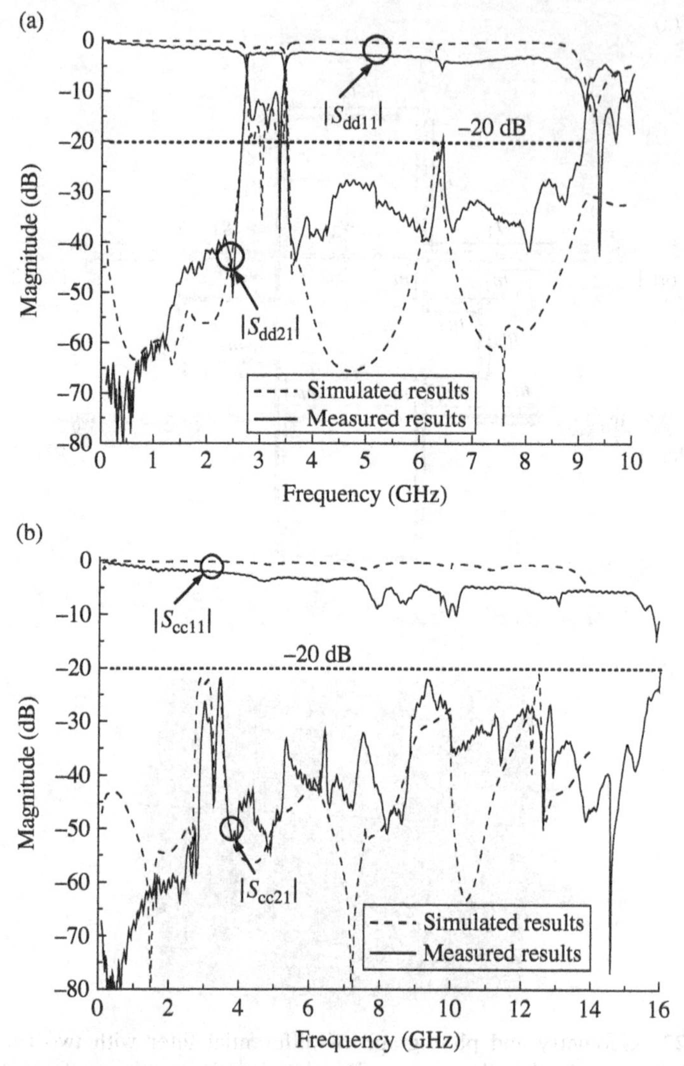

Figure 6.28 Measured and simulated results of the balanced filter with two transmission zeros. (a) Differential mode. (b) Common mode. Reprinted with permission from Ref. [25]; copyright 2015 IEEE.

appears along the symmetric line in Figure 6.29(a), as shown in Figure 6.29(b) and (c). The matrices of the differential-mode/common-mode circuit can be defined as $M_{os} \times M_{cs1} \times M_s \times M_t \times M_s \times M_{cs2} \times M_{os}$ (M_{os}, open/shorted coupled lines). After matrix conversion, when $S_{dd21}/S_{cc21} = 0$, from 0 to $2f_0$, besides the two transmission zeros (f_{tz1}, f_{tz2}) realized by the dual-mode resonator, another two transmission zeros can be also obtained as

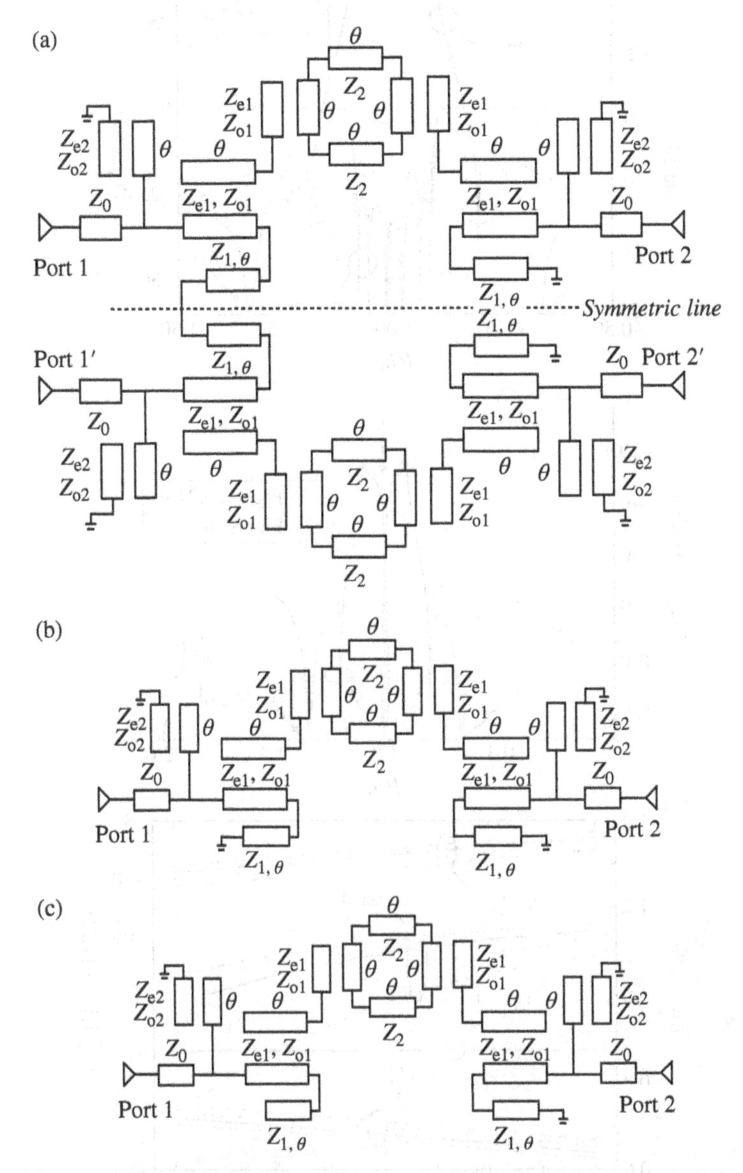

Figure 6.29 (a) The ideal circuit of the balanced filter with four transmission zeros. (b) Equivalent circuit for the differential mode. (c) Equivalent circuit for the common mode. Reprinted with permission from Ref. [25]; copyright 2015 IEEE.

$$\theta_{tz3} = \arccos\frac{Z_{e2} - Z_{o2}}{Z_{e2} + Z_{o2}}, \quad \theta_{tz4} = \pi - \theta_{tz3} \tag{6.28}$$

The two transmission zeros f_{tz3}, f_{tz4} are realized by the four open/shorted coupled lines. The simulated frequency responses of Figure 6.29(b) and (c) are shown in Figure 6.30 (simulated with ANSYS

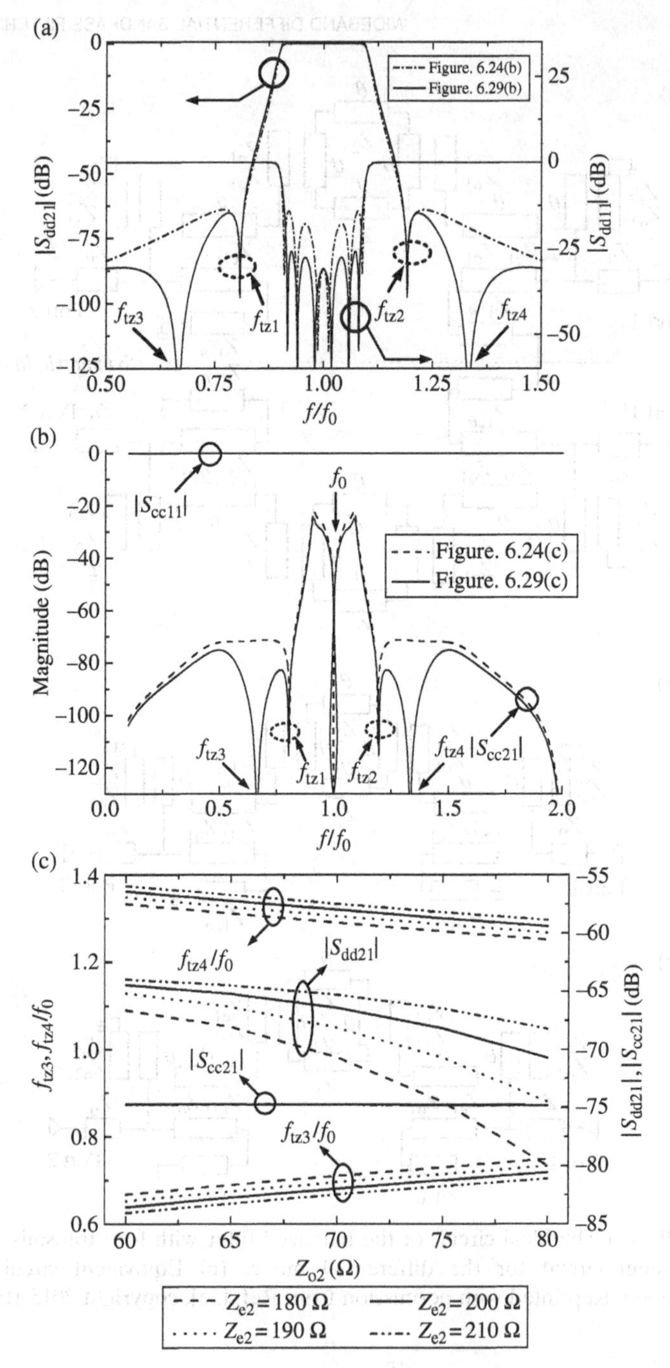

Figure 6.30 Simulated frequency responses of Figure 6.29. (a) $|S_{dd21}|$ and $|S_{dd11}|$, $Z_0 = 50\ \Omega$, $Z_1 = 90\ \Omega$, $Z_2 = 120\ \Omega$, $Z_{e1} = 180\ \Omega$, $Z_{o1} = 110\ \Omega$, $Z_{e2} = 200\ \Omega$, $Z_{o2} = 70\ \Omega$. (b) $|S_{cc21}|$ and $|S_{cc11}|$, $Z_0 = 50\ \Omega$, $Z_1 = 90\ \Omega$, $Z_2 = 120 = \Omega$, $Z_{e1} = 180\ \Omega$, $Z_{o1} = 110\ \Omega$, $Z_{e2} = 200\ \Omega$, $Z_{o2} = 70\ \Omega$. (c) f_{tz3}, f_{tz4}/f_0, $|S_{dd21}|$, $|S_{cc21}|$ versus Z_{e2}, Z_{o2}, $(0 - f_{tz1}, f_{tz2} - 2f_0)$, $Z_0 = 50\ \Omega$, $Z_1 = 90\ \Omega$, $Z_2 = 120\ \Omega$, $Z_{e1} = 180\ \Omega$, $Z_{o1} = 110\ \Omega$. Reprinted with permission from Ref. [25]; copyright 2015 IEEE.

Designer v3.0). Compared with the balanced filter of Figure 6.24(a), the two transmission zeros f_{tz3}, f_{tz4} can be used to further improve the differential-mode passband selectivity and the common-mode suppression level from 0 to f_{tz1}, f_{tz2} to $2f_o$. In addition, the locations of two transmission zeros f_{tz1}, f_{tz2} do not change with Z_1, Z_{e2}, and Z_{o2}, and the locations of two transmission zeros f_{tz3}, f_{tz4} do not change with Z_1, Z_2, Z_{e1}, and Z_{o1}. Therefore, besides the advantage for balanced filter with two transmission zeros, the two transmission zeros (f_{tz3}, f_{tz4}) can be seen as another two independently adjustable parameters to improve the transmission characteristic for the balanced filter with four transmission zeros.

The 3-dB bandwidth of the balanced filter with four transmission zeros is chosen as 22.5%, and the final parameters for the filter of Figure 6.29 are listed as follows: $Z_0 = 50\ \Omega$, $Z_1 = 90\ \Omega$, $Z_2 = 123\ \Omega$, $Z_{e1} = 181\ \Omega$, $Z_{o1} = 109\ \Omega$, $Z_{e2} = 210\ \Omega$, $Z_{o2} = 70\ \Omega$. The structure parameters for the balanced filter (80 mm × 70 mm, on the 1-mm-thick substrate with $\varepsilon_r = 2.65$, and tan $\delta = 0.002$.) shown in Figure 6.31(a) and (b) are $l_1 = 17.2$ mm, $l_2 = 17$ mm, $l_3 = 17.1$ mm, $l_4 = 17$ mm, $m_1 = 17.2$ mm, $m_2 = 5.5$ mm, $m_3 = 7.1$ mm, $m_4 = 10$ mm, $m_5 = 12.4$ mm, $w_0 = 2.7$ mm, $w_1 = 0.2$ mm, $w_2 = 0.42$ mm, $w_3 = 0.47$ mm, $w_4 = 0.26$ mm, $g_1 = 0.57$ mm, $g_2 = 0.25$ mm, $d_1 = 0.7$ mm, $d_2 = 0.7$ mm.

The measured results of the balanced filter with four transmission zeros are shown in Figure 6.32. Good agreements can be observed between the simulation and the experiments. For the differential mode of the balanced filter in Figure 6.32(a), five transmission zeros are located at 1.91, 2.50, 3.58, 3.71, and 4.37 GHz; the 3-dB bandwidth is 21.9% (2.76–3.44 GHz) with return loss greater than 12.5 dB; and the insertion loss is greater than 20 dB from 3.54 to 9.11 GHz ($2.94f_0$); for the common mode of Figure 6.32(b), the insertion loss is greater than 20 dB from 0 to 16.0 GHz ($5.16f_0$). The second harmonic around 6.35 GHz is mainly due to the desynchronization between the quarter-wavelength lines in the resonator [22]. In addition, the slight frequency discrepancies of the transmission zeros in the upper stopband and larger insertion loss for the passband between the measured and simulated results are mainly caused by the imperfect soldering skill of the shorted stubs and folded transmission lines of the balanced filters, and the even-/odd-mode phase velocities of the microstrip coupled lines also affect the positioning of the transmission zeros of the filters. Some slot etched in the shorted/coupled lines can be considered to extend the electrical path of the odd mode, and the effective phase velocity of odd mode will be reduced [28].

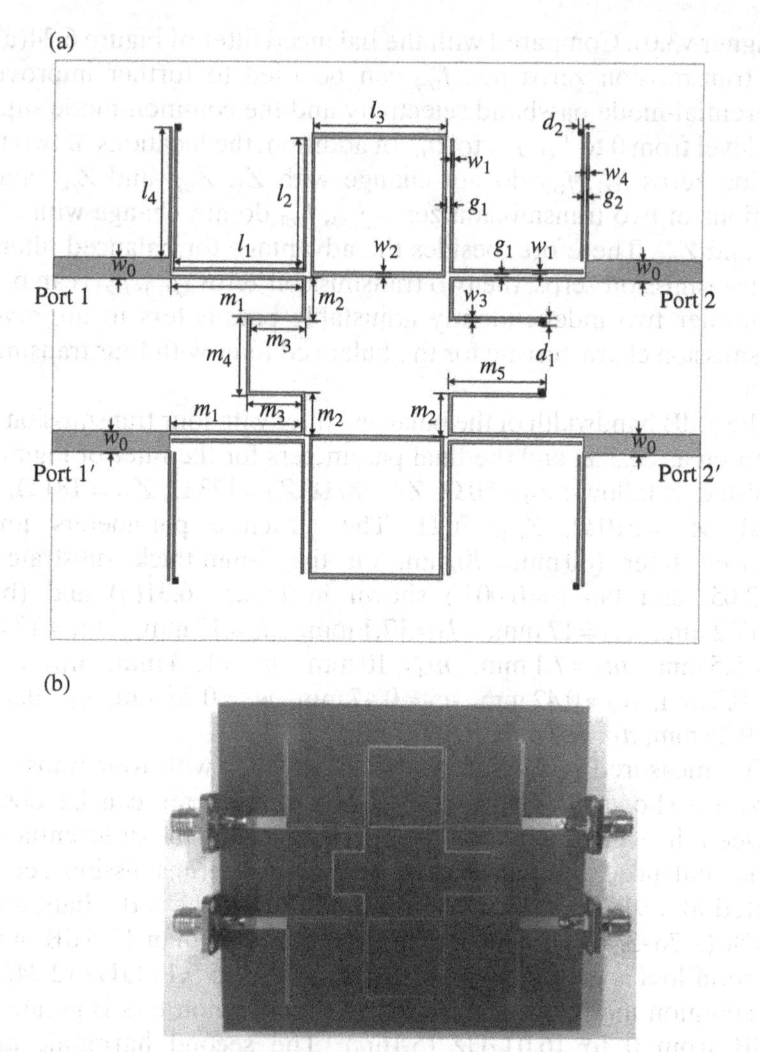

Figure 6.31 Geometry and photograph of the proposed balanced filter with four transmission zeros. (a) Geometry. (b) Photograph. Reprinted with permission from Ref. [25]; copyright 2015 IEEE.

6.3.4 Comparisons of Several Wideband Balanced Filters Based on Different Techniques

In this chapter, recently proposed wideband balanced filters based on T-shaped structures and half-/full-wavelength ring resonators are discussed. Detailed comparisons of effective circuit size, bandwidth and upper stopband for differential mode, out-of-band transmission zeros, and common-mode suppression for wideband balanced filters

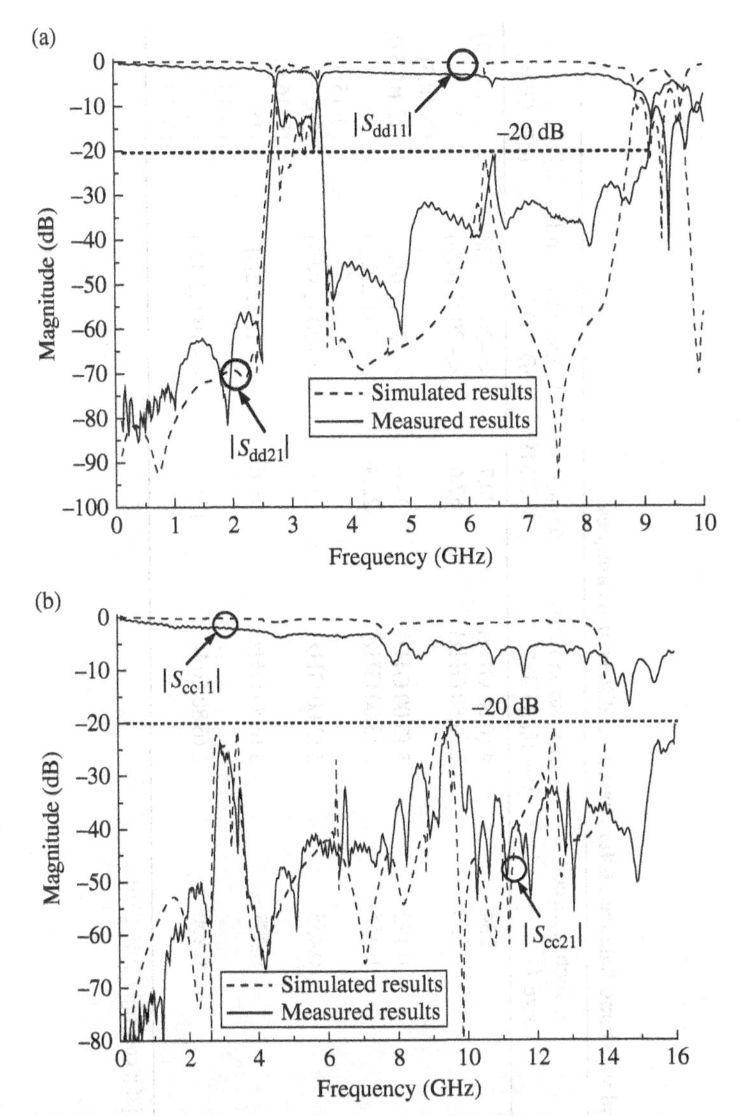

Figure 6.32 Measured and simulated results of the balanced filter with four transmission zeros. (a) Differential mode. (b) Common mode. Reprinted with permission from Ref. [25]; copyright 2015 IEEE.

introduced in this chapter are presented in Table 6.2. The filter structures using microstrip-slot-microstrip structure and DSPSL 180° phase shifting structure can meet the UWD bandwidth/band demand, while the selectivity for the differential-mode and common-mode suppression could be further improved. The T-shaped structure, half/full-wavelength ring resonator, and cascaded T-shaped structures can increase the

Table 6.2 Comparisons of several wideband balanced filters based on different techniques

Balanced filter structures	Effective circuit size (λ_0^2)	Transmission zeros ($0-2f_0$, GHz)	Δf_{3dB} (%, DM)	Upper stopband (dB, DM)	$\|S_{cc21}\|$, (dB, GHz)
Ref. [12], T-shaped structure	0.234	4 (6.85 GHz)	70.7	<-20 ($2.77f_0$)	<-14 (0–19)
Ref. [17], cascaded T-shaped structure	0.250	2 (6.80 GHz)	67.6	<-20 ($2.58f_0$)	<-15 (0–18.5)
Ref. [24], half-wavelength ring	0.192	5 (3.00 GHz)	79.0	<-25 ($2.6f_0$)	<-13 (0–8.0)
Ref. [25]-I, full-wavelength ring	0.455	3 (3.00 GHz)	22.9	<-20 ($2.92f_0$)	<-20 (0–15.9)
Ref. [25]-II, full-wavelength ring	0.455	5 (3.00 GHz)	21.9	<-20 ($2.94f_0$)	<-20 (0–16.0)
Ref. [29], DSPSL phase inverter	0.397	3 (6.80 GHz)	75.0	<-15 ($2.2f_0$)	<-15 (0–13.5)
Ref. [30], Marchand balun	0.250	2 (6.80 GHz)	111	<-100 ($2.1f_0$)	<-13 (0–10.0)

passband order and transmission zeros of the differential mode and extend the upper stopband of the differential mode, while the common mode can be extended to over $5f_0$.

6.4 WIDEBAND DIFFERENTIAL NETWORKS USING MARCHAND BALUN

Besides wideband differential bandpass filters, it is also valuable to develop a balanced-to-balanced or differential-mode power divider/combiner for RF/microwave front-end systems [1]. In former works, the bandwidths of the balanced power dividing/combining networks [31, 32] are only 28 and 13% ($|S_{dd21,31}| < 4$ dB), respectively, and the bandwidth of the common-mode suppression should be further extended also. The Marchand balun can realize a wideband out-of-phase power division (over 50% bandwidth) and can thus be used to implement wideband balanced networks with common-mode suppression. In this section, two new wideband in-phase/out-of-phase balanced power dividing/combining networks with wideband common-mode suppression using Marchand balun are proposed [33], as shown in Figure 6.33(a) and (b). The main advantages of the two balanced networks are the wideband for the differential-mode power division (bandwidth over 50%, $|S_{dd21,31}| < 4$ dB) and wideband common-mode suppression (bandwidth over 100%, $|S_{cc11}| > 15$ dB).

6.4.1 *S*-Parameter for Six-Port Differential Network

For a six-port balanced power dividing/combining network, the mixed-mode scattering matrix ($\mathbf{S}_{mm}$) can be extracted from the standard matrix ($\mathbf{S}_{std}$) using the matrix transformation [1, 31], and the mixed-mode scattering matrix ($\mathbf{S}_{mm}$) can be defined as

$$\mathbf{S}_{mm} = \begin{bmatrix} S_{dd} & S_{dc} \\ S_{cd} & S_{cc} \end{bmatrix} = \begin{bmatrix} S_{dd11} & S_{dd12} & S_{dd13} & S_{dc11} & S_{dc12} & S_{dc13} \\ S_{dd21} & S_{dd22} & S_{dd23} & S_{dc21} & S_{dc22} & S_{dc23} \\ S_{dd31} & S_{dd32} & S_{dd33} & S_{dc31} & S_{dc32} & S_{dc33} \\ S_{cd11} & S_{cd12} & S_{cd13} & S_{cc11} & S_{cc12} & S_{cc13} \\ S_{cd21} & S_{cd22} & S_{cd23} & S_{cc21} & S_{cc22} & S_{cc23} \\ S_{cd31} & S_{cd32} & S_{cd33} & S_{cc31} & S_{cc32} & S_{cc33} \end{bmatrix} \quad (6.29)$$

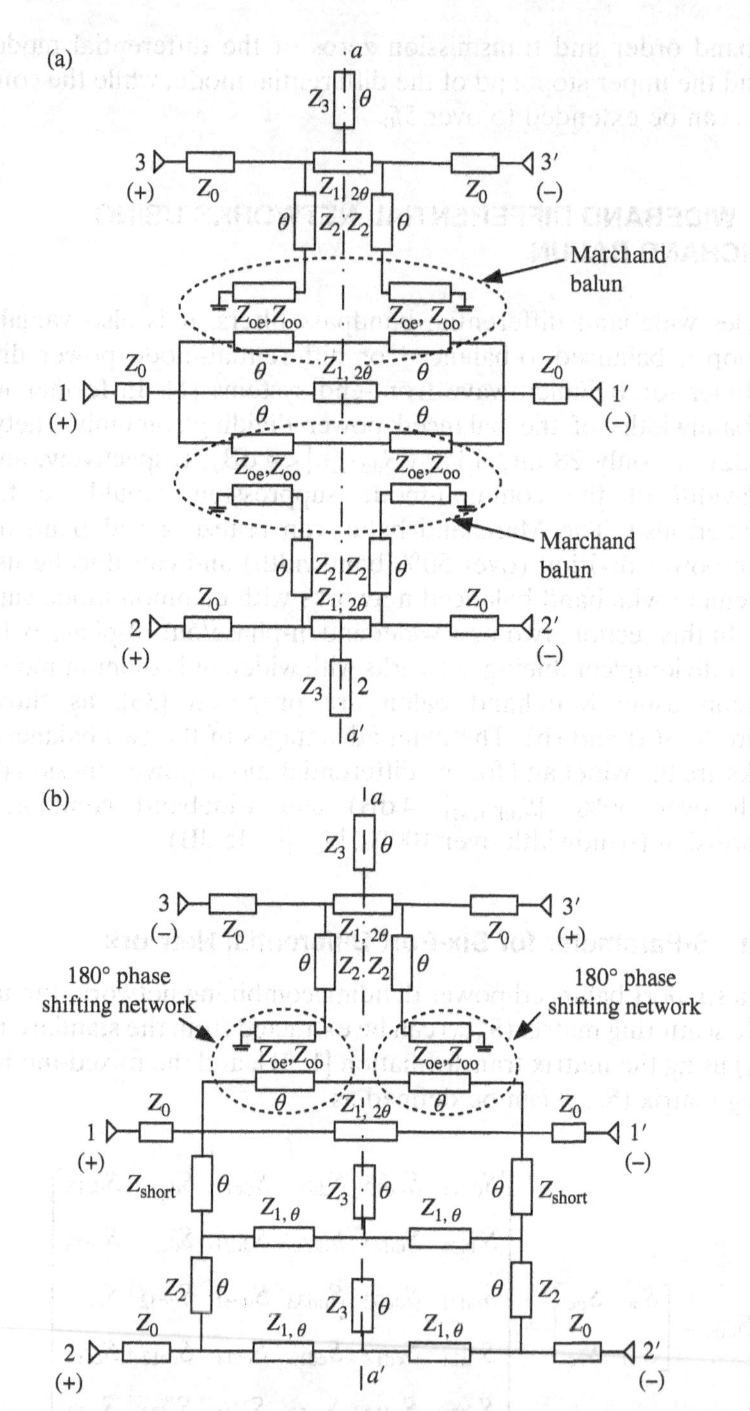

Figure 6.33 Circuits of two wideband balanced power dividing/combining networks. (a) In-phase. (b) Out-of-phase. Reprinted with permission from Ref. [33]; copyright 2014 IEEE.

and

$$S_{dd}=\begin{bmatrix} S_{dd11} & S_{dd12} & S_{dd13} \\ S_{dd21} & S_{dd22} & S_{dd23} \\ S_{dd31} & S_{dd32} & S_{dd33} \end{bmatrix},\quad S_{dc}=\begin{bmatrix} S_{dc11} & S_{dc12} & S_{dc13} \\ S_{dc21} & S_{dc22} & S_{dc23} \\ S_{dc31} & S_{dc32} & S_{dc33} \end{bmatrix} \tag{6.30}$$

$$S_{cd}=\begin{bmatrix} S_{cd11} & S_{cd12} & S_{cd13} \\ S_{cd21} & S_{cd22} & S_{cd23} \\ S_{cd31} & S_{cd32} & S_{cd33} \end{bmatrix},\quad S_{cc}=\begin{bmatrix} S_{cc11} & S_{cc12} & S_{cc13} \\ S_{cc21} & S_{cc22} & S_{cc23} \\ S_{cc31} & S_{cc32} & S_{cc33} \end{bmatrix} \tag{6.31}$$

$$S_{dd}=(S_{dd})^{T}, S_{dc}=(S_{dc})^{T}, S_{cd}=(S_{cd})^{T}, S_{cc}=(S_{cc})^{T} \tag{6.32}$$

where the third-order submatrices S_{dd} and S_{cc} represent the differential-mode and common-mode scattering matrices, respectively, and the third-order submatrices S_{dc} and S_{cd} indicate the conversion between the differential mode and common mode [31]. The mixed-mode S-parameters ($\mathbf{S}_{mm}$) of the wideband in-phase balanced power dividing/combining network can be obtained as

$$\left.\begin{cases} S_{dd11}=0.5\times(S_{11}-S_{1'1}-S_{11'}+S_{1'1'}) \\ S_{dd21,31}=0.5\times(S_{21}-S_{2'1}-S_{3'1}+S_{31}) \\ S_{dd22,33}=0.5\times(S_{22}-S_{2'2}-S_{22'}+S_{2'2'}) \\ S_{dd23,32}=0.5\times(-S_{23'}+S_{23}+S_{32}-S_{2'3}) \end{cases}\right\} S_{dd},$$

$$\left.\begin{cases} S_{dc11}=0 \\ S_{dc21,31}=0.5\times(S_{21}-S_{2'1}+S_{3'1}-S_{31}) \\ S_{dc22,33}=0.25\times(S_{22}-S_{2'2'}+S_{3'3'}-S_{33}) \\ S_{dc23,32}=0.25\times(S_{23'}+S_{2'3}+S_{32'}+S_{3'2}) \end{cases}\right\} S_{dc} \tag{6.33}$$

$$\left.\begin{cases} S_{cd11}=0 \\ S_{cd21,31}=0.5\times(S_{21}+S_{2'1}-S_{3'1}-S_{31}) \\ S_{cd22,33}=0.25\times(-S_{22}-S_{2'2'}+S_{3'3'}+S_{33}) \\ S_{cd23,32}=0.25\times(S_{23'}-S_{2'3}+S_{32'}-S_{3'2}) \end{cases}\right\} S_{cd},$$

$$\left.\begin{cases} S_{cc11}=0.5\times(S_{11}+S_{1'1}+S_{11'}+S_{1'1'}) \\ S_{cc21,31}=0.5\times(S_{21}+S_{2'1}+S_{3'1}+S_{31}) \\ S_{cc22,33}=0.5\times(S_{22}+S_{2'2}+S_{22'}+S_{2'2'}) \\ S_{cc23,32}=0.5\times(S_{23'}+S_{23}+S_{32}+S_{2'3}) \end{cases}\right\} S_{cc} \tag{6.34}$$

and for the wideband out-of-phase balanced power dividing/combining network, $S_{dd21} = -S_{dd31}$, $S_{dd22} = -S_{dd33}$, $S_{dc21} = -S_{dc31}$, $S_{dc22} = -S_{dc33}$, $S_{cd21} = -S_{cd31}$, $S_{cd22} = -S_{cd33}$, $S_{cc21} = -S_{cc31}$, $S_{cc22} = -S_{cc33}$. As an ideal wideband in-phase or out-of-phase balanced power dividing/combining network, the mixed-mode S-parameters ($\mathbf{S}_{mm}$) can be illustrated as

$$\mathbf{S}_{\text{mm-in,out}} = \begin{bmatrix} 0 & j\frac{\sqrt{2}}{2} & \pm j\frac{\sqrt{2}}{2} & 0 & 0 & 0 \\ j\frac{\sqrt{2}}{2} & 0 & 0 & 0 & 0 & 0 \\ \pm j\frac{\sqrt{2}}{2} & 0 & 0 & 0 & 0 & 0 \\ 0 & 0 & 0 & -1 & 0 & 0 \\ 0 & 0 & 0 & 0 & -1 & 0 \\ 0 & 0 & 0 & 0 & 0 & \mp 1 \end{bmatrix} \tag{6.35}$$

To meet the mixed-mode S-parameters (S_{mm}) of the in-phase and out-of-phase power dividing/combining networks, the following equations can be obtained as [31]

$$S_{11} = S_{11'}, S_{21} = -S_{2'1}, S_{31} = -S_{3'1}, |S_{21}| = |S_{31}| = \frac{\sqrt{2}}{4} \tag{6.36}$$

$$S_{22} = S_{2'2} = S_{2'2'}, S_{22'} = S_{23} = S_{33'} = S_{2'3'} = 0 \tag{6.37}$$

In this way, for the wideband in-phase balanced power dividing/combining network, when the differential modes are excited from the port 1 (port 1′), ports 2, 3 (ports 2′, 3′) can be seen as an equal in-phase power divider; from the constraint conditions of (6.21) and (6.22), out-of-phase should be also realized for ports 2, 2′ and ports 3, 3′, respectively; when the common mode are excited from the port 1 (port 1′), stopband structures should be realized to ports 2, 3 (ports 2′, 3′) for wideband common-mode suppression. For the wideband out-of-phase balanced power dividing/combining network, when the differential mode transmit from port 1 (port 1′) to ports 2, 3 (ports 2′, 3′), a wideband balun power divider can be realized for ports 2, 3 (ports 2′, 3′), respectively, and the other conditions are the same as the in-phase balanced power dividing/combining network. Next, the analysis and design of the wideband in-phase and out-of-phase balanced power dividing/combining networks will be given in detail.

6.4.2 Wideband In-Phase Differential Network

The equivalent circuit of the wideband in-phase balanced power dividing/combining network is shown in Figure 6.33(a), two Marchand baluns are connected between ports 1, 1′, and three transmission lines with characteristic impedance of Z_1 and electrical length of 2θ are located in the center of ports 1, 1′ and ports 2, 2′ (ports 3, 3′), while two quarter-wavelength open stubs (characteristic impedance Z_3, electrical length θ, $\theta = \pi/2$ at f_0, f_0 is the center frequency of the two networks) are introduced to improve the common-mode suppression. The characteristic impedances of the microstrip lines at the input/output ports are $Z_0 = 50\ \Omega$.

In Figure 6.33(a), to realize good impedance matching for each port, the input impedance ($Z_{in} = Z_0$) and output impedance ($Z_{out} = Z_2$) of the Marchand balun should meet the following conditions [34, 35]:

$$k = \frac{Z_{oe} - Z_{oo}}{Z_{oe} + Z_{oo}} = \frac{1}{\sqrt{1 + 2Z_2/Z_0}} \tag{6.38}$$

For a given set of balun impedances at input and output ports, the coupled line parameters (Z_{oe} and Z_{oo}) are not unique, and the bandwidth for the Marchand balun increases as the coupling coefficient k increases [34]. In addition, from the Marchand balun to the output ports 2, 2′ and ports 3, 3′, the quarter-wavelength transmission line can be seen as an impedance transformer, and the input impedance of the shorted coupled lines is

$$Z_{short} = \frac{2Z_{oe}Z_{oo}}{Z_{oe} - Z_{oo}} \tag{6.39}$$

and

$$Z_2 = \sqrt{Z_o \times Z_{short}} \tag{6.40}$$

From (6.38) to (6.40), we can get the following relationships:

$$\left(\frac{Z_{oe} - Z_{oo}}{Z_{oe} + Z_{oo}}\right)^2 = \frac{\sqrt{Z_0(Z_{oe} - Z_{oo})}}{\sqrt{Z_0(Z_{oe} - Z_{oo})} + 2\sqrt{2Z_{oe}Z_{oo}}} \tag{6.41}$$

$$Z_{oe} = \frac{Z_2^2/\sqrt{Z_0(Z_o + 2Z2)}}{1 - \left(1/\sqrt{1 + 2Z2/Z0}\right)}, \quad Z_{oo} = \frac{Z_2^2/\sqrt{Z_0(Z_o + 2Z_2)}}{1 + \left(1/\sqrt{1 + 2Z_2/Z_0}\right)}. \tag{6.42}$$

Referring to (6.38)–(6.42), the desired relationship between Z_{oe} and Z_{oo} for different combinations of Z_2 is plotted in Figure 6.34(a). For a given set of characteristic impedance of Z_2, the coupled line parameters (Z_{oe} and Z_{oo}) are unique, and the coupling coefficient k decreases as Z_2 increases. In this way, by properly choosing the even-/odd-mode characteristic impedance of the shorted coupled line and the characteristic impedance of Z_2, good impedance matching can be realized for each port of Figure 6.34(a).

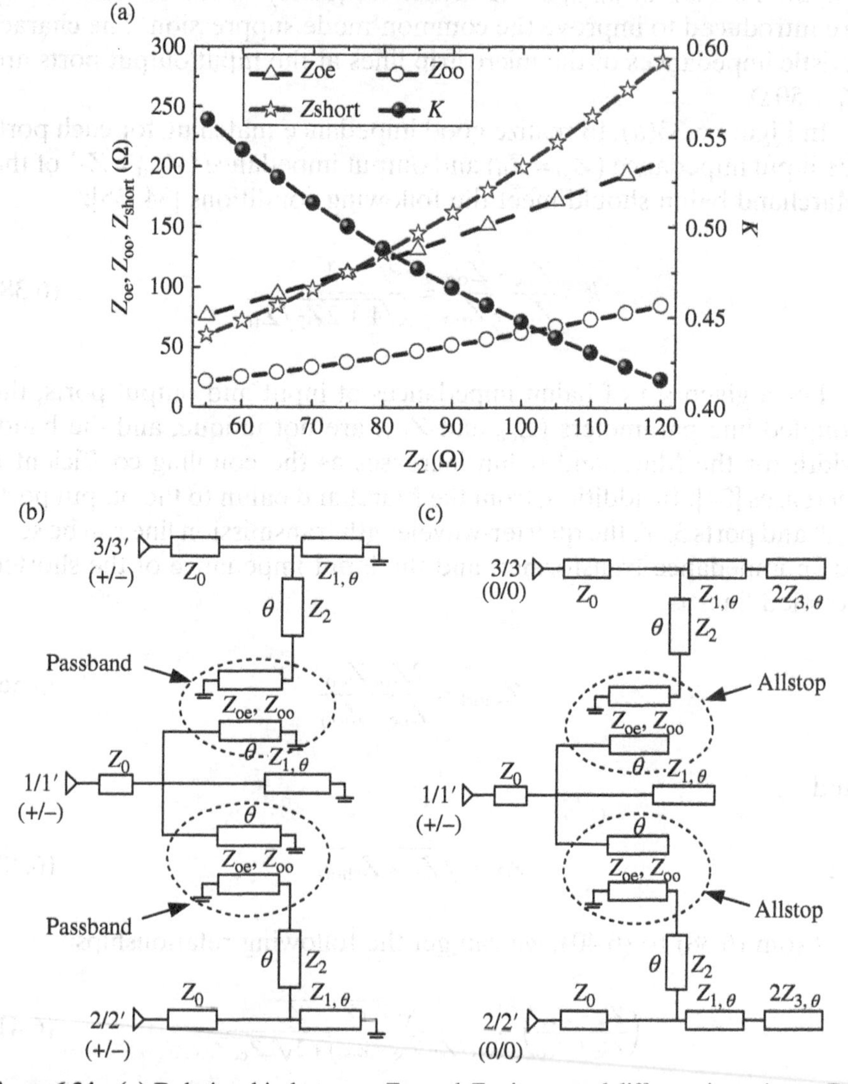

Figure 6.34 (a) Relationship between Z_{oe} and Z_{oo} in case of different impedance Z_2, $Z_0 = 50\,\Omega$. (b) Differential-mode circuit. (c) Common-mode circuit. Reprinted with permission from Ref. [33]; copyright 2014 IEEE.

Moreover, when the differential-mode and common-mode signals are excited from ports 1, 1′ in Figure 6.34(a), a virtual short/open appears along the symmetric line of *a*–*a*′, as shown in Figure 6.34(b) and (c). For the differential-mode circuit of Figure 6.34(b), the shorted coupled lines and the three shorted stubs are all wide passband structures [36]. Based on the relationships of (6.38)–(6.42), the simulated results of Figure 6.34(b) versus different Z_{oe}, Z_{oo}, Z_1, and Z_2 are shown in Figure 6.35(a) and (b). The bandwidth of the differential mode

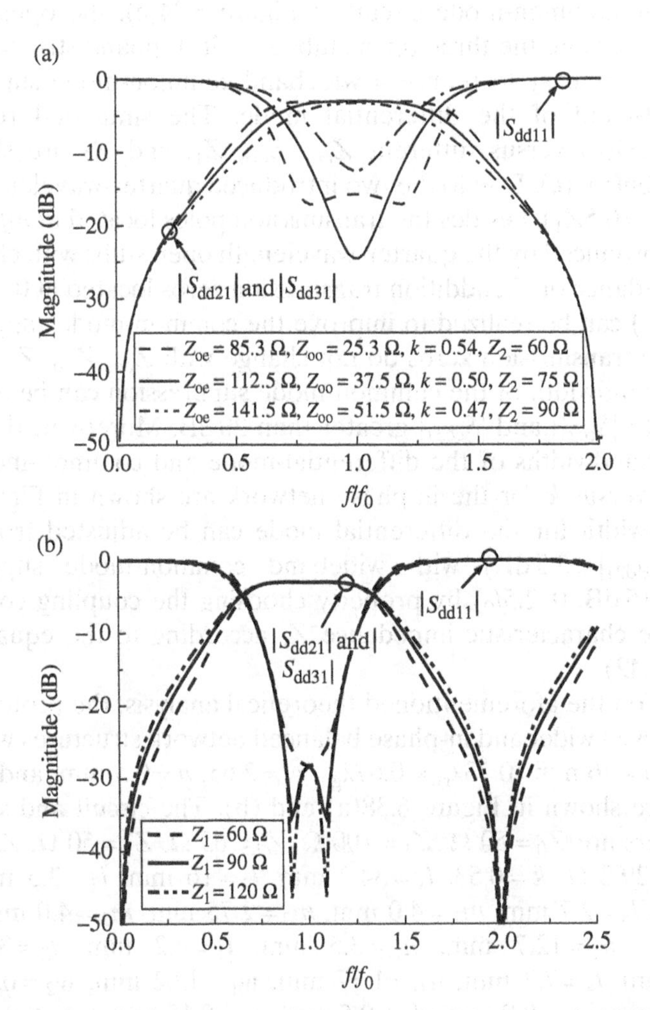

Figure 6.35 Simulated frequency responses of the differential mode for the in-phase balanced network. (a) Versus Z_{oe}, Z_{oo}, Z_2, $Z_1 = 100\ \Omega$, $Z_0 = 50\ \Omega$. (b) Versus Z_1, $Z_{oe} = 103.2\ \Omega$, $Z_{oo} = 33.2\ \Omega$, $Z_2 = 70\ \Omega$, $Z_0 = 50\ \Omega$. Reprinted with permission from Ref. [33]; copyright 2014 IEEE.

increases as the coupling coefficient k increases and decreases as Z_1 decreases. As a matter of fact, when the characteristic impedance Z_1 of the shorted stub becomes smaller, the capacitance of the shorted stub increases, while the inductance decreases, and the shorted stub can be seen as an ideal lossless parallel resonance circuit in the center frequency f_0. Because the Q factor of the passband for the shorted stub is proportional to the susceptance slope parameter $\sqrt{C_S/L_S}$, when the Q factor increases, the bandwidth for the passband will become narrower [12].

For the common-mode circuit of Figure 6.34(c), the open/shorted coupled lines and the three open stubs are all stopband structures [36], and it is very easy to realize a wideband common-mode suppression in-/out-of-band of the differential mode. The simulated results of Figure 6.34(c) versus different Z_{oe}, Z_{oo}, Z_1, and Z_2 are shown in Figure 6.36(a)–(c). Due to the two introduced quarter-wavelength open stubs ($Z_3 = 0.5Z_1$), besides the transmission poles located at mf_0 ($m = 0$, $1, 2, \ldots$) produced by the quarter-wavelength open stubs with characteristic impedance of Z_1, addition transmission zeros located at $0.5nf_0$ ($n =$ $1, 3, 5, \ldots$) can be realized to improve the common-mode suppression, and these transmission zeros do not change with Z_{oe}, Z_{oo}, Z_1, and Z_2, and the bandwidth of the common-mode suppression can be extended to $3f_0$ with $|S_{cc21}|$ and $|S_{cc31}|$ greater than 20 dB. Moreover, the power division bandwidths of the differential-mode and common-mode suppression versus k for the in-phase network are shown in Figure 6.37; the bandwidth for the differential mode can be adjusted from 20 to 70% ($|S_{dd21}| < 3.5$ dB) with wideband common-mode suppression ($|S_{cc21}| > 15$ dB, 0–$2.5f_0$) by properly choosing the coupling coefficient k and the characteristic impedance Z_1 according to the equations of (6.29)–(6.42).

Based on the aforementioned theoretical analysis, the prototypes of the proposed wideband in-phase balanced network structures with sizes of 50 mm × 46 mm ($0.73\lambda_{g0} \times 0.67\lambda_{g0}$, $\varepsilon_r = 2.65$, $h = 0.5$ mm, and $\tan\delta =$ 0.002.) are shown in Figure 6.38(a) and (b). The circuit and structure parameters are $Z_0 = 50$ Ω, $Z_1 = 100$ Ω, $Z_2 = 65$ Ω, $Z_3 = 50$ Ω, $Z_{oe} = 94.2$ Ω, $Z_{oo} = 29.2$ Ω, $k = 0.53$, $l_1 = 34.2$ mm, $l_2 = 16$ mm, $l_3 = 3.5$ mm, $l_4 =$ 10.2 mm, $l_5 = 2.7$ mm, $m_1 = 4.0$ mm, $m_2 = 2.75$ mm, $m_3 = 4.0$ mm, $m_4 =$ 2.95 mm, $n_1 = 12.7$ mm, $n_2 = 3.5$ mm, $t_1 = 4.2$ mm, $t_2 = 3.0$ mm, $t_3 = 3.9$ mm, $t_4 = 7.3$ mm, $w_0 = 1.37$ mm, $w_1 = 1.62$ mm, $w_2 = 0.68$ mm, $w_3 = 0.3$ mm, $w_4 = 0.9$ mm, $d_1 = 0.5$ mm, $g_1 = 0.15$ mm, $g_2 = 0.8$ mm, g_3 $= 0.3$ mm, $g_4 = 0.3$ mm, $s_1 = 1.91$ mm, $s_2 = 1.13$ mm. To realize the high impedance ratio of the shorted/open coupled line for single-layer PCB structure, patterned ground-plane technique proposed in Refs.

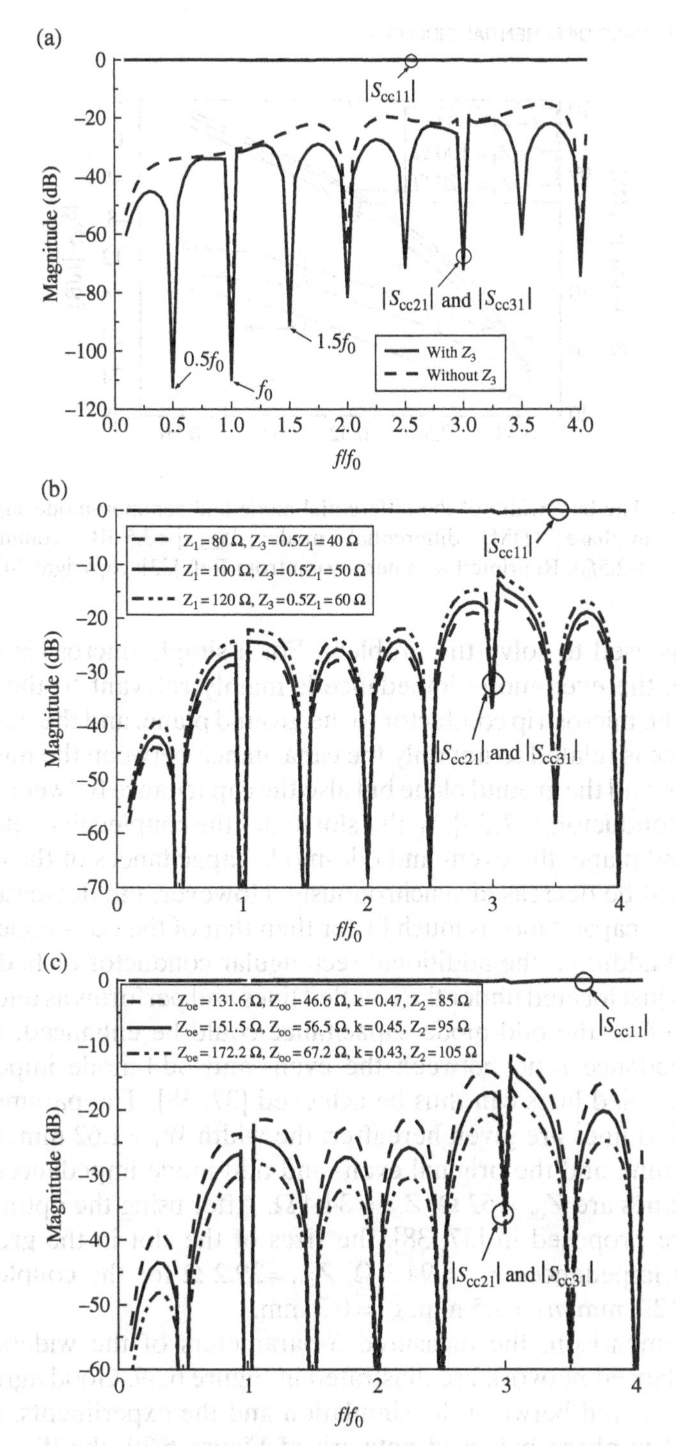

Figure 6.36 Simulated frequency responses of the common mode for the in-phase balanced network. (a) With/without Z_3. (b) Versus Z_1, $Z_1 = 2Z_3$, $Z_{oe} = 131.6\,\Omega$, $Z_{oo} = 46.6\,\Omega$, $Z_2 = 85\,\Omega$, $Z_0 = 50\,\Omega$. (c) Versus Z_{oe}, Z_{oo}, Z_2, $Z_1 = 100\,\Omega$, $Z_3 = 50\,\Omega$, $Z_0 = 50\,\Omega$. Reprinted with permission from Ref. [33]; copyright 2014 IEEE.

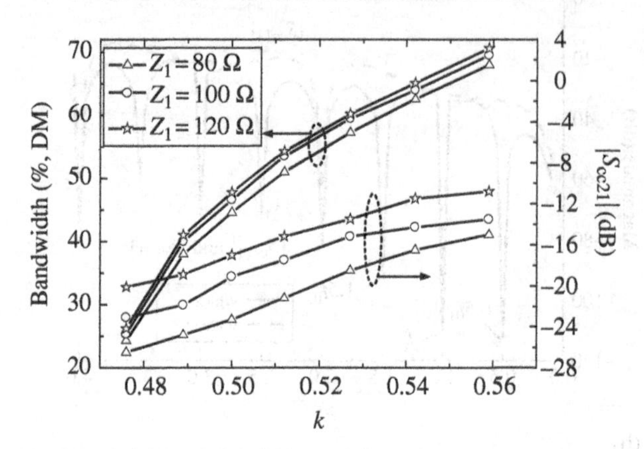

Figure 6.37 The bandwidth of the differential-mode and common-mode suppression versus k, in-phase (DM, differential mode: $|S_{dd21}| < 3.5$ dB; common-mode suppression: 0–2.5f_0). Reprinted with permission from Ref. [33]; copyright 2014 IEEE.

[37, 38] is used to solve the problem. For a simple microstrip coupler structure, the even-mode impedance is mainly relevant to the capacitance of the microstrip conductor to the ground plane, and the odd-mode impedance is related to not only the capacitance between the microstrip conductor and the ground plane but also the capacitance between the two coupled conductors [37, 38]. With a slot under the coupled lines etched on the ground plane, the even- and odd-mode capacitances of the coupled lines would be decreased synchronously. However, the decrease of the even-mode capacitance is much faster than that of the odd-mode capacitance. In addition, the additional rectangular conductor etched on the ground is just located under the coupled lines and performs as one capacitor; therefore the odd-mode capacitance could be enhanced, and the high impedance ratio between the even- and odd-mode impedances for the coupled lines can thus be achieved [37, 38]. The parameters of the coupled lines are given hereafter: the width $W_1 = 1.62$ mm, the gap $g_1 = 0.15$ mm, and the original even- and odd-mode impedances of the coupled lines are $Z_{oe} = 52\ \Omega$, $Z_{oo} = 34.4\ \Omega$. After using the optimization procedure proposed in [37, 38], the sizes of the slot in the ground to meet the impedances $Z_{oe} = 94.2\ \Omega$, $Z_{oo} = 29.2\ \Omega$ for the coupled lines are $n_1 = 12.7$ mm, $n_2 = 3.5$ mm, $g_3 = 0.3$ mm.

For comparison, the measured S-parameters of the wideband in-phase balanced network are illustrated in Figure 6.39. Good agreement can be observed between the simulation and the experiments. For the wideband in-phase balanced network of Figure 6.39, the $|S_{dd21}|$ and $|S_{dd31}|$ of the differential mode are less than 3.85 dB from 2.0 to 3.55 GHz (bandwidth 55.3%) with $|S_{dd11}|$ greater than 12.5 dB (2.15–3.6

(a)

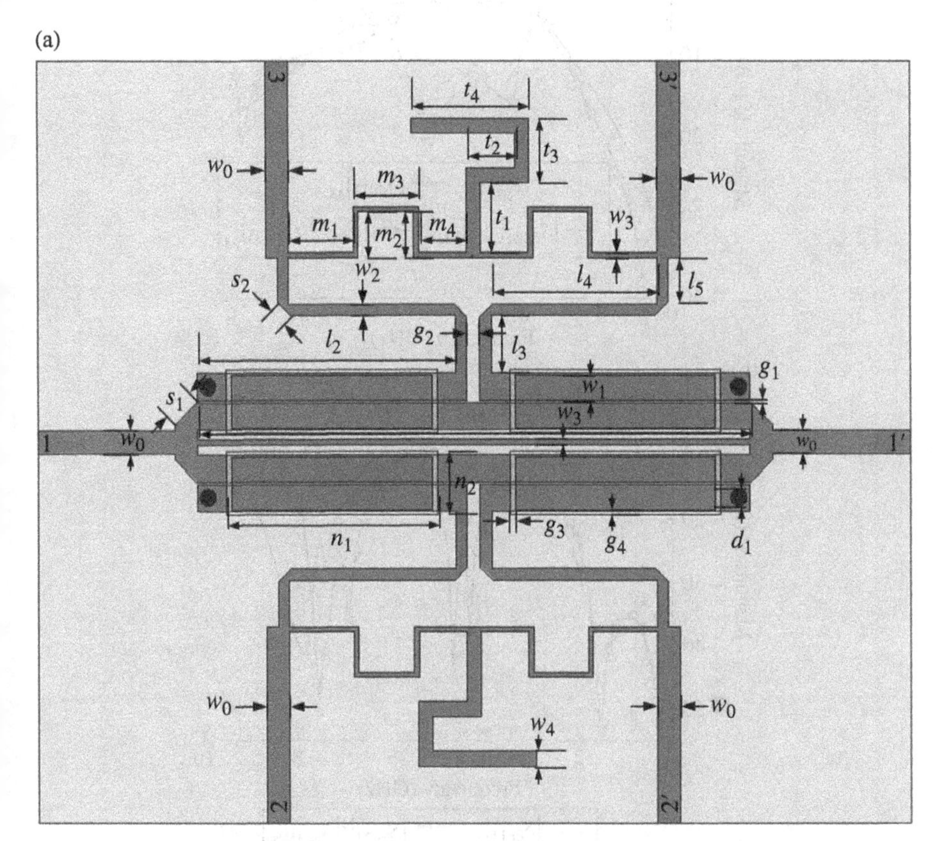

(b)

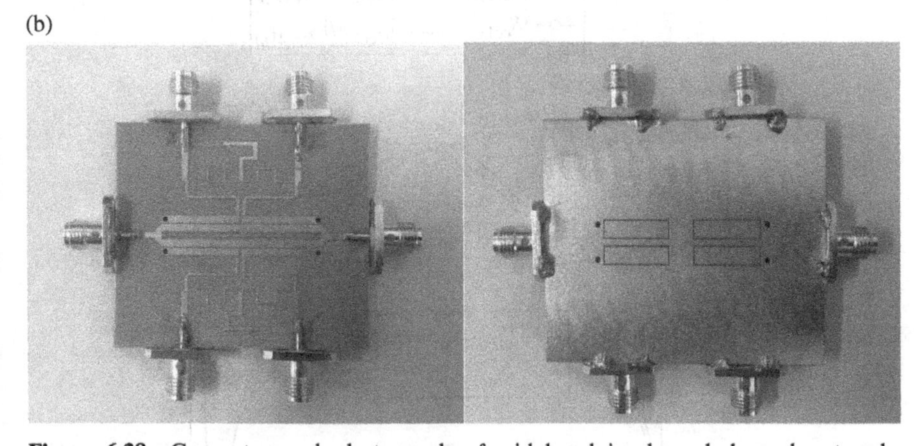

Figure 6.38 Geometry and photograph of wideband in-phase balanced network. (a) Geometry. (b) Photograph. Reprinted with permission from Ref. [33]; copyright 2014 IEEE.

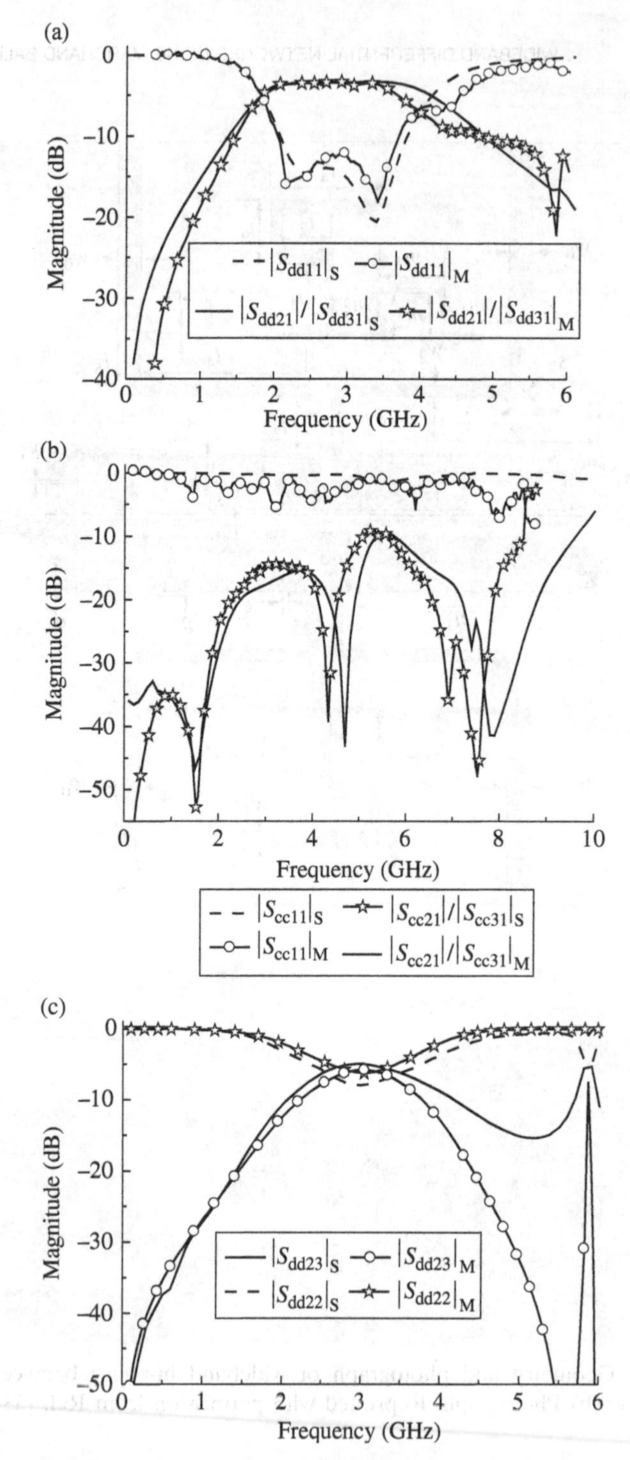

Figure 6.39 Measured and simulated results of the wideband in-phase balanced network (a) $|S_{dd11}|/|S_{dd21}|$. (b) $|S_{cc11}|/|S_{cc21}|$. (c) $|S_{dd22}|/|S_{dd23}|$. (d) $|S_{cc22}|/|S_{cc23}|$. (e) $|S_{cd}|/|S_{dc}|$ (M, measurement; S, simulation). Reprinted with permission from Ref. [33]; copyright 2014 IEEE.

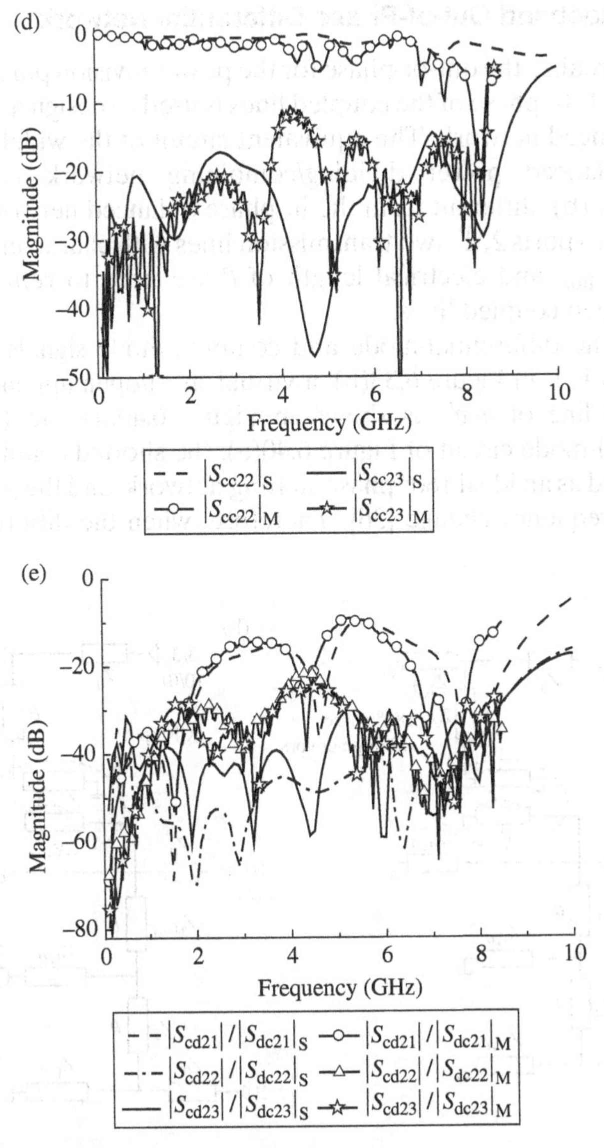

Figure 6.39 (Continued)

GHz), the $|S_{cc21}|$ and $|S_{cc31}|$ of the common mode are greater than 15 dB from 0 to 4.5 GHz ($1.5f_0$) and 10 dB to 8.6 GHz ($2.87f_0$), and the $|S_{cc23}|$ is over 10 dB from 0 to 8.5 GHz ($2.8f_0$); in addition, the $|S_{cd21}|/|S_{dc21}|$ and $|S_{cd31}|/|S_{dc31}|$ between the differential-mode and common-mode conversion are greater than 15 dB from 0 to 4.8 GHz ($1.6f_0$) and 10 dB from 0 to 9 GHz ($3f_0$); the $|S_{cd22}|/|S_{dc22}|$ and $|S_{cd23}|/|S_{dc23}|$ are over 25 dB from 0 to 8.3 GHz ($2.77f_0$).

6.4.3 Wideband Out-of-Phase Differential Network

To better realize the out-of-phase for the power division ports (ports 2/2′, 3/3′), here 180° phase of the coupled lines is used to design another wideband balanced network. The equivalent circuit of the wideband out-of-phase balanced power dividing/combining network is shown as Figure 6.33(b), different from the in-phase balanced network; from the ports 1, 1′ to ports 2, 2′, two transmission lines with characteristic impedance of Z_{short} and electrical length of θ are used to replace the two shorted/open coupled lines.

When the differential-mode and common-mode signals are excited from ports 1, 1′ in Figure 6.33(b), a virtual short/open appears along the symmetric line of *a*–*a*′, as shown in Figure 6.40(a) and (b). For the differential-mode circuit of Figure 6.40(a), the shorted coupled lines can be regarded as an ideal 180° phase shifting network, and the phase is fixed with the frequency change [36]. Therefore, when the differential-mode

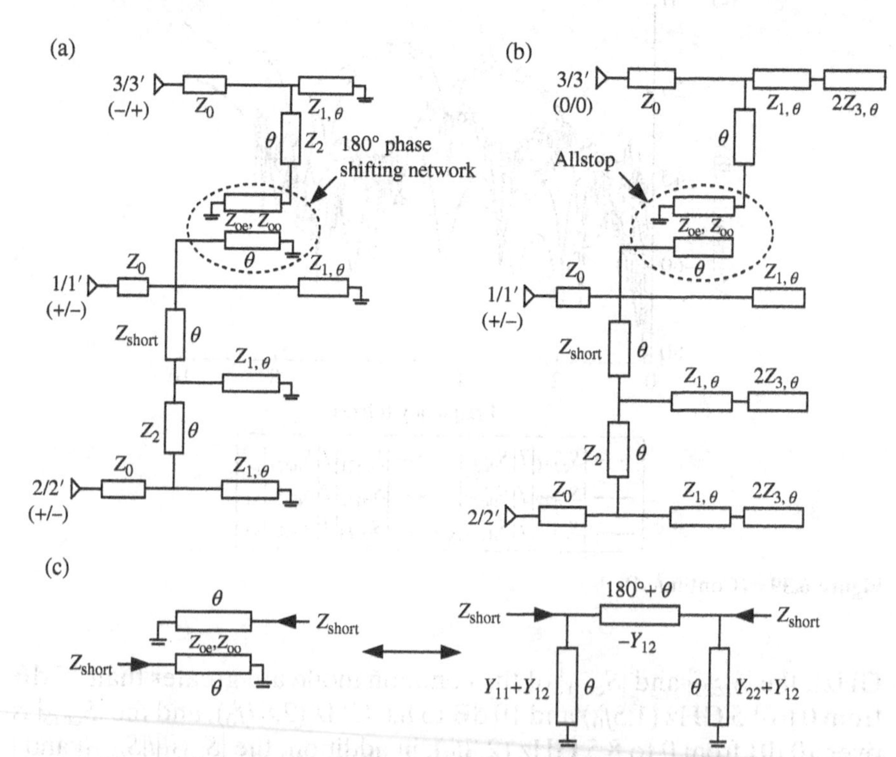

Figure 6.40 (a) Differential-mode circuit of the wideband out-of-phase balanced power dividing/combining network. (b) Common-mode circuit of the wideband out-of-phase balanced power dividing/combining network. (c) Equivalent circuit of the shorted couple lines. Reprinted with permission from Ref. [33]; copyright 2014 IEEE.

signals transmit from port 1 (1′) to ports 2, 3 (2′, 3′), an out-of-phase power division can be realized. In addition, to realize good impedance matching for each port, the characteristic impedance of transmission line instead of the shorted coupled line should be equal to the input impedance of the shorted coupled line as $Z_{short} = 2Z_{oe}Z_{oo}/(Z_{oe} - Z_{oo})$.

The simulated results of Figure 6.40(a) versus different Z_{oe}, Z_{oo}, Z_1, Z_{short}, and Z_2 and the phases of S_{dd21} and S_{dd31} are shown in Figure 6.41 (a)–(c), and similar as the in-phase balanced network, the bandwidth of differential mode for the out-of-phase network increases as the coupling coefficient k increases and decreases as Z_1 decreases. In addition, for the high input impedance (Z_{short}) of the shorted coupled lines, the difference between the even and odd modes cannot enlarge infinitely, and due to the precision limitations in practical fabrication, here the impedances for high input impedance (Z_{short}) should be chosen below 120 Ω.

For the common-mode circuit of Figure 6.40(b), the open/shorted coupled lines and the open stubs are also stopband structure. The simulated results of Figure 6.40(b) versus different Z_{oe}, Z_{oo}, Z_1, and Z_2 are shown in Figure 6.42(a) and (b). Similarly, besides the transmission poles located at mf_0 ($m = 1, 3, \ldots$) produced by the quarter-wavelength open stubs with characteristic impedance of Z_1, addition transmission zeros located at $0.5nf_0$ ($n = 1, 3, 5, \ldots$) can be realized to improve the common-mode suppression, and these transmission zeros do not change with Z_{oe}, Z_{oo}, Z_1, and Z_2. However, since there is no DC isolation from ports 1, 1′ to ports 2, 2′, the common-mode suppression cannot be extended to out of band of the differential mode, while in-band common-mode suppression with $|S_{cc21}|$ and $|S_{cc31}|$ greater than 20 dB can be realized from $0.5f_0$ to $1.5f_0$. The power division bandwidth of the differential-mode and common-mode suppression versus k for the out-of-phase network is also shown in Figure 6.43; the bandwidth for the differential mode can be also adjusted from 40 to 76% ($|S_{dd21}| < 3.5$ dB) with wideband common-mode suppression ($|S_{cc21}| > 17$ dB, $1.5f_0 - 2.5f_0$) by properly choosing the coupling coefficient k and the characteristic impedance Z_1.

Based on the aforementioned theoretical analysis, the proposed out-of-phase balanced network structures can be designed; the prototype and photograph of the network with size of 60 mm × 50 mm ($0.88\lambda_{g0} \times 0.73\lambda_{g0}$, $\varepsilon_r = 2.65$, $h = 0.5$ mm, and $\tan\delta = 0.002$.) are shown in Figure 6.44(a) and (b). The circuit and structure parameters are $Z_0 = 50$ Ω, $Z_1 = 115$ Ω, $Z_{short} = 84.5$ Ω, $Z_2 = 65$ Ω, $Z_3 = 67$ Ω, $Z_{oe} = 94.2$ Ω, $Z_{oo} = 29.2$ Ω, $k = 0.53$, $l_1 = 34.2$ mm, $l_2 = 15.7$ mm, $l_3 = 3.5$ mm, $l_4 = 10.2$ mm, $l_5 = 2.7$ mm, $m_1 = 4.0$ mm, $m_2 = 2.75$ mm, $m_3 = 4.0$ mm, $m_4 = 2.95$ mm, $n_1 = 13.5$ mm, $n_2 = 3.5$ mm, $t_1 = 4.2$ mm, $t_2 = 3.0$ mm, $t_3 = 3.9$ mm,

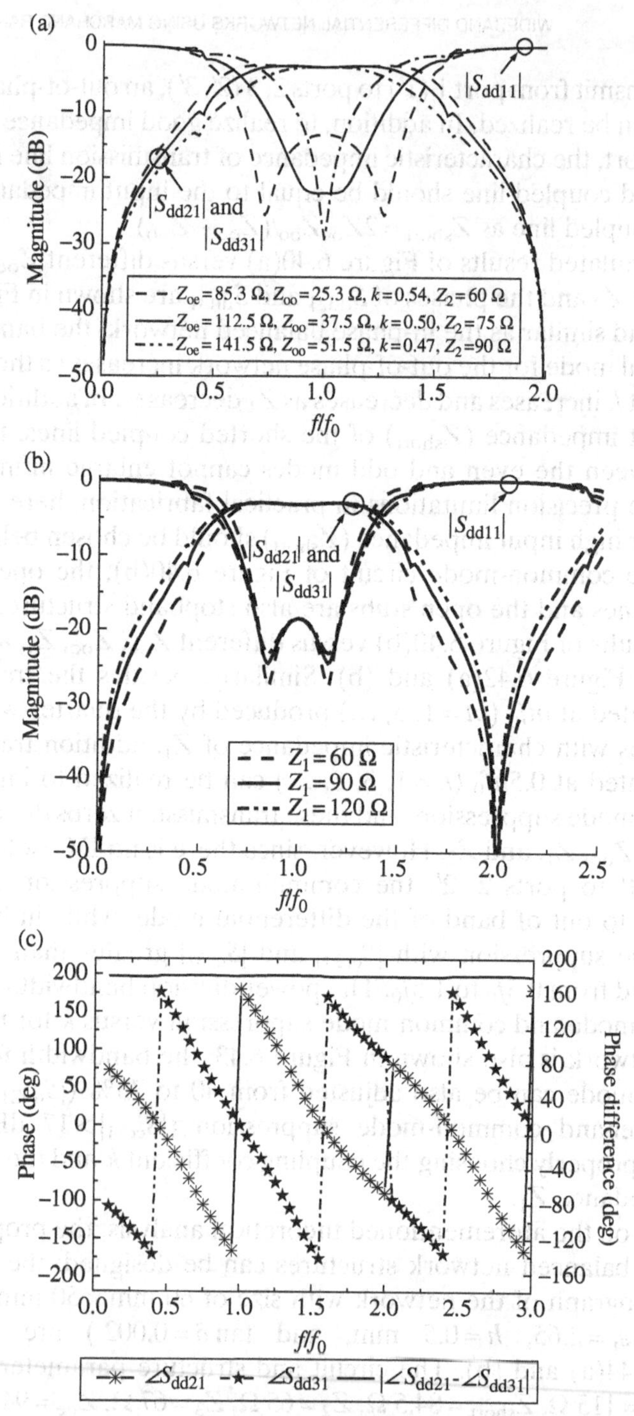

Figure 6.41 Simulated frequency responses of the differential mode for the out-of-phase balanced network. (a) Versus Z_{oe}, Z_{oo}, Z_2, $Z_1 = 115\,\Omega$, $Z_0 = 50\,\Omega$. (b) Versus Z_1, $Z_{oe} = 103.2\,\Omega$, $Z_{oo} = 33.2\,\Omega$, $Z_{short} = 98\,\Omega$, $Z_2 = 70\,\Omega$, $Z_0 = 50\,\Omega$. (c) Phase of S_{dd21} and S_{dd31}. Reprinted with permission from Ref. [33]; copyright 2014 IEEE.

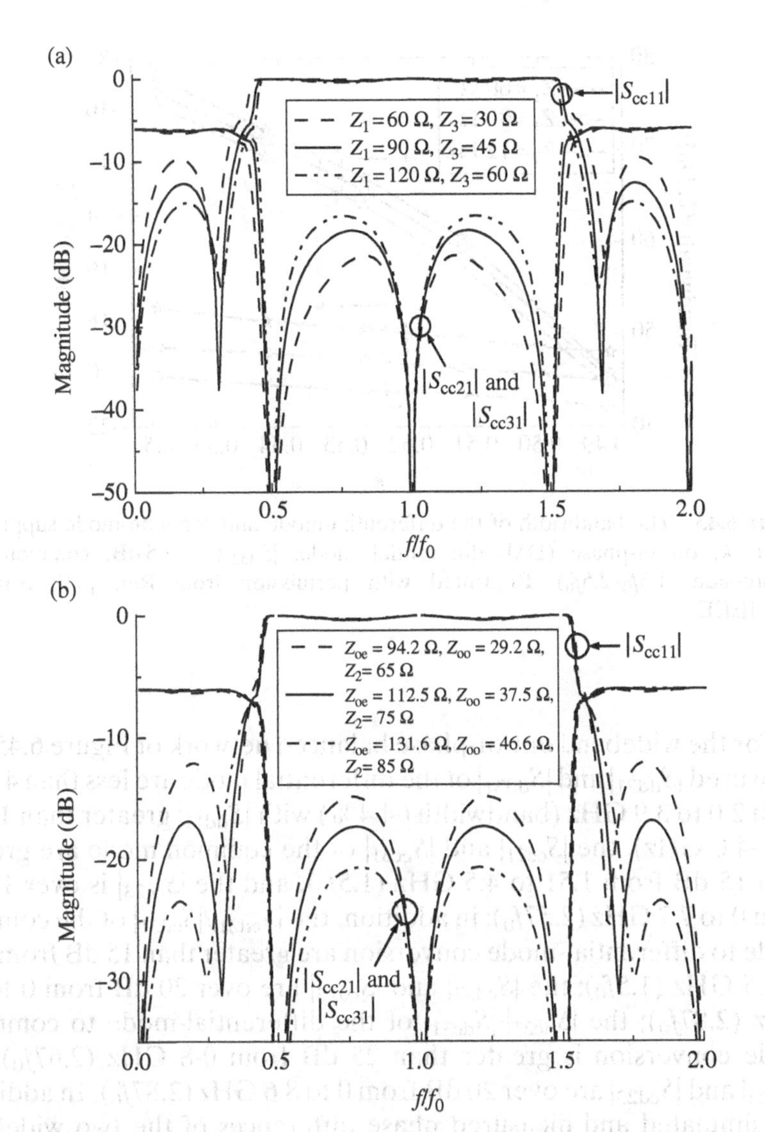

Figure 6.42 Simulated frequency responses of the common mode for the out-of-phase balanced network. (a) Versus Z_1, $Z_1 = 2Z_3$, $Z_{oe} = 103.2\ \Omega$, $Z_{oo} = 33.2\ \Omega$, $Z_{short} = 98\ \Omega$, $Z_2 = 70\ \Omega$, $Z_0 = 50\ \Omega$. (b) Versus Z_{oe}, Z_{oo}, Z_2, $Z_1 = 110\ \Omega$, $Z_3 = 55\ \Omega$, $Z_0 = 50\ \Omega$. Reprinted with permission from Ref. [33]; copyright 2014 IEEE.

$t_4 = 7.3$ mm, $t_5 = 3.5$ mm, $t_6 = 13.2$ mm, $p_1 = 4.6$ mm, $p_2 = 2.6$ mm, $p_3 = 3.3$ mm, $p_4 = 4.6$ mm, $p_5 = 3.2$ mm, $p_6 = 2.4$ mm, $p_7 = 2.5$ mm, $w_0 = 1.37$ mm, $w_1 = 1.62$ mm, $w_2 = 0.68$ mm, $w_3 = 0.3$ mm, $w_4 = 0.9$ mm, $w_5 = 0.6$ mm, $w_6 = 0.9$ mm, $d_1 = 0.5$ mm, $g_1 = 0.15$ mm, $g_2 = 1.5$ mm, $g_3 = 0.3$ mm, $g_4 = 0.3$ mm, $s_1 = 1.4$ mm, $s_2 = 1.13$ mm, $s_3 = 1.0$ mm, $s_4 = 1.13$ mm.

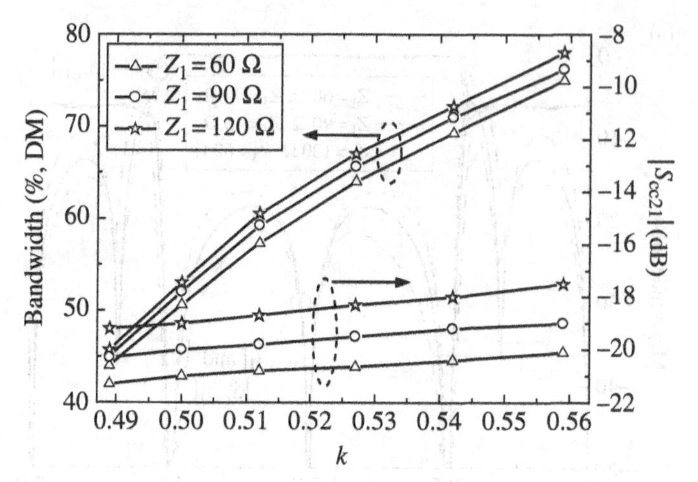

Figure 6.43 The bandwidth of the differential-mode and common-mode suppression versus k, out-of-phase (DM, differential mode: $|S_{dd21}| < 3.5$ dB; common-mode suppression: $1.5f_0$–$2.5f_0$). Reprinted with permission from Ref. [33]; copyright 2014 IEEE.

For the wideband out-of-phase balanced network of Figure 6.45, the measured $|S_{dd21}|$ and $|S_{dd31}|$ of the differential mode are less than 4.0 dB from 2.0 to 3.9 GHz (bandwidth 64.4%) with $|S_{dd11}|$ greater than 10 dB (2.2–4.0 GHz), the $|S_{cc21}|$ and $|S_{cc31}|$ of the common mode are greater than 15 dB from 1.51 to 4.5 GHz ($1.5f_0$), and the $|S_{cc23}|$ is over 15 dB from 0 to 7.7 GHz ($2.57f_0$); in addition, the $|S_{cd21}|/|S_{cd31}|$ of the common mode to differential-mode conversion are greater than 15 dB from 1.51 to 4.5 GHz ($1.5f_0$); the $|S_{cd22}|$ and $|S_{cd23}|$ are over 20 dB from 0 to 8.6 GHz ($2.87f_0$); the $|S_{dc21}|/|S_{dc31}|$ of the differential-mode to common-mode conversion is greater than 25 dB from 0-8 GHz ($2.67f_0$); the $|S_{cd22}|$ and $|S_{cd23}|$ are over 20 dB from 0 to 8.6 GHz ($2.87f_0$). In addition, the simulated and measured phase differences of the two wideband balanced networks are also plotted in Figure 6.46, which are better than 0.85° (in-phase) and 1.1° (out-of-phase) within the operating band, respectively. The slight frequency discrepancies between the measured and simulated results are mainly caused by the limited fabrication precision and measurement errors. However, since no isolation resistors are used for the two networks, the isolation ($|S_{dd23}|$) of the differential-mode power division is not as good as the ideal mixed-mode S-parameters ($\mathbf{S}_{mm}$). To overcome this drawback, wideband Gysel power dividers [39] can be used to realize wideband differential network with high isolation [40].

(a)

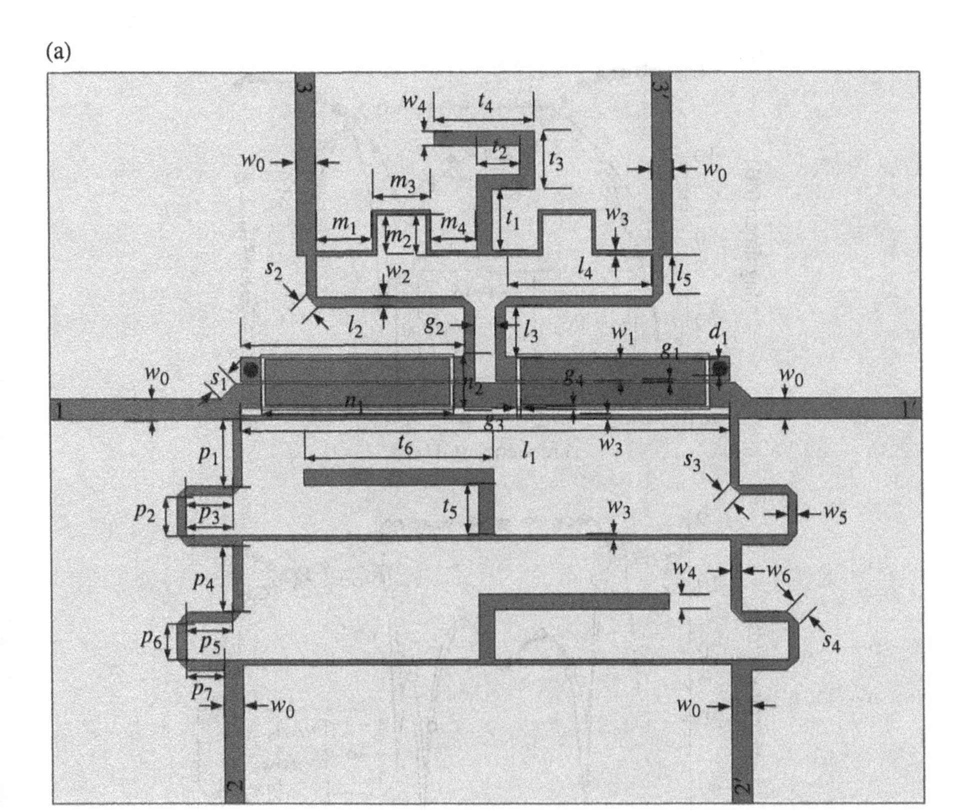

(b)

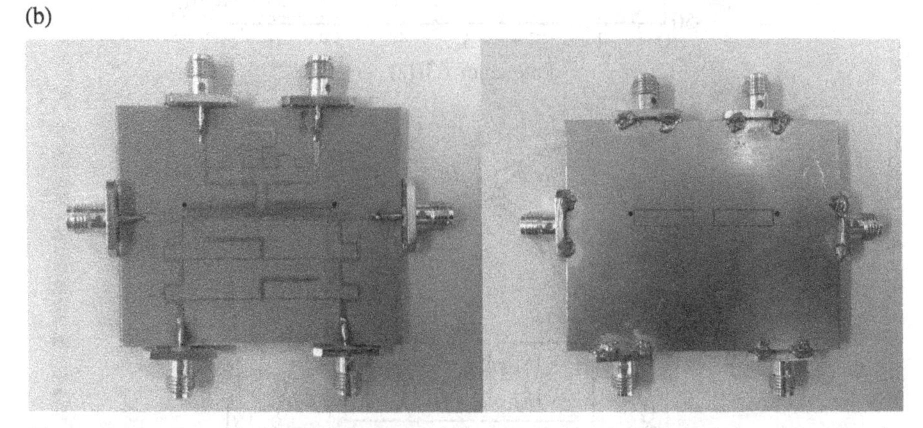

Figure 6.44 Geometry and photograph of wideband out-of-phase balanced network. (a) Geometry. (b) Photograph. Reprinted with permission from Ref. [33]; copyright 2014 IEEE.

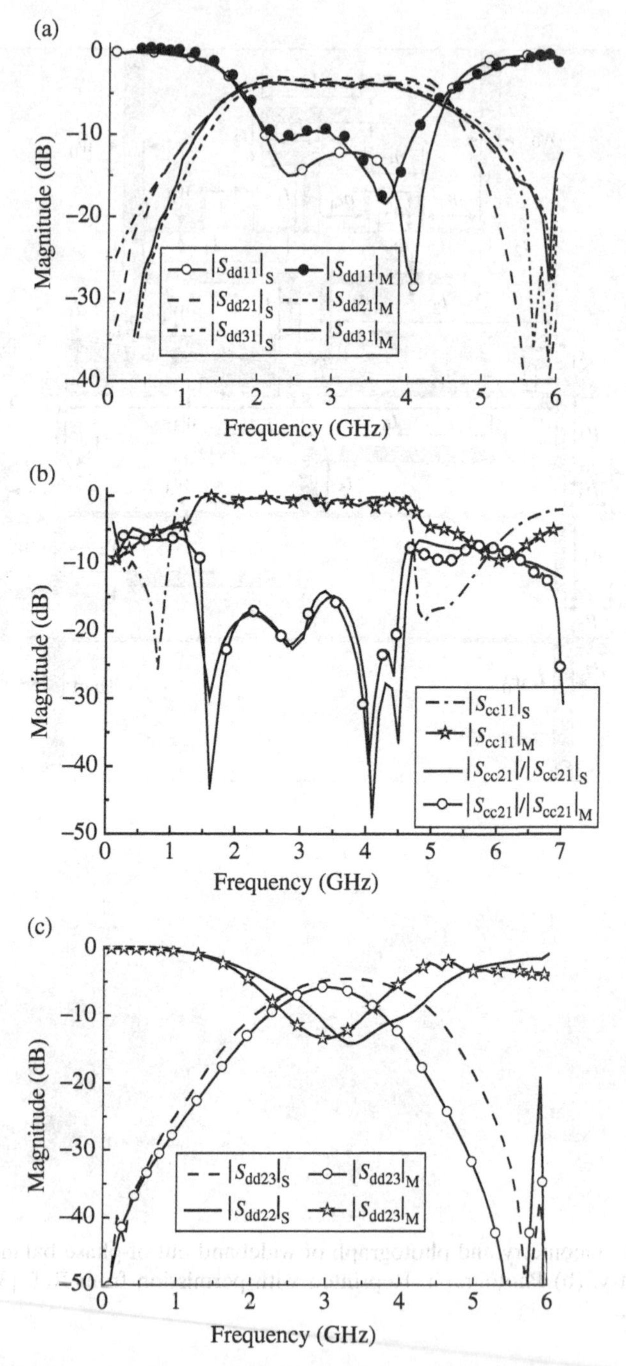

Figure 6.45 Measured and simulated results of the wideband out-of-phase balanced network. (a) $|S_{dd11}|/|S_{dd21}|/|S_{dd31}|$. (b) $|S_{cc11}|/|S_{cc21}|$. (c) $|S_{dd22}|/|S_{dd23}|$. (d) $|S_{cc22}|/|S_{cc23}|$. (e) $|S_{cd21}|/|S_{dc31}|$. (f) $|S_{cd22}|/|S_{dc23}|$ (M, measurement; S, simulation). Reprinted with permission from Ref. 6.45; copyright 2014 IEEE.

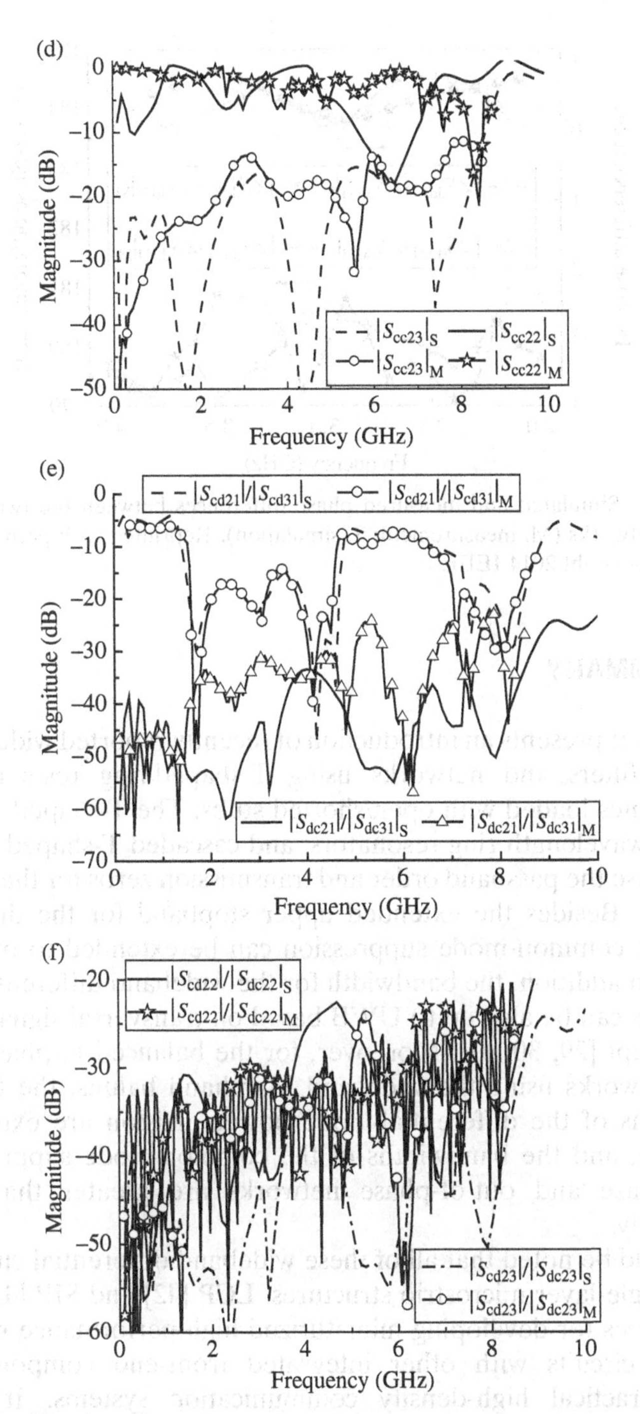

Figure 6.45 (Continued)

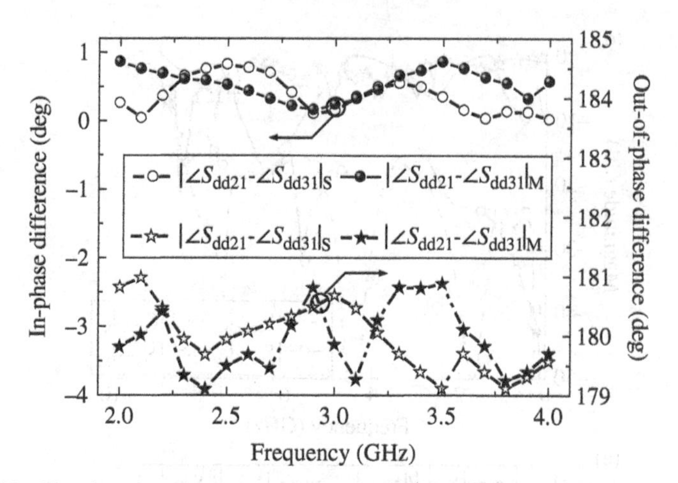

Figure 6.46 Simulated and measured phase differences between the two wideband balanced networks (M, measurement; S, simulation). Reprinted with permission from Ref. [33]; copyright 2014 IEEE.

6.5 SUMMARY

This chapter presents an introduction of recently reported wideband differential filters and networks using T-shaped/ring resonators and coupled lines loaded with open/shorted stubs. The T-shaped structure, half-/full-wavelength ring resonators, and cascaded T-shaped structure can increase the passband order and transmission zeros for the differential mode. Besides the extended upper stopband for the differential mode, the common-mode suppression can be extended to over three octaves. In addition, the bandwidth for the wideband differential bandpass filters can be also up to UWB based on transversal signal interaction concept [29, 30, 41]. Moreover, for the balanced in-phase/out-of-phase networks using the wideband Marchand baluns, the fractional bandwidths of the differential-mode power division are extended to over 50%, and the bandwidths of the common-mode suppression of the in-phase and out-of-phase networks are greater than 100%, respectively.

It should be noted that all of these wideband differential circuits are nearly single-layer microstrip structures. LCP [42] and SIP [43] can be good choices for developing miniaturized high-performance wideband balanced circuits with other integrated front-end components for further practical high-density communication systems. It can be expected that more and more novel fully differential microwave circuits and systems will be presented and put into practical applications in the near further.

REFERENCES

1. W. R. Eisenstant, B. Stengel, and B. M. Thompson, *Microwave differential circuit design using mixed-mode S-parameters*, Artech House, Boston, MA, 2006.
2. D. Bockelman and W. Eisenstadt, "Combined differential and common mode scattering parameters: Theory and simulation," *IEEE Trans. Microwave Theory Tech.*, vol. **43**, no. 7, pp. 1530–1539, Jul. 1995.
3. W. J. Feng, W. Q. Che, Q. Xue, "The proper balance: Overview of microstrip wideband balanced circuits with wideband common mode suppression," *IEEE Microwave Mag.*, vol. **16**, no. 5, pp. 55–68, Jun. 2015.
4. FCC, "Revision of part 15 of the commission's rules regarding ultra-wideband transmission system," Federal Communications Commission, Washington, DC, Technical Report ET-Docket 98-153 FCC02-48, Apr. 2002.
5. P. S. Kshetrimayum, "An introduction to UWB communication systems," *IEEE Potentials*, vol. **28**, pp. 9–13, Mar. 2009.
6. S. Sun and L. Zhu, "Multiple-resonator-based bandpass filters," *IEEE Microwave Mag.*, vol. **10**, no. 2, pp. 88–98, Apr. 2009.
7. Z. C. Hao and J. S. Hong, "Ultrawideband filter technologies," *IEEE Microwave Mag.*, vol. **11**, no. 4, pp. 56–68, Jun. 2010.
8. W. J. Feng, W. Q. Che, and Q. Xue, "Transversal signal interaction: Overview of high-performance wideband bandpass filters," *IEEE Microwave Mag.*, vol. **15**, no. 2, pp. 84–96, Mar. 2014.
9. W. H. Tu and K. Chang, "Compact second harmonic-suppressed bandstop and bandpass filters using open stubs," *IEEE Trans. Microwave Theory Tech.*, vol. **54**, no. 6, pp. 2497–2502, Jun. 2006.
10. X. Y. Zhang, J.-X. Chen, Q. Xue, and S.-M. Li, "Dual-band bandpass filters using stub-loaded resonators," *IEEE Microwave Wireless Compon. Lett.*, vol. **17**, no. 8, pp. 583–585, Aug. 2007.
11. Y. L. Wu, Y.A. Liu, Y. Zhang, J. Gao, and H. Zhou, "A dual-band unequal Wilkinson power divider without reactive components," *IEEE Trans. Microwave Theory Tech.*, vol. **57**, no. 1, pp. 216–222, Jan. 2009.
12. W. J. Feng and W. Q. Che, "Novel wideband differential bandpass filters based on T-shaped structure," *IEEE Trans. Microwave Theory Tech.*, vol. **60**, no. 6, pp. 1560–1568, Jun. 2012.
13. L. Zhu, S. Sun, and W. Menzel, "Ultra-wideband (UWB) bandpass filters using multiple-mode resonator," *IEEE Microwave Wireless Compon. Lett.*, vol. **15**, no. 11, pp. 796–798, Nov. 2005.
14. Y.-C. Chiou, J.-T. Kuo, and E. Cheng, "Broadband quasi-Chebyshev bandpass filters with multimode stepped-impedance resonators (SIRs)," *IEEE Trans. Microwave Theory Tech.*, vol. **54**, no. 8, pp. 3352–3358, Aug. 2006.
15. K. J. Song and Q. Xue, "Inductance-loaded Y-shaped resonators and their applications to filters," *IEEE Trans. Microwave Theory Tech.*, vol. **58**, no. 4, pp. 978–984, Apr. 2010.

16. D. M. Pozar, *Microwave Engineering*, 2nd ed. John Wiley & Sons, Inc., New York, 1998.
17. W. J. Feng, W. Q. Che, L. M. Gu, and Q. Xue, "High selectivity wideband balanced bandpass filters using symmetrical multi-mode resonators," *IET Microwave Antennas Propag.*, vol. **7**, no. 12, pp. 1005–1015, Nov. 2013.
18. W. J. Feng, X. Gao, W. Q. Che, W. C. Yang, and Q. Xue, "High selectivity wideband balanced filters with multiple transmission zeros," *IEEE Trans. Circuits Syst. II: Express Briefs*, vol. **64**, no. 10, pp. 1182–1186, Oct. 2017.
19. I. Wolff, "Microstrip bandpass filters using degenerate modes of a microstrip ring resonators," *IET Electron. Lett.*, vol. **8**, no. 12, pp. 163–164, Jun. 1972.
20. L. H. Hsieh and K. Chang, "Compact, low insertion-loss, sharp-rejection, and wide-band microstrip bandpass filters," *IEEE Trans. Microwave Theory Tech.*, vol. **51**, no. 4, pp. 1241–1246, Apr. 2003.
21. J. S. Hong and S. Li, "Theory and experiment of dual-mode microstrip triangular patch resonators and filters," *IEEE Trans. Microwave Theory Tech.*, vol. **52**, no. 4, pp. 1237–1243, Apr. 2004.
22. M. Salleh, G. Prigent, O. Pigaglio, and R. Crampagne, "Quarter-wavelength side-coupled ring resonator for bandpass filters," *IEEE Trans. Microwave Theory Tech.*, vol. **56**, no. 1, pp. 156–162, Jan. 2008.
23. W. J. Feng, W. Q. Che, and Q. Xue, "Compact ultra-wideband bandpass filters with notched bands based on transversal signal-interaction concepts," *IET Microwave Antennas Propag.*, vol. **7**, no. 12, pp. 961–969, Nov. 2013.
24. W. J. Feng, W. Q. Che, Y. L. Ma, and Q. Xue, "Compact wideband differential bandpass filter using half-wavelength ring resonator," *IEEE Microwave Wireless Compon. Lett.*, vol. **23**, no. 2, pp. 81–83, Feb. 2013.
25. W. J. Feng, W. Q. Che, and Q. Xue, "Balanced filters with wideband common mode suppression using dual-mode ring resonators," *IEEE Trans. Circuits Syst. I: Regular Papers*, vol. **62**, no. 6, pp. 1499–1507, Jun. 2015.
26. G. Matthaei, L. Young, and E. M. T. Jones, *Microwave Filters, Impedance Matching Networks and Coupling Structures*, Artech House Inc., Norwood, MA, Section 5, pp. 222, 224, 1985.
27. W. J. Feng, W. Q. Che, and H. D. Chen, "Balanced filter circuit based on open/shorted coupled lines," *2015 International Microwave Symposium*, Phoenix, AZ, USA, pp. 1–3, 17–22 May 2015.
28. M. A. Sánchez-Soriano, E. Bronchalo, and G. Torregrosa-Penalva, "Compact UWB bandpass filter based on signal interference techniques," *IEEE Microwave Wireless Compon. Lett.*, vol. **19**, no. 11, pp. 692–694, Nov. 2009.
29. W. J. Feng, W. Q. Che, T. F. Eibert, and Q. Xue, "Compact wideband differential bandpass filter based on the double-sided parallel-strip line and transversal signal-interaction concepts," *IET Microwave Antennas Propag.*, vol. **6**, no. 2, pp. 186–195, Apr. 2012.

30. H. T. Zhu, W. J. Feng, W. Q. Che, and Q. Xue, "Ultra-wideband differential bandpass filter based on transversal signal-interference concept," *IET Electron. Lett.*, vol. **47** no. 18, pp. 1033–1034, Sep. 2011.
31. B. Xia, L.-S. Wu, and J. F. Mao, "A new balanced-to-balanced power divider/combiner," *IEEE Trans. Microwave Theory Tech.*, vol. **60**, no. 9, pp. 287–295, Sep. 2012.
32. L. S. Wu, B. Xia, and J. F. Mao, "A half-mode substrate integrated waveguide ring for two-way power division of balanced circuit," *IEEE Microwave Wireless Compon. Lett.*, vol. **22**, no. 7, pp. 333–335, Jul. 2012.
33. W. J. Feng, H. T. Zhu, W. Q. Che, and Q. Xue, "Wideband in-phase and out-of-phase balanced power dividing and combining networks," *IEEE Trans. Microwave Theory Tech.*, vol. **62**, no. 5, pp. 1192–1202, May 2014.
34. K. S. Ang and I. D. Robertson, "Analysis and design of impedance transforming planar Marchand baluns," *IEEE Trans. Microwave Theory Tech.*, vol. **49**, no. 2, pp. 402–406, Feb. 2001.
35. W. J. Feng and W. Q. Che, "Ultra-wideband bandpass filter using broadband planar Marchand balun," *IET Electron. Lett.*, vol. **47** no. 3, pp. 198–199, Feb. 2011.
36. J. S. Hong and M. J. Lancaster, *Microstrip Filters for RF/Microwave Applications*, John Wiley & Sons, Inc., New York, 2001.
37. M. C. Velazquez, J. Martel, and F. Medina, "Parallel coupled microstrip filters with floating ground-plane conductor for spurious-band suppression," *IEEE Trans. Microwave Theory Tech.*, vol. **53**, no. 5, pp. 1823–1828, May 2005.
38. Y. X. Guo, Z. Y. Zhang, and L. C. Ong, "Improved wide-band schiffman phase shifter," *IEEE Trans. Microwave Theory Tech.*, vol. **54**, no. 3, pp. 2412–2418, Mar. 2006.
39. L. Chiu and Q. Xue, "A parallel-strip ring power divider with high isolation and arbitrary power-dividing ratio," *IEEE Trans. Microwave Theory Tech.*, vol. **55**, no. 11, pp. 2419–2426, Nov. 2007.
40. W. J. Feng, C. Y. Zhao, W. Q. Che, and Q. Xue, "Wideband balanced network with high isolation using double-sided parallel-strip line," *IEEE Trans. Microwave Theory Tech.*, vol. **63**, no. 12, pp. 4113–4118, Dec. 2015.
41. S. Y. Shi, W. W. Choi, W. Q. Che, K. W. Tam, and Q. Xue, "Ultra-wideband differential bandpass filter with narrow notched band and improved common-mode suppression by DGS," *IEEE Microwave Wireless Compon. Lett.*, vol. **22**, no. 4, pp. 185–187, Apr. 2012.
42. Z.-C. Hao and J.-S. Hong, "Compact UWB filter with double notch-bands using multilayer LCP technology," *IEEE Microwave Wireless Compon. Lett.*, vol. **19**, no. 8, pp. 500–502, 2009.
43. H. C. Chen, C. H. Tsai, and T. L. Wu, "A compact and embedded balanced bandpass filter with wideband common-mode suppression on wireless SiP," *IEEE Trans. Compon., Packag. Manuf. Technol.*, vol. **2**, no. 6, pp. 1030–1038, Jun. 2012.

CHAPTER 7

UWB AND NOTCHED-BAND UWB DIFFERENTIAL FILTERS USING MULTILAYER AND DEFECTED GROUND STRUCTURES (DGSs)

Jian-Xin Chen,[1] Li-Heng Zhou,[1] and Quan Xue[2]

[1]School of Electronics and Information, Nantong University, Nantong, China
[2]School of Electronic and Information Engineering, South China University of Technology, Guangzhou, China

The development of UWB differential filters is an attractive topic in microwave circuit research because of the capacity of these filters for high data rate transmission, high immunity to environmental noise, and low electromagnetic interference (EMI) [1–8]. The mechanism of the multilayer microstrip-to-slotline transition (MST) and its applications in differential topology are analyzed, with the slotline defected ground structure (DGS) pattern etched in the ground. In addition to allowing differential signal transmission, the MST can also realize multimode characteristics with a controllable differential passband by

Balanced Microwave Filters, First Edition. Edited by Ferran Martín, Lei Zhu, Jiasheng Hong, and Francisco Medina.

adjusting the impedance of the slotline. Additionally, the MST has inherent wideband CM suppression characteristics and thus has the potential to be used in UWB differential filter designs. In this chapter, several typical types of UWB and notched-band UWB differential filters based on the MST are presented and described.

7.1 CONVENTIONAL MULTILAYER MICROSTRIP-TO-SLOTLINE TRANSITION (MST)

Figure 7.1(a) depicts the layout of a conventional MST: the slotline is etched on the ground plane (bottom layer) and is crossed at a right angle by the microstrip on the top layer [9]. The microstrip extends approximately a quarter of a wavelength beyond the slot, and the transition is achieved through magnetic coupling. The extended portions of the microstrip and slotline are θ_m and θ_s, respectively, which represent their quarter-wavelength electrical lengths. The corresponding equivalent circuit and its simplified counterpart are shown in Figure 7.1(b) and (c), respectively, where the transformer turns ratio, n, describes the magnitude of the coupling between the microstrip and slotline. Here,

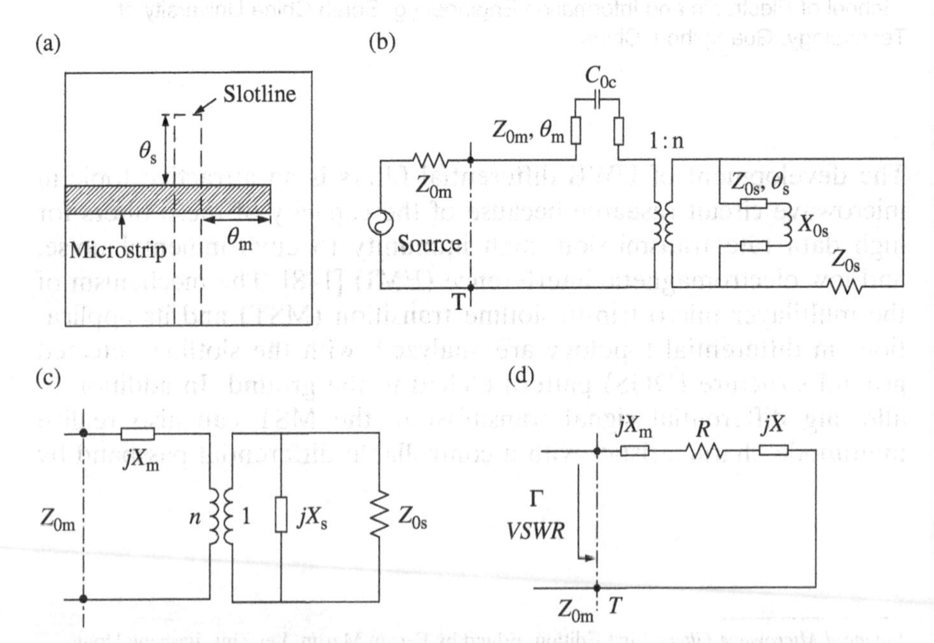

Figure 7.1 (a) Conventional MST. (b) Transmission-line equivalent circuit for the MST. (c) Reduced equivalent circuit. (d) Transformed equivalent circuit.

$$jX_s = Z_{0s}\frac{jX_{0s} + jZ_{0s}\tan\theta_s}{Z_{0s} - X_{0s}\tan\theta_s} \tag{7.1}$$

$$jX_m = Z_{0m}\frac{1/j\omega C_{0c} + jZ_{0m}\tan\theta_m}{Z_{0m} + \tan\theta_m/\omega C_{0c}} \tag{7.2}$$

After transformation to the microstrip side, the equivalent circuit in Figure 7.1(c) can be further reduced, as shown in Figure 7.1(d). In this circuit,

$$R = n^2\frac{Z_{0s}X_s^2}{Z_{0s}^2 + X_s^2} \tag{7.3}$$

$$X = n^2\frac{Z_{0s}^2 X_s}{Z_{0s}^2 + X_s^2} \tag{7.4}$$

Finally, the reflection coefficient Γ is given by

$$\Gamma = \frac{R - Z_{0m} + j(X_m + X)}{R + Z_{0m} + j(X_m + X)} \tag{7.5}$$

According to the aforementioned analysis, the characteristic impedance of the slotline, Z_{0s}, must be optimized to match the microstrip impedance Z_{0m}. The bandwidth of the cross-junction transition is determined primarily by the characteristic impedances of the microstrip line and slotline.

7.2 DIFFERENTIAL MST

7.2.1 Differential MST with a Two-Layer Structure

Based on the conventional MST shown in Figure 7.1, transformations have been developed for applications in a differential topology. Figure 7.2 shows a simple model in which a microstrip line is deployed on the top layer with a vertical arrangement, while the slotline is etched on the ground horizontally [10]. The corresponding charge distributions under DM and CM excitations are also portrayed in this figure. Under the DM excitation, a virtual electrical wall is formed along the symmetrical line, and the currents on the two sides of the slotline are opposite, as shown in Figure 7.2(a). This matches the features of the slotline resonator. Thus, the DM signal can be successfully coupled to the slotline

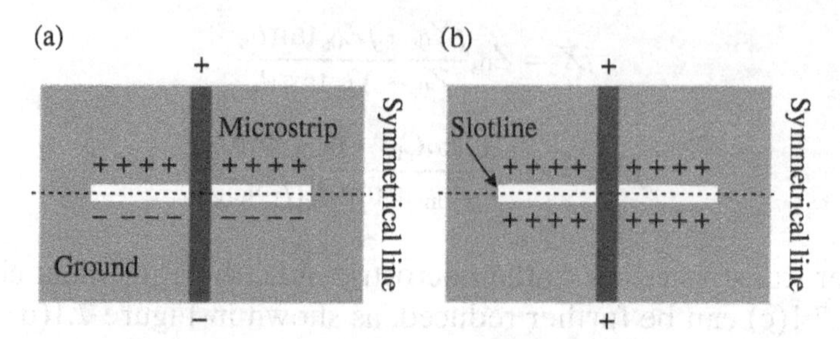

Figure 7.2 Geometry and charge distribution of the differential MST with two-layer structure: (a) Differential mode and (b) common mode.

resonator through strong magnetic coupling and then converted into the slotline mode propagating along the slotline. The virtual ground halves the width of the slotline, changing its impedance. This operation only affects the bandwidth of the differential passband. In contrast to the traditional MST, with the differential MST, there is no need to use a quarter-wavelength microstrip open stub to achieve strong magnetic coupling, since the virtual electrical wall provides complete mode conversion to the slotline mode.

When the CM signal is excited, currents with identical polarity appear on the two sides of the slotline, as shown in Figure 7.2(b). This situation is different from the electric field characteristics of the slotline. Under such circumstances, a virtual magnetic wall exists along the symmetrical line, and the electric field of the slotline mode is perpendicular to this magnetic wall, which conflicts with the boundary condition of the magnetic wall. The CM signal cannot be converted to the slotline mode and is totally reflected, leading to inherently high CM suppression.

7.2.2 Differential MST with Three-Layer Structure

Figure 7.3(a) shows the top view of the three-layer differential MST. The slotline resonator is located on the middle ground plane, while the differential I/O ports are placed on the top and bottom layers, respectively. The reasons for using three-layer substrates are summarized as follows: (i) the slotline portion with the desired impedance on the two-layered substrate is too narrow to be fabricated and (ii) the three-layer substrate can effectively reduce undesired radiation in the slotline portion. Under DM operation, electric fields in the slotline are excited out of phase by a pair of strip conductors, resulting in a faithful signal transmission, as illustrated in Figure 7.3(b). Under CM operation, electric fields in the slotline are excited in phase, as shown in

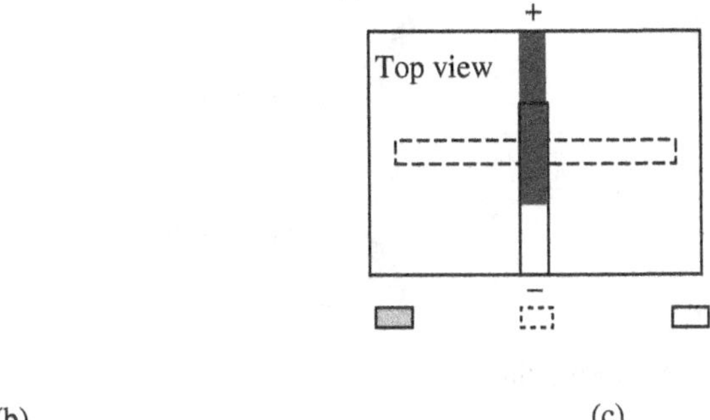

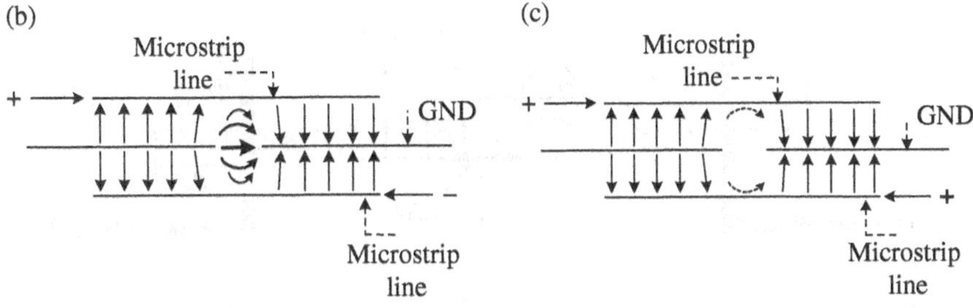

Figure 7.3 (a) Top view of the three-layer differential MST. (b) Cross view of electrical field distribution under DM feeding. (c) Cross view of electrical field distribution under CM feeding.

Figure 7.3(c), thus canceling each other and resulting in excellent CM rejection.

To achieve UWB differential filters based on the MST, many investigators have presented a variety of designs with satisfactory performance, which are analyzed in detail in the following section.

7.3 UWB DIFFERENTIAL FILTERS BASED ON THE MST

7.3.1 Differential Wideband Filters Based on the Conventional MST

The structure of the filter is shown in Figure 7.4(a). It is based on the conventional MST structure shown in Figure 7.1, where the four-port circuit is ideally symmetric with respect to the horizontal central plane, that is, A–A′ [11]. Its half bisections under DM and CM excitations are drawn in Figure 7.4(b) and (c), respectively. For the DM excitation, the central plane (A–A′) acts as a perfectly electric wall. The uniform-impedance slotline resonator can be replaced by a stepped-impedance slotline multimode resonator to expand the bandwidth. In this design, the fractional bandwidth of the filter can be flexibly adjusted by

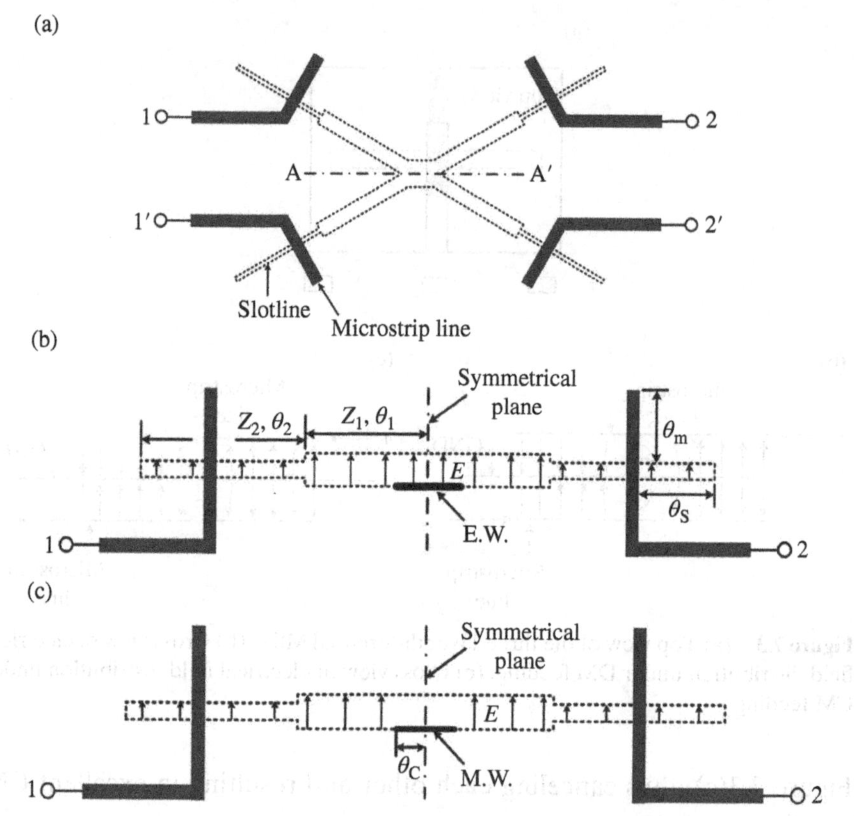

Figure 7.4 (a) Schematic of the developed wideband DM bandpass filter. (b) Half of a two-port bisection under DM excitation (E.W., electric wall). (c) Half of a two-port bisection under CM excitation (M.W., magnetic wall). Reprinted with permission from Ref. [11]; copyright 2015 IEEE.

choosing the different impedance ratio Z_1/Z_2 of the stepped-impedance slotline resonator and electrical lengths θ_m and θ_s, as listed in Table 7.1. For the CM operation, a virtual magnetic wall appears along the central plane (A–A′), as depicted in Figure 7.4(c). The structure of the slotline is destroyed, and the CM signals cannot propagate through the middle portion of the slotline resonator, resulting in favorable CM suppression.

A template is designed and implemented on the Rogers 6010 with a thickness of 0.635 mm and dielectric constant of 10.8, as shown in Figure 7.5. Figure 7.6(a) and (b) shows the simulated and the measured transmission coefficients under DM and CM operations, respectively. The measured DM passband achieves a maximum insertion loss of 0.5 dB. The 3 dB fractional bandwidth is 107.7% over the frequency range of 1.2–4.0 GHz, while the CM signal is suppressed lower than −20 dB in the range of 0.5–5.0 GHz.

Table 7.1 Comparison in Fraction Bandwidth (FBW) among Three Cases

Parameters	Case I	Case II	Case III
Z_1 (Ω)	170	159	90
Z_2 (Ω)	68	68	68
θ_1 at 2 GHz	45°	45°	45°
θ_2 at 2 GHz	45°	45°	45°
θ_m at 3 GHz	63°	83°	102°
θ_S at 3 GHz	43°	91°	110°
FBW (%)	98.6	103.8	109

Reprinted with permission from Ref. [11]; copyright 2015 IEEE.

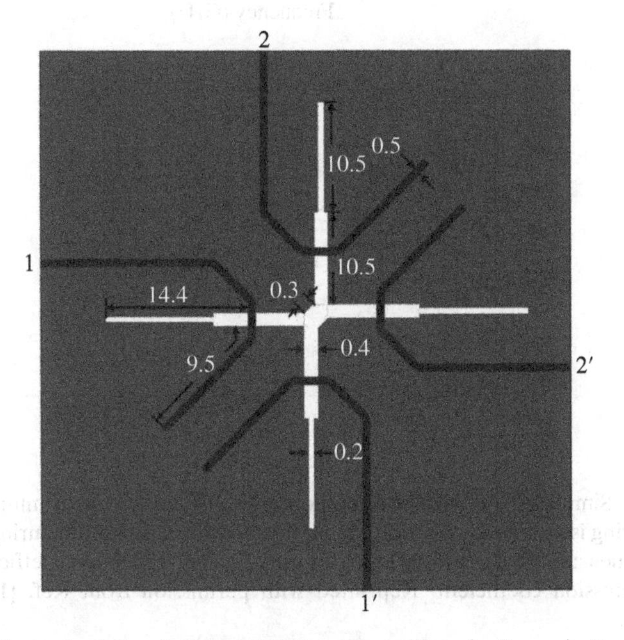

Figure 7.5 Layout of the designed DM bandpass filter (unit: mm). Reprinted with permission from Ref. [11]; copyright 2015 IEEE.

7.3.2 Differential Wideband Filters Based on the Differential MST

Filter 1, depicted in Figure 7.7(a), is based on the differential MST with two layers (as shown in Figure 7.2). Here, the microstrip lines with a pair of differential ports on the two sides are directly tapped on the slotline resonator [12]. Figure 7.7(b) shows that Filter 2, with the three-layer differential MST topology (as shown in Figure 7.4), is a transformation of Filter 1. In Filter 2, the slotline resonator is now connected with two pairs of open-ended microstrip stubs on both top and bottom layers.

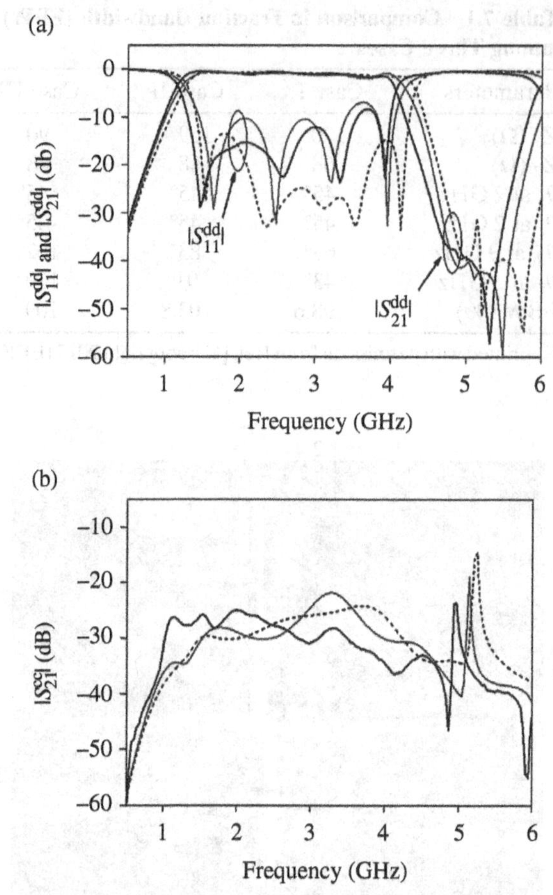

Figure 7.6 Simulated and measured responses (short dashed line: simulation ignoring manufacturing issues; solid gray line: simulation considering manufacturing issues; solid black line: measurement). (a) DM case: transmission and reflection coefficients. (b) CM case: transmission coefficient. Reprinted with permission from Ref. [11]; copyright 2015 IEEE.

The first three or four resonant modes are depicted by their corresponding electric field distributions, shown in Figure 7.8. For Filter 1, the resonator is fed approximately $L_{slot}/3$ (L_{slot} is the total length of the slotline resonator). The first two resonant modes, shown by the solid line in Figure 7.8(a), can be effectively excited. The third mode, shown by the dashed line, cannot be excited because the electric field along the slotline resonator becomes null at the coupling position. Similarly, the slotline resonator for Filter 2 is fed approximately $L_{slot}/4$ to excite the three resonant modes. In this case, the first three modes in

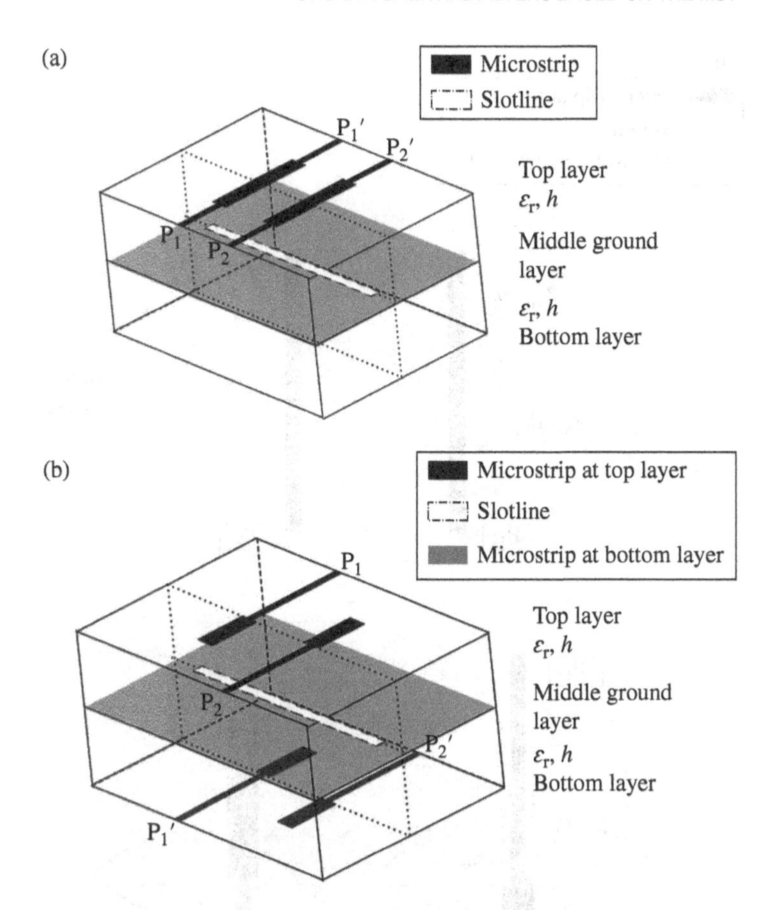

Figure 7.7 Schematics of two developed differential filters: (a) Filter 1 and (b) Filter 2. Reprinted with permission from Ref. [12]; copyright 2015 IEEE.

the solid line can be excited because of the non-null electric field intensity at the feeding position, while the fourth mode in the dashed line cannot emerge.

To design these differential bandpass filters, the initial circuit models deduced from the geometrical schemes in Figure 7.8 are displayed in Figure 7.9. These models are directly under DM operation and are composed of the microstrips or microstrip stubs, slotline shorted stubs, and connecting slotline portions. The transformers with their turn ratios, shown in Figure 7.9, represent the degree of coupling between the microstrip and slotline. The electrical lengths of the microstrip portions, θ_{MS} or θ_{ML}, are equal to 90° at the middle frequency of the filter (f_{mid}), where the strongest coupling is achieved. All the slotlines or slot stubs

(a)

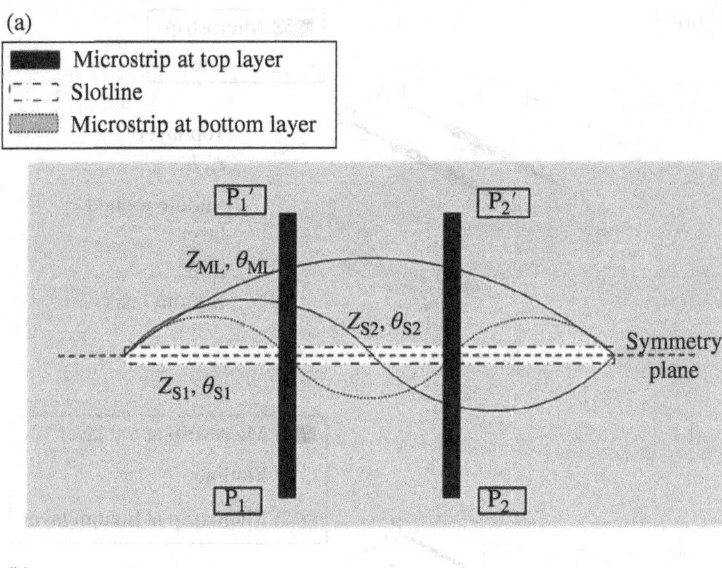

(b)

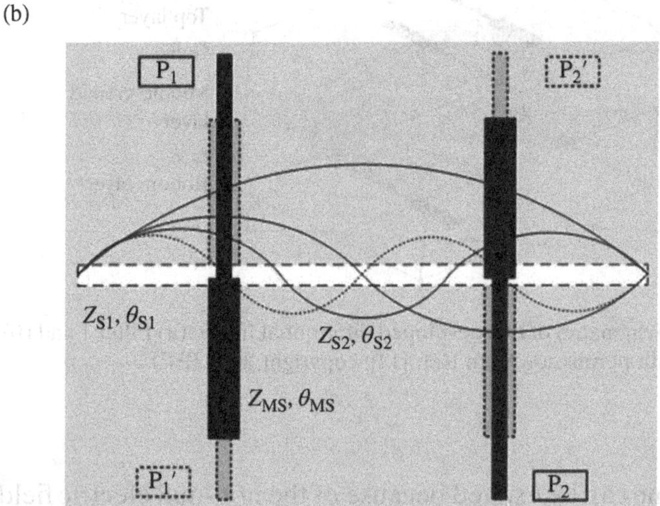

Figure 7.8 Top view of two developed differential filters and electric field distribution of respective resonant mode in the slotline resonator: (a) Filter 1 and (b) Filter 2. Reprinted with permission from Ref. [12]; copyright 2015 IEEE.

are determined at the first resonant frequency, f_1, with the electrical lengths θ_{SS} and θ_{CS}, using (7.6) for two and three resonant modes:

$$\left.\begin{aligned} &2\times\theta_{SS}+\theta_{CS}=180^\circ \\ &\theta_{SS}=\frac{1}{3}\times 180^\circ(\text{two-mode case}) \\ &\theta_{SS}=\frac{1}{4}\times 180^\circ(\text{three-mode case}) \end{aligned}\right\} \tag{7.6}$$

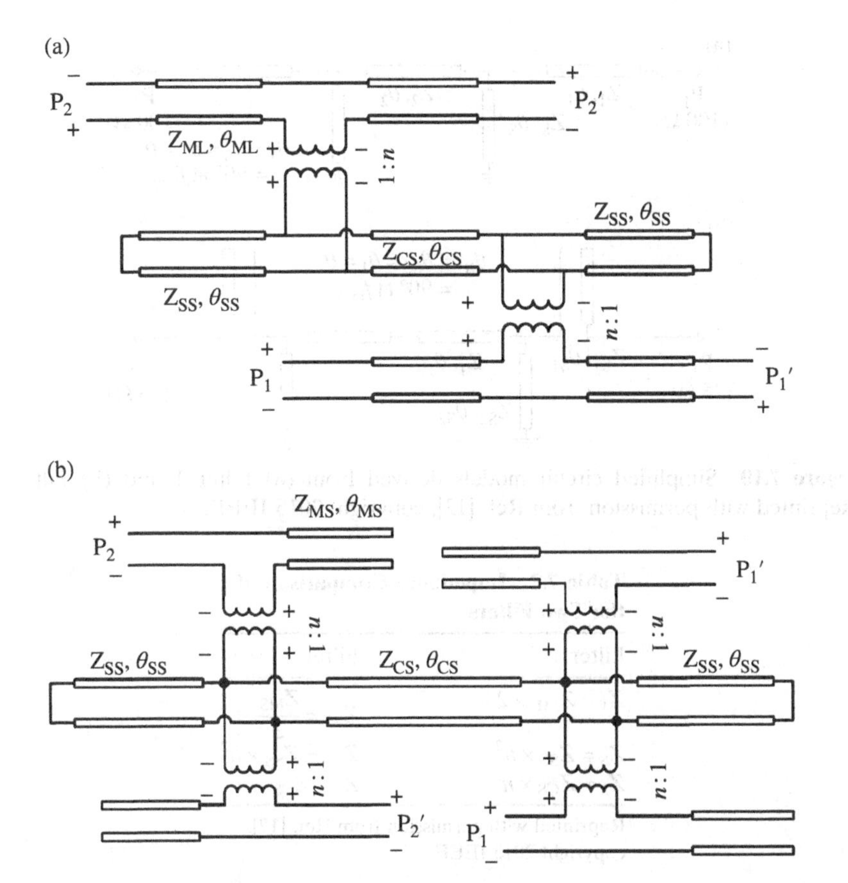

Figure 7.9 Equivalent DM circuit models derived from (a) Filter 1 and (b) Filter 2. Reprinted with permission from Ref. [12]; copyright 2015 IEEE.

It should be noted that the pairs of differential ports in Filter 1 are series connected, whereas they are shunt connected in Filter 2. Therefore, the reference differential-port impedances in the following synthetic process are markedly different.

The initial circuit models can be further simplified into the ones shown in Figure 7.10. While the four-port circuit models are converted into two-port ones, all electrical lengths are converted by referring to 90° at the center frequency of the filter. Impedance transformers n (shown in Figure 7.9) are absorbed into the characteristic impedances of the related transmission-line sections, as listed in Table 7.2.

This procedure yields the simplified circuit models, which represent the general high-pass prototypes. All transmission-line cells involved have the same electrical length of θ, which is 90° at f_{mid}.

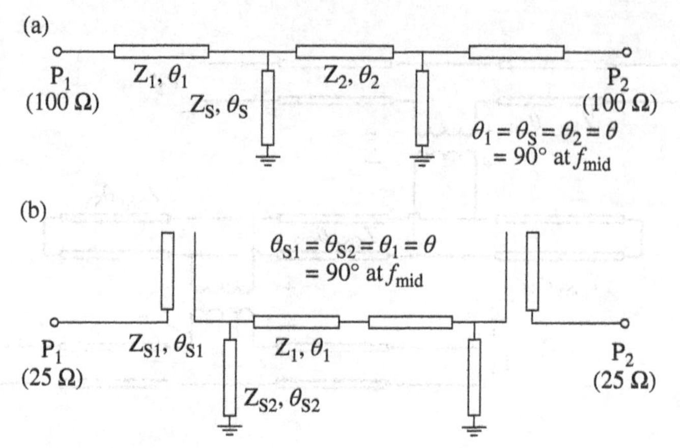

Figure 7.10 Simplified circuit models derived from (a) Filter 1 and (b) Filter 2. Reprinted with permission from Ref. [12]; copyright 2015 IEEE.

Table 7.2 Impedance Comparison of the Two Filters

Filter 1	Filter 2
$Z_1 = Z_{ML} \times 2$	$Z_{S1} = \frac{Z_{MS}}{2}$
$Z_S = Z_{SS} \times n^2$	$Z_{S2} = Z_{SS} \times n^2$
$Z_2 = Z_{CS} \times n^2$	$Z_1 = Z_{CS} \times n^2$

Reprinted with permission from Ref. [12]; copyright 2015 IEEE.

The prototypes can now be synthesized easily, following the method in Ref. [13]. In theory, filters with any desired bandwidths and ripple constants can be achieved by this method. For conciseness, the design and synthesis of only one of the two filters, that is, Filter 1, is discussed in this section.

The complete matrix for the topology in Figure 7.14(a) is

$$\begin{pmatrix} A & B \\ C & D \end{pmatrix} = (M_1)(M_s)(M_2)(M_s)(M_1) \tag{7.7}$$

where

$$(M_n) = \begin{pmatrix} \cos\theta & jz_n \sin\theta \\ j\sin\theta/z_n & \cos\theta \end{pmatrix},\ n=1,2;\ (M_s) = \begin{pmatrix} 1 & 0 \\ 1/(jz_s \tan\theta) & 1 \end{pmatrix} \tag{7.8}$$

z_1, z_2, and z_s are the unit normalized impedances. With

$$|S_{21}|^2 = \frac{1}{1+|F|^2} \text{ and } F = \frac{B-C}{2} \tag{7.9}$$

F can be calculated from the aforementioned relations and rearranged as

$$F = j\left[k_1 \frac{\cos^4\theta}{\sin\theta} + k_2 \frac{\cos^2\theta}{\sin\theta} + k_3 \frac{1}{\sin\theta}\right] \tag{7.10}$$

where

$$k_1 = \frac{z_2}{2z_1^2} - \frac{z_2}{2} - \frac{z_1^2-1}{2z_2} - z_1 - \frac{z_1^2+z_1z_2-1}{z_s} + \frac{z_2 - z_1^2 z_2}{2z_s^2} + \frac{1}{z_1} + \frac{z_2}{z_1 z_s} \tag{7.11a}$$

$$k_2 = z_1 + \frac{z_2}{2} - \frac{z_2}{z_1^2} + \frac{z_1^2+z_1z_2}{z_s} + \frac{2z_1^2-1}{2z_2} - \frac{1}{z_1} - \frac{z_2}{z_1 z_s} + \frac{z_1^2 z_2}{2z_s^2} \tag{7.11b}$$

$$k_3 = -\frac{z_1^4 - z_2^2}{2z_1^2 z_2} \tag{7.11c}$$

To exhibit the Chebyshev responses with equal-ripple in-band behavior, the squared magnitude of the filters is expressed as

$$|S_{21}|^2 = \frac{1}{1+\varepsilon^2\cos^2(n\phi_{\text{u.e.}} + m\phi_{\text{L}})} = \frac{1}{1+\varepsilon^2[T_n(x)T_m(y) - U_n(x)U_m(y)]^2} \tag{7.12}$$

where $n = 3$ and $m = 1$. In (7.12), ε is the specified equal-ripple constant in the passband. $T_n(x)$ and $U_n(x)$ are the Chebyshev polynomial functions of the first and second types of degree n, respectively, and

$$x = \cos\phi_{\text{u.e.}} = \alpha\cos\theta \quad \alpha = \frac{1}{\cos\theta_c} \tag{7.13a}$$

$$y = \cos\phi_{\text{L}} = \alpha\sqrt{\frac{\alpha^2-1}{\alpha^2-x^2}} \tag{7.13b}$$

where θ_c is the phase at the lower cutoff frequency. Based on the relations between these variables, the condition in (7.14) should be enforced as

$$\varepsilon\cos(n\phi_{\text{u.e.}}+m\phi_{\text{L}})=K_1\frac{\cos^4\theta}{\sin\theta}+K_2\frac{\cos^2\theta}{\sin\theta}+K_3\frac{1}{\sin\theta} \tag{7.14}$$

$$K_1=-\varepsilon\left(4\alpha^4+4\alpha^3\sqrt{\alpha^2-1}\right) \tag{7.15a}$$

$$K_2=-\varepsilon\left(5\alpha^2+3\alpha\sqrt{\alpha^2-1}\right) \tag{7.15b}$$

$$K_3=\varepsilon \tag{7.15c}$$

After solving $k_1=K_1$, $k_2=K_2$, and $k_3=K_3$, the three normalized impedances, z_1, z_2, and z_s, are all determined by any specified ripple constant ε and bandwidth reflected in α. The impedances of all stubs and connecting lines Z_1, Z_2, and Z_s are calculated by multiplying them by the corresponding differential-port impedances.

Two differential wideband filters are built on the substrate, namely, Roger's RT/Duriod 6010LM with a permittivity of 10.7 and thickness of 0.635 mm in each layer. Figures 7.11(a) and 7.12(a) depict the physical dimensions of the designed filters, while Figures 7.11(b) and 7.12(b) display the top/bottom view photographs of the assembled filters. Figures 7.11 and 7.12 show plots of both DM and CM frequency responses of these two filters from synthesis, simulation, and measurement. These figures indicate that the three sets of DM frequency responses match well with each other, especially the synthesized and simulated ones. The 3 dB fractional bandwidths of the two filters are equal to approximately 120 and 123.2%, respectively. For Filter 1, CM rejection is realized by the unmatched field, which turns out to produce an all-stop performance. The CM rejection is higher than 25.8 dB within the realized DM passband and higher than 22.7 dB over the frequency range of 0–18.5 GHz. Additionally, the CM insertion loss in Filter 2 is higher than 21.6 dB over the frequency range of 0–10.45 GHz, which covers the entire DM passband.

7.4 DIFFERENTIAL WIDEBAND FILTERS BASED ON THE STRIP-LOADED SLOTLINE RESONATOR

In this section, the strip-loaded slotline resonator is introduced and applied to differential filters with multimode operation. The employment of back-to-back microstrip resonators enables easy control of the resonant modes for constructing wideband differential filters.

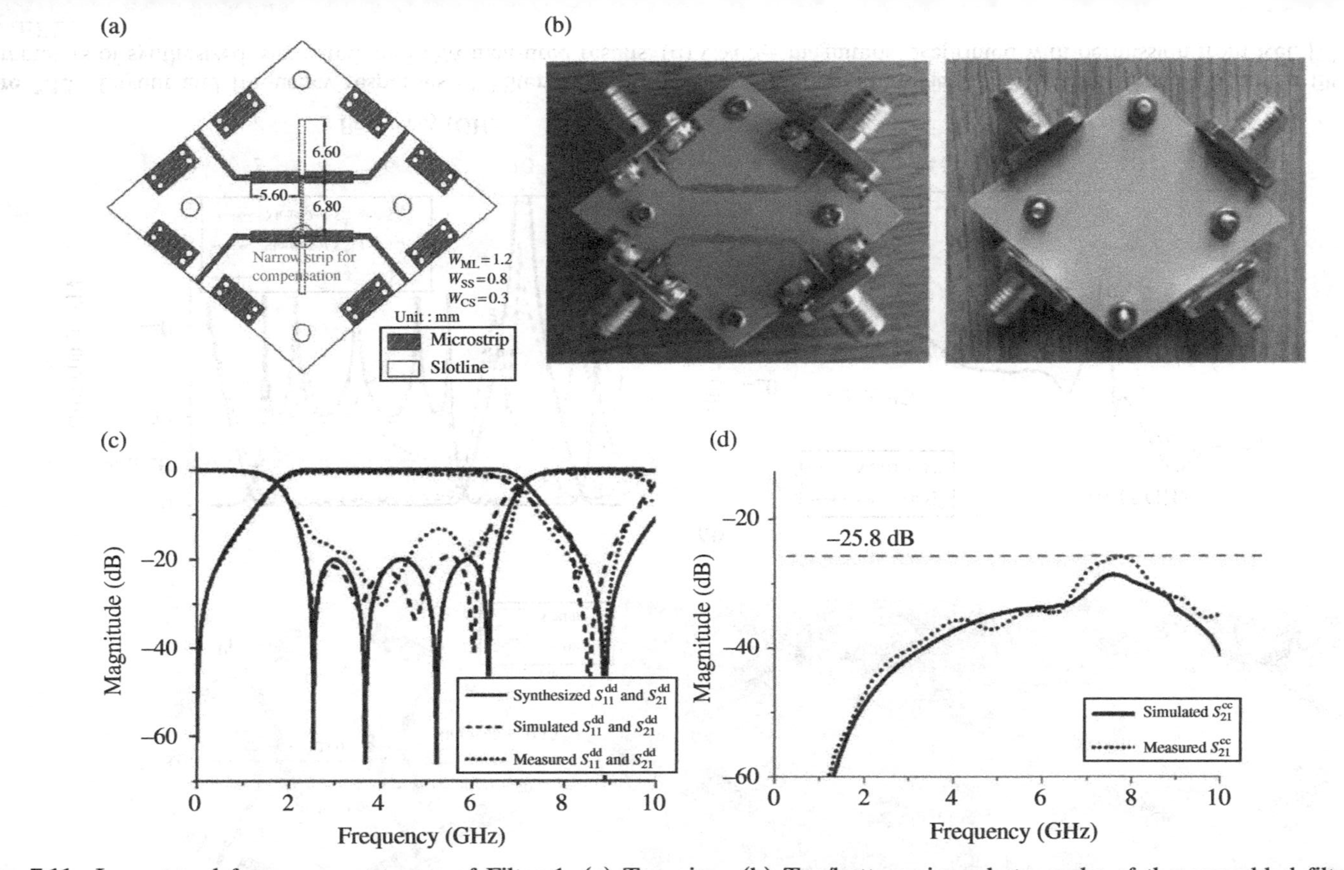

Figure 7.11 Layout and frequency responses of Filter 1. (a) Top view. (b) Top/bottom-view photographs of the assembled filter. (c) DM S-parameters for synthesized, simulated, and DM-measured results. (d) CM S_{21} magnitude. Reprinted with permission from Ref. [12]; copyright 2015 IEEE.

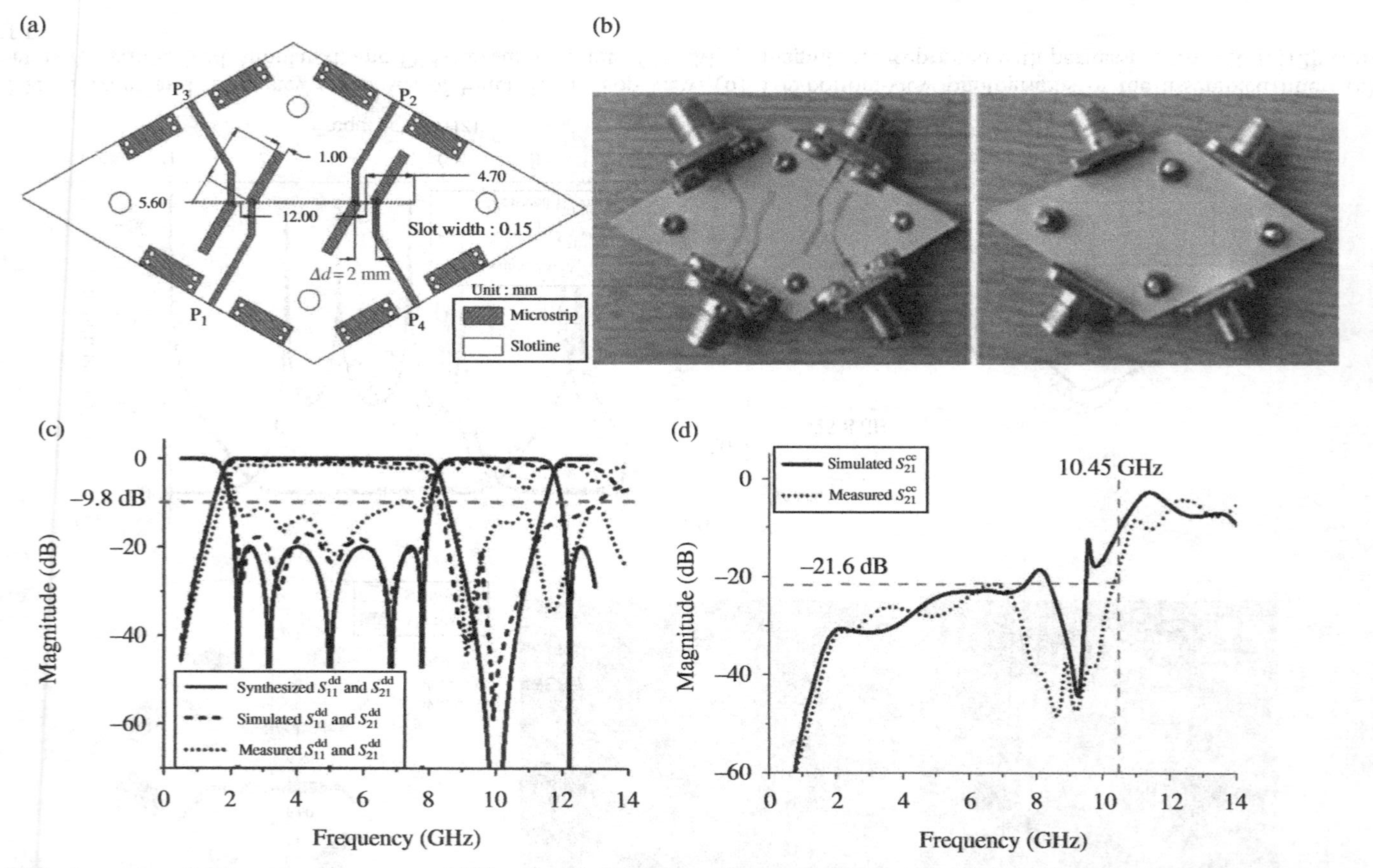

Figure 7.12 Layout and frequency responses of Filter 2. (a) Top view. (b) Top/bottom-view photographs of the assembled filter. (c) DM *S*-parameters of synthesized, simulated, and DM-measured results. (d) CM S_{21} magnitude. Reprinted with permission from Ref. [12]; copyright 2015 IEEE.

7.4.1 Differential Wideband Filters Using Triple-Mode Slotline Resonator

A novel strip-loaded slotline resonator is proposed by perpendicular and symmetrical loading of a pair of identical strip conductors, as illustrated in Figure 7.13. By virtue of the location and length of these paired strips, three and four resonant modes in the proposed slotline resonator can be properly excited and employed to realize a wide differential BPF with desirable CM rejection. The proposed filter, which is based on a strip-loaded resonator fed or tightly coupled with differential I/O ports placed on the top and bottom layers, is shown in Figure 7.14. By virtue of the widely used frequency-dispersive property of parallel coupled lines, two additional transmission poles can be produced within this DM passband to achieve satisfactory in-band flatness. Figure 7.15(a) depicts the variation of resonances in the developed resonator, with $\theta_M = 0°$, 15°, and 90°. The unloaded slotline resonator has a virtual-short-circuit condition at the middle position in even-order modes according to the electric field distribution in the slotline. When the strips are centrally loaded to excite the resonator, these even-order modes are not excited at all. However, the odd-order modes, which are under the virtual open-circuit condition at the middle position, tend to be affected if these strips are introduced. The resonator under the resonance of odd-order modes can be viewed as two individual $\lambda/4$ resonators. As the initial virtual-open-circuit end is moved from the middle position to the open-circuit end of the loaded strips, a frequency shift emerges. At the same time, the loaded strips themselves may induce additional resonance due to their role as a $\lambda/2$ resonator. In Figure 7.15, the first resonant frequency is

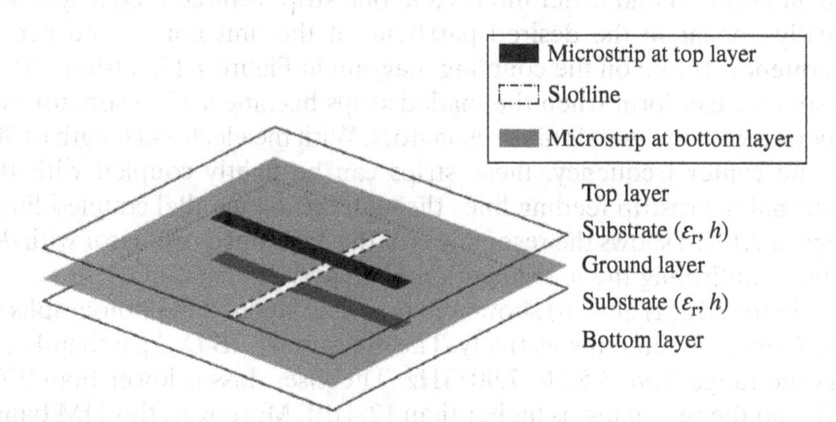

Figure 7.13 Proposed strip-loaded slotline resonator. Reprinted with permission from Ref. [14]; copyright 2016 IEEE.

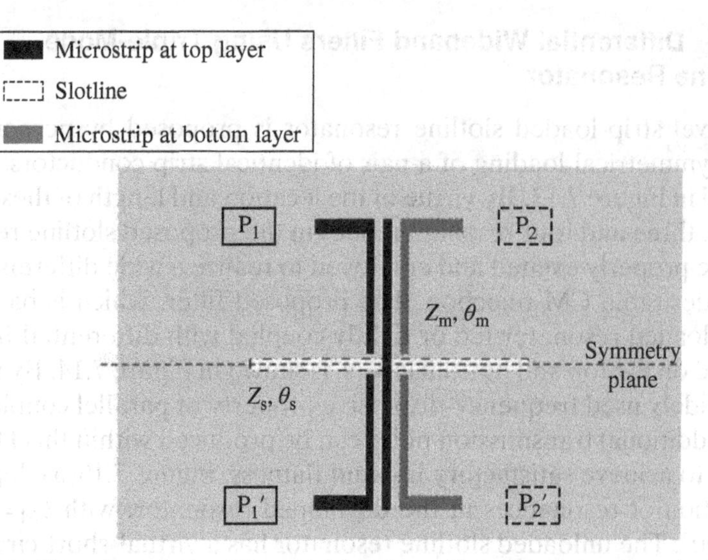

Figure 7.14 Top view of the differential bandpass filter based on the strip-loaded slotline resonator. Reprinted with permission from Ref. [14]; copyright 2016 IEEE.

labeled f_{addi}. Importantly, the short-ended slotline, labeled Z_s and θ_s in Figure 7.16, will definitely produce multiple transmission zeros at the frequencies for which the electrical length from the short-circuited ends to the strips is equal to $\theta_s = n \times 180°$ $(n = 1,2,3,\ldots)$. At these frequencies, the short-circuited ends cause the emergence of a virtual-short-circuit condition at the center, thus generating transmission zeros. In Figure 7.15, these zeros appear at normalized frequencies of 2, 4, and 6.

When θ_M is lengthened to 90°, three resonant modes, that is, two frequency-shifted odd-order modes and one strip-inspired mode, symmetrically appear in the desired passband at the unit normalized center frequency. Based on the coupling diagram in Figure 7.17, a triple-mode resonator can form when the loaded strips become a $\lambda/2$ resonator and operate with the two slotline resonators. With the electrical length of 90° at the center frequency, these strips can be tightly coupled with the external microstrip feeding lines through the $\lambda/4$ parallel coupled lines. Figure 7.15(b) shows the resonances of the developed resonator with $\theta_M = 90°$, confirming the aforementioned analysis.

Figure 7.18(a) and (b) shows the top-view layout and photographs of the fabricated filter, respectively. The measured 3 dB DM passband covers the range from 3.58 to 7.90 GHz. The insert loss is lower than 0.77 dB, and the return loss is higher than 12.4 dB. Moreover, this DM bandpass filter has a widened upper stopband, with attenuation higher than 20 dB up to 14.5 GHz. The measured CM attenuation is better than 38.5

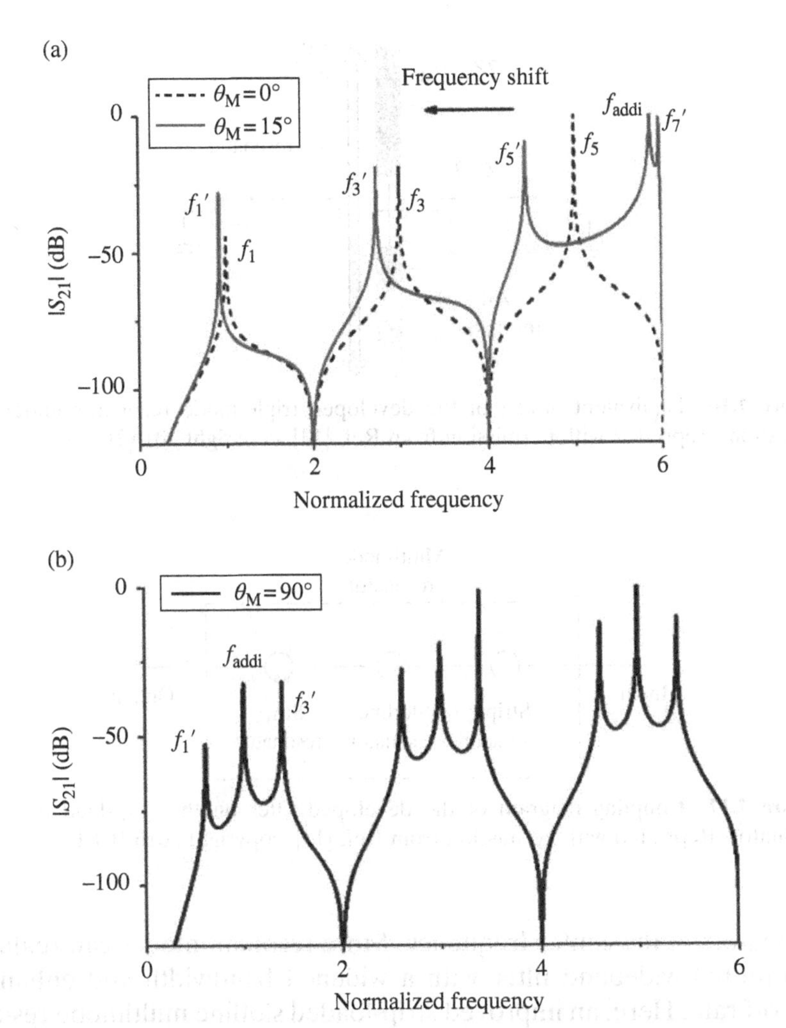

Figure 7.15 Resonances of the triple-mode resonator with different electrical lengths of loaded strips. (a) $\theta_M = 0°$ and 15° and (b) 90°. Reprinted with permission from Ref. [14]; copyright 2016 IEEE.

dB over the entire DM passband, and better than 20 dB over the plotted frequency range of 0–14.9 GHz. The bandwidth of the realizable triple-mode filter is in the range of approximately 67–87%.

7.4.2 Differential Wideband Filters Using Quadruple-Mode Slotline Resonator

A quadruple-mode resonator can be formed if the slotline resonator is under dual-mode operation, and the two loaded strip lines act as two

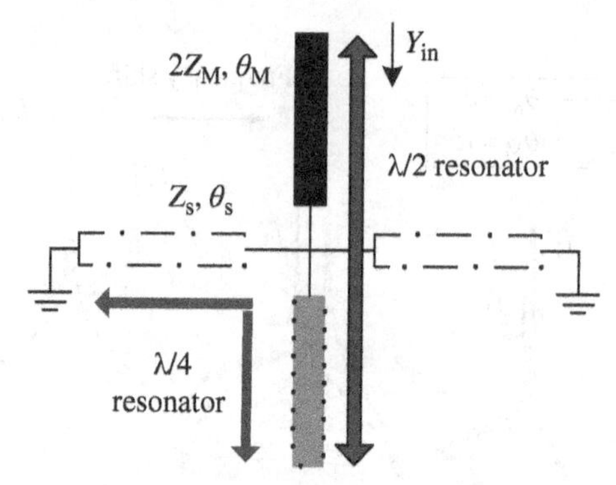

Figure 7.16 Equivalent model of the developed triple-mode resonator under DM operation. Reprinted with permission from Ref. [14]; copyright 2016 IEEE.

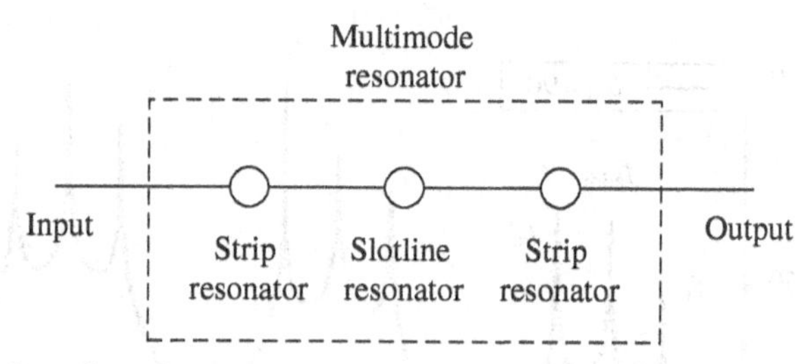

Figure 7.17 Coupling diagram of the developed filter on the strip-loaded slotline resonator. Reprinted with permission from Ref. [14]; copyright 2016 IEEE.

resonators at the center frequency. More resonant modes can realize a differential wideband filter with a widened bandwidth and enhanced roll-off rate. Here, an improved strip-loaded slotline multimode resonator is presented in which a pair of identical strips is symmetrically placed along the slotline resonator, with an offset distance from the center position. These strips are fed by parallel coupled lines at their two sides, as illustrated in Figure 7.19 [14]. Figure 7.20(a) and (b) shows the physical layout and photographs of the fabricated filter, respectively. Figure 7.20 (c) indicates the simulated and measured frequency responses. The measured DM 3 dB passband covers the range of 3.45–8.45 GHz, and the in-band CM rejection is higher than 24.5 dB. The results demonstrate satisfactory out-of-band performance, with harmonic suppression up to 16.0 GHz. The bandwidth of the realizable quadruple-mode filter ranges from 82 to 94%. Compared with the design in Section 7.4.1, this filter contains an additional transmission pole, which is beneficial for a wider DM passband.

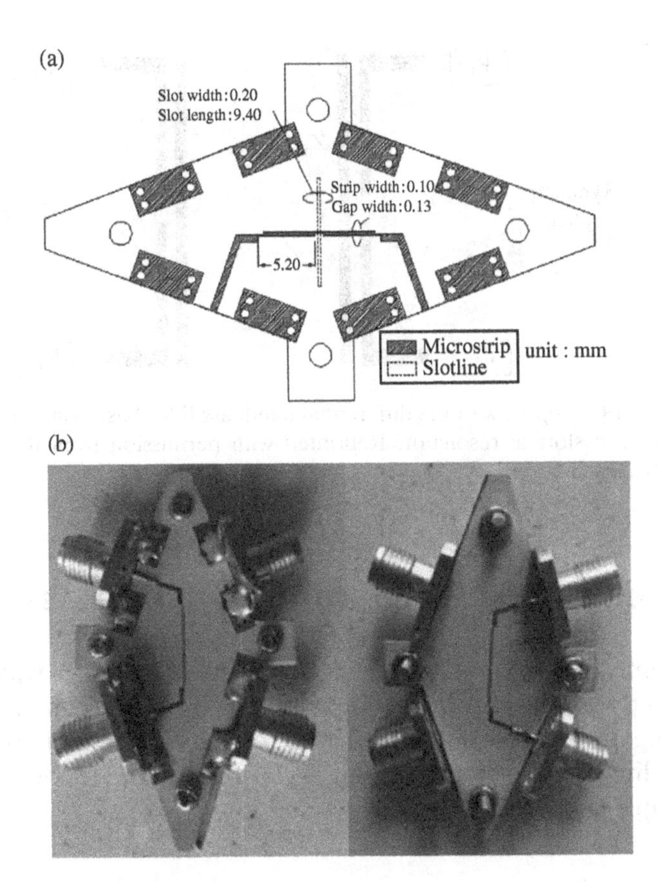

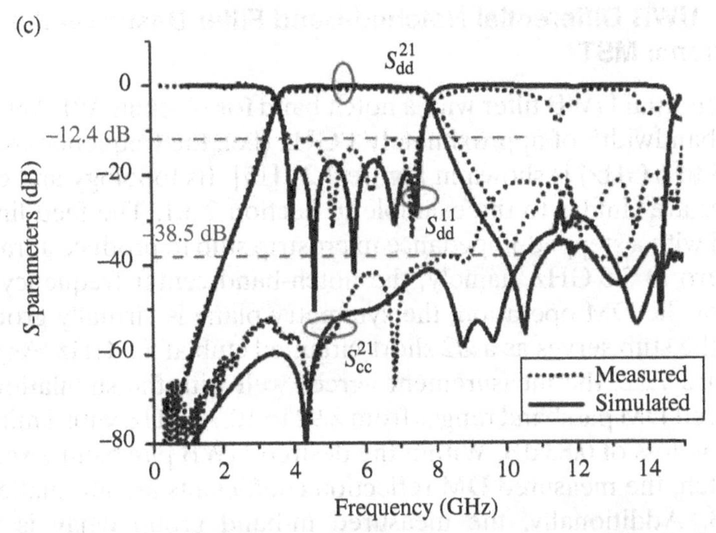

Figure 7.18 Physical layout and frequency responses of the developed triple-mode filter. (a) Top view with all dimensions labeled. (b) Top/bottom-view photographs of the assembled filter. (c) Simulated and measured results. Reprinted with permission from Ref. [14]; copyright 2016 IEEE.

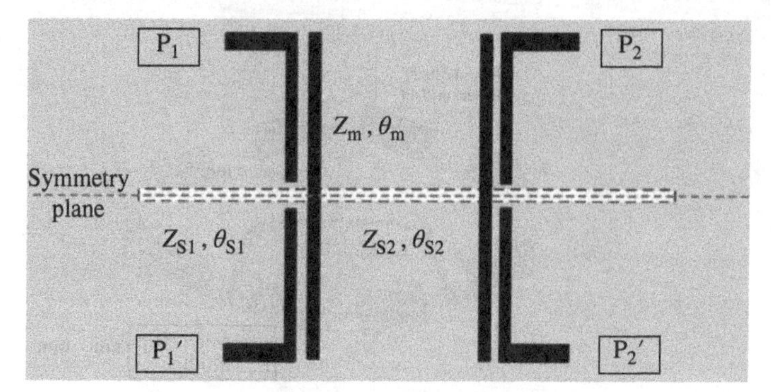

Figure 7.19 Top view of the differential bandpass filter based on the quadruple-mode strip-loaded slotline resonator. Reprinted with permission from Ref. [14]; copyright 2016 IEEE.

7.5 UWB DIFFERENTIAL NOTCHED-BAND FILTER

In some applications, notched bands are desired in wideband filters for suppressing unwanted signals. These bands can be realized by adding an extra stopband structure to block transmissions at the corresponding frequency. This point is clarified by the following two simple examples.

7.5.1 UWB Differential Notched-Band Filter Based on the Traditional MST

A differential UWB filter with a notch band for blocking WLAN signals and a bandwidth of approximately 1 GHz (i.e., the frequency spectrum from 5 to 6 GHz) is shown in Figure 7.21 [15]. Its topology and characteristics are similar to the example in Section 7.3.1. The feed lines are loaded with a stepped-impedance microstrip stub to produce a transmission zero at 5.5 GHz, namely, the notch-band center frequency. Note that, for the DM operation, the symmetry plane is virtually grounded. Thus, the stub serves as a $\lambda/2$ short-circuited stub at 5.5 GHz. As shown in Figure 7.22, the measurement agrees well with the simulation. The measured DM passband ranges from 2.92 to 10.73 GHz, with a minimum insertion loss of 0.83 dB. Within the desired UWB passband, except for the notch, the measured DM reflection coefficients are all smaller than −10 dB. Additionally, the measured in-band group delay is within 0.31–0.6 ns, with a maximum variation of 0.29 ns. The measured CM rejection levels are greater than 18.85 dB over the entire UWB

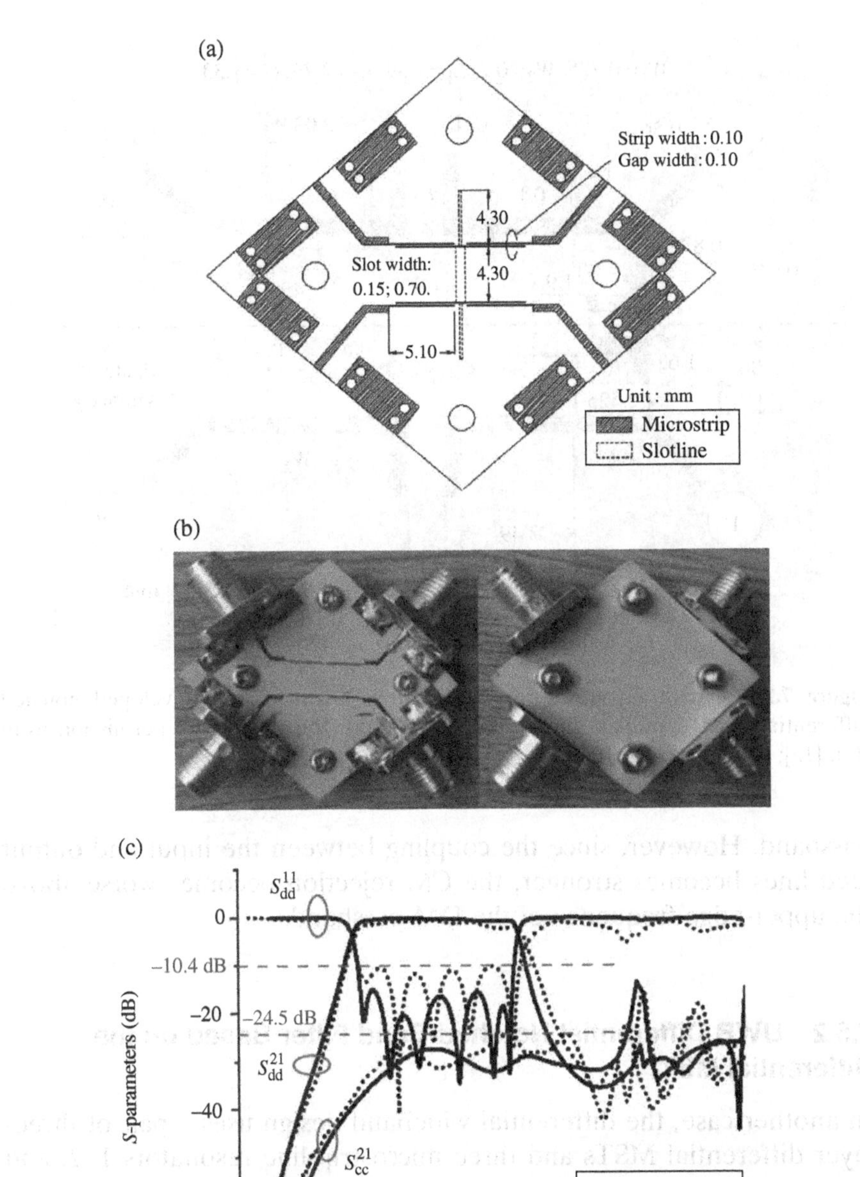

Figure 7.20 Physical layout and frequency responses of the developed quadruple-mode filter. (a) Top view with all the dimensions labeled. (b) Top/bottom-view photographs of the assembled filter. (c) Simulated and measured results. Reprinted with permission from Ref. [14]; copyright 2016 IEEE.

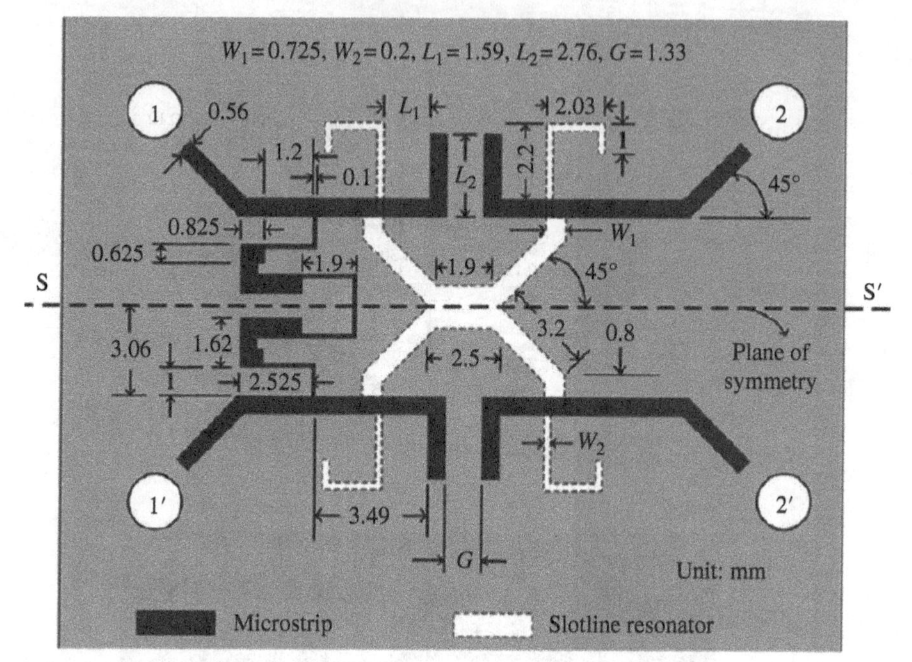

Figure 7.21 Circuit layout of top and bottom layers of the developed notched differential UWB bandpass filter with 5 GHz band. Reprinted with permission from Ref. [15]; copyright 2012 IEEE.

passband. However, since the coupling between the input and output feed lines becomes stronger, the CM rejection becomes worse above the upper-edge frequency of the DM passband.

7.5.2 UWB Differential Notched-Band Filter Based on the Differential MST

In another case, the differential wideband design uses a pair of three-layer differential MSTs and three microstrip-line resonators 1, 2, and 3 on the bottom layer [16], as shown in Figure 7.23. The proposed wide-band differential BPF is fabricated on two layers of RO4003C substrate (dielectric constant of 3.38, loss tangent of 0.0027, and $h = 0.813$ mm) and one layer of prepreg (dielectric constant of 3.48, loss tangent of 0.003, $h_p = 0.2$ mm). The simulated results for a pair of multilayer differential MSTs are shown in Figure 7.24 by the line marked as "without resonator 1." For a pair of multilayer differential MSTs, a wideband DM response with two transmission poles can be realized. Moreover, the two transmission poles can be controlled by the length of the

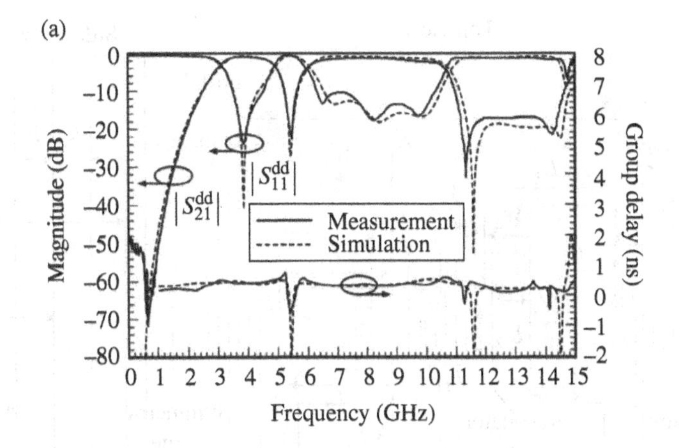

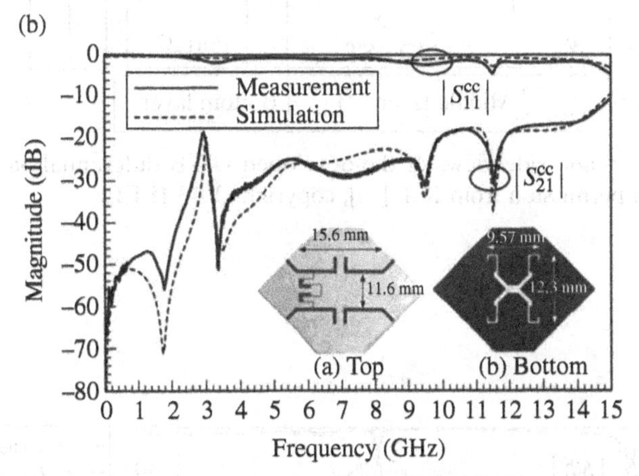

Figure 7.22 Simulated and measured *S*-parameters for (a) DM and (b) CM operations of the notched differential UWB bandpass filter with 5 GHz band. Reprinted with permission from Ref. [15]; copyright 2012 IEEE.

short-circuit slotline stub L_2 and the length of the open circuit microstrip stub L_3 when the distance between the pair of multilayer differential MSTs is fixed. L_2 and L_3 can be roughly obtained by [17]

$$L_2 = \lambda_s \arctan(Z_s/2\pi f_{0t} L_{sc})/2\pi \tag{7.16}$$

$$L_3 = \lambda_m \frac{\arctan(2\pi f_{0t} Z_m C_{oc})^{-1}}{2\pi} \tag{7.17}$$

where λ_s and λ_m represent the guided wavelengths of the microstrip line and slotline at the center frequency (f_{0t}) of the multilayer differential MST, respectively. Z_s and Z_m are the characteristic impedances of the

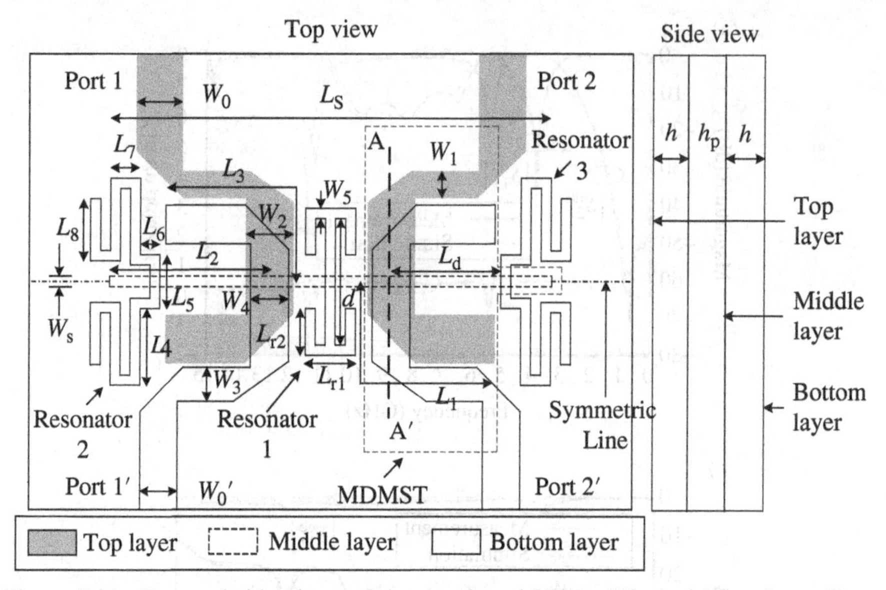

Figure 7.23 Top and side views of the developed UWB differential bandpass filter. Reprinted with permission from Ref. [16]; copyright 2013 IEEE.

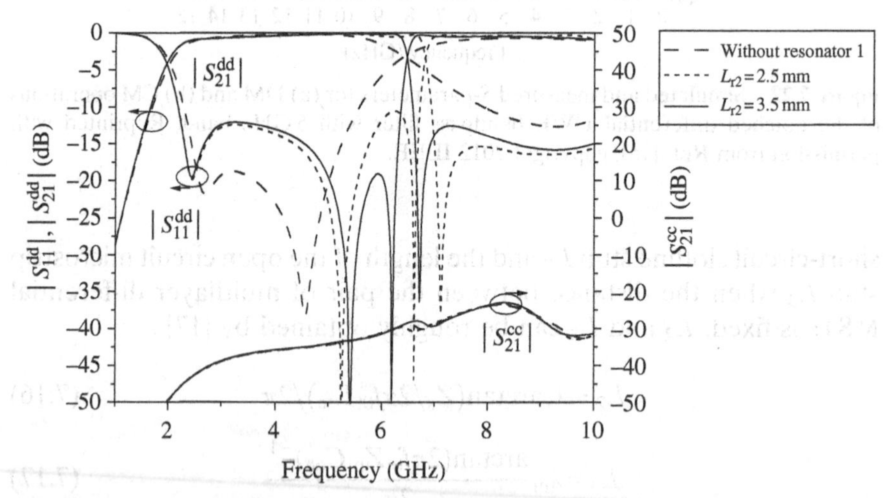

Figure 7.24 Simulated results of the developed wideband differential bandpass filter with or without resonator 1. ($W_0 = 2.36$ mm, $W_0' = 1.86$ mm, $W_1 = 1.3$ mm, $W_2 = 2.36$ mm, $W_3 = 1.65$ mm, $W_4 = 1.86$ mm, $W_5 = 0.5$ mm, $L_s = 22$ mm, $W_s = 0.5$ mm, $L_1 = 7.85$ mm, $L_2 = 8$ mm, $L_3 = 7.85$ mm, $d = 6.5$ mm, and $L_{r1} = 2.5$ mm.) Reprinted with permission from Ref. [16]; copyright 2013 IEEE.

microstrip line and slotline, respectively. L_{sc} and C_{oc} are the equivalent inductance and capacitance for the shorted and open ends of the stubs, respectively. To further improve the DM bandwidth, a microstrip-line resonator (resonator 1) is added to the middle of the pair of multilayer differential MSTs. An additional transmission pole and transmission zero are produced in the DM response after adding resonator 1. To simplify the design, the resonant frequency of resonator 1 is used to control the upper fringe of the DM passband while keeping the other dimensions unchanged. The resonant frequency of resonator 1 can be obtained when B_d equals zero, where B_d represents the normalized DM input susceptance of resonator 1 and can be evaluated by

$$B_d = \frac{-1/\tan\beta(L_{r1}/2 + L_{r2}) + 2\tan\beta(d/2)}{1 + (-1/\tan\beta(L_{r1}/2 + L_{r2}) - \tan\beta(d/2))\tan\beta(d/2)} \tag{7.18}$$

where β is the phase constant of the microstrip line. Figure 7.24 displays the DM responses at different L_{r2}. It can be seen from Figure 7.24 and (7.18) that the higher the resonant frequency of resonator 1, the wider the DM bandwidth. The shape factor at the upper side of the passband is improved because of the new transmission zero created by resonator 1, while the lower part of the DM passband generated by the multilayer differential MSTs remains almost the same.

The electric fields on each side of the slotline can be transferred to the slotline only when they are out of phase, so that only DM signals can pass, while the CM signal is reflected naturally. This is also why CM suppression remains nearly unchanged when tuning resonator 1, as shown in Figure 7.23. This result leads to a simple design procedure, because only the DM response needs to be addressed for the developed filter. The simulated and measured results for the wideband differential bandpass filter without a notched band are shown in Figure 7.25. The filter centered at 4.25 GHz possesses a fractional bandwidth of 115%, with a minimum insertion loss of 0.43 and 20 dB CM suppression from 1 to 10 GHz. The remarkable variation in $|S_{11}^{dd}|$ arises because of the fabrication tolerance of the multilayer structure.

The notched band inside the differential wideband passband can be realized using two microstrip-line resonators (resonator 2/3) above the slotline stubs to break the operation of transitions at the corresponding frequency. The position of the notched band is decided by the resonant frequency of the resonators. The attenuation of the notched band can be controlled by changing the distance (L_d) between resonator 2/3 and the multilayer differential MST, as shown in Figure 7.26. Clearly, a smaller

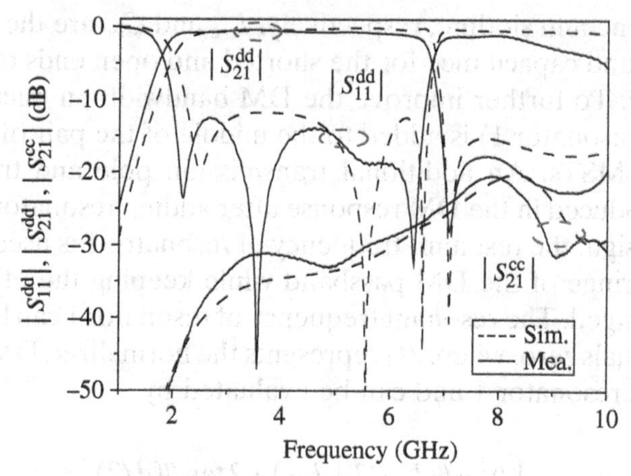

Figure 7.25 Simulated and measured results of the wideband differential bandpass filter ($L_{r2} = 2.5$ mm). Reprinted with permission from Ref. [16]; copyright 2013 IEEE.

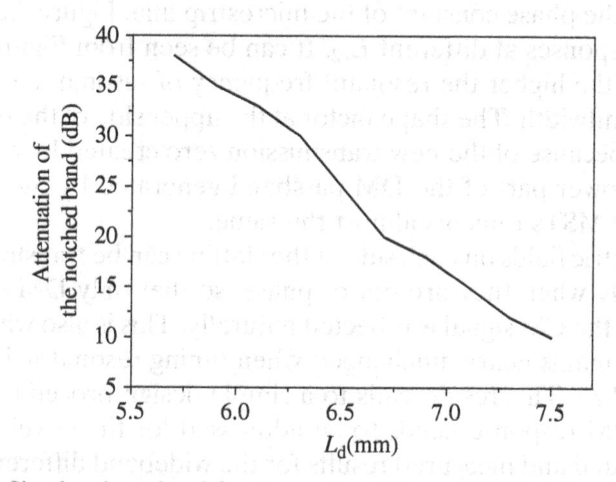

Figure 7.26 Simulated results of the attenuation of the notched band against L_d. ($L_4 = 3.7$ mm, $L_5 = 3.7$ mm, $L_6 = 1$ mm, $L_7 = 1.5$ mm, $L_8 = 3$ mm, and the other dimensions are the same as those in Figure 7.4.) Reprinted with permission from Ref. [16]; copyright 2013 IEEE.

L_d will cause larger attenuation of the notched band because of stronger coupling between resonator 2/3 and the shorted slot stub. To preserve the compact size and high attenuation of the notched band, L_d is selected as 5.7 mm. The measured and simulated results of the differential wideband filter with a notched band are shown in Figure 7.27. The minimum insertion loss, bandwidth, center frequency, and CM suppression remain unchanged in comparison to the filter without the notched

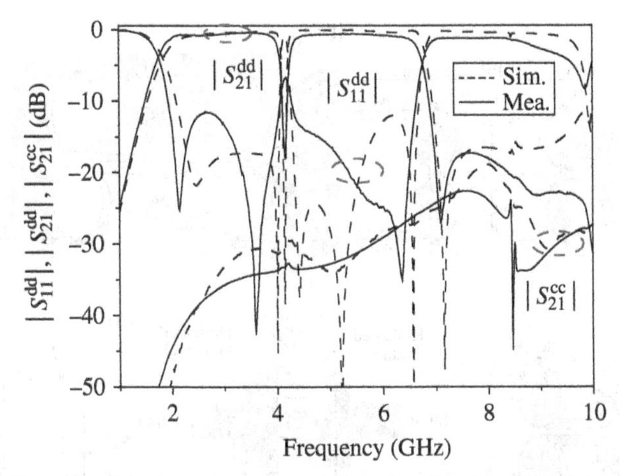

Figure 7.27 Simulated and measured results of the developed wideband differential bandpass filter with notched band. Reprinted with permission from Ref. [16]; copyright 2013 IEEE.

band. The measurement shows a sharp, notched band positioned at 4.15 GHz, with 18 dB attenuation.

Furthermore, a dual notched-band UWB differential bandpass filter is designed, using a folded triple-mode slotline resonator and the MST [18]. Here, two groups of inverse $\lambda/4$ resonators with different sizes are separately applied to realize two notched bands at 5.2 and 5.8 GHz.

7.6 DIFFERENTIAL UWB FILTERS WITH ENHANCED STOPBAND SUPPRESSION

The dumbbell-shaped DGS at the ground plane of the middle layer can be connected with an H-shaped shunt with open-ended stubs on the bottom layer, which functions as a low-pass filter for widening the upper stopband of the UWB differential bandpass filter [19]. This structure includes three conductive layers interleaved by two substrates; two pairs of similar tapered microstrip patches at the top and bottom layers are coupled through a pair of tapered slots at the middle layer of the structure, which also contains the ground plane. There are two vias connecting the far ends of the coupled microstrip patches to the ground plane, as shown in Figure 7.28.

It has been shown that a DM wideband can be easily designed, based on a broadside-coupled structure. However, such a design has a narrow upper stopband with a slow cutoff rate [20]. Thus, to improve the sharpness of the stopbands and to make the filter ports uniplanar, two sections

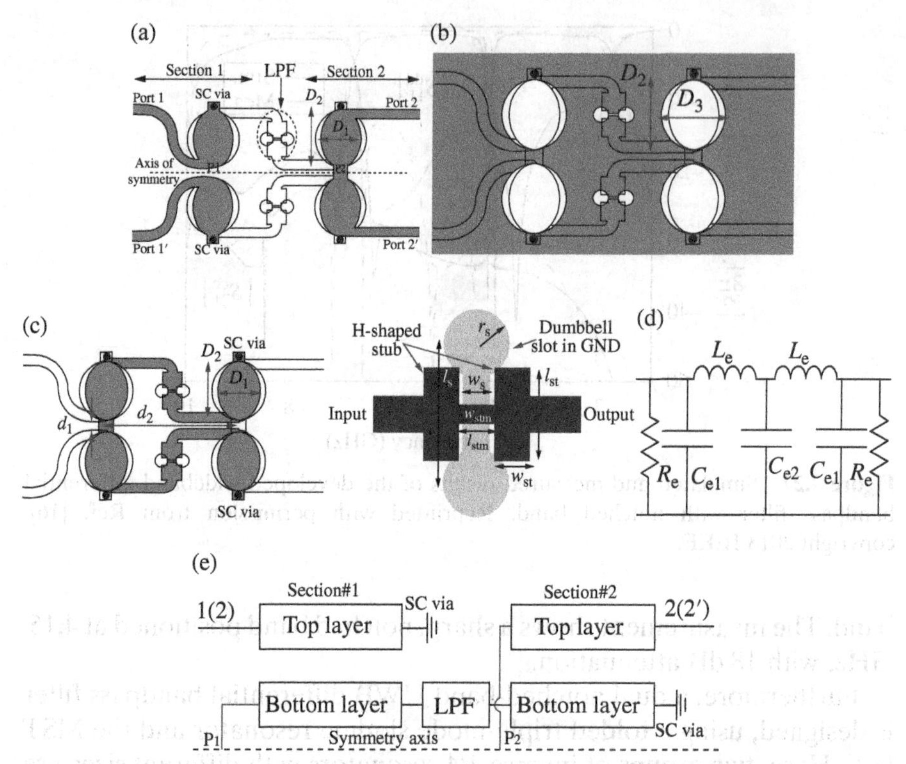

Figure 7.28 Configuration of the developed differential bandpass filter with the following layers highlighted: (a) top layer, (b) middle layer, and (c) bottom layer. (d) Details of the embedded low-pass filter and its equivalent circuit. (e) Simplified diagram of the device. Reprinted with permission from Ref. [19]; copyright 2011 IEEE.

are employed, as illustrated in Figure 7.28. A simplified diagram for the device is depicted in Figure 7.28(a)–(e). In the DM operation, a virtual short circuit appears along the axis of symmetry. In this case, section 1 and section 2 are connected virtually to the ground at points P_1 and P_2, respectively. Using four-port network analysis, the reflection (Γ) and transmission (T) coefficients for each coupled section in this mode are

$$\Gamma = \frac{1 - C^2\left(1 + \sin^2(\beta_{ef} D_2)\right)}{\left[\sqrt{1 - C^2}\cos(\beta_{ef} D_2) + j\sin(\beta_{ef} D_2)\right]^2} \tag{7.19}$$

$$T = \frac{j2C\sqrt{1 - C^2}\sin(\beta_{ef} D_2)}{\left[\sqrt{1 - C^2}\cos(\beta_{ef} D_2) + j\sin(\beta_{ef} D_2)\right]^2} \tag{7.20}$$

where C is the coupling factor between the top and bottom layers, β_{ef} is the effective phase constant at the center of the coupled structure, and

D_2 is the physical length of the coupled structure (Figure 7.28a), which is equal to $\lambda/4$ at the center frequency of the passband. In CM operation, a virtual open circuit appears along the axis of symmetry, and, thus, at P_1 and P_2 marked in Figure 7.28(e). Γ and T for each of the coupled sections in this mode can be found using four-port network analysis, that is, $\Gamma = 1$ and $T = 0$. Thus, the structure in the CM behaves as an all-stop filter. This is the preferred behavior of the differential circuit in the CM, compared to having a bandstop filter, because the all-stop filter secures the rejection of CM signals across the entire DM passband of the bandpass filter, whereas a bandstop filter usually has a low rejection capability at the lower and upper ends of its stopband. The embedded low-pass filter can be represented by the equivalent RLC circuit in Figure 7.28(d). The stubs with length l_{st} and the slot with the length l_s in the ground plane are designed to be slightly below $\lambda/4$ at the required cutoff frequency. These lengths ensure that the stubs and slot add effectively, thus representing pure reactive elements to the microstrip line.

The designed filter was fabricated using two layers of Rogers TMM4 as the substrate, and the final dimensions (in mm) are as follows: $D_1 = 4.16$, $D_2 = 5.86$, $D_3 = 5.28$, $l_s = 2.14$, $d_1 = 0.9$, $d_2 = 13$, $l_{st} = 1.94$, $l_{stm} = 0.75$, $w_s = 0.14$, $w_{st} = 1.4$, $w_{stm} = 0.86$, and $r_s = 0.65$. The simulated and measured results are depicted in Figure 7.29. The filter has a 3 dB DM

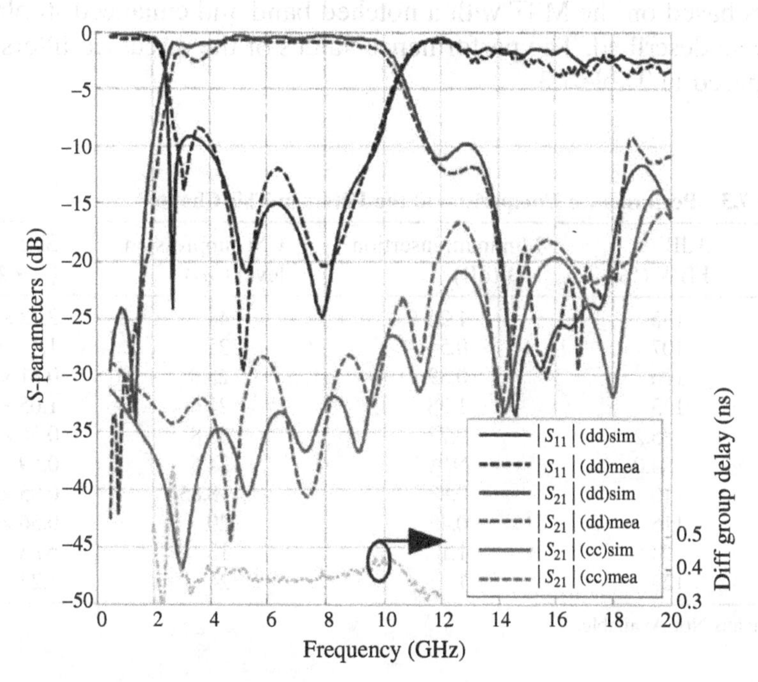

Figure 7.29 Differential and CM performance of the filter. Reprinted with permission from Ref. [19]; copyright 2011 IEEE.

passband that extends from 2.5 to 10.6 GHz, which is equivalent to a 123% fractional bandwidth centered at 6.55 GHz. The insertion loss is less than 1 dB across the band from 2.5 to 10 GHz according to the simulations and from 4.2 to 9.9 GHz in the measured results. The return loss is more than 14 dB across the band from 4.3 to 9.2 GHz. As required by UWB regulations, the filter has sharp lower and upper cutoff bands and a wide upper stopband that extends beyond 20 GHz. The developed filter suppresses CM signals across the entire investigated band, that is, up to 20 GHz. With regard to the UWB range (3.1–10.6 GHz), the CM signal is attenuated by more than 27 dB according to the simulations and by 24 dB in the measurements.

7.7 SUMMARY

This chapter introduces a variety of UWB differential bandpass filters based on the conventional or differential MST, which combines the merits of a microstrip line and a slotline to achieve wideband characteristics. The MST-based differential filter with two- and three-layer structures realizes wide DM passband and high CM suppression due to the inherent immunity of the slotline to the CM noise. Finally, UWB differential filters based on the MST with a notched band and enhanced stopband are also described. The performance values of the discussed filters are compared in Table 7.3.

Table 7.3 Performance Comparison of the Filters in This Chapter

	3 dB FBW (%)	Minimum insertion loss (dB)	CM suppression level (dB)	Size ($\lambda_g \times \lambda_g$)
[10]	105	1.22	44.7	2.33 × 0.39
[11]	107	0.5	25	1.4 × 1.4
[12]	120	0.59	25.8	0.81 × 0.48
	123	1.28	21.6	1.05 × 0.51
[14]	75.2	0.77	35.8	0.56 × 0.48
	84.0	N/A	24.5	0.59 × 0.7
[15]	110	0.83	18.85	0.95 × 0.75
[16]	115	0.43	20	0.56 × 0.28
[18]	94.8	1.1	13	N/A
[19]	123	1	24	1.23 × 0.87

N/A means Not Available.

REFERENCES

1. T. B. Lim and L. Zhu, "A DM wideband bandpass filter on microstrip line for UWB application," *IEEE Microwave Wireless Compon. Lett.*, vol. **19**, no. 10, pp. 632–634, Oct. 2009.
2. T. B. Lim and L. Zhu, "Highly selective DM wideband bandpass filter for UWB application," *IEEE Microwave Wireless Compon. Lett.*, vol. **21**, no. 3, pp. 133–135, Mar. 2011.
3. W. J. Feng and W. Q. Che, "Novel wideband differential bandpass filters based on T-shaped structure," *IEEE Trans. Microwave Theory Tech.*, vol. **60**, no. 6, pp. 1560–1568, Jun. 2012.
4. W. J. Feng, W. Q. Che, Y. L. Ma, and Q. Xue, "Compact wideband differential bandpass filters using half-wavelength ring resonator," *IEEE Microwave Wireless Compon. Lett.*, vol. **23**, no. 2, pp. 81–83, Feb. 2013.
5. X.-H. Wang, H. L. Zhang, and B.-Z. Wang, "A novel ultra-wideband differential filter based on microstrip line structures," *IEEE Microwave Wireless Compon. Lett.*, vol. **23**, no. 3, pp. 128–130, Mar. 2013.
6. X. Wu, Q. X. Chu, "Compact differential ultra-wideband bandpass filter with CM suppression," *IEEE Microwave Wireless Compon. Lett.*, vol. **22**, no. 9, pp. 456–458, Sep. 2012.
7. X. H. Wu, Q. X. Chu, and L. L. Qiu, "Differential wideband bandpass filter with high-selectivity and CM suppression," *IEEE Microwave Wireless Compon. Lett.*, vol. **23**, no. 12, pp. 644–646, Dec. 2013.
8. P. Vélez *et al.*, "Ultra-compact (80 mm^2) DM ultra-wideband (UWB) bandpass filters with CM noise suppression," *IEEE Trans. Microwave Theory Tech.*, vol. **63**, no. 4, pp. 1272–1280, Apr. 2015.
9. K. C. Gupta, *Microstrip lines and slotlines*, Artech House, Inc., Norwood, MA, 1996.
10. Y.-J. Lu, S.-Y. Chen, and P. Hsu, "A DM wideband bandpass filter with enhanced CM suppression using slotline resonator," *IEEE Microwave Wireless Compon. Lett.*, vol. **22**, no. 10, pp. 503–505, Oct. 2012.
11. D. Chen, H. Bu, L. Zhu, and C. Cheng, "A DM wideband bandpass filter on slotline multi-mode resonator with controllable bandwidth," *IEEE Microwave Wireless Compon. Lett.*, vol. **25**, no. 1, pp. 28–30, Jan. 2015.
12. X. Guo, L. Zhu, K.-W. Tam, and W. Wen, "Wideband differential bandpass filters on multimode slotline resonator with intrinsic CM rejection," *IEEE Trans. Microwave Theory Tech.*, vol. **63**, no. 5, pp. 1587–1594, May 2015.
13. R. Li, S. Sun, and L. Zhu, "Synthesis design of ultra-wideband bandpass filters with composite series and shunt stubs," *IEEE Trans. Microwave Theory Tech.*, vol. **57**, no. 3, pp. 684–692, Mar. 2009.

14. X. Guo, L. Zhu, and W. Wen, "Strip-loaded slotline resonators for differential wideband bandpass filters with intrinsic CM rejection," *IEEE Trans. Microwave Theory Tech.*, vol. **62**, no. 4, pp. 450–458, Feb. 2016.
15. C.-H. Lee, C.-I. G. Hsu, and C.-J. Chen, "Band-notched differential UWB bandpass filter with stepped-impedance slotline multi-mode resonator," *IEEE Microwave Wireless Compon. Lett.*, vol. **22**, no. 4, pp. 182–184, Apr. 2012.
16. J. Shi, C. Shao, J.-X. Chen, Q.-Y. Lu, Y. Peng, and Z.-H. Bao, "Compact low-loss wideband differential bandpass filter with high CM suppression," *IEEE Microwave Wireless Compon. Lett.*, vol. **23**, no. 9, pp. 480–482, Sep. 2013.
17. A. Podcameni and M. L. Coimbra, "Slotline-microstrip transition on iso/anisotropic substrate: A more accurate design," *Electron. Lett.*, vol. **16**, no. 20, pp. 780–781, 1980.
18. H.-W. Deng, Y. Zhao, Y. He, S.-L. Jia, and M. Wang, "Compact dual-notched balanced UWB bandpass filter with folded triple-mode slotline resonator," *Electron. Lett.*, vol. **50**, no. 6, pp. 447–449, Mar. 2014.
19. A. M. Abbosh,"Ultra wideband balanced bandpass filter," *IEEE Microwave Wireless Compon. Lett.*, vol. **21**, no. 9, pp. 480–482, Sep. 2011.
20. A. M. Abbosh, "Planar bandpass filters for ultra-wideband applications," *IEEE Trans. Microwave Theory Tech.*, vol. **55**, no. 10, pp. 2262–2269, Oct. 2007.

CHAPTER 8

APPLICATION OF SIGNAL INTERFERENCE TECHNIQUE TO THE IMPLEMENTATION OF WIDEBAND DIFFERENTIAL FILTERS

Wei Qin[1] and Quan Xue[2]

[1]School of Electronics and Information, Nantong University, Nantong, China
[2]School of Electronic and Information Engineering, South China University of Technology, Guangzhou, China

8.1 BASIC CONCEPT OF THE SIGNAL INTERFERENCE TECHNIQUE

In wireless communications, sources of interference are ubiquitous and can be classified into various types, such as same-frequency interference, intermodulation interference, out-of-band interference, spurious emission interference, and image interference. In general, interference is undesirable in wireless communications. However, the signal interference technique, which makes use of interference, has been widely

Balanced Microwave Filters, First Edition. Edited by Ferran Martín, Lei Zhu, Jiasheng Hong, and Francisco Medina.

applied to improve the quality and security of modern wireless communication systems, especially in digital signal processing. For example, factitious interference can be used to modulate transmitted signals, thus providing an effective method of enhancing the anti-jamming ability of the communication system.

8.1.1 Fundamental Theory

The signal interference technique has been applied in the microwave/radio-frequency (RF) field since 1985 when Professor C. Rauscher proposed and investigated this concept in the design of distributed microwave filters [1]. Two different types of circuit models are available. One type of the models is the transversal structure, which uses the feedforward technique as shown in Figure 8.1(a). In this transversal filter model, the filtering responses are obtained by adding the amplitude and phase of each feedforward component. The other type is the

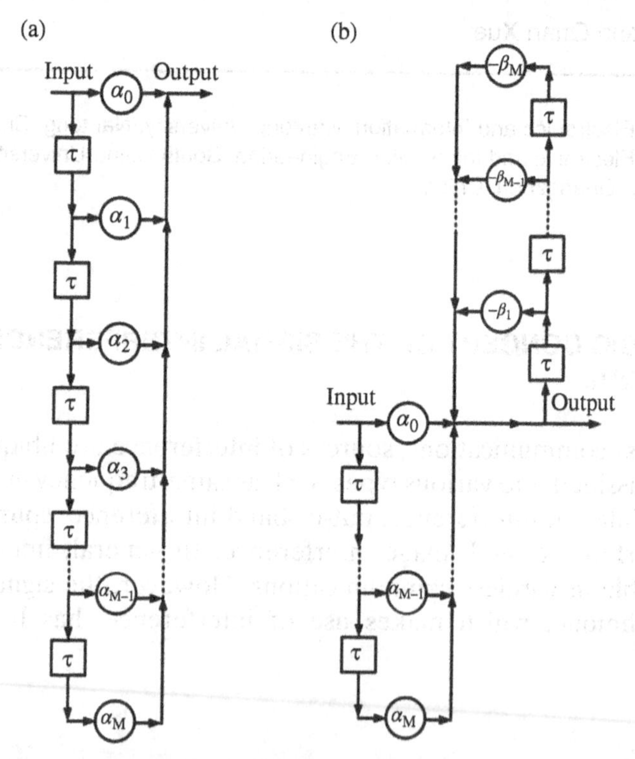

Figure 8.1 Signal flow graphs for distributed microwave filters in (a) a transversal structure and (b) a recursive structure. Reprinted with permission from Ref. [1]; copyright 1985 IEEE.

recursive structure, which is illustrated in Figure 8.1(b). In the recursive filter model, the feedback technique is applied based on the transversal filter structure. The filtering responses are determined by the amplitude and phase of each feedforward and feedback component. With the feedback technique, the recursive filter structure shows increasingly stable characteristics. In practice, however, parasitic feedback cannot be allowed in most microwave active circuits. Therefore, the transversal-type filter is commonly selected as the theoretical model for the design of microwave filters. Moreover, transmission poles and zeros can be generated by in-phase superposition and out-of-phase elimination, respectively, to improve the selectivity and harmonic suppression of the filters.

For an $(M+1)$-order transversal filter, the transfer function can be defined using the following equation:

$$H_{\mathrm{T}}(j\omega)=\sum_{m=0}^{M}\alpha_m \cdot e^{-j2\pi m\omega/\omega_{\mathrm{S}}} \tag{8.1}$$

where the angular frequency $\omega_{\mathrm{S}}=2\pi/\tau$ is the independent variable (τ is the time-delay constant in Figure 8.1a) and α_m is the amplitude-weighting factor of each feedforward component.

In a practical microwave filter, the bandwidth of the transversal filter is restricted to a certain extent because of the feedforward components. In addition, the influence of the conventional filter segments on the transversal structure should also be considered. These additional filtering effects can be represented using a supplementary transfer function $H_{\mathrm{S}}(j\omega)$, and the overall filtering response can be written as follows:

$$H(j\omega)=H_{\mathrm{S}}(j\omega)\cdot H_{\mathrm{T}}(j\omega) \tag{8.2}$$

which implies the additional influence from all transversal components. Therefore, this decomposition tends to be most suitable for practical implementations of microwave filters.

To determine the key parameters ω_{s} and α_m ($m=0, 1, \ldots, M$) in (8.2), a prescribed target response function $G(j\omega)$ must be selected to approximate the function $H(j\omega)$. To construct an approach applicable to most transversal filters, $G(j\omega)$ is assumed to represent a bandpass/bandstop filter for which only the magnitude response matters. The Fourier expansion of (8.1) leads to

$$\alpha_m=\frac{1}{\omega_{\mathrm{S}}}\cdot\int_{\omega_{\mathrm{A}}}^{\omega_{\mathrm{B}}}\frac{|G(j\omega)|}{|H_{\mathrm{S}}(j\omega)|}\cdot e^{+j2\pi\left(m-\frac{M}{2}\right)\cdot\frac{\omega}{\omega_{\mathrm{S}}}}\cdot d\omega,\quad m=0,1,\ldots,M \tag{8.3}$$

where ω_A and ω_B represent the upper and lower cutoff frequencies of the bandpass/bandstop filter, respectively. Beyond the passband, $|G(j\omega)|$ is set to zero. By defining $(M+1)$ uniformly spaced frequency points within the interval $0<\omega<\omega_s$, expression (8.3) can be approximated by the following summation form:

$$\alpha_m = \frac{1}{M+1}\cdot\sum_{k=K_A}^{K_B}\frac{\left|G\left(j\frac{k}{M+1}\omega_s\right)\right|}{\left|H_S\left(j\frac{k}{M+1}\omega_s\right)\right|}\cdot e^{+j2\pi k\cdot\frac{m-M/2}{M+1}},\quad m=0,1,\ldots,M \qquad (8.4)$$

where K_A and K_B are the index limits corresponding to ω_A and ω_B in (8.3), respectively. Generally, the values of α_m are complex numbers. However, for even periodic transfer functions, all coefficients are real. To implement wideband bandpass/bandstop filters from the distributed low-pass/high-pass prototypes, the complementary responses exhibit double symmetry with the passband/stopband centered at multiples of $\omega_s/2$. Because of the symmetry feature, the amplitude coefficients also satisfy $\alpha_m=\alpha_{M-m}$. For practicality, the bandpass/bandstop filters are generally centered near $\omega_s/2$. The negative terms in (8.1) are approximately equal to

$$\alpha_m\cdot e^{-j2\pi m\omega/\omega_s}\approx-\alpha_m\cdot\gamma\cdot\left\{e^{-j2\pi(m-1)\omega/\omega_s}+e^{-j2\pi(m+1)\omega/\omega_s}\right\},$$
$$\text{for }\ \alpha_m/\alpha_{M/2}<0\text{ and }m>0 \qquad (8.5)$$

Without a loss of generality, M is assumed to be an even number, and a new set of amplitude coefficients $\bar{\alpha}_m$ is produced as follows:

$$\begin{aligned}
\bar{\alpha}_0&=\bar{\alpha}_M\\
&=\alpha_{M/2}\cdot\left\{\frac{1}{2}\cdot\left(\left|\frac{\alpha_0}{\alpha_{M/2}}\right|+\frac{\alpha_0}{\alpha_{M/2}}\right)+\frac{\gamma}{2}\cdot\left(\left|\frac{\alpha_1}{\alpha_{M/2}}\right|-\frac{\alpha_1}{\alpha_{M/2}}\right)\right\},\\
\bar{\alpha}_1&=\bar{\alpha}_{M-1}\\
&=\alpha_{M/2}\cdot\left\{\frac{1}{2}\cdot\left(\left|\frac{\alpha_1}{\alpha_{M/2}}\right|+\frac{\alpha_1}{\alpha_{M/2}}\right)+\frac{1}{2}\cdot\left(\left|\frac{\alpha_0}{\alpha_{M/2}}\right|-\frac{\alpha_0}{\alpha_{M/2}}\right)+\frac{\gamma}{2}\cdot\left(\left|\frac{\alpha_2}{\alpha_{M/2}}\right|-\frac{\alpha_2}{\alpha_{M/2}}\right)\right\},\\
\bar{\alpha}_m&=\bar{\alpha}_{M-m}\\
&=\alpha_{M/2}\cdot\left\{\frac{1}{2}\cdot\left(\left|\frac{\alpha_m}{\alpha_{M/2}}\right|+\frac{\alpha_m}{\alpha_{M/2}}\right)+\frac{\gamma}{2}\cdot\left(\left|\frac{\alpha_{m-1}}{\alpha_{M/2}}\right|-\frac{\alpha_{m-1}}{\alpha_{M/2}}\right)+\frac{\gamma}{2}\cdot\left(\left|\frac{\alpha_{m+1}}{\alpha_{M/2}}\right|-\frac{\alpha_{m+1}}{\alpha_{M/2}}\right)\right\},\\
&\quad m=2,3,\cdots,\frac{M}{2}-1\\
\bar{\alpha}_{M/2}&=\alpha_{M/2}
\end{aligned} \qquad (8.6)$$

An empirical value for γ is approximately 0.6, and the aforementioned substitutions are applicable for bandwidths at up to an octave of the center frequency.

8.1.2 One Filter Example Based on Ring Resonator

Since the concept of designing microwave distributed filters based on the transversal signal interference technique was first proposed, a large number of corresponding designs have been tested and reported [2–9]. However, bandpass filters based on the transversal signal interference technique present a primary disadvantage because multiple transmission paths must be involved to realize a high-order filtering response, which enlarges the physical size of the filter and increases the difficulty of tuning the impedance matching among multiple transmission paths. A simple structure has been proposed to realize a transversal microwave filter that presents multiple transmission zeros and consists of a one-wavelength (λ) ring resonator with direct-connected orthogonal feeders and only two feedforward paths. Nevertheless, because of the resonant characteristic of the ring resonator, the bandwidth is too limited to support wideband applications. To overcome this drawback, Chang *et al.* [2] proposed to attach two $\lambda/4$ open-ended stubs orthogonally to the ring resonator to increase the number of resonant modes. A small piece of square tuning stub was also added to the ring resonator to create good passband performance. Gómez-García *et al.* [3] explored the application of a branch-line directional coupler to the design of microwave bandpass filters. The general concept used a branch-line coupler as a transversal filtering section by loading the coupled ports of the coupler with suitable open-ended stubs and using the isolated port as the output port. Based on the transversal signal interference technique, the wideband transfer functions of the proposed bandpass filters with perceptible stopbands and sharp cutoff slopes were derived, and six out-of-band transmission zeros were realized. Following this concept, a series of wideband single-/dual-band bandpass filters were designed and implemented [4–6]. The disadvantages of the aforementioned bandpass filters include the inefficient resolution of DC blocking and circuit-size miniaturization and an insufficient bandwidth for wideband/ultra-wideband (UWB) applications. Furthermore, an interdigital feeding structure was used in the ring resonator by Zhu *et al.* [7, 8] to create a wideband bandpass filter with excellent DC-blocking characteristics. Moreover, additional transmission zeros were generated because of the interdigital feeding structure, which improved selectivity. In 2010, parallel stepped-impedance stubs were proposed by Chang *et al.* [9] for attachment to the ring resonator, thereby resulting in an increased number of resonant modes and improved harmonic suppression.

The aforementioned bandpass filters based on the signal interference technique all use ring resonators, and the weak points of wideband bandpass filter designs are described as follows:

1. Simplex structure—The primary resonant cells are the ring resonators and traditional coupled line and open-ended/short-ended stub structures, among others, are not well utilized.
2. Imperfect theory–Only the passband characteristics are considered, while the stopband characteristics caused by elimination between signals through different paths are not well addressed.
3. Limited bandwidth—The bandwidth must be further extended for applications that require wider bandwidths, such as UWB systems, although the ring resonator allows the structure to realize a fractional bandwidth (*FBW*) as large as 80% as reported in Ref. [9].

8.1.3 Simplified Circuit Model

The aforementioned microwave bandpass filters based on ring resonators use ordinary transmission lines to create a time delay for the transversal signal interference technique. The passband performance can be sufficiently controlled and optimized, while the stopband performance is insufficient. Let us consider whether the control of both the passband and stopband performance can be realized if other structures, such as the ideal/wideband 180° phase shift and traditional parallel-coupled lines, were utilized for the time delay. As specified in Ref. [10], a parallel-coupled line usually introduces a 90° phase difference.

Based on (8.1), a simplified circuit model is presented in this work to analyze the phase characteristics of the transversal-type filter with only two signal paths as shown in Figure 8.2. The signal transfer relationship can be expressed as follows:

$$\begin{aligned} y(\theta) &= [h_1(\theta_1) + h_2(\theta_2)] \cdot x(\theta) \\ h_1(\theta_1) &= A_1 \cdot e^{-j\theta_1} \\ h_2(\theta_2) &= A_2 \cdot e^{-j\theta_2} \end{aligned} \tag{8.7}$$

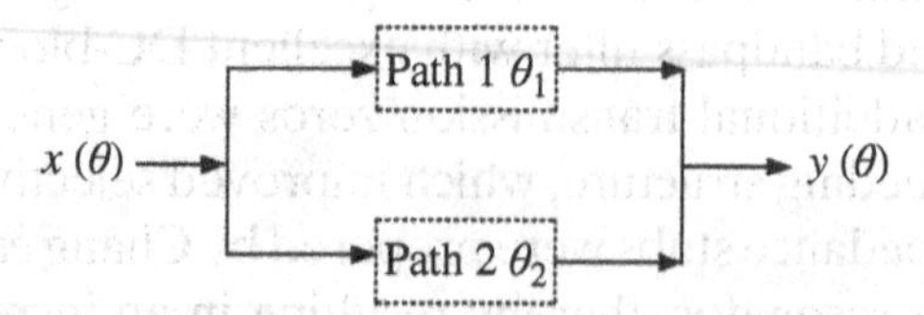

Figure 8.2 Simplified circuit model for filters using the signal interference technique.

It is rational to assume that $A_1 = A_2 = 0.5A$. The passband characteristics can be realized if the signal phases of the two paths satisfy
$\theta_1 = \theta_2 \pm 2n\pi,\ (n = 0, 1, 2, \ldots),\ (0 < \theta_2 < 2\pi)$ (8.8)
Similarly, the stopband characteristics are obtained if the phases satisfy

$$\theta_1 = \theta_2 \pm n\pi,\ (n = 1, 3, 5, \ldots),\ (0 < \theta_2 < 2\pi) \tag{8.9}$$

In the filter design, the concrete realization of the aforementioned simplified circuit model should be considered in detail, and methods of controlling the phases of the passband and stopband must be determined. Generally, to construct a wideband bandpass filter, the signals of the two paths must have equal amplitudes and phases at the center frequency (f_0) of the passband, and the harmonic frequencies must have an equal amplitude and 180° phase difference. If an ideal 180° phase shifter is introduced into one of the two paths and a 180° transmission line (at f_0) is added in another path, the signals maintain an equal amplitude and phase at f_0 and each have an equal amplitude and a 180° phase difference at $2f_0$. In this manner, the designed bandpass filter offers preferable wideband characteristics and a wider range of harmonic suppression. Thus, the selectivity and harmonic suppression are effectively improved without cascading low-pass filtering structures.

The structure in Figure 8.3 is used as an example of a wideband bandpass filter with second harmonic suppression. Without the ideal 180° phase shifter and the transmission line with an electrical length of 180° at f_0, the second harmonic cannot be suppressed. However, when these two components are introduced, the total phase of path 1 at f_0 is $(180° + \theta_1 + 360°)$ and that of path 2 is $(180° + \theta_1)$. The phase difference between the two paths is 360° and thus forms a passband at f_0. At the second harmonic $2f_0$, the total phase of path 1 is $(180° + 2\theta_1 + 720°)$ and that of path 2 is $(360° + 2\theta_1)$. Thus, the phase difference at $2f_0$ is

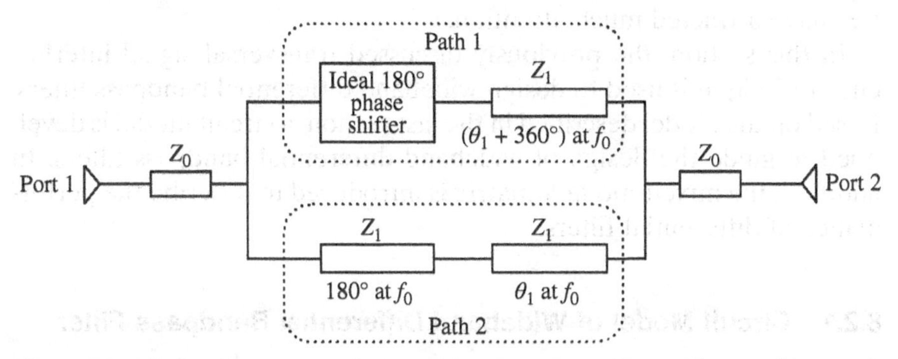

Figure 8.3 Illustrative structure for a wideband bandpass filter with second harmonic suppression.

540°, which results in a stopband. It should be stated that the key element in this example is the ideal 180° phase shifter, which does not occur in reality but fortunately can be replaced with a wideband phase shifter or a wideband balun.

The basic concept and theories in the signal interference techniques have been introduced and discussed in this section and are used to design microwave filters with different performance requirements by applying different circuit topologies along the two signal paths.

8.2 SIGNAL INTERFERENCE TECHNIQUE FOR WIDEBAND DIFFERENTIAL FILTERS

The construction of one miniaturized communication system must highly integrate various RF components, analogue and digital circuits. Currently, the problems of interference and crosstalk are becoming more serious, especially if the frequencies move toward higher bands. Because of the advantages of differential systems, such as higher immunity to these interferences compared with that of single-ended systems, various differential components have been developed in recent years. In these components, differential bandpass filters play an important role in modern wireless communication systems. For a well-designed differential bandpass filter, only the differential-mode signals are desired and allowed to pass through the filter, whereas the common-mode signals must be suppressed. In addition, the differential-mode frequency responses should possess excellent out-of-band rejection and high selectivity.

Several approaches have been proposed to design narrowband differential bandpass filters with good common-mode noise suppression; however, these filters suffer from the disadvantage of high insertion losses. Additionally, wideband and UWB circuits can satisfy the requirement of high data rate transmissions. Therefore, wideband/UWB differential filters have attracted much attention.

In this section, the previously discussed transversal signal interference technique is used to design wideband differential bandpass filters. Based on the model described in the last section, a circuit model is developed to guide the design of wideband differential bandpass filters. In addition, the mixed-mode *S*-matrix is introduced to describe the performance of differential filters.

8.2.1 Circuit Model of Wideband Differential Bandpass Filter

Similar to its applications in microwave wideband single-ended bandpass filters, the signal interference technique can also be applied to the design of wideband differential bandpass filters. Figure 8.4 shows a simple circuit model of a type of wideband differential filter in which the signal interference technique is used. A two-port differential filter is also a four-port network with two input ports and two output ports, and its transfer relationships can be expressed as shown in Equations (8.10) and (8.11):

$$\begin{aligned} y(\theta) &= h_1(\theta_1)\cdot x(\theta) + h_2(\theta_2)\cdot x(\theta') \\ h_1(\theta_1) &= A_1 e^{-j\theta_1} \\ h_2(\theta_2) &= A_2 e^{-j\theta_2} \end{aligned} \tag{8.10}$$

$$\begin{aligned} y(\theta') &= h_1(\theta'_1)\cdot x(\theta) + h_2(\theta'_2)\cdot x(\theta') \\ h_1(\theta'_1) &= A_1 e^{-j\theta'_1} \\ h_2(\theta'_2) &= A_2 e^{-j\theta'_2} \end{aligned} \tag{8.11}$$

It is also assumed that $A_1 = A_2 = 0.5A$. In addition, the phases of the four paths can be stated as shown in the following equations for circuit symmetry:

$$\theta_1 = \theta'_1 \text{ and } \theta_2 = \theta'_2 \tag{8.12}$$

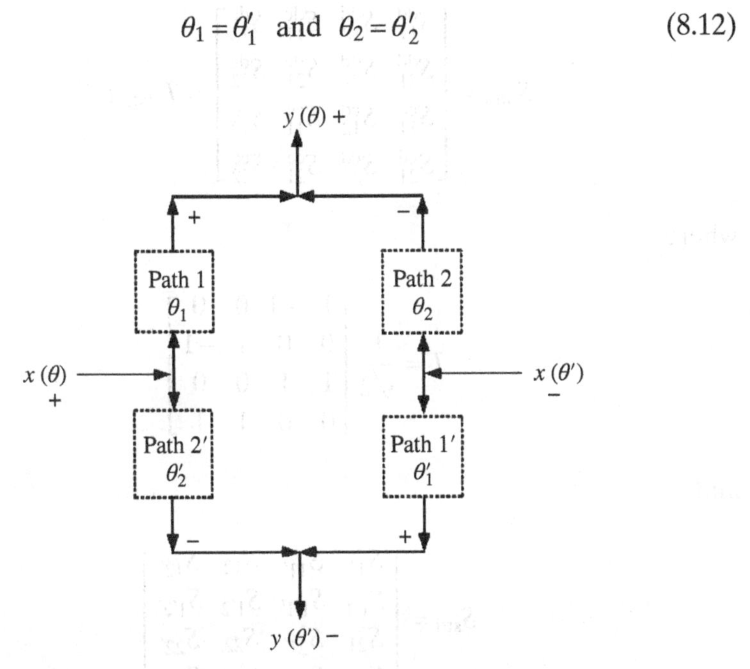

Figure 8.4 Circuit model for differential bandpass filter using the signal interference technique.

Under the phase characteristics for the stopband stated in (8.9), the following relationships can be obtained:

$$y(\theta) = -y(\theta'), \text{ if } x(\theta) = -x(\theta') \tag{8.13}$$

$$y(\theta) = y(\theta') = 0, \text{ if } x(\theta) = x(\theta') \tag{8.14}$$

According to (8.13), a passband is constructed for the differential mode at the desired frequency f_0. Moreover, (8.14) indicates that a common-mode stopband is formed at f_0.

With the circuit model in Figure 8.4, the transversal signal interference technique can be flexibly expanded to the design of differential bandpass filters.

8.2.2 S-Matrix for Differential Bandpass Filters

The standard four-port *S*-matrix (S_{std}) cannot directly exhibit the performance of differential bandpass filters. Fortunately, the mixed-mode *S*-matrix (S_{mm}) has been widely applied in differential filter designs, which increases the simplicity and facilitates the implementation of the design procedure. The design can be calculated from S_{std} and is expressed as follows:

$$S_{mm} = \begin{bmatrix} S_{11}^{dd} & S_{12}^{dd} & S_{11}^{dc} & S_{12}^{dc} \\ S_{21}^{dd} & S_{22}^{dd} & S_{21}^{dc} & S_{22}^{dc} \\ S_{11}^{cd} & S_{12}^{cd} & S_{11}^{cc} & S_{12}^{cc} \\ S_{21}^{cd} & S_{22}^{cd} & S_{21}^{cc} & S_{22}^{cc} \end{bmatrix} = TS_{std}T^{-1} \tag{8.15}$$

where

$$T = \frac{1}{\sqrt{2}} \begin{bmatrix} 1 & -1 & 0 & 0 \\ 0 & 0 & 1 & -1 \\ 1 & 1 & 0 & 0 \\ 0 & 0 & 1 & 1 \end{bmatrix} \tag{8.16}$$

and

$$S_{std} = \begin{bmatrix} S_{11} & S_{11'} & S_{12} & S_{12'} \\ S_{1'1} & S_{1'1'} & S_{1'2} & S_{1'2'} \\ S_{21} & S_{21'} & S_{22} & S_{22'} \\ S_{2'1} & S_{2'1'} & S_{2'2} & S_{2'2'} \end{bmatrix} \tag{8.17}$$

After simplification, the differential-/common-mode S-parameters of interest are expressed as follows:

$$\begin{aligned} S_{21}^{dd} &= (S_{21} - S_{2'1} - S_{21'} + S_{2'1'})/2 \\ S_{11}^{dd} &= (S_{11} - S_{1'1} - S_{11'} + S_{1'1'})/2 \end{aligned} \tag{8.18}$$

and

$$\begin{aligned} S_{21}^{cc} &= (S_{21} + S_{2'1} + S_{21'} + S_{2'1'})/2 \\ S_{11}^{cc} &= (S_{11} + S_{1'1} + S_{11'} + S_{1'1'})/2 \end{aligned} \tag{8.19}$$

These equations can be applied to better investigate the transmission characteristics of the differential and common modes.

8.3 SEVERAL DESIGNS OF WIDEBAND DIFFERENTIAL BANDPASS FILTERS

In the previous section, a simple circuit model is proposed for wideband differential bandpass filters, and based on this model, several design examples are provided in this section to verify the convenience and efficiency of the circuit model. Marchand balun, π-type UWB 180° phase shifter, and double-sided parallel-strip line (DSPSL) UWB 180° phase inverter will be applied to create the 180° phase difference in the different signal paths of the circuit model.

8.3.1 Differential Bandpass Filter Based on Wideband Marchand Baluns

The Marchand balun is a well-known balun structure composed of two pairs of $\lambda/4$ parallel-coupled lines as shown in Figure 8.5 [11, 12]. The working conditions for a balun are as follows:

$$S_{11} = 0 \text{ and } S_{31} = -S_{21} \tag{8.20}$$

Therefore, the following relationship between the input-/output-port impedances and the even-/odd-mode impedances of the parallel-coupled lines can be obtained as follows:

$$\frac{Z_{oe} - Z_{oo}}{Z_{oe} + Z_{oo}} = \frac{1}{\sqrt{2Z_{out}/Z_{in} + 1}} \tag{8.21}$$

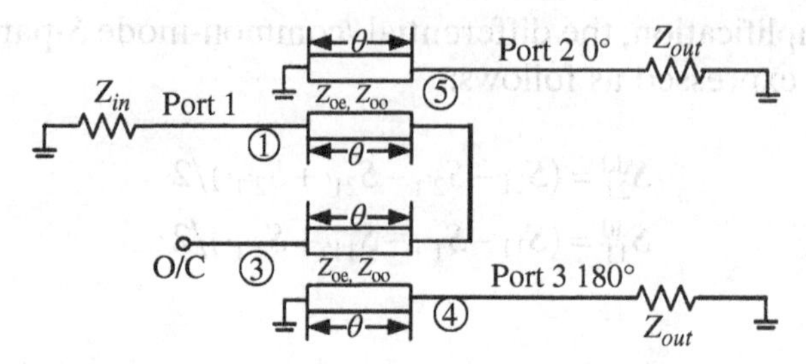

Figure 8.5 Circuit model for the Marchand balun. Reproduced by permission of the Institution of Engineering & Technology.

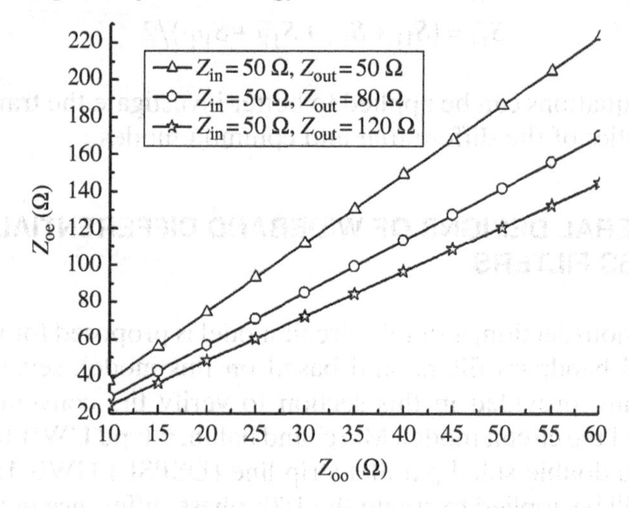

Figure 8.6 Relationship between the input-/output-port impedances and the even-/odd-mode impedances of parallel-coupled lines. Reproduced by permission of the Institution of Engineering & Technology.

In the case of different input-/output-port impedances, the relationship between the even- and odd-mode impedances is illustrated in Figure 8.6, and the proper values can be selected to design Marchand baluns with different bandwidths. Figure 8.7 shows the simulation results for the Marchand balun under different input-/output-port impedances. The simulation results indicate that the bandwidth of the Marchand balun exhibits a wide adjustable range from 12 to 76% with $|S_{11}|$ greater than 10 dB. Therefore, wideband differential bandpass filters with various bandwidths can be designed and implemented if the 180° phase difference between the paths (path 1/1′ and 2/2′) in Figure 8.4 is realized using two Marchand baluns.

Based on this information, a new UWB differential bandpass filter that uses two planar Marchand baluns is presented [13]. Two different transmission paths are introduced to realize the signal transmission from ports 1 and 1′ to ports 2 and 2′, and four microstrip lines with a

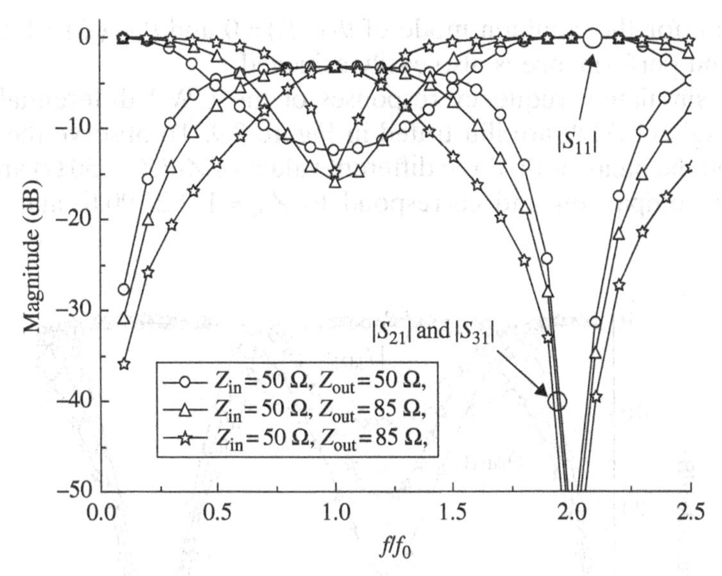

Figure 8.7 Simulation results for the Marchand balun under different input-/output-port impedances. Reproduced by permission of the Institution of Engineering & Technology.

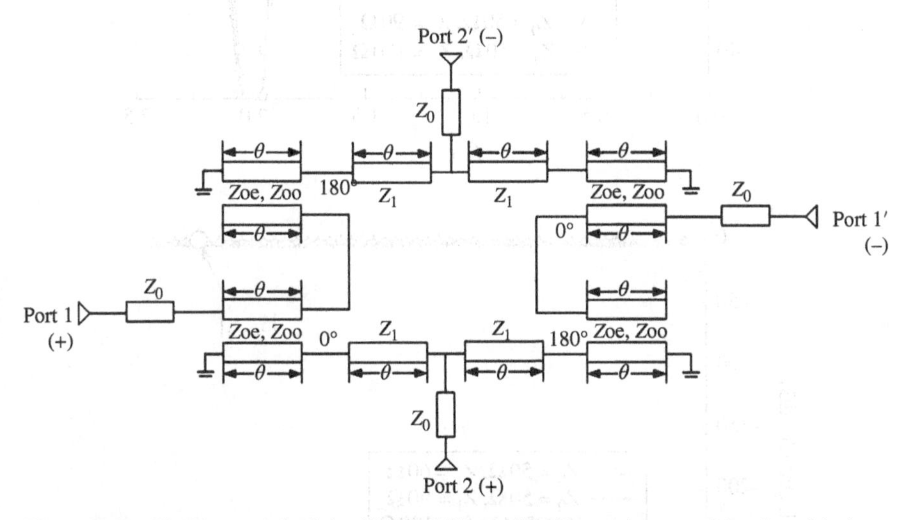

Figure 8.8 Circuit model of the UWB differential filter using two Marchand baluns. Reproduced by permission of the Institution of Engineering & Technology.

characteristic impedance of $Z_0 = 50\ \Omega$ are used as input/output ports as shown in Figure 8.8.

If the differential-mode signals are excited to transmit from ports 1 and 1′ to ports 2 and 2′, the passband performance is easily achieved because of the equal 180° phase difference power division of the Marchand balun; thus, $\theta_{12}\ (f_0) = 0°$ and $\theta_{1'2}\ (f_0) = 180° + 180° = 360°$. In addition, for the common mode of θ_{12} $(f_0) = 0$ and $\theta_{1'2}$ $(f_0) = 180°$, the stopband performance is also easily achieved.

The simulated frequency responses of the UWB differential filter using Agilent ADS are illustrated in Figure 8.9. To observe the effect of Z_1 on the bandwidth, three different values of Z_1 ($Z_0 = 50\ \Omega$) are chosen for comparison and correspond to $Z_{oe} = 111.2$, 90.6, and 79.5 Ω

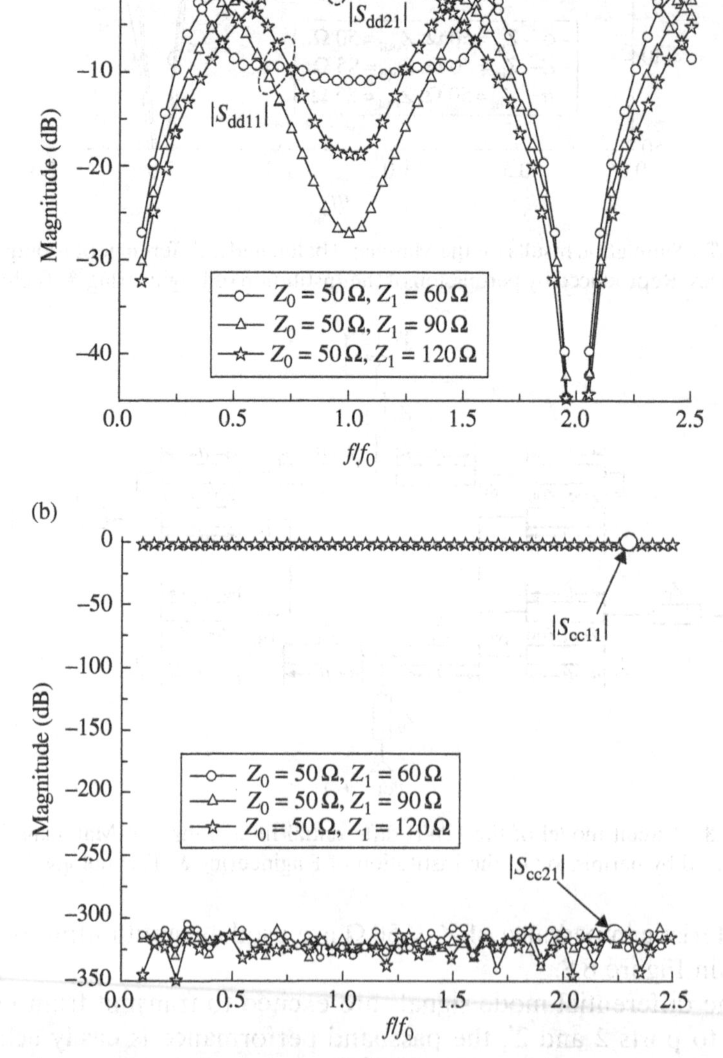

Figure 8.9 Simulated responses of the UWB differential bandpass filter: (a) differential mode and (b) common mode. Reproduced by permission of the Institution of Engineering & Technology.

($Z_{oo} = 33\ \Omega$) [14]. The 3-dB *FBW* for the differential mode ranges from 80 to 124%. In addition, the common mode is easily suppressed over the entire frequency band because of the 180° phase difference power division of the Marchand balun.

The proposed UWB differential bandpass filter is constructed on the substrate with $\varepsilon_r = 2.65$, $h = 0.5$ mm, and $\tan\delta = 0.002$ as shown in Figure 8.10. To satisfy the bandwidth requirement of the UWB band

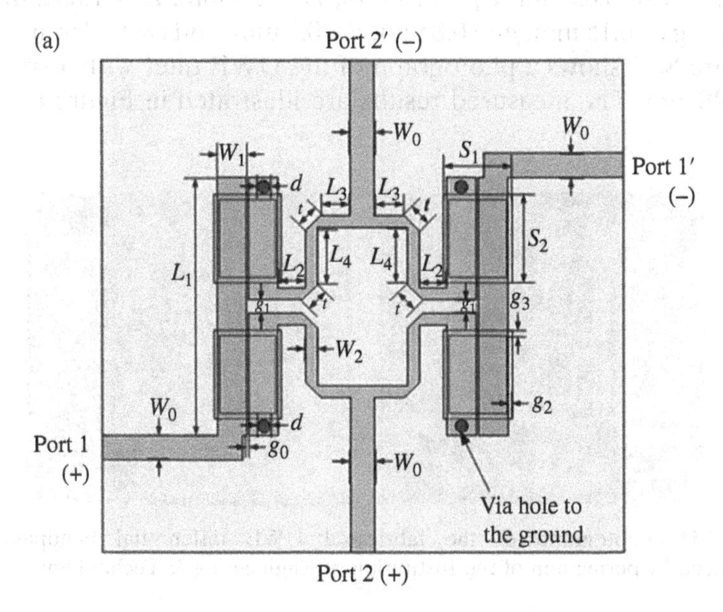

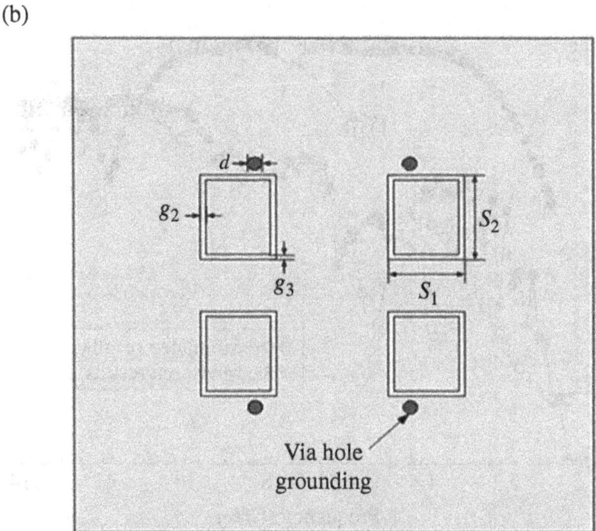

Figure 8.10 (a) Top view and (b) bottom view of the UWB differential bandpass filter. Reproduced by permission of the Institution of Engineering & Technology.

(3.1–10.6 GHz), the parameters for the UWB differential filter are chosen as follows: $Z_0 = 50\ \Omega$ and $Z_1 = 68\ \Omega$ ($Z_{oe} = 104\ \Omega$ and $Z_{oo} = 33\ \Omega$). The patterned ground-plane technique in Ref. [15] is used to obtain the desired even-/odd-mode values. After using the optimization procedure in Ref. [15], the slot sizes in the case of $Z_{oe} = 104\ \Omega$ and $Z_{oo} = 33\ \Omega$ for the coupled lines are $S_1 = 3.8$ mm, $S_2 = 5.05$ mm, and $g_2 = g_3 = 0.3$ mm. Other physical dimensions in Figure 8.10(a) are finally defined as $W_0 = 1.37$ mm, $W_1 = 1.62$ mm, $W_2 = 0.65$ mm, $L_1 = 14.7$ mm, $L_2 = 1.5$ mm, $L_3 = 1.63$ mm, $L_4 = 3.27$ mm, $g_0 = 0.15$ mm, $g_1 = 0.8$ mm, $d = 0.7$ mm, and $t = 1.17$ mm.

Figure 8.11 shows a photograph of the UWB filter with a size of 28 mm × 28 mm. The measured results are illustrated in Figure 8.12. The

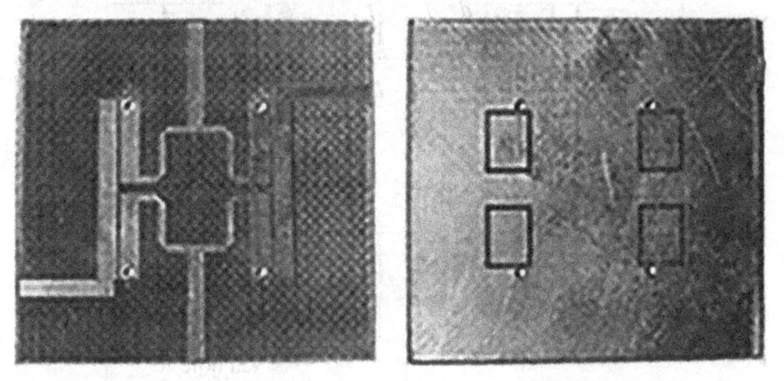

Figure 8.11 Photograph of the fabricated UWB differential bandpass filter. Reproduced by permission of the Institution of Engineering & Technology.

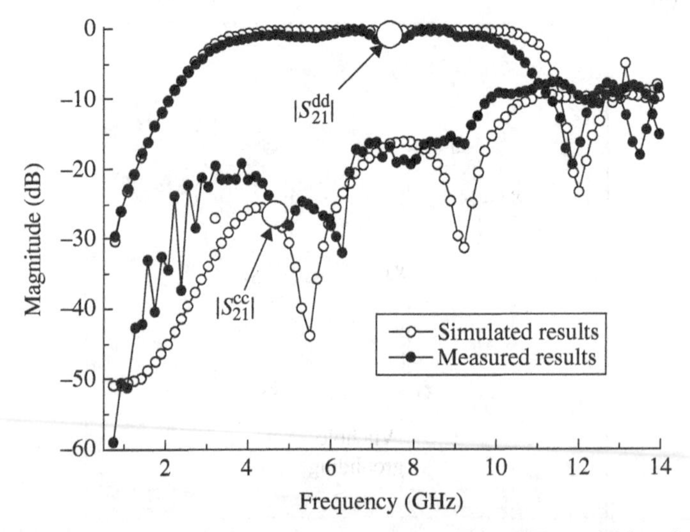

Figure 8.12 Measured and simulated results of the UWB differential bandpass filter. Reproduced by permission of the Institution of Engineering & Technology.

insertion loss of the differential mode is less than 1.75 dB within the passband (the 3-dB *FBW* is approximately 111% and 3–10.6 GHz). Broadband rejection (0–10 GHz) of the common mode greater than 13 dB is achieved, which is so limited because the performance of the realized Marchand balun is suboptimal in amplitude and phase balances.

8.3.2 Differential Bandpass Filter Based on π-Type UWB 180° Phase Shifters

In this section, a wideband differential bandpass filter is presented using two π-type UWB 180° phase shifters [16]. Figure 8.13 shows the schematic of the presented differential filter. Four paths with two 180° phase shifters and two 360° transmission lines are present. For the differential-mode signals in Figure 8.13(a), the input out-of-phase signals are converted to in-phase signals as they pass through the first 180° phase shifter and the 360° transmission line. These in-phase signals propagate to the other two paths connected with the output ports. The second 180° phase shifter and 360° transmission line convert these in-phase signals back to differential-mode signals with out-of-phase responses. Therefore, the differential-mode signals propagate well in the proposed schematic. However, the common-mode noise signals in Figure 8.13(b) are first converted to out-of-phase signals. These signals meet and cancel each other out at the center point C of the filter because they are out of phase.

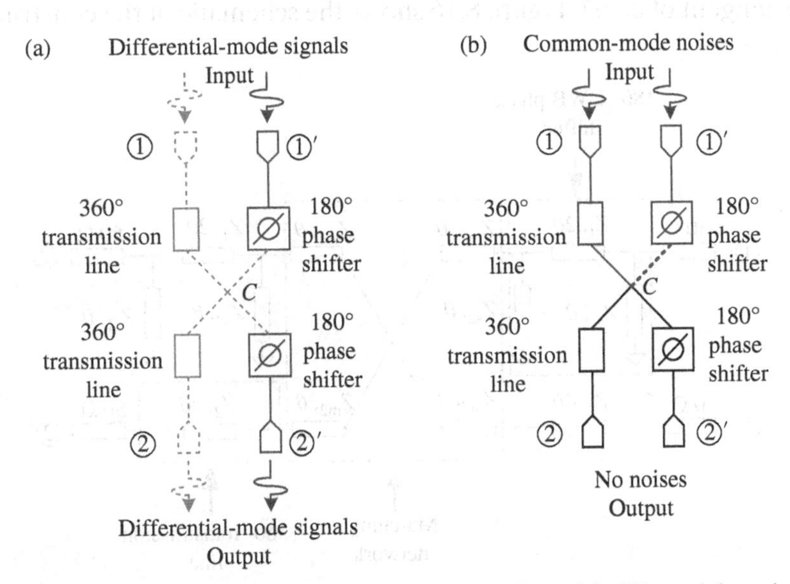

Figure 8.13 Schematic of the differential bandpass filter: (a) differential mode and (b) common mode. Reprinted with permission from Ref. [16]; copyright 2013 IEEE.

In this case, the center point is equivalent to a short-circuit point, and the output ports do not receive any common-mode signals. In Figure 8.13, the solid lines represent the in-phase signals, and the dashed lines represent the out-of-phase signals. This figure clearly shows that the common-mode noises are canceled out and the differential-mode signals pass through the proposed filter. Therefore, any type of differential filter can be created based on the 180° phase shifters and the 360° transmission lines. Furthermore, if the 180° phase shifter features a UWB response, then a UWB differential filter can be realized.

Figure 8.14 shows the transmission-line model of the proposed differential UWB filter, which consists of three components: π-type 180°UWB phase shifters, 360° transmission lines, and a matching network. For the phase shifter, the 180° phase difference can be realized by adjusting Z_1, Z_{s1}, and Z_{s2} via circuit simulations (assuming $Z_2 = 50\ \Omega$ and $\theta = 90°$). In the proposed filter, $Z_1 = 62\ \Omega$, $Z_2 = Z_{m1} = 50\ \Omega$, $Z_{s1} = Z_{m2} = 76\ \Omega$, and $Z_{s2} = 60.5\ \Omega$ are eventually determined based on simulations with an Agilent ADS. Figure 8.15 shows the S-parameter calculated using the method described in Ref. [17], in which the center frequency is 6.85 GHz. The differential-mode signals propagate well in ultrawide frequency band, and the common-mode noise is well suppressed. These simulation results validate the proposed design method for differential filters.

For further demonstration, the proposed UWB differential filter was designed and fabricated on a Rogers RT/duroid 6006 substrate with a thickness of 0.635 mm, a relative dielectric constant of $\varepsilon_r = 6.15$, and a loss tangent of 0.002. Figure 8.16 shows the schematic of the constructed

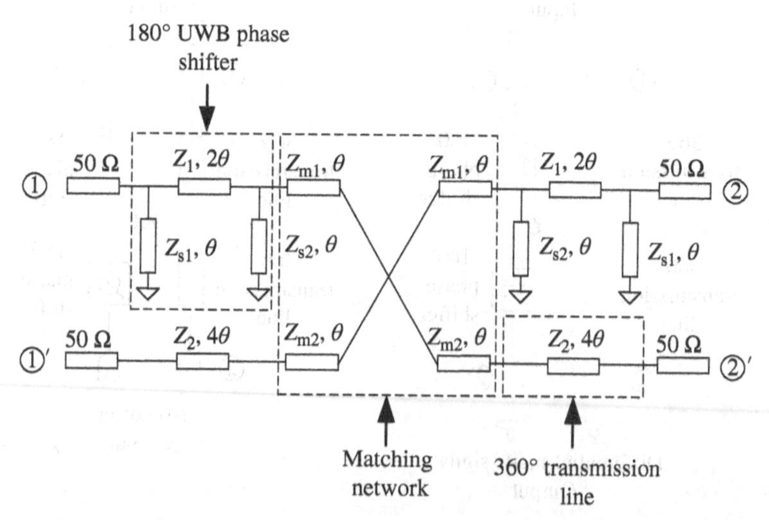

Figure 8.14 Transmission-line model of the proposed filter. Reprinted with permission from Ref. [16]; copyright 2013 IEEE.

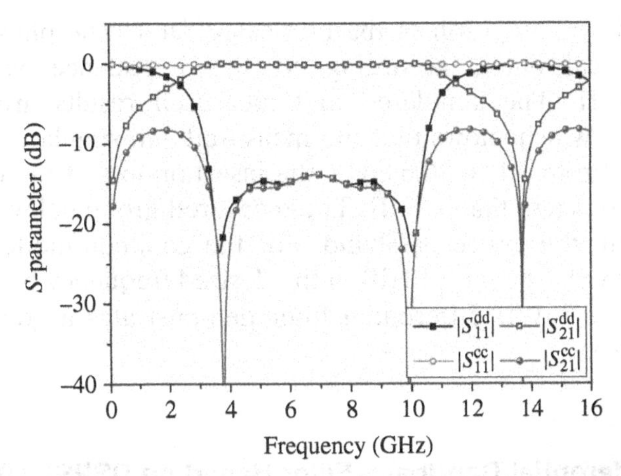

Figure 8.15 Calculated *S*-parameter curves from the equivalent transmission-line model. Reprinted with permission from Ref. [16]; copyright 2013 IEEE.

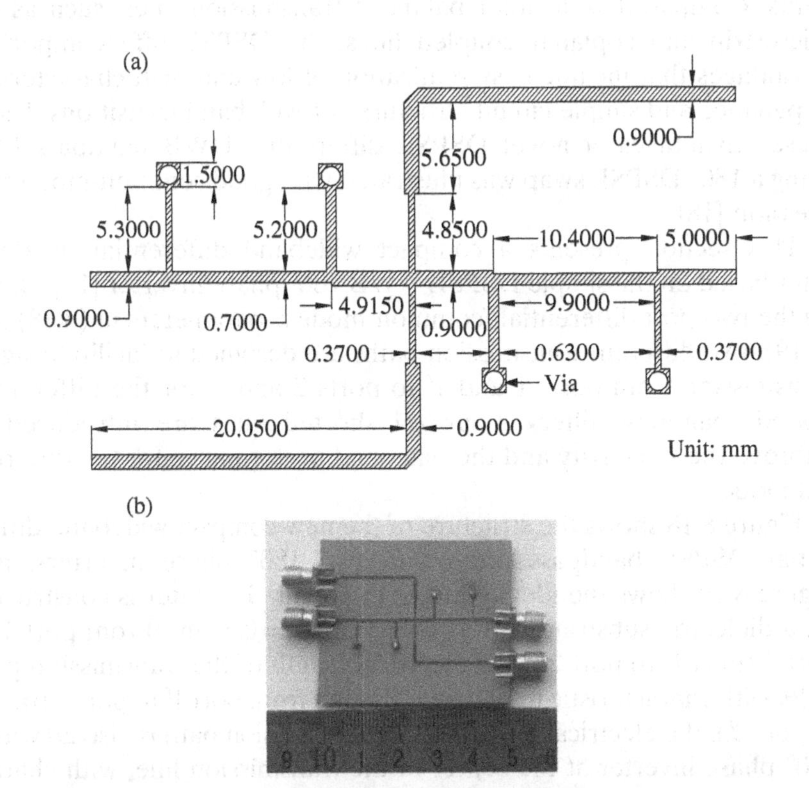

Figure 8.16 (a) Zeland IE3D model and (b) photograph of the proposed filter. Reprinted with permission from Ref. [16]; copyright 2013 IEEE.

model and a photograph of the fabricated filter. The physical dimensions are calculated according to the impedance values given in Section II. The simulated and measured results are shown in Figure 8.17, which shows that the measured and simulated results are consistent. From 3.1 to 10.6 GHz, the insertion loss of the differential-mode signal is less than –1 dB. The measured group delay is as low as 1.5 ns within the entire passband. For the common-mode signal, the measured $|S_{21}^{cc}|$ is below –10 dB in the desired frequency band. Overall, the proposed UWB differential filter demonstrates a good electrical performance.

8.3.3 Differential Bandpass Filter Based on DSPSL UWB 180° Phase Inverter

The DSPSL is a well-known type of balanced transmission line that is useful and convenient for the design of balanced microwave components. Compared with other balanced transmission lines, such as the microstrip and coplanar coupled lines, the DSPSL offers important advantages that include easy realization of low and high characteristic impedance and simple circuit structures of wideband transitions. Using these advantages, a novel DSPSL differential UWB bandpass filter using a 180° DSPSL swap was illustrated with good common-mode suppression [18].

This section presents a compact wideband differential bandpass filter based on the simple DSPSL UWB 180° phase inverter [19]. Based on the two-port differential-/common-mode S-parameters in (8.18) and (8.19), two different transmission paths are designed to facilitate signal transmission from ports 1 and 1′ to ports 2 and 2′ for the differential DSPSL bandpass filters. Four $\lambda/4$ shorted lines are introduced to improve the selectivity and the passband performance for the differential mode.

Figure 8.18 shows the structure of the new compact wideband differential DSPSL bandpass filter with two 180° phase inverters, and Figure 8.19 shows the ideal equivalent circuit. The filter is constructed on a dielectric substrate with $\varepsilon_r = 2.65$ and $h = 0.5$ mm. From port 1 to port 2 (port 1′ to port 2′), the electrical length of the transmission path is 2θ, with characteristic impedance Z_1, and from port 1′ to port 2 (port 1 to port 2′), the electrical length of the transmission path is also 2θ with a 180° phase inverter at the center of the transmission line, with characteristic impedance Z_2. To facilitate the filter design, differential-mode and common-mode analyses are conducted.

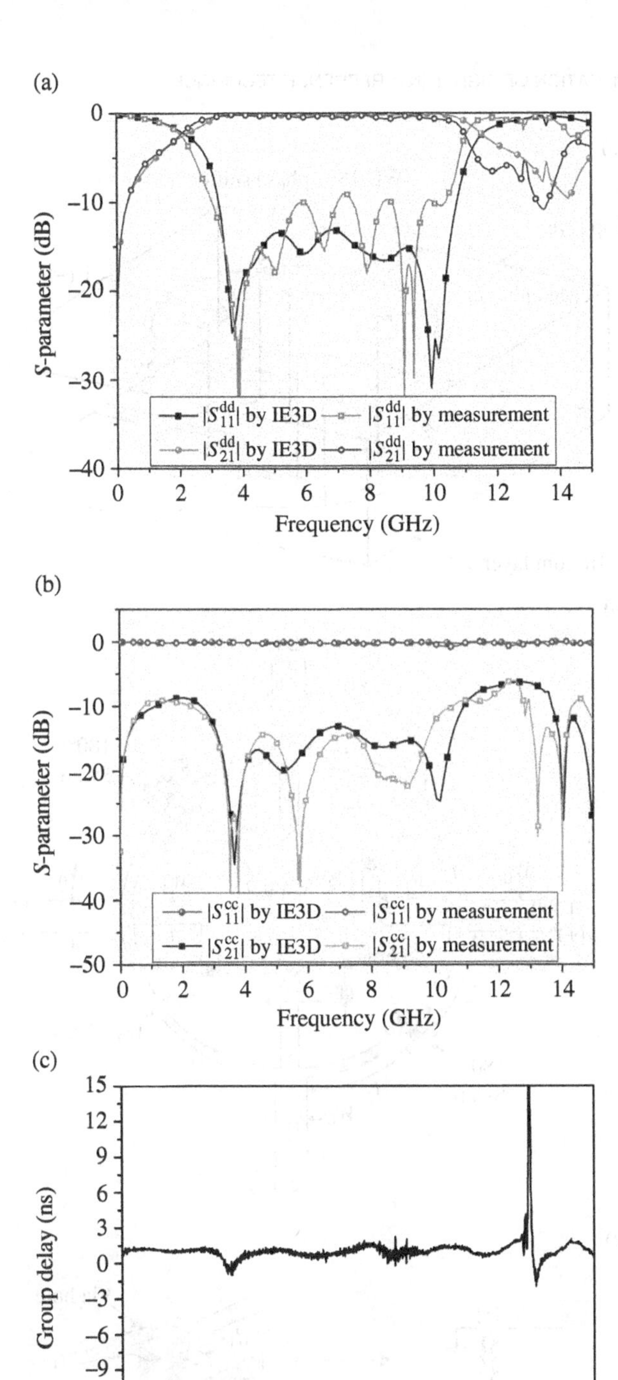

Figure 8.17 Simulated and measured results: (a) differential-mode response, (b) common-mode response, and (c) differential-mode to common-mode response. Reprinted with permission from Ref. [16]; copyright 2013 IEEE.

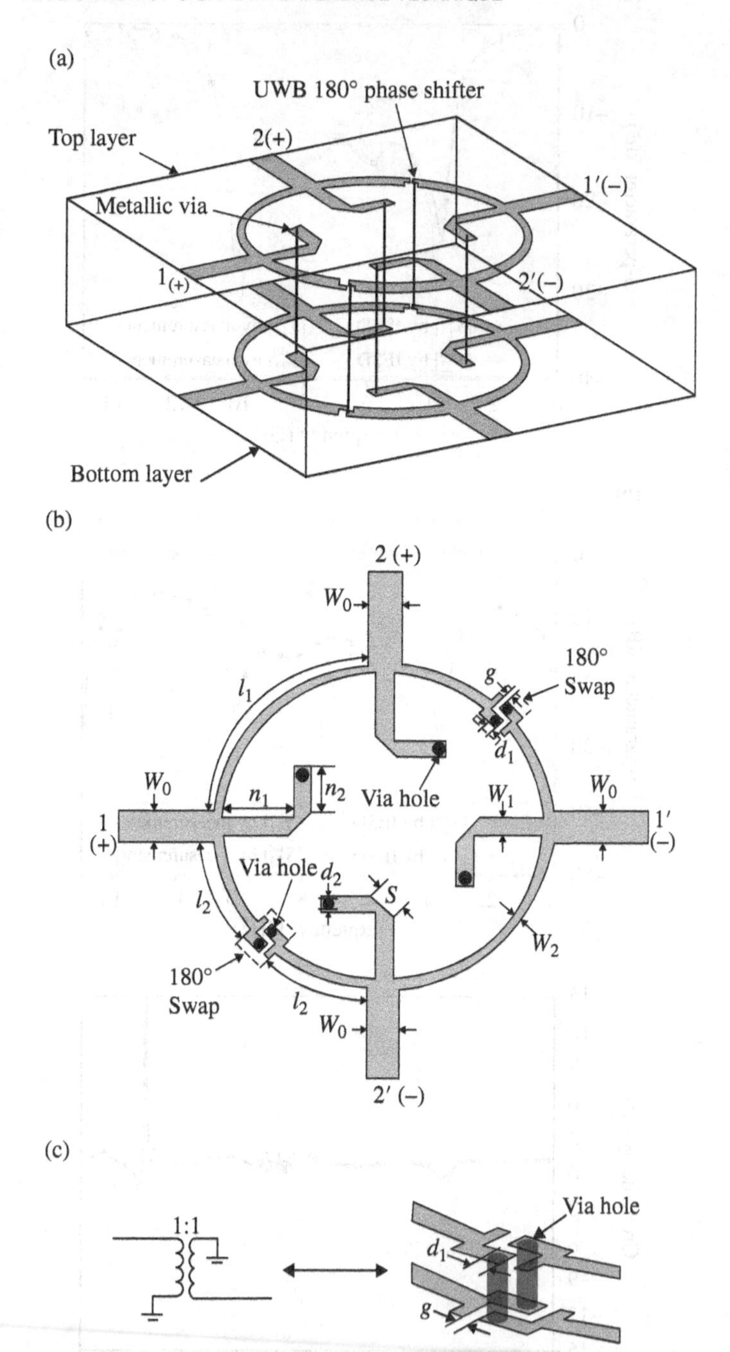

Figure 8.18 Structure of the wideband differential bandpass filter using the DSPSL UWB 180° phase shifter: (a) 3D layout, (b) top view, and (c) DSPSL UWB 180° phase inverter layout. Reproduced by permission of the Institution of Engineering & Technology.

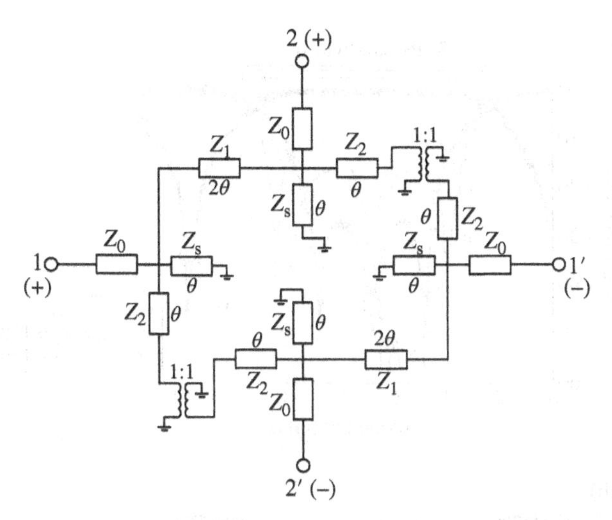

Figure 8.19 Circuit model of the wideband differential bandpass filter using the DSPSL UWB 180° phase inverter. Reproduced by permission of the Institution of Engineering & Technology.

8.3.3.1 Differential-Mode Analysis From (8.9), when the differential-mode signals are excited to transmit from ports 1 and 1′ to ports 2 and 2′, because of the two different transmission paths with electrical lengths of 2θ ($\theta = 90°$ at the center frequency f_0) and a 180° phase inverter, $\theta_{12}(f_0) = 180°$ and $\theta_{1'2}(f_0) = 540°$ (including the 180° phase difference of the differential-mode signals), a passband performance for the differential mode can be realized from ports 1 and 1′ to ports 2 and 2′. Figure 8.20 (a) shows three different frequency responses of the differential mode, and the bandwidth for the differential mode is decreased with increases of Z_1 and Z_2. Figure 8.20(b) illustrates the simulated results of the transmission poles for the differential mode in two different cases.

Moreover, introducing four shorted lines with Z_s causes several transmission poles and helps improve the passband transmission characteristic for the differential mode.

8.3.3.2 Common-Mode Analysis When the common-mode signals are excited to transmit from ports 1 and 1′ to ports 2 and 2′, because of the two transmission paths and the two 180° phase inverters, $\theta_{12}(f_0) = 180°$ and $\theta_{1'2}(f_0) = 360°$, the stopband performance for the common mode can be easily realized from ports 1 and 1′ to ports 2 and 2′. In addition, because of the ideal 180° phase difference of the two 180° phase inverters as discussed in Ref. [20], the common mode can be easily suppressed in the passband and out of the band for the differential mode;

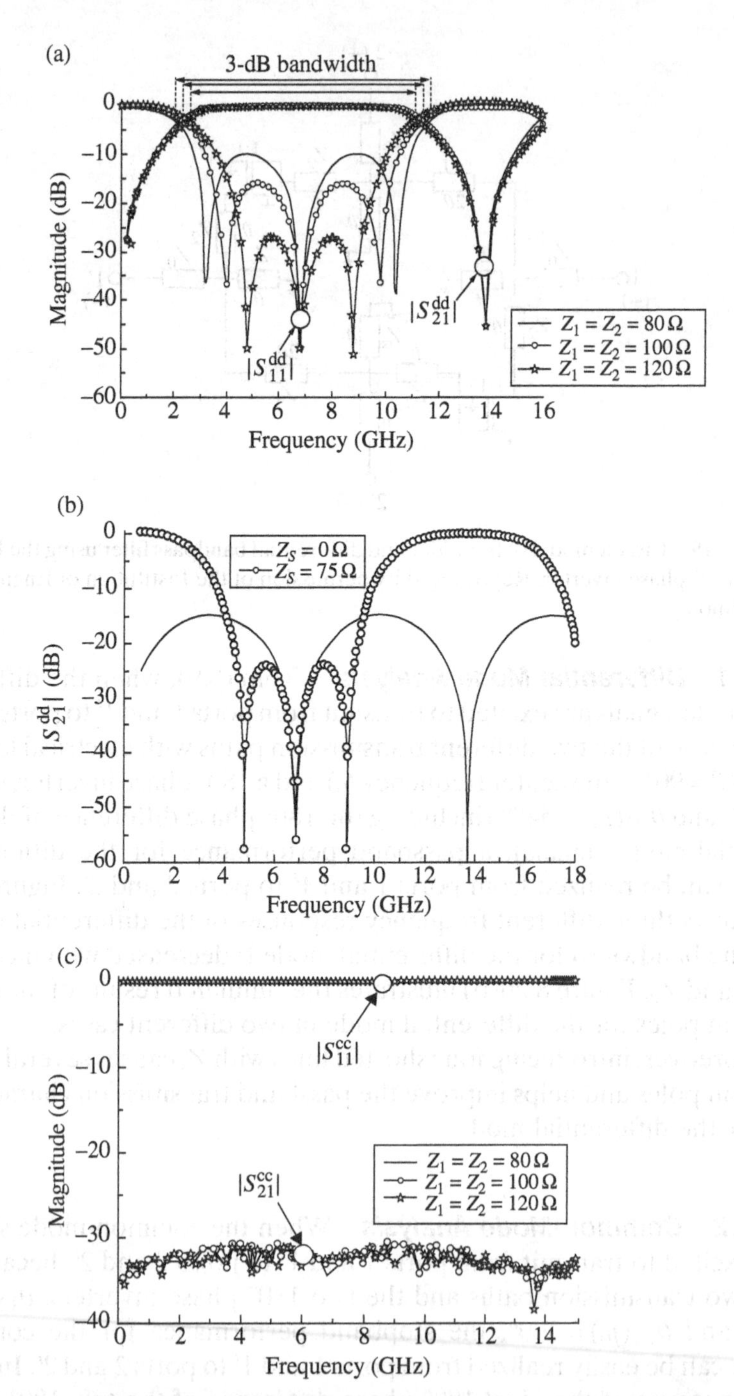

Figure 8.20 Differential-mode and common-mode frequency responses of the wideband differential bandpass filter using the DSPSL UWB 180° phase inverter: (a) $Z_s = 75\,\Omega$, (b) $Z_1 = Z_2 = 120\,\Omega$, and (c) $Z_s = 75\,\Omega$. Reproduced by permission of the Institution of Engineering & Technology.

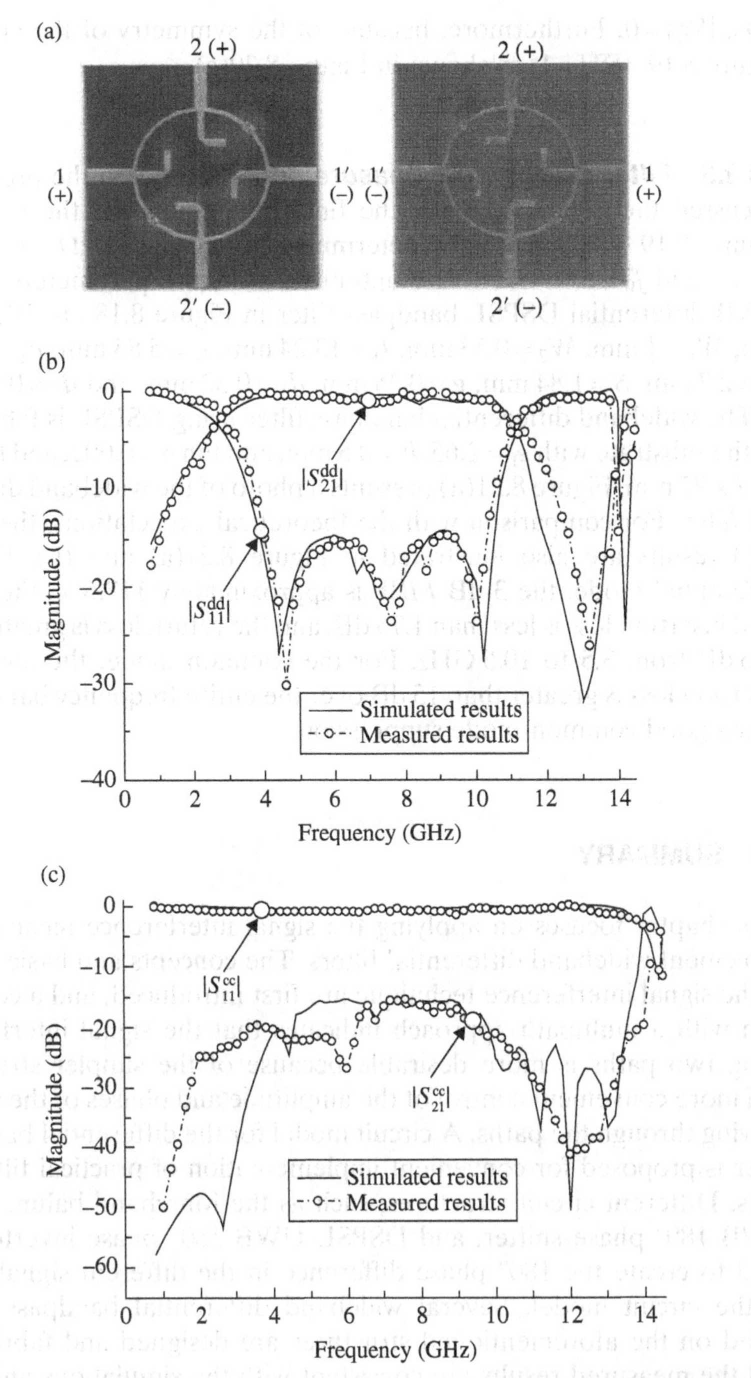

Figure 8.21 Measured and simulated results of the wideband differential bandpass filter using the DSPSL UWB 180° phase inverter: (a) photograph, (b) differential mode, and (c) common mode. Reproduced by permission of the Institution of Engineering & Technology.

thus, $|S_{21}^{cc}| = 0$. Furthermore, because of the symmetry of the circuit in Figure 8.19, $|S_{11}^{cc}| = 1$, as shown in Figure 8.20(c).

8.3.3.3 Filter Design and Measurement

Based on the previously discussed theoretical analysis, the final parameters for the circuit of Figure 8.19 are eventually determined to be $Z_s = 80\,\Omega$, $Z_1 = Z_2 = 120\,\Omega$, and $f_0 = 6.85$ GHz. The optimized structure parameters for the UWB differential DSPSL bandpass filter in Figure 8.18 are $W_0 = 1.85$ mm, $W_1 = 1$ mm, $W_2 = 0.53$ mm, $l_1 = 13.24$ mm, $l_2 = 5.83$ mm, $n_1 = 4$ mm, $n_2 = 2.7$ mm, $S = 1.84$ mm, $g = 0.25$ mm, $d_1 = 0.32$ mm, and $d_2 = 0.6$ mm.

The wideband differential bandpass filter using DSPSL is fabricated on the substrate with $\varepsilon_r = 2.65$, $h = 0.5$ mm, and $\tan\delta = 0.002$, and the size is 27 × 27 mm. Figure 8.21(a) presents a photo of the wideband differential filter. For comparison with the theoretical expectations, the measured results are also illustrated in Figure 8.21(a) and (b). For the differential mode, the 3-dB *FBW* is approximately 117.6%, the measured insertion loss is less than 1.75 dB, and the return loss is greater than 15.5 dB from 3.5 to 10.5 GHz. For the common mode, the measured insertion loss is greater than 15 dB over the entire frequency band, indicating good common-mode suppression.

8.4 SUMMARY

This chapter focuses on applying the signal interference technique to implement wideband differential filters. The concepts and basic theory of the signal interference technique are first introduced, and a comparison with a multipath approach indicates that the signal interference using two paths is more desirable because of the simpler structures and more convenient control of the amplitude and phases of the signals flowing through the paths. A circuit model for the differential bandpass filter is proposed for convenient implementation of practical filter circuits. Different circuit structures, such as the Marchand balun, π-type UWB 180° phase shifter, and DSPSL UWB 180° phase inverter, are used to create the 180° phase difference in the different signal paths of the circuit model. Several wideband differential bandpass filters based on the aforementioned structures are designed and fabricated, and the measured results are consistent with the simulations and indicate good bandpass performance for the differential mode and high suppression for the common mode.

REFERENCES

1. C. Rauscher, "Microwave active filters based on transversal and recursive principles", *IEEE Trans. Microw. Theory Tech.*, vol. **33**, pp. 1350–1360, Dec. 1985.
2. L. H. Hsieh and K. Chang, "Compact, low insertion-loss, sharp-rejection, and wide-band microstrip bandpass filters," *IEEE Trans. Microw. Theory Tech.*, vol. **51**, pp. 1241–1246, Apr. 2003.
3. R. Gómez-García, J. I. Alonso, and D. Amor-Martn, "Using the branchline directional coupler in the design of microwave bandpass filters," *IEEE Trans. Microw. Theory Tech.*, vol. **53**, pp. 3221–3229, Oct. 2005.
4. R. Gómez-García and J. I. Alonso, "Design of sharp-rejection and low-loss wideband planar filters using signal-interference techniques," *IEEE Microw. Wireless Compon. Lett.*, vol. **15**, pp. 530–532, Aug. 2005.
5. R. Gómez-García, M. Sánchez-Renedo, B. Jarry, J. Lintignat, and B. Barelaud, "A class of microwave transversal signal-interference dual-passband planar filters," *IEEE Microw. Wireless Compon. Lett.*, vol. **19**, pp. 158–160, Mar. 2009.
6. R. Gómez-García and M. Sánchez-Renedo, "Microwave dual-band bandpass planar filters based on generalized branch-line hybrids," *IEEE Trans. Microw. Theory Tech.*, vol. **58**, pp. 3760–3769, Dec. 2010.
7. S. Sun and L. Zhu, "Wideband microstrip ring resonator bandpass filters under multiple resonances," *IEEE Trans. Microw. Theory Tech.*, vol. **55**, pp. 2176–2182, Apr. 2007.
8. S. Sun, L. Zhu, and H. H. Tian, "A compact wideband bandpass filter using transversal resonator and asymmetrical interdigital coupled lines," *IEEE Microw. Wireless Compon. Lett.*, vol. **18**, pp. 1753–1755, Mar. 2008.
9. C. H. Kim and K. Chang, "*Wideband ring resonator bandpass filter with dual stepped impedance stubs*," IEEE MTT-S International Microwave Symposium Digest, California, USA, May 2010, pp. 229–232.
10. Y. C. Li, Q. Xue, and X. Y. Zhang, "Single- and dual-band power dividers integrated with bandpass filters," *IEEE Trans. Microw. Theory Tech.*, vol. **61**, pp. 69–76, Jan. 2013.
11. K. S. Ang and I. D. Robertson, "Analysis and design of impedance-transforming planar Marchand baluns," *IEEE Trans. Microw. Theory Tech.*, vol. **49**, pp. 402–406, Mar. 2001.
12. K. S. Ang and Y. C. Leong, "Converting baluns into broad-band impedance-transforming 180° hybrids," *IEEE Trans. Microw. Theory Tech.*, vol. **50**, pp. 1990–1995, Aug. 2002.
13. H. T. Zhu, W. J. Feng, W. Q. Che, and Q. Xue, "Ultra-wideband differential bandpass filter based on transversal signal-interference concept", *Electron. Lett.*, vol. **47**, no. 18, pp. 1033–1035, Sept. 2011.

14. W. J. Feng and W. Q. Che, "Ultra-wideband bandpass filter using broadband planar Marchand balun," *Electron. Lett.*, vol. **47**, no. 3, pp. 198–199, Feb. 2011.
15. Z. Y. Zhang, Y. X. Guo, L. C. Ong, and M. Y. W. Chia, "Improved planar Marchand balun with a patterned ground plane," *Int. J. RF Microw. Comput.-Aided Eng.*, vol. **15**, no. 3, pp. 307–316, Mar. 2005.
16. X. H. Wang, H. Zhang, and B. Z. Wang, "A novel ultra-wideband differential filter based on microstrip line structures," *IEEE Microw. Wireless Compon. Lett.*, vol. **23**, no. 3, pp. 128–130, Mar. 2013.
17. W. R. Eisenstant, B. Stengel, and B. M. Thompson, *Microwave Differential Circuit Design Using Mixed-Mode S-Parameters*, Artech House, Norwood, MA, 2006.
18. X. H. Wang, Q. Xue", and W. W. Choi, "A novel ultra-wideband differential filter based on double-sided parallel-strip line, " *IEEE Microw. Wireless Compon. Lett.*, vol. **20**, no. 8, pp. 471–473, Aug. 2010.
19. W. J. Feng, W. Q. Che, T. F. Eibert, and Q. Xue, "Compact wideband differential bandpass filter based on the double-sided parallel-strip line and transversal signal-interaction concepts," *IET Microw. Antennas Propag.*, vol. **6**, no. 2, pp. 186–195, Apr. 2012.
20. K. W. Wong, L. Chiu, and Q. Xue, "Wideband parallel-strip bandpass filter using phase inverter," *IEEE Microw. Wireless Compon. Lett.*, vol. **18**, no. 8, pp. 503–505, Aug. 2008.

CHAPTER 9

WIDEBAND BALANCED FILTERS BASED ON MULTI-SECTION MIRRORED STEPPED IMPEDANCE RESONATORS (SIRs)

Ferran Martín,[1] Jordi Selga,[1] Paris Vélez,[1] Marc Sans,[1] Jordi Bonache,[1] Ana Rodríguez,[2] Vicente E. Boria,[2] Armando Fernández-Prieto,[3] and Francisco Medina[3]

[1]CIMITEC, Departament d'Enginyeria Electrònica, Universitat Autònoma de Barcelona, Bellaterra, Spain
[2]Departamento de Comunicaciones-iTEAM, Universitat Politècnica de València, Valencia, Spain
[3]Departamento de Electrónica y Electromagnetismo, Universidad de Sevilla, Sevilla, Spain

9.1 INTRODUCTION

The purpose of this chapter is to study some implementations of wideband balanced bandpass filters based on multi-section mirrored stepped impedance resonators (SIRs)[1] and to provide guidelines for their design.

[1] The stepped impedance resonator (SIR) concept was first introduced in Ref. [1], where such resonators were analyzed and applied to the implementation of high performance UHF filters.

Balanced Microwave Filters, First Edition. Edited by Ferran Martín, Lei Zhu, Jiasheng Hong, and Francisco Medina.

By multi-section mirrored SIRs, we refer to balanced SIR structures that exhibit a symmetry plane along the axial direction, four single-ended ports (or two differential ports), and at least three different SIR sections, where each section is characterized by a certain line impedance (which should be high or low) and must be electrically short. Thanks to the small electrical lengths and extreme impedances of the SIR sections, the balanced SIR can be described by a lumped element circuit (as will be seen) useful for design purposes. The specific considered topologies of the mirrored SIRs introduce transmission zeros for the common mode, which are fundamental for the suppression of that mode in the differential filter passband. Two types of mirrored SIRs are considered: 5-section [2] and 7-section [3, 4] SIRs. In the latter case, additional transmission zeros for the differential mode, which can be exploited to improve the out-of-band performance for that mode, are also generated. In order to implement these balanced filters, the mirrored SIRs are alternated either with balanced quarter-wavelength transmission-line sections (acting as admittance inverters) [2–4] or with series resonators (in balanced configuration) implemented by means of lumped elements (interdigital or patch capacitances and inductive strips) [5, 6]. It will be shown that it is possible to achieve quasi-Chebyshev responses through this approach. A method to compensate for the bandwidth shrinkage (differential mode) related to the limited functionality of the quarter-wavelength transmission lines as admittance inverters is reported and applied to the design of an illustrative prototype of a wideband balanced filter based on 7-section SIRs. The second example corresponds to an ultra-wideband (UWB) balanced filter, in this case fully implemented by semi-lumped (i.e., electrically small) elements.

9.2 THE MULTI-SECTION MIRRORED STEPPED IMPEDANCE RESONATOR (SIR)

Figure 9.1 depicts three topologies corresponding to balanced SIRs with different number of sections and the corresponding lumped element equivalent circuits (losses are ignored). The narrow sections are described by inductances, whereas the capacitors account for the wide sections of the SIR. Note that the considered balanced SIRs are actually bi-symmetric since there is also a symmetry plane at the midplane between the differential input (1–1′) and output (2–2′) ports (besides the symmetry plane in the axial direction, A–A′). This bi-symmetry is not actually necessary but convenient in order to simplify the circuit model (i.e., to reduce the number of elements). For the implementation

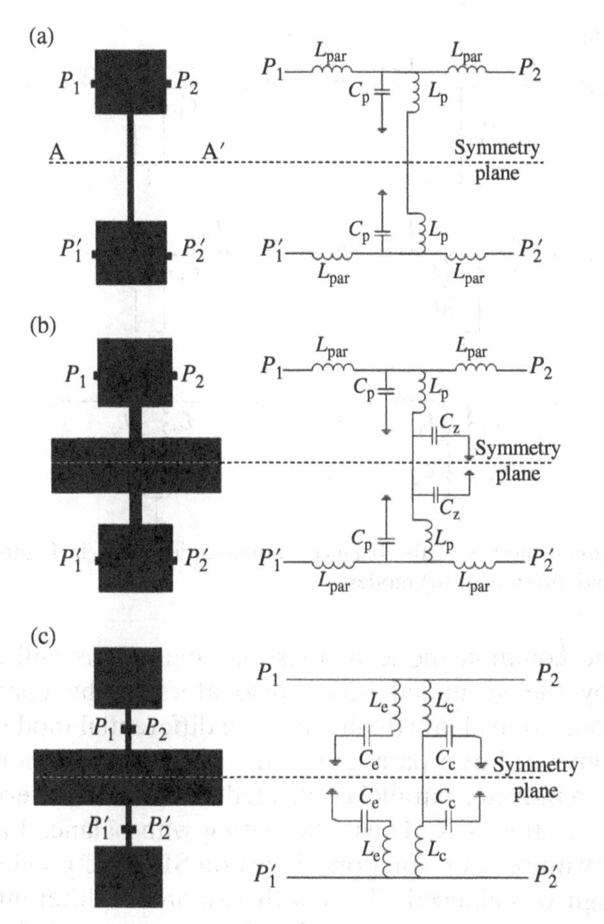

Figure 9.1 Typical topologies of balanced 3-section (a), 5-section (b), and 7-section (c) SIRs and the corresponding lumped element equivalent circuit models.

of common-mode suppressed balanced filters, the topology in Figure 9.1 (a) is not of interest since it is not possible to generate transmission zeros, necessary to reject that mode. To gain insight on this, let us consider the lumped element equivalent circuits of the balanced SIRs in Figure 9.1 for the differential and common modes (see Figure 9.2). These models are inferred from those in Figure 9.1 by simply grounding (differential mode) or opening (common mode) the nodes intersecting the symmetry plane A–A′ (an electric and magnetic wall for the differential and common mode, respectively). In view of the resulting circuits, it follows that for the configuration in Figure 9.1(a), common-mode transmission zeros are not expected. However, by virtue of the central capacitive patch in the 5- and 7-section balanced SIRs, transmission

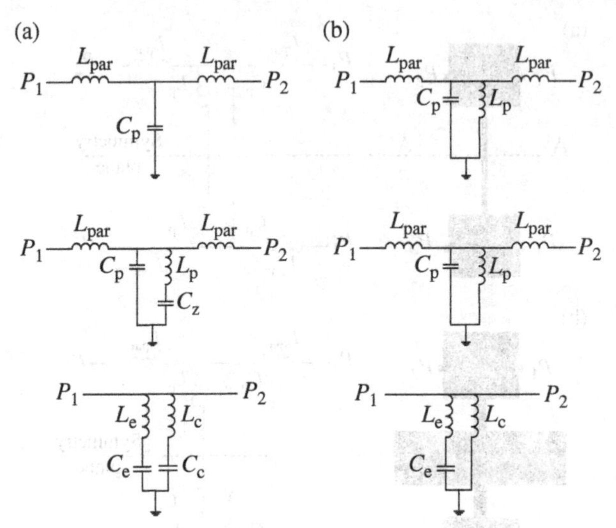

Figure 9.2 Circuit models of the balanced multi-section SIRs in Figure 9.1 for the common (a) and differential (b) modes.

zeros for the common mode in these configurations will appear, as explained by the grounded series resonators in the corresponding common-mode equivalent circuits. For the differential mode, the shunt branch is composed of a parallel resonator for the 5-section balanced SIR and by an inductor parallel connected to a grounded series resonator for the 7-section SIR. Thus, alternating with balanced admittance inverters or with series resonators, 5-section SIRs are good candidates for the design of balanced filters with responses (differential mode) roughly identical to those achievable with the canonical circuit model of a bandpass filter (e.g., a Chebyshev response) [7–9]. For 7-section SIRs, the presence of the grounded series resonator prevents from obtaining the canonical circuit of a bandpass filter for the differential mode, but such resonator can be tailored (through the inductance L_e) in order to generate differential-mode transmission zeros above the differential filter pass band and thus improve the stop band response for that mode, as will be shown later. It is worth mentioning that there is a parasitic inductance (series branch) in the circuit model of the 5-section SIR. The origin of such inductance is the non-negligible length of the external square patches of the balanced SIR. For an accurate description of the structure through a lumped element equivalent circuit, in general such inductance cannot be ignored. However, in certain configurations, such as filters based on balanced SIRs coupled through quarter-wavelength transmission lines, these parasitic inductances can be disregarded.

Let us now validate the previous circuit models on the basis of the 5-section balanced SIR. To this end, let us consider the specific topology depicted in Figure 9.3 and substrate parameters corresponding to the *Rogers RO3010* substrate with thickness $h = 0.635$ mm and dielectric constant $\varepsilon_r = 10.2$. The lossless electromagnetic simulations of the differential- and common-mode responses, inferred by the *Keysight Momentum* commercial software, are depicted in Figure 9.4. In order to extract the circuit elements in Figure 9.1(b), the reactances (or susceptances) of the series and shunt branches of the equivalent T-circuits for both modes are obtained (see Figure 9.5). This is done from the simulated *S*-parameters of both modes, according to well-known transformations reported in several textbooks [8,9]. Once the reactances are known, the parasitic inductance L_{par} is inferred from the slope of the series reactance (identical for the two modes). The elements L_p and C_p are obtained from the shunt susceptance of the differential mode. One condition is inferred from the resonance frequency,

$$\omega_0 = \frac{1}{\sqrt{L_p C_p}} \tag{9.1}$$

whereas the second one is inferred from the susceptance slope,

$$b = \frac{\omega_0}{2} \left. \frac{dB(\omega)}{d\omega} \right|_{\omega = \omega_0} = \omega_0 C_p \tag{9.2}$$

Finally, the capacitance C_z is determined from the transmission zero frequency for the common mode, given by

$$\omega_z = \frac{1}{\sqrt{L_p C_z}} \tag{9.3}$$

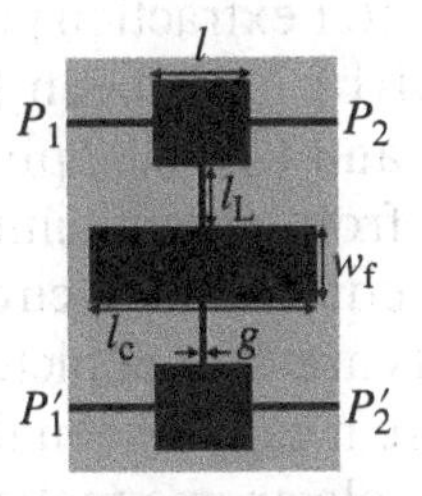

Figure 9.3 Topology used to validate the circuit model of the balanced 5-section SIR. The upper metal level is indicated in black color, whereas the ground plane is corresponded by the grey color. Dimensions are $l = 4$ mm, $l_L = 1.32$ mm, $g = 0.2$ mm, $w_f = 3.4$ mm, and $l_c = 9.6$ mm.

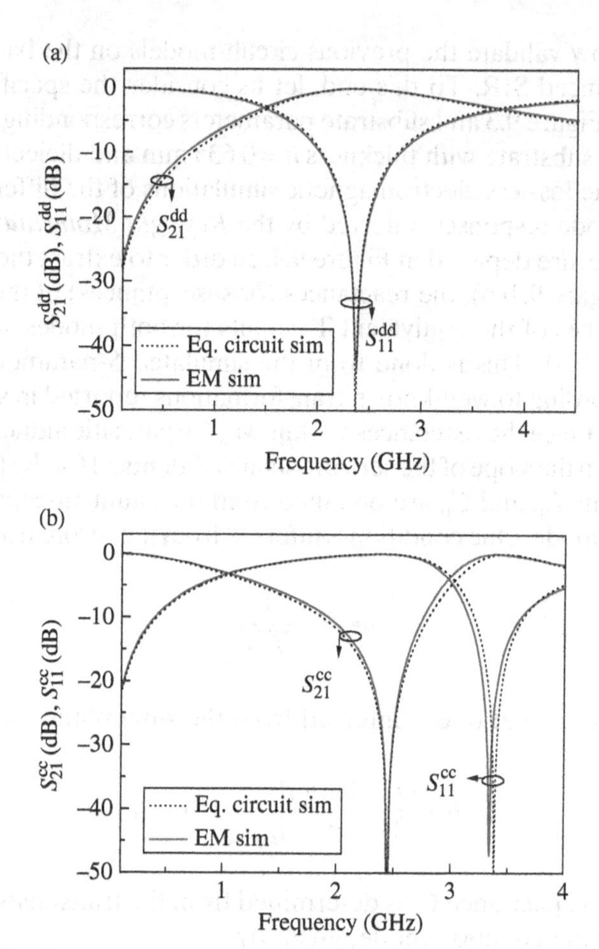

Figure 9.4 Lossless electromagnetic simulation and circuit simulation (with the extracted parameters indicated in the text) of the balanced SIR in Figure 9.3 for differential (a) and common (b) modes.

With the previous parameter extraction procedure, the elements of the circuit model in Figure 9.1(b) have been found to be $L_{par} = 0.19$ nH, $L_p = 1.54$ nH, $C_p = 3.00$ pF, and $C_z = 3.02$ pF. With these element values, the *S*-parameters inferred from circuit simulation, as well as the reactances/susceptances of the equivalent T-circuit models for both modes, can be inferred. The results are also depicted in Figures 9.4 and 9.5 for comparison purposes, where it can be seen the excellent agreement with the results inferred from electromagnetic simulation. Hence, these results validate the circuit model of the 5-section balanced SIR. Note that the accuracy is very good in the considered frequency range (roughly $2f_0$). Obviously, it is expected that the predictions of the circuit model progressively fail as frequency increases, but it has been shown

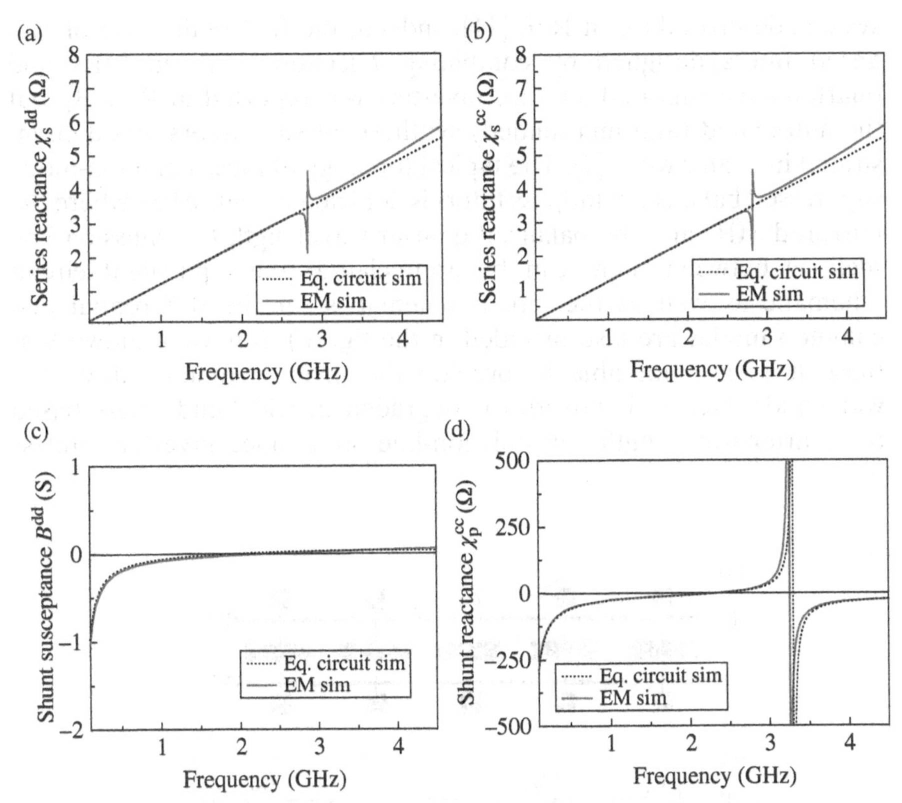

Figure 9.5 Reactances/susceptances of the equivalent T-circuit model (common and differential mode) of the balanced SIR in Figure 9.3. (a) Reactance of the series branch for the differential mode, (b) reactance of the series branch for the common mode, (c) susceptance of the shunt branch for the differential mode, and (d) reactance of the shunt branch for the common mode.

that the reported model of the 5-section balanced SIR is useful for the design of UWB balanced bandpass filters [5].

A similar parameter extraction procedure can be applied to the circuit model of the 7-section SIR. Nevertheless, the validity of this model will be pointed out later by comparing the electromagnetic responses of designed filters with the responses of the circuit schematics. Note that in this case, the series parasitic inductance is not present due to the proximity of the differential input and output ports.

9.3 WIDEBAND BALANCED BANDPASS FILTERS BASED ON 7-SECTION MIRRORED SIRs COUPLED THROUGH ADMITTANCE INVERTERS

The filters reported in this section are inspired by the single-ended filters first reported in Ref. [10] and subsequently designed following a systematic procedure in Ref. [11]. Indeed, the first realization of balanced filters designed by combining 7-section mirrored SIRs and quarter-wavelength admittance inverters was reported in Ref. [3], but the automated (and unattended) synthesis of such filters was demonstrated in a later work [4]. The typical topology of these common-mode suppressed balanced bandpass filters is depicted in Figure 9.6, where the mirrored SIRs and the balanced quarter-wavelength transmission line sections between them can be appreciated (the equivalent circuit schematic as well as the circuit schematics for the differential and common modes are also included in the figure). It is well known that these lines are not able to provide the inverter functionality over widebands. Hence, bandwidth is degraded in wideband filters based on quarter-wavelength transmission-line admittance inverters, unless

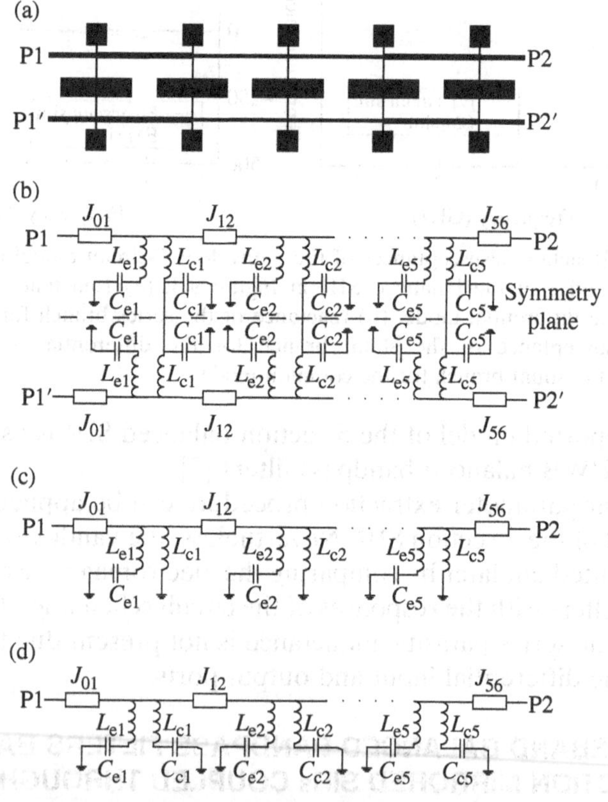

Figure 9.6 Typical topology (order 5) of a wideband balanced bandpass filter based on 7-section SIRs coupled through admittance inverters (a), circuit schematic (b), and circuit schematic for the differential (c) and common (d) modes.

a compensation procedure at the design level is introduced.[2] Such compensation method (first reported in single-ended filters [11] and then applied to balanced filters [4]), based on aggressive space mapping (ASM) optimization [12–14], is included in this chapter. This method provides the filter schematic (optimum schematic) able to satisfy the specifications for the differential-mode and common-mode responses. Then, once the schematic is determined, a second ASM algorithm is applied for the determination of the filter layout. Hence, the unattended design tool for these balanced filters follows a two-step ASM process.

9.3.1 Finding the Optimum Filter Schematic

Obviously, the inherent bandwidth degradation in wideband filters based on quarter-wavelength admittance inverters can be alleviated by over-dimensioning the filter bandwidth. However, the central frequency and in-band return loss level (or ripple) are also modified as a consequence of the limited functionality of the inverters. Thus, a systematic procedure to guarantee that the required filter specifications (central frequency, f_0, fractional bandwidth, FBW, and ripple, L_{Ar}) can be satisfied is needed.

The main hypothesis of the method is that there exists a set of filter specifications (f_0, FBW, and L_{Ar}), different than the target, that provides a filter response (after application of the well-known synthesis formulas [7–9] and replacement of the inverters with quarter-wavelength transmission lines), satisfying the target specifications. If these specifications (different than the target) are known, the resulting filter schematic (composed of lumped elements, i.e., the resonators, plus distributed elements, viz., the quarter-wavelength transmission lines) is the one that must be synthesized by the considered layout. Thus, filter design is a two-step process (as mentioned before), where first the filter schematic providing the required specifications (optimum filter schematic) is determined, and then the layout is generated. For the two design steps, an ASM algorithm is developed. Let us now detail the first ASM algorithm and leave the second one for the next subsection. For better understanding of both ASM algorithms, the general formulation of ASM is included in Appendix 9.A.

The first ASM applies only to the schematic corresponding to the differential mode (Figure 9.6c). Hence, the capacitances C_{ci} are

[2] Indeed, not only the fractional bandwidth but also the filter central frequency and the in-band return loss level are affected by the narrow band functionality of the admittance inverters.

independently determined in a later stage in order to set the common-mode transmission zeros to certain values. Such values are chosen with an eye toward achieving efficient common-mode suppression in the region of interest (differential-mode passband). Nevertheless, as will be shown later, the second ASM involves the whole filter cell, hence including the patches corresponding to the capacitances C_{ci}.

Actually, the inverters (with admittance J_{ij}) of the circuits in Figure 9.6 are implemented by means of transmission-line sections. In the single-ended filters reported in Ref. [11], such transmission-line sections were forced to be identical (i.e., with a characteristic admittance of 0.02 S), resulting in different resonators from stage to stage. However, for balanced filters it is convenient to consider identical resonators and different admittances of the inverters.[3] By considering identical resonators, the synthesis of the layout is simpler since it is guaranteed that the distance between the pair of lines is uniform along the whole filter. Otherwise, if different resonators are considered, the inductances L_{ci} may be different, resulting in diverse lengths if the widths are considered to be the same, as is the case (see Section 9.4). Note that these widths should be identical in order to reduce the number of geometrical parameters in the second ASM. Hence different length means that the distance between the bisection plane and the lines is not uniform unless meanders are used, which is not considered to be the optimum solution.

Once the filter order, n, is set to a certain value that suffices to achieve the required filter selectivity, the filter specifications (differential mode) are the central frequency, f_0, the FBW, and the in-band ripple level L_{Ar} (or minimum return loss level). The transmission zero frequencies provided by the resonators L_{ei}–C_{ei} are all set to $f_z = 2f_0$, since this provides spurious suppression and good filter selectivity above the upper band edge [11]. The elements L_e, C_e, and L_c (identical for all filter stages) are inferred from the following three conditions: (i) the filter central frequency, given by

$$f_0 = \frac{1}{2\pi\sqrt{(L_e + L_c)C_e}} \tag{9.4}$$

[3] In the design of filters based on resonant elements coupled through admittance/impedance inverters, the resonator element values and the admittance/impedance of the inverters are not univocally determined [9]. Therefore, it is possible to force the admittances/impedances of the inverters to the same value, to force identical resonators, or to consider different resonators and inverters from stage to stage.

(ii) the transmission zero frequency:

$$f_z = \frac{1}{2\pi\sqrt{L_e C_e}} \tag{9.5}$$

and (iii) the susceptance slope at f_0:

$$b = 2\pi f_0 \frac{C_e(L_e + L_c)^2}{L_c^2} \tag{9.6}$$

Let us now consider the following target specifications, corresponding to the example used to illustrate the automated design of the considered balanced filters: order $n = 5$, Chebyshev response with $f_0 = 2.4$ GHz, FBW = 40% (corresponding to a 43.91% –3-dB FBW), and $L_{Ar} = 0.2$ dB. By setting the susceptance slope to $b = 0.067$ S, the element values of the shunt resonators are found to be $L_e = 0.4401$ nH, $C_e = 2.4983$ pF, and $L_c = 1.3202$ nH, and the admittance of the inverters (obtained by the formulas that can be found in Ref. [9]) are $J_{0,1} = J_{5,6} = 0.0200$ S, $J_{1,2} = J_{4,5} = 0.0200$ S, and $J_{2,3} = J_{3,4} = 0.0157$ S. This susceptance slope value has been chosen in order to obtain an admittance value of 0.02 S for the inverters of the extremes of the device (so that the corresponding transmission-line sections can be elongated at wish for connector soldering).

It is worth to mention that for the Chebyshev bandpass filters, the FBW is given by the ripple level and is hence smaller than the –3-dB FBW. However, it is convenient to deal with the –3-dB FBW since the ripple level is not constant in the optimization process (to be described). From now on, this –3-dB FBW is designated as FBW, rather than $\text{FBW}_{-3\text{dB}}$ (as usual), for simplicity, and to avoid an excess of subscripts in the formulation.

The quasi-Chebyshev filter response (i.e., the one inferred from the schematic in Figure 9.6(c), but with ideal admittance inverters), depicted in Figure 9.7, is similar to the ideal (target) Chebyshev response in the passband region, and it progressively deviates from it as frequency approaches f_z, as expected. The discrepancies are due to the fact that the shunt resonator is actually a combination of a grounded series resonator (providing the transmission zero) and a grounded inductor. The quasi-Chebyshev response satisfies the specifications to a rough approximation. Hence the target is considered to be the ideal Chebyshev response (except for the transmission zero frequency), also included in the figure. If the ideal admittance inverters are replaced with quarter-wavelength transmission lines, the response is further modified.

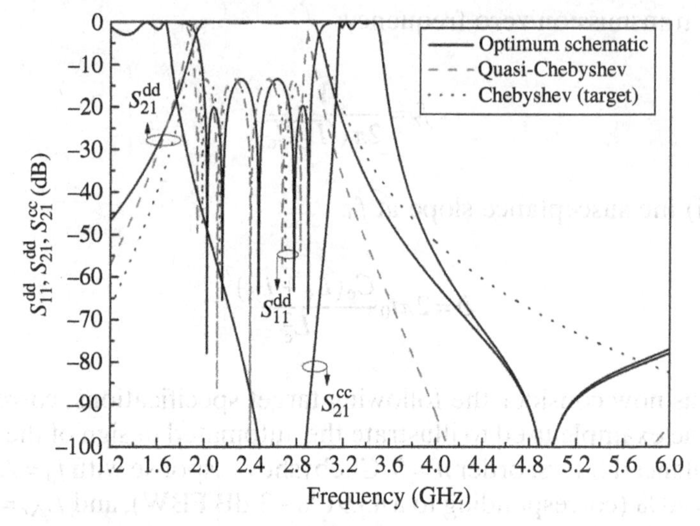

Figure 9.7 Differential-mode quasi-Chebyshev response of the filter that results by using the element values indicated in the text and ideal admittance inverters with the indicated admittances, compared with the ideal Chebyshev (target) response and the response of the optimum filter schematic. The response of the optimum filter schematic to the common mode is also included. Reprinted with permission from Ref. [4]; copyright 2015 IEEE.

Thus, the aim is to find the filter schematic for the differential mode (Figure 9.6c) able to satisfy the specifications. To this end, an ASM tool that carries out the optimization at the schematic level is applied.

As mentioned before, the key point in the development of this first iterative ASM algorithm is to assume that there exists a set of filter specifications, different than the target, that leads to a filter schematic (inferred by substituting the ideal admittance inverters with quarter-wavelength transmission lines), whose response satisfies the target specifications. In this first ASM algorithm (see Appendix 9.A), the optimization (coarse model) space is constituted by the set of specifications, f_0, FBW, and L_{Ar}, being its response the ideal Chebyshev response (target response) depicted in Figure 9.7. The validation (fine model) space is constituted by the same variables, but their response is inferred from the schematic in Figure 9.6(c), with element values calculated as specified above, and quarter-wavelength transmission lines at f_0, where f_0 is the considered value of this element in the validation space (not necessarily the target filter central frequency). The variables of each space are differentiated by a subscript. Thus, the corresponding vectors in the coarse and fine models are written as $\mathbf{x_c} = [f_{0c}, \text{FBW}_c, L_{Arc}]$ and $\mathbf{x_f} = [f_{0f}, \text{FBW}_f, L_{Arf}]$, respectively. The coarse model solution (target specifications) is expressed as $\mathbf{x}_c^* = [f_{0c}^*, \text{FBW}_c^*, L_{\text{Arc}}^*]$. Note that the transmission zero frequency, set to $f_z = 2f_0$, as indicated before, is not a variable in the optimization process.

The first vector in the validation space is set to $\mathbf{x}_f^{(1)} = \mathbf{x}_c^*$. From $\mathbf{x}_f^{(1)}$, the response of the fine model space is obtained (using the schematic with quarter-wavelength transmission lines), and from it we directly extract the parameters of the coarse model by direct inspection of that response, that is, $\mathbf{x}_c^{(1)} = \mathbf{P}\left(\mathbf{x}_f^{(1)}\right)$. Applying (9.11) (see Appendix 9.A), we can thus obtain the first error function. The Jacobian matrix is initiated by slightly perturbing the parameters of the fine model, f_{0f}, FBW_f, and L_{Arf}, and inferring the effects of such perturbations on the coarse model parameters, f_{0c}, FBW_c, and L_{Arc}. Thus, the first Jacobian matrix is given by

$$\mathbf{B} = \begin{pmatrix} \dfrac{\delta f_{0c}}{\delta f_{0f}} & \dfrac{\delta f_{0c}}{\delta \text{FBW}_f} & \dfrac{\delta f_{0c}}{\delta L_{\text{Arf}}} \\ \dfrac{\delta \text{FBW}_c}{\delta f_{0f}} & \dfrac{\delta \text{FBW}_c}{\delta \text{FBW}_f} & \dfrac{\delta \text{FBW}_c}{\delta L_{\text{Arf}}} \\ \dfrac{\delta L_{\text{Arc}}}{\delta f_{0f}} & \dfrac{\delta L_{\text{Arc}}}{\delta \text{FBW}_f} & \dfrac{\delta L_{\text{Arc}}}{\delta L_{\text{Arf}}} \end{pmatrix} \tag{9.7}$$

Once the first Jacobian matrix is obtained, the process is iterated (obtaining $\mathbf{x}_f^{(2)}$ from (9.12), using (9.13), etc.) until convergence is obtained. At each iteration, the elements of the coarse space vector, $\mathbf{x}_c^{(j)}$, are compared with the target (filter specifications), $\mathbf{x}_c^*$, and the error function is obtained according to

$$\|f_{\text{norm}}\| = \sqrt{\left(1 - \frac{f_{0c}}{f_{0c}^*}\right)^2 + \left(1 - \frac{\text{FBW}_c}{\text{FBW}_c^*}\right)^2 + \left(1 - \frac{L_{\text{Arc}}}{L_{\text{Arc}}^*}\right)^2} \tag{9.8}$$

The flow diagram of this first ASM algorithm, able to provide the optimum filter schematic, is depicted in Figure 9.8. Applying the developed ASM algorithm to the considered example ($\mathbf{x}_c^* = [f_{0c}^*, \text{FBW}_c^*, L_{\text{Arc}}^*]$ = [2.4 GHz, 43.91%, 0.2 dB]), the error function rapidly decreases, being the error smaller than 0.02% after $N = 3$ iterations. The fine model parameters for the last iteration ($N = 3$) are $\mathbf{x}_f^{(3)} = \left[f_{0f}^{(3)}, \text{FBW}_f^{(3)}, L_{\text{Arf}}^{(3)}\right]$ = [2.4703 GHz, 65.3%, 0.2786 dB] and the coarse model parameters are $\mathbf{x}_c^{(3)} = \left[f_{0c}^{(3)}, \text{FBW}_c^{(3)}, L_{\text{Arc}}^{(3)}\right]$ = [2.400 GHz, 43.91%, 0.199987 dB]. Note that

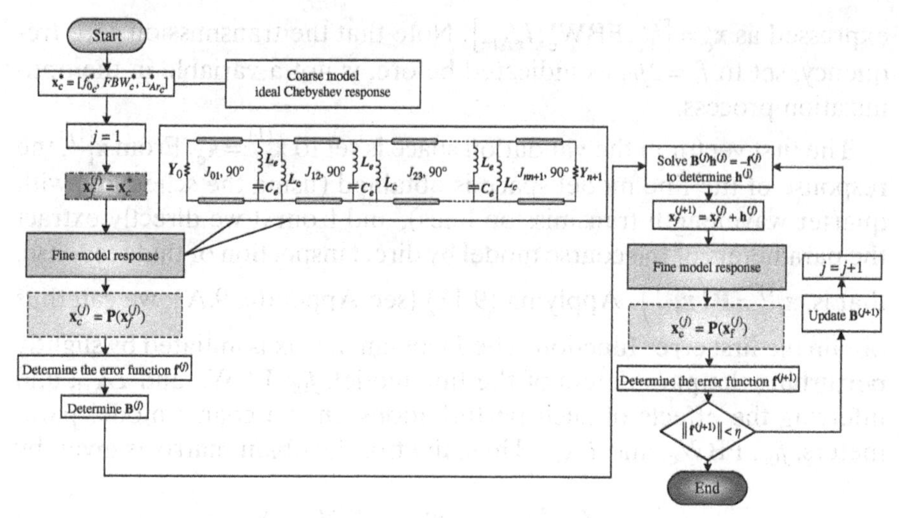

Figure 9.8 Flow diagram of the first ASM. Reprinted with permission from Ref. [11]; copyright 2014 IEEE.

$\mathbf{x_f}$ is appreciably different from $\mathbf{x_c^*}$. The optimum filter schematic is the one that gives the last fine model response (which provides an error below a predefined value). The elements of the shunt resonators for this optimum filter schematic are $L_e = 0.5935$ nH, $C_e = 1.7486$ pF, and $L_c = 1.7804$ nH, whereas the admittances of the inverters (quarter-wavelength transmission-line sections at $f_{0f}^{(3)} = 2.4703\text{ GHz} \neq f_0^*$) are $J_{0,1} = J_{5,6} = 0.02$ S, $J_{1,2} = J_{4,5} = 0.0211$ S, and $J_{2,3} = J_{3,4} = 0.0168$ S.

The response of the optimum schematic is compared with the target response in Figure 9.7. The agreement in terms of central frequency, bandwidth, and in-band ripple is very good, as expected on account of the small value of the error function that results after $N = 3$ iterations. However, the positions of the reflection zero frequencies are different in both responses.[4] The reason is that these frequency positions are not goals in the optimization process. Nevertheless, the synthesized circuit fulfills the target specifications for the differential mode. To complete the circuit schematic of Figure 9.6(b), valid for both modes, the capacitances C_{ci} need to be determined. As mentioned before, such capacitances are determined by the position of the transmission zeros for the common mode according to

[4] Note also that the response of the optimum schematic is not equiripple. Nevertheless, the in-band ripple level is better than the target.

$$f_{zi}^{cc} = \frac{1}{2\pi\sqrt{L_c C_{ci}}} \tag{9.9}$$

where the super-index indicates that these transmission zeros correspond to the common mode, and the subindex i indicates the filter stage. Note that there is no reason, a priori, to set the transmission zeros to the same value. Nevertheless, for this guide example, all the transmission zeros are set to 2.717 GHz (i.e., 1.1 $f_{0f}^{(3)}$), and hence $C_{ci} = 1.9268$ pF (the resulting response for the common mode is also depicted in Figure 9.7). Thus, the schematic resulting from this first ASM process, including C_{ci}, is the optimum filter schematic used as the starting point in the ASM algorithm developed to obtain the filter layout, to be described in the next subsection.

9.3.2 Layout Synthesis

The layout synthesis involves the determination of (i) the dimensions of the resonant elements (7-section mirrored SIRs), (ii) the width of the transmission line sections (inverters), and (iii) their lengths. Hence, three specific ASM subprocesses are developed for the automated synthesis of the filter layout, following a scheme similar to that reported in Ref. [11] for the synthesis of single-ended filters. Since the resonant elements are all identical (for the reasons explained before), the ASM devoted to the determination of resonator dimensions is applied only once. Let us now discuss in detail these three independent ASM subprocesses.

9.3.2.1 Resonator Synthesis In the ASM process devoted to the resonator synthesis, the variables in the optimization space are the resonator elements, that is, $\mathbf{x_c} = [L_e, L_c, C_e, C_c]$, and the coarse model response is obtained through circuit simulation. The validation space is constituted by a set of four geometrical variables. The other geometrical variables necessary to completely define the resonator layout are set to fixed values and are not variables of the optimization process. By this means, we deal with the same number of variables in both spaces, necessary for the inversion of the Jacobian matrix. Specifically, the variables in the validation space are the lengths of the narrow (inductive) and wide (capacitive) sections of the 7-section mirrored SIRs, that is, $\mathbf{x_f} = [l_{L_e}, l_{L_c}, l_{C_e}, l_{C_c}]$. The fine model response is obtained through electromagnetic simulation of the layout, inferred from the fine model variables plus the fixed dimensions, namely, the widths of the narrow and wide sections of the mirrored SIRs, and substrate parameters. The considered substrate parameters are those of the *Rogers RO3010* with thickness h = 635 μm and dielectric constant $\varepsilon_r = 10.2$. Concerning the fixed dimensions, the values are set to $W_{L_e} = W_{L_c} = 0.2$ mm, and there are two bounded values, $W_{C_e} = l_{C_e}$ (i.e., a square shaped geometry for the external patch capacitors is chosen), and $W_{C_c} = 0.75 \cdot \lambda/4$ mm, where λ is the guided wavelength at the central frequency of the optimum filter schematic. The value of 0.2 mm for the narrow inductive strips is slightly above the critical dimensions that are realizable with most available technologies. Concerning the square geometry of the external capacitive patches, with this shape factor the patches are described by a lumped capacitance to a very good approximation. Finally, the width of the central patches, W_{C_c}, has been chosen with the above criterion in order to avoid overlapping between adjacent patches.

In order to initiate the ASM algorithm, it is necessary to obtain an initial layout for the multi-section SIR. This is obtained from the following approximate formulas [8, 15]:

$$l_{L_e} = \frac{L_e v_{ph}}{Z_h} \tag{9.10a}$$

$$l_{C_e} = C_e v_{pl} Z_l \tag{9.10b}$$

$$l_{L_c} = \frac{L_c v_{ph}}{Z_h} \tag{9.10c}$$

$$l_{C_c} = C_c v_{pl} Z_l \tag{9.10d}$$

where v_{ph} and v_{pl} are the phase velocities of the high- and low-impedance transmission-line sections, respectively, and Z_h and Z_l are the corresponding characteristic impedances.

Once the initial layout (i.e., $\mathbf{x}_f^{(1)}$) is determined, the four circuit elements can be extracted from the electromagnetic response using (9.4)–(9.6) and (9.9). The specific procedure is as follows: The 4-port *S*-parameters (considering 50 Ω ports) of the 7-section SIR are obtained (e.g., by means of the *Keysight Momentum* electromagnetic solver). From these results, the *S*-parameters corresponding to the differential and common mode are inferred from formulas given in Chapter 1 and [16]. Then, from f_0, f_z and b (expressions 9.4–9.6) of the differential-mode response, the element values L_e and C_e and L_c are extracted, whereas C_c is determined from the transmission zero (expression 9.9) corresponding to the common-mode response. This provides

$\mathbf{x}_c^{(1)} = \mathbf{P}\left(\mathbf{x}_f^{(1)}\right)$, and, using (9.11), the first error function can be inferred. To iterate the process using (9.12), with $\mathbf{h}^{(1)}$ derived from (9.13), a first approximation of the Jacobian matrix is needed. Following a similar approach to the one explained in Section 9.3.1, the lengths l_{L_e}, l_{L_c}, l_{C_e}, and l_{C_c} are slightly perturbed, and the values of L_e, L_c, C_e, and C_c resulting after each perturbation from parameter extraction are then obtained. This provides the first order-4 Jacobian matrix. By means of this procedure, the layouts of the 7-section mirrored SIRs are determined.

9.3.2.2 Determination of the Line Width The widths of the quarter-wavelength (at $f_{0f}^{(3)}$) transmission lines are determined through a one-variable ASM procedure, where the fine model variable is the line width, W, whereas the variable of the coarse model is the characteristic impedance. However, it has to be taken into account that this ASM must be repeated as many times as different admittance inverters are present in the filter. It is also important to bear in mind that the pair of differential lines are widely separated, so that the differential- and common-mode impedances take the same value, that is, identical to that of the isolated line.[5]

9.3.2.3 Optimization of the Line Length (Filter Cell Synthesis) To determine the length of the inverters, the procedure is to consider the whole filter cell, consisting of the resonator cascaded in between the inverter halves (not necessarily of the same width or admittance). As it was pointed out in Ref. [11], optimization of the whole filter cell is necessary since the resonators may introduce some (although small) phase shift. In Ref. [11], the whole filter cell was forced to exhibit a phase shift of 90° at the central frequency of the optimum schematic. However, the fact that the inverters at both sides of the resonator have different admittance means that the phase of S_{21} is no longer 90° at the central frequency of the optimum schematic. Nevertheless, the phase shift of the cell can be easily inferred from circuit simulation, and the resulting value is the goal of this third ASM subprocess. Thus, the ASM optimization consists of varying the length of the lines cascaded to the resonator until the required phase per filter cell is achieved (the other geometrical

[5] Actually, according to the definition of characteristic impedance for the differential/common mode and even/odd mode given in Chapter 1, for uncoupled lines the even- and odd-mode impedances, rather than the differential and common mode impedances, are identical to the one of the isolated lines. However, sometimes these impedances (even/common and odd/differential) are indistinctly used as synonymous.

parameters of the cell are kept unaltered). The phase is directly inferred from the frequency response of the cell obtained from electromagnetic simulation at each iteration step.

Once each filter cell is synthesized, the cells are cascaded and no further optimization is required. The flow diagram of the complete ASM process able to automatically provide the layout from the optimum filter schematic, consisting of the three independent quasi-Newton iterative algorithms described, is depicted in Figure 9.9.

Using the mirrored SIR element values and inverter admittances corresponding to the optimum filter schematic of the example reported in the previous section, the developed ASM algorithm has been used to automatically generate the filter layout (depicted in Figure 9.10). Resonator dimensions are $l_{L_e} = 0.9075$ mm, $l_{L_c} = 2.4136$ mm, $l_{C_e} = 2.5262$ mm, and $l_{C_c} = 1.0986$ mm, and the lengths of the filter cells give

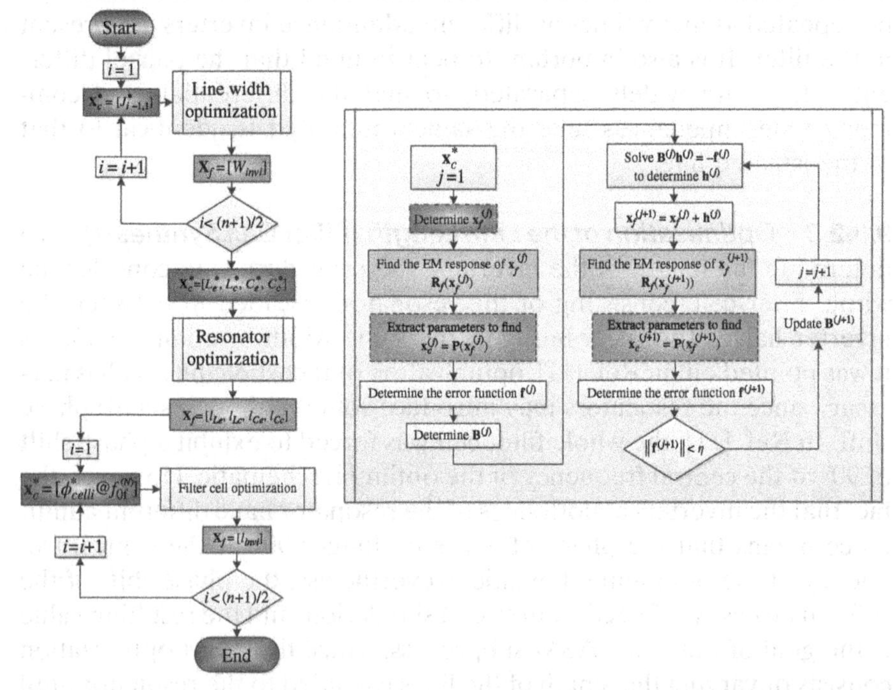

Figure 9.9 Flow diagram of the second ASM algorithm. The subprocess depicted at the right of the figure represents a typical ASM algorithm used in each optimization process (particularly the indicated one is for the resonator optimization). Notice that the loop must be executed $(n+1)/2$ times, n being the filter order (which is assumed to be odd), since for odd Chebyshev response, the cells i and $n+1-i$ are identical. However, this does not affect resonator optimization since resonators are identical for all filter stages.

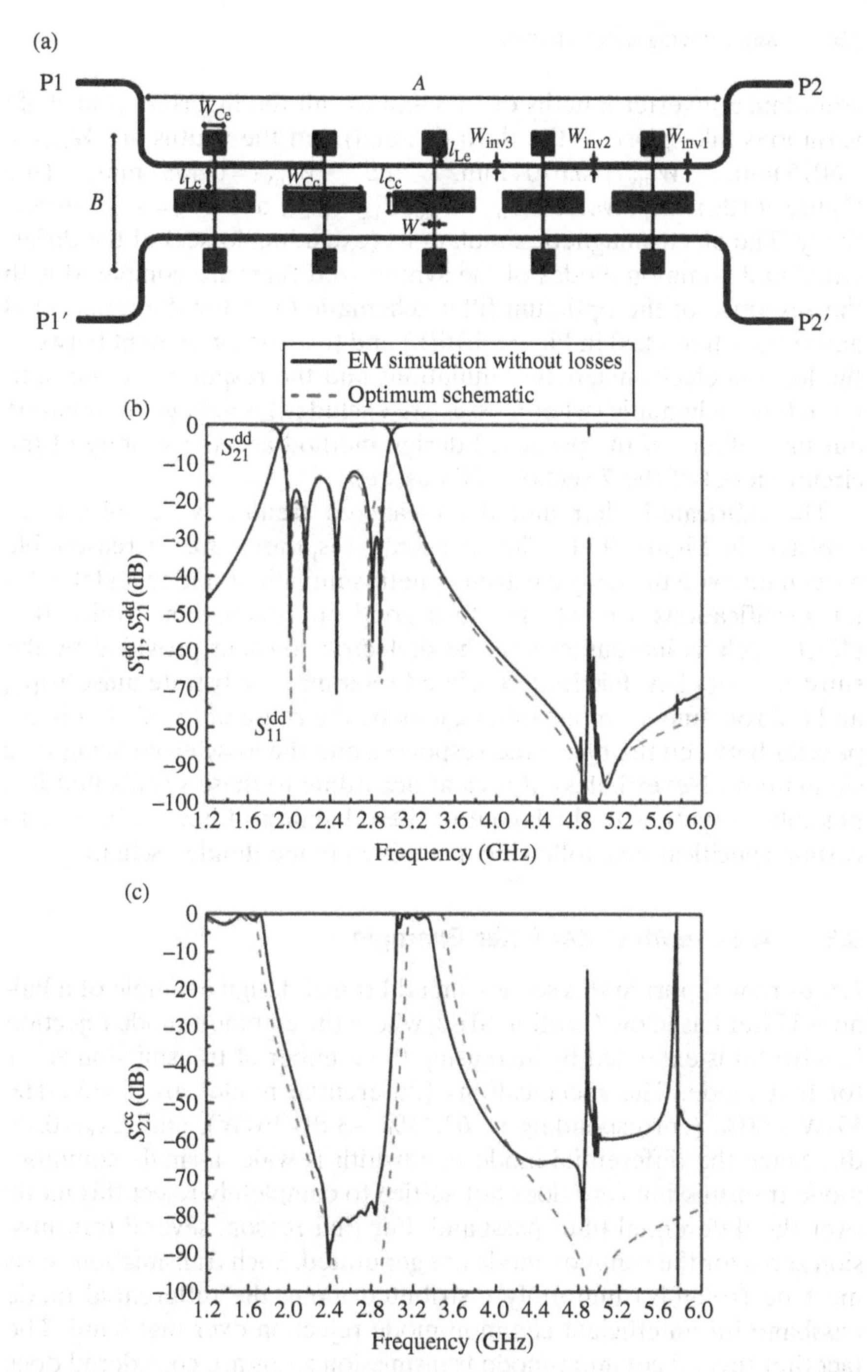

Figure 9.10 Layout of the synthesized order-5 filter (a), differential-mode response (b), and common-mode response (c). In (b) and (c), the lossless electromagnetic simulations of the synthesized layout are compared with the circuit simulations of the optimum filter schematic. The relevant dimensions in (a) are $A = 63$ mm and $B = 15$ mm. Reprinted with permission from Ref. [4]; copyright 2015 IEEE.

admittance inverter lengths of 11.4 mm for all the inverters (the slight variations take place at the third decimal) and the widths are W_{inv1} = 0.6015 mm, W_{inv2} = 0.6704 mm, and W_{inv3} = 0.4087 mm (see Figure 9.10a) for inverters $J_{0,1} = J_{5,6}$, $J_{1,2} = J_{4,5}$, and $J_{2,3} = J_{3,4}$, respectively. The electromagnetic simulations (excluding losses) of the differential and common modes of the synthesized filter are compared with the response of the optimum filter schematic (also for the differential and common modes) in Figure 9.10(b) and (c). The agreement between the lossless electromagnetic simulations and the responses of the optimum filter schematic (where losses are excluded) is very good, pointing out the validity of the proposed design method and the validity of the circuit model of the 7-section SIRs as well.

The fabricated filter and the measured frequency responses are depicted in Figure 9.11. The measured responses are in reasonable agreement with the lossy electromagnetic simulations and reveal that filter specifications are satisfied to a good approximation. Notice that effects such as inaccuracies in the dielectric constant provided by the substrate supplier, fabrication related tolerances, substrate anisotropy, and foil roughness, among others, may be the cause of the slight discrepancies between the measured responses and the lossy electromagnetic simulations. Nevertheless, it is clear according to these results that it is possible to synthesize the layout of these balanced filters, subjected to certain specifications, following a completely unattended scheme.

9.3.3 A Seventh-Order Filter Example

Let us now report in this section an additional design example of a balanced filter based on 7-section SIRs, where the common-mode rejection bandwidth is extended by increasing the number of transmission zeros for that mode. The specifications (differential mode) are f_0 = 3 GHz, FBW = 60% (corresponding to 63.43% −3-dB FBW), and L_{Ar} = 0.15 dB. Since the differential-mode bandwidth is wide, a single common-mode transmission zero does not suffice to completely reject this mode over the differential filter passband. For that reason, several transmission zeros for the common mode are generated. Such transmission zeros must be (roughly) uniformly distributed along the differential mode passband for an efficient common-mode rejection over that band. The fact that several common-mode transmission zeros are considered does not affect the first ASM algorithm. However, as many different capacitances C_{ci} as transmission zeros must be calculated by means of expression (9.9) to completely determine the elements of the optimum filter schematic. Since the capacitances C_{ci} determine the area of the central

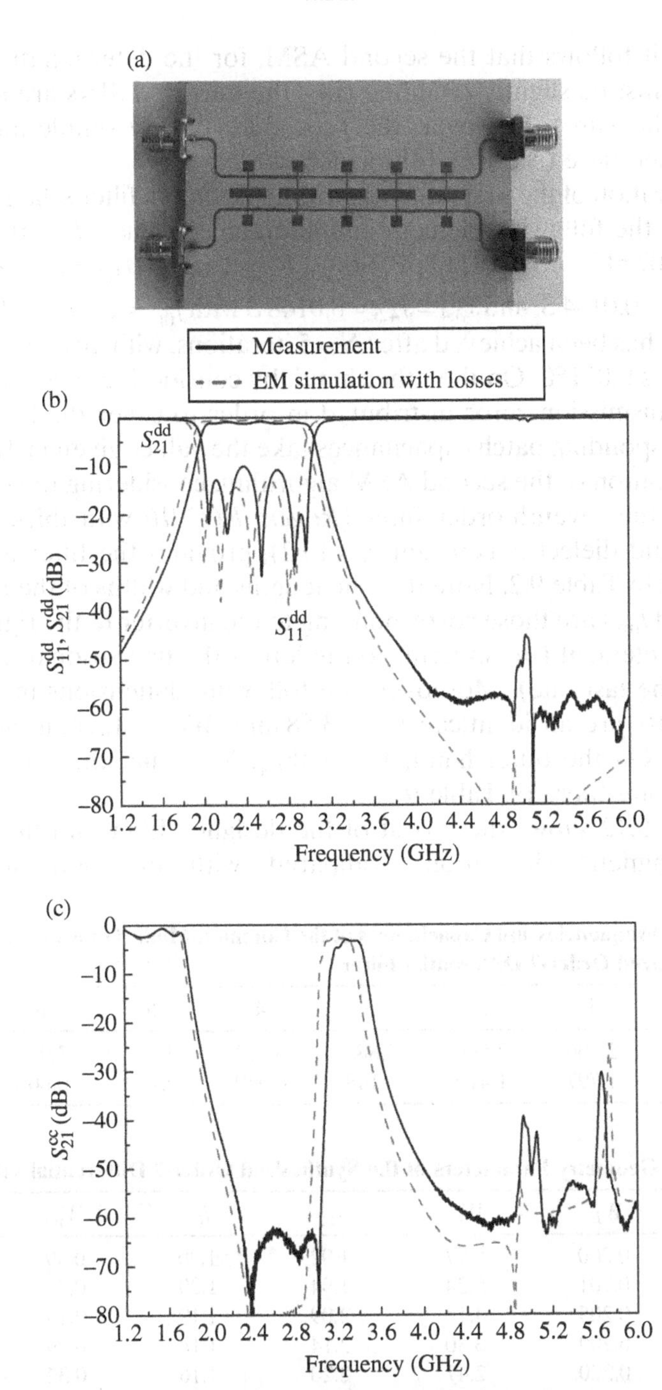

Figure 9.11 Photograph of the fabricated order-5 filter (a), differential-mode response (b), and common-mode response (c). In (b) and (c), the measured responses are compared to the lossy electromagnetic simulations of the synthesized layout. Reprinted with permission from Ref. [4]; copyright 2015 IEEE.

patches, it follows that the second ASM, for the determination of the layout, must be slightly modified (i.e., the mirrored SIRs are not identical in this case). However, the procedure is very simple and is not reproduced here (see Ref. [4] for details).

Application of the first ASM algorithm (optimum filter schematic) has provided the following element and admittance values: $L_e = 0.8836$ nH, $L_c = 2.6507$ nH, $C_e = 0.7114$ pF, $J_{0,1} = J_{7,8} = 0.02$ S, $J_{1,2} = J_{6,7} = 0.0199$ S, $J_{2,3} = J_{5,6} = 0.0154$ S, and $J_{3,4} = J_{4,5} = 0.0148$ S with $f_{0f}^{(5)} = 3.1739$ GHz. Convergence has been achieved after $N = 5$ iterations, with an error function as small as 0.021%. On the other hand, by considering seven common-mode transmission zeros distributed in order to cover the bandwidth, the corresponding patch capacitances take the values given in Table 9.1.

Application of the second ASM algorithm, considering the substrate used for the seventh-order filter (*Rogers RO3010* with thickness $h =$ 635 μm and dielectric constant $\varepsilon_r = 10.2$), provides the filter geometry indicated in Table 9.2. Note that the lengths and widths of the inverters (W_{inv} and l_{inv}) are those corresponding to the inverter to the right of the resonant element (the inverter to the left of the first resonator is identical to the last one). Moreover, the following dimensions in the mirrored SIRs are all identical: $l_{L_c} = 3.58$ mm, $W_{C_e} = 1.2$ mm, and $l_{C_c} =$ 2.92 mm. On the other hand, $W_{L_{ei}} = W_{L_{ci}}$. Note that the optimization variables are those of Table 9.2.

Figure 9.12 shows the layout of the designed filter and the lossless electromagnetic simulation, compared with the optimum filter

Table 9.1 Frequencies and Capacitances of the Common-Mode Transmission Zeros of the Synthesized Order-7 Differential Filter

Stage	1	2	3	4	5	6	7
f_z^{cc} (GHz)	2.306	2.596	2.885	3.174	3.572	3.970	4.369
C_c (pF)	1.797	1.419	1.148	0.949	0.749	0.606	0.501

Table 9.2 Geometry Parameters of the Synthesized Order-7 Differential Filter

Stage	W_{L_c}	W_{C_c}	l_{L_e}	l_{C_e}	W_{inv}	l_{inv}
1	0.200	6.77	1.92	1.20	0.59	8.87
2	0.201	5.24	1.94	1.20	0.32	8.87
3	0.205	4.14	2.09	1.19	0.29	8.86
4	0.209	3.30	2.14	1.18	0.29	8.86
5	0.220	2.47	2.20	1.16	0.32	8.85
6	0.233	1.86	2.19	1.12	0.59	8.84
7	0.253	1.40	2.38	1.06	0.59	8.76

All dimensions in millimeters.

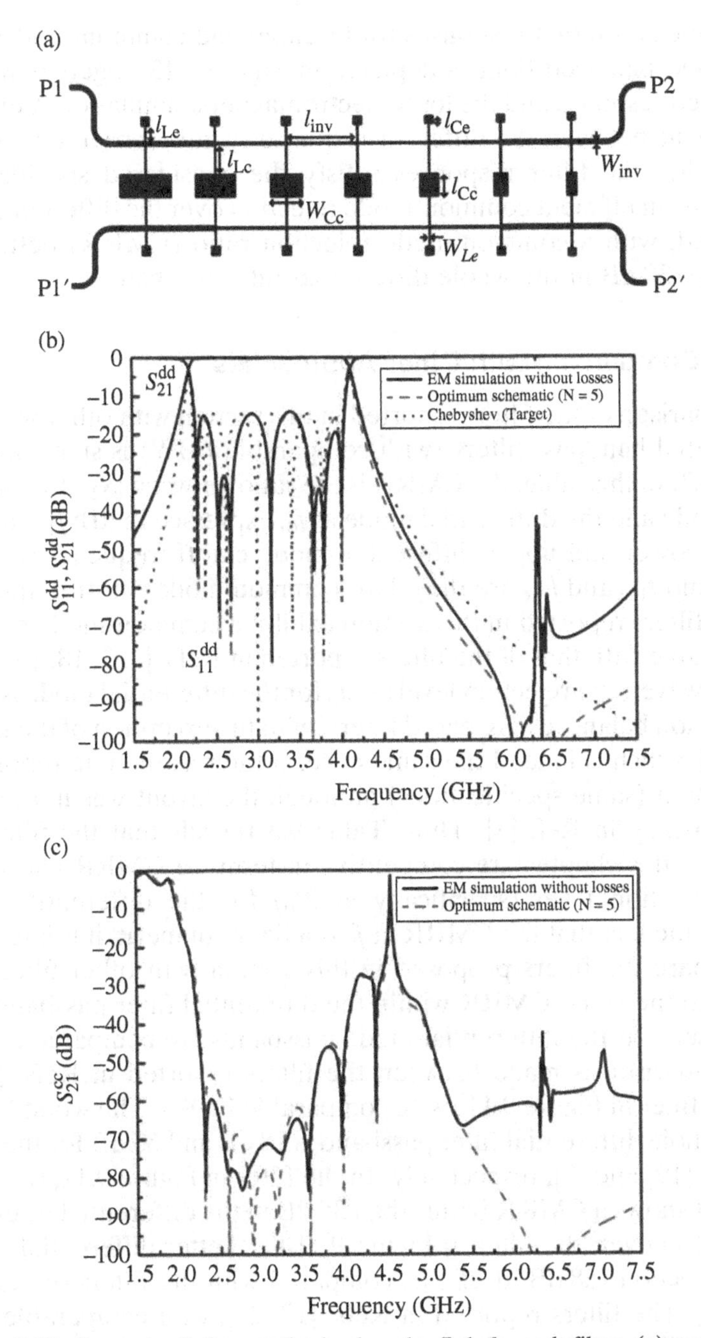

Figure 9.12 Layout of the synthesized order-7 balanced filter (a) and lossless electromagnetic simulation compared to the response of the optimum filter schematic and target response for the differential (b) and common mode (c). Reprinted with permission from Ref. [4]; copyright 2015 IEEE.

schematic and target responses (differential and common-modes). The fabricated balanced filter is depicted in Figure 9.13, together with the measured response and the lossy electromagnetic simulation. Very good agreement between the different responses can be observed, and it is found that the filter responses satisfy the considered specifications, including an efficient common-mode rejection over the differential filter passband, with a common-mode rejection ratio (CMRR) better than CMRR = 30 dB in the whole differential filter passband.

9.3.4 Comparison with Other Approaches

A comparison of the filters reported in this section with other wideband differential bandpass filters (with comparable FBW) is summarized in Table 9.3. In this table, the CMRR is the ratio between $|S_{21}|$ for the common mode and the differential mode at f_0, expressed in dB; f_{1dd} and f_{2dd} are the lower and upper differential-mode cutoff frequencies, respectively, and f_{1cc} and f_{2cc} are the –3 dB common-mode cutoff frequencies.

The filters reported in this section exhibit a common-mode rejection comparable with that of the filters reported in Refs. [2, 3, 18, 19, 23, 25, 26]. However, the rejection level at $2f_0$ for the differential mode is larger in 7-section balanced SIR-based filters, with the exception of the filter of Ref. [3], which is indeed the same order-5 filter as the one reported in this section (same specifications) although the layout was not inferred automatically in Ref. [3]. Thus, Table 9.3 reveals that the filters discussed in this chapter are competitive in terms of CMRR and out-of-band rejection level (specifically at $2f_0$) for the differential mode. Despite the fact that the CMRR at f_0 is a figure of merit, it is interesting to compare the filters proposed in this section with other filters with regard to the worst CMRR within the differential filter passband. This makes sense if the differential-mode passbands are comparable. Thus, the comparison is made between the filters reported in Refs. [2, 19] and the filter in Figure 9.11 (with comparable FBW). The worst CMRR in the whole differential filter passband is 18 dB and 63 dB for the filters of Refs. [19] and [2], respectively. In the filter in Figure 9.11, the measurement shows a CMRR better than 35 dB in the differential filter passband. Moreover, the filter in Figure 9.11 has better differential out-of-band rejection (58 dB at $2f_0$) as compared with the filters of Refs. [2] and [19]. The filters reported in Refs. [17, 22], with comparable FBW with the filter in Figure 9.13, have a CMRR in the whole differential passband better than 22 dB and 14.5 dB, respectively. In the filter in Figure 9.13, the measurement shows a CMRR better than 30 dB in the differential filter passband. Moreover, the filter in Figure 9.13 has

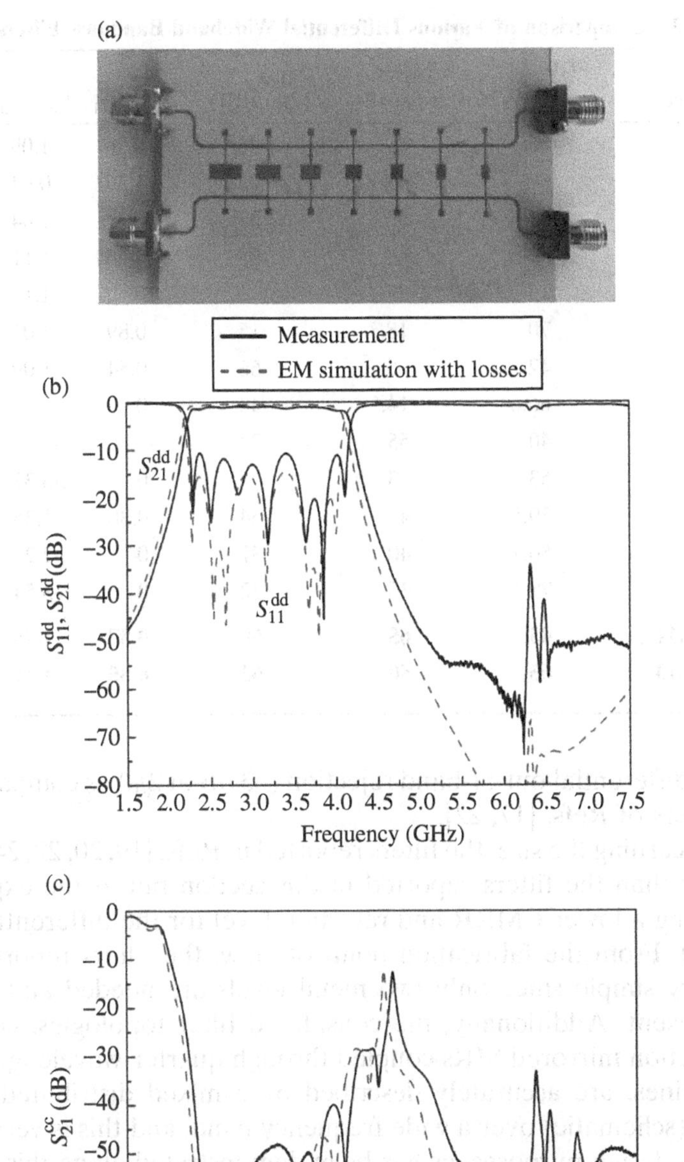

Figure 9.13 Photograph of the fabricated order-7 filter (a) and measured response compared to the lossy electromagnetic simulation for the differential (b) and common mode (c). Reprinted with permission from Ref. [4]; copyright 2015 IEEE.

Table 9.3 Comparison of Various Differential Wideband Bandpass Filters

Reference	FBW (−3 dB) (%)	CMRR at f_0 (dB)	S_{21dd} ($2f_0$) (dB)	f_{1cc}/f_{1dd}	f_{2cc}/f_{2dd}	Effective area
[17]	65	29	46	0.96	1.09	$0.25\lambda_g^2$
[18]	50	48	23	1.25	0.94	$0.25\lambda_g^2$
[19]	45	53	35	0	1.04	$0.045\lambda_g^2$
[2]	43	63	38	0.66	1.11	$0.6\lambda_g^2$
[20]	15	27	9	0	1.15	$0.022\lambda_g^2$
[21]	70	18.7	15	0.89	1.01	$0.52\lambda_g^2$
[3]	42	65	66	0.54	1.09	$0.37\lambda_g^2$
[22]	66.6	14.5	40	0	2.21	$0.163\lambda_g^2$
[23]	40	55	22	–	–	$0.25\lambda_g^2$
[24]	53	13	7	0	>3.37	$0.175\lambda_g^2$
[25]	59.5	47	54	0.38	1.35	$0.39\lambda_g^2$
[26]	56.7	40	37	0	>2.2	$0.2\lambda_g^2$
[27]	79	35	32	0	>3.54	$0.19\lambda_g^2$
Figure 9.11	**43**	**65**	**58**	**0.87**	**1.05**	$\mathbf{0.36\lambda_g^2}$
Figure 9.13	**65**	**50**	**63**	**0.85**	**1.11**	$\mathbf{0.28\lambda_g^2}$

better differential out-of-band rejection (63 dB at $2f_0$) as compared with the filters of Refs. [17, 22].

Concerning the size, the filters reported in Refs. [19, 20, 22, 24, 27] are smaller than the filters reported in this section but at the expense of obtaining a lower CMRR and rejection level for the differential mode (at $2f_0$). From the fabrication point of view, the filters reported here are very simple since only two metal levels are needed and vias are not present. Additionally, the considered filter topologies, consisting of 7-section mirrored SIRs coupled through quarter-wavelength differential lines, are accurately described by a mixed distributed-lumped model (schematic) over a wide frequency band, and this is very important for design purposes, as has been demonstrated along this section.

9.4 COMPACT ULTRA-WIDEBAND (UWB) BALANCED BANDPASS FILTERS BASED ON 5-SECTION MIRRORED SIRs AND PATCH CAPACITORS

To reduce filter size, multi-section mirrored SIRs should be combined with semi-lumped series resonators, rather than with quarter-wavelength admittance inverters. In this section, we present a topology for the series resonators useful to achieve UWB responses for the differential mode. Note also that by avoiding distributed elements, the ASM-based compensation method reported in the previous section is not needed. As a reference, we will consider the implementation of balanced filters covering the UWB defined by the Federal Communications Commission (FCC), that is, 3.1–10.6 GHz.

9.4.1 Topology and Circuit Model of the Series Resonators

To a rough approximation, the considered UWB balanced filters are described by the canonical circuit of a bandpass filter, consisting of a ladder network with series resonators in the series branches and parallel resonators in the shunt branches. As it is well known from filter theory, the capacitances of the series branches and the inductances of the shunt branches increase with filter bandwidth. Conversely, the inductances of the series branches and the capacitances of the shunt branches are small for bandpass filters with very wide bandwidths. This means that large series capacitances are needed in order to achieve the required bandwidth (3.1–10.6 GHz). Hence, series gaps or interdigital capacitors do not seem to be the optimum solution, on account of the limited capacitance values achievable with these topologies.[6]

Indeed, the typical required values force us to implement such series capacitors by means of patch capacitances (i.e., broadside elements, exhibiting large capacitance per area unit). Therefore, ground plane etching is necessary to implement such capacitors. Metallic vias can in principle be avoided by series-connecting two identical patch capacitances. However, this strategy increases the required capacitor area since each individual capacitance value is twice the nominal one. The best option (in terms of miniaturization) is thus to implement the series capacitances through an optimized topology that uses a metallic connection between the upper and lower metal levels (via). The series inductances of the series branches (of small value for UWB filters) are simply implemented by means of a pair of short inductive strips cascaded to the capacitor terminals. Nevertheless, the presence of the vias may contribute to the series inductance as well. The top and 3D views of a typical series resonator are depicted in Figure 9.14(a) and (b), respectively.

The lumped element equivalent circuit model of the series branch (Figure 9.14c) is not simple, as compared with the one of the 5-section

[6] Nevertheless, interdigital capacitors have been applied to the implementation of balanced filters with smaller bandwidths [6].

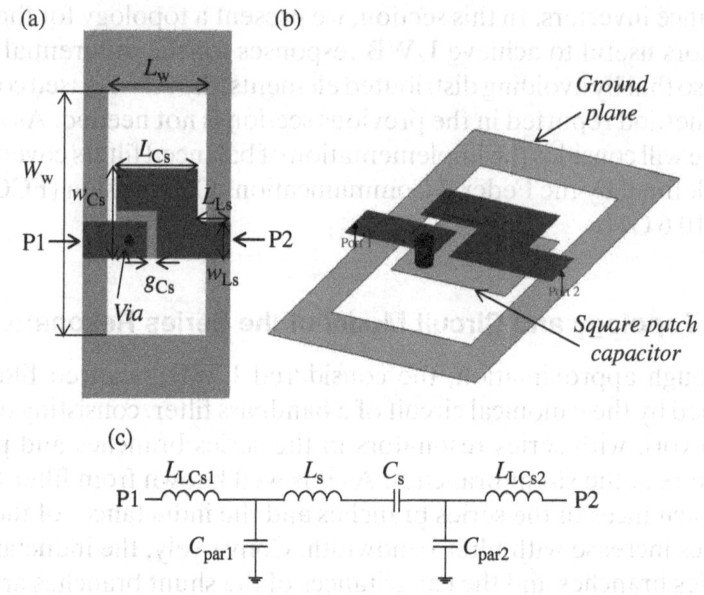

Figure 9.14 Top view (a), 3D view (b), and lumped element equivalent circuit model (c) of the series branches of the balanced UWB bandpass filter. Dimensions are W_w = 3.3 mm, L_w = 1.46 mm, w_{Ls} = 0.5 mm, L_{Ls} = 0.49 mm, w_{Cs} = 1.22 mm, L_{Cs} = 1.2 mm, and g_{Cs} = 0.144 mm.

balanced SIR. Note that since the pairs of series resonators replacing the admittance inverters are very distant, the differential- and common-mode circuits describing such structures are undistinguishable and given by the circuit model describing the isolated resonators (Figure 9.14c). Note also that the considered resonator is asymmetric with regard to the ports. Such asymmetry is accounted for by the circuit model by considering different parasitic inductances (L_{LCs1} and L_{LCs2}) and capacitances (C_{par1} and C_{par2}).

To validate the circuit model, the *S*-parameters of the structure depicted in Figure 9.14 (with dimensions indicated in the caption and substrate parameters corresponding to *Rogers RO3010* with dielectric constant $\varepsilon_r = 10.2$ and thickness $h = 254$ μm) have been obtained from full wave electromagnetic simulation (see Figure 9.15). From these parameters, the series reactance and the shunt reactances of the equivalent π-circuit model are inferred. Such reactances are depicted in Figure 9.16, where they are compared with the reactances derived from the transformation of the model in Figure 9.14(c) to its equivalent π-circuit model. It is remarkable that the shunt reactances of the equivalent π-circuit model are unequal, confirming the need to consider an asymmetric circuit model (i.e., such reactances are significantly different) to describe the

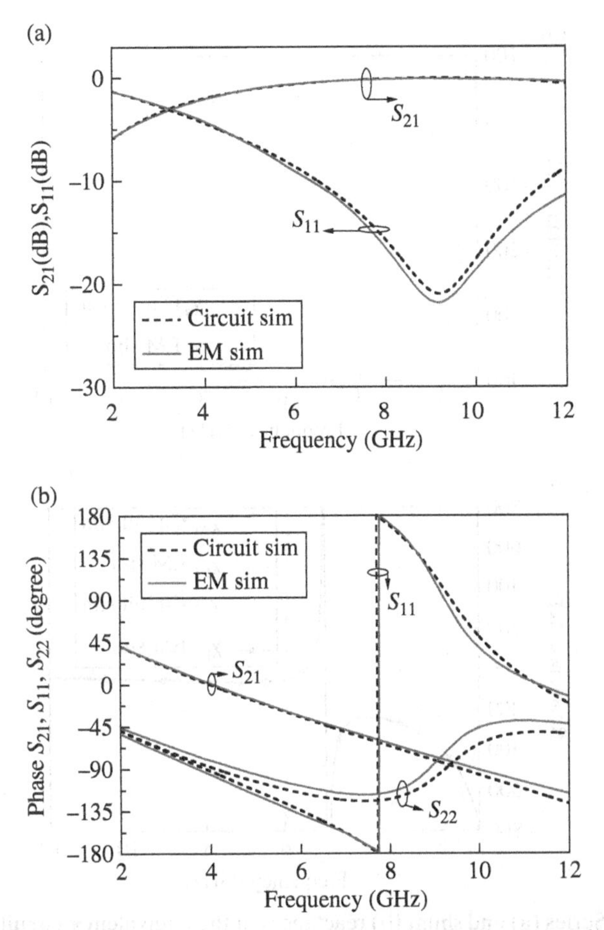

Figure 9.15 Magnitude (a) and phase (b) of the *S*-parameters for the structure in Figure 9.14 inferred from electromagnetic and circuit simulation. The element values of the circuit model in Figure 9.14(c) are $C_s = 0.453$ pF, $L_s = 1.39$ nH, $C_{par1} = 0.33$ pF, $L_{LCs1} = 0.31$ nH, $C_{par2} = 0.18$ pF, and $L_{LCs2} = 0.123$ nH. The magnitude of S_{22} is roughly undistinguishable from that of S_{11} and is not depicted. Reprinted with permission from Ref. [5]; copyright 2015 IEEE.

series resonator. Note also that in certain frequency regions the right-hand side shunt reactance exhibits negative slope. This does not contradict the Foster reactance theorem [28], since the curves depicted in Figure 9.16, rather than corresponding to reactances of circuits with identifiable reactive elements, are the reactances of an equivalent circuit (π-circuit model), and such a circuit may not be described by lumped elements (this is indeed the case).

The agreement between the results inferred from electromagnetic and circuit simulation in Figures 9.15 and 9.16 is very reasonable and

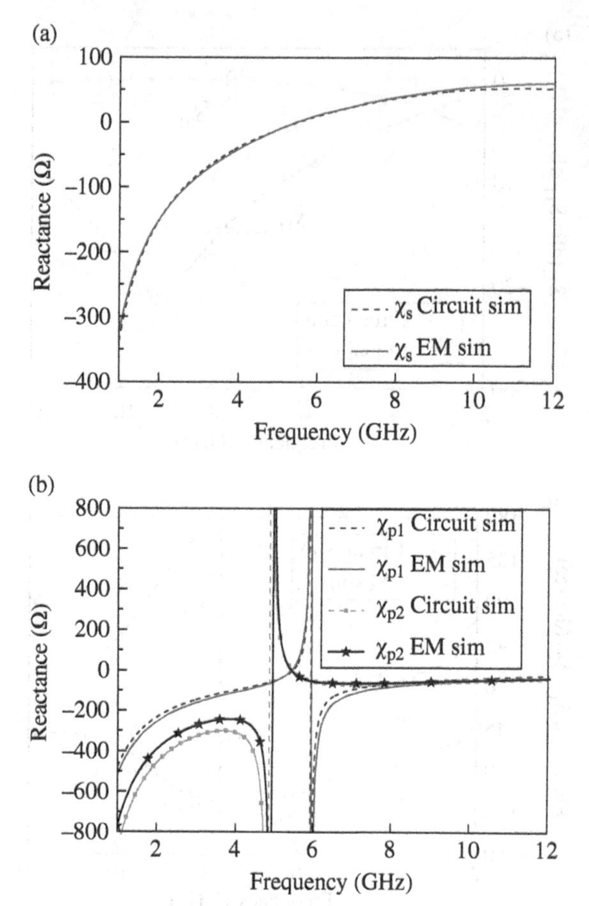

Figure 9.16 Series (a) and shunt (b) reactances of the equivalent π-circuit model of the structure in Figure 9.14(a) inferred from electromagnetic and circuit simulation. The element values are those indicated in the caption in Figure 9.15. Reprinted with permission from Ref. [5]; copyright 2015 IEEE.

validates the circuit model in Figure 9.14(c). The determination of the circuit parameters, necessary to obtain the equivalent π-circuit model, is not so straightforward. First of all, in the low frequency limit, the inductances can be neglected, and the resulting model is a capacitive asymmetric π-circuit model. Thus by representing the susceptances of the series and shunt branches, the capacitors C_{par1}, C_{par2}, and C_s can be inferred from the susceptance slope in the DC limit. Once these element values are known, the inductances are determined by curve fitting the reactances of the three branches of the equivalent circuit model (the element values are indicated in the caption in Figure 9.15). That is, curve fitting is necessary for parameter extraction due to the high number of independent elements (6) of the circuit model. Conversely, for the 5-section or 7-section balanced SIRs, parameter extraction is based on analytical expressions (see Sections 9.2 or 9.3).

9.4.2 Filter Design

The filters reported in this section are based on the previous series resonators, combined with 5-section mirrored SIRs. In these UWB balanced filters, it is necessary to distribute the common-mode transmission zeros, provided by the mirrored SIRs, at different locations over the differential filter passband (notice that the series branches do not have any influence on the position of these transmission zeros). Since the inductances of the narrow strips of the SIRs are determined from the specifications of the differential-mode filter response, it follows that the transmission zero locations must be controlled through the dimensions of the central (wide) capacitive sections of the mirrored SIRs, and each balanced SIR provides one transmission zero for the common mode. Typically, three transmission zeros are enough in order to achieve a significant common-mode rejection over the regulated UWB band (this is the case in the proposed example, as will be later shown). Due to the higher complexity of the circuit model of these filters, as compared with the model of the filters discussed in the previous section, the layout is not automatically synthesized by means of an ASM-based tool. Filter design in this case is based on parameter extraction and tuning, as will be shown in brief. First the layout that provides the required differential-mode response is obtained. This excludes the determination of the central patch capacitors of the 5-section SIRs. Then, the dimensions of these patch capacitances are determined in order to set the transmission zeros for the common mode and thus achieve a stop band for that mode over the differential filter passband.

Let us now focus on the design of the balanced filter in order to cover the regulated UWB band for the differential mode. The intended filter response for the differential mode is an order-5 Chebyshev response covering the band from 3.1 GHz up to 10.6 GHz with a ripple of 0.15 dB. By considering the canonical order-5 bandpass filter circuit model with shunt connected parallel resonators at the first, third, and fifth stages, and series connected series resonators at the second and fourth stages, the element values giving such a response are $C_1 = C_5 = 0.535$ pF, $L_1 = L_5 = 1.44$ nH, $C_2 = C_4 = 0.532$ pF, $L_2 = L_4 = 1.448$ nH, $C_3 = 0.888$ pF, and $L_3 = 0.868$ nH, where the subindex denotes the filter stage. Once these values are known, the next step is to independently synthesize the layouts for each filter stage (shunt and series branches) with an eye toward achieving the aforementioned element values for the active elements of the filter model for the differential mode (active elements are those reactive elements in the models in Figures 9.1(b) and 9.14 (c) not being the parasitic elements, viz., C_{pi}, L_{pi}, C_{si}, and L_{si}). The parameter extraction procedures explained before are used for that purpose, and optimization at the layout level is done as specified next. For the shunt branches, the dimensions of the patch capacitors (C_{pi}) and inductive strips (L_{pi}) are estimated from the formulas providing the dimensions of the low and high impedance transmission-line sections of stepped-impedance low pass filters (expressions 9.10). Then, optimization is carried out at layout level in order to obtain the required element values for C_{pi} and L_{pi} from the parameter extraction procedure detailed before. Notice that parameter extraction gives also the value of the parasitic inductance L_{par}. For the series branches, a square geometry for the patch capacitor and a via with a diameter of 0.1 mm, and 0.5 mm wide access lines are considered. Then the area of the square capacitor and the length of the access lines are tuned until the active elements C_s and L_s, inferred from parameter extraction and curve fitting as specified before, are those corresponding to the ideal Chebyshev response. This procedure provides also the parasitic elements of the series branch.

Obviously, due to the presence of the parasitic elements, we do not expect that the filter response for the differential mode agrees with the ideal Chebyshev response. Thus, the filter layout must be modified in order to obtain a better approximation to the required Chebyshev response. The key aspect in the design process is to assume that layout tuning does not substantially modify the parasitic elements. Therefore, the next step is to set the parasitic elements in the circuit model to those values inferred from parameter extraction and curve fitting, and tune the active elements until the filter response for the differential mode agrees with the intended Chebyshev response to a good approximation in the band of interest. Notice that this is a tuning at the circuit level. Nevertheless, since there are too many degrees of freedom, tuning may be complex and time-consuming, unless an automatic process is considered. Therefore, an automatic optimization routine in *Keysight ADS* was implemented in order to find the element values that provide a minimum in-band return loss level (10 dB) with the frequency positions of the five matching points (reflection zeros) as close as possible to those of the ideal Chebyshev response. This routine gives the active element values of the filter circuit model. Once these elements are known, the next step is to modify (tune) the layout in order to infer these element values from parameter extraction and curve fitting for either filter stage. It is worth mentioning that synthesis is not excessively complex since the number of geometrical free parameters per filter stage is small. The element values, including parasitics, are given in Table 9.4 (notice that the first and fifth stages are slightly different).

As it was already mentioned, the wide patch capacitances of SIRs determine the position of the transmission zeros for the common mode, but they do not have any influence on the differential filter response. Thus, with exception of such capacitances, filter layout is determined from the design method detailed earlier. For which concerns the patches corresponding to the grounded capacitances, C_z, the areas have been determined in order to obtain transmission zeros for the common mode at 5.5 GHz, 7 GHz, and 10 GHz.

The layout and photograph of the designed filter are depicted in Figure 9.17. The filter has been fabricated on the *Rogers RO3010* substrate with dielectric constant $\varepsilon_r = 10.2$ and thickness $h = 254$ μm. The differential and common-mode filter responses are depicted in Figure 9.18. Due to the presence of parasitics, the measured and electromagnetically

Table 9.4 Circuit Elements for the Equivalent Circuit of the Differential UWB Filter

i	L_p	C_p	L_{par}	L_s	C_s	L_{LCs1}	L_{LCs2}	C_{par1}	C_{par2}
1	1.44	0.38	0.06						
2,4				1.25	0.48	0.31	0.09	0.30	0.23
3	0.87	0.47	0.16						
5	1.46	0.40	0.11						

The units are nH for the inductances and pF for the capacitances.

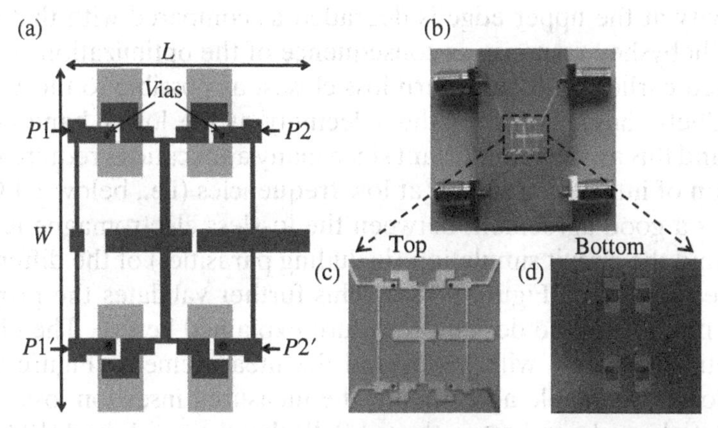

Figure 9.17 Layout (a) and photograph (b) of the fabricated filter. A zoom of the photograph is shown in (c) top view and (d) bottom view. Dimensions are $W = 10.5$ mm and $L = 7.6$ mm. Reprinted with permission from Ref. [5]; copyright 2015 IEEE.

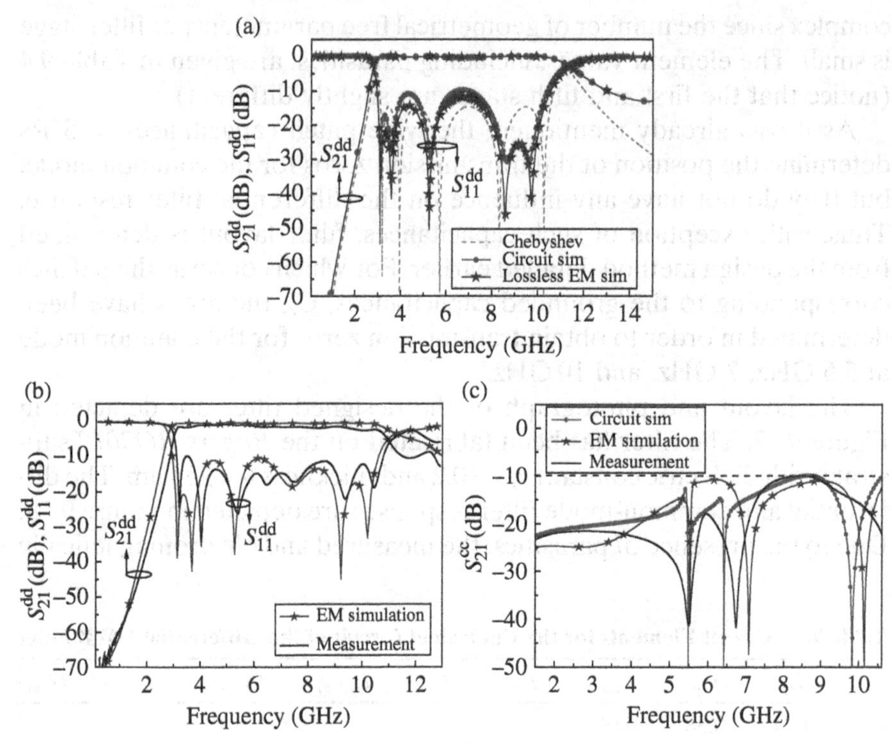

Figure 9.18 Differential (a)–(b) and common-mode (c) filter response. Reprinted with permission from Ref. [5]; copyright 2015 IEEE.

simulated differential filter responses are a rough approximation of the ideal Chebyshev response (also included in the figure). Indeed filter selectivity at the upper edge is degraded as compared with that of the ideal Chebyshev response as consequence of the optimization criterion indicated earlier (in-band return loss closest as possible to those of the ideal Chebyshev). However the selectivity at the lower band edge is good, and this aspect is important since many applications require strong rejection of interfering signals at low frequencies (i.e., below 3.1 GHz). There is a good agreement between the lossless electromagnetic simulation and the circuit simulation (including parasitics) of the differential filter response (see Figure 9.18a). This further validates the proposed circuit model and the design procedure explained before. The electromagnetic simulation with losses and the measurement (Figure 9.18b) are also in reasonable agreement. The measured insertion loss for the differential mode is better than 1.9 dB between 3.1 and 10.6 GHz, whereas the return loss is higher than 10 dB within the same frequency interval. The slight discrepancies in the insertion loss are mainly

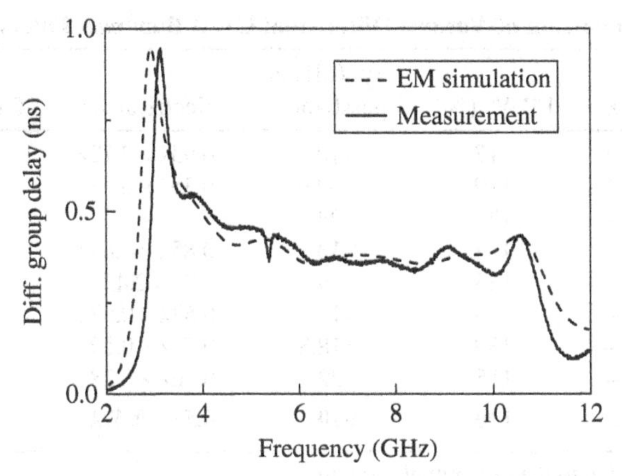

Figure 9.19 Differential group delay of the filter in Figure 9.17. Reprinted with permission from Ref. [5]; copyright 2015 IEEE.

attributed to the effects of the connectors and soldering. The common-mode response is reasonably predicted by the circuit model (common mode), as depicted in Figure 9.18(c). Thanks to the presence of three transmission zeros, a common-mode rejection better than 10 dB over the whole regulated UWB is achieved. Concerning dimensions, these are as small as 10.5 mm × 7.6 mm, that is, $0.5\lambda \times 0.37\lambda$ (excluding the tapered access lines), λ being the guided wavelength at the central filter frequency. The differential group delay, shown in Figure 9.19, exhibits a value smaller than 0.5 ns in the whole differential pass band and a variation smaller than 0.28 ns.

9.4.3 Comparison with Other Approaches

In order to appreciate the competitiveness, in terms of performance and dimensions, of the balanced UWB bandpass filters presented in this section with other similar filters reported in the literature [29–36], Table 9.5 summarizes some relevant parameters of such filters. The filters reported in Refs. [31, 34, 35] exhibit strong common-mode rejection in the differential filter passband, but size is large as compared with the filter in Figure 9.17. The other filters (including the one proposed in this section) exhibit comparable common-mode rejection, but the filter proposed here is much smaller. Exception is the filter proposed in Ref. [36], with strong common-mode rejection in the differential filter passband and small size. This filter is very competitive, although the reported band (with and without notched band) covers a smaller frequency range (between roughly 2 and 7 GHz). By increasing the filter order, more

Table 9.5 Comparison of Various Differential UWB Bandpass Filters

Reference	n	FBW[a] (%)	S_{21}^{cc} (dB) at passband	Electrical size	Dimension (mm)
[29]	6	117	>10	$0.93\lambda \times 2.12\lambda$	18.1×41.3
[30]	4	119	>9.6	$0.35\lambda \times 0.7\lambda$	10.4×20.4
[31]	—	123	>24	—	20×30
[32]	—	124	>13	$0.85\lambda \times 0.85\lambda$	28×28
[33]	—	135	>9	$1.9\lambda \times 2.4\lambda$	10.26×41
[34]	—	113	>13	$0.57\lambda \times 0.57\lambda$	—
[35]	—	139	>18.8	$0.75\lambda \times 0.59\lambda$	15.6×12.3
[36]	—	115	>22	$0.56\lambda \times 0.28\lambda$	24×12
Figure 9.17	**5**	**130**	**>10**	$\mathbf{0.5\lambda \times 0.37\lambda}$	$\mathbf{10.5 \times 7.6}$

[a]FBW corresponding to the differential response.

common-mode transmission zeros in the proposed balanced filters can be generated, and this can potentially improve the rejection level of that mode. Another possibility would be to load the central patch capacitances with additional SIRs oriented parallel to the filter axis (with the result of transmission zero splitting).

Concerning the differential-mode response, the combination of in-band insertion and return loss and selectivity of the filter proposed in this section is competitive, as compared with the other filters. The filter reported in Ref. [29] exhibits comparable selectivity at the lower band edge and better selectivity at the upper band edge, but the measured in-band insertion loss is better in the prototype reported in this section. In terms of in-band insertion losses, the filter reported in Ref. [33] is very competitive, but this filter is large and selectivity is limited. The filter reported in Ref. [30] exhibits a symmetric differential-mode response with very reasonable in-band insertion loss and significant selectivity at the upper transition band, but the size is large, as compared to the prototype in Figure 9.17. Moreover, selectivity at the lower transition band is much better in the filter in Figure 9.17. The filter proposed in Ref. [36] (without notched band) is very competitive concerning the differential mode in-band response and selectivity, but it is implemented in multilayer technology, and it does not cover the UWB regulated band from 3.1 GHz up to 10.6 GHz.

9.5 SUMMARY

Wideband balanced bandpass filters with inherent common-mode suppression based on multi-section mirrored SIRs have been discussed in the present chapter. Through the use of such resonant elements, it is possible to achieve wideband differential-mode responses and an efficient common-mode rejection by the selective introduction of transmission zeros for the common mode. Three filter examples are reported in the chapter, including an order-5 and an order-7 wideband filter, both based on 7-section SIRs coupled through admittance inverters, and an order-5 UWB balanced filter based on 5-section SIRs and patch capacitors. For the two former filters, an automated synthesis scheme, based on ASM has been proposed.

APPENDIX 9.A: GENERAL FORMULATION OF AGGRESSIVE SPACE MAPPING (ASM)

ASM [13] is an optimization technique used to find an approximation of the optimal design of a computationally expensive model, called fine model. To this end, a fast but inaccurate surrogate of the original fine model, designated as coarse model, is exploited. In a typical ASM problem, a coarse model optimal design whose coarse model response satisfies certain specifications is known, and hence this coarse model response is considered to be the target response of the fine model. Two simulation spaces are thus used in ASM: (i) the optimization space, $\mathbf{X_c}$, where the variables are linked to those of the coarse model, which is simple and computationally efficient, although not accurate (as mentioned earlier), and (ii) the validation space, $\mathbf{X_f}$, where the variables are linked to the fine model, typically more complex and CPU intensive, but significantly more precise. In each space, a vector containing the different model parameters can be defined. Such vectors are denoted as $\mathbf{x_f}$ and $\mathbf{x_c}$ for the fine and coarse model spaces, respectively, and their corresponding responses are $\mathbf{R_f}(\mathbf{x_f})$ and $\mathbf{R_c}(\mathbf{x_c})$. The variables of the optimization and validation spaces may be very diverse. It is possible that both spaces involve the same variables, as is the case of the ASM used in Section 9.3.1 to find the optimum filter schematic. Typically, in optimization problems involving planar circuits described by a lumped element circuit model, the variables of the optimization space are the set of lumped elements, and the response in this space is inferred from the circuit simulation of the lumped element model. The variables of the validation space are the set of dimensions that define the circuit layout (the substrate parameters are usually fixed and hence they are not optimization variables), and the response in this space is obtained from the electromagnetic simulation of the layout. In the determination of the filter layout discussed in Section 9.3.2, it was the considered scenario.

In ASM, the goal is to minimize the following error function:

$$\mathbf{f}(\mathbf{x_f}) = \mathbf{P}(\mathbf{x_f}) - \mathbf{x_c^*} \tag{9.11}$$

where $\mathbf{x_c^*}$ is the coarse model solution that gives the target response, $\mathbf{R_c}(\mathbf{x_c^*})$, and $\mathbf{P}(\mathbf{x_f})$ is a parameter transformation mapping the fine model parameter space to the coarse model parameter space. In reference to the two spaces considered earlier, $\mathbf{P}(\mathbf{x_f})$ provides the coarse model parameters from the fine model parameters typically by means of a parameter extraction procedure.

Let us assume that $\mathbf{x_f}^{(j)}$ is the jth approximation to the solution in the validation space and $\mathbf{f}^{(j)}$ the error function corresponding to $\mathbf{f}(\mathbf{x_f}^{(j)})$. The next vector of the iterative process $\mathbf{x_f}^{(j+1)}$ is obtained by a quasi-Newton iteration according to

$$\mathbf{x_f}^{(j+1)} = \mathbf{x_f}^{(j)} + \mathbf{h}^{(j)} \tag{9.12}$$

where $\mathbf{h}^{(j)}$ is given by

$$\mathbf{h}^{(j)} = -\left(\mathbf{B}^{(j)}\right)^{-1} \mathbf{f}^{(j)} \tag{9.13}$$

and $\mathbf{B}^{(j)}$ is an approach to the Jacobian matrix, which is updated according to the Broyden formula [13]:

$$\mathbf{B}^{(j+1)} = \mathbf{B}^{(j)} + \frac{\mathbf{f}^{(j+1)}\mathbf{h}^{(j)T}}{\mathbf{h}^{(j)T}\mathbf{h}^{(j)}} \tag{9.14}$$

In (9.14), $\mathbf{f}^{(j+1)}$ is obtained by evaluating (9.11), and the super-index T stands for transpose.

For the application of the previous algorithm, it is necessary to provide an initial vector in the fine model, $\mathbf{x_f}^{(1)}$, and to initiate the Jacobian matrix. If the variables in both spaces are the same, a simple solution is $\mathbf{x_f}^{(1)} = \mathbf{x_c^*}$, and $\mathbf{B}^{(1)} = \mathbf{I}$, where $\mathbf{I}$ is the identity matrix. However, for which concerns $\mathbf{B}^{(1)}$, a better choice is to slightly perturb the parameters of the fine model, infer the effects of such perturbations on the coarse model parameters, and construct the first approximation to the Jacobian matrix from the first-order partial derivatives, in practice expressed as finite differences (see expression 9.7, corresponding to $\mathbf{B}^{(1)}$ for the ASM algorithm used to determine the optimum schematic of the filters reported in Section 9.3).

Finally, the ASM algorithm ends once the norm of the error function is smaller than a predefined value. Typically, this norm is expressed as the relative error between the extracted parameters of the coarse model and the target values (see, e.g., expression 9.8).

REFERENCES

1. M. Makimoto and S. Yamashita, "Compact bandpass filters using stepped impedance resonators," *Proc. IEEE*, vol. **67** (1), pp. 16–19, Jan. 1979.
2. P. Vélez, M. Durán-Sindreu, J. Bonache, A. Fernández Prieto, J. Martel, F. Medina and F. Martín, "Differential bandpass filters with common-mode suppression based on stepped impedance resonators (SIRs)," *IEEE MTT-S International Microwave Symposium*, Seattle, WA, USA, 2–7 Jun. 2013.
3. P. Velez, J. Selga, M. Sans, J. Bonache, and F. Martin, "Design of differential-mode wideband bandpass filters with wide stop band and common-mode suppression by means of multisection mirrored stepped impedance resonators (SIRs)," *IEEE MTT-S International Microwave Symposium*, Phoenix, AZ, USA, 22–27 May 2015.
4. M. Sans, J. Selga, P. Vélez, A. Rodríguez, J. Bonache, V.E. Boria, and F. Martín, "Automated design of common-mode suppressed balanced wideband bandpass filters by means of aggressive space mapping (ASM)," *IEEE Trans. Microw. Theory Tech.*, vol. **63**, no. 12, pp. 3896–3908, Dec. 2015.
5. P. Vélez, J. Naqui, A. Fernández-Prieto, J. Bonache, J. Mata-Contreras, J. Martel, F. Medina, and F. Martín "Ultra-compact (80 mm^2) differential-mode ultra-wideband (UWB) bandpass filters with common-mode noise suppression," *IEEE Trans. Microw. Theory Tech.*, vol. **63**, pp. 1272–1280, Apr. 2015.
6. M. Sans, J. Selga, P. Vélez, A. Rodríguez, J. Bonache, V.E. Boria, and F. Martín, "Automated design of balanced wideband bandpass filters based on mirrored stepped impedance resonators (SIRs) and interdigital capacitors," *Int. J. Microw. Wireless Technol.*, Vol. **8**, Issue 4–5, pp. 731–740. June 2016.
7. G. Matthaei, L. Young, and E. M. T. Jones, *Microwave Filters, Impedance-Matching Networks, and Coupling Structures*, Artech House, Norwood, MA, 1980.
8. D. M. Pozar, *Microwave Engineering*, Addison Wesley, Reading, MA, 1990.
9. J. S. Hong and M. J. Lancaster, *Microstrip Filters for RF/Microwave Applications*, John Wiley & Sonc Inc., New York, 2001.
10. J. Bonache, I. Gil, J. García-García, and F. Martín, "Compact microstrip band-pass filters based on semi-lumped resonators," *IET Microw. Antennas Propag.*, vol. **1**, pp. 932–936, Aug. 2007.

11. M. Sans, J. Selga, A. Rodríguez, J. Bonache, V.E. Boria, and F. Martín, "Design of planar wideband bandpass filters from specifications using a two-step aggressive space mapping (ASM) optimization algorithm," *IEEE Trans. Microw. Theory Tech.*, vol. **62**, pp. 3341–3350, Dec. 2014.
12. J. W. Bandler, R. M. Biernacki, S. H. Chen, P. A. Grobelny, and R.H. Hemmers, "Space mapping technique for electromagnetic optimization," *IEEE Trans. Microw. Theory Tech.*, vol. **42**, pp. 2536–2544, Dec. 1994.
13. J. W. Bandler, R. M. Biernacki, S. H. Chen, R. H. Hemmers, and K. Madsen, "Electromagnetic optimization exploiting aggressive space mapping," *IEEE Trans. Microw. Theory Tech.*, vol. **43**, pp. 2874–2882, Dec. 1995.
14. S. Koziel, Q.S. Cheng, and J.W. Bandler, "Space mapping," *IEEE Microw. Mag.*, vol. **9**, pp. 105–122, Dec. 2008.
15. I. Bahl and P. Barthia, *Microwave Solid State Circuit Design*, John Wiley & Sons, Inc., New York, 1998.
16. W. R. Eisenstadt, B. Stengel, and B. M. Thompson, *Microwave Differential Circuit Desing Using Mixed Mode S-Parameters*, Artech House, Boston, MA, 2006.
17. T. B. Lim and L. Zhu, "A differential-mode wideband bandpass filter on microstrip line for UWB applications," *IEEE Microw. Wireless Compon. Lett.*, vol. **19**, pp. 632–634, Oct. 2009.
18. T. B. Lim and L. Zhu, "Highly selective differential-mode wideband bandpass filter for UWB application," *IEEE Microw. Wireless Compon. Lett.*, vol. **21**, no. 3, pp. 133–135, Mar. 2011.
19. P. Vélez, J. Naqui, A. Fernández-Prieto, M. Durán-Sindreu, J. Bonache, J. Martel, F. Medina, and F. Martín, "Differential bandpass filter with common mode suppression based on open split ring resonators and open complementary split ring resonators," *IEEE Microw. Wireless Compon. Lett.*, vol. **23**, no. 1, pp. 22–24, Jan. 2013.
20. A. K. Horestani, M. Durán-Sindreu, J. Naqui, C. Fumeaux, and F. Martín, "S-shaped complementary split ring resonators and application to compact differential bandpass filters with common-mode suppression," *IEEE Microw. Wireless Compon. Lett.*, vol. **24**, no. 3, pp. 150–152, Mar. 2014.
21. X.-H. Wangand and H. Zhang, "Novel balanced wideband filters using microstrip coupled lines," *Microw. Opt. Technol. Lett.*, vol. **56**, pp. 1139–1141, May 2014.
22. L. Li, J. Bao, J.-J. Du, and Y.-M. Wang, "Compact differential wideband bandpass filters with wide common-mode suppression," *IEEE Microw. Wireless Compon. Lett.*, vol. **24**, no. 3, pp. 164–166, Mar. 2014.
23. H. Wang, L.-M. Gao, K.-W. Tam, W. Kang, and W. Wu, "A wideband differential BPF with multiple differential- and common-mode transmission zeros using cross-shaped resonator," *IEEE Microw. Wireless Compon. Lett.*, vol. **24**, no. 12, pp. 854–856, Oct. 2014.

24. W. Feng, W. Che, and Q. Xue, "High selectivity wideband differential bandpass filter with wideband common mode suppression using marchand balun," *IEEE International Wireless Symposium*, Xian, China, Mar. 2014.
25. L. Li, J. Bao, J.-J. Du, and Y.-M. Wang, "Differential wideband bandpass filters with enhanced common-mode suppression using internal coupling technique," *IEEE Microw. Wireless Compon. Lett.*, vol. **24**, no. 5, pp. 300–302, Feb. 2014.
26. J. G. Zhou, Y.-C. Chiang, and W. Che, "Compact wideband balanced bandpass filter with high common-mode suppression based on cascade parallel coupled lines," *IET Microw. Antennas Propag.*, vol. **8**, no. 8, pp. 564–570, June 2014.
27. W. Feng, W. Che, Y. Ma, and Q. Xue, "Compact wideband differential bandpass filters using half-wavelength ring resonator," *IEEE Microw. Wireless Compon. Lett.*, vol. **23**, no. 2, pp. 81–83, Feb. 2013.
28. R. A. Foster, "A reactance theorem," *Bell Syst. Tech. J.*, vol. **3**, pp. 259–267, Apr. 1924.
29. T. B. Lim and L. Zhu, "Differential-mode ultra-wideband bandpass filter on microstrip line," *Electron. Lett.*, vol. **45**, no. 22, pp. 1124–1125, Oct. 2009.
30. X.-H. Wu, Q.-X. Chu, "Compact differential ultra-wideband bandpass filter with common-mode suppression," *IEEE Microw. Wireless Compon. Lett.*, vol. **22**, pp. 456–458, Sep. 2012.
31. A. M. Abbosh, "Ultrawideband balanced bandpass filter," *IEEE Microw. Wireless Compon. Lett.*, vol. **21**, pp. 480–482, Sep. 2011.
32. H. T. Zhu, W. J. Feng, W. Q. Che, and Q. Xue, "Ultra-wideband differential bandpass filter based on transversal signal-interference concept," *Electron. Lett.*, vol. **47**, no. 18, pp. 1033–1035, Sep. 2011.
33. X.-H. Wang, H. Zhang, and B.-Z. Wang, "A novel ultra-wideband differential filter based on microstrip line structures," *IEEE Microw. Wireless Compon. Lett.*, vol. **23**, pp. 128–130, Mar. 2013.
34. S. Shi, W.-W. Choi, W. Che, K.-W. Tam, and Q. Xue, "Ultra-wide-band differential bandpass filter with narrow notched band and improved common-mode suppression by DGS," *IEEE Microw. Wireless Compon. Lett.*, vol. **22**, no. 4, pp. 185–187, Apr. 2012.
35. C-H Lee, C.-I.G. Hsu, and C.-J. Chen, "Band-notched balanced UWB BPF with stepped-impedance slotline multi-mode resonator," *IEEE Microw. Wireless Compon. Lett.*, vol. **22**, no. 4, pp. 182–184, Apr. 2012.
36. J. Shi, C. Shao, J.-X. Chen, Q.-Y. Lu, Y. Peng, and Z.-H. Bao, "Compact low-loss wideband differential bandpass filter with high common-mode suppression," *IEEE Microw. Wireless Compon. Lett.*, vol. **23**, no. 9, pp. 480–482, Sep. 2013.

CHAPTER 10

METAMATERIAL-INSPIRED BALANCED FILTERS

Ferran Martín,[1] Paris Vélez,[1] Ali Karami-Horestani,[2] Francisco Medina,[3] and Christophe Fumeaux[2]

[1]CIMITEC, Departament d'Enginyeria Electrònica, Universitat Autònoma de Barcelona, Bellaterra, Spain
[2]School of Electrical and Electronic Engineering, The University of Adelaide, Adelaide, SA, Australia
[3]Departamento de Electrónica y Electromagnetismo, Universidad de Sevilla, Sevilla, Spain

10.1 INTRODUCTION

This chapter is focused on two different strategies for the design of balanced bandpass filters with common-mode suppression, where the relevant aspect is the small size of the resulting devices. Miniaturization is achieved by using electrically small resonators, formerly applied to the design of planar metamaterial structures and devices. In particular, it will be shown that the combination of open split ring resonators (OSRRs) and open complementary split ring resonators (OCSRRs) is useful to implement compact balanced filters with high common-mode

Balanced Microwave Filters, First Edition. Edited by Ferran Martín, Lei Zhu, Jiasheng Hong, and Francisco Medina.

rejection ratio (CMRR) in the differential-mode passband. Then, balanced filters based on S-shaped complementary split ring resonators (S-CSRRs) are discussed. Since both filter types are based on metamaterial resonators (all of them inspired by the split ring resonator (SRR) [1]), such filters are designated in this book as metamaterial-inspired balanced filters.

10.2 BALANCED BANDPASS FILTERS BASED ON OPEN SPLIT RING RESONATORs (OSRRs) AND OPEN COMPLEMENTARY SPLIT RING RESONATORS (OCSRRs)

The OSRR and OCSRR are open resonators that can be excited not only by external magnetic or electric fields but also by means of a current or voltage source. Let us first briefly review the topology of these resonant elements and justify their small electrical size, and then let us study their application to the design of compact balanced filters.

10.2.1 Topology of the OSRR and OCSRR

The OSRR and OCSRR were first proposed in Refs. [2] and [3], respectively. The topology of the OSRR is depicted in Figure 10.1, where it is compared with the topology of the SRR [1], from which it is inspired. The OSRRs consist of a pair of open and concentric metallic rings, like the SRR, but unlike the SRR, the apertures in the OSRR are located at the same angular position, and two crossed apertures are extended outward, resulting in a two-terminal (open) resonator. Note that the topology of the OSRR resembles a pair of concentric hooks. In the SRR, terminals do exist as well, but this particle is typically driven by means of an axial magnetic field or by means of an electric field applied in the orthogonal direction to the symmetry plane. In the SRR, the inductance

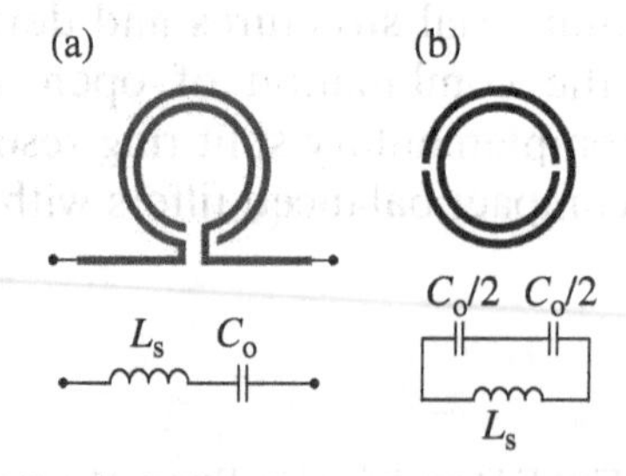

Figure 10.1 Typical topology of the OSRR (a) and SRR (b). The lumped element equivalent circuits are also included.

is equivalent to the inductance of a single ring with the same width and average radius [4, 5], whereas the capacitance is given by the series connection of the edge capacitance of the two halves. The reason is that in one of the halves, the displacement current "circulates" from the inner to the outer ring, whereas it flows from the outer to the inner ring in the other half. Thus, if the edge capacitance of the whole circumference is called C_o, it follows that the capacitance of the SRR is $C_o/4$. By contrast, in the OSRR the displacement current flows from one ring to the other along the whole circumference, and for that reason the capacitance of the OSRR is simply C_o. The inductance of the OSRR is, however, identical to the one of the SRR (provided ring dimensions are identical). Therefore, from the previous words it follows that the resonance frequency of the OSRR is half the resonance frequency of the SRR or, equivalently, the electrical size of the OSRR is half the electrical size of the SRR. It is worth mentioning that if the terminals of the OSRR are connected, the resulting closed particle is the two-turn spiral resonator (2-SR) [6]. Thus, the inductance and capacitance of the OSRR and 2-SR are identical [5].

The topology of the OCSRR is depicted in Figure 10.2. In this case, the comparison to the topology of the complementary split ring resonator (CSRR), formerly proposed in Ref. [7], is most appropriate. Indeed, the OCSRR topology can be inferred from the topology of the CSRR following a procedure similar to that applied to the SRR in order to obtain the topology of the OSRR (as presented in the previous paragraph). Alternatively, the OCSRR can be inferred from the topology of the OSRR by applying duality. Namely, the OCSRR is the negative (or complementary) image of the OSRR, where metallic strips are

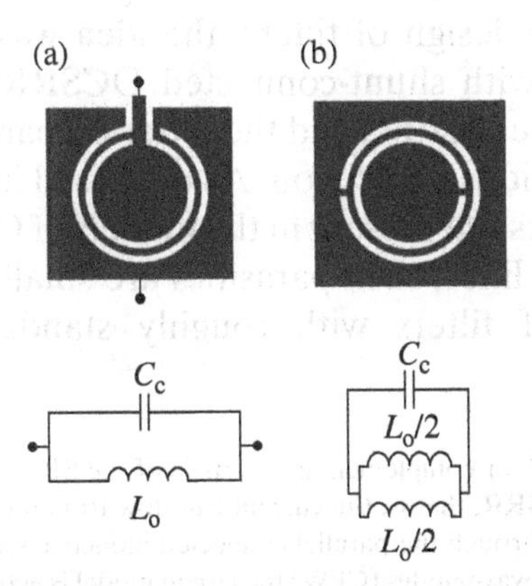

Figure 10.2 Typical topology of the OCSRR (a) and CSRR (b). The lumped element equivalent circuits are also included.

replaced by slots and vice versa.[1] From duality arguments, it can be inferred that the electrical size of the OCSRR is half the electrical size of the CSRR and identical to the electrical size of the OSRR (assuming that perfect duality applies) [4]. Nevertheless, let us demonstrate this property from similar considerations as those used in reference to the SRR and OSRR. In the case of the CSRR and OCSRR, the capacitances of both resonators are identical and given by the whole capacitance between the inner and outer metallic regions. The difference concerns the inductance. In the CSRR, it is given by the parallel connection between the inductances of the pair of strips that connect the inner and outer metallic regions of the particle. If we call L_o the inductance of the whole circumference, the equivalent inductance of the CSRR is $L_o/4$. Conversely, in the OCSRR the current can flow from one terminal to the other not only across the capacitance between the inner and outer metallic regions but also along the circular strip present between the two slot rings.[2] The inductance of this strip is L_o, and hence this is the inductance of the OCSRR. Therefore, it follows that the electrical size of the OCSRR is half the electrical size of the CSRR.

It should be taken into account that the previous analysis is valid as long as the electrical size of the particles is small. To this end it is necessary that the distance between the rings (SRR and OSRR) or between the slot rings (CSRR and OCSRR) is very small [4, 5].

10.2.2 Filter Design and Illustrative Example

OSRRs and OCSRRs combined were first applied to the design of dual-band microwave components based on composite right-/left-handed (CRLH) transmission lines and to the design of single-ended filters in Refs. [8, 9]. For the design of filters, the idea was to alternate series-connected OSRRs with shunt-connected OCSRRs, the former being described by series resonators and the latter by parallel resonant tanks, as shown in the previous subsection. As discussed in Refs. [8, 9], despite the fact that parasitics are present in the models of OSRR- and OCSRR-loaded transmission lines, such parasitics are small and do not prevent from the design of filters with roughly standard responses (e.g., Chebyshev).[3]

[1] The CSRR is also the dual, or complementary, particle of the SRR.
[2] This also applies to the CSRR, that is, the current can flow from the inner to the outer metallic regions across the slots or through the parallel connected inductive strips.
[3] In OSRR-loaded coplanar waveguides (CPW) the circuit model is actually a π-circuit with a series LC resonator in the series branch and parasitic capacitances in the shunt branches. OCSRR-loaded CPWs are described by a T-circuit with a parallel resonator in the shunt branch and parasitic

Since the considered balanced filters of this chapter are implemented in microstrip technology, let us first present the topology of the microstrip single-ended bandpass filter based on OSRRs and OCSRRs [8]. Such topology, corresponding to an order-3 filter, is depicted in Figure 10.3. The first and third stages are implemented by means of series OSRRs, whereas the central (second) stage uses a shunt OCSRR connected to the ground plane by means of a pair of vias. Alternatively, the filter can be implemented with two OCSRRs (first and third stages) and one OSRR (second stage). With the topology of Figure 10.3, the filter is roughly described by the canonical circuit of an order-3 bandpass filter with series-connected series resonators in the first and third stages and a shunt parallel resonator in the central stage. Actually, as mentioned before, parasitics are present, and circuit design is not as simple as if parasitics were not present in the model (the detailed procedure is described in Ref. [8]). Nevertheless, the OCSRR stage is described by a simple shunt-connected parallel resonator. It is not necessary to include series inductances at both sides of the resonant element in a T-circuit configuration, such as is needed in CPW technology [8]. However, the accurate description of the OSRR-loaded microstrip line section (first

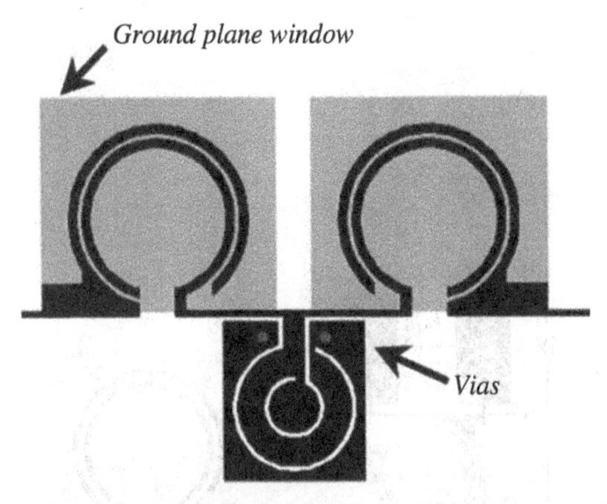

Figure 10.3 Typical topology of a single-ended microstrip order-3 bandpass filter implemented by means of OSRRs and OCSRRs. The upper metal level is indicated in black, whereas the ground plane windows are depicted in gray.

---

inductors in the series branches. Conversely, in microstrip lines loaded with shunt-connected OCSRRs (connected to ground through vias), the parasitics can be neglected, and the cell is described simply by a shunt parallel resonator, but in OSRR-loaded microstrip lines, the circuit model cannot be described by the canonical circuit of Figure 10.1, and parasitics must be accounted for, like in OSRR-loaded CPWs.

and third stages) requires a relatively complex model composed of an asymmetric π-circuit cascaded in between series-connected inductances (usually of small value) [8].[4] With this circuit model, the response of the filter is accurately described, as demonstrated in Ref. [8] in reference to an order-3 Chebyshev bandpass filter with 0.1 dB ripple, central frequency $f_c = 1.3$ GHz, and fractional bandwidth of 50%.

The OSRR/OCSRR balanced bandpass filters are implemented by mirroring the topology of Figure 10.3 and excluding the vias. A typical layout of the proposed differential-mode bandpass filters is depicted in Figure 10.4. The structure is symmetric with respect to the indicated plane (dashed line). For the differential mode, where there is a virtual ground in that plane (electric wall), the OCSRRs are grounded (without the presence of vias), and the structure exhibits a passband. If the distance between the host lines is large, no coupling effects take place between mirror elements, and filter design is as simple as designing a single-ended bandpass filter.

The proposed strategy to suppress the common mode in the region of interest (which must cover at least the differential-mode passband) consists of tailoring the metallic region surrounding the OCSRRs. To gain insight on filter design, let us consider the lumped element equivalent circuit model of the four-port section corresponding to one pair of

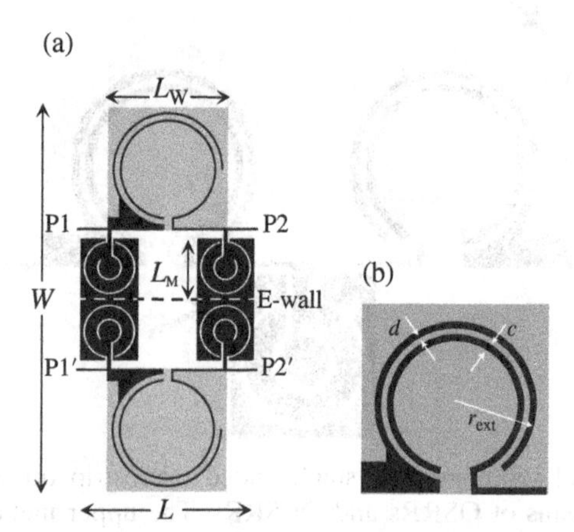

Figure 10.4 Typical layout of the OSRR-/OCSRR-based differential-mode bandpass filter (order-3) with common-mode suppression (a) and detail of the OSRR (b). Relevant dimensions are indicated.

[4] Note that the ground plane is windowed in the regions beneath the OSRRs in order to obtain a more accurate description of the particle by means of a series resonator, despite that parasitics are present.

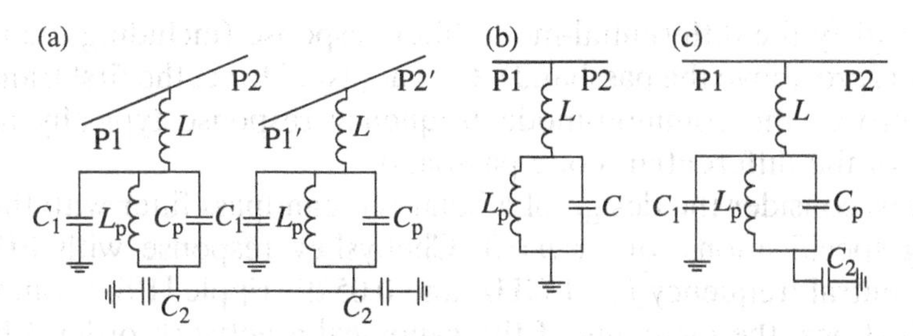

Figure 10.5 Equivalent circuit model of the mirrored OCSRRs section. (a) Complete model; (b) differential-mode model; (c) common-mode model.

mirrored OCSRRs (Figure 10.5a). L_p and C_p model the OCSRR, L accounts for the inductive strip present between the microstrip lines and the center part of the OCSRRs, C_1 is the capacitance between the central strip of the OCSRR and the ground plane, and C_2 is the patch capacitance corresponding to the metallic region surrounding the OCSRRs. Since this circuit model is by far more complex than a simple pair of parallel resonators, a detailed explanation is necessary. First of all, the inductance L can be neglected if the strip that connects the central metallic region of the OCSRRs and the host line is short and wide. However, this inductance is useful since it introduces a transmission zero for the differential mode above the passband, and selectivity for that mode can be improved [9]. Thus, we consider this inductance as a design parameter, rather than a parasitic element. For which concern the capacitance C_2, it is grounded by the presence of vias in single-ended filters, hence forcing the capacitance C_1 to be parallel connected to the capacitance C_p in those filters. This explains that the circuit model describing single-ended OCSRR-loaded microstrip lines with grounded OCSRR (through vias) is as simple as a shunt-connected parallel resonator, except by the presence of L if the strip connecting the line and the OCSRR provides enough inductance. However, for balanced OCSRR-loaded microstrip structures, the accurate circuit model is the one depicted in Figure 10.5(a).

The models for the differential and common modes are depicted in Figure 10.5(b) and (c), respectively. Notice that the capacitance C_2 is grounded for the differential mode and does not play any role. Thus, for this mode, the shunt OCSRR is described by a parallel resonator in series with an inductor (useful to introduce a transmission zero above the differential filter passband, as mentioned before). However, for the common mode, the symmetry plane is an open circuit, and the effect of the capacitance C_2 is the presence of two transmission zeroes in the common-mode frequency response. Once the elements L, L_p, C_p, and C_1 are set to satisfy the differential-mode filter response (including the transmission zero above the passband), C_2 is adjusted to set the first transmission zero of the common-mode frequency response typically in the center of the differential-mode passband.

Let us consider the design of a balanced bandpass filter with the following specifications: order $n = 3$, Chebyshev response with FBW = 45%, central frequency $f_o = 1$ GHz, and 0.05 dB ripple [10]. From these specifications, the elements of the canonical π-network order-3 bandpass filter can be obtained through well-known transformations from the low-pass filter prototype [11]. Once these elements are known, the topology of the series-connected OSRRs is obtained by curve fitting the response of the series LC resonator, giving the ideal Chebyshev response in the region of interest. For the OCSRRs, the elements of the model of Figure 10.5(b) (L, L_p, and $C = C_1 + C_p$) are derived from the susceptance slope at f_o (i.e., by forcing it to be equal to that of the LC tank giving the ideal Chebyshev response) and from the differential-mode transmission zero, given by

$$f_Z^{dd} = \frac{1}{2\pi}\sqrt{\frac{1}{C}\left(\frac{1}{L} + \frac{1}{L_p}\right)} \tag{10.1}$$

The parasitic capacitances of the π-model of the OSRR (which can be easily extracted from the electromagnetic simulation of the isolated particle) [8] are small and have negligible effect on the differential filter response. Once L, L_p, and C are known, the OCSRRs are synthesized with the help of the model reported in Ref. [3] and a parameter extraction procedure similar to that reported in Ref. [9]. From this model, the OCSRR capacitance, C_p, can be estimated, and hence C_1 can be derived. Finally, C_2 is adjusted to the required value to force the common-mode transmission zero at f_o. The metallic region surrounding the OCSRRs is then expanded or contracted to adjust the common-mode transmission zero to that value (the initial size is inferred from the parallel plate capacitor formula). Following this procedure, the element values and the layout of the shunt branch are inferred. The differential- and common-mode frequency responses of the OCSRR section are shown in Figure 10.6. The good agreement between circuit and electromagnetic simulations in the region of interest is appreciable and validates the proposed model of Figure 10.5.

The photograph of the designed and fabricated filter is shown in Figure 10.7(a) (dimensions are $0.15\lambda_g \times 0.30\lambda_g$, λ_g being the guided wavelength at the filter central frequency). Figure 10.7(b) shows the simulated and measured insertion and return loss for the differential

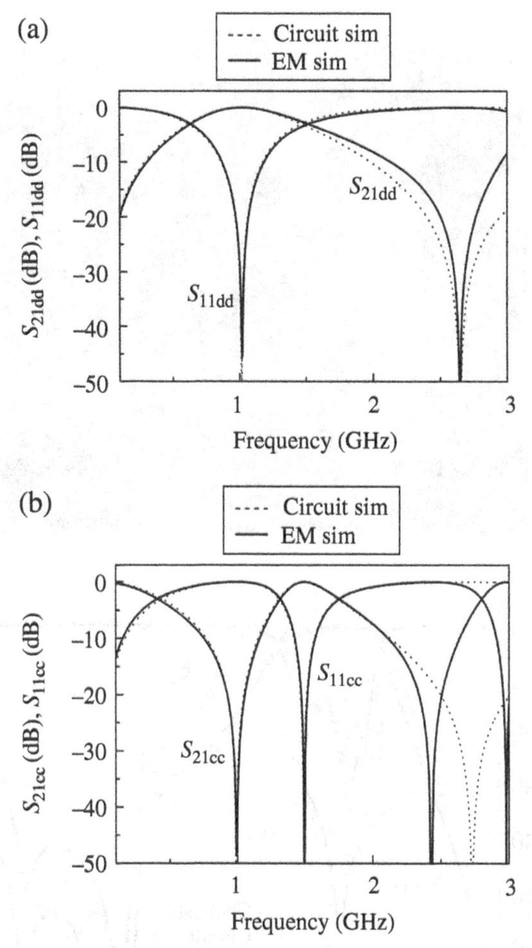

Figure 10.6 Circuit and electromagnetic simulation of the differential- (a) and common- (b) mode frequency response of the OCSRR pair section. Element values are: $L = 0.628$ nH, $L_p = 3.675$ nH, $C_p = 0.1$ pF, $C_1 = 6.85$ pF, $C_2 = 5.65$ pF. Reprinted with permission from Ref. [10]; copyright 2013 IEEE.

mode, as well as the insertion loss for the common mode. The circuit simulations for the differential- and common-mode models are also included in the figure. Both the differential- and common-mode filter responses are in good agreement with the circuit simulations up to roughly 2 GHz. The agreement with the Chebyshev response is also good, although the selectivity and stopband rejection above the passband of the designed balanced filter are better due to the effects of the transmission zero (the first spurious appears at roughly $3f_o$). As expected, there is a transmission zero at f_o for the common mode (the

(a)

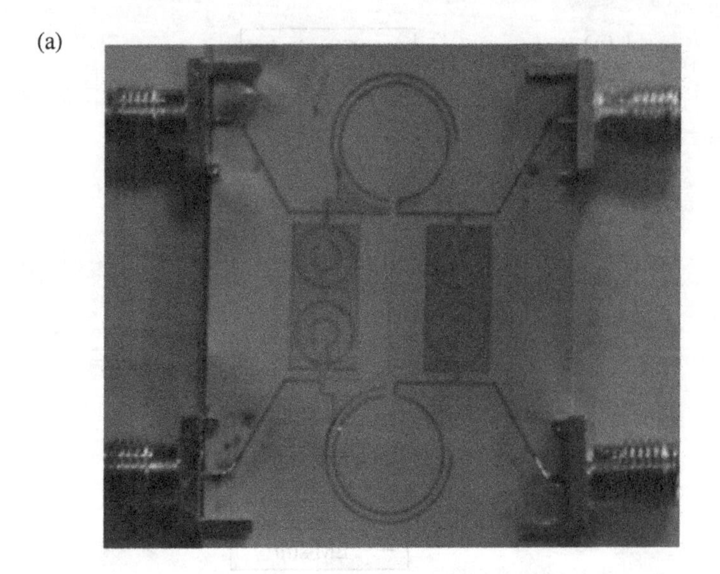

(b)

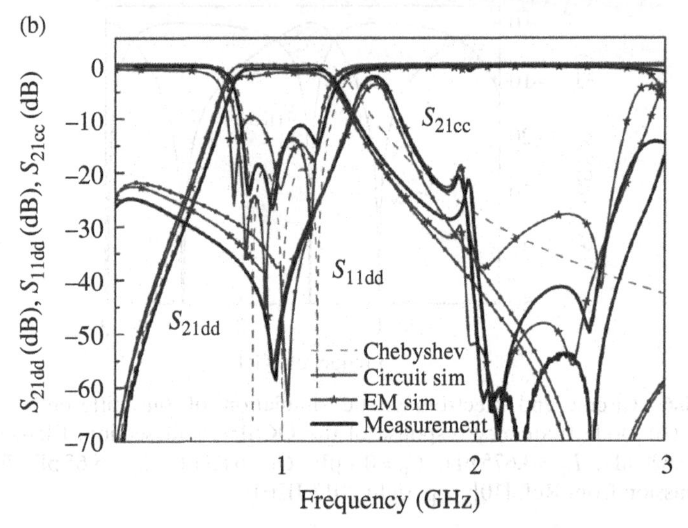

Figure 10.7 Photograph (a) and frequency response (b) of the fabricated balanced OSRR-/OCSRR-based bandpass filter. The considered substrate is the *Rogers RO3010* with dielectric constant $\varepsilon_r = 10.2$ and thickness $h = 0.254$ mm. Dimensions (in reference to Figure 10.4) are $L = 18.9$ mm, $W = 37.8$ mm, $L_W = 12.6$ mm, and $L_M = 6$ mm. For the OCSRR: $r_{ext} = 2.7$ mm, $c = 0.2$ mm, and $d = 1.2$ mm. For the OSRR, $r_{ext} = 5.8$ mm, $c = 0.2$ mm, and $d = 0.55$ mm. The 50-Ω microstrip lines have a width of 0.21 mm. For the π-model of the OSRR, the elements of the series LC tank are 22.28 nH and 1.22 pF, and the values of the shunt capacitors are 1.35 pF (left) and 0.74 pF (right). The elements of the shunt branch are $L = 0.628$ nH, $L_p = 3.675$ nH, $C_p = 0.1$ pF, $C_1 = 6.85$ pF, and $C_2 = 5.65$ pF. Reprinted with permission from Ref. [10]; copyright 2013 IEEE.

CMRR at f_0 being 53 dB), and the common-mode rejection within the differential filter passband is better than 20 dB. The combination of filter size and common-mode rejection is very competitive, and this type of filter is especially suitable for wideband applications. However, ultra-wideband responses cannot be obtained with these resonant elements. Other balanced filters based on OCSRRs coupled through admittance inverters can be found in Ref. [12]. Such filters are implemented by means of a combination of semi-lumped (OCSRRs) and distributed (quarter-wavelength transmission-line sections) components, but due to the absence of OSRRs, the ground plane is kept unaltered.

10.3 BALANCED FILTERS BASED ON S-SHAPED COMPLEMENTARY SPLIT RING RESONATORS (S-CSRRs)

The S-CSRR is the complementary version of the S-shaped split ring resonator (S-SRR). This later resonator (S-SRR), with typical topology depicted in Figure 10.8, has been used for the implementation of metamaterials [13, 14], microwave filters [15, 16], and sensors [17]. At the fundamental resonance frequency, the current flow in the two loops of the S-SRR is opposite, that is, it is clockwise in one loop and counterclockwise in the other one. Therefore, the S-SRR cannot be excited (fundamental resonance) by means of a uniform time-varying magnetic field applied orthogonally to the plane of the particle (as usually done to drive an SRR). However, it can be excited by a nonuniform magnetic field with counter components in the orthogonal plane of the particle in each loop. From duality considerations, it follows that the complementary counterpart, the S-CSRR, can be excited by a nonuniform electric field with components in the orthogonal plane of the particle of opposite directions in each loop.

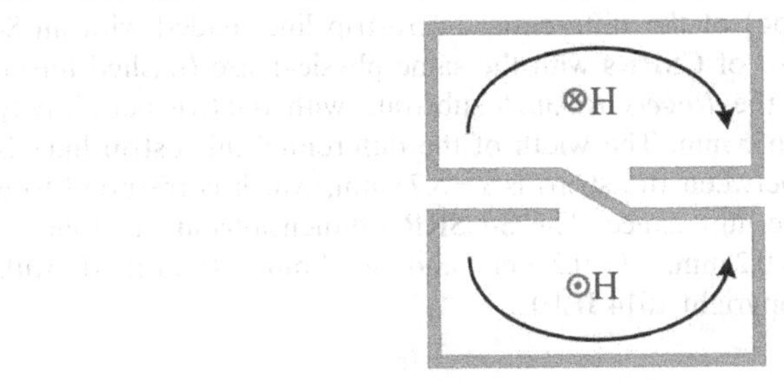

Figure 10.8 Typical topology of the S-SRR with indication of current flow at the fundamental resonance frequency.

Let us now consider a differential line loaded with an S-CSRR, as indicated in Figure 10.9(a) [18]. According to the previous paragraph, the structure should be roughly transparent to the common mode at the fundamental resonance frequency of the S-CSRR, but it should prevent the propagation of differential signals in the vicinity of that frequency, as Figure 10.10 illustrates. This figure also shows the response of a differential line loaded with a pair of CSRRs with identical overall dimensions to the S-CSRR (see Figure 10.9b).[5] As can be seen,

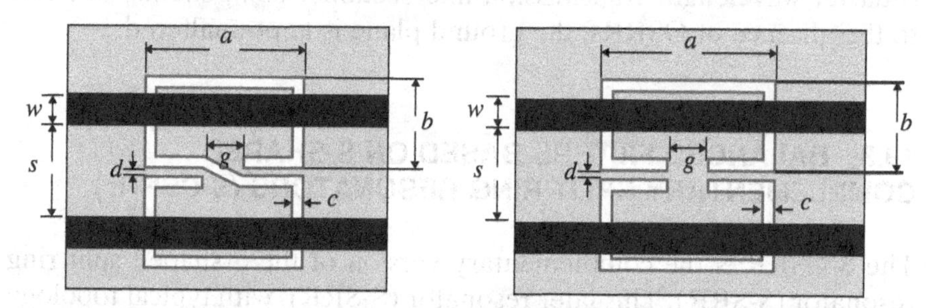

Figure 10.9 A differential microstrip line loaded with (a) an S-CSRR and (b) a pair of rectangular single ring CSRRs. Relevant dimensions are indicated. Reprinted with permission from Ref. [18]; copyright 2014 IEEE.

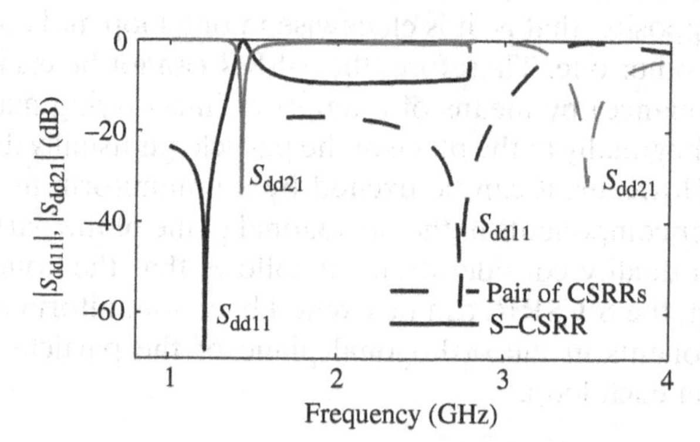

Figure 10.10 A comparison between the simulated transmission and reflection coefficients (differential mode) of the differential microstrip line loaded with an S-CSRR (solid lines) and a pair of CSRRs with the same physical size (dashed lines). The considered substrate is the *Rogers RO3010* substrate with relative permittivity $\varepsilon_r = 10.2$ and thickness $h = 0.635$ mm. The width of the differential microstrip lines is $w = 0.5$ mm, and the space between the strips is $s = 5.75$ mm, which correspond to a 50-Ω odd-mode characteristic impedance. The S-CSRR's dimensions are as follows: $a = 6.2$ mm, $b = 5.7$ mm, $c = 0.2$ mm, $d = 0.2$ mm, and $g = 1$ mm. Reprinted with permission from Ref. [18]; copyright 2014 IEEE.

[5] In the CSRR-loaded differential line of Figure 10.9(b), both the differential and common modes are rejected at the fundamental resonance frequency.

the notch frequency appears at much higher frequency in the CSRR-loaded structure. This indicates the potential of S-CSRRs for size reduction. Namely, the S-CSRR, like the S-SRR, is a very small resonator in terms of electrical size.

10.3.1 Principle for Balanced Bandpass Filter Design and Modeling

The bandstop functionality of the S-CSRR for the differential mode can be switched to a bandpass behavior by simply introducing series capacitive gaps in the strips of the differential line, as shown in Figure 10.11. Note that in this structure, inner split slot rings are added to the S-CSRR in order to achieve more compactness. The response of the structure of Figure 10.11 for the differential mode is depicted in Figure 10.12, where the bandpass behavior can be appreciated. The lumped element equivalent circuit model for this mode is depicted in Figure 10.13. It is identical to the circuit model of a single-ended microstrip line loaded with a CSRR and a series gap [19]. In this model, L and C are the per-section equivalent inductance and capacitance of the differential line with the presence of the S-CSRR, and C_g models the capacitive gaps. The S-CSRR is modeled with an equivalent capacitance C_c and an equivalent inductance L_c. Using the procedure explained in Ref. [20], the parameters of the circuit model of Figure 10.13 can be extracted from the

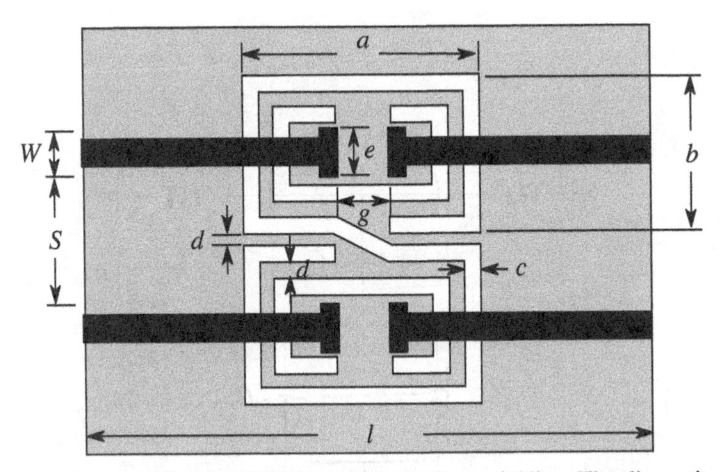

Figure 10.11 Layout of an S-CSRR/gap loaded differential line. The dimensions of the structure are as follows: $a = 9.8$ mm, $b = 7$ mm, $c = 0.8$ mm, $d = 0.2$ mm, $g = 0.6$ mm, $w = 0.5$ mm, $s = 5.75$ mm, $e = 2$ mm, and $l = 22$ mm. Reprinted with permission from Ref. [18]; copyright 2014 IEEE.

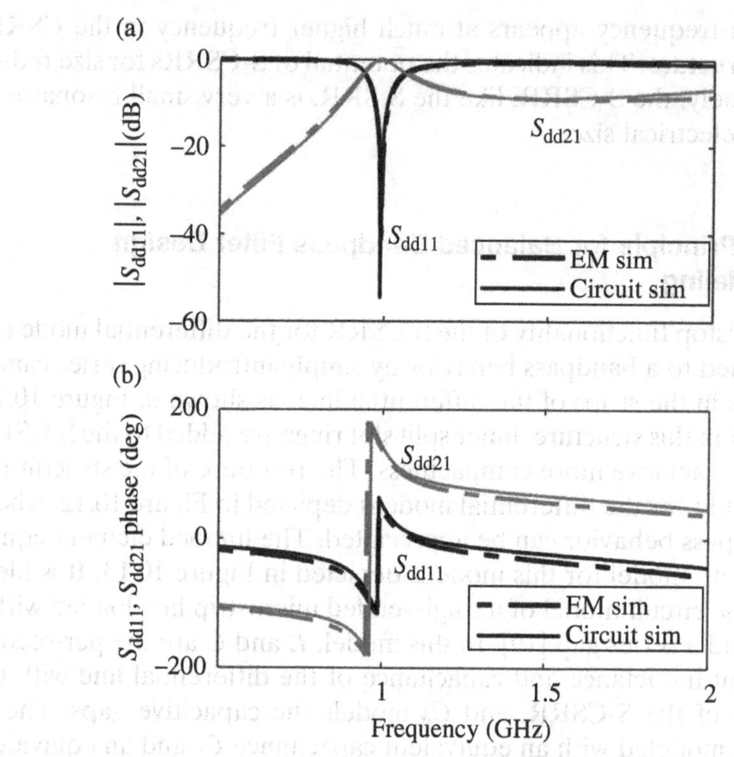

Figure 10.12 Comparison between the amplitude (a) and phase (b) of the differential-mode transmission and reflection coefficients from the electromagnetic simulation of the structure of Figure 10.11 and those from the proposed circuit model. Reprinted with permission from Ref. [18]; copyright 2014 IEEE.

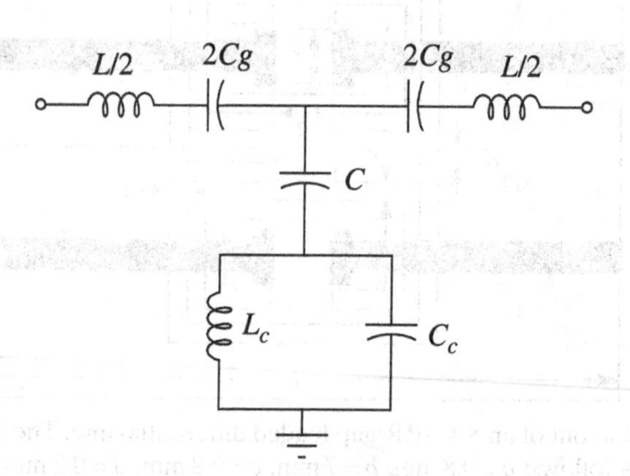

Figure 10.13 Lumped element equivalent circuit model of the unit cell for the differential mode.

electromagnetic simulations of the structure of Figure 10.11 for the differential mode. The dimensions of the structure are indicated in the caption of Figure 10.11, and the *Rogers RO3010* substrate with relative permittivity $\varepsilon_r = 10.2$ and thickness $h = 0.635$ mm has been considered. The extracted parameters for differential mode are $L = 8$ nH, $C = 70$ pF, $C_g = 0.33$ pF, $C_c = 1.2$ pF, and $L_c = 10$ nH. The good agreement between the circuit and electromagnetic simulation of Figure 10.12 validates the model for the differential mode.

The structure of Figure 10.11 is opaque for the common mode in the vicinity of the fundamental resonance of the S-CSRR. This is desirable since the common mode must be rejected in the frequency region where the passband for the differential mode arises. However, the common mode exhibits a passband at higher frequency, due to excitation of the second resonance, as reported in Ref. [18].

10.3.2 Illustrative Example

To illustrate the potential of S-CSRRs in the design of differential-mode bandpass filters with common-mode suppression, a third-order periodic (although periodicity is not mandatory from the design point of view) bandpass filter with fractional bandwidth of 10%, 0.16 dB in-band ripple and central frequency of 1 GHz [18] is presented. The focus of the design procedure is simply placed on the differential characteristics of the filter, since the common mode is intrinsically suppressed by the proposed structure, as justified in the previous subsection. The structure of Figure 10.11, which can be considered as a distributed resonator with two $\lambda/8$ feed lines, is used as the filter's unit cell. The feed lines of each unit cell in conjunction with those of the adjacent cells act as 90° impedance inverters with normalized impedance of $K = 1$. Based on the theory of impedance inverters, the fractional bandwidth of the filter can be controlled by the reactance slope of the unit cell, excluding the feed lines [11], which is in turn controlled by gap dimensions. Thus, a parametric analysis is applied in the design process in order to determine the optimum gap dimensions. The next step is to optimize the length of the feed lines to achieve a 90° phase shift from port one to port two of the unit cell at resonance frequency. Finally, the filter is realized by cascading the optimized unit cells. The photograph of the fabricated filter, using the *Rogers RO3010* substrate with relative permittivity $\varepsilon_r = 10.2$ and thickness $h = 0.635$ mm and the dimensions indicated in the caption of Figure 10.11, is depicted in Figure 10.14. The impedance inverters are meandered to achieve further compactness. The filter is as small as $0.09\lambda_g \times 0.25\lambda_g$, where λ_g is the guided

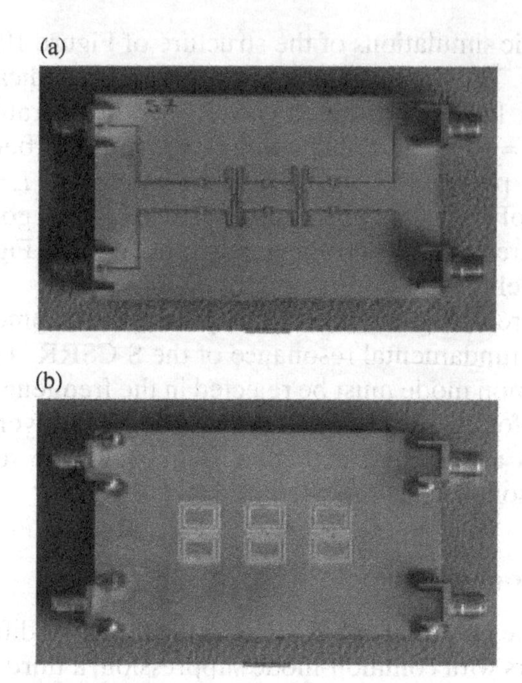

Figure 10.14 (a) Top and (b) bottom photograph of the fabricated third-order differential bandpass filter with common-mode suppression based on S-CSRRs. Reprinted with permission from Ref. [18]; copyright 2014 IEEE.

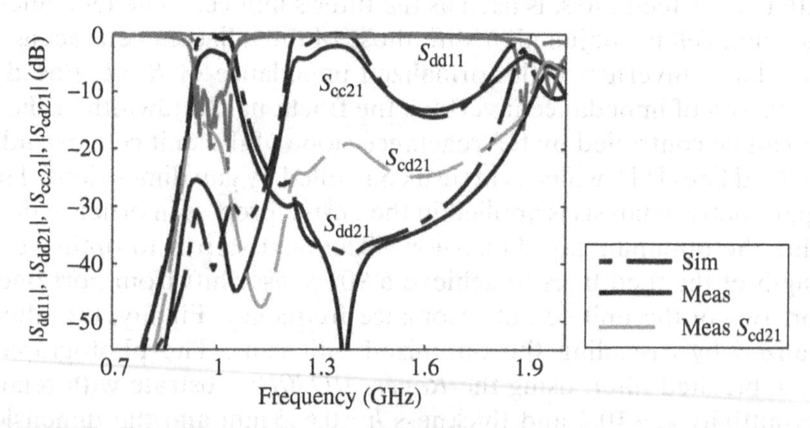

Figure 10.15 Comparison between simulated (lossless) and measured transmission and reflection coefficients of the third-order differential filter with common-mode suppression of Figure 10.14. Reprinted with permission from Ref. [18]; copyright 2014 IEEE.

wavelength at $f = 1$ GHz. A comparison between the electromagnetic simulation and the measured frequency response of the filter in differential and common modes is depicted in Figure 10.15. The figure shows that the structure acts as a differential bandpass filter with 2.8 dB insertion loss at central frequency, common-mode suppression better than −25 dB, and common-to-differential mode conversion better than −16 dB.

10.4 SUMMARY

Compact balanced filters based on electrically small resonators have been reviewed in this chapter. As long as the considered resonant elements have been key building blocks for the implementation of metamaterial structures or are based on the main resonator topology used for that purpose (SRRs), such balanced filters have been designated as metamaterial-inspired balanced filters in this book. The two design approaches considered, one of them based on a combination of OSRRs and OCSRRs and the other one based on S-CSRRs, provide compact filter size and efficient common-mode suppression in the region of interest.

REFERENCES

1. J. B. Pendry, A. J. Holden, D. J. Robbins, and W. J. Stewart, "Magnetism from conductors and enhanced nonlinear phenomena," *IEEE Trans. Microw. Theory Tech.*, vol. **47**, pp. 2075–2084, 1999.
2. J. Martel, R. Marqués, F. Falcone, J. D. Baena, F. Medina, F. Martín, and M. Sorolla, "A new LC series element for compact band pass filter design," *IEEE Microw. Wireless Compon. Lett.*, vol. **14**, pp. 210–212, 2004.
3. A. Velez, F. Aznar, J. Bonache, M. C. Velázquez-Ahumada, J. Martel, and F. Martín, "Open complementary split ring resonators (OCSRRs) and their application to wideband CPW band pass filters," *IEEE Microw. Wireless Compon. Lett.*, vol. **19**, pp. 197–199, 2009.
4. R. Marqués, F. Martín, and M. Sorolla, *Metamaterials with Negative Parameters: Theory, Design and Microwave Applications*, John Wiley & Sons, Inc., Hoboken, NJ, 2007.
5. F. Martín, *Artificial Transmission Lines for RF and Microwave Applications*, John Wiley & Sons, Inc., Hoboken, NJ, 2015.
6. J. D. Baena, R. Marqués, F. Medina, and J. Martel, "Artificial magnetic metamaterial design by using spiral resonators," *Phys. Rev. B*, vol. **69**, paper 014402, 2004.

7. F. Falcone, T. Lopetegi, J. D. Baena, R. Marqués, F. Martín, and M. Sorolla, "Effective negative-ε stop-band microstrip lines based on complementary split ring resonators," *IEEE Microw. Wireless Compon. Lett.*, vol. **14**, pp. 280–282, 2004.
8. M. Durán-Sindreu, A. Vélez, F. Aznar, G. Sisó, J. Bonache, and F. Martín, "Application of open split ring resonators and open complementary split ring resonators to the synthesis of artificial transmission lines and microwave passive components," *IEEE Trans. Microw. Theory Tech.*, vol. **57**, pp. 3395–3403, Dec. 2009.
9. M. Durán-Sindreu, P. Vélez, J. Bonache, and F. Martín, "Broadband microwave filters based on open split ring resonators (OSRRs) and open complementary split ring resonators (OCSRRs): improved models and design optimization," *Radioengineering*, vol. **20**, pp. 775–783, Dec. 2011.
10. P. Vélez, J. Naqui, A. Fernández-Prieto, M. Durán-Sindreu, J. Bonache, J. Martel, F. Medina, and F. Martín, "Differential bandpass filter with common mode suppression based on open split ring resonators and open complementary split ring resonators," *IEEE Microw. Wireless Compon. Lett.*, vol. **23**, pp. 22–24, 2013.
11. J.-S. Hong, and M. J. Lancaster, *Microstrip Filters for RF/Microwave Applications*, John Wiley & Sons, Inc., New York, 2001.
12. P. Vélez, M. Durán-Sindreu, J. Naqui, J. Bonache, and F. Martín, "Common-mode suppressed differential bandpass filter based on open complementary split ring resonators fabricated in microstrip technology without ground plane etching," *Microw. Opt. Technol. Lett.*, vol. **56**, pp. 910–916, Apr. 2014.
13. H. Chen, L. Ran, J. Huangfu, X. Zhang, K. Chen, T. M. Grzegorczyk, and J. Au Kong, "Left-handed materials composed of only S-shaped resonators," *Phys. Rev. E*, vol. **70**, paper 057605, 2004.
14. H. Chen, L. Ran, J. Huangfu, X. Zhang, K. Chen, T. M. Grzegorczyk, and J. A. Kong, "Negative refraction of a combined double S-shaped metamaterial," *Appl. Phys. Lett.*, vol. **86**, paper 151909, 2005.
15. A. K. Horestani, J. Naqui, M. Durán-Sindreu, C. Fumeaux, and F. Martín, "Coplanar waveguides loaded with S-Shaped split ring resonators (S-SRRs): modeling and application to compact microwave filters," *IEEE Antennas Wireless Propag. Lett.*, vol. **13**, pp. 1349–1352, 2014.
16. A. K. Horestani, M. Durán-Sindreu, J. Naqui, C. Fumeaux, and F. Martín, "Compact coplanar waveguide band-pass filter based on coupled S-shaped split ring resonators," *Microw. Opt. Technol. Lett.*, vol. **57**, pp. 1113–1116, May 2015.
17. J. Naqui, J. Coromina, A. Karami-Horestani, C. Fumeaux, and F. Martín, "Angular displacement and velocity sensors based on coplanar waveguides (CPWs) loaded with S-shaped split ring resonator (S-SRR)," *Sensors*, vol. **15**, pp. 9628–9650, 2015.

18. A. K. Horestani, M. Durán-Sindreu, J. Naqui, C. Fumeaux, and F. Martín, "S-shaped complementary split ring resonators and application to compact differential bandpass filters with common-mode suppression," *IEEE Microw. Wireless Compon. Lett.*, vol. **24**, no. 3, pp. 150–152, Mar. 2014.
19. J. D. Baena, J. Bonache, F. Martín, R. Marqués, F. Falcone, T. Lopetegi, M. A. G. Laso, J. García, I. Gil, M. Flores-Portillo, and M. Sorolla, "Equivalent circuit models for split ring resonators and complementary split rings resonators coupled to planar transmission lines," *IEEE Trans. Microw. Theory Tech.*, vol. **53**, pp. 1451–1461, Apr. 2005.
20. J. Bonache, M. Gil, I. Gil, J. Garcia-García, and F. Martín, "On the electrical characteristics of complementary metamaterial resonators," *IEEE Microw. Wireless Compon. Lett.*, vol. **16**, pp. 543.545, Oct. 2006.

CHAPTER 11

WIDEBAND BALANCED FILTERS ON SLOTLINE RESONATOR WITH INTRINSIC COMMON-MODE REJECTION

Xin Guo,[1,2] Lei Zhu,[1] and Wen Wu[2]

[1]Department of Electrical and Computer Engineering, Faculty of Science and Technology, University of Macau, Macau SAR, China
[2]Ministerial Key Laboratory, JGMT, Nanjing University of Science and Technology, Nanjing, China

11.1 INTRODUCTION

As it has been discussed in the previous chapters, balanced circuits have significant advantages over their single-ended counterparts. Thus, balanced or differential-mode bandpass filters (BPFs) have been extensively studied and developed as one of the key circuit blocks in modern communication systems with low noise, small interference, and little crosstalk, thanks to their excellent property of common-mode (CM) rejection. With the wideband or ultra-wideband single-ended BPFs being used widely, there is an increasing demand for exploration

Balanced Microwave Filters, First Edition. Edited by Ferran Martín, Lei Zhu, Jiasheng Hong, and Francisco Medina.

of wideband balanced BPFs. In addition to excellent differential-mode (DM) passbands, these balanced filters should have good CM rejections within their designated wide DM passbands. Thus, the design of wideband balanced BPFs is more challenging than the design of wideband single-ended BPFs or narrowband balanced BPFs.

Similarly to narrowband balanced BPFs, a typical wideband balanced BPF is usually constructed by mirroring two wideband single-ended BPFs with some auxiliary components added at the symmetrical axis in order to highly suppress CM signals [1–4]. Meanwhile, wideband balanced BPFs based on the transversal signal interference concept are developed as reported in Refs. [5, 6]. But the dimensions of these filters are all decided by parametric analysis and full-wave optimization at the cost of intensive computation, due to the shortage of systematic synthesis approaches. What's more, some trade-offs are usually inevitable between their DM and CM performances, thus bringing out extra problematic issues in the entire optimization procedure.

A brand new wideband balanced BPF is proposed in Ref. [7] by making use of three transmission lines. It features its simple configuration and especially an intrinsic CM rejection property during the electromagnetic field conversion between two different kinds of transmission lines. Other similar researches display a few wideband balanced BPFs by using slotlines in certain parts [8, 9], verifying the intrinsic CM rejection property. However, these works not only have missed reasonable description on the working mechanisms of the wide DM passbands but also have not provided any systematic synthesis or design methods to determine all the physical dimensions of the filters. In this aspect, only the time-intensive cut-and-try approach has been implemented relying on intensive multiple-round simulations over the entire layouts of these filters.

Herein, this chapter displays several outcomes of our extensive researches on the wideband balanced or DM BPFs based on the technology of slotline multimode resonator (MMR) with intrinsic CM rejection [10–14]. In Section 11.2, two kinds of wideband balanced BPFs evolved from previous researches [7–9] based on slotline MMR are deeply studied not only on the working mechanism but also on the synthesis method; in Section 11.3, a novel kind of strip-loaded slotline resonator with intrinsic CM rejection is proposed and utilized to design a class of wideband balanced BPFs; in Section 11.4, a more generalized hybrid MMR is proposed by two kinds of transmission line, that is, microstrip line and slotline, and it is utilized to design an alternative wideband balanced BPF with intrinsic CM rejection.

11.2 WIDEBAND BALANCED BANDPASS FILTER ON SLOTLINE MMR

11.2.1 Working Mechanism

The three-dimensional view and top view of the two proposed balanced BPFs on slotline MMR [10] are depicted in Figures 11.1 and 11.2, respectively. In each filter, a slotline resonator is formed on the middle ground plane of a two-layer substrate, and its multimode resonance is utilized to build up a wideband balanced BPF. Our first effort is made to investigate the excitation of multi-resonant modes in a slotline resonator. In this context, the first three or four resonant modes are depicted by their electric field distribution as illustrated in Figure 11.2. For filter 1, the

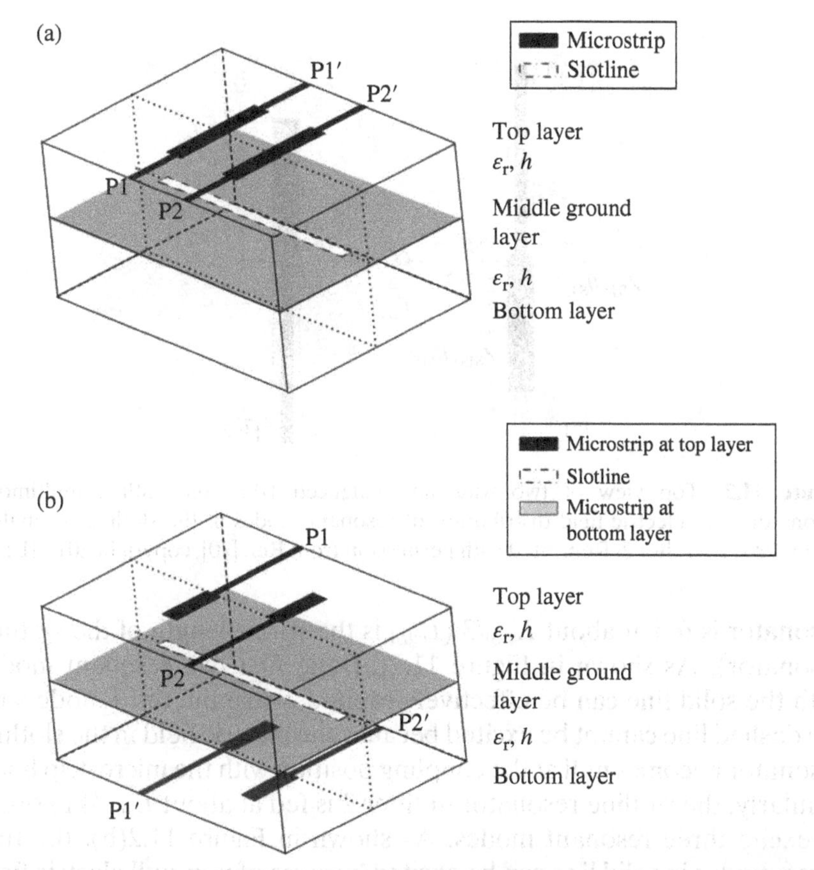

Figure 11.1 Three-dimensional view of two wideband balanced BPFs on slotline multimode resonators. (a) Filter 1. (b) Filter 2. Reprinted with permission from Ref. [10]; copyright 2015 IEEE.

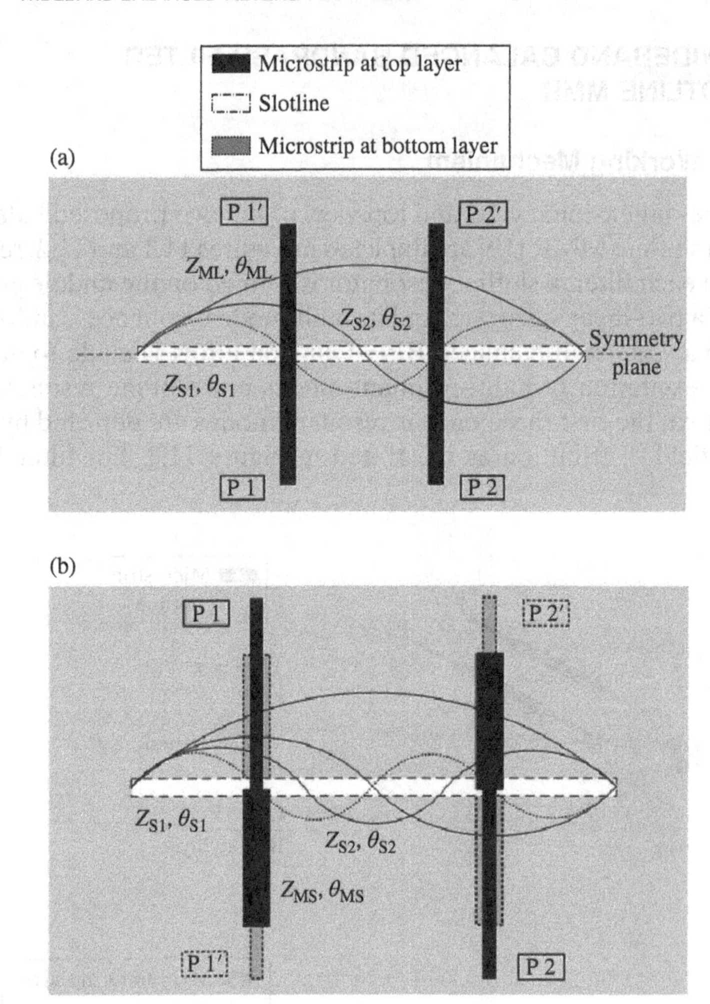

Figure 11.2 Top view of two wideband balanced BPFs on slotline multimode resonators and electric field distribution of resonant modes in the slotline resonator. (a) Filter 1. (b) Filter 2. Reprinted with permission from Ref. [10]; copyright 2015 IEEE.

resonator is fed at about $L_{slot}/3$ (L_{slot} is the whole length of the slotline resonator). As shown in Figure 11.2(a), the first two resonant modes with the solid line can be effectively excited, while the third mode with the dashed line cannot be excited because the electric field in the slotline resonator becomes null at the coupling position with the microstrip lines. Similarly, the slotline resonator in filter 2 is fed at about $L_{slot}/4$ in order to excite three resonant modes. As shown in Figure 11.2(b), the first three modes in solid line can be excited because of non-null electric field intensity at the feeding position, while the fourth mode in dashed line cannot emerge.

Next, two kinds of feeding approaches are proposed. The available feeding scheme should provide wideband coupling capability after the desired resonant modes are excited for good DM signal transmission over a specific frequency band, but reject or non-excite all the CM signals within the DM operating band. Figures 11.1(a) and 11.2(a) show the geometrical layout and equivalent representation of the first filter, that is, filter 1, where the microstrip lines with paired differential ports at their two sides are directly tapped on the slotline resonator. The strips with differential ports symmetrically connected make the symmetrical plane along the slotline to be a virtual electric wall under DM operation; thus the DM signals can be transmitted along the slotline portion. However, under CM operation, this symmetrical plane becomes a virtual magnetic wall, thereby decaying all the CM signals along the slotline. For filter 2 as shown in Figures 11.1(b) and 11.2(b), the slotline resonator is now connected with paired microstrip open-ended stubs. Two microstrip open-ended stubs are loaded at the same position of the slotline but on the top or bottom layer, respectively, and they are referred to as one paired stubs. Figure 11.3 shows the cross-view of the electrical field distribution in the 3-metal-layer structure, and it is herein prepared to demonstrate its working mechanism. Under DM operation, the

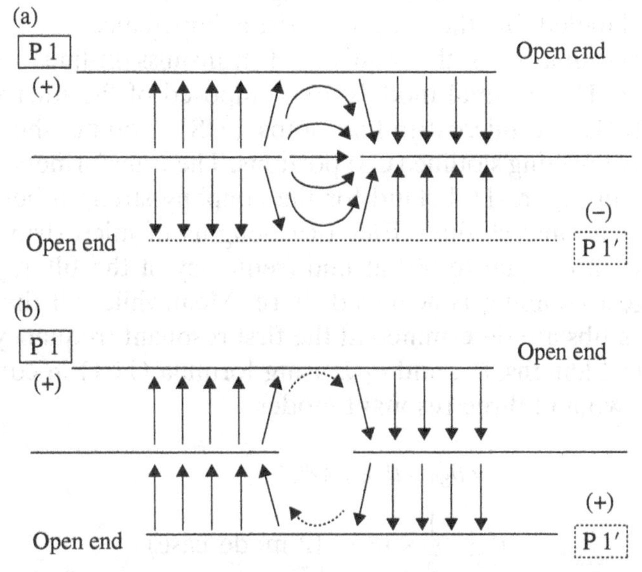

Figure 11.3 Cross-view of electrical field distribution in the 3-metal-layer structure for filter 2. (a) DM feeding. (b) CM feeding. Reprinted with permission from Ref. [10]; copyright 2015 IEEE.

electric fields in the slotline are excited in-phase by a pair of strip conductors, resulting in good signal transmission, as shown in Figure 11.3 (a). While in CM, the electric fields in the slotline are excited with an out-of-phase signal as illustrated in Figure 11.3(b) and they are canceled out with each other, thus gaining excellent CM rejection.

In conclusion, filter 1 makes use of its symmetrical characteristic in geometry to get a field mismatching situation in CM, whereas filter 2 achieves good CM rejection using the transversal signal cancelation concept. Thus the CM signal can be well rejected during field conversion by using the described feeding approaches. In DM, the two feeding schemes are developed based on variations of wideband microstrip–slotline transitions, so they have wideband coupling property. Thereafter, the desired DM wideband passband could be satisfactorily achieved after a considered design is executed.

11.2.2 Synthesis Method

As the CM signals have been intrinsically rejected, all the efforts are made to synthesize these balanced BPFs with the desired DM passband performance. The initial DM circuit models are displayed in Figure 11.4, and they are directly deduced from the geometrical schemes in Figure 11.2 under DM operation. In Figure 11.2, transmission-line portions are labeled by their characteristic impedances and electrical lengths that link with the equivalent transmission-line sections in Figure 11.4. These initial models are composed of the microstrip-line portions (ML) or microstrip-line stubs (MS), slotline shorted-stubs (SS), and connecting slotline (CS) portions. The transformers with their turn ratios in Figure 11.4 stand for the coupling strength between the microstrip line and slotline. Electrical lengths of microstrip portions, θ_{MS} or θ_{ML}, are equal to 90° at mid-frequency of the filter, f_{mid}; thus the strongest coupling is achieved there. Meanwhile, all the slotlines or slotline stubs are determined at the first resonant frequency, f_1, with the electrical lengths, θ_{SS} and θ_{CS}, using formula (11.1) in correspondence with two and three resonant modes:

$$
\begin{aligned}
&2\times\theta_{SS}+\theta_{CS}=180^\circ \\
&\theta_{SS}=\frac{1}{3}\times 180^\circ \ \text{(2-mode case)} \\
&\theta_{SS}=\frac{1}{4}\times 180^\circ \ \text{(3-mode case)}
\end{aligned}
\tag{11.1}
$$

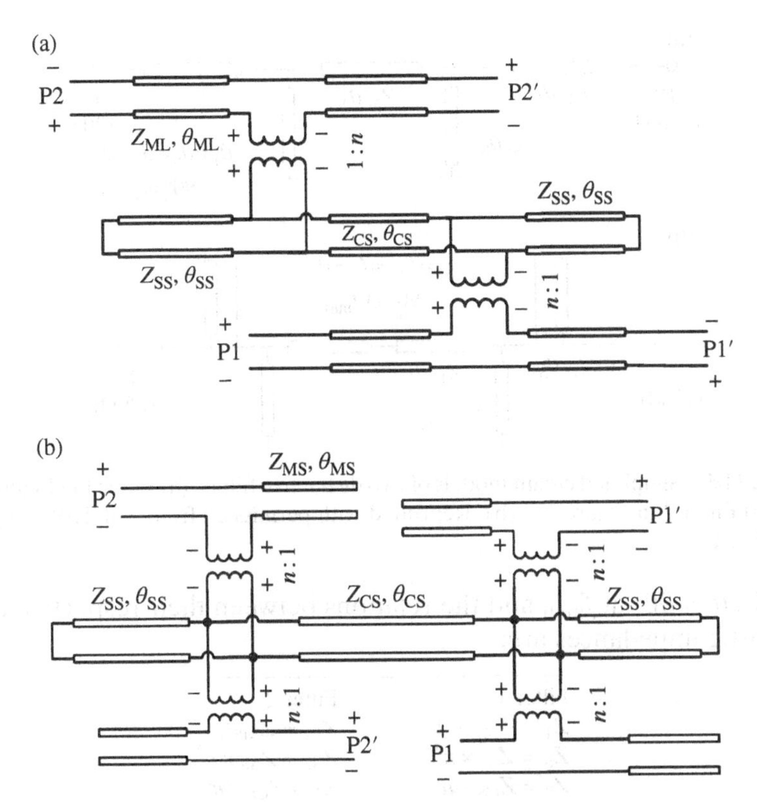

Figure 11.4 Initial DM equivalent circuit models directly deduced from the two geometrical schemes. (a) Filter 1. (b) Filter 2. Reprinted with permission from Ref. [10]; copyright 2015 IEEE.

Interestingly, it is noticeable from the two initial circuit models that the differential-port pairs in filter 1 are series connected, whereas they in filter 2 are shunt connected. This property makes the differential-port impedances distinctively different in the following synthesis process.

The initial circuit models are further simplified into the ones shown in Figure 11.5 in order to design the balanced BPFs using a unified synthesis approach. In this process of simplification, the four-port circuit models are converted into two-port ones in which series or shunt portions are merged. Then, all the electrical lengths are converted with respect to f_{mid}, and impedance transformers are absorbed into the characteristic impedances of the related transmission-line sections. As such, the simplified circuit models come out as general high-pass prototypes. All the involved transmission-line cells have the same electrical length,

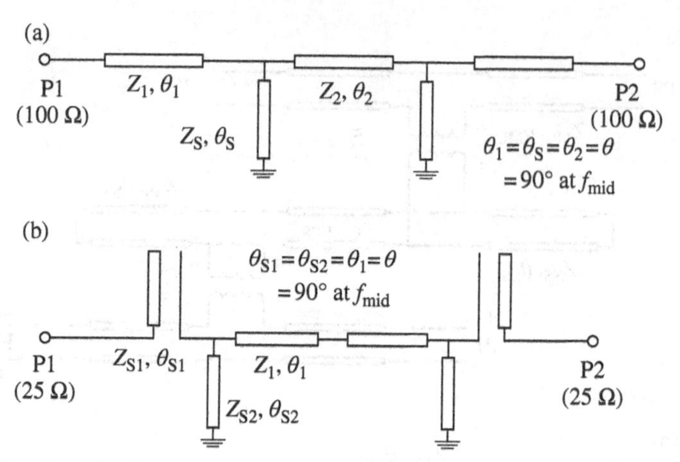

Figure 11.5 Simplified circuit models of two distinctive filters. (a) Filter 1 in Figure 11.4(a). (b) Filter 2 in Figure 11.4(b). Reprinted with permission from Ref. [10]; copyright 2015 IEEE.

that is, $\theta = 90°$ at f_{mid}, and the relations between their respective characteristic impedances are:

Filter 1	Filter 2
$Z_1 = Z_{ML} \times 2$	$Z_{S1} = Z_{MS}/2$
$Z_S = Z_{SS} \times n^2$	$Z_{S2} = Z_{SS} \times n^2$
$Z_2 = Z_{CS} \times n^2$	$Z_1 = Z_{CS} \times n^2$

The prototypes in Figure 11.5 can be synthesized following the method in Ref. [15]. By this method, filtering performances of these prototypes can be designed with any desired bandwidth (BW) and ripple constant in theory. In order to make the description concise, only one of the presented filters, that is, filter 1, is selected for description of synthesis design in this subsection.

The whole **ABCD** matrix of the topology in Figure 11.5(a) is

$$\begin{bmatrix} \mathbf{A} & \mathbf{B} \\ \mathbf{C} & \mathbf{D} \end{bmatrix} = M_1 M_S M_2 M_S M_1 \tag{11.2}$$

In (11.2), all the **M** matrices can be theoretically derived as

$$M_n = \begin{bmatrix} \cos(\theta) & jz_n \sin(\theta) \\ j\left(\dfrac{1}{z_n}\right)\sin(\theta) & \cos(\theta) \end{bmatrix}, \quad n = 1,2; \; M_S = \begin{bmatrix} 1 & 0 \\ \dfrac{1}{jz_S \tan(\theta)} & 1 \end{bmatrix} \tag{11.3}$$

where z_1, z_S, and z_2 are the unit or normalized impedances.

With the **ABCD** matrix in (11.2) known, the transmission coefficient of the two-port topology in Figure 11.5(a) can be derived as

$$|S_{21}|^2 = \frac{1}{1+|F|^2} \text{ and } F = \frac{B-C}{2} \tag{11.4}$$

Using (11.2) and (11.3), F can be deduced and properly rearranged as

$$F = j\left[k_1\frac{\cos^4(\theta)}{\sin(\theta)} + k_2\frac{\cos^2(\theta)}{\sin(\theta)} + k_3\frac{1}{\sin(\theta)}\right] \tag{11.5}$$

where

$$k_1 = \frac{z_2}{2{z_1}^2} - \frac{z_2}{2} - \frac{{z_1}^2-1}{2z_2} - z_1 - \frac{{z_1}^2+z_2z_1-1}{z_S} + \frac{z_2-{z_1}^2z_2}{2{z_S}^2} + \frac{1}{z_1} + \frac{z_2}{z_1z_S} \tag{11.6a}$$

$$k_2 = z_1 + \frac{z_2}{2} - \frac{z_2}{{z_1}^2} + \frac{{z_1}^2+z_2z_1}{z_S} + \frac{2{z_1}^2-1}{2z_2} - \frac{1}{z_1} - \frac{z_2}{z_1z_S} + \frac{{z_1}^2z_2}{2{z_S}^2} \tag{11.6b}$$

$$k_3 = -\frac{{z_1}^4-{z_2}^2}{2{z_1}^2z_2} \tag{11.6c}$$

To exhibit the Chebyshev responses with equal-ripple in-band behavior using the topology in Figure 11.5(a), the squared magnitude of the filters is expressed as

$$|S_{21}|^2 = \frac{1}{1+\varepsilon^2\cos^2(n\phi_{\text{u.e.}}+m\phi_{\text{L}})} = \frac{1}{1+\varepsilon^2[T_n(x)T_m(y)-U_n(x)U_m(y)]^2} \tag{11.7}$$

where $n=3$, $m=1$.

In (11.7), ε is the specified equal-ripple constant in the passband. $T_n(x)$ and $U_n(x)$ are the Chebyshev polynomial functions of the first and second kind of degree n, respectively. And

$$x = \cos(\phi_{\text{u.e.}}) = \alpha\cos(\theta), \quad \alpha = \frac{1}{\cos(\theta_c)}, \tag{11.8a}$$

$$y=\cos(\phi_{\rm L})=\alpha\sqrt{\frac{\alpha^2-1}{\alpha^2-x^2}} \tag{11.8b}$$

where θ_c is the electrical length or phase at the lower cutoff frequency.

Based on the relations among those variables, one can establish the condition as

$$\varepsilon\cos(n\phi_{\rm u.e.}+m\phi_{\rm L})=K_1\frac{\cos^4(\theta)}{\sin(\theta)}+K_2\frac{\cos^2(\theta)}{\sin(\theta)}+K_3\frac{1}{\sin(\theta)} \tag{11.9}$$

$$K_1=-\varepsilon\left(4\alpha^4+4\alpha^3\sqrt{\alpha^2-1}\right) \tag{11.10a}$$

$$K_2=-\varepsilon\left(5\alpha^2+3\alpha\sqrt{\alpha^2-1}\right) \tag{11.10b}$$

$$K_3=\varepsilon \tag{11.10c}$$

After solving $k_1=K_1$, $k_2=K_2$, and $k_3=K_3$, three normalized impedances, z_1, z_S, and z_2, are all determined by any specified ripple constant ε and BW reflected in α. The characteristic impedances of all stubs and connecting lines, Z_1, Z_S, and Z_2, are calculated by multiplying them with the corresponding differential-port impedances. It needs to be pointed out that different feeding approaches make the differential-port impedances different. In filter 1, the differential-port impedance equals to 100 Ω due to the series connection; and in filter 2, the differential-port impedance becomes 25 Ω due to the shunt connection.

After the filtering responses of the simplified models are synthesized, the process of mapping them into the final filters with physical dimensions is executed to design the two wideband balanced BPFs as will be discussed in the following subsections.

11.2.3 Geometry and Layout

At the beginning, all the dimensions of the microstrip lines and slotlines involved could crudely be determined under the assumption of the unity turn ratio for the impedance transformers as done in many reported works, for example, Ref. [15]. Afterward, an effective method needs to be employed to calculate the real values of these turn ratios in filter 1, helping get final exact dimensions or to find proper values of turn ratios in filter 2 to realize the desired impedance values in the simplified circuits.

As for filter 1, the turn ratios are calculated as follows. At mid-frequency, f_{mid}, the **ABCD** matrix of the shunt shorted stub can be calculated as

$$M_{SS} = \begin{bmatrix} 1 & 0 \\ -j\frac{1}{Z_{SS}}\cot(\theta) & 1 \end{bmatrix} = \begin{bmatrix} 1 & 0 \\ 0 & 1 \end{bmatrix} \text{ at } f_{mid} \tag{11.11}$$

Thus, the frequency response of this filter at around f_{mid} should be equal (or very approximate) to that of its respective circuit model by taking out the aforementioned portions shown in Figure 11.6(a). Compared with its simplified format in Figure 11.6(b), the turn ratios of these transformers can be expressed as

$$n = \sqrt{\frac{Z_2'}{Z_{CS} \text{ in the crude dimension}}} = \sqrt{\frac{Z_2'}{Z_2}} \tag{11.12}$$

where Z_2' stands for the impedance in Figure 11.6(b) causing the same return loss (RL) as the simulated RL of the crude dimension at f_{mid} and Z_2 is the impedance of the connecting slotline in the synthesized circuit model.

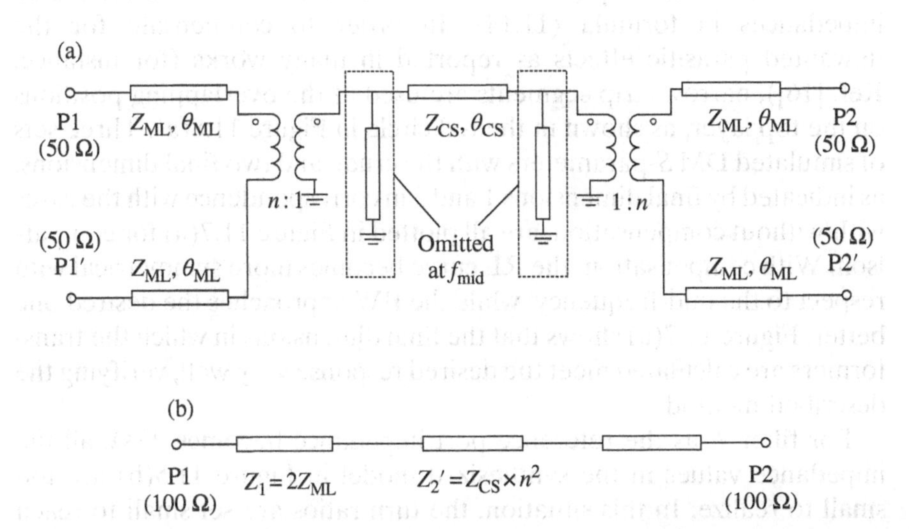

Figure 11.6 (a) Direct DM circuit model of filter 1 at f_{mid} for calculating turn ratios. (b) Its simplified DM circuit model. Reprinted with permission from Ref. [10]; copyright 2015 IEEE.

In our design, the circuit model with maximum RL of 20 dB and BW of 2.75 : 1 are first synthesized as $Z_{ML} = 65.8\,\Omega$; $Z_{SS} = 68.5\,\Omega$; $Z_{CS} = 47.9\,\Omega$. The crude dimensions are simulated using the strip/slot widths with the previously derived impedances, and its simulated RL at f_{mid} is 17.1 dB as shown in Figure 11.7(a). Z_2' in Figure 11.6(b) causing the same RL at f_{mid} is 38.3 Ω; thus

$$n = \sqrt{\frac{Z_2'}{Z_2}} = \sqrt{\frac{38.3}{47.8}} = 0.895 \tag{11.13}$$

During the design process, the turn ratio n is assumed to be invariable, since the crude dimensions do not appear to be much different from the final dimensions. After n is calculated, the final exact impedances can be calculated as

$$Z_{ML\text{final}} = Z_{ML} = 65.8 \tag{11.14a}$$

$$Z_{SS\,\text{final}} = \frac{Z_{SS}}{n^2} = \frac{68.5}{0.895^2} = 85.6 \tag{11.14b}$$

$$Z_{CS\,\text{final}} = \frac{Z_{CS}}{n^2} = \frac{47.9}{0.895^2} = 59.8(\text{unit}: \Omega) \tag{11.14c}$$

In final, all the strip/slot widths are derived from these calculated impedances in formula (11.14). In order to compensate for the unwanted parasitic effects as reported in many works (for instance, Ref. [16]), narrow strip segments are used at the overlapping positions on the top layer, as shown in the red circle in Figure 11.8(a). Three sets of simulated DM S-parameters with the crude and two final dimensions, as indicated by final dimensions 1 and 2 in correspondence with the cases with/without compensation, are all plotted in Figure 11.7(a) for comparison. With compensation, the RL curve becomes more symmetrical with respect to the mid-frequency, while the BW approaches the desired one better. Figure 11.7(a) shows that the final dimensions in which the transformers are calculated meet the desired response very well, verifying the described method.

For filter 2, as the reference port impedance becomes 25 Ω, all the impedance values in the synthesized model in Figure 11.5(b) are too small to realize. In this situation, the turn ratios are set small to reach the desired small impedance in the simplified model. In our design, the paired feeding strips on the top and bottom layers are located with an offset distance, Δd, from each other along the direction of slotline,

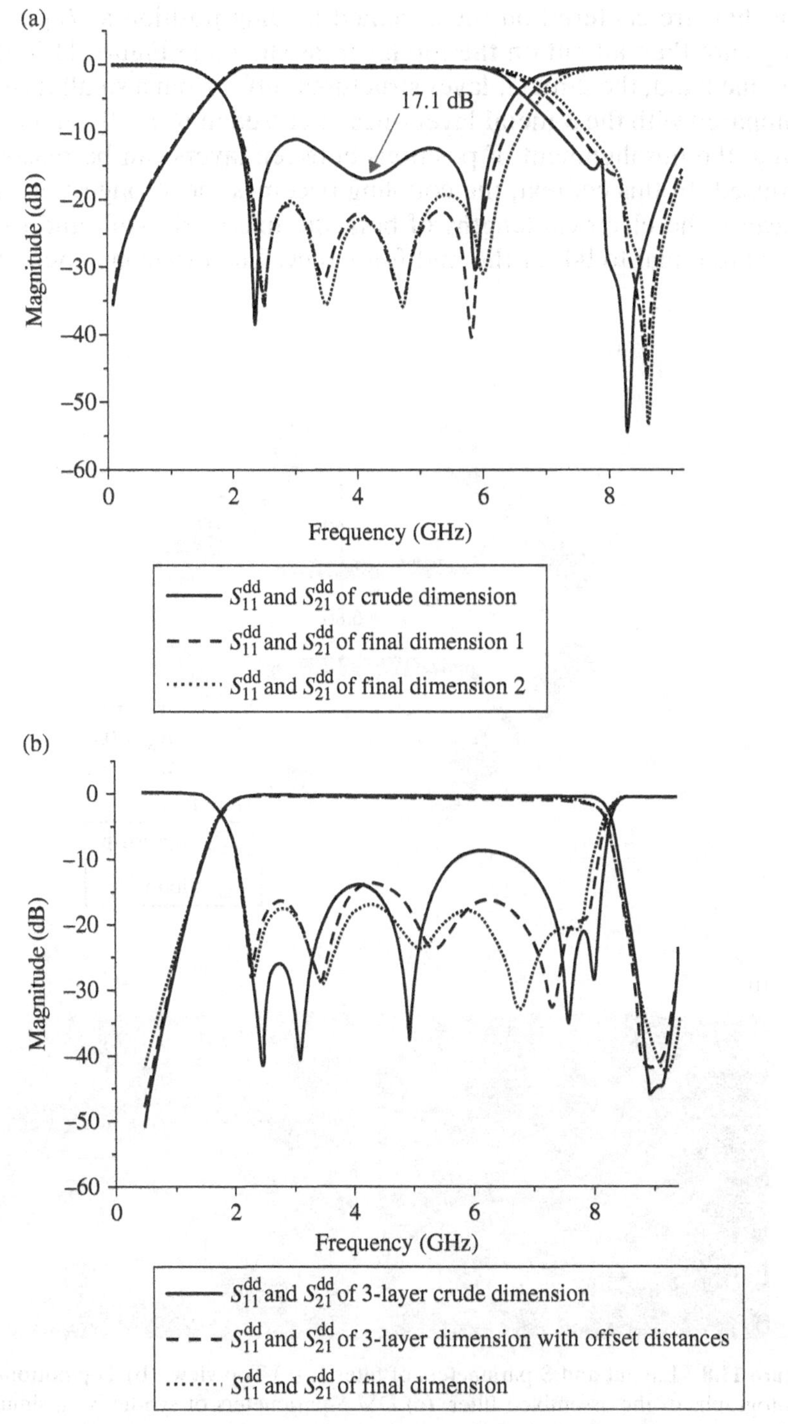

Figure 11.7 Simulated DM *S*-parameters after different design procedures. (a) Filter 1. (b) Filter 2. Reprinted with permission from Ref. [10]; copyright 2015 IEEE.

but they are centered on the specified feeding position at $L_{slot}/4$. The strips are then all put on the top layer, as shown in Figure 11.9(a). On the one hand, the 2-metal-layer structures suffer from a small turn ratio compared with the 3-metal-layer ones discussed in Ref. [9]; on the other hand, the misalignment of positions between layers can be reasonably reduced. In this context, the coupling becomes the strongest at $L_{slot}/4$, because the electrical lengths of both the microstrip stub and slotline stub are equal to 90° at the mid-frequency. The electromagnetic fields

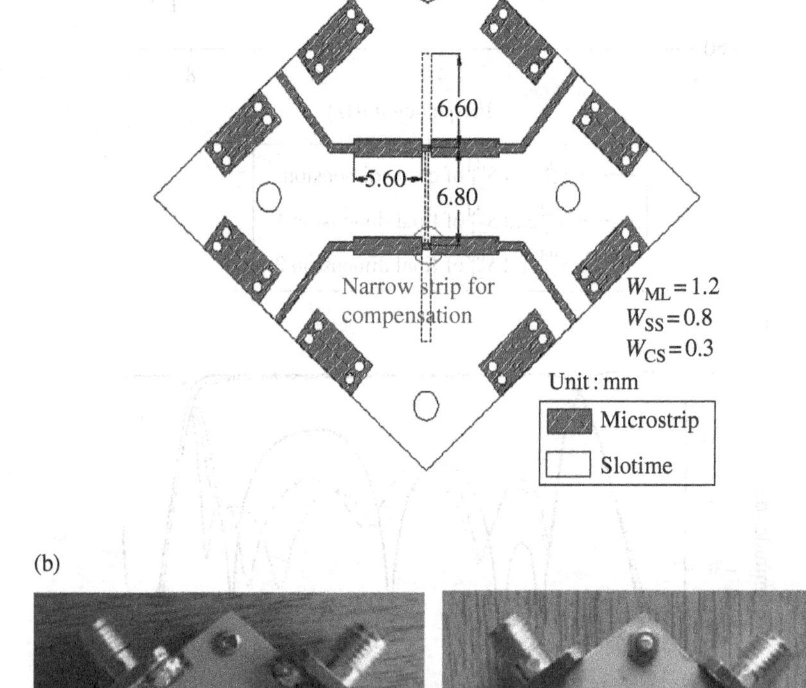

Figure 11.8 Layout and S-parameters of filter 1. (a) Top view. (b) Top/bottom view photographs of the assembled filter. (c) DM *S*-parameters of synthesized, simulated, and measured results. (d) CM S_{21}-magnitude. Reprinted with permission from Ref. 11.8; copyright 2015 IEEE.

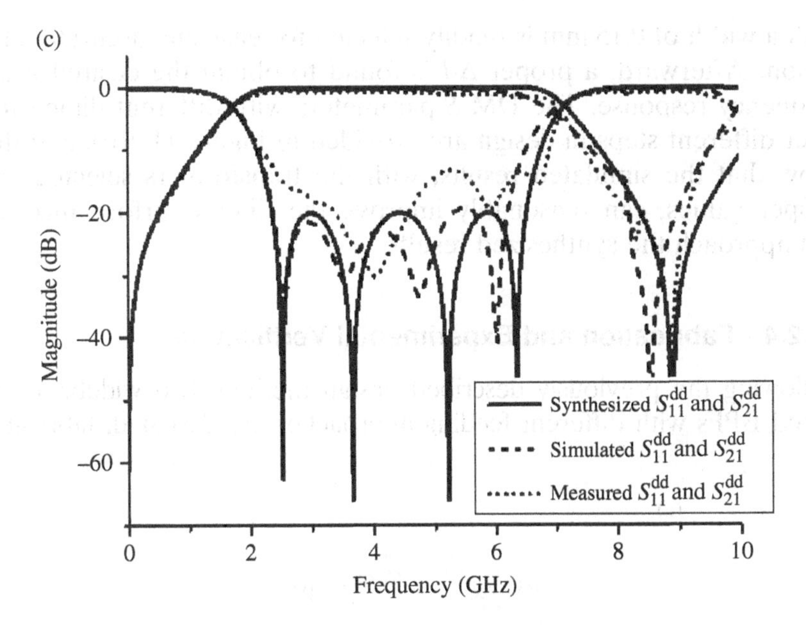

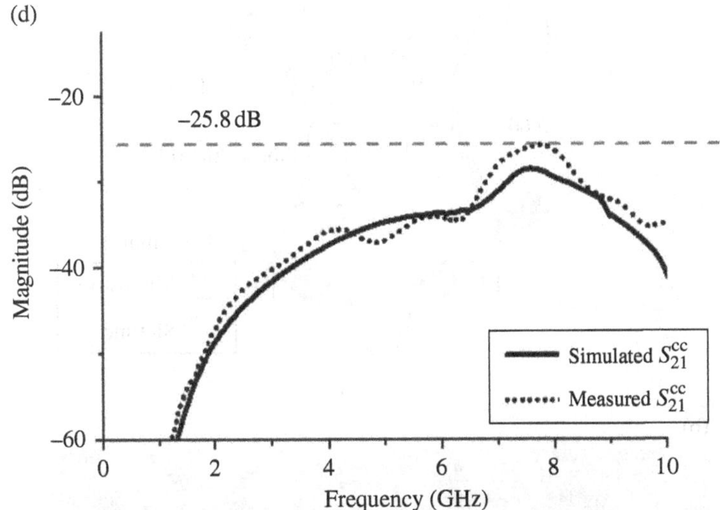

Figure 11.8 (Continued)

induced by these two kinds of lines at this position reach their maximal values simultaneously. With the offset distance, the slotline stubs are either longer or shorter than 90° at the feeding positions. Since the electric field in the slotline is not maximal at the mid-frequency, the coupling degree reflected in n tends to be accordingly decreased.

In the design of filter 2, the BW of 3.75 : 1 is chosen and achieved using a slotline uniform impedance resonator (UIR). To realize the small impedances as desired in the synthesized results, a narrow slotline with a width of 0.15 mm is readily selected for easy and accurate fabrication. Afterward, a proper Δd is found to obtain the desired n and frequency response. The DM S-parameters with different dimensions after different steps in design are provided in Figure 11.7(b), and they show that the simulated results, with the transformers selected with proper values, can reasonably improve the filter's performance and can approach the synthesized results.

11.2.4 Fabrication and Experimental Verification

Following the previously described design method, two wideband balanced BPFs with different feeding approaches are designed, fabricated,

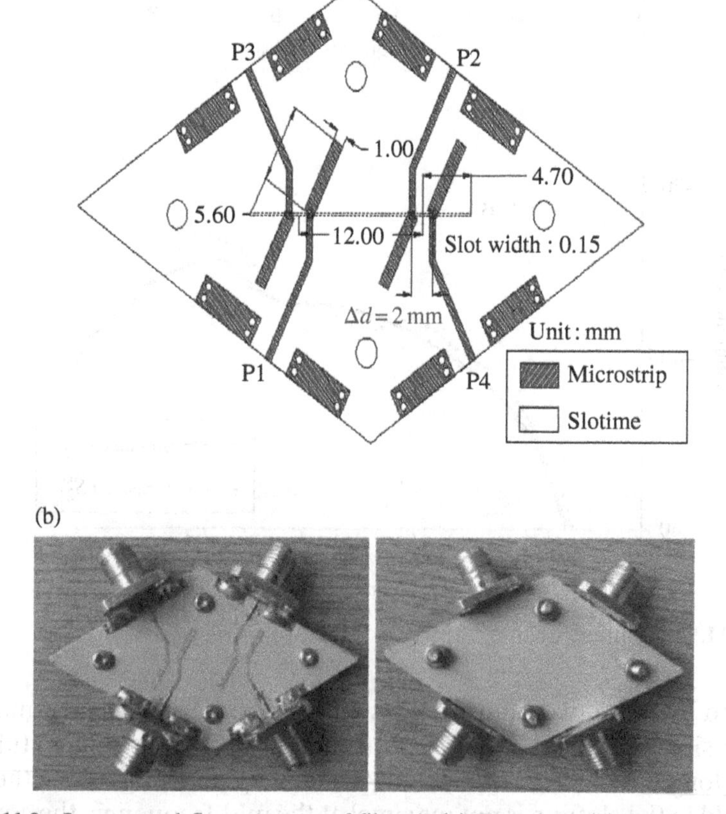

Figure 11.9 Layout and S-parameters of filter 1. (a) Top view. (b) Top/bottom view photographs of the assembled filter. (c) DM S-parameters of synthesized, simulated, and measured results. (d) CM S_{21}-magnitude. Reprinted with permission from Ref. 11.9; copyright 2015 IEEE.

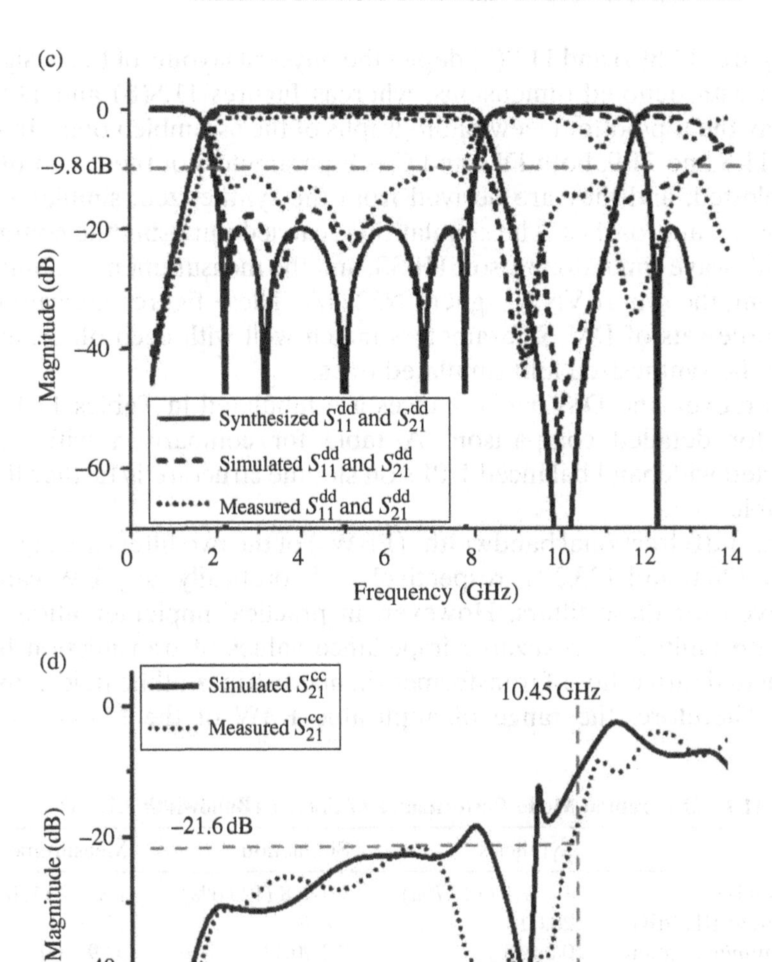

Figure 11.9 (Continued)

and measured. The filters are built on a 2-layer substrate, that is, Rogers RT/duroid 6010LM with a permittivity of 10.7 and thickness of 0.635 mm in each layer. The reasons of using 2-layer substrates are given: on the one hand, the slotline portions with desired impedances on the single-layer substrate are too narrow to be fabricated; on the other hand, the 2-layer substrate can effectively reduce undesired radiation in the slotline portion.

Figures 11.8(a) and 11.9(a) depict the physical layouts of the designed filters with denoted dimensions, whereas Figures 11.8(b) and 11.9(b) display the top/bottom view photographs of the assembled ones. In Figures 11.8 and 11.9, both DM and CM S-parameters of these two filters are plotted, and they are derived from the synthesized, simulated, or measured approaches. The simulation is carried out using the commercial full-wave simulator Ansoft HFSS, and the measurement is obtained by using the 4-port VNA Agilent N5224A. These figures indicate that the three sets of DM *S*-parameters match well with each other, especially the synthesized and simulated ones.

Moreover, the DM performances are tabulated in Tables 11.1 and 11.2 for detailed comparison. A table for comparison with other reported wideband balanced BPFs on slotline structure is further listed in Table 11.3.

The 3-dB fractional bandwidths (FBWs) of the two filters are equal to about 120.0 and 123.2%, respectively. Theoretically, any BW can be achieved for these filters. However, in practical implementation, the BW are limited by realizable impedance values of transmission lines and actual turn ratios of transformers in microstrip–slotline field conversion. Therefore, the range of applicable FBW of the filters can be

Table 11.1 Differential-Mode Performance of Filter 1 (Bandwidth 2.75 : 1)

	Synthesis	Simulation	Measurement
BW (GHz)	1.7–7.1 (122.7%)	1.7–6.8 (120.0%)	1.8–7.2 (120.0%)
Maximum RL (dB)	20.00	19.18	13.41
Minimum/maximum IL (dB)	0/0.043	0.23/0.43	0.59/1.60
Impedance of each section (Ω)	$Z_1/Z_S/Z_2$ 65.8/68.5/47.8	$Z_{ML}/Z_{SS}/Z_{CS}$ 33.2/84.8/55.9	
Turn ratio	0.895		

IL, insertion loss; RL, return loss.

Table 11.2 Differential-Mode Performance of Filter 2 (Bandwidth 3.75 : 1)

	Synthesis	Simulation	Measurement
BW (GHz)	1.7–8.3 (132%)	1.7–7.9 (129.2%)	1.9–8.0 (123.2%)
Maximum RL (dB)	20.00	17.08	9.80
Minimum/maximum IL (dB)	0/0.043	0.37/1.10	1.28/2.12
Impedance of each section (Ω)	$Z_{S1}/Z_{S2}/Z_1$ 18.4/33.0/32.3	$Z_{MS}/Z_{SS}/Z_{CS}$ 37.0/49.2/49.2	
Turn ratio	0.819		

Table 11.3 Comparison with Other Wideband Balanced Bandpass Filters on Slotline

	DM performance						
	Central frequency (GHz)	3-dB fractional bandwidth (%)	Maximum RL (dB)	Minimum IL (dB)	In-band CM rejection (dB)	CM rejection bandwidth	Effective circuit size
[7]	3.60	105.0	10.0	1.22	47.5	0.3–8.0 GHz (minimum 44.7 dB)	$0.67\ \lambda_0 \times 0.20\ \lambda_0$
[8]	6.83	114.3	10.1	0.83	18.9	Not given	$0.36\ \lambda_0 \times 0.28\ \lambda_0$
[9]	4.25	115.0	12.0	0.43	20.0	0–10 GHz (minimum 20.0 dB)	$0.56\ \lambda_0 \times 0.28\ \lambda_0$
Filter 1	4.50	120.0	13.4	0.59	25.8	0–18.5 GHz (minimum 22.7 dB)	$0.30\ \lambda_0 \times 0.18\ \lambda_0$
Filter 2	5.45	123.2	9.8	1.28	21.6	0–10.6 GHz (minimum 21.6 dB)	$0.39\ \lambda_0 \times 0.19\ \lambda_0$

estimated as about 100–140% under the required in-band ripple level of 0.043 dB and the preselected substrate. Regarding the DM passband, the simulated and measured results are well matched with each other for filter 1, but they are not in such good agreement with each other for filter 2. It may be primarily caused by the fact that the unexpected tolerance error in fabricating the narrow slotline in filter 2 is more serious than that in fabricating the wide slotline in filter 1.

Besides, the intrinsic CM rejection can be satisfactorily realized as confirmed in the simulation and measurement. In filter 1, the CM rejection is realized by unmatched field, which turns out to be an all-stop performance. The CM rejection is higher than 25.8 dB within the realized DM passband and higher than 22.7 dB over the frequency range of 0–18.5 GHz. In filter 2, good CM rejection is realized by the signal cancelation technique as discussed previously. In fact, the CM performance of filter 2 represents a bandpass response with weak input/output coupling strength. The CM insertion loss is higher than 21.6 dB over the frequency range of 0–10.45 GHz, which covers the entire DM passband. In short, the CM signals have been effectively rejected within the desired DM passband in the two proposed filters, which is fully consistent with our prediction in synthesis and simulation.

11.3 WIDEBAND BALANCED BPF ON STRIP-LOADED SLOTLINE RESONATOR

11.3.1 Strip-Loaded Slotline Resonator

Wideband BPFs based on MMRs have been widely studied and developed in modern microwave systems by virtue of many attractive features, for instance, a straightforward design procedure. In categories of MMRs, the stub-loaded resonator is among the simplest and widespread ones [17–24]. Compared with stepped-impedance MMRs, stub-loaded MMRs usually have more flexible abilities in mode reallocation, more compact sizes, better out-of-band properties, and so forth. However, the stub-loaded resonator has been rarely applied on design of wideband balanced BPFs with high CM rejection properties. The reported ones suffer from either low CM suppression level [25] or limited CM suppression BW [26]. Thus, a kind of strip-loaded resonator is proposed here and utilized to design a class of wideband balanced BPFs with intrinsic CM rejection [12].

Figure 11.10 depicts the scenarios of two distinct stub-loaded resonators, namely, the traditional microstrip stub-loaded resonator and the strip-loaded slotline resonator. The traditional stub-loaded resonator

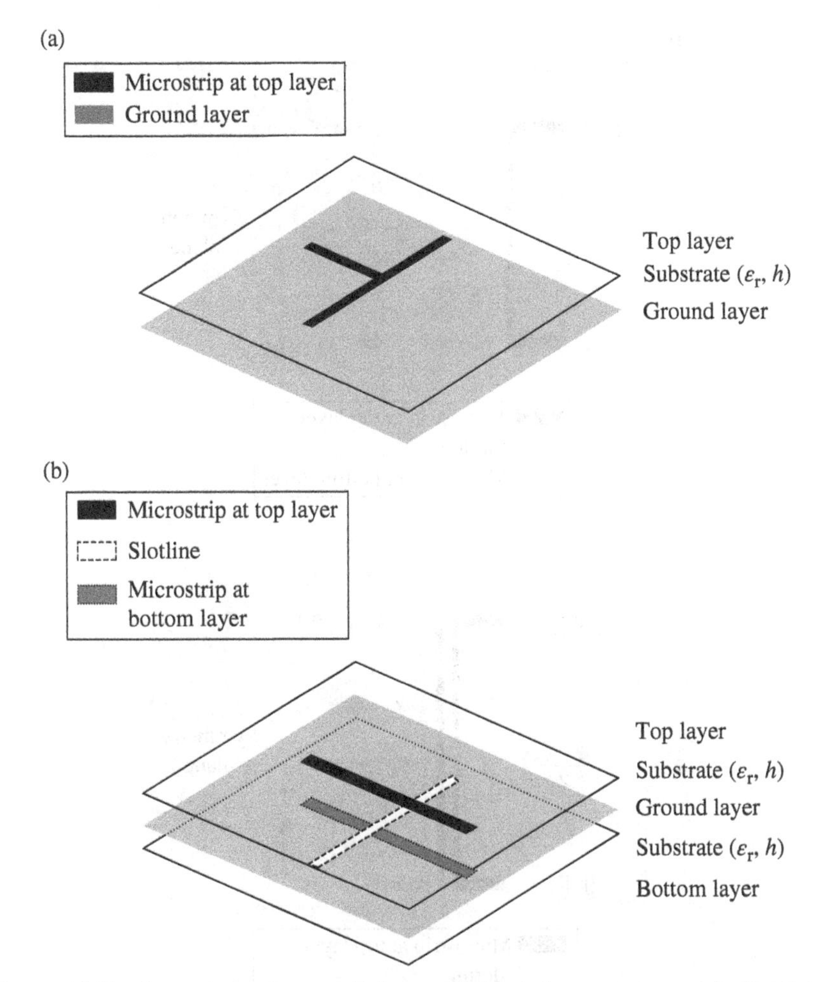

Figure 11.10 Scenarios of two distinct stub-loaded resonators. (a) Traditional microstrip stub-loaded resonator. (b) Proposed strip-loaded slotline resonator. Reprinted with permission from Ref. [12]; copyright 2016 IEEE.

is commonly formed on the same interface of a dielectric substrate in the format of either microstrip line or slotline. A strip-loaded slotline resonator is formed by loading a pair of identical strip conductors on the slot portion perpendicularly and symmetrically, as illustrated in Figure 11.10 (b). With proper use of the location and length of the loaded strips, three or four resonant modes of a strip-loaded slotline resonator can be excited and employed for realization of wideband balanced BPFs with good CM rejection.

Figure 11.11 indicates the top view of two types of wideband balanced BPFs based on a general slotline resonator in Section 11.2 and the

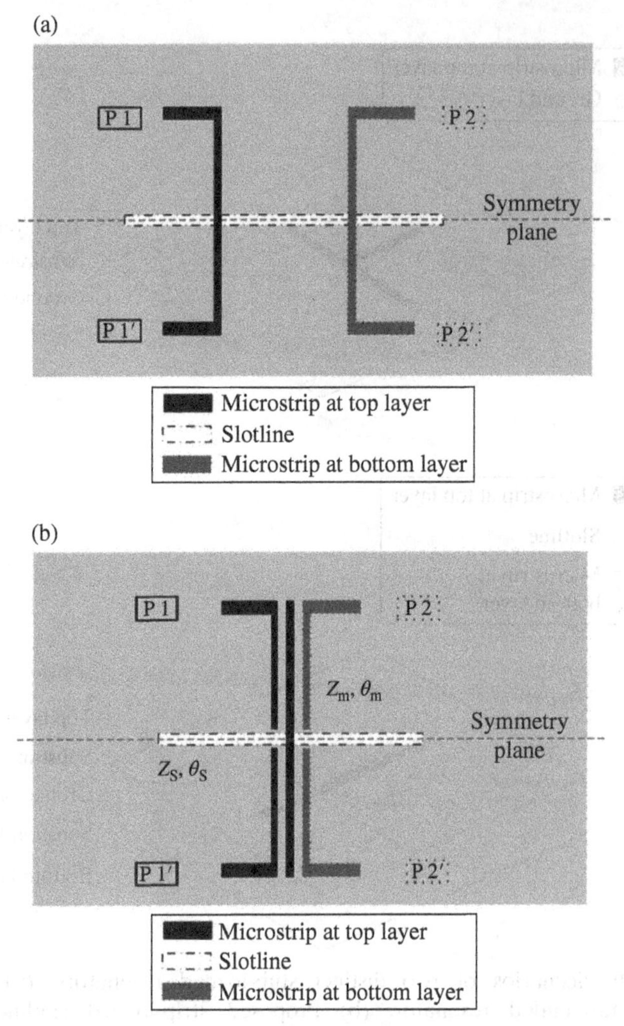

Figure 11.11 Top view of two types of wideband balanced BPFs based on different slotline resonators. (a) Initial slotline resonator in Section 11.2. (b) Proposed strip-loaded slotline resonator. Reprinted with permission from Ref. [12]; copyright 2016 IEEE.

strip-loaded slotline resonator developed herein. As shown in Figure 11.11(a), the balanced BPF, as investigated in Section 11.2, utilizes two microstrip–slotline transitions to feed the slotline resonator so that a wide DM passband with good CM suppression could be reasonably achieved. However, under the restriction of inevitable orthogonal transitions, the balanced BPFs on slotline MMRs cannot be implemented well in the format of stub-loaded resonator. By loading microstrip lines on the slotline resonator as depicted in Figure 11.11(b), there are two types of lines in the resonator, and the resonator is tightly coupled from the loaded strip stubs using the parallel-coupled lines [27]. In this aspect, the design of such a wideband balanced BPF can effectively follow the previous work on design of a wideband single-ended multimode BPF using the stub-loaded resonator [22].

As a balanced BPF, the working mechanism of its DM transmission and CM rejection should be qualitatively demonstrated. Figure 11.12(a) and (b) depicts the cross-view of the electrical field distribution in the two-layer structure under DM and CM excitations, respectively. Under DM operation in Figure 11.12(a), the symmetrical plane becomes a virtual electric wall so that the electric field in the DM feeding line is analogous to that in the slotline. Thus, it can be highly expected that a DM passband can be achieved. The symmetrical plane is converted into a virtual magnetic wall under CM operation as shown in Figure 11.12(b). In this case, the even-symmetrical field in the feeding lines is orthogonal with the odd-symmetrical field in the slotline, thereby intrinsically rejecting the transmission of all the CM signals along the slotline.

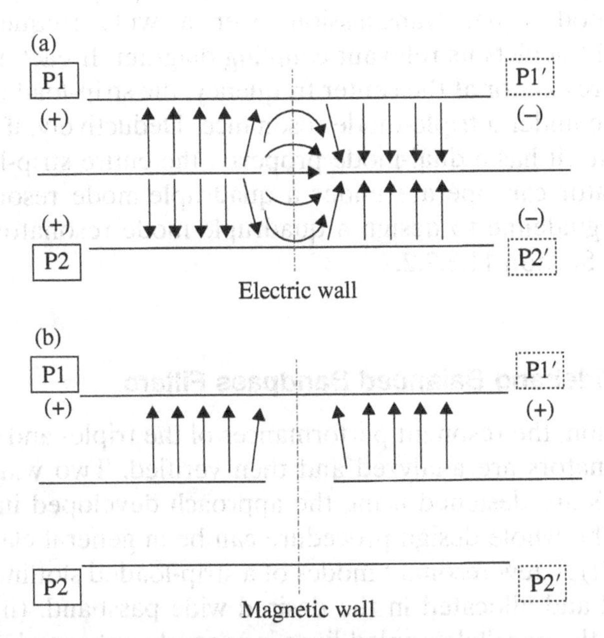

Figure 11.12 Cross-view of electrical fields in the two-layer structure under distinct feeding schemes. (a) DM feeding. (b) CM feeding. Reprinted with permission from Ref. [12]; copyright 2016 IEEE.

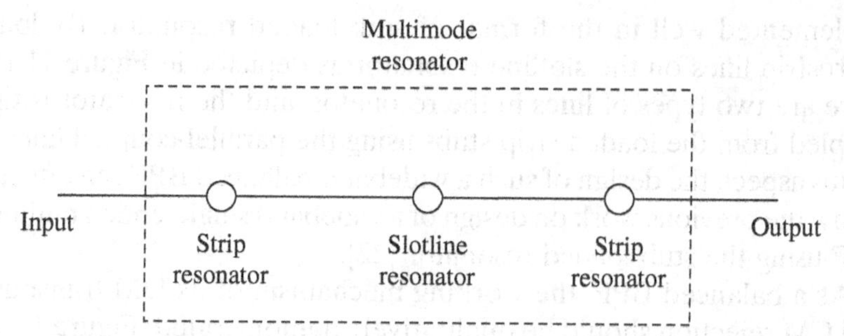

Figure 11.13 Coupling diagram of the wideband balanced BPF on the strip-loaded slotline resonator. Reprinted with permission from Ref. [12]; copyright 2016 IEEE.

From the other perspective, the wideband balanced BPF on the strip-loaded slotline resonator can be interpreted by a coupling diagram, as displayed in Figure 11.13. In Figure 11.13, each node represents either strip resonator or slotline resonator, and each solid line segment indicates the coupling between two adjacent resonators. Under CM operation, the coupling between the strip resonator and slotline resonator is tiny; thus CM signals are hardly transmitted. Under DM operation, all the coupling paths are excited with strong and proper degree so as to achieve good signal transmission over a wide frequency range. Figure 11.13 depicts its relevant coupling diagram. If each node represents a $\lambda/2$ resonator at the center frequency, the strip-loaded resonator can operate under a triple-mode resonance. Deductively, if the slotline resonator itself has a dual-mode property, the entire strip-loaded slotline resonator can operate under a quadruple-mode resonance. This gives us a guideline to design a quadruple-mode resonator as will be detailed in Section 11.3.2.2.

11.3.2 Wideband Balanced Bandpass Filters

In this section, the resonant performances of the triple- and quadruple-mode resonators are analyzed and then verified. Two wideband balanced BPFs are designed using the approach developed in Refs. [22] and [28]. The whole design procedure can be in general classified into two steps: (i) A few resonant modes of a strip-loaded slotline resonator are excited and allocated in the desired wide passband. (ii) Coupling strength in the parallel-coupled lines is properly enhanced, and its frequency dispersion needs to be regulated by placing its coupling peak near the center of the desired DM passband.

11.3.2.1 *Wideband Balanced BPF on Strip-Loaded Triple-Mode Slotline Resonator* The strip-loaded triple-mode slotline resonator is depicted in Figure 11.11(b). Its strips are centrally placed above and beneath the slotline. Figure 11.14 shows its equivalent circuit model under DM operation. In Figure 11.14, all the involved transmission lines are represented by their characteristic impedances and electrical lengths. For clear demonstration, each portion in Figure 11.14(b) uses the same line style or format as that in Figure 11.11. The portion with $2 \times \theta_S = 180°$ at the first resonant frequency (normalized to 1) represents the $\lambda/2$ slotline resonator. The phase θ_M stands for the electrical length of the loaded strip at the unity or normalized frequency. Initially, all the impedance values are set as 50 Ω. Under DM operation, this resonator can be referred to as a typical stub-loaded resonator. When the strip stubs are loaded on slotline and the strip-loaded slotline resonator is excited from the strips, all the even-order modes in this slotline resonator can be hardly excited at all. With proper extension of loaded stubs, the odd-order modes in the slotline resonator are gradually moved down in frequency, while an extra strip-induced resonance is accompanied. By properly locating these multiple resonant modes, a wide DM passband can be formed up as well studied in Refs. [22] and [28].

Figure 11.15(a) depicts the variation in resonant frequencies of this resonator with $\theta_M = 0°$, 15°, and 90°. As for the unloaded slotline resonator, it turns out to be a virtual-short-circuit condition at the middle position for its even-order modes, according to the electric field distribution in the slotline in Section 11.2. When the stubs are loaded centrally to launch the resonator, the even-order modes cannot be excited at all. But the odd-order modes with virtual-open-circuit condition at the middle position tend to be certainly affected, as these stubs are introduced. The resonator under the operation of odd-order modes can be viewed as

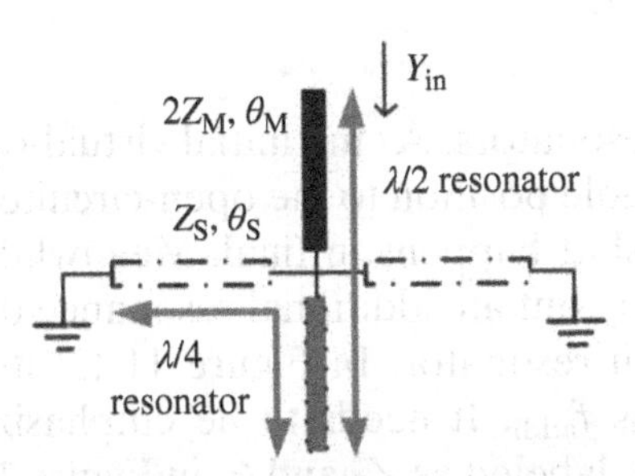

Figure 11.14 Equivalent circuit model of the triple-mode strip-loaded slotline resonator under DM operation. Reprinted with permission from Ref. [12]; copyright 2016 IEEE.

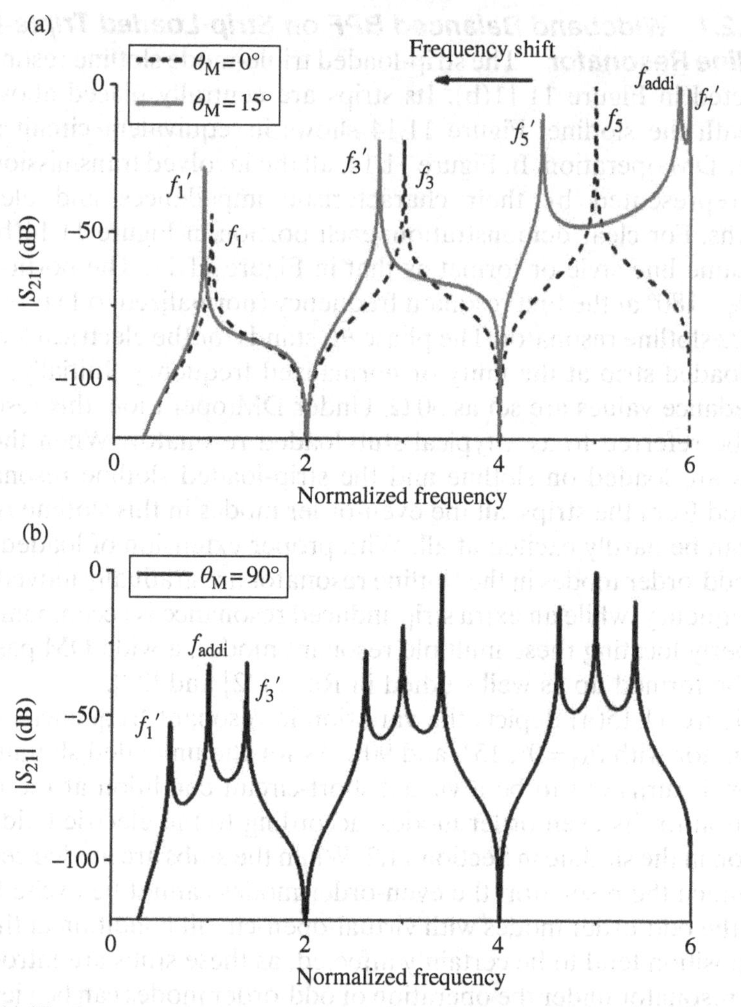

Figure 11.15 Resonances of the triple-mode resonator with different electrical lengths of loaded stubs. (a) $\theta_M = 0°$ and 15°. (b) $\theta_M = 90°$. Reprinted with permission from Ref. [12]; copyright 2016 IEEE.

two individual $\lambda/4$ resonators. As the initial virtual-open-circuited end is moved from the middle position to the open-circuited end of the loaded stubs, a frequency shift happens in final. Meanwhile, the loaded stubs themselves may bring out an additional resonance due to their function of a half-wavelength resonator. In Figure 11.15, its first resonant frequency is labeled as f_{addi}. It needs to be emphasized herein that the short-ended slotline, labeled as Z_S and θ_S in Figure 11.14, will definitely bring out a few transmission zeros at the frequencies where the electrical length from the short-circuited ends to the strips is equal to $\theta_S = n \times 180°$

($n = 1, 2, 3, \ldots$). At these frequencies, the short-circuited ends can cause the emergence of virtual-short-circuit condition at the center of the resonator, thus generating a number of transmission zeros. In Figure 11.15, these zeros usually appear at the normalized frequencies of 2, 4, 6, ...

When θ_M is lengthened to 90°, three resonant modes, that is, two frequency-shifted odd-order modes and one strip-induced mode, symmetrically appear in the desired passband at the normalized center frequency. This phenomenon can be verified by the coupling diagram in Figure 11.13: a triple-mode resonator can be constituted when the two loaded strips both become $\lambda/2$ resonators ($2 \times \theta_M = 180°$) and cooperate with the $\lambda/2$ slotline resonator. Additionally, with the electrical length θ_M of 90° at the center frequency, these stubs can be tightly coupled with the external microstrip feeding lines through the $\lambda/4$ parallel-coupled lines. Figure 11.15(b) shows the resonances of the strip-loaded slotline resonator with $\theta_M = 90°$, which proves the previous analysis.

To further verify the triple-mode property, the resonant frequencies of the circuit model in Figure 11.14 are calculated with $\theta_M = 90°$. Herein, all the electrical lengths, that is, θ_M and θ_S, are set to be the same, so they are both labeled as θ. As shown in Figure 11.14, the input resistance of the resonator, Y_{in}, is derived as

$$Y_{in} = j\frac{\tan\theta - (2Z_M/Z_S\tan\theta)}{Z_M(1 + (4Z_M/Z_S) - \tan^2\theta)} \tag{11.15}$$

When θ moves to 90°, Y_{in} gradually approaches 0, thus producing a resonance at the center frequency where $\theta = 90°$. Under $Y_{in} = 0$, two more resonant frequencies can be generated if the electrical length θ satisfies

$$\theta = \arctan\left(\sqrt{\frac{2Z_M}{Z_S}}\right) \tag{11.16a}$$

$$\theta = \pi - \arctan\left(\sqrt{\frac{2Z_M}{Z_S}}\right) \tag{11.16b}$$

In practical design, the circuit model is not strictly accurate because of the existing nonideal voltage transformers at the microstrip–slotline coupling position, as discussed in Section 11.2. Moreover, the design of a multimode BPF is usually carried out based on the procedure of mode allocation and coupling enhancement. Based on the previous discussion, Figure 11.16 is plotted to indicate the variation of resonant frequencies with respect to W_S under fixed $W_M = 0.10$ mm as derived in full-wave simulation. Herein, W_S and W_M are the physical widths of the slotline and microstrip-line stubs corresponding with the portions labeled by Z_S and Z_M in the circuit model in Figure 11.14, respectively. $W_M = 0.10$ mm is readily chosen to realize the tight coupling under the restriction of accuracy in etching fabrication process. It can be seen that the modes are further separated as W_S becomes larger. When

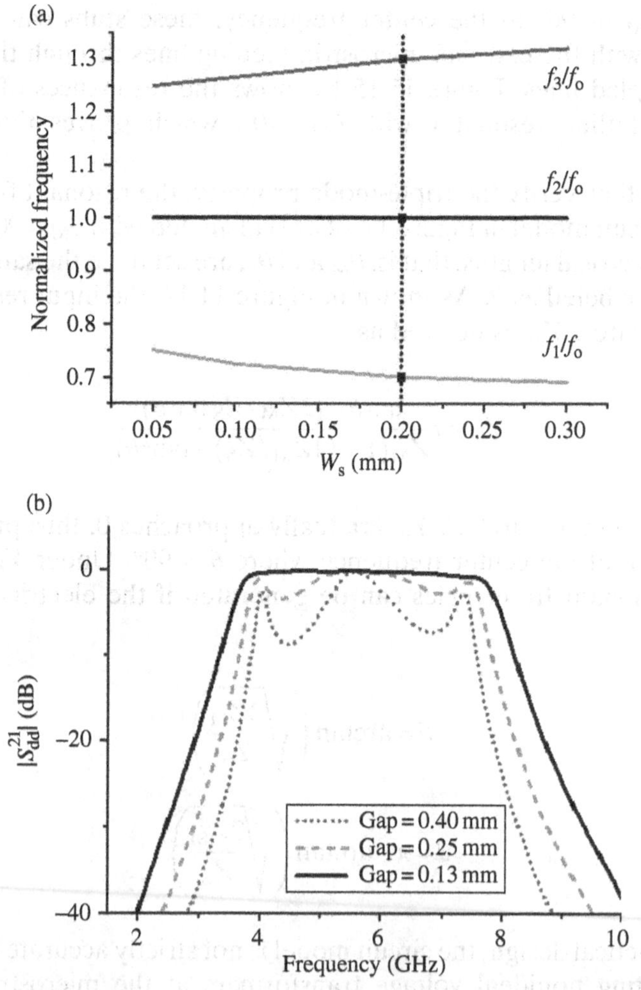

Figure 11.16 (a) Variation of resonant frequencies in the triple-mode resonator with respect to W_S when $W_M = 0.10$ mm. (b) Improved in-band insertion loss of the triple-mode filter with decreased coupling gaps. Reprinted with permission from Ref. [12]; copyright 2016 IEEE.

$W_S = 0.20$ mm, the mode distribution is consistent with what we desire as marked in the figure for design of a quasi-Chebyshev BPF with an FBW of 70.0% at 5.6 GHz. Herein, the theoretical values for these modes are calculated [29]. For a kth-order Chebyshev BPF with an FBW, Δ, at center frequency f_o, the pole frequency f_n can be calculated as

$$\frac{f_n}{f_o} = 1 + x_n \times \frac{\Delta}{2} \tag{11.17a}$$

$$x_n = \cos\left(\frac{2k+1-2n}{2k}\pi\right), \quad n = 1,2,3,\ldots,k \tag{11.17b}$$

In this design, the normalized theoretical pole frequencies are calculated $f_1 = 0.697, f_2 = 1, f_3 = 1.303$; and the designed resonator achieves three normalized resonant peaks at $f_1 = 0.705$, $f_2 = 1$, $f_3 = 1.298$. Afterward, the parallel-coupled lines are introduced at two sides with proper frequency-dispersive and tightened coupling strength. Figure 11.16(b) displays the improved in-band insertion loss of the filter with decreased coupling gaps. The in-band insertion loss of the filter is gradually flattened as the gaps are decreased. When the gap between two coupled lines is 0.13 mm, the required in-band flatness can be satisfactorily obtained in final.

The previously designed wideband balanced BPF is fabricated on a 2-layer substrate, that is, Rogers RT/duroid 6010LM with a permittivity of 10.7 and thickness of 0.635 mm. Figure 11.17(a) shows its top view layout with all the critical dimensions labeled, whereas Figure 11.17(b) gives the top/bottom view photographs of the fabricated balanced BPF with four SMA connectors installed. The simulated and measured frequency responses are both plotted in Figure 11.17(c) for comparison. The simulated results are derived by the commercial full-wave simulator, Ansoft HFSS, and the measured ones are obtained by using the 4-port VNA Agilent N5224A.

The two sets of results are found in good agreement with each other over a wide frequency range up to 15 GHz. On the one hand, the measured 3-dB DM passband covers a wide range of 3.58–7.90 GHz. Over this passband, the measured insertion loss is lower than 0.77 dB and the RL is higher than 12.4 dB. Comparatively, they are reasonably close to 0.2 and 16.9 dB in simulation. Moreover, this DM BPF has a widened upper stopband with attenuation higher than 20 dB in a frequency range up to 14.5 GHz. On the other hand, the measured CM attenuation is better than 38.5 dB in the entire DM passband and better than 20 dB over the plotted frequency range of 0–14.9 GHz. It has been confirmed that the CM signals in the proposed filter have been highly suppressed. The FBW of the realizable triple-mode filter is in a range of 67–87% with

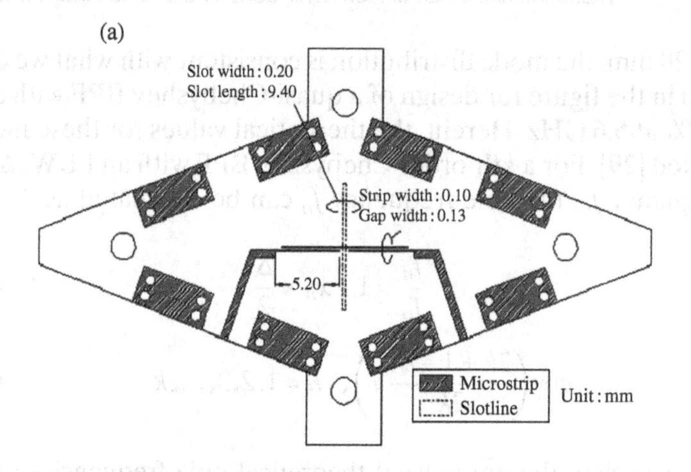

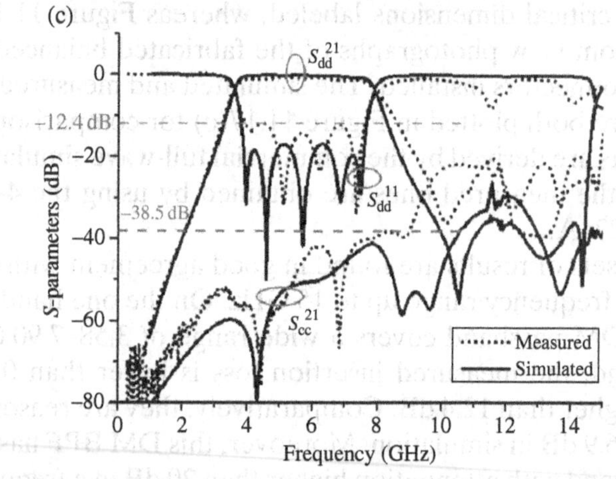

Figure 11.17 Physical layout and *S*-parameters of the proposed triple-mode filter. (a) Top view with all the dimensions labeled. (b) Top/bottom view photographs of the assembled filter. (c) Simulated and measured results. Reprinted with permission from Ref. [12]; copyright 2016 IEEE.

several practical restrictions considered, such as the limited coupling strength or line width.

11.3.2.2 Wideband Balanced BPF on Strip-Loaded Quadruple-Mode Slotline Resonator

In order to excite and use more resonant modes in realization of a balanced BPF with widened BW and enhanced roll-off rate, a strip-loaded quadruple-mode slotline resonator is presented, and its geometrical structure is depicted in Figure 11.18(a).

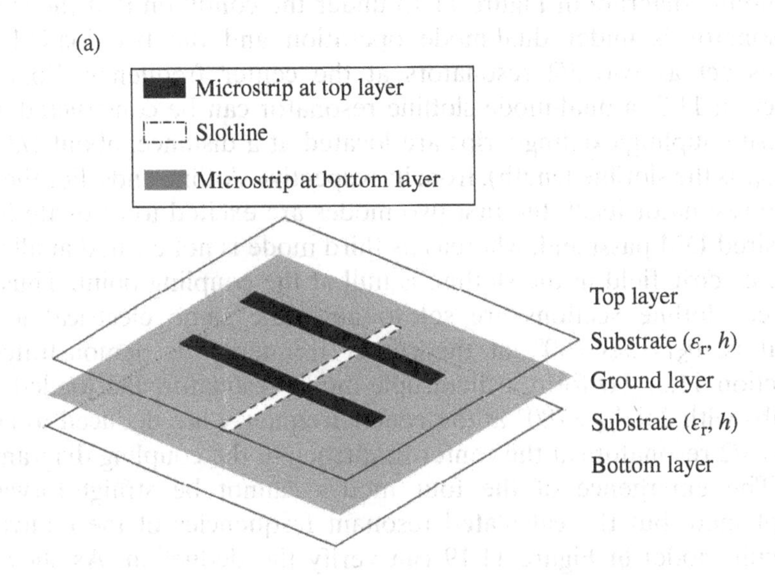

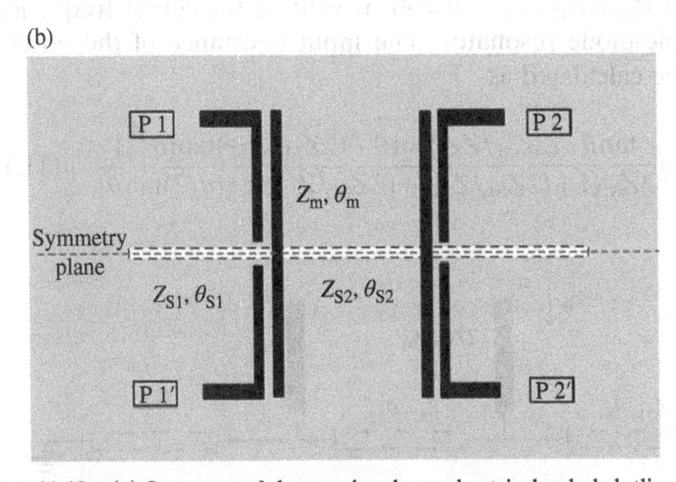

Figure 11.18 (a) Structure of the quadruple-mode strip-loaded slotline resonator. (b) Top view of wideband balanced bandpass filter on the resonator in (a). Reprinted with permission from Ref. [12]; copyright 2016 IEEE.

A pair of identical strip stubs is symmetrically placed along the slotline resonator with an offset distance from the middle position. By introducing the parallel-coupled lines at its two sides, a wideband balanced BPF can be formulated, and its geometry is illustrated in Figure 11.18(b). The equivalent circuit model of the quadruple-mode resonator is presented in Figure 11.19 with all the transmission lines represented by their characteristic impedances and electrical lengths. Each portion in Figure 11.19 uses the same line style or format as that in Figure 11.18(b).

This quadruple-mode resonator is derived from the deduction of the coupling diagram in Figure 11.13 under the condition that the slotline resonator is under dual-mode operation and the two loaded strip lines act as two $\lambda/2$ resonators at the center frequency. Following Section 11.2, a dual-mode slotline resonator can be constructed when these coupling/exciting strips are located at a distance, about 1/3 L_{slot} (L_{slot} is the slotline length), from its respective slotline ends. For the slotline resonator itself, the first two modes are excited to resonate in the desired DM passband, whereas its third mode is not excited at all since the electric field in the slotline is null at the coupling point. Thus, the three slotline sections are set to have the same electrical length, that is, $\theta_{S1} = \theta_{S2} = 90°$, at the center frequency as demonstrated in Section 11.2. To form a quadruple-mode resonator, the loaded strip stubs with $2 \times \theta_M = 180°$ at the center frequency are deduced to act as two $\lambda/2$ resonators at the center frequency in the coupling diagram.

The emergence of the four modes cannot be straightforwardly explained, but the calculated resonant frequencies of the equivalent circuit model in Figure 11.19 can verify the deduction. As shown in Figure 11.19, $\theta_{S1} = \theta_{S2} = \theta_M = \theta = 90°$ is valid at the center frequency of the quadruple-mode resonator. The input resistance of the odd/even modes can be calculated as

$$Y_{\text{in\_even}} = j\frac{\tan\theta - (2Z_M/Z_{S1})\cot\theta + (2Z_M/Z_{S2})\tan(\theta/2)}{2Z_M(1 + (2Z_M/Z_{S1}) - (2Z_M/Z_{S2})\tan(\theta/2)\tan\theta)} \tag{11.18a}$$

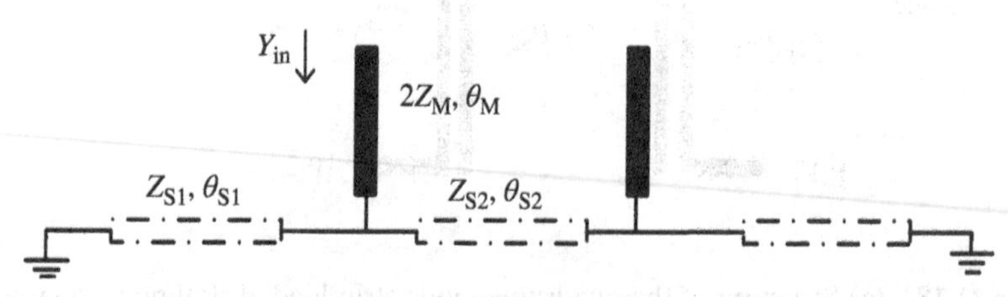

Figure 11.19 Equivalent circuit model of the quadruple-mode resonator under DM operation. Reprinted with permission from Ref. [12]; copyright 2016 IEEE.

$$Y_{in\_odd} = j\frac{\tan\theta - (2Z_M/Z_{S1})\cot\theta - (2Z_M/Z_{S2})\cot(\theta/2)}{2Z_M(1 + (2Z_M/Z_{S1}) + (2Z_M/Z_{S2})\cot(\theta/2)\tan\theta)} \quad (11.18b)$$

At the resonant frequencies, the input resistance of each mode is definitely equal to zero, that is, $Y_{in\_even} = 0$ for even-order modes and $Y_{in\_odd}=0$ for odd-order modes. Four resonances with electrical length θ are calculated from (11.19a) and (11.19b). It is obvious that the roots satisfying (11.19a) and (11.19b) are symmetric with respect to $\theta = 90°$, thereby indicating that the four resonances symmetrically occur with respect to the center frequency:

$$\left(1 + 2\frac{Z_{S1}}{Z_{S2}}\right)\tan^4\frac{\theta}{2} - 2\left(\frac{Z_{S1}}{Z_M} + \frac{Z_{S1}}{Z_{S2}} + 1\right)\tan^2\frac{\theta}{2} + 1 = 0 \quad (11.19a)$$

$$\left(1 + 2\frac{Z_{S1}}{Z_{S2}}\right)\cot^4\frac{\theta}{2} - 2\left(\frac{Z_{S1}}{Z_M} + \frac{Z_{S1}}{Z_{S2}} + 1\right)\cot^2\frac{\theta}{2} + 1 = 0 \quad (11.19b)$$

In practical design, the procedure based on the widely used mode allocation and coupling enhancement is implemented by full-wave simulation. Figure 11.20 shows the variation of resonant modes with varied W_{S1} when $W_{S2} = 0.70$ mm or varied W_{S2} when $W_{S1} = 0.15$ mm. Herein, W_{S1} and W_{S2} are the physical widths of the slotline portions labeled by Z_{S1} and Z_{S2} in Figure 11.19, respectively. It is found that W_{S1} has little influence on f_2 and f_3, so the mode allocation can be implemented as follows: (i) Set the width of loaded strips as 0.10 mm to obtain the tight coupling easily. (ii) Choose the proper value of W_{S2} to match the theoretical values of f_2 and f_3. (iii) Determine the proper value of W_{S1} to match the theoretical values of f_1 and f_4 with the determined W_{S2}. The theoretical positions of these resonant modes can be also calculated from (11.17). In this situation, $W_{S1} = 0.15$ mm and $W_{S2} = 0.70$ mm are finally decided for a quadruple-mode (fourth-order Chebyshev) filter with FBW of 80.0% at the center frequency of 5.8 GHz. The theoretical values of normalized pole frequencies are $f_1 = 0.630$, $f_2 = 0.847$, $f_3 = 1.153$, $f_4 = 1.370$, while the simulated normalized resonant peaks of the designed resonators are $f_1 = 0.634$, $f_2 = 0.846$, $f_3 = 1.156$, $f_4 = 1.366$.

In final, the designed resonator is coupled in parallel with two external $\lambda/4$ lines for achieving the dispersive coupling strength with the peak at the central frequency. Herein, the wideband balanced BPF on the quadruple-mode stub-loaded resonator can be realized with a gap of 0.10 mm.

Figure 11.21(a) and (b) depicts the physical layout of the designed quadruple-mode balanced BPF with all the dimensions labeled and

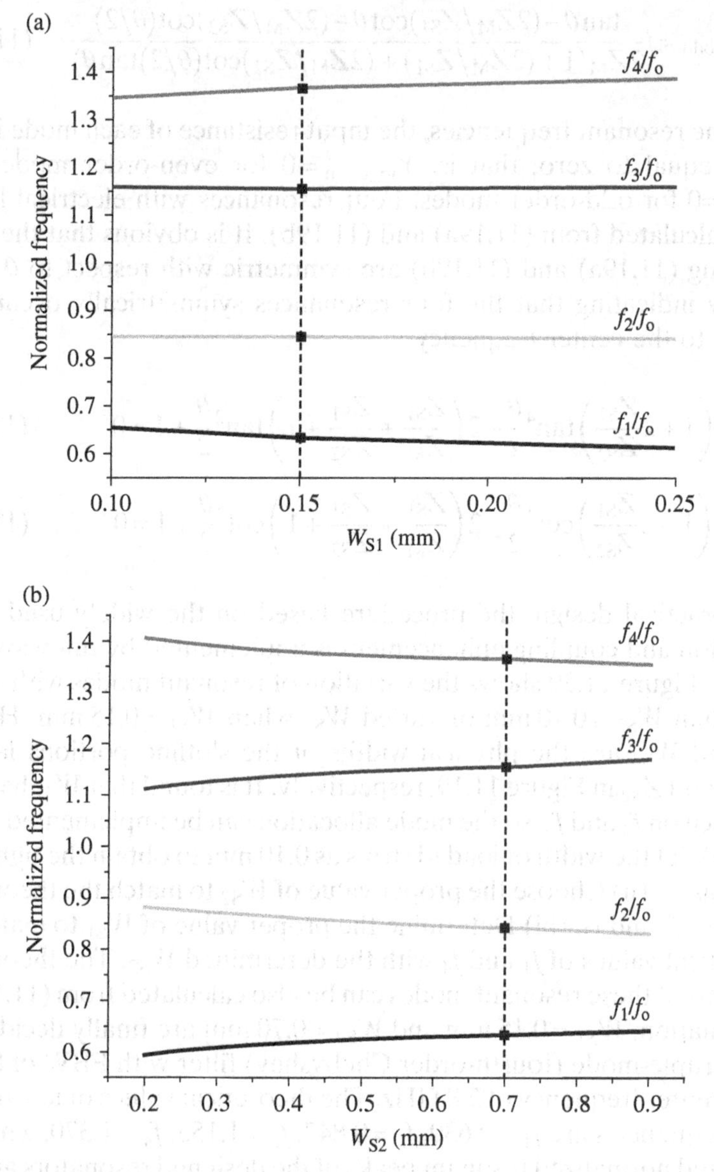

Figure 11.20 Variation of resonant frequencies of the quadruple-mode resonator. (a) Varied W_{S1} when $W_{S2} = 0.70$ mm. (b) Varied W_{S2} when $W_{S1} = 0.15$ mm.

the top/bottom view photographs of the fabricated filter, respectively. Figure 11.21(c) indicates the simulated and measured *S*-parameters. The simulated results achieve a 3-dB passband of 3.42–8.30 GHz in DM, within which the CM rejection is better than 27.4 dB. Meanwhile, the measured results reach a DM 3-dB passband of 3.45–8.45 GHz with

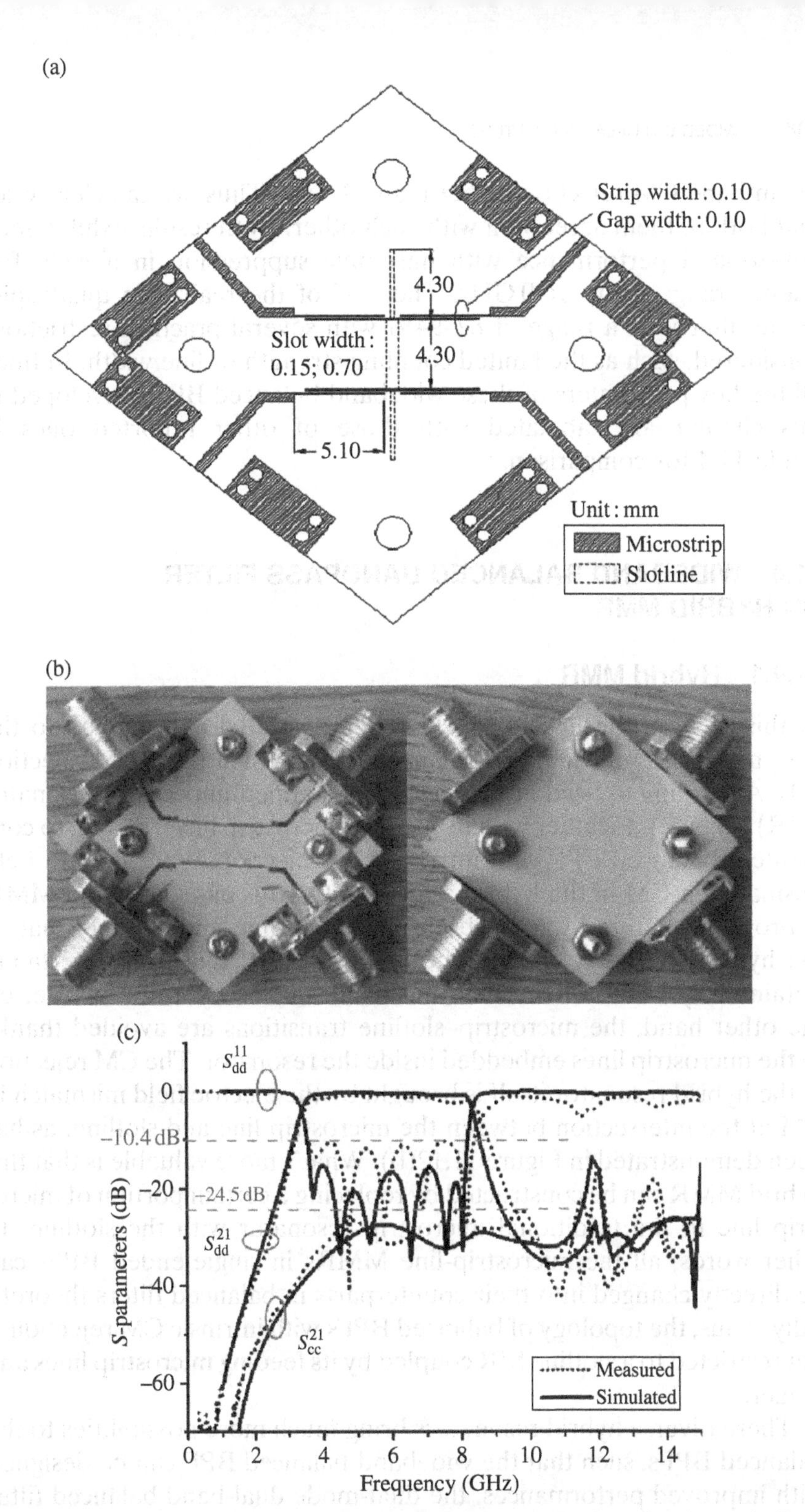

Figure 11.21 Physical layout and *S*-parameters of the proposed quadruple-mode filter. (a) Top view with all the dimensions labeled. (b) Top/bottom view photographs of the assembled filter. (c) Simulated and measured results. Reprinted with permission from Ref. [12]; copyright 2016 IEEE.

the in-band CM rejection higher than 24.5 dB. Thus, we can clearly see that both of them agree well with each other. Both results exhibit good out-of-band performance with harmonic suppression in a wide frequency range up to 16.0 GHz. The BW of the realizable quadruple-mode filter is in a range of 82–94% with several practical restrictions considered, such as the limited coupling strength or line width. In final, all the key parameters of these wideband balanced BPFs developed in this chapter are tabulated with those of other reported ones in Table 11.4 for comparison.

11.4 WIDEBAND BALANCED BANDPASS FILTER ON HYBRID MMR

11.4.1 Hybrid MMR

In this section, the hybrid MMR is demonstrated and applied to the design of a class of wideband balanced BPFs with intrinsic CM rejection [14]. According to Section 11.2, a slotline stepped-impedance resonator (SIR) can work together with its feeding microstrip lines, aiming to constitute a balanced BPF with intrinsic CM rejection by virtue of a field mismatch in CM in the hybrid structure. On this basis, a hybrid MMR is proposed herein by using both microstrip line and slotline, that is, the hybrid structure. On the one hand, the CM rejection is certainly obtained by virtue of the field mismatch in this hybrid structure; on the other hand, the microstrip–slotline transitions are avoided thanks to the microstrip lines embedded inside the resonator. The CM rejection in the hybrid resonator itself is brought by the electric field mismatch in CM at the intersection between the microstrip line and slotline, as has been demonstrated in Figure 11.12(b). What's more valuable is that this hybrid MMR can be constructed by replacing a certain portion of microstrip line in the traditional microstrip resonator with the slotline. In other words, all the microstrip-line MMRs in single-ended BPFs can be directly changed into their counterparts in balanced filters theoretically. Thus, the topology of balanced BPFs with intrinsic CM rejection is not restricted to a slotline SIR coupled by its feeding microstrip lines any longer.

These diverse hybrid resonators bring much more possibilities to the balanced BPFs, such that the wideband balanced BPF can be designed with improved performances, the dual-mode dual-band balanced filter can be constituted, and so on. Figure 11.22 shows the previously stated comparison between the traditional slotline resonator and the hybrid

Table 11.4 Comparison with Other Similar or Latest Balanced Wideband Filters

Filters		DM performance		CM performance			
		Center frequency (GHz)	Fractional bandwidth (%)	In-band rejection (dB)	Rejection bandwidth	Effective circuit size	Topology type
[26]		3.5	40	40	2.8–4.2 GHz (minimum 40 dB)	$0.25\lambda_0^2$	Cross-shaped resonator
[25]	Without cross coupling	6.85	70	14.5	0–19.0 GHz (minimum 14.5 dB)	$0.54\ \lambda_0 \times 0.52\ \lambda_0$	T-shaped resonator
	With cross coupling		70.7	13.5	0–19.5 GHz (minimum 13.5 dB)	$0.90\ \lambda_0 \times 0.26\ \lambda_0$	
Proposed filters	Triple-mode	5.74	75.2	38.5	0–14.9 GHz (minimum 20 dB)	$0.21\ \lambda_0 \times 0.18\ \lambda_0$	Strip-loaded slotline resonator with intrinsic CM rejection
	Quadruple-mode	5.95	84.0	24.5	0–11.8 GHz (minimum 20 dB)	$0.22\ \lambda_0 \times 0.26\ \lambda_0$	
[8]		2.60	107	25.0	0.5–5 GHz (minimum 20 dB)	$0.43\ \lambda_0 \times 0.43\ \lambda_0$	Slotline stepped-impedance resonator

(*Continued*)

Table 11.4 (Continued)

Filters		DM performance		CM performance			
		Center frequency (GHz)	Fractional bandwidth (%)	In-band rejection (dB)	Rejection bandwidth	Effective circuit size	Topology type
[10]	Dual-mode	4.50	120.0	25.8	0–18.5 GHz (minimum 22.7 dB)	$0.30\ \lambda_0 \times 0.18\ \lambda_0$	Slotline multimode resonator
	Triple-mode	5.45	123.2	21.6	0–10.6 GHz (minimum 21.6 dB)	$0.39\ \lambda_0 \times 0.19\ \lambda_0$	Slotline multimode resonator
[30]		Not mentioned	130	10	1–10.6 GHz (minimum 10 dB)	$0.50\ \lambda_0 \times 0.37\ \lambda_0$	Planar semi-lumped components
[31]	Single section	3.5	145	20	Not mentioned	Not mentioned	Wire-bonded multiconductor transmission lines
	Double section		97		Not mentioned	Not mentioned	
[32]		1.85	59.5	42	0.82–2.8 GHz (minimum 40 dB)	$0.60\ \lambda_0 \times 0.30\ \lambda_0$	Modified branch-line structure
[33]		1.62	66.6	14.5	0–4.8 GHz (minimum 14.5 dB)	$0.27\ \lambda_0 \times 0.27\ \lambda_0$	Coupled full-wavelength loop

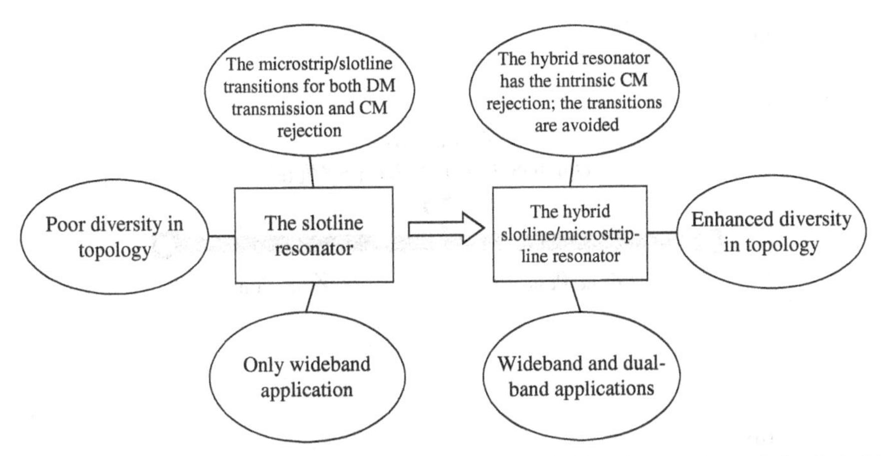

Figure 11.22 Comparison between the traditional slotline resonator and the hybrid microstrip-line/slotline resonators used in balanced BPFs with intrinsic CM rejection. Reprinted with permission from Ref. [14]; copyright 2016 IEEE.

microstrip-line/slotline resonator used in balanced BPFs with intrinsic CM rejection.

To verify the aforementioned hypothesis, a complete design of a wideband balanced BPF based on the hybrid MMR will be demonstrated herein. Initially, the single-ended counterpart of this hybrid MMR is chosen as the resonator in Ref. [34]. The conceptual model of this hybrid resonator can be directly derived from its single-ended counterpart. An improvement is followed so that the resonances can be reallocated to enhance the desired performances. In this context, the CM intrinsic rejection does not need to be taken into consideration at all.

Now, the conceiving procedure of this hybrid MMR is illustrated at first. Figure 11.23(a) shows the single-ended resonator that is a simplified sketch of the resonator in Ref. [34]. The SIR in Ref. [34] is simplified into the UIR in Figure 11.23(a) for easy analysis later on. This simplification does hardly affect its resonant property qualitatively. In its derived hybrid MMR as shown in Figure 11.23(b), the format of this loaded short-ended stub is transferred from the microstrip line into the slotline. Since the short-ended SIR remains in the format of microstrip line, a hybrid resonator can be preliminarily achieved. Figure 11.23 (b) represents the conceptual sketch of this hybrid MMR, and it is convenient to analyze the resonant property qualitatively.

Then, the resonant property of this hybrid MMR can be investigated by using Figure 11.23(b) to decide its basic configuration. An improvement is made from Figure 11.23(a) to Figure 11.23(b) so that the

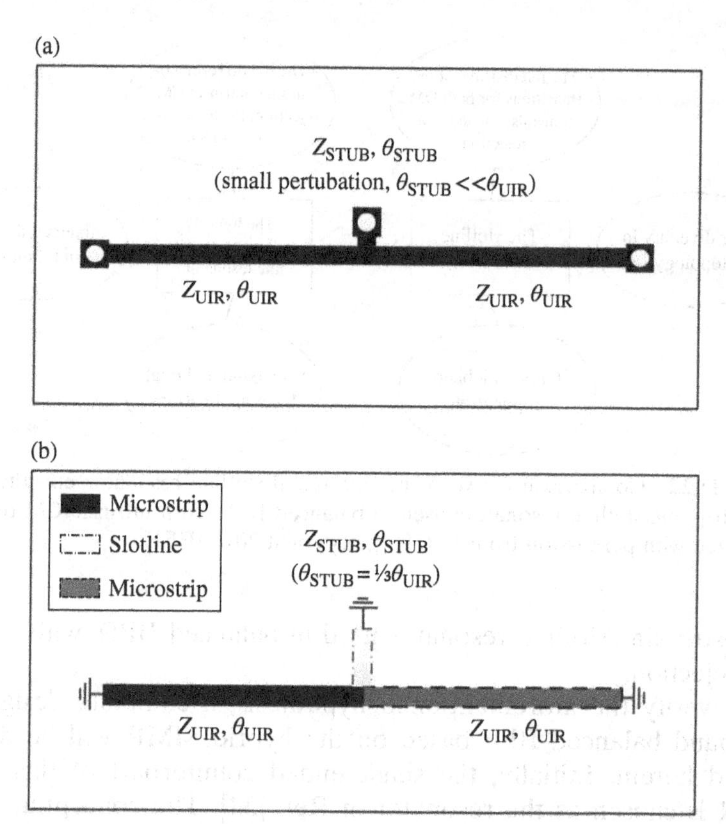

Figure 11.23 Conceptual circuit models of two distinct resonators. (a) Single-ended resonator in Ref. [34]. (b) Proposed hybrid resonator improved from (a). Reprinted with permission from Ref. [14]; copyright 2016 IEEE.

resonances can be reallocated to improve the desired performances. Herein, a comparative research is conducted to clearly demonstrate the enhancement. To simplify the analysis, the characteristic impedance of the loaded stubs, Z_{STUB}, is set as one half of the impedance of the UIR, Z_{STUB}.

Figure 11.24(a) and (b) is prepared to provide the comparative study on the resonances in Figure 11.23(a) and the resonances in its original unloaded resonator. The stub used in Ref. [34] is short in length, occupying less than 10% of the entire UIR. All the original odd-order modes in this unloaded UIR, that is, $f_1, f_3, f_5 \ldots$, are moved upward, because the loaded stub alters the boundary condition at the middle position. The unmoved even-order modes, that is, $f_2, f_4, f_6 \ldots$, are referred to as the odd modes, and the moved odd-order modes are referred to as the even modes, according to their field distributions along this symmetric

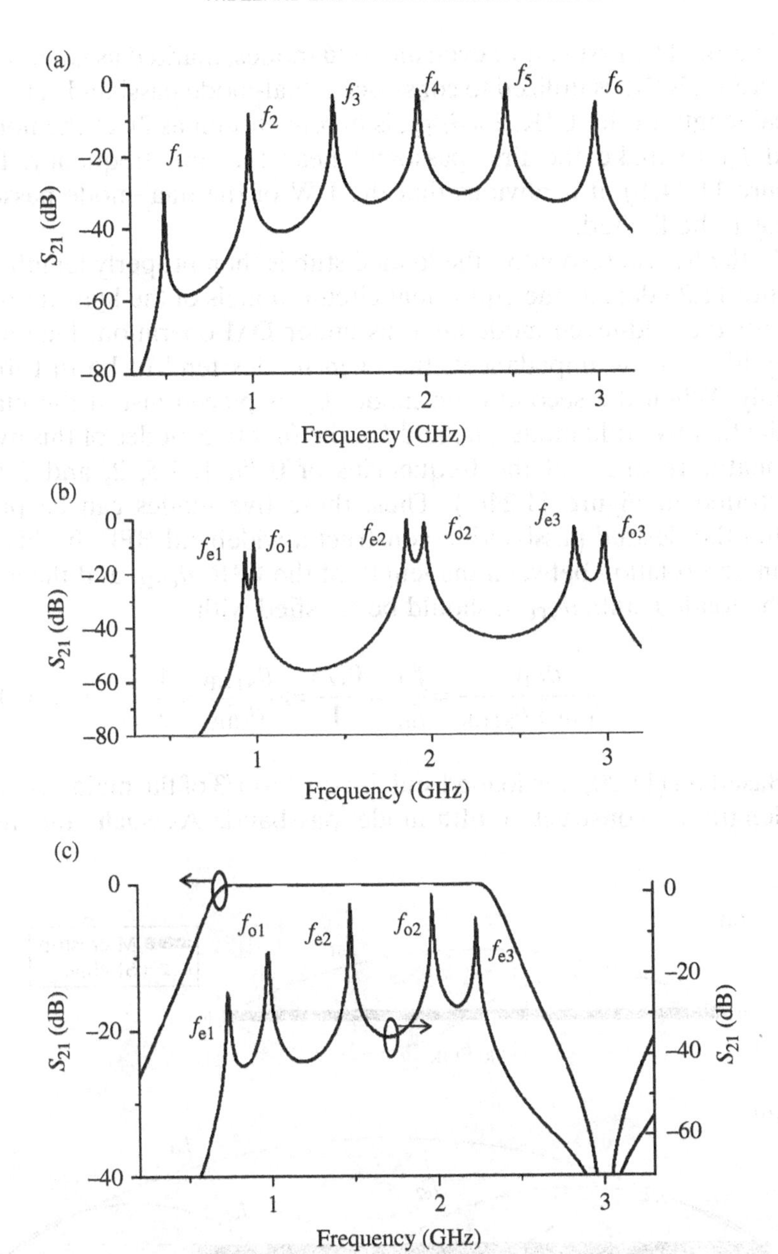

Figure 11.24 Graphical expression of resonant properties of different resonators and expected wideband frequency responses: (a) unloaded UIR, (b) short stub-loaded UIR [35], (c) proposed hybrid resonator with expected wideband DM frequency response. Reprinted with permission from Ref. [14]; copyright 2016 IEEE.

resonator. The first pair of even and odd modes, marked as f_{e1} and f_{o1} in Figure 11.24(b), is utilized to construct a dual-mode passband. The electrical length of the UIR, $2 \times \theta_{UIR}$, is herein chosen as 2π at the normalized f_{o1} to make the first passband near the unit frequency. From Figure 11.24(b), it is obvious that the BW of the dual-mode passband tends to be limited.

In the hybrid resonator, the loaded stub is then properly lengthened. Figure 11.25 depicts the equivalent circuit models of the hybrid resonator for the odd-/even-mode analysis under DM operation. Due to the simplification on impedances, the even modes tend to be distributed evenly. When the second even mode, f_{e2}, is placed just at the middle of the first two odd modes, f_{o1} and f_{o2}, the first five modes of this hybrid resonator resonate at the frequencies of 0.75, 1, 1.5, 2, and 2.25 as illustrated in Figure 11.24(c). Thus, these five modes can be placed within the desired passband to construct a wideband BPF. In this situation, the relation between the length of the UIR, θ_{UIR}, and the length of the loaded stub, θ_{STUB}, should be satisfied with

$$\frac{\theta_{UIR}}{\theta_{UIR}+\theta_{STUB}}=\frac{f_{e1}}{f_{o1}}=\frac{0.75}{1}\Rightarrow\frac{\theta_{STUB}}{\theta_{UIR}}=\frac{1}{3} \tag{11.20}$$

Based on (11.20), the loaded stub is equal to 1/3 of the main resonator in length to construct a fifth-mode passband. As such, the initial

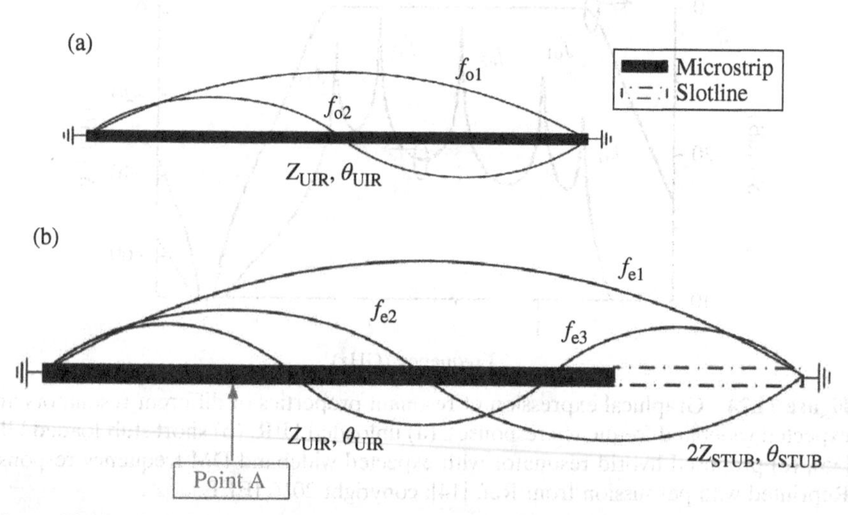

Figure 11.25 DM equivalent circuit models of the hybrid resonator with electric field distribution with respect to its multiple resonant modes. (a) Odd mode. (b) Even mode. Reprinted with permission from Ref. [14]; copyright 2016 IEEE.

configuration of the hybrid resonator is determined to contain a main microstrip-line UIR with the loading of a slotline stub whose length is equal to 1/3 length of the main resonator.

Next, the hybrid resonator with its sketch in Figure 11.23(b) should be converted into its physical layout for development of a wideband balanced BPF. On the whole, the practical hybrid resonator and paired differential ports, Pn and Pn' (n stands for the port number), are symmetric with respect to the slotline so as to achieve the intrinsic CM rejection. But, once one pair of ports (P1 and P1′) is placed at two sides of the microstrip line symmetrically in Figure 11.23(b), there is insufficient space for implementation of the other pair of ports (P2 and P2′). In order to fit the four-port framework, an additional microstrip line is loaded at the bottom layer, in parallel with that on the top layer, as shown in Figure 11.26. In Figure 11.26, points A and A′ stand for the paired ports' positions in the wideband balanced BPF, as will be explained in Section 11.4.2. For better understanding, the dot lines are plotted on the ground plane to reflect the relative location between the slotline and microstrip lines on the top and bottom layers. Interestingly, we can find that the inconvenient and inevitable via holes in the single-ended resonator are all avoided in the balanced one. The short-circuited

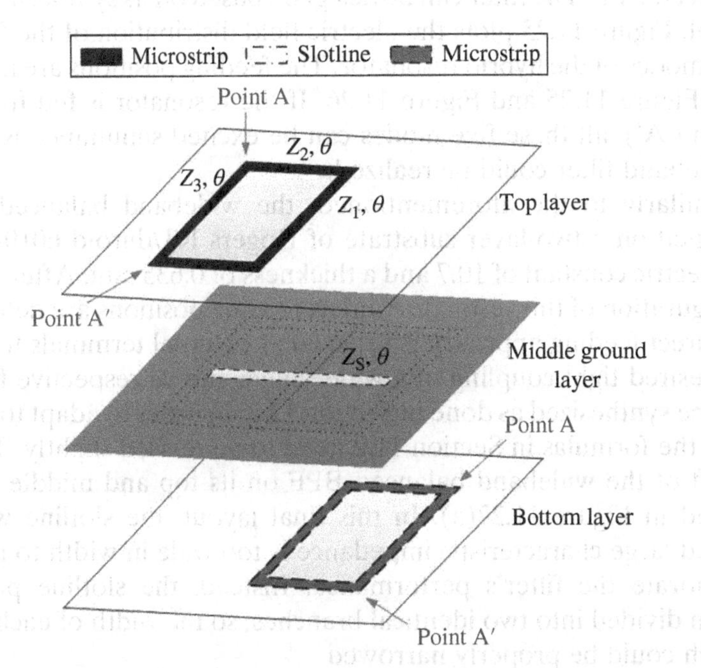

Figure 11.26 Three-dimensional view of the hybrid resonator on microstrip line and slotline. Reprinted with permission from Ref. [14]; copyright 2016 IEEE.

ends in the single-ended UIR in Figure 11.23(a) are craftily implemented in its balanced counterpart. They are allocated at the symmetrical plane by making good use of the virtual electric wall under DM operation. Meanwhile, the other short-circuited end in the loaded stub is naturally completed by the slotline itself.

As a result of a comparison between the layout under DM operation in Figure 11.26 and the sketch in Figure 11.23(b), the microstrip line on the top layer is equivalent to the left part of the hybrid resonator in Figure 11.23(b), whereas the right part is placed on the bottom layer. The slotline section on the middle ground plane functions as a loaded stub. The resultant resonator is not a UIR anymore so as to achieve the good in-band performance. Its configuration further gets certain modifications by using different characteristic impedances of all the involved sections in Figure 11.26. The impedances, Z_1, Z_2, Z_3, and Z_S, are only used to indicate different values rather than those used in the following paragraphs.

11.4.2 Wideband Balanced Bandpass Filters

In this section, the hybrid resonator is applied to construct a wideband balanced BPF. The filter can be designed based on its synthesized circuit model. Figure 11.25 plots the electric field distribution of the five resonant modes of the hybrid resonator. The feeding positions are marked in both Figure 11.25 and Figure 11.26. If the resonator is fed from point A (and A′), all these five modes can be excited simultaneously. Thus, a wideband filter could be realized.

Similarly to the aforementioned, the wideband balanced BPF is designed on a two-layer substrate of Rogers RT/duroid 6010LM with a dielectric constant of 10.7 and a thickness of 0.635 mm. After the basic configuration of this resonator and its feeding positions are determined, the direct feeding approach is selected at external terminals to achieve the desired tight coupling in a wide range, and its respective filter can then be synthesized as done in Section 11.2. In order to adapt to the case here, the formulas in Section 11.2 need to be revised slightly. The final layout of the wideband balanced BPF on its top and middle layers is plotted in Figure 11.27(a). In this final layout, the slotline with prescribed large characteristic impedance is too wide in width to probably deteriorate the filter's performance. Instead, the slotline portion is herein divided into two identical branches, so the width of each slotline branch could be properly narrowed.

During synthesizing the frequency response of the wideband balanced BPF, its DM circuit model with each portion represented by its

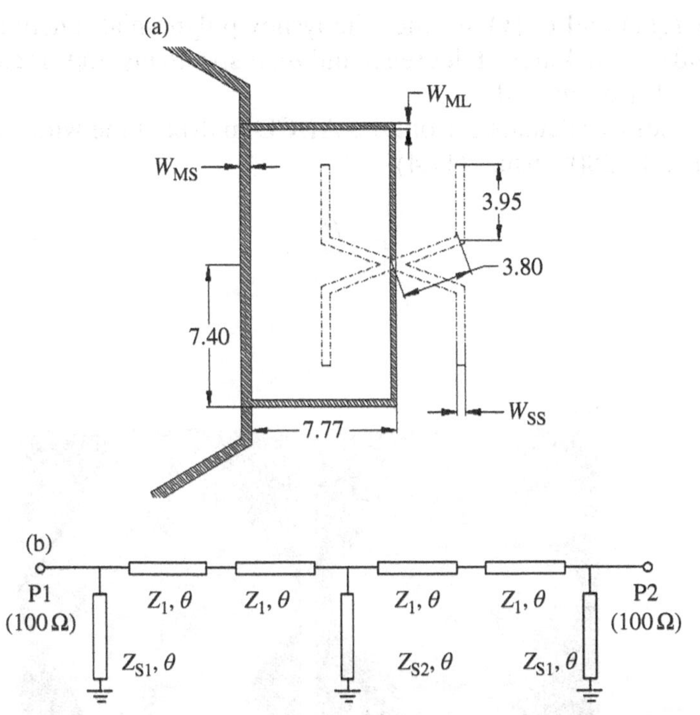

Figure 11.27 (a) Final layout of the wideband balanced BPF on its top and middle layers with all the key dimensions labeled (unit: mm). (b) DM equivalent circuit model of proposed wideband balanced BPF. Reprinted with permission from Ref. [14]; copyright 2016 IEEE.

equivalent impedance and electrical length is firstly derived as depicted in Figure 11.27(b). Ideally, the relations between characteristic impedances of practical components and equivalent impedances in the circuit model are readily set as $Z_{S1} = 2 \times Z_{MS}$, $Z_1 = 2 \times Z_{ML}$ and $Z_{S2} = Z_{SS}$. Z_{MS}, Z_{ML}, and Z_{SS} are the characteristic impedances of transmission lines with the same subscripts in Figure 11.27(a). Due to its equivalence with a typical optimum high-pass filter, the filtering characteristics of the network in Figure 11.27(b) can be in general described by a transfer (insertion) function:

$$|S_{21}|^2 = \frac{1}{1 + \varepsilon^2 F^2} \tag{11.21}$$

where ε is the specified equal-ripple constant in the passband and F is the filtering function as given in (11.7):

$$F = T_n(x)T_m(y) - U_n(x)U_m(y) \tag{11.22}$$

where $T_n(x)$ and $U_n(x)$ are the Chebyshev polynomial functions of the first and second kinds of degree n and m, respectively. In this case, $n = 4$ and $m = 1$ are selected.

Now, let us calculate F from the ABCD matrix of the whole network in Figure 11.28(b) using (11.4):

$$F = \frac{B - C}{2} \tag{11.23}$$

(a)

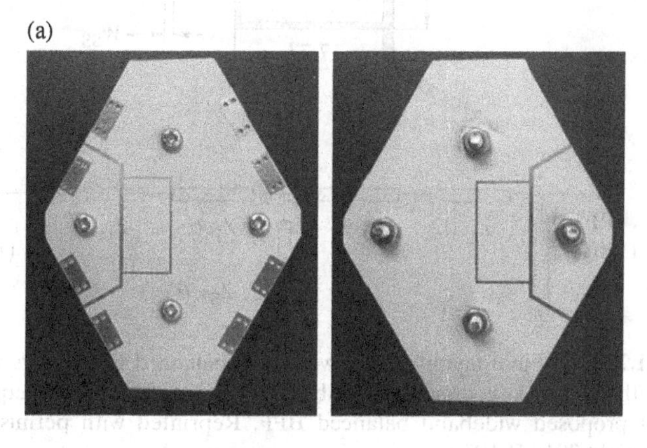

(b)

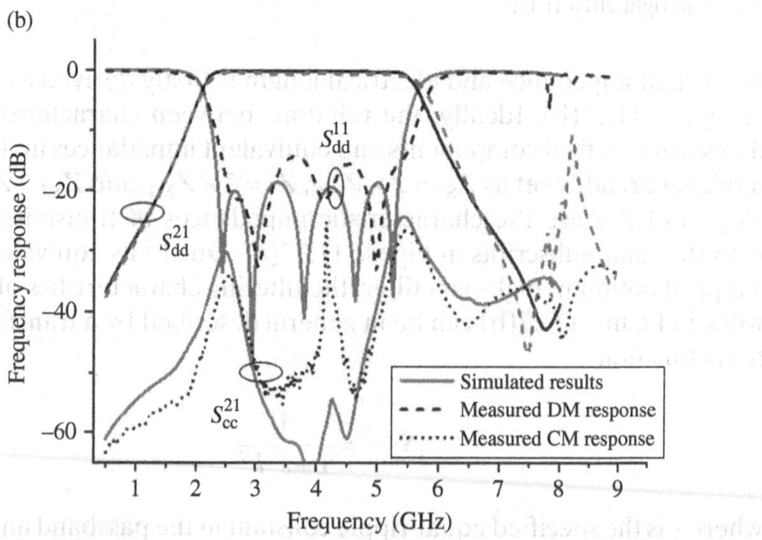

Figure 11.28 (a) Top/bottom view photographs of the fabricated wideband balanced BPF. (b) Simulated and measured S-parameters. Reprinted with permission from Ref. [14]; copyright 2016 IEEE.

Under the exact equalization of the two F in (11.22) and (11.23), all the three normalized impedances, z_{S1}, z_1, and z_{S2}, can be determined by the specified ripple constant and BW. As an example, the filter can be synthesized according to the prescribed specifications, that is, 0.043 dB ripple level and lower cutoff frequency of $\theta_c = 55°$ in the desired passband. The synthesized impedance values with reference to the port impedance of 100 Ω are equal to $Z_{S1} = 99.1\ \Omega$, $Z_1 = 128.3\ \Omega$, $Z_{S2} = 59.5\ \Omega$. The initial dimensions of this filter can be reasonably estimated by using the geometrical parameters of microstrip lines or slotline with prescribed characteristic impedances. In this wideband BPF, which is analogous to the high-pass prototype, all the electrical lengths are equal to 90° at the center frequency. The widths of microstrip lines are calculated as $W_{MS} = 0.57$ mm and $W_{ML} = 0.31$ mm. However, there are unexpectedly nonideal voltage transformers at the microstrip–slotline intersection portions, thus causing certain deviation in design if this equivalent circuit model is directly used. In order to compensate for this effect, the slotline's characteristic impedance needs to be selected slightly larger than $Z_{SS} = Z_{S2}$. After all the microstrip portions are determined by synthesized impedances, the slotline portion should be slightly widened, and it is finally chosen in full-wave simulation as $W_{SS} = 0.48$ mm with the length plotted in Figure 11.28(a).

Figure 11.28(a) gives the top/bottom view photographs of the fabricated wideband balanced filter. The simulated and measured S-parameters are both plotted in Figure 11.28(b) for comparison. The simulated results are derived by the commercial full-wave simulator Ansoft HFSS, and the measured ones are obtained by using the 4-port VNA Anritsu MS4624D. The filter is expected to operate at the central frequency of 4.0 GHz with a 3-dB FBW of 93.3% under the lower cutoff frequency selected in the synthesis design. The simulated and measured results are highly consistent with each other. The measured balanced filter has the insertion loss of 0.45–0.78 dB and RL higher than 14.2 dB within the entire DM passband from 2.09 to 6.09 GHz with an FBW of 97.8%. The CM signals in the desired DM passband remain satisfactorily suppressed to a level higher than 26.7 dB.

Compared with other reported slotline balanced BPFs, this wideband balanced filter has made effective use of multiple resonances. It is for the first time reported that a five-mode resonator can be used in the design of a balanced BPF with widened BW and good CM rejection. Moreover, the developed filter has better in-band flatness and sharper roll-off near the upper and lower cutoff frequencies. Additionally, its measured in-band RL is found to match well with the simulated one.

11.5 SUMMARY

This chapter has proposed several novel wideband balanced BPFs with intrinsic CM rejection. The hybrid microstrip–slotline structure has been used throughout the chapter in order to maintain the intrinsic CM rejection. On this basis, MMRs of various configurations have been exploited to design these wideband balanced BPFs, including slotline MMR, strip-loaded slotline resonator, and hybrid resonator. Working mechanisms and design methods of these wideband balanced filters have been intuitively explained, theoretically demonstrated and experimentally validated over a wide frequency range for both DM and CM frequency responses.

REFERENCES

1. T. B. Lim and L. Zhu, “A differential-mode wideband bandpass filter on microstrip line for UWB application,” *IEEE Microw. Wireless Compon. Lett.*, vol. **19**, no. 10, pp. 632–634, Oct. 2009.
2. T. B. Lim and L. Zhu, “Highly selective differential-mode wideband bandpass filter for UWB application,” *IEEE Microw. Wireless Compon. Lett.*, vol. **21**, no. 3, pp. 133–135, Mar. 2011.
3. X.-H. Wu and Q.-X. Chu, “Compact differential ultra-wideband bandpass filter with common-mode suppression,” *IEEE Microw. Wireless Compon. Lett.*, vol. **22**, no. 9, pp. 456–458, Sep. 2012.
4. X.-H. Wu, Q.-X. Chu, and L.-L. Qiu, “Differential wideband bandpass filter with high-selectivity and common-mode suppression,” *IEEE Microw. Wireless Compon. Lett.*, vol. **23**, no. 12, pp. 644–646, Dec. 2013.
5. X.-H. Wang, Q. Xue, and W.-W. Choi, “A novel ultra-wideband differential filter based on double-sided parallel-strip line,” *IEEE Microw. Wireless Compon. Lett.*, vol. **20**, no. 8, pp. 471–473, Aug. 2010.
6. W. Feng, W. Che, and Q. Xue, “The proper balance: overview of microstrip wideband balance circuits with wideband common mode suppression,” *IEEE Microw. Mag.*, vol. **16**, no. 5, pp. 55–68, Jun. 2015.
7. Y.-J. Lu, S.-Y. Chen, and P. Hsu, “A differential-mode wideband bandpass filter with enhanced common-mode suppression using slotline resonator,” *IEEE Microw. Wireless Compon. Lett.*, vol. **22**, no. 10, pp. 503–505, Oct. 2012.
8. C.-H. Lee, C. I. G. Hsu, and C.-J. Chen, “Band-notched balanced UWB BPF with stepped-impedance slotline multi-mode resonator,” *IEEE Microw. Wireless Compon. Lett.*, vol. **22**, no. 4, pp. 182–184, Apr. 2012.
9. J. Shi, C. Shao, J.-X. Chen, et al., “Compact low-loss wideband differential bandpass filter with high common-mode suppression,” *IEEE Microw. Wireless Compon. Lett.*, vol. **23**, no. 9, pp. 480–482, Sep. 2013.

10. X. Guo, L. Zhu, K.-W. Tam, and W. Wu, "Wideband differential bandpass filters on multimode slotline resonator with intrinsic common-mode rejection," *IEEE Trans. Microw. Theory Tech.*, vol. **63**, no. 5, pp. 1587–1594, May 2015.
11. X. Guo, W. Wu, and L. Zhu, "A novel design method of wideband differential bandpass filters on the multimode slotline resonator," in *Asia–Pacific Microwave Conference*, Nanjing, China, vol. **3**, pp. 1–3, 6–9 Dec. 2015.
12. X. Guo, L. Zhu, and W. Wu, "Strip-loaded slotline resonators for differential wideband bandpass filters with intrinsic common-mode rejection," *IEEE Trans. Microwave Theory Tech.*, vol. **64**, no. 2, pp. 450–458, Feb. 2016.
13. X. Guo, W. Wu, and L. Zhu, "Research on slotline stub-loaded resonator and hybrid stub-loaded resonator for differential wideband filters," in *IEEE International Workshop on Electromagnetics*, Nanjing, China, 2016, pp. 1–3, 16–18 May 2016.
14. X. Guo, L. Zhu, and W. Wu, "Balanced wide-/dual-band BPFs on a hybrid multimode resonator with intrinsic common-mode rejection," *IEEE Trans. Microw. Theory Tech.*, vol. **64**, no. 7, pp. 1991–2005, Jul. 2016.
15. R. Li, S. Sun, and L. Zhu, "Synthesis design of ultra-wideband bandpass filters with composite series and shunt stubs," *IEEE Trans. Microw. Theory Tech.*, vol. **57**, no. 3, pp. 684–692, Mar. 2009.
16. R. Li, S. Sun, and L. Zhu, "Direct synthesis of transmission line low-/high-pass filters with series stubs," *IET Microw. Antennas Propag.*, vol. **3**, no. 4, pp. 654–662, Jun. 2009.
17. L. Zhu and W. Menzel, "Compact microstrip bandpass filter with two transmission zeros using a stub-tapped half-wavelength line resonator," *IEEE Microw. Wireless Compon. Lett.*, vol. **13**, no. 1, pp. 16–18, Jan. 2003.
18. H. Ishii, T. Kimura, N. Kobayashi, A. Saito, Z. Ma, and S. Ohshima, "Development of UWB HTS bandpass filters with microstrip stubs-loaded three-mode resonator," *IEEE Trans. Appl. Supercond.*, vol. **23**, no. 3, Art. no. 1500204, Jun. 2013.
19. L. Han, K. Wu, and X. Zhang, "Development of packaged ultra-wideband bandpass filters," *IEEE Trans. Microw. Theory Tech.*, vol. **58**, no. 1, pp. 220–228, Jan. 2010.
20. K. Song and Q. Xue, "Inductance-loaded Y-shaped resonators and their applications to filters," *IEEE Trans. Microw. Theory Tech.*, vol. **58**, no. 4, pp. 978–984, Apr. 2010.
21. W.-H. Tu, "Broadband microstrip bandpass filters using triple-mode resonator," *IET Microw. Antennas Propag.*, vol. **4**, no. 9, pp. 1275–1282, Sep. 2010.
22. R. Li and L. Zhu, "Compact UWB bandpass filter using stub-loaded multiple-mode resonator," *IEEE Microw. Wireless Compon. Lett.*, vol. **17**, no. 1, pp. 40–42, Jan. 2007.
23. Z. Shang, X. Guo, B. Cao, B. Wei, X. Zhang, Y. Heng, G. Suo, and X. Song, "Design of a superconducting ultra-wideband (UWB) bandpass filter with

sharp rejection skirts and miniaturized size," *IEEE Microw. Wireless Compon. Lett.*, vol. **23**, no. 2, pp. 72–74, Feb. 2013.

24. Q.-X. Chu and X.-K. Tian, "Design of UWB bandpass filter using stepped-impedance stub-loaded resonator," *IEEE Microw. Wireless Compon. Lett.*, vol. **23**, no. 12, pp. 644–646, Dec. 2013.
25. W. Feng and W. Che, "Novel wideband differential bandpass filters based on T-shaped structure," *IEEE Trans. Microw. Theory Tech.*, vol. **60**, no. 6, pp. 1560–1568, Jun. 2012.
26. H. Wang, L.-M. Gao, K.-W. Tam, W. Kang, and W. Wu, "A wideband differential BPF with multiple differential- and common-mode transmission zeros using cross-shaped resonator," *IEEE Microw. Wireless Compon. Lett.*, vol. **24**, no. 12, pp. 854–856, Dec. 2014.
27. K. Song and X. Quan, "Novel broadband bandpass filters using Y-shaped dual-mode microstrip resonators," *IEEE Microw. Wireless Compon. Lett.*, vol. **19**, no. 9, pp. 548–550, Sep. 2009.
28. L. Zhu, S. Sun, and W. Menzel, "Ultra-wideband (UWB) bandpass filters using multiple-mode resonator," *IEEE Microw. Wireless Compon. Lett.*, vol. **15**, no. 11, pp. 796–798, Nov. 2005.
29. Y.-C. Chiou, J.-T. Kuo, and E. Cheng, "Broadband quasi-Chebyshev bandpass filters with multimode stepped-impedance resonators (SIRs)," *IEEE Trans. Microw. Theory Tech.*, vol. **54**, no. 8, pp. 3352–3358, Aug. 2015.
30. P. Vélez, J. Naqui, A. Fernández-Prieto, J. Bonache, J. Mata-Contreras, J. Martel, F. Medina, and F. Martín, "Ultra-compact (80 mm^2) differential-mode ultra-wideband (UWB) bandpass filters with common-mode noise suppression," *IEEE Trans. Microw. Theory Tech.*, vol. **63**, no. 4, pp. 1272–1280, Apr. 2015.
31. J. J. Sánchez-Martínez and E. Márquez-Segura, "Analytical design of wire-bonded multiconductor transmission-line-based ultra-wideband differential bandpass filters," *IEEE Trans. Microw. Theory Tech.*, vol. **62**, no. 10, pp. 2308–2315, Oct. 2014.
32. L. Li, J. Bao, J.-J. Du, and Y.-M. Wang, "Differential wideband bandpass filters with enhanced common-mode suppression using internal coupling technique," *IEEE Microw. Wireless Compon. Lett.*, vol. **24**, no. 5, pp. 300–302, May 2014.
33. L. Li, J. Bao, J.-J. Du, and Y.-M. Wang, "Compact differential wideband bandpass filters with wide common-mode suppression," *IEEE Microw. Wireless Compon. Lett.*, vol. **24**, no. 3, pp. 164–166, Mar. 2014.
34. S.-J. Sun, B. Wu, T. Su, K. Deng, and C.-H. Liang, "Wideband dual-mode microstrip filter using shorted-ended resonator with centrally loaded inductive stub," *IEEE Trans. Microw. Theory Tech.*, vol. **60**, no. 12, pp. 3667–3673, Dec. 2013.

PART 4

Narrowband and Dual-Band Balanced Bandpass Filters with Intrinsic Common-Mode Suppression

CHAPTER 12

NARROWBAND COUPLED-RESONATOR BALANCED BANDPASS FILTERS AND DIPLEXERS

Armando Fernández-Prieto,[1] Francisco Medina,[1] and Jesús Martel[2]

[1]Departamento de Electrónica y Electromagnetismo, Universidad de Sevilla, Sevilla, Spain
[2]Departamento de Física Aplicada II, Universidad de Sevilla, Sevilla, Spain

12.1 INTRODUCTION

Narrowband bandpass filters (BPFs) are key components in modern communication systems and other applications of wireless technologies. Obviously, this kind of filters is also required in its balanced differential form. The frontier between narrow- and wideband devices is rather diffuse, but, in common practice, narrowband refers, at least in this context, to fractional bandwidths (FBW) up to around 10% (in some cases, bandwidths as high as 20% can be considered narrowband implementations [1]). In planar technology, such FBW can easily be achieved by employing coupled resonators, whereas other approaches must be used to reach larger bandwidths. Probably owing to this reason, coupled resonators are, by far, behind the most common implementations of relatively

Balanced Microwave Filters, First Edition. Edited by Ferran Martín, Lei Zhu, Jiasheng Hong, and Francisco Medina.

narrow bandwidth balanced filters. Thus, loop-shaped uniform impedance resonators (UIRs) [2–4], stepped-impedance resonators (SIRs) [3, 5–12], interdigital line resonators (ILRs) [13], dual-mode resonators (DMRs) [14, 15], complementary split ring resonators (CSRRs) [16], multilayer resonators [17, 18], square patch resonators [19], and even edge-coupled line resonators [20–22] have been used to design narrowband balanced BPFs, yielding good differential-mode (DM) performance and, simultaneously, effective common-mode (CM) noise suppression. This chapter offers a review of the landscape of available solutions for the appropriate design of differential filters with FBW below 20%.

Diplexers are also important components in modern dual-band wireless communication systems. The diplexer is the most basic passive device intended for frequency-domain multiplexing operation. A wide frequency band can be divided into two sub-bands (or, on the contrary, two narrow sub-bands can be combined into a single wideband signal) using diplexers. This functionality is required, for instance, in satellite communications and radio-frequency (RF) front ends of cellular radio base stations. In these two applications the separation of the transmit and receive channels is obviously necessary. In comparison with single-ended diplexers, very few works can be found in the literature related to balanced or differential diplexers. This chapter devotes its last section to the discussion of some contributions to this topic.

This chapter is organized as follows: Section 12.2 discusses balanced filter topologies where CM is inherently rejected due to the geometric properties (including symmetry) of the proposed structures. Several examples of this kind of narrowband filter implementation strategy are discussed in that section. Section 12.3 reports on filters where the involved resonators require electrical loading at some point located in their symmetry planes in order to improve CM rejection, since the unloaded structure does not provide enough common-mode rejection ratio (CMRR). Balanced filters based on the use of coupled line structures are contemplated in Section 12.4. Finally, Section 12.5 presents a brief discussion on some proposed balanced diplexers made in microstrip technology.

12.2 COUPLED-RESONATOR BALANCED FILTERS WITH INTRINSIC COMMON-MODE REJECTION

Filters based on coupled resonators can be designed making use of different coupling mechanisms, namely, electric (capacitive), magnetic (inductive), or mixed (both electric and magnetic) coupling. As it will become apparent along this chapter, mixed and magnetic coupling provide inherent CM rejection. On the contrary, electric coupling implementations usually demand the introduction of additional elements placed along the filter symmetry plane[1] in order to improve the CM response (this usually means to improve the CM rejection level and/or rejection bandwidth). In the next subsections some examples of filters involving mixed or magnetic coupling will be discussed. Filters with electric coupling will be considered in Section 12.3.

12.2.1 Loop and SIR Resonator Filters with Mixed Coupling

Open- and closed loop-shaped UIRs [2–4] and SIRs [3, 5–10, 20] have been extensively used in the implementation of balanced BPFs with a good intrinsic CM rejection level. These resonators are, basically, TEM or quasi-TEM resonant printed transmission-line sections with the shape of rectangular (or square) closed (or open) rings [23]. In the case of UIRs, the characteristic impedance of the transmission-line section used to implement the resonant structure is uniform (i.e., the microstrip line width is the same all around the perimeter of the ring). Conversely, SIRs are formed by, at least, two different transmission-line sections having different characteristic impedances. Several examples of the layout of this kind of resonators in microstrip technology are depicted in Figure 12.1(a) and (b). SIRs offer more degrees of freedom for the design and are more suitable for miniaturization. Besides, their successive resonances are not integer multiples of the fundamental resonance frequency, thus mitigating the problem of spurious band response associated with UIRs.

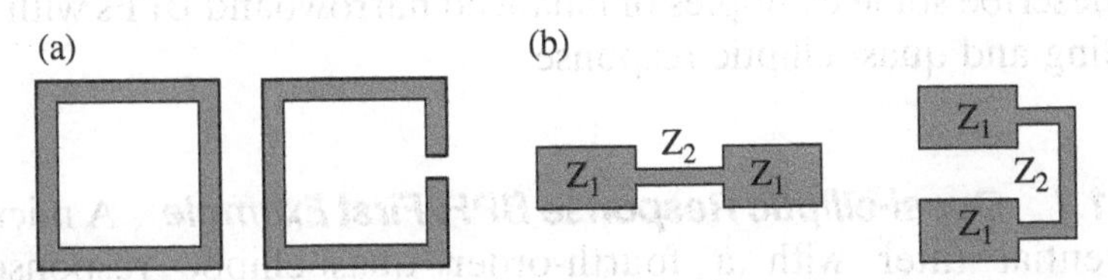

Figure 12.1 (a) Two UIRs in the form of closed (left) and open (right) square loops. (b) Basic structure of two three-section SIRs. The one to the rightmost side corresponds to the SIR folded version (FSIR). Gray regions represent the metalized parts of the microstrip implementation.

[1] Differential balanced structures use to be symmetrical with respect to a certain symmetry plane. This ensures proper balanced operation of the device, although structural symmetry is not strictly necessary if the equivalent circuit has the suitable symmetry properties.

The electromagnetic response of a SIR is basically controlled by the ratio of the characteristic impedances and lengths of the sections composing the resonator. Thus, the following non-dimensional parameters control the electrical response of a two-stage or of a symmetrical three-stage SIR [23]:

$$R = \frac{Z_2}{Z_1}, \quad \alpha = \frac{L_2}{L_1 + L_2} \tag{12.1}$$

where Z_1 and Z_2 are the characteristic impedances of the different sections and L_1 and L_2 their respective lengths. The resonance frequency of the SIR can be adjusted by tuning the values of R and α. Note that the resonance frequency of UIRs is almost exclusively controlled by their physical length. An extra degree of freedom, in terms of design flexibility, is therefore offered by SIRs.

The use of mixed couplings in the design of balanced BPFs based on both UIRs and SIRs allows for the implementation of differential filters with good DM selectivity and strong CM rejection. Several standard transfer functions can be implemented by means of coupled resonators with mixed coupling, such as Butterworth (maximally flat), Chebyshev, quasi-elliptic, or elliptic responses. Elliptic and quasi-elliptic transfer functions offer the best performance in terms of selectivity and out-of-band rejection thanks to the possibility of introducing a finite number of transmission zeros (TZs) outside the passband. As a disadvantage, it can be mentioned that they tend to give place to more complex physical implementations, thus requiring more sophisticated layouts than those demanded to achieve Butterworth or Chebyshev responses. Let us now describe some examples of balanced narrowband BPFs with mixed coupling and quasi-elliptic response.

12.2.1.1 Quasi-elliptic Response BPF: First Example A microstrip differential filter with a fourth-order quasi-elliptic response was reported in Ref. [5]. The layout of such filter is depicted in Figure 12.2. The filter consists of two pairs of symmetrically located folded SIRs (FSIRs) of the half-wavelength type. Note that SIRs of this type are not necessarily $\lambda/2$ long, since the true length of the resonator at the resonance frequency depends on the impedance ratio. The use of this terminology refers to the fact that no vias to ground are used, in such a way that, if the resonator were a UIR, its length at resonance would be $\lambda/2$.

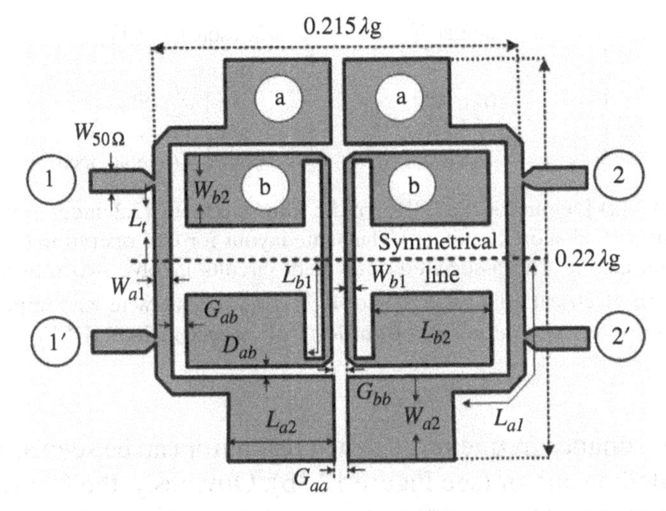

Figure 12.2 Layout of the fourth-order balanced filter proposed in Ref. [5] using symmetric $\lambda/2$ folded SIRs. Physical dimensions (in mm): $W_{50\Omega}=1.9$, $W_{a1}=1.5$, $W_{a2}=7.6$, $W_{b1}=0.8$, $W_{b2}=6.8$, $L_{a1}=17.1$, $L_{a2}=9.5$, $L_{b1}=11.9$, $L_{b2}=11.3$, $L_t=7.2$, $G_{aa}=1.2$, $G_{bb}=1.4$, $G_{ab}=0.8$, and $D_{ab}=0.4$. Reprinted with permission from Ref. [5]; copyright 2007 IEEE.

Since two different FSIRs are involved in this design, two different sets of characteristic impedances and lengths (one for each FSIR) are required: Z_{i1}, Z_{i2}, L_{i1}, and L_{i2}. The subscript i denotes resonators a and b (see Figure 12.2). The corresponding impedance and length ratios are then defined as follows:

$$R_i=\frac{Z_{i2}}{Z_{i1}},\quad \alpha_i=\frac{L_{i2}}{L_{i1}+L_{i2}};\; i=a,b \tag{12.2}$$

The configuration in Figure 12.2 allows for the implementation of a quasi-elliptic response thanks to the presence of cross coupling between the different resonators. The cross coupling introduces a couple of TZs at the lower and upper sides of the differential passband.

The existence of a horizontal symmetry plane clearly determines the different boundary conditions that apply for the DM and the CM excitations. Under DM operation (odd excitation), the symmetry plane behaves as a virtual perfect electric wall (PEW). This means that each FSIR can be treated as a shorted SIR for such specific excitation. At the resonance frequency, f_0^{d}, the FSIR is a $\lambda/4$-like SIR, as depicted in Figure 12.3(a). On the other hand, for CM (even) excitation, the symmetry plane behaves as a virtual perfect magnetic wall (PMW). At

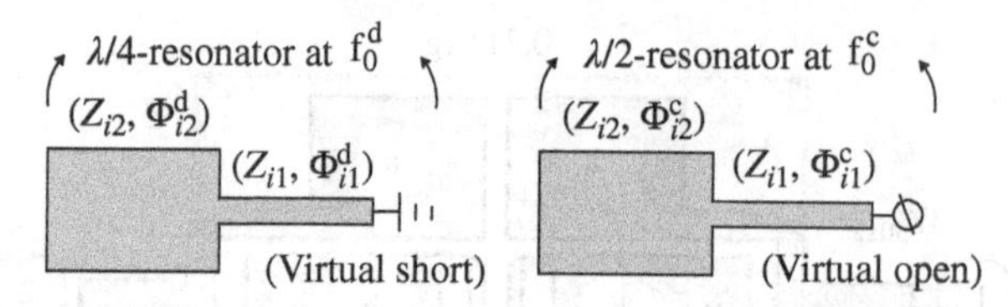

Figure 12.3 (a) Layout for the FSIRs of the filter in Figure 12.2 under DM operation, showing the virtual short circuit. (b) The same layout for CM operation exhibiting the virtual open circuit. The associated equivalent circuits involve two transmission-line sections with electrical lengths $\Phi_{i1}^{c,d}$ and $\Phi_{i2}^{c,d}$ and with characteristic impedances Z_{i1} and Z_{i2}. Reprinted with permission from Ref. [5]; copyright 2007 IEEE.

the CM resonance frequency, f_0^c, each resonator can be seen as a λ/2-like open-ended resonator (see Figure 12.3b). Obviously, the CM resonance frequency, f_0^c, will be significantly greater than f_0^d [5, 23]. This fact will provide inherent poor CM transmission at f_0^d, at least over a certain frequency range including the whole differential passband. This feature is important concerning the achievable CMRR.

The fourth-order quasi-elliptic balanced BPF depicted in Figure 12.2 was designed with the following specifications: DM center frequency $f_0^d = 1.0\text{GHz}$, 3-dB fractional bandwidth FBW = 10%, return loss $L_R = 20\text{dB}$, and $\Omega_a = 2.00$.[2] The filter was implemented on an FR4 substrate ($\varepsilon_r = 4.4$) having a thickness of $h = 1\text{mm}$. The values of the corresponding elements for the quasi-elliptic low-pass response are $g_1 = 0.95449$, $g_2 = 1.38235$, $J_1 = -0.16271$, and $J_2 = 1.06062$. Note that this type of filters can be designed by using the well-known coupled-resonator filter design procedure thoroughly explained in Ref. [24]. Such technique makes use of the coupling coefficients, $M_{i,j}^{dd}$ ($i = a, b$, $j = a, b$), between the resonators and the external quality factor, Q_e^{dd}. The superscript "dd" in these expressions stands for DM operation. These parameters can be expressed in terms of the elements of the low-pass prototype filter as follows:

$$Q_e^{dd} = \frac{g_0 g_1}{\text{FBW}} = 9.55, \quad M_{ab}^{dd} = \frac{\text{FBW}}{\sqrt{g_1 g_2}} = 0.087 \tag{12.3}$$

[2] Ω_a is a normalized frequency defined in the context of quasi-elliptic filter design theory [24]. It basically controls the location of the out-of-band TZs of the desired quasi-elliptic response. In this particular case Ω_a has been chosen to be 2.

$$M_{bb}^{\mathrm{dd}} = \frac{\mathrm{FBW} \cdot J_2}{g_2} = 0.0767 \tag{12.4}$$

$$M_{aa}^{\mathrm{dd}} = \frac{\mathrm{FBW} \cdot J_1}{g_1} = -0.017 \tag{12.5}$$

Note that only the differential response is designed, thus determining the final physical dimensions. The CM response is simply expected to be good enough for practical purposes.

The first step in the design process consists in determining the physical dimensions of the resonators in order to set the desired center frequency (in our case, $f_0^{\mathrm{d}} = 1\,\mathrm{GHz}$). As any other resonator used in microwave filter design, SIRs present spurious frequency responses. However, conversely to what happens with UIRs, the spurious resonance frequencies of SIRs can be adjusted by tuning R_i and α_i [23]. It can be demonstrated that the DM stopband for the topology shown in Figure 12.2 is mainly controlled by the inter-coupled resonators (resonators labeled as b) [5, 21]. This means that, by properly designing the b resonators, it is possible to push up the first DM spurious harmonic. The specific choice of the SIR parameters $R_b = 0.25$ and $\alpha_b = 0.5$ moves the first DM spurious resonance to 5.5 f_0^{d}, while the fundamental DM resonance is kept at the desired value ($f_0^{\mathrm{d}} = 1\,\mathrm{GHz}$) [5]. On the other hand, resonator a is designed with the highest level of feasible compactness compatible with the level of cross coupling between the input and output resonators required for the introduction of a TZ. Taking into consideration all what has been commented so far, the final dimensions of the resonators providing the required center frequency and the extended stopband for the DM are the ones given in the caption of Figure 12.2.

The next step in the design process is the realization of the required external quality factor, $Q_{\mathrm{e}}^{\mathrm{dd}}$, and coupling coefficients, M_{ab}^{dd}, M_{aa}^{dd}, and M_{bb}^{dd}. $Q_{\mathrm{e}}^{\mathrm{dd}}$ depends on the tapping distance, L_t, as illustrated in Figure 12.4(a). In this figure, the CM external quality factor ($Q_{\mathrm{e}}^{\mathrm{cc}}$) is also shown to illustrate the poor matching level at f_0^{d}. The DM coupling coefficients, M_{ab}^{dd}, M_{aa}^{dd}, and M_{bb}^{dd}, are mainly controlled by the gaps G_{ab}, G_{aa}, and G_{bb}, respectively (see Figure 12.4(b)–(d)). The final obtained values of tapping distance and gaps are given in the caption to Figure 12.2.

The measured and simulated DM and CM responses of the designed filter are shown in Figure 12.5(a) and (b), respectively. Good agreement between simulations and measurements is found for both excitations. The measured center frequency is $f_0^{\mathrm{d}} = 1.02\,\mathrm{GHz}$, the measured insertion loss is 3.51 dB, and the experimental fractional bandwidth FBW = 12%.

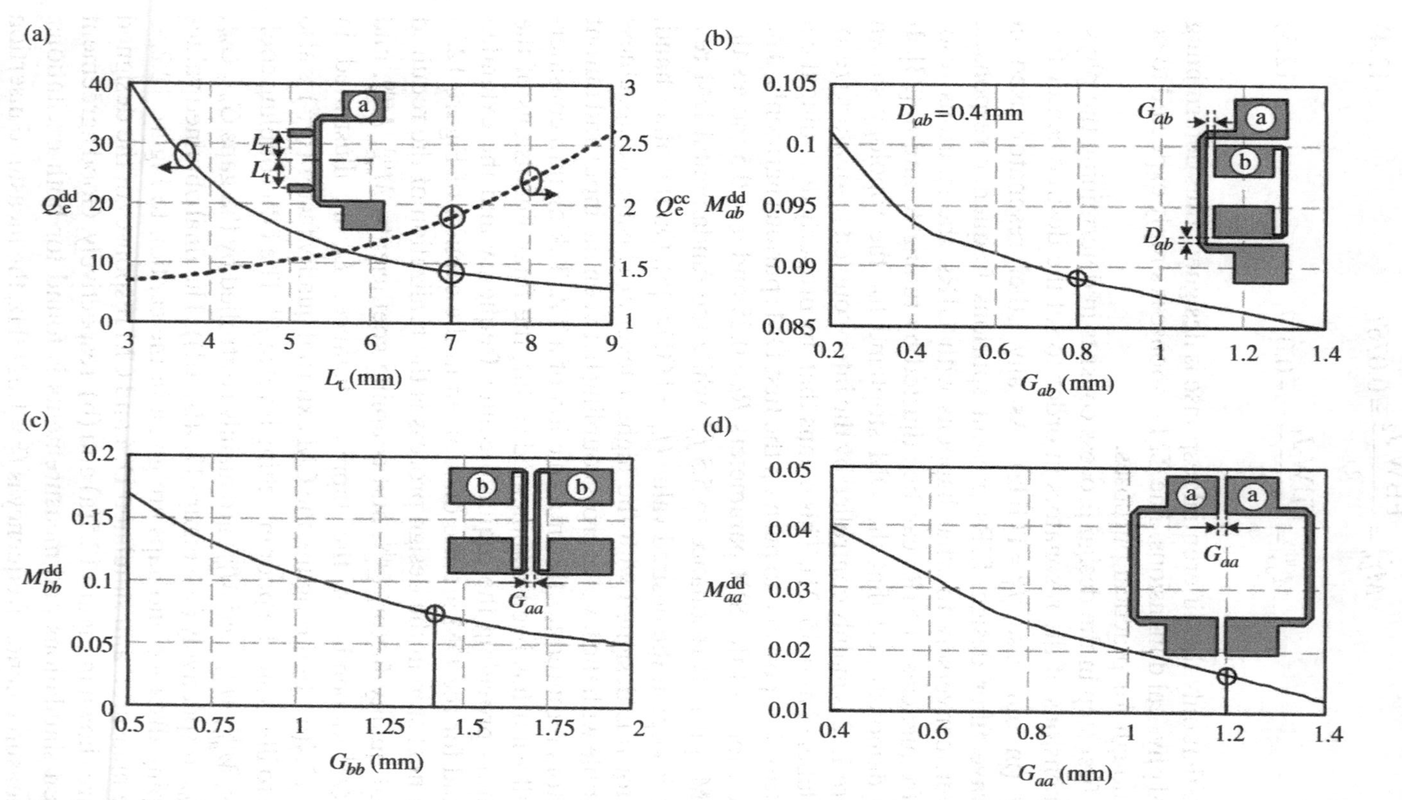

Figure 12.4 (a) External quality factors Q_e^{dd} and Q_e^{cc} for differential- and common-mode excitations as a function of L_t. (b) M_{ab}^{dd}, (c) M_{bb}^{dd}, and (d) M_{aa}^{dd} are the relevant coupling coefficients, controlled by G_{ab}, G_{aa}, and G_{bb}, respectively. Reprinted with permission from Ref. [5]; copyright 2007 IEEE.

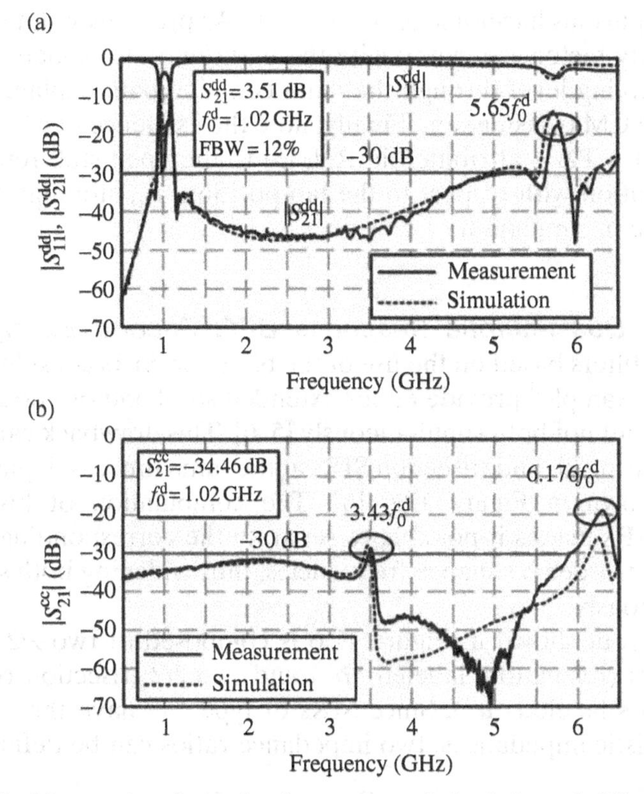

Figure 12.5 Measured and simulated responses of the balanced differential filter reported in Ref. [5]. (a) Differential-mode response; (b) common-mode response. Reprinted with permission from Ref. [5]; copyright 2007 IEEE.

The spurious passband has been pushed up to $5.65f_0^d$. The filter has a very compact size ($0.215\lambda_g \times 0.22\lambda_g$, where λ_g is the microstrip guided wavelength at the center frequency). A suppression level of 34.46 dB at f_0^d is obtained for the CM. This is a more than acceptable achievement (for narrowband applications the CMRR is typically required to be better than 20 dB). The reasons behind the good CMRR is the poor CM external quality factor at f_0^d (poor matching) and the weak coupling level that CM presents through the main coupling path of the filter (from input resonator, *a*, through inter-coupled resonators, *b*, and then to output resonator, *a*). The CM transmission response is mainly determined by the direct coupling path between input/output resonators (from input resonator *a* directly to output resonator *a*). Note that the CM rejection level is better than 30 dB from 0.5 to 6 GHz. The spurious CM response at $3.43f_0^d$ has been suppressed. This spurious response is associated with the first spurious harmonic of resonator *b*. As previously stated, the low CM quality factor associated with the input/output resonators and the weak coupling level through the main coupling path explain the weak wideband CM transmission. Finally, note the existence of a TZ at about 5 GHz. This TZ is attributed in Ref. [5] to the open stub remaining in CM operation with respect to the tap position, L_t, although this TZ is not visible in simulation.

12.2.1.2 *Quasi-elliptic Response BPF: Second Example* The balanced filters based on the use of the bisection SIRs considered in the previous example[3] provide either extended stopband or good CM suppression, but not both simultaneously [5, 6]. This drawback can be overcome by using bi- and trisection SIRs at the same time, as depicted in the layout shown in Figure 12.6 [6]. The combination of bi- and trisection SIRs makes it possible to separate the corresponding DM and CM higher-order resonance frequencies, thus widening both stopbands simultaneously.

The layout shown in Figure 12.6 is composed of two $\lambda/2$ bisection resonators (resonators labeled "*b*") and two $\lambda/2$ trisection resonators (resonators labeled "*a*"). Since SIRs of type "*a*" have three different characteristic impedances, two impedance ratios can be defined as

$$R_{a2} = \frac{Z_{a2}}{Z_{a1}}, \quad R_{a3} = \frac{Z_{a3}}{Z_{a1}} \tag{12.6}$$

In order to simplify the design process, the lengths L_{a2} and L_{a3} are initially set to be identical ($L_{a2} = L_{a3}$). Besides, the width W_{a1} is fixed to a specific value ($W_{a1} = 0.3$ mm), with the aim of reducing the number of parameters to be calculated. Therefore, the remaining parameters to be determined are W_{a2}, W_{a3}, L_{a1}, and L_{a2}.

Let us consider the equivalent circuits for the input/output resonators (resonators "*a*") under DM and CM operation conditions shown in Figure 12.7. In this figure, $(\phi^{d}_{a1}, \phi^{d}_{a2}, \phi^{d}_{a3})$ and $(\phi^{c}_{a1}, \phi^{c}_{a2}, \phi^{c}_{a3})$ are the corresponding DM and CM electrical lengths, whereas γ is a parameter associated with the tap position. If CM input impedance is set to zero

[3] Actually, the SIRs used in the previous filter design consist of three transmission line sections. However, these SIRs are symmetrical, in such a way that only two different characteristic impedances appear in the DM and CM equivalent structures. This is the reason to call those resonators bisection SIRs. Following this criterion, a trisection SIR will be made of five transmission-line sections having a symmetry plane, thus leading to three-section structures for the DM and CM equivalent circuits.

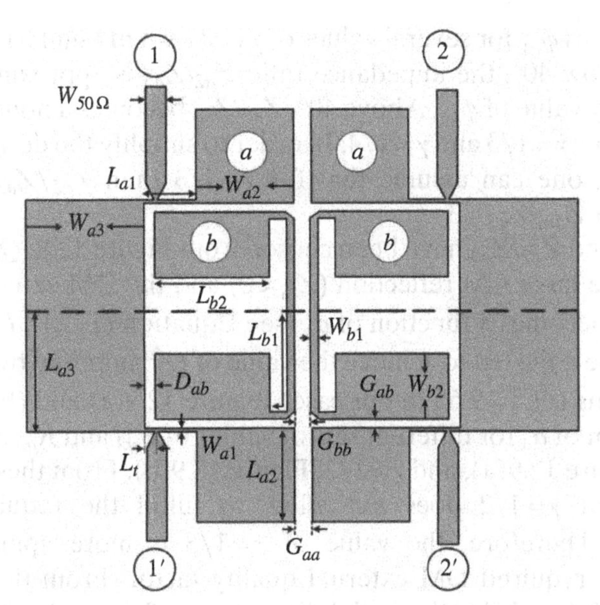

Figure 12.6 Layout of a fourth-order balanced filter using symmetric bi-/trisection SIRs (see Ref. [6]). Physical dimensions are (in mm) $W_{50\Omega} = 1.9$, $W_{a1} = 0.3$, $W_{a2} = 9.3$, $W_{a3} = 13$, $W_{b1} = 0.8$, $W_{b2} = 6.6$, $L_{a1} = 4.9$, $L_{a2} = 8.7$, $L_{a3} = 10.8$, $L_{b1} = 11.9$, $L_{b2} = 11.3$, $L_t = 1.3$, $G_{aa} = 1.6$, $G_{bb} = 1.4$, $G_{ab} = 0.4$, and $D_{ab} = 0.5$. Reprinted with permission from Ref. [6]; copyright 2007 IEEE.

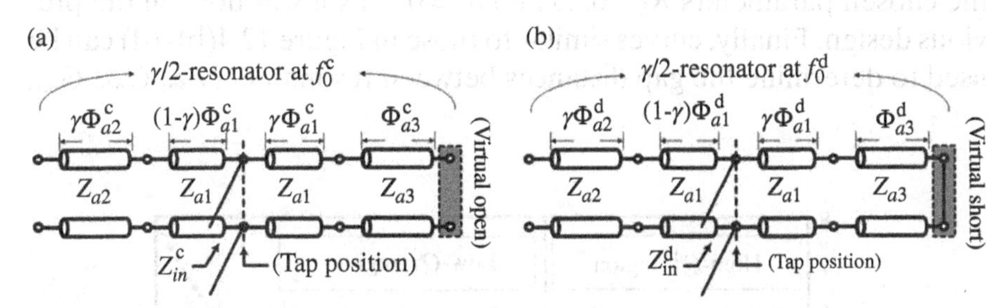

Figure 12.7 (a) Equivalent half circuit for resonator "*a*" under CM excitation. (b) Equivalent half circuit for the same resonator under DM operation. Reprinted with permission from Ref. [6]; copyright 2007 IEEE.

in Figure 12.7(a), that is, $Z_{\text{in}}^{\text{c}} = 0$ ($Q_{\text{e}}^{\text{cc}} = \infty$), the condition for total reflection of the CM can be found to be

$$Z_{a2} \cdot \tan \gamma \phi_{a1}^{\text{c}} = Z_{a3} \cdot \tan(1-\gamma) \phi_{a1}^{\text{c}} \tag{12.7}$$

which is a very helpful expression to design the input/output resonators in order to achieve good CM rejection. A plot of the impedance ratio

Z_{a2}/Z_{a3} versus ϕ^{c}_{a1} for several values of γ is shown in Figure 12.8. For ϕ^{c}_{a1} roughly below 40°, the impedance ratio Z_{a2}/Z_{a3} is approximately constant for any value of ϕ^{c}_{a1}. Above 40°, Z_{a2}/Z_{a3} becomes a nonlinear function of ϕ^{c}_{a1} for $\gamma = 1/3$ and $\gamma = 1/4$. In order to simplify the design process, once again, one can assume that for $\gamma = 1/3$ then $Z_{a2}/Z_{a3} = 2$ and if $\gamma = 1/4$ then $Z_{a2}/Z_{a3} = 3$.

Once γ and Z_{a2}/Z_{a3} have been chosen from Figure 12.8, Q^{cc}_{e} is almost fixed to give large CM reflection ($Z^{c}_{in} \approx 0$) and the DM external quality factor Q^{dd}_{e} becomes a function of α_a (see Equation (12.2)), R_{a2} and R_{a3}, which can be adjusted to achieve the value of Q^{dd}_{e} imposed by the design specifications ($Q^{dd}_{e} = 9.55$ in our case). Figure 12.9(a) and (b) plots Q^{dd}_{e} as a function of α_a for different sets of values of R_{a2} and R_{a3} considering $\gamma = 1/2$, Figure 12.9(a), and $\gamma = 1/3$, Figure 12.9(b). From these figures it is clear that $\gamma = 1/2$ does not allow to fulfill the requirement of $Q^{dd}_{e} = 9.55$. Therefore, the value of $\gamma = 1/3$ is more appropriate to achieve the required DM external quality factor. From the curves in Figure 12.9(a) and (b), the initial dimensions of resonator "*a*" and the tap position L_t can be determined. Thus, the initial values chosen for the aforementioned parameters result to be $\gamma = 1/3$, $\alpha_a = 0.8$, $R_{a2} = 1/6$, and $R_{a3} = 1/12$.

The determination of the dimensions for resonator "*b*" is based on the chosen parameters $R_b = 0.25$ and $\alpha_b = 0.5$, as it was done in the previous design. Finally, curves similar to those in Figure 12.4(b)–(d) can be used to determine the gap distances between resonators (viz., G_{ab}, G_{aa},

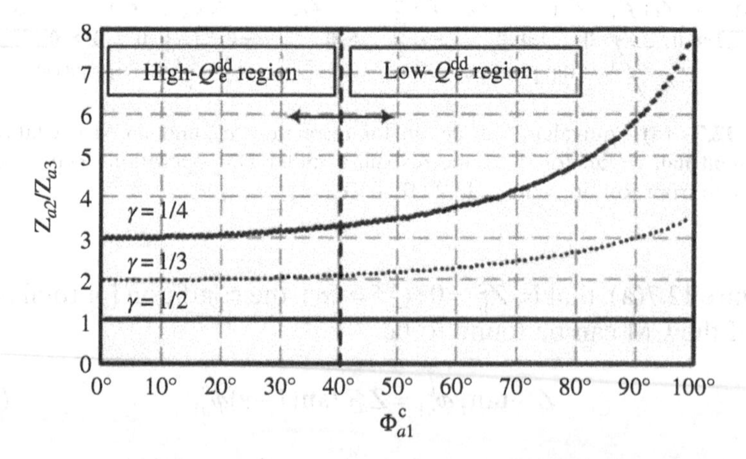

Figure 12.8 Parameters leading to perfect reflection condition under common-mode operation. Reprinted with permission from Ref. [6]; copyright 2007 IEEE.

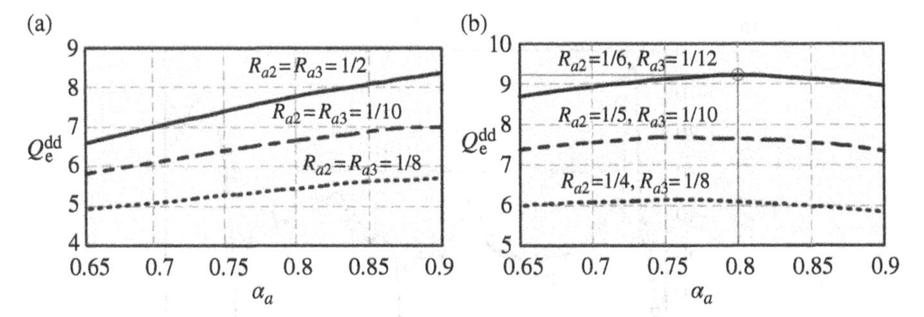

Figure 12.9 Simulated DM external quality factor, Q_e^{dd}, versus α_a for different values of R_{a2} and R_{a3} (see Ref. [6]). (a) $\gamma = 1/2$ and (b) $\gamma = 1/3$. Reprinted with permission from Ref. [6]; copyright 2007 IEEE.

G_{bb}, and D_{ab}) in order to achieve the coupling coefficients imposed by the design specifications (Equations (12.3)–(12.5)). Since the design is complex and involves many parameters, a fine-tune process based on full-wave simulation is required to obtain the final dimensions, which are specified in the caption to Figure 12.6. This final tuning process makes that $L_{a2} \neq L_{a3}$, with a slight difference between them of about 10%. Simulated and measured results for a fabricated prototype are shown in Figure 12.10. The measured 3-dB FBW of the DM is 11.5%, with a minimum level of insertion loss of 3.88 dB. The physical size is $0.233\lambda_g \times 0.34\lambda_g$ and the out-of-band transmission is below −30 dB between 1.15 and 7.5 GHz. This represents a clear improvement when compared with the results obtained with the filter prototype I. For CM operation, the measured insertion loss within the differential passband is 52.7 dB (about 18 dB improvement when compared with the filter prototype I). In addition, the CM stopband is below −30 dB up to $5f_0^d$. For this filter, the CMRR is better than 45 dB from 0.966 to 1.084 GHz. These results demonstrate that both bi- and tri-section resonators provide good CM suppression, wide CM stopband with good rejection level, and a differential response with high selectivity and wide stopband.

Butterworth balanced BPFs using resonators with mixed coupling have also been developed for SIRs and loop-shaped UIRs [2, 7]. The design process for Butterworth filters is easier than for the examples considered earlier. Unfortunately, the prize to pay is the degradation of the out-of-band performance for both DM and CM responses.

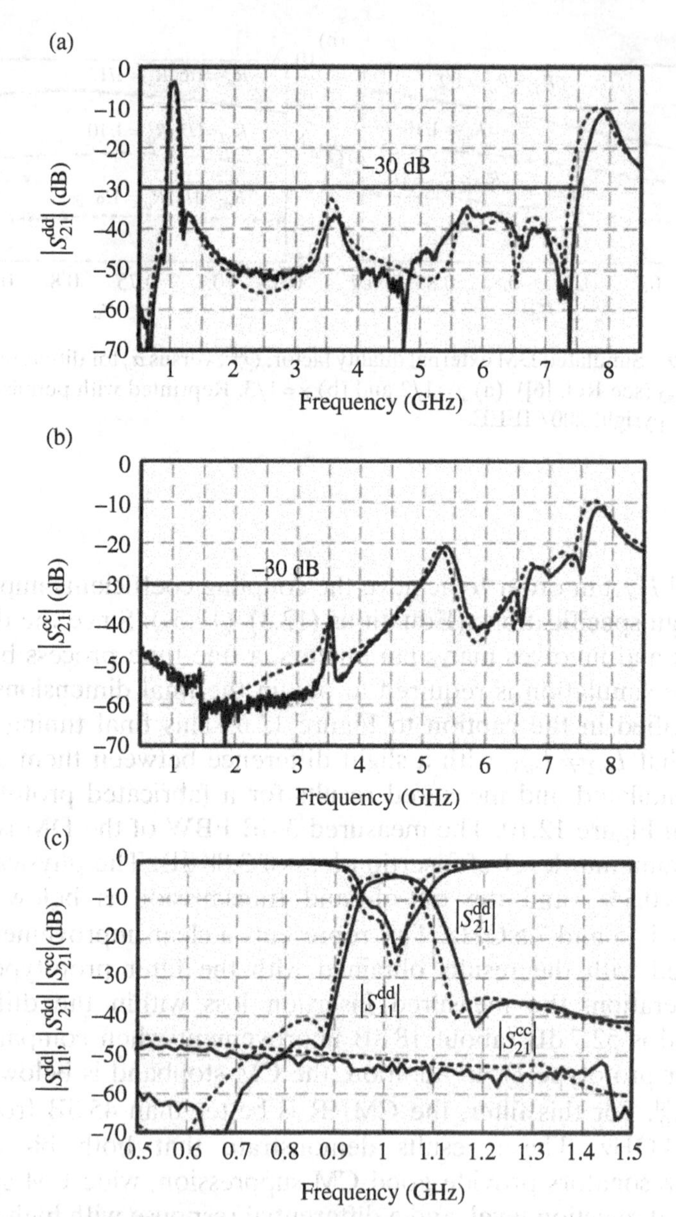

Figure 12.10 Simulated (dashed lines) and measured (solid lines) results for the differential filter proposed in Ref. [6]. (a) Differential-mode and (b) common-mode responses. (c) Detail of the differential passband. Reprinted with permission from Ref. [6]; copyright 2007 IEEE.

12.2.2 Magnetically Coupled Open-Loop and FSIR Balanced Filters

In previous section it has been demonstrated that mixed-coupled resonators are very useful for designing narrowband balanced BPFs with good selectivity and good inherent CM rejection. In this section it will be shown that the use of pure magnetic coupling [3, 12] is a very interesting alternative to mixed coupling to attain the same goal. Two key features of the magnetic coupling approach are as follows: (i) no additional tuning is required after the design of the DM response to improve the CM rejection, and (ii) increasing the order of the filter is a trivial task in comparison with implementations based on the use of resonators with relatively complicated geometries (such as the ones studied until this point in this chapter). The basic resonator used in this section is geometrically very simple and allows for the implementation of high-order filters in a very easy manner. Let us now examine some examples of differential coupled-resonator filters with magnetically coupled stages.

12.2.2.1 Filters with Magnetic Coupling: First Example In this subsection a filter prototype based on open-loop resonators is designed for demonstration of the positive features of the magnetic coupling alternative. Filter specifications are: second-order Butterworth type, center frequency $f_0^{\mathrm{d}} = 2.45\,\mathrm{GHz}$, and fractional bandwidth FBW = 10%. The dielectric constant of the substrate used for the implementation of this filter is $\varepsilon_r = 3.0$, and its thickness is $h = 1.016\,\mathrm{mm}$. The design process starts by choosing the resonator dimensions, leading to the required resonance frequency. Basically, f_0^{d} is mainly controlled by the length of the open-loop resonator, which has to be close to half the guided wavelength at such frequency. The resonator dimensions are given in the caption of Figure 12.11, where two magnetically coupled open-loop resonators are depicted.

Next, as it was done in previous sections, the filter is designed (DM response) by following the classical design procedure described in Ref. [24]. Recall that such procedure makes use of the inter-resonator coupling factor, M, and the external quality factor, Q_e. In order to obtain the desired FBW, the values for M and Q_e have to be calculated by using the following expressions [24]:

$$M_{i,i+1} = \frac{\mathrm{FBW}}{\sqrt{g_i g_{i+1}}}, \quad \text{for } i = 1,\ldots,n-1 \tag{12.8}$$

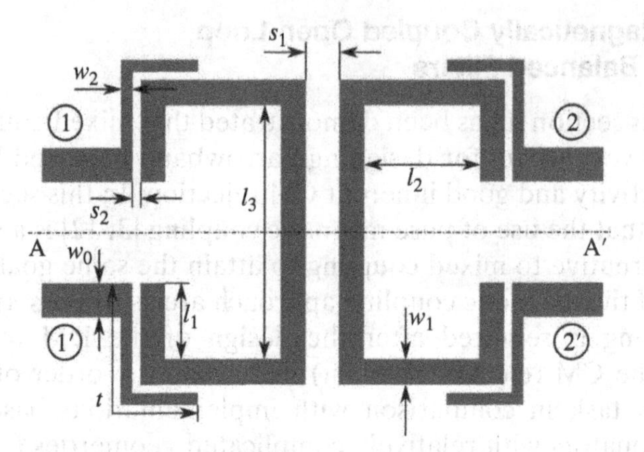

Figure 12.11 Layout of a basic magnetically coupled open-loop balanced bandpass filter of second-order [3]. The final dimensions (in mm) of the designed filter are $l_1 = 4.0$, $l_2 = 4.9$, $l_3 = 16.9$, $w_0 = 2.53$, $w_1 = 0.8$, $w_2 = 0.3$, $t = 9$, $s_1 = 1.6$, and $s_2 = 0.1$. Reprinted with permission from Ref. [3]; copyright 2015 IEEE.

$$Q_{e1} = \frac{g_0 g_1}{\text{FBW}} \quad Q_{en} = \frac{g_n g_{n+1}}{\text{FBW}} \tag{12.9}$$

where n is the filter order and g_i $(i = 1,\ldots,n-1)$ are the Butterworth low-pass prototype element values for the filter response to be implemented. In the case at hand, $g_0 = g_3 = 1$ and $g_1 = g_2 = 1.4142$, in such a way that the values $M_{1,2} = 0.071$ and $Q_{e1} = Q_{e2} = Q_e = 14$ are required. Design curves similar to those shown in Figure 12.4 can be plotted to extract the values of the geometrical parameters, yielding the desired Q_e and $M_{1,2}$. These new curves are depicted in Figure 12.12.

The parameters now controlling Q_e are the width, w_2, and length, t, of the capacitively coupled feeding lines and the separation between these lines and the resonator, s_2. For the sake of simplicity, the width has been fixed to $w_2 = 0.3$mm, and the other two dimensional parameters have been adjusted to obtain the design curves in Figure 12.12. From these figures, the extracted values of the physical dimension fulfilling the filter specs turn out to be $s_1 = 1.6$mm, $s_2 = 0.1$mm, and $t = 9.0$mm. Note that the feeding lines and the resonators must be closely spaced to reach the required value of Q_e. A wide range of values of w_2 are, in principle, acceptable. However, w_2 has been chosen so that t is short enough to make negligible the input/output coupling and also to avoid fabrication tolerance troubles associated with a too small value of s_2.

This filter has been simulated with a full-wave simulator (*ADS Momentum* [25]) and then fabricated and measured. Simulations and measurements have been included in Figure 12.13(a) and (b). From

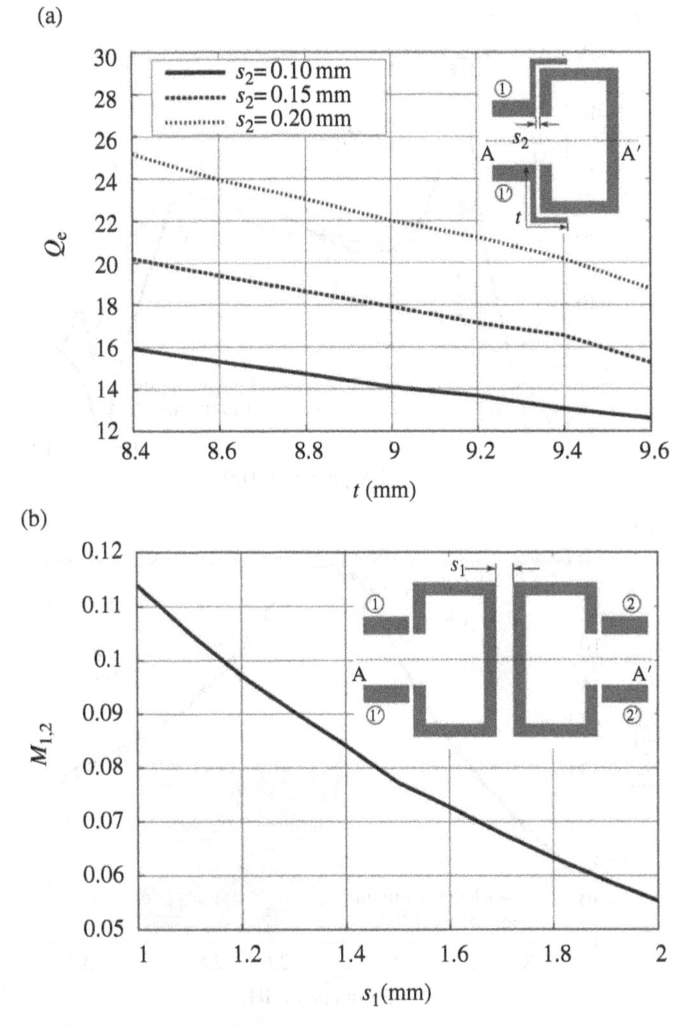

Figure 12.12 (a) Differential-mode external quality factor, Q_e, as a function of the length of the capacitive feeding lines, t, for several values of the separation s_2. (b) Simulated differential-mode coupling coefficient, $M_{1,2}$, versus the gap between the resonators, s_1, for the magnetically coupled open-loop resonators used in the design. Reprinted with permission from Ref. [3]; copyright 2015 IEEE.

these figures it is clear that CM noise has a measured suppression level of about 40 dB. The DM also exhibits a very good performance in terms of out-of-band rejection level, matching, selectivity, and insertion loss (1.28 dB, measured). Note that in contrast with the case of quasi-elliptic response, no TZs appear now at the lower and upper bounds of the pass-band (note that a Butterworth transfer function has been implemented).

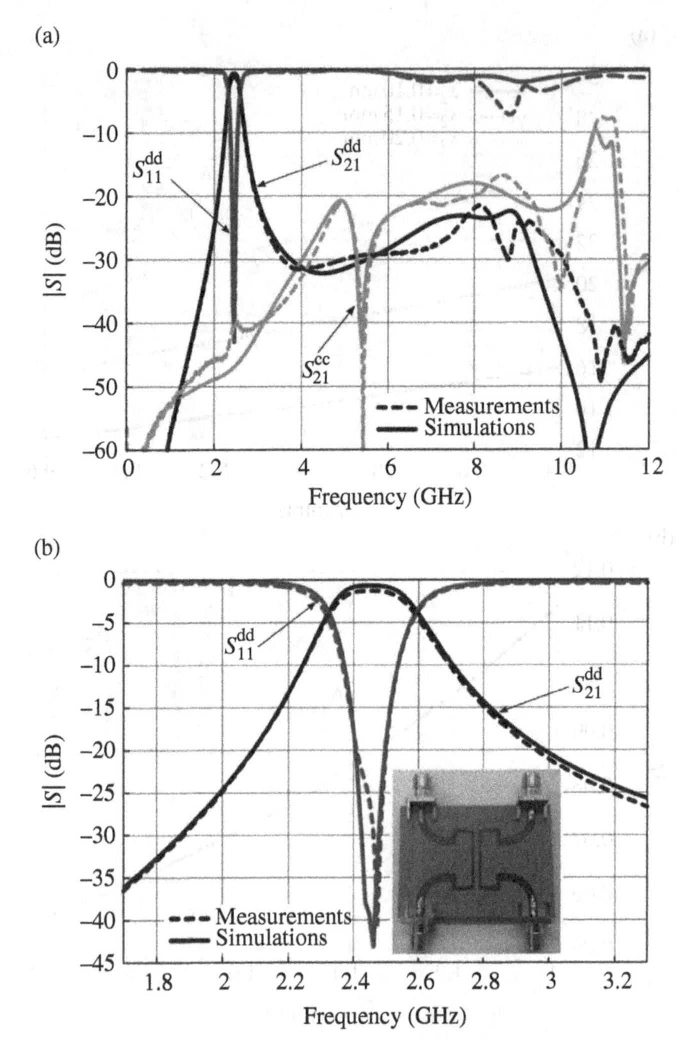

Figure 12.13 (a) Simulated and measured DM and CM responses of the open-loop balanced bandpass filter in Figure 12.11. (b) Detail of the differential passband including a photograph of the fabricated filter. Reprinted with permission from Ref. [3]; copyright 2015 IEEE.

It is very interesting to compare the performance of this filter with its electrically coupled resonators version. The layout of the latter filter is shown in Figure 12.14. Note that the resonator dimensions are different from those in Figure 12.11. This is due to the fact that the excitation of the resonators is different in both designs (capacitive for the magnetic coupling case and inductive for the electric coupling case). The excitation type modifies the resonance frequency, thus resulting in different

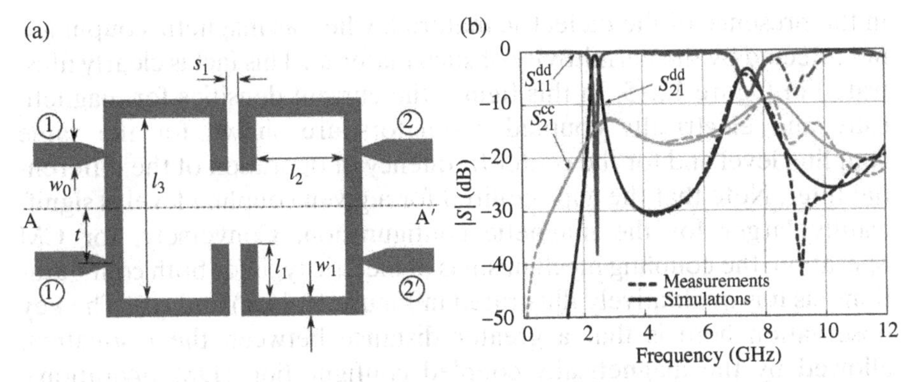

Figure 12.14 (a) Layout of a coupled open-loop balanced bandpass filter with the same DM response of the filter in Figures 12.11 and 12.13 but based on electrical coupling. The final dimensions of the designed 2nd-order Butterworth filter ($f_0^d = 2.45$ GHz, $\Delta = 10\%$) are $l_1 = 2.5$mm, $l_2 = 8.75$mm, $l_3 = 16.9$mm, $w_0 = 2.53$mm, $w_1 = 0.8$mm, $t = 2.435$mm, and $s_1 = 0.53$mm. (b) Simulated and measured DM and CM responses. Reprinted with permission from Ref. [3]; copyright 2015 IEEE.

physical dimensions for the two coupling schemes. Since the design process for the filter with electric coupling is the same used in the previous designs, we will limit our discussion to the obtained performances and to the role played in CM cancelation by both types of coupling. The simulated and measured response of the filter based on electric coupling is shown in Figure 12.14(b).

By comparing Figures 12.13(a) and 12.14(b), it is clear that the CM rejection for the filter based on electrically coupled resonators within the differential passband is very poor (only 15 dB) when compared with the rejection provided by the equivalent filter based on magnetic coupling. Indeed, a 26 dB improvement of CMRR is measured at the center frequency. Moreover, DM and CM out-of-band behavior is better for the filter based on magnetic coupling. In order to qualitatively explain why magnetic coupling prevents CM transmission (at least, better than electric coupling), it is necessary to consider the effect of the different boundary conditions imposed at the symmetry plane AA′ for DM and CM excitations in the two filter implementations. For the DM, the coupling between resonators in the configuration of Figure 12.11 is of magnetic type, while for the topology in Figure 12.14(a) is of electric type. In general, for this kind of geometries, and for a given separation between resonators, magnetic coupling is stronger than electric coupling. This is because the presence of the dielectric substrate makes the electric coupling to be quite different from the magnetic one. The electric coupling for a given separation between microstrips is smaller in the presence of the dielectric material, whereas magnetic coupling is not affected by the permittivity of such material. This fact is clearly illustrated in Figure 12.15. In this figure, the current densities for magnetically and electrically coupled resonators are shown for the same coupling level and for the center frequency of operation of the differential filter. Note that the gap required for a given coupling level is significantly larger for the magnetic configuration. Conversely, for CM operation, the coupling mechanism is of electric type for both configurations, as it is qualitatively illustrated in Figure 12.15(c) and (d). The key observation here is that a greater distance between the resonators, allowed by the magnetically coupled configuration (DM operation), helps to improve the CM rejection level. It can be clearly appreciated that the current level in the output ports (under CM excitation) is much smaller for the magnetically coupled implementation.

Another point worth to be mentioned is the appearance of a CM TZ at $f = 5.3$GHz for the case of magnetically coupled resonators (see

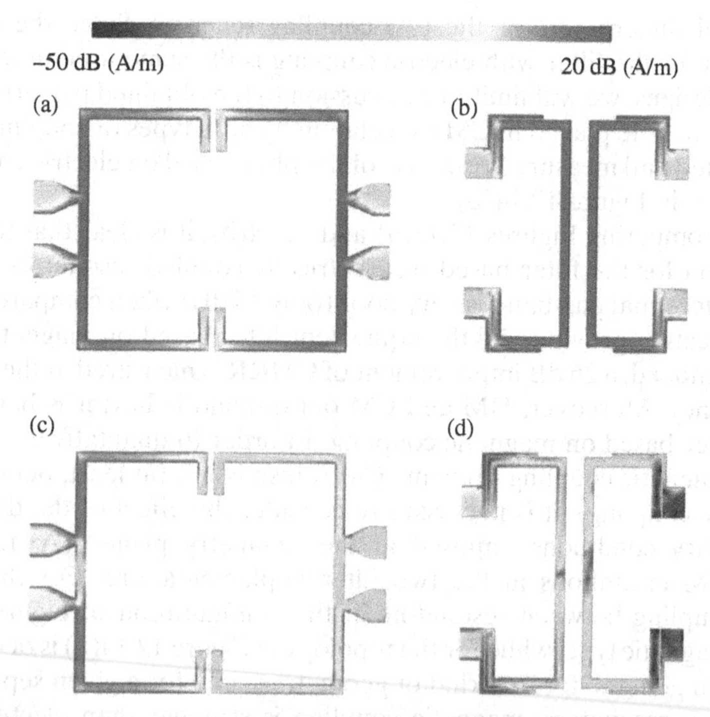

Figure 12.15 Surface current distributions at $f_0^{\rm d}$ for the electrically coupled (a) and (c) and the magnetically coupled (b) and (d) configurations. The top figures correspond to DM operation. The bottom figures correspond to CM operation. Reprinted with permission from Ref. [3]; copyright 2015 IEEE.

Figure 12.13(a)). This TZ does not appear when the resonators are electrically coupled (see Figure 12.14(a)). Its existence is associated with the coupled section of the resonators in Figure 12.11, which basically corresponds to a pair of coupled microstrip lines (see Figure 12.16). It is well known from coupled transmission-line theory that when two edge-coupled lines are implemented in a nonhomogeneous substrate (such as microstrip), the two fundamental propagating modes (viz., odd and even modes) have different modal phase velocities. The coupled section of length l_c is characterized by its even- and odd-mode characteristic impedances, Z_{0e}, Z_{0o}, respectively, and their even- and odd-mode electrical lengths, θ_e, θ_o. Since the structure is symmetric with respect to the plane BB′, even and odd excitation analysis can be employed [26] to obtain the CM and DM equivalent circuits (see Figure 12.16(b) and (c), respectively). According to the aforementioned symmetry, it is possible to define the following transmission coefficient, S_{21}, for the coupled line section:

$$S_{21} = \frac{1}{2}\left[S_{11}^{e} - S_{11}^{o}\right] \tag{12.10}$$

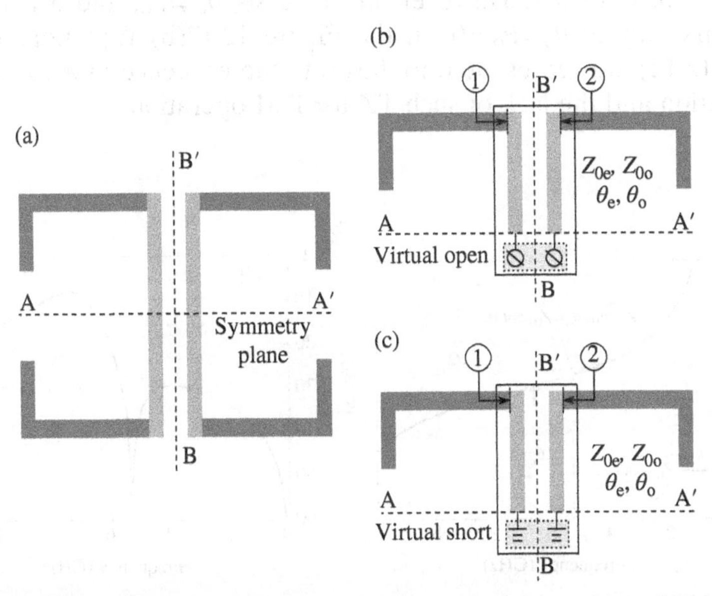

Figure 12.16 (a) Basic structure of magnetically coupled $\lambda_g/2$-resonators; (b) common-mode equivalent half circuit, and (c) differential-mode equivalent half circuit. Reprinted with permission from Ref. [3]; copyright 2015 IEEE.

where S_{11}^{e} and S_{11}^{o} are the reflection coefficients for even and odd excitations. This results into the following expressions for CM and DM transmission coefficients:

$$S_{21}^{cc} = jZ_0 \frac{Z_{0o}\cot\theta_o - Z_{0e}\cot\theta_e}{(jZ_{0e}\cot\theta_e - Z_0)(jZ_{0o}\cot\theta_o - Z_0)} \tag{12.11}$$

$$S_{21}^{dd} = jZ_0 \frac{Z_{0o}\tan\theta_o - Z_{0e}\tan\theta_e}{(jZ_{0e}\tan\theta_e - Z_0)(jZ_{0o}\tan\theta_o - Z_0)} \tag{12.12}$$

where $Z_0 = 25\Omega$ for CM and $Z_0 = 100\Omega$ for DM, respectively.

A simple analysis of (12.11) and (12.12) shows that for an inhomogeneous transmission medium such as microstrip ($\theta_e \neq \theta_o$), a CM TZ ($S_{21}^{cc} = 0$) occurs at the frequency f_{TZ} at which $Z_{0o}\cot\theta_o = Z_{0e}\cot\theta_e$. This equation can be graphically solved as shown in Figure 12.17(a). Note that the crossing point is located at a frequency corresponding to values of θ_e and θ_o lower than $\pi/2$. If the medium were homogeneous (i.e., $\theta_e = \theta_o$), a CM TZ occurs when $\cot\theta_e = \cot\theta_o = 0$ or $\cot\theta_e = \cot\theta_o = \infty$. The first two zeros are then found when $\theta_{e,o} = 0$ or $\theta_{e,o} = \pi/2$. For the DM, Equation (12.12) shows that for a homogeneous medium ($\theta_e = \theta_o$), the TZs would appear at frequencies corresponding to $\theta_e = \theta_o = 0$, π, or $\pi/2$. However, in our case, $\theta_e \neq \theta_o$, and a trivial TZ appears only at $\theta_e = \theta_o = 0$. In the Figure 12.17(b) Equations (12.11) and (12.12) are represented to illustrate the existence of a TZ for CM operation and the lack of such TZ for DM operation.

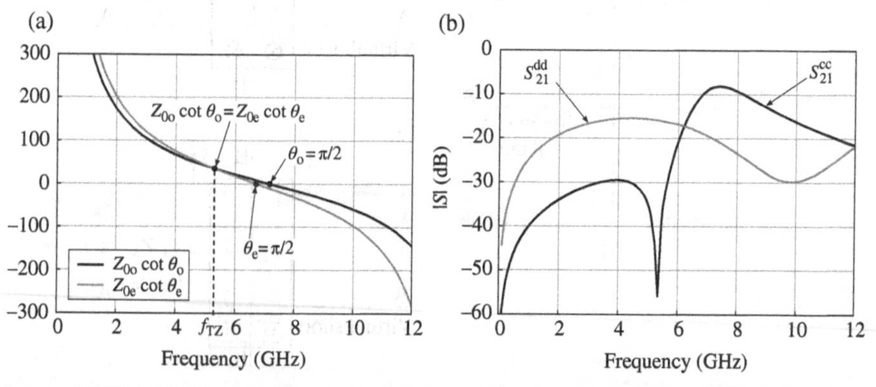

Figure 12.17 (a) Plots of $Z_{0o}\cot\theta_o$ and $Z_{0e}\cot\theta_e$ versus frequency. (b) Transmission coefficients for the coupled lines depicted in Figure 12.16. Reprinted with permission from Ref. [3]; copyright 2015 IEEE.

12.2.2.2 Filters with Magnetic Coupling: Second Example It has been shown in the previous section that the use of magnetic coupling is a simple but efficient strategy for designing balanced BPFs with high CM suppression and good DM performance. This fact has been illustrated with the simplest implementation, consisting of the design of a second-order ($n = 2$) Butterworth filter based on open-loop resonators. However this concept can be extended to an arbitrary number of resonators, thus allowing higher-order filters. As an example, a fourth-order ($n = 4$) Butterworth filter based on FSIRs using an *in-line* configuration is designed now. The following specifications are chosen: center frequency $f_0^d = 2.45\,\text{GHz}$ and fractional bandwidth FBW = 14%. A substrate of relative permittivity $\varepsilon_r = 3$ and thickness $h = 1.016\,\text{mm}$ is used. The required coupling coefficients and external quality factors for these specs can be calculated by using Equations (12.8)–(12.9), and they turn to be $Q_{e1} = Q_{e4} = 5.47$, $M_{12} = M_{34} = 0.117$, and $M_{23} = 0.076$. Design curves for $M_{i,i+1}$ and Q_e are similar to those of the previously designed filters. The filter layout and its final dimensions are shown in Figure 12.18. Note that the implementation of a higher-order filter using an in-line configuration results into alternating coupling schemes: electric coupling exists between resonators 1–2 and 3–4, whereas magnetic coupling applies to resonators 2–3. Therefore, this topology ensures the existence of at least a pair of

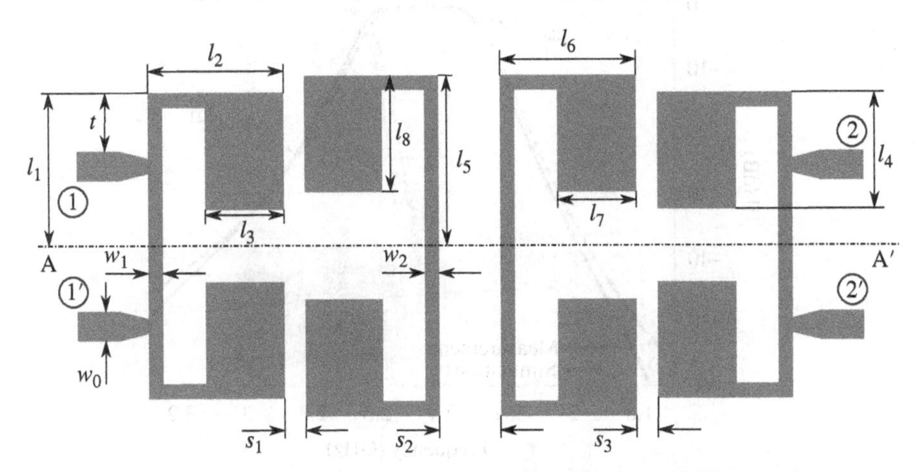

Figure 12.18 Layout of a fourth-order balanced BPF based on in-line coupled folded SIRs. The final dimensions of the designed filter are (in mm) $l_1 = 8.4$, $l_2 = 7.32$, $l_3 = 2.98$, $l_4 = 4.86$, $l_5 = 8.7$, $l_6 = 6.35$, $l_7 = 3.1$, $l_8 = 4.26$, $w_0 = 2.53$, $w_1 = w_2 = 1$, $t = 2.93$, $s_1 = s_3 = 0.21$, and $s_2 = 1.54$. Reprinted with permission from Ref. [3]; copyright 2015 IEEE.

magnetically coupled resonators. This should lead to a good CM noise suppression level.

When comparing the fourth-order filter with the second-order one, two differences are expected: (i) the higher-order filter should provide better DM selectivity, and (ii) better CM suppression should be observed. To corroborate these expectations, the simulated and measured results for this design are depicted in Figure 12.19(a) and (b). These

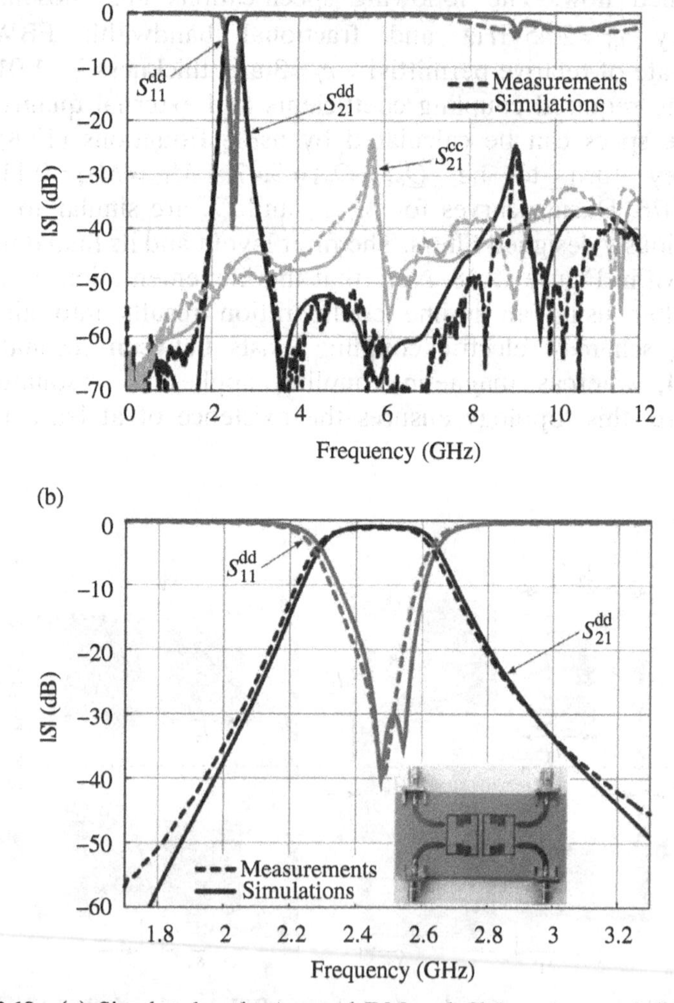

Figure 12.19 (a) Simulated and measured DM and CM responses of the designed fourth-order FSIR balanced bandpass filter in Figure 12.18. (b) Detail of the differential passband including a photograph of the fabricated prototype. Reprinted with permission from Ref. [3]; copyright 2015 IEEE.

figures show a CM rejection of approximately −50 dB within the differential passband, which has a measured insertion loss at f_0^{d} of 1.2 dB. In addition, both DM and CM out-of-band regions have been improved when compared with the filter shown in Figure 12.11. More specifically, $|S_{21}^{\mathrm{dd}}|$ is below −35 dB in most of the band until 12 GHz and no spurious passbands are observed. Selectivity has also been improved due to the increasing of the order of the filter.

Before ending this subsection, it should be pointed out that resonators 1 and 4 are different from resonators 2 and 3. As it has been explained before, the resonance frequency of the resonators is affected by the type of excitation. Hence, the dimensions of the resonators must be adjusted to ensure that all of them resonate at the same frequency.

12.2.3 Interdigital Line Resonators Filters

Another type of resonator that can be used to design narrowband balanced BPFs is the interdigital line resonator (ILR) [13], whose layout has been drawn in Figure 12.20. This resonator allows for the implementation of compact filters that offer good DM behavior and weak CM transmission. Figure 12.20(b) and (c) depicts the equivalent circuits for an ILR under DM and CM excitations, respectively. When operated in DM, the symmetry plane of the resonator is a virtual short circuit. On the other hand, for CM excitation, the symmetry plane is a virtual open circuit. A different response is then expected, as usual, for each mode of operation. For DM operation the structure can be seen as a hybrid resonator whose resonance frequency is f_0^{d}. For CM operation it can be seen as a multistub-loaded resonator with a resonance frequency f_0^{c}.

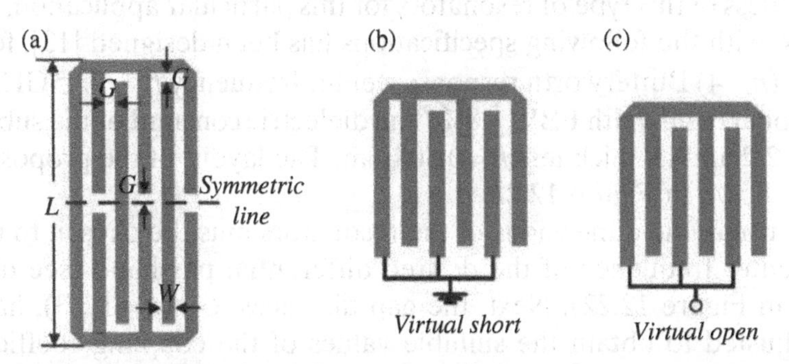

Figure 12.20 (a) Configuration of an interdigital line resonator (ILR). (b) Equivalent circuit for differential-mode operation. (c) Equivalent circuit for common-mode operation. Reprinted with permission from Ref. [13]; copyright 2015 John Wiley & Sons.

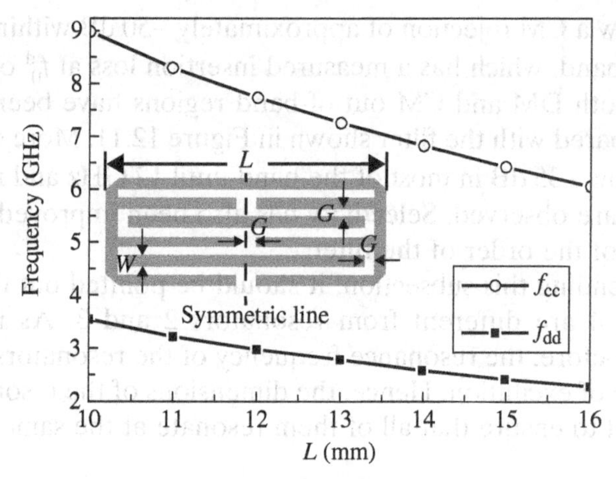

Figure 12.21 DM and CM resonance frequencies as a function of L for $W = 0.5$ and $G = 0.2$. Substrate parameters are dielectric constant $\varepsilon_r = 2.2$ and thickness $h = 0.508$. Reprinted with permission from Ref. [13]; copyright 2015 John Wiley & Sons.

In order to design a balanced filter, the dimensions of the resonators must be chosen in such a manner that their fundamental resonance frequency for DM excitation corresponds to the center frequency of the differential passband. If the gap size (G) and the strip width (W) are fixed, the resonance frequency of the resonator will be a function of the total length, L. The resonance frequency for both DM and CM excitations are plotted in Figure 12.21. According to this figure, $f_0^c >> f_0^d$. This means that within the differential passband, CM will be barely excited, ensuring in this way a good CM rejection at f_0^d.

12.2.3.1 ILR Filter Design Example In order to demonstrate the usefulness of this type of resonators for this particular application, a prototype with the following specifications has been designed [13]: fourth-order ($n = 4$) Butterworth response, center frequency $f_0^d = 2.45\,\text{GHz}$, and fractional bandwidth FBW = 8%. The dielectric constant of the substrate is $\varepsilon_r = 2.2$ and its thickness $h = 0.508\,\text{mm}$. The layout of the proposed filter is shown in Figure 12.22.

As usual, the dimensions of the resonators must be chosen to match the center frequency of the desired differential passband (see dimensions in Figure 12.22). Next, the gap distances, G_k ($k = 1\ldots3$), have to be adjusted to obtain the suitable values of the coupling coefficients. Finally, the tapping position, H, is adjusted to meet the required theoretical external quality factor. For this specific design, the values of the aforementioned electrical parameters are $Q_{e1} = Q_{e4} = 11.67$,

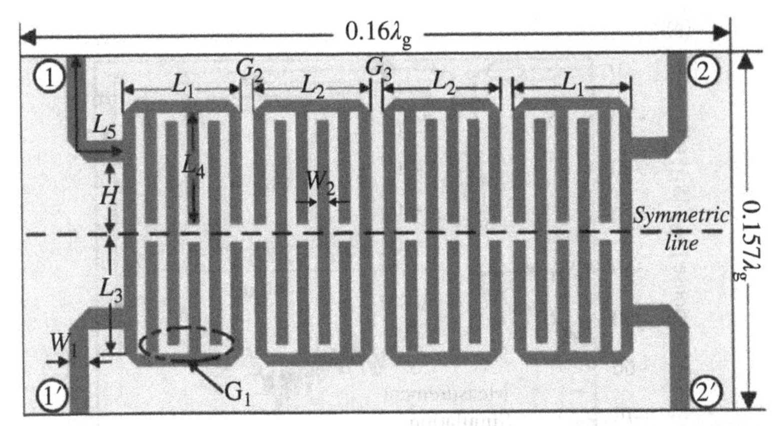

Figure 12.22 Balanced bandpass filter based on symmetric ILRs proposed in Ref. [13]. Dimensions (in mm) are $W_1 = 1.5$, $W_2 = 0.5$, $L_1 = 4$, $L_2 = 4.1$, $L_3 = 7.2$, $L_4 = 6.75$, $L_5 = 11.5$, $G_1 = 0.2$, $G_2 = 0.2$, $G_3 = 0.8$, and $H = 1.74$. Reprinted with permission from Ref. [13]; copyright 2015 John Wiley & Sons.

$M_{12} = M_{34} = 0.0728$, and $M_{23} = 0.056$. Once again, design curves for $M_{i,i+1}$ and Q_e, similar to those used in previous designs, can be plotted to extract the proper gap values and tapping positions. The filter layout and its final dimensions are shown in Figure 12.22. Simulated and measured results for both DM and CM responses are shown in Figure 12.23. The measured DM passband FBW is 8.5%, with a minimum insertion loss of 1.84 dB. It can be observed how CM is below −50 dB from DC to about 6.4 GHz. It is worth mentioning again that the CM has been suppressed thanks to its weak coupling value and poor excitation associated with the input/output resonators. It has been demonstrated in this section, following Ref. [13], that ILRs offer the possibility of designing balanced filters with good DM and CM performance and, at the same time, with very compact size.

12.2.4 Dual-Mode and Dual-Behavior Resonators for Balanced Filter Design

A different class of resonators commonly used in single-ended filters is dual-mode and dual-behavior resonators (DMRs and DBRs). The purpose of this subsection is to explore the possibility of applying such kind of resonators to balanced differential filter design. DMRs consist of electromagnetic resonant structures with two degenerated modes (same frequency of resonance) whose field distributions are orthogonal to each other [24]. Since no coupling exists between orthogonal modes, some

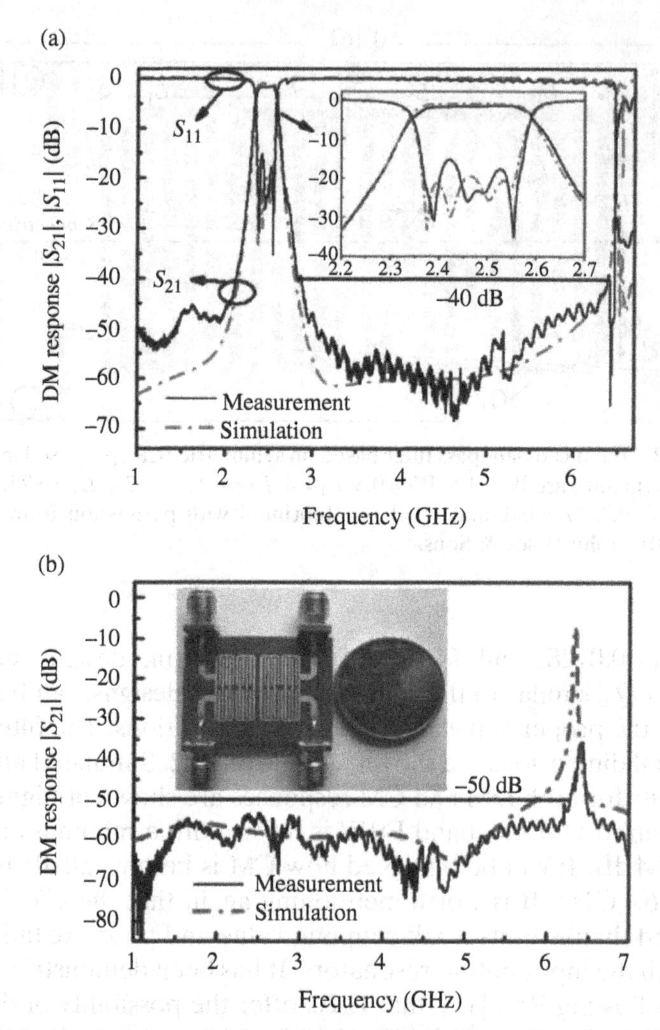

Figure 12.23 Simulations and measurements for the filter whose layout is shown in Figure 12.22 [13]. (a) Differential-mode response and (b) common-mode response. Reprinted with permission from Ref. [13]; copyright 2015 John Wiley & Sons.

perturbation must be introduced in the symmetry plane of the structure to give place to the required level of coupling. Note that, although a single physical structure is present, from the electrical point of view, the structure behaves as a pair of coupled resonators. This feature is, obviously, very useful for filter miniaturization. Planar DBRs result from the association of two different parallel bandstop structures. Each of them introduces a TZ at its own resonance frequency, but, at the same time, this association will be transparent along a narrow frequency range located between the two stopbands. In this way, DBR results in a high selectivity bandpass response created between the lower and upper stopbands. Let us give next some examples of balanced filters designed using DMRs and DBRs.

12.2.4.1 Dual-Mode Square Patch Resonator Filters

A very well-known, simple, and commonly used DMR is the square patch resonator. This resonator can be viewed as a resonant cavity by using the magnetic wall model depicted in Figure 12.24 [24]. The physical dimensions of the cavity are slightly larger than those of the physical printed patch in order to account for edge stray fields. In the frame of this model, this resonator is, basically, a parallelepiped-shaped microwave cavity with PEW (top and bottom walls) and PMW (lateral walls). This cavity has two degenerated fundamental modes that can be used for filter design purposes. It is possible to express the electromagnetic fields inside the cavity in terms of TM^{z}_{mn0} modes as follows:

$$E_z = \sum_{m=0}^{\infty} \sum_{n=0}^{\infty} A_{mn} \cos\left(\frac{m\pi}{a}x\right) \cos\left(\frac{n\pi}{a}y\right) \tag{12.13}$$

$$H_x = \left(\frac{j\omega\varepsilon_{\text{eff}}}{k_c^2}\right)\left(\frac{\partial E_z}{\partial y}\right), \quad H_y = \left(\frac{j\omega\varepsilon_{\text{eff}}}{k_c^2}\right)\left(\frac{\partial E_z}{\partial x}\right) \tag{12.14}$$

$$k_c^2 = \left(\frac{m\pi}{a}\right)^2 + \left(\frac{n\pi}{a}\right)^2 \tag{12.15}$$

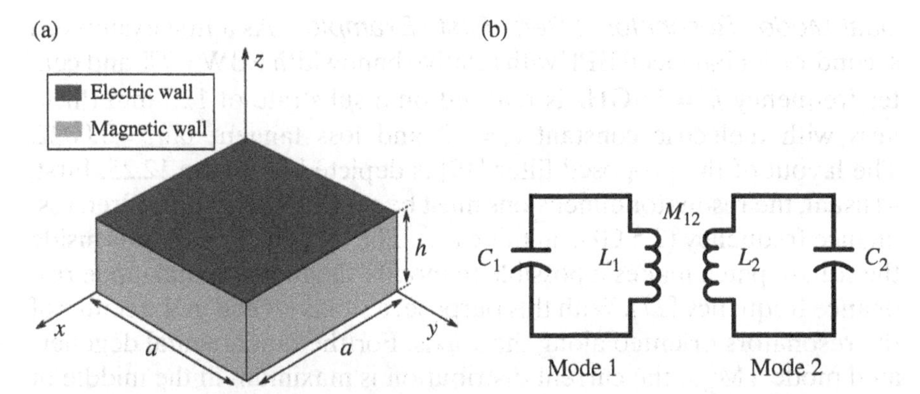

Figure 12.24 (a) Magnetic wall model of a square patch resonator (cavity model) and (b) its equivalent circuit [24].

where A_{mn} is the amplitude of the mnth mode, ω the angular frequency, and a and ε_{eff} the effective width and permittivity, respectively, of the equivalent magnetic wall cavity model. The resonance frequencies of the cavity can be calculated through the following expressions:

$$f_{mn0} = \frac{1}{2\pi\sqrt{\mu\varepsilon_{\text{eff}}}}\sqrt{\left(\frac{m\pi}{a}\right)^2 + \left(\frac{n\pi}{a}\right)^2} \tag{12.16}$$

From Equations (12.13), (12.14), and (12.15), it is clear that two degenerated modes, namely, TM^z_{100} and TM^z_{010}, exist. Both modes have orthogonal field distributions, being the electric field maximum at the edges of the patch and minimum in the middle of the cavity. From Equation (12.16), the resonance frequency of the two degenerated fundamental modes is

$$f_{100} = f_{010} = \frac{1}{2a\sqrt{\mu\varepsilon_{\text{eff}}}} \tag{12.17}$$

As previously mentioned, by perturbing the resonator symmetry plane, the two degenerated modes will be coupled to each other. This fact has been used in the implementation of compact single-ended filters since, compared with filters based on single-mode resonators, the total number of resonators required to implement the filtering function is, obviously, halved. Not much research has been carried out regarding DMR balanced BPFs. However, there are some contributions in the literature where square patch resonators are used with this purpose. We will briefly describe this work in the next paragraphs.

Dual-Mode Resonator Filter: First Example As a first example, a second-order balanced BPF with relative bandwidth FBW = 7% and center frequency $f_0^{\text{d}} = 2.5\,\text{GHz}$ is realized on a substrate of 1.27 mm thickness with dielectric constant $\varepsilon_r = 9.2$ and loss tangent $\tan\delta = 0.0022$. The layout of the proposed filter [19] is depicted in Figure 12.25. First, as usual, the resonator dimensions must be set to achieve the desired resonance frequency (2.5 GHz in this case). The introduction of slots inside the square patch makes it possible to modify the fundamental mode resonance frequency [27]. With this purpose, a slot is etched in the center of the resonators oriented along the x axis. For the fundamental degenerated mode TM^z_{100}, the current distribution is maximum in the middle of the patch, along the y axis. On the other hand, due to the orthogonality of the two degenerated modes, for the other fundamental mode TM^z_{010},

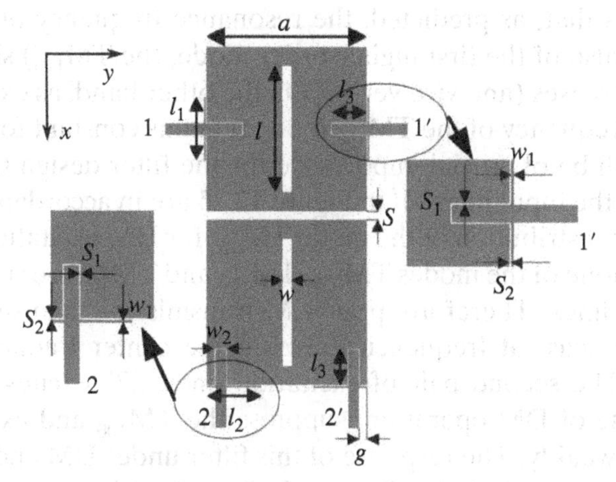

Figure 12.25 Layout of the second-order square patch resonator balanced filter proposed in Ref. [19]. Dimensions (in mm) are $l = 15.4$, $a = 20$, $w = 1.1$, $w_1 = 0.2$, $w_2 = 1.5$, $l_1 = 7.8$, $l_2 = 6.8$, $l_3 = 4.5$, $S = 0.6$, $S_1 = 0.1$, $S_2 = 0.1$, and $g = 1$. Reprinted with permission from Ref. [19]; copyright 2015 IEEE.

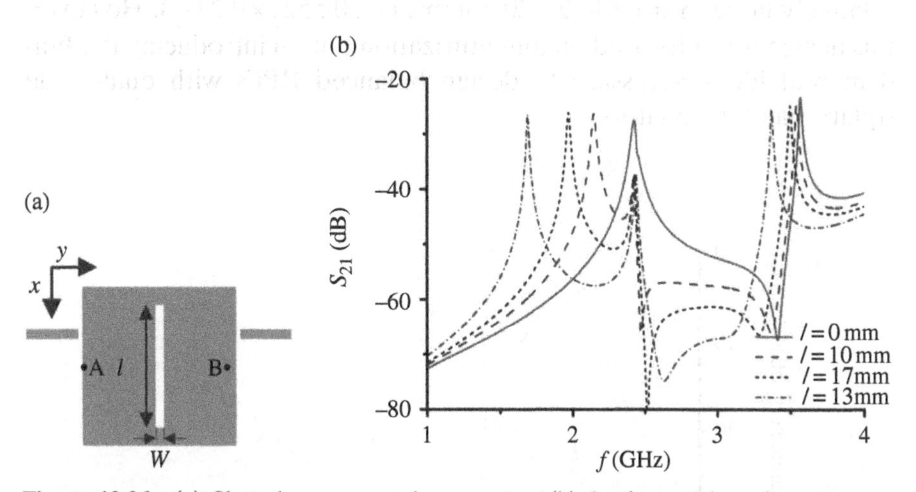

Figure 12.26 (a) Slotted square patch resonator; (b) fundamental mode resonance frequency of this resonator as a function of l. Reprinted with permission from Ref. [19]; copyright 2015 IEEE.

the current distribution will be maximum in the middle and along the x direction.

Since the slot is oriented along the x axis, the mode TM^z_{010} will be strongly affected by its presence. On the contrary, the TM^z_{100} mode will remain almost unperturbed. The influence of the slot length on the resonance frequency of the perturbed mode is shown in Figure 12.26. This figure shows that, as predicted, the resonance frequency of the TM^z_{010} mode (and also of the first higher-order mode, the TM^z_{110}) shifts downward as l increases (and vice versa). On the other hand, as expected, the resonance frequency of the TM^z_{100} mode remains constant for any value of l. This will be of crucial importance for the filter design because the positions of the input lines 11′ in Figure 12.25 are in accordance with the electric field distribution of the mode TM^z_{010} for DM excitation. For CM operation, none of the modes TM^z_{010}, TM^z_{110} and TM^z_{100} are supported by the feeding lines. Therefore, poor CM transmission is ensured by the structure, at least at frequencies around the center frequency of the DM filter. The second pair of excitation lines, 22′, excites the TM^z_{010} mode in case of DM operation, suppress the TM^z_{100} and excites mode TM^z_{110} very weakly. The response of this filter under DM and CM operation conditions is plotted in Figure 12.27. DM exhibits a passband centered at 1.8 GHz, with FBW of 7% and minimum level of insertion loss of 1.3 dB. CM is below −40 dB within the differential passband. In comparison with the filters considered previously in this chapter, this filter is relatively large in size (40.9×20.6 mm^2, i.e., $0.55\lambda_g \times 0.27\lambda_g$). However, this design is not focused on miniaturization but on introducing the fundamental ideas necessary to design balanced BPFs with dual-mode square patch resonators.

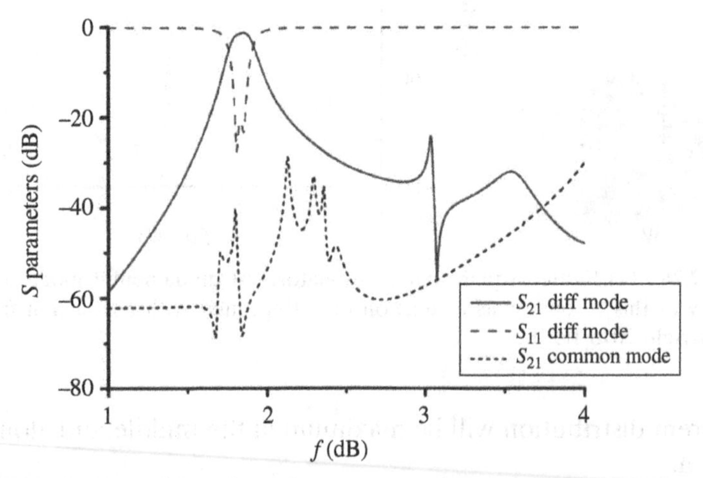

Figure 12.27 Differential-mode and common-mode responses of the second-order square patch resonator balanced filter reported in Ref. [19]. Dimensions (in mm) are (see Figure 12.25 for notation) $l = 15.4$, $a = 20$, $w = 1.1$, $w_1 = 0.2$, $w_2 = 1.5$, $l_1 = 7.8$, $l_2 = 6.8$, $l_3 = 4.5$, $s = 0.6$, $s_1 = 0.1$, $s_2 = 0.1$, and $g = 1$. Reprinted with permission from Ref. [19]; copyright 2015 IEEE.

Dual-Mode Resonator Filter: Second Example The schematics of a balanced BPF based on double-sided parallel-strip line (DSPSL) and dual-mode square patch resonators [14] are depicted in Figure 12.28. Figure 12.28(a) is a 3D view of the structure and Figure 12.28(b) and (c) shows the top and bottom layouts, respectively. The structure consists of a DSPSL dual-mode square patch resonator etched in a substrate with thickness h and two transformers located between the DSPSL and the microstrip feeding pair. The ground plane is inserted in the middle of the substrate, beneath the two circuits located at the top and bottom sides, which are identical in size. The layout of the transformer between the DSPSL and the microstrip feeding lines is depicted in Figure 12.29(a).

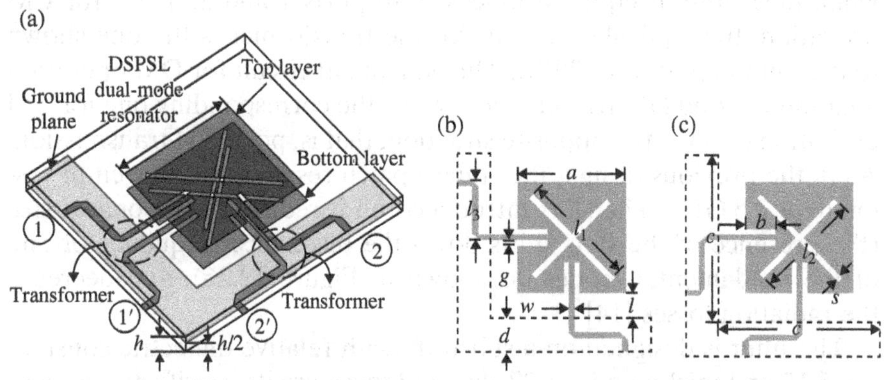

Figure 12.28 (a) 3D view of the DSPSL dual-mode square patch resonator filter described in Ref. [14]. (b) Top layer and (c) bottom layer of such filter. Reprinted with permission from Ref. [14]; copyright 2007 John Wiley & Sons.

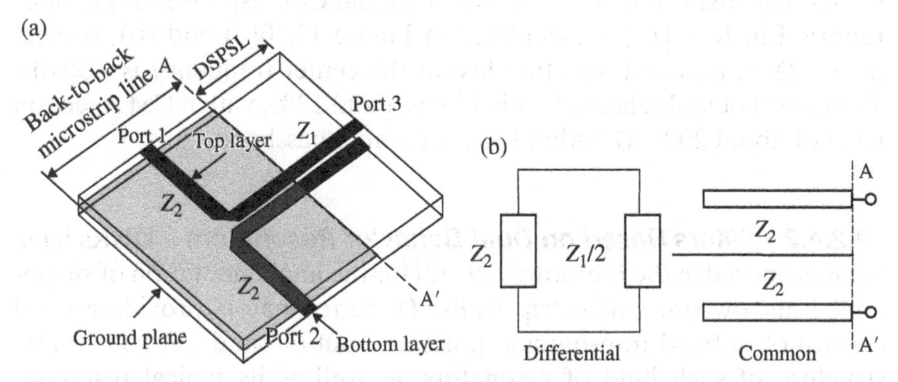

Figure 12.29 Layout of the transformer between DSPSL and microstrip feeding pair and its equivalent circuits [14]. (a) 3D view and (b) differential-mode (left) and common-mode (right) equivalent circuits. Reprinted with permission from Ref. [14]; copyright 2007 John Wiley & Sons.

The ground plane introduced in the middle of the structure transforms the DSPSL into two back-to-back microstrip lines, as it is shown in Figure 12.29(a). If both strips have the same width, it can be shown that $Z_1 = 2Z_2$ [28], Z_1 being the characteristic impedance of the DSPSL and Z_2 the characteristic impedance of the microstrip line on a substrate with thickness $h/2$. Figure 12.29(b) and (c) shows the DM and CM equivalent circuits of the structure, respectively. Under DM operation all the power is transferred from the differential port 1–2 to port 3, and then the transformer can be modeled as the left circuit figure in Figure 12.29(b). For CM operation the reference plane AA′ in Figure 12.29(a) can be treated as an open circuit, since the voltage distribution at the top and bottom strips in the back-to-back microstrip region are identical when measured at equal distances from ports 1 and 2. Thus, for CM operation, the equivalent circuit for the transformer is the one shown to the right in Figure 12.29(b). The equivalent circuit for DM operation guarantees good DM transmission, while the corresponding one for CM excitation ensures the opposite situation, that is, poor CM transmission. As in the previous design, the square patch resonator is chosen to resonate at 2.5 GHz. The slots introduced on it reduce the resonator size (the presence of the slots shifts down the resonance frequency of one of the fundamental modes, as shown in Figure 12.26), and decrease the radiation losses [14].

The filter is designed on a substrate with relative dielectric constant $\varepsilon_r = 6.15$ and thickness $h = 1.27$mm, and these are its specifications: center frequency $f_0^d = 2.5$GHz and fractional bandwidth FBW = 12%. The final physical dimensions are (in mm; see Figure 12.28) $g = 0.6$, $w = 0.9$, $d = 10$, $a = 16$, $b = 4.5$, $c = 39$, $l = 15$, $l_1 = 17$, $l_2 = 15.5$, $l_3 = 8.9$, and $s = 0.3$. The measured and simulated DM and CM responses of the filter reported in Ref. [14] are depicted in Figure 12.30(a) and (b), respectively. The measured insertion loss at the center frequency is 1.55 dB. The experimental relative bandwidth is about 12%, with a CM rejection level of about 29.5 dB within the differential passband.

12.2.4.2 Filters Based on Dual-Behavior Resonators DBRs have been proposed in the literature [29–32] for the implementation of single-ended narrowband microstrip BPFs. These resonators provide a good control of in-band transmission poles and out-of-band TZs. The basic structure of such kind of resonators, as well as its typical response, can be seen in Figure 12.31. Note that a DBR results from the association of two parallel resonant structures, which will be characterized by the impedances Z_{S1} and Z_{S2}, exhibiting a bandstop frequency response. At resonance, each branch introduces a TZ. Once the two stopbands

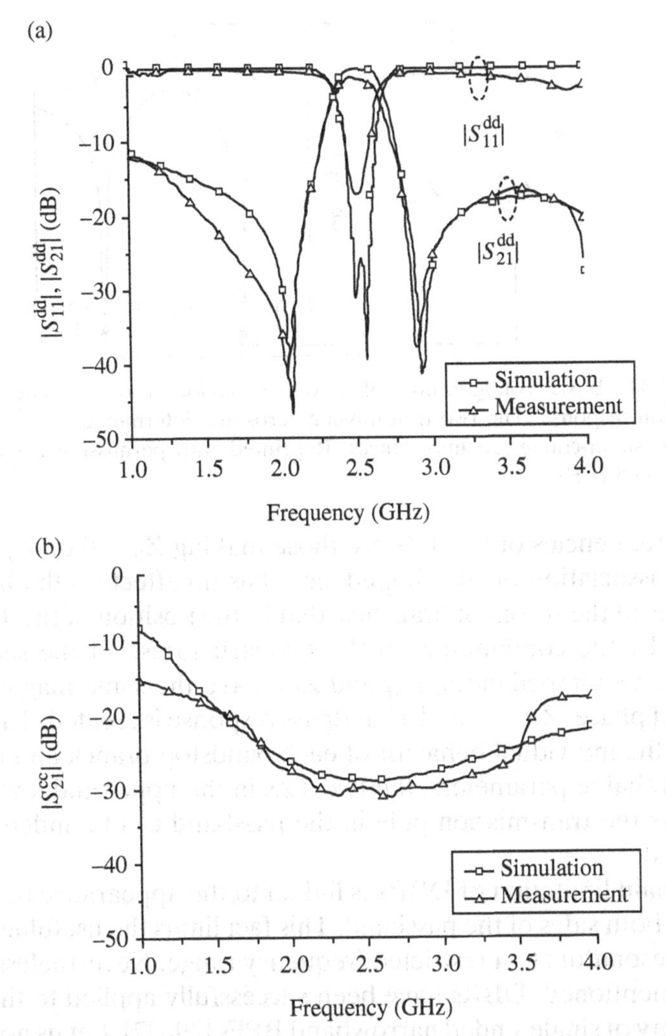

Figure 12.30 Measured and simulated responses of the balanced BPF reported in Ref. [14]: (a) differential-mode and (b) common-mode operation. Reprinted with permission from Ref. [14]; copyright 2007 John Wiley & Sons.

have been properly allocated, the shunt association of the two branches gives place to the appearance of a passband somewhere between the upper and lower rejected bands. To understand why this association is transparent within a given frequency range, it is convenient to consider the frequency behavior of the total impedance of the shunt structure, given by

$$Z = \frac{Z_{S1} Z_{S2}}{Z_{S1} + Z_{S2}} \tag{12.18}$$

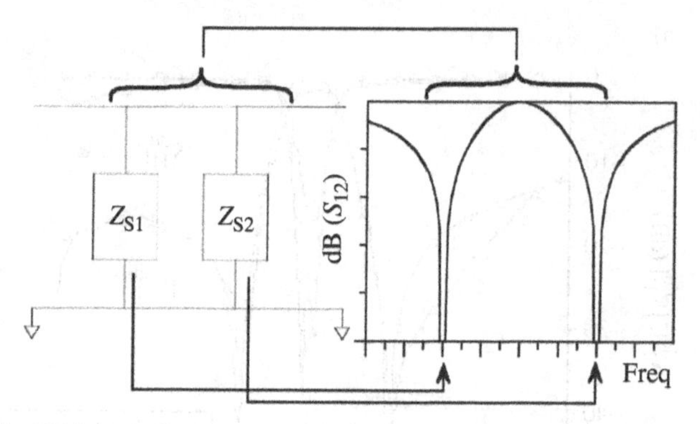

Figure 12.31 Basic configuration of a dual-behavior resonator and its typical transmission response. The two transmission zeros are determined by the resonances of the two shunt-connected impedances. Reprinted with permission from Ref. [31]; copyright 2005 IEEE.

The frequencies of the TZs are those making $Z_{S1} = 0$ or $Z_{S2} = 0$. The parallel association of such impedances has no effect on the individual responses of the resonant branches, that is, the position of the TZs is not affected by the combination of the two structures. At the same time, when the two impedances, Z_{S1} and Z_{S2}, have the same magnitude but are out of phase, $Z \rightarrow \infty$ and a bandpass response is created. Taking into account the individual behavior of each bandstop branch and the number of available parameters, the two TZs in the upper and lower bands as well as the transmission pole in the passband can be independently controlled.

The main limitation of DBRs is linked to the appearance of spurious bands at both sides of the passband. This fact limits the usefulness of this kind of resonators to a restricted frequency range. Nevertheless, as previously mentioned, DBRs have been successfully applied to the design of a variety of single-ended narrowband BPFs [29–32]. Let us now see an example of how this type of resonators can also be used to design BPFs with differential response. Figure 12.32(a) shows the schematic circuit representation of the balanced BPF based on DBRs proposed in Ref. [15]. The filter consists of two pairs of DBRs with one short-circuited stub (Z_3, θ_3) and one open-circuited stub (Z_4, θ_4). Each DBR is composed of two coupled lines that are short-circuited (Z_1, θ_1), whose even- and odd-mode characteristic impedances are Z_{oe} and Z_{oo}, respectively, plus two pairs of transmission lines (Z_1, θ_5, Z_2, θ_2). The equivalent circuits for DM and CM operation are depicted in Figure 12.32(b) and (c), respectively. Under DM excitation, the symmetry plane behaves as a virtual short circuit. The external quality factor and input admittance for the circuit in Figure 12.32(b) can be calculated as

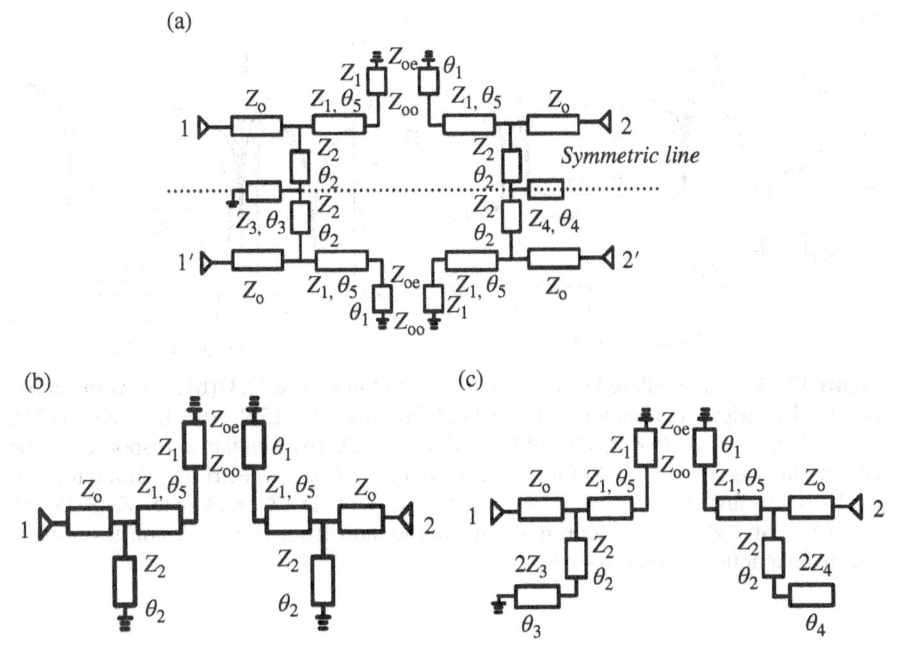

Figure 12.32 (a) Schematic representation of the differential filter based on DBRs proposed in Ref. [15]. (b) Differential-mode equivalent circuit and (c) common-mode equivalent circuit. Reprinted with permission from Ref. [15]; courtesy of the Electromagnetics Academy 2015.

$$Y_{\text{in1}} = -j\cot(\theta_1 + \theta_5)/Z_1 - j\cot\theta_2/Z_2 \tag{12.19}$$

$$Q_e = 0.5R_L\left((\theta_1 + \theta_5)\csc^2(\theta_1 + \theta_5)/Z_1 + \theta_2\csc^2\theta_2/Z_2\right) \tag{12.20}$$

where R_L is the load impedance for Z_1 and Z_2. The TZs appear when the condition $\theta_1 + \theta_5 = \pi$ is fulfilled [33]. The resonance condition for the two shorted stubs can be obtained by imposing $Y_{\text{in1}} = 0$. This leads to a band-pass response whose quality factor and coupling coefficient can be extracted from the filter specifications using the standard procedure used before in this chapter [24]. The simulated response of the circuit in Figure 12.32(b) (DM operation) is plotted in Figure 12.33(a). The TZs f_{tz1} and f_{tz2} are realized by the short-circuited stub (Z_1, θ_1), and the other two TZs, f_{ctz1} and f_{ctz2}, are generated due to the bandstop response of the coupled transmission lines. When the circuit is excited by a CM signal, the symmetry plane behaves as a virtual open circuit. In such case, the input impedance can be written as

$$Y_{\text{in2}} = -j\cot(\theta_1 + \theta_5)/Z_1 + j\frac{Z_2 + 2Z_4\cot\theta_4\tan\theta_2}{2Z_2Z_4\cot\theta_4 - Z_2^2\tan\theta_2} \tag{12.21}$$

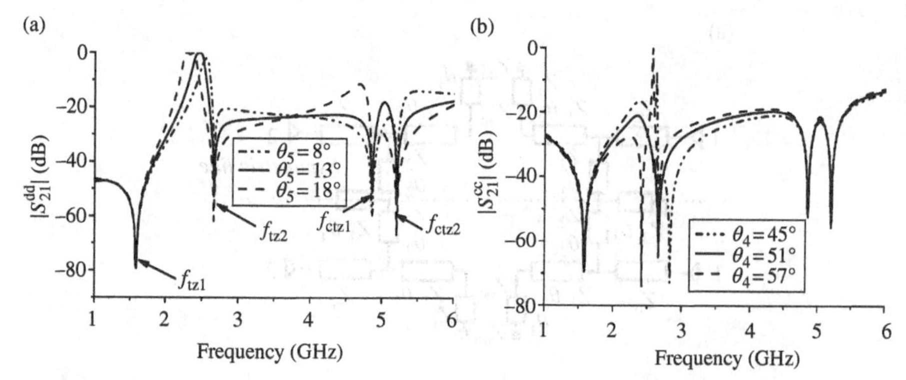

Figure 12.33 (a) Simulated response of the circuit in Figure 12.32(b) for several values of θ_5. Electrical parameters at $f=2.45\,\text{GHz}$ are $\theta_1=159°$, $\theta_2=13°$, $Z_1=117\,\Omega$, $Z_{oe}=147.6\,\Omega$, $Z_{oo}=75.8\,\Omega$, $Z_2=65\,\Omega$, and $Z_0=50\,\Omega$. (b) Simulated response of the circuit in Figure 12.32(c) for several values of θ_4. Electrical parameters at $f=2.45\,\text{GHz}$ are $\theta_1=159°$, $\theta_2=13°$, $\theta_5=13°$, $Z_1=117\,\Omega$, $Z_{oe}=147.6\,\Omega$, $Z_{oo}=75.8\,\Omega$, $Z_2=65\,\Omega$, and $Z_0=50\,\Omega$. Reprinted with permission from Ref. [15]; courtesy of the Electromagnetics Academy 2015.

It is clear that a CM TZ occurs when $Z_{in2}=1/Y_{in2}=0$. The CM frequency response for several values of θ_4 is represented in Figure 12.33(b). This figure shows that although the CM response has several TZs, a transmission peak appears close to the upper differential passband for $\theta_4=45°$ and $\theta_4=57°$. This transmission maximum is associated with the contribution of the open stub with impedance Z_2 and electrical length θ_2. This is the reason for the inclusion of the open stub (Z_4, θ_4), whose aim is to suppress this unwanted CM resonance. From the observation of Figure 12.33(b), it can be noticed that a TZ is introduced at the frequency at which the CM appears by choosing $\theta_4=51°$. In this way, good CM rejection is obtained over a wide frequency band. The authors of this study designed and fabricated a filter prototype [15] with $f_0^d=2.45\,\text{GHz}$, FBW = 4.9%, and Butterworth transfer function. They used a Rogers 5880 substrate ($\varepsilon_r=2.2$, loss tangent $\tan\delta=0.0009$, and thickness $h=0.508$ mm). The layout of the filter can be seen in Figure 12.34(a). The coupled-resonator procedure that has been extensively used in previous sections is employed here again to extract the required external quality factor $Q_e=28.6$ and coupling coefficient $M=0.049$. First, by means of (12.19) and (12.20), the desired values of $\theta_1+\theta_5$ and θ_2 are obtained by properly adjusting the ratio Z_1/Z_2. Then, the separation distance between the two shorted coupled lines and θ_1 is selected so that the DM transmission has a good performance. Finally, CM suppression is maximized by finding the better values of the electrical parameters Z_3, θ_3, Z_4, and θ_4. The measured and simulated DM and CM responses of the designed filter are reproduced in Figure 12.34(b) and (c), respectively. Good agreement between simulation and measurements is appreciated. Note that the differential response exhibits good selectivity, with four TZs (simulation) located

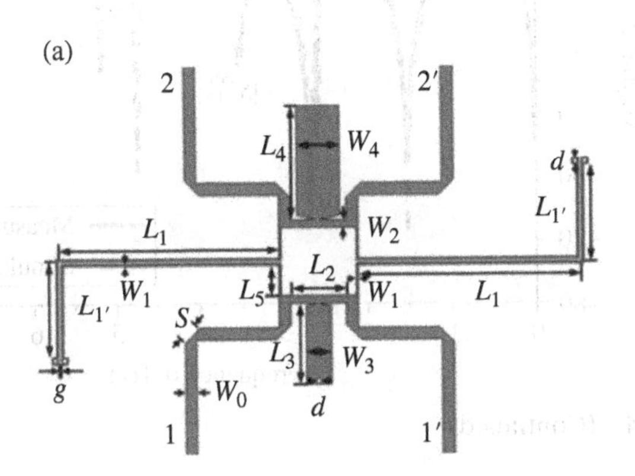

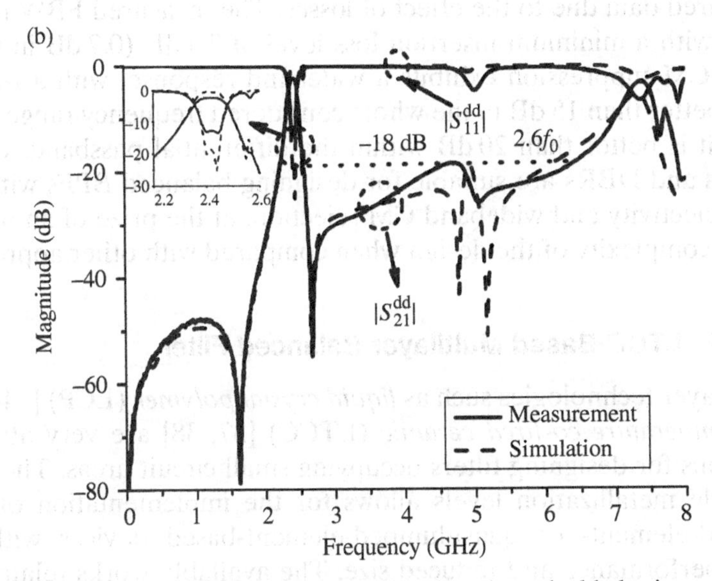

Figure 12.34 (a) Layout of the balanced filter based on dual-behavior resonators proposed in Ref [15]. Dimensions are (in mm) $L_1 = 28$, $L_{1'} = 11.9$, $L_2 = 6.8$, $L_3 = 10$, $L_4 = 14$, $L_5 = 3.7$, $W_0 = 1.53$, $W_1 = 0.3$, $W_2 = 1$, $W_3 = 3.1$, $W_4 = 5.3$, d = 0.6, $g = 0.3$, and S = 2.63. (b) Measured and simulated data for the DM response; (c) measured and simulated data for the CM response. Reprinted with permission from Ref. [15]; courtesy of the Electromagnetics Academy 2015.

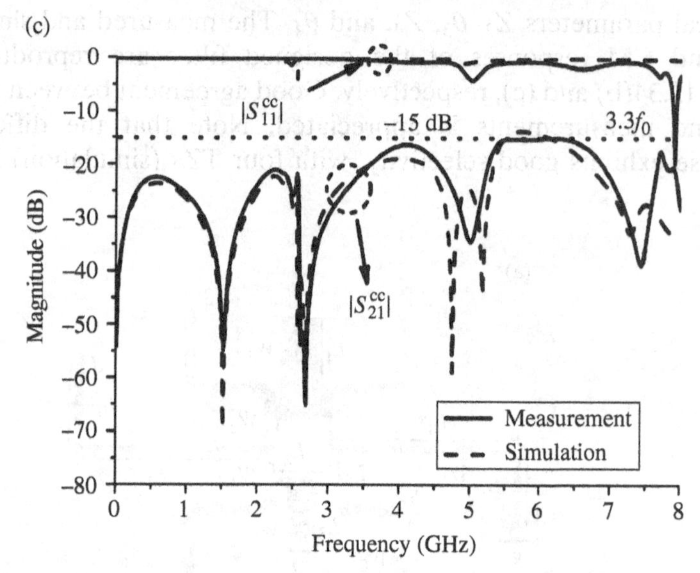

Figure 12.34 (Continued)

at 1.62, 2.67, 4.78, and 5.18 GHz. One of the TZs does not appear in the measured data due to the effect of losses. The measured FBW is about 5.2% with a minimum insertion loss level of 2.4 dB (0.7 dB in simulation). CM suppression exhibits a wideband response, with a rejection level better than 15 dB in the whole considered frequency range. Moreover, it is better than 20 dB within the differential passband. Clearly, DMRs and DBRs are suitable for designing balanced BPFs with good DM selectivity and wideband CM rejection, at the prize of an increase in the complexity of the design when compared with other approaches.

12.2.5 LTCC-Based Multilayer Balanced Filter

Multilayer technologies such as *liquid crystal polymer* (LCP) [34–36] or *low-temperature co-fired ceramic* (LTCC) [37, 38] are very attractive solutions for designing filters occupying small circuit areas. The use of multiple metallization levels allows for the implementation of many lumped-element- or quasi-lumped-element-based devices with very good performance and reduced size. The available works related with the application of these multilayer compact technologies to the implementation of differential narrowband filters with inherent CM suppression are very scarce, and more research is probably required in the years to come. In this subsection an example of differential filter based on the use of LTCC technology is discussed [17].

A balanced BPF on LTCC technology can be obtained if, for instance, a classical folded $\lambda_g/2$ resonator is fabricated using an LTCC multilayer structure. This was suggested in Ref. [17]. The implementation of this idea is shown in Figure 12.35. The open-ended $\lambda_g/2$ resonator has been divided into six parts (asymmetrical strip lines), as depicted in Figure 12.35(b). The four middle layers are all connected by means of via holes. The filter based on this resonator, proposed in Ref. [17], is shown in Figure 12.36.

For the resonator in Figure 12.35, the fundamental resonance frequency corresponds to an odd-symmetry resonance. The normalized voltage distribution for this resonator at the fundamental resonance frequency is the one shown in Figure 12.36(b). From this figure it is clear that the symmetry plane can be treated as a virtual ground plane, resulting in a multilayered structure where each layer has its own ground plane (actual or virtual), thus maintaining the stripline

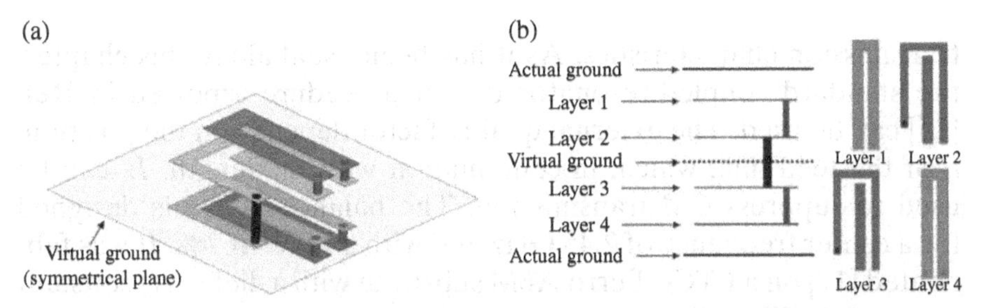

Figure 12.35 (a) 3D view of the distributed-element resonator (ground not shown) proposed in Ref. [17]. (b) Cross-sectional view and four layer configuration. Reprinted with permission from Ref. [17]; copyright 2013 Institution of Engineering & Technology.

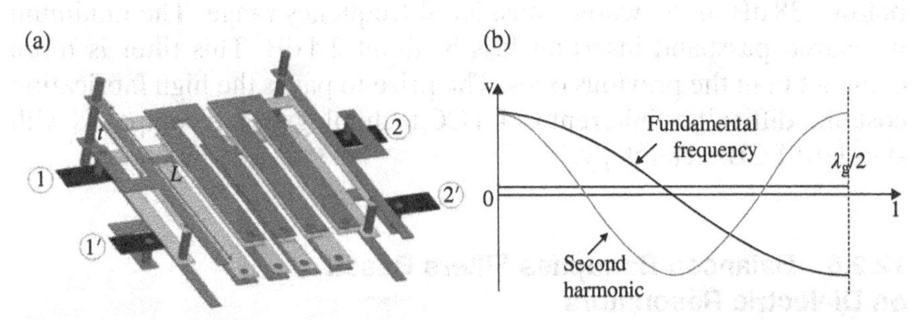

Figure 12.36 (a) Structure of an LTCC balanced bandpass filter proposed in Ref. [17]. (b) Normalized voltage distribution of the resonator in Figure 12.35. Reprinted with permission from Ref. [17]; copyright 2013 Institution of Engineering & Technology.

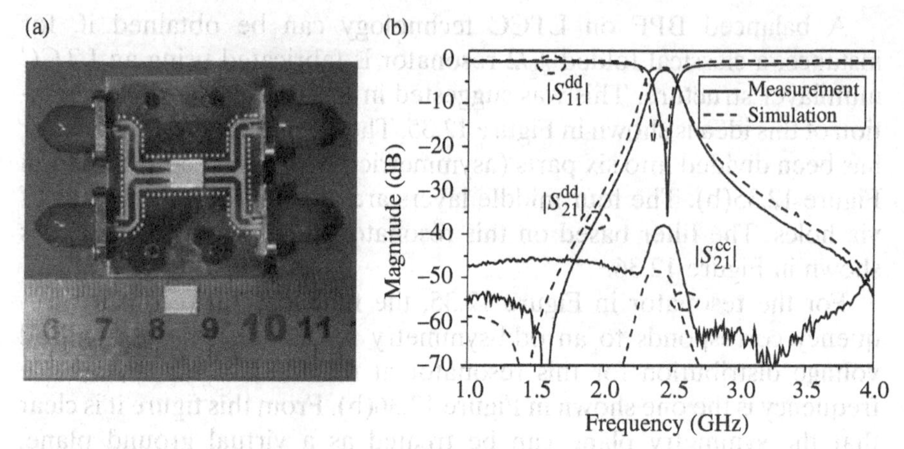

Figure 12.37 (a) Photograph of the LTCC balanced filter fabricated in Ref. [17]. (b) Measured and simulated results for its differential- and common-mode responses. Reprinted with permission from Ref. [17]; copyright 2013 Institution of Engineering & Technology.

transmission characteristics. As it has been usual along this chapter, the standard coupled-resonator design procedure reported in Ref. [24] can be used. The external quality factor depends on the position, t, of the feed line, which, in combination with the length, L, can be used to suppress CM transmission. The balanced filter is designed for a center frequency of 2.45 GHz and with a FBW of 7%. It was fabricated [17] on a LTCC Ferro A6M substrate with a dielectric constant of $\varepsilon_r = 5.9$. The filter is compact in size, occupying an area of $0.119\lambda_g \times 0.099\lambda_g$ at f_0^d. A photograph of the fabricated prototype mounted on a PCB for measurement purposes is depicted in Figure 12.37(a). Simulated and measured performance for both DM and CM operations is plotted in Figure 12.37(b). CM transmission is below −38 dB in the whole considered frequency range. The minimum measured passband insertion loss is about 2.4 dB. This filter is more compact than the previous ones. The prize to pay is the high fabrication cost and difficulties inherent to LTCC technology when compared with standard PCB technology.

12.2.6 Balanced Bandpass Filters Based on Dielectric Resonators

The technologies used in the preceding sections are, in general, suitable for relative FBW larger than, at least, 1%. For FBW < 1%, the low quality factor of such technologies causes a significant increase in the insertion loss within the passband, resulting in an overall degradation of the filter performance. A method to overcome this problem is based on the use of high-Q dielectric resonators (DRs). This kind of resonator has widely been used in single-ended circuits [39, 40], but it has not been until very recently that DRs have been employed in the implementation of differential filters [41–43]. DRs are, basically, resonant structures made of a low-loss high-dielectric-constant material. They use to be mounted inside a metal enclosure using a low-dielectric-constant support [44]. Most of the electromagnetic field is concentrated inside the DR, although a significant portion of the energy is confined in the close proximity of the dielectric structure in the form of evanescent-like waves. The low-loss material used to fabricate the DR provides the required high Q factor (greater than 10 000 at 4 GHz). This feature makes this type of resonators very suitable for very narrowband filter applications. Typically, DRs have cylindrical shape although other geometrical forms are also possible. Besides, the metal enclosure is not always employed. It should be mentioned that, in contrast with other kind of frequently employed resonators, it is difficult to find analytical or approximate expressions for the resonance frequencies and the obtaining of accurate enough solutions requires the use of computationally intensive numerical methods [44].

Rectangular DRs have very recently been used to design narrowband balanced BPFs [41]. In that reference, the modes $TE_{11\delta}$ and $TM_{11\delta}$, which have the lowest resonance frequencies, are utilized to control the DM and CM responses. The resonance frequency of both modes can be adjusted by means of the resonator dimensions, as depicted in Figure 12.38(a). In this figure, A and B have been chosen to be $A = 20$mm and $B = 16$mm, while L is varied. The dielectric constant of the resonator is $\varepsilon_r = 36.5$ and its loss tangent is 2.5×10^{-4}. To understand how this resonator can be used to design a filter with differential operation, it is convenient to analyze the field distribution for the fundamental mode ($TE_{11\delta}$). Figure 12.38(b) shows a top view of the pair of coupled rectangular DRs. Taking into account the field pattern of the $TE_{11\delta}$ mode (see Ref. [41]), it can be easily appreciated that this mode is excited under differential excitation of the structure. On the other hand, if a CM excitation is applied, the $TE_{11\delta}$ will not be generated, since the electric field distribution for CM is not consistent with the electric field of such mode. In this way, a differential filter with no CM transmission within the passband is expected to be obtained using the configuration in Figure 12.38(b). A second-order balanced BPF with $f_0^{\mathrm{d}} = 3.966$GHz, 0.13 dB-ripple (Chebyshev response), and a small fractional bandwidth

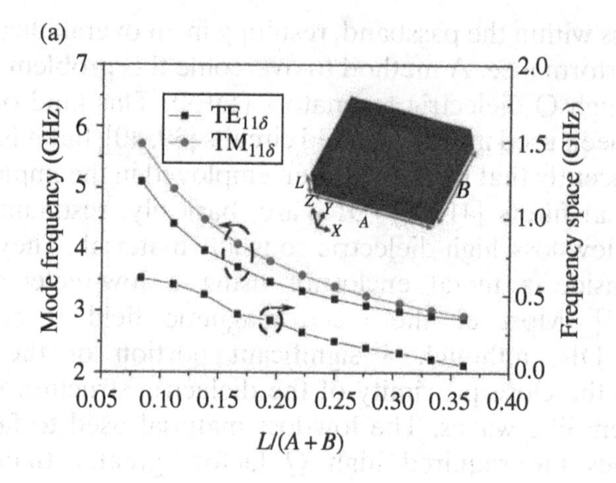

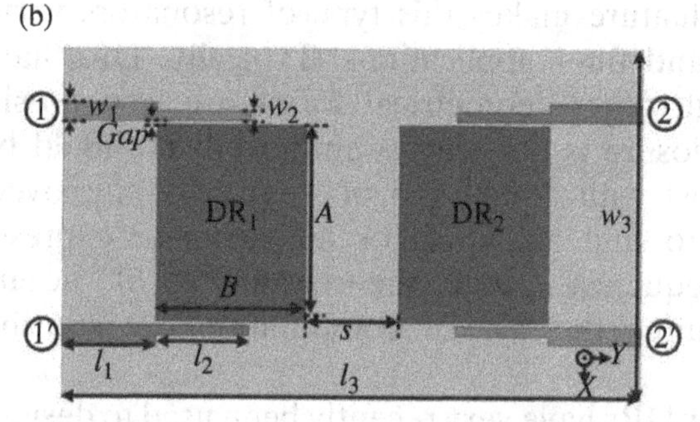

Figure 12.38 (a) TE$_{11\delta}$ and TM$_{11\delta}$ resonance frequencies and frequency spacing between them as a function of $L/(A+B)$. Parameters: $\varepsilon_r=36.5$, $A=20$mm, and $B=16$mm (L is the changing variable). (b) Top view of the filter configuration showing the DRs without shielding cavity. Reprinted with permission from Ref. [41]; copyright 2015 IEEE.

of FBW = 0.265% was designed, fabricated, and measured in Ref. [41]. The PCB substrate parameters are $\varepsilon_r=3.38$, thickness $h=0.8182$mm, and loss tangent $\tan\delta=0.0027$. The low-pass prototype element values are $g_0=1$, $g_1=0.9107$, and $g_2=0.6437$. For these specifications, the required external quality factor and coupling coefficient result to be $Q_e=344$ and $M_{12}=0.0035$. The external quality factor depends on the length l_2 (see Figure 12.38b), while the coupling coefficient is a function of the separation between resonators. Using a full-wave electromagnetic simulator, the values of Q_e and M_{12} can be represented as functions of the relevant dimensional parameters, and, from those curves, the physical dimensions of the filter are obtained as usual. In this case, the appropriate values of the physical dimensions are (in mm) $A=20$, $B=16$, $L=5$, $s=10$, gap = 0.4, $w_1=1.86$, $w_2=1$, $w_3=40.8$, $l_1=10$, $l_2=10$,

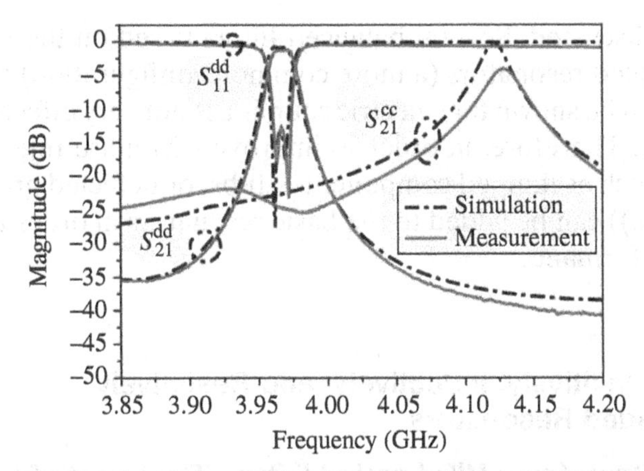

Figure 12.39 Simulated and measured results for the balanced filter reported in Ref. [41]. Reprinted with permission from Ref. [41]; copyright 2015 IEEE.

$l_3 = 62$. The height of the cavity is $h_1 = 12.3$mm. The simulated and measured results for this filter are depicted in Figure 12.39. CM suppression is better than 20 dB within the differential passband. DM offers very good selectivity and low insertion loss in the passband in spite of the very small FBW. This is the advantage of using high Q value DRs.

Another design similar to the one discussed in the previous paragraphs can be found in another recent publication [42]. In this case, cylindrical resonators are used instead of the rectangular ones employed in Ref. [41]. The main difference between cylindrical and rectangular resonators relies in the resonance modes. Whereas the modes with the lowest resonance frequencies in rectangular resonators are $TE_{11\delta}$ and $TM_{11\delta}$, for cylindrical resonators the modes with the lowest resonance frequencies are $TE_{01\delta}$ and $TM_{01\delta}$. Otherwise, the design process is exactly the same.

12.3 LOADED RESONATORS FOR COMMON-MODE SUPPRESSION IMPROVEMENT

In the previous sections of this chapter, several balanced BPFs with intrinsic CM rejection have been reported. They make use of magnetic (or mixed) coupled printed resonators or more complex configurations (such as DMRs, DBRs, multilayer LTCC resonators, or DRs). Despite the unquestionable advantages of having intrinsic CM rejection features, the developed filters require the use of relatively complicated designs to implement the desired DM response (with the exception of the case treated in Section 12.2.2). In the present section a different strategy is discussed. Several balanced filters based on the use of electrically coupled resonators (a more common configuration) will be presented. It will be shown that electric coupling is not very effective for CM suppression. Therefore, in order to improve CM noise rejection, extra elements (such as lumped components, stubs, or defected ground structures (DGSs)) can be added to the basic resonators in order to improve the CM performance.

12.3.1 Capacitively, Inductively, and Resistively Center-Loaded Resonators

12.3.1.1 Open-Loop UIR-Loaded Filter The layout of two electrically coupled open-loop UIRs is depicted in Figure 12.40(a). This configuration was used in Ref. [4] as the basic building block to implement a balanced BPF. In order to improve the CM response, the two resonators are proposed to be loaded with lumped elements placed along the symmetry plane. Under DM excitation, the symmetry plane is a virtual short circuit. In such case, the loading lumped elements will have no influence on the observed electrical response. On the contrary, for CM excitation the symmetry plane behaves as a virtual open circuit, in such a way that the CM transmission is expected to be strongly affected by the loading components. The DM and CM equivalent circuits for the center-loaded half-wavelength open-loop UIR are shown in Figure 12.40(b). In the following paragraphs two different cases will be analyzed: (i) the loading components are a capacitor and a resistor, and (ii) the loading elements are an inductance and a resistor.

Common-Mode Response for Capacitor–Resistor Loading Let us consider the equivalent circuit, for CM operation, depicted in Figure 12.40(b) (bottom right) assuming that the loading elements are a capacitor (C) and a resistor (R). The input admittance, Y_{in}^{c}, for this circuit is given by

$$Y_{\text{in}}^{\text{c}} = \frac{Y_{\text{L}} + j2Y\tan\theta}{2Y + jY_{\text{L}}\tan\theta}; \quad Y_{\text{L}} = \frac{j\omega C}{jR\omega C + 1} \tag{12.22}$$

where θ is the electrical length of the microstrip line section and ω the angular operation frequency. The input admittance as a function of R and C is found then to be

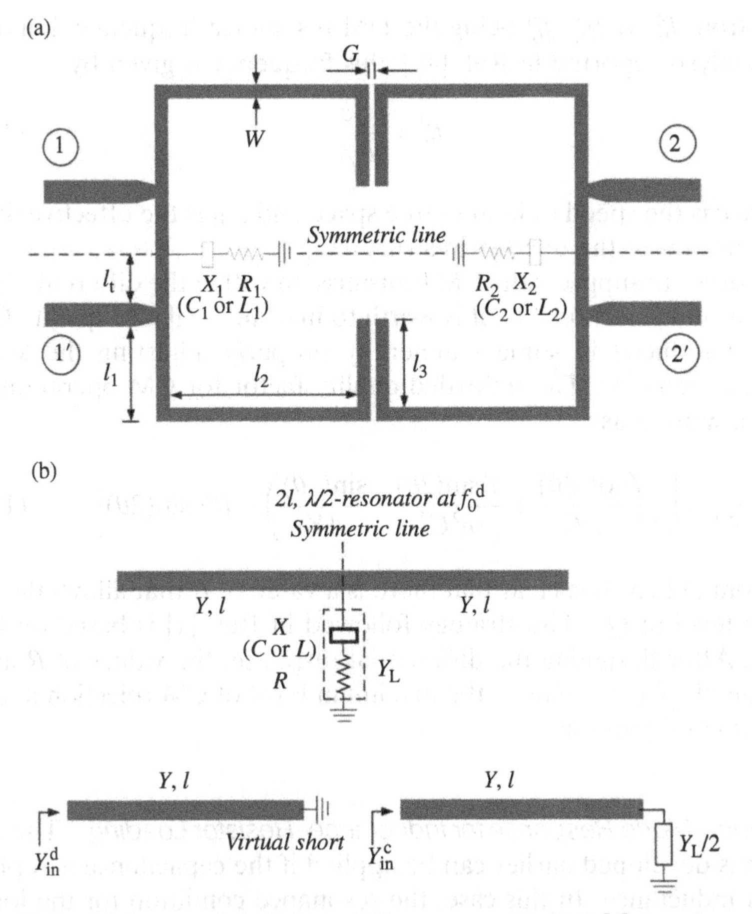

Figure 12.40 (a) Layout of the balanced filter proposed in Ref. [4], based on center-loaded open-loop uniform impedance resonators. (b) A half-wavelength center-loaded resonator (top) and its differential-mode (bottom left) and common-mode (bottom right) equivalent circuits. Reprinted with permission from Ref. [4]; copyright 2010 IEEE.

$$Y_{in}^{c} = \frac{Y}{(2Y - \omega C \tan\theta)^2 + (2YR\omega C)^2} \times \{2YR\omega^2 C^2 (1 + \tan^2\theta) + j[2Y\omega C(1 - \tan^2\theta) + (4Y^2R^2\omega^2C^2 + 4Y^2 - \omega^2C^2)\tan\theta]\} \quad (12.23)$$

The resonance condition is given by $Y_{in}^{c} = 0$. An inspection of (12.23) when the resonance condition is enforced reveals that by changing C it is possible to tune the CM resonance frequency. For $R = 0$ the tuning range goes from f_0^{d} to $2f_0^{\mathrm{d}}$, f_0^{d} being the DM resonance frequency. Following the analysis reported in Ref. [45], this frequency is given by

$$f_0^{\mathrm{d}} = \frac{c}{4l\sqrt{\varepsilon_{\mathrm{eff}}}} \tag{12.24}$$

where c is the speed of light in free space and $\varepsilon_{\mathrm{eff}}$ is the effective dielectric constant of the microstrip structure.

In order to suppress the CM transmission within the differential passband as much as possible, it is worth to investigate if CM quality factor can be reduced in some manner by properly adjusting the loading lumped elements. The unloaded quality factor for CM operation, $Q_{\mathrm{u}}^{\mathrm{c}}$, can be written as

$$Q_{\mathrm{u}}^{\mathrm{c}} = \left| \frac{1}{R}\left(\frac{\cos(2\theta)}{\omega C} + \frac{Y\sin(2\theta)}{\omega^2 C^2} - \frac{\sin(2\theta)}{4Y} \right) + RY\sin(2\theta) \right| \tag{12.25}$$

From (12.25) it is clear that there is a value of R that allows the minimum level of $Q_{\mathrm{u}}^{\mathrm{c}}$. The strategy followed in Ref. [4] is based on these ideas. After designing the differential response, the values of R and C are tuned so as to achieve the maximum level of CM rejection allowed by this configuration.

Common-Mode Response for Inductance–Resistor Loading The same analysis developed earlier can be applied if the capacitance is replaced by an inductance. In this case, the resonance condition for the loaded transmission line is given by

$$\begin{aligned} 0 = & \frac{Y}{(2Y\omega L + \tan\theta)^2 + (2YR)^2} \\ & \times \left[2Y\omega L(\tan^2\theta - 1) + (4Y^2R^2 + 4Y^2\omega^2L^2 - 1)\tan\theta \right] \end{aligned} \tag{12.26}$$

From (12.26), the CM resonance frequency fulfills the condition $f_0^{\mathrm{c}} < f_0^{\mathrm{d}}$, and it can be tuned by varying the inductance, L. The unloaded quality factor for the CM resonance is now expressed as

$$Q_{\mathrm{u}}^{\mathrm{c}} = \left| \frac{1}{R}\left(Y\omega^2 L^2 \sin(2\theta) - \omega L\cos(2\theta) - \frac{\sin(2\theta)}{4Y} \right) + RY\sin(2\theta) \right| \tag{12.27}$$

Once again, it is possible to find a value of R providing the minimum value of $Q_{\mathrm{u}}^{\mathrm{c}}$.

In Ref. [4] the filter whose layout is shown in Figure 12.40(a) was implemented in a Taconic RF-60A-0310 substrate with a thickness of 0.82 mm, $\varepsilon_r = 6.03$, and $\tan\delta = 0.0038$. The DM design is not affected by the presence of the lumped components located at the symmetry plane since, as it was explained before, this plane is virtually short-circuited. The standard procedure used in the previous sections for the synthesis of filters based on coupled resonators can be then applied. The resonator dimensions are adjusted in order to obtain the desired center frequency passband (in this case $f_0^{\mathrm{d}} = 1.57\,\mathrm{GHz}$). Then, the tapping position of the feeding lines and the gap distance are set to attain the required external quality factor, $Q_e = 17.67$, and coupling coefficient, $M_{1,2} = 0.056$ (the chosen fractional bandwidth is FBW = 8%). The final dimensions for the filter in Figure 12.40(a) are (all in mm) $l_t = 3.1$, $G = 0.45$, $l_1 = 5.05$, $l_2 = 10$, $l_3 = 4.6$, and $W = 0.7$. The simulated DM and CM responses of this filter without any extra added lumped elements are shown in Figure 12.41(a). As it can be seen in this figure, the CM rejection level is rather poor (around 17 dB within the differential passband and less than 15 dB from 0.5 to 3.5 GHz) when compared with the results obtained with the filters described in previous sections.

As it has been commented before, loading with appropriate lumped elements, the resonators in the symmetry plane can significantly enhance the CMRR. In Ref. [4] it has been proven that the CM quality factor can be controlled through the values of R, C, and L in a manner that, ideally, total reflection could be achieved at a desired frequency. In Figure 12.42 we plot the simulation results for CM transmission ($|S_{21}^{\mathrm{cc}}|$) when the resonator is loaded with a capacitor and a resistor in its symmetry plane. In Figure 12.42(a) several plots of $|S_{21}^{\mathrm{cc}}|$ as a function of the frequency are given for a set of values of the loading capacitance. The value of the resistance is kept fixed ($R = 0.5\,\Omega$). On the contrary, in Figure 12.42(b) the value of the capacitance is kept invariable ($C = 11.6\,\mathrm{pF}$), whereas the resistance is changed. It can clearly be observed that the CM response at the center frequency of the differential filter (f_0^{d}) considerably changes when C and/or R are modified. From the inspection of Figure 12.42(a), it can be seen that there is a value of the capacitance providing the best CM response. For this value ($C = 11.6\,\mathrm{pF}$), the CM external quality factor tends to infinite. Once the capacitor is fixed, the curves in Figure 12.42(b) can be used to find the best value of R in order to reach the minimum value of the unloaded quality factor, Q_u^{c}. This ensures maximum CM rejection. The optimum value is found to be $R = 0.5\,\Omega$ in this example.

Similar curves for the CM response versus L and R are shown in Figure 12.43(a) and (b), respectively. When these responses are

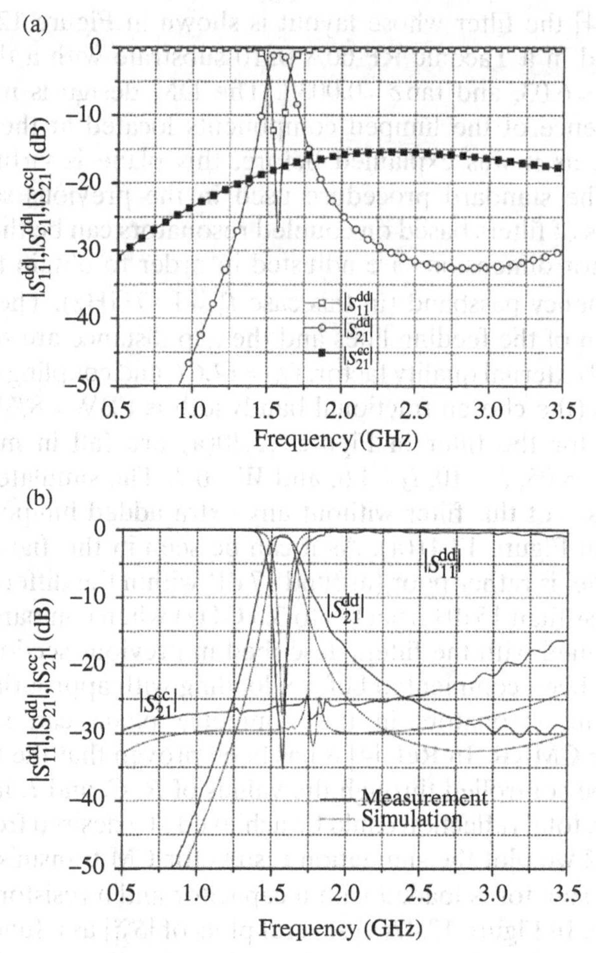

Figure 12.41 (a) Simulated response of the balanced filter in Figure 12.40(a) (differential- and common-mode responses) without adding extra loading elements. Dimensions are in the main text. (b) Measured and simulated DM and CM responses for the fabricated filter with RC loading lumped elements. Reprinted with permission from Ref. [4]; copyright 2010 IEEE.

compared with the one shown in Figure 12.41(a) (conventional filter without lumped elements), no special improvement is appreciated. This is because the inductance cannot sufficiently increase the external quality factor of the loaded resonator. It is then clear that the best way to reduce CM noise with this configuration is to use the series-connected capacitor–resistor set. Figure 12.41(b) shows the measured and simulated DM and CM responses for the filter fabricated in Ref. [4] using the geometry depicted filter in Figure 12.40(a) and lumped capacitors

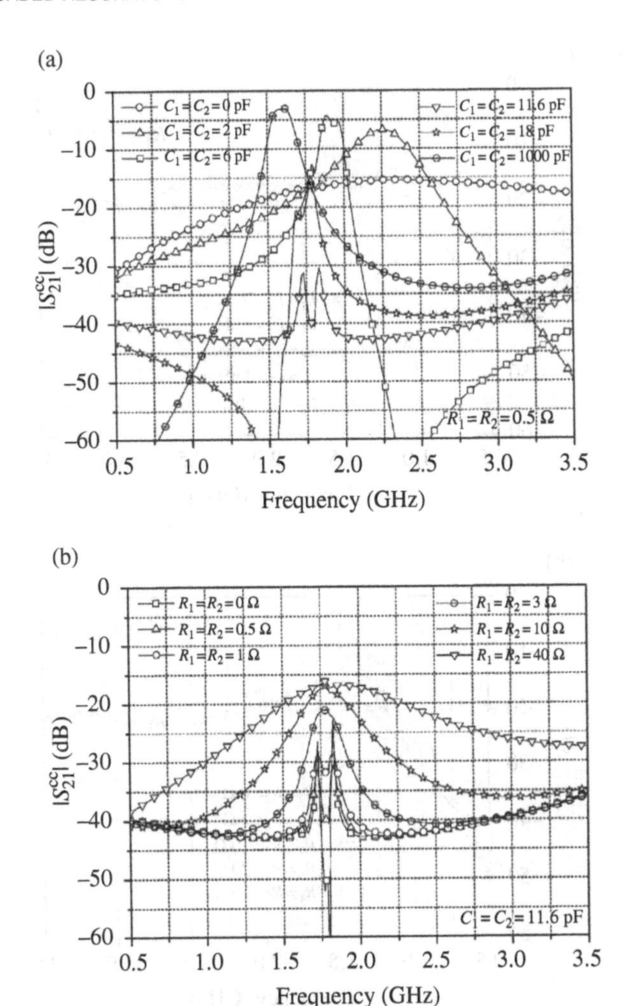

Figure 12.42 Simulated common-mode response for the RC loaded version of the filter in Figures 12.40(a) and 12.41. (a) Response curves for several values of C with fixed $R=0.5\,\Omega$. (b) Response curves for several values of R with fixed $C=11.6$pF. Reprinted with permission from Ref. [4]; copyright 2010 IEEE.

for loading. In the fabricated filter the chosen values for the capacitances and resistances were $C_1=3.7$pF, $C_2=3.8$pF, and $R_1=R_2=0\,\Omega$. These values differ from the ones obtained in the previous analysis because the real capacitors used in the fabrication have parasitic inductances and resistances that cannot be ignored [46, p. 18]. In order to minimize the effect of such parasitic elements, the capacitors and resistors must be chosen with the given values. From Figure 12.41(b) it can be noted that the CM rejection level has been improved in almost 10 dB when

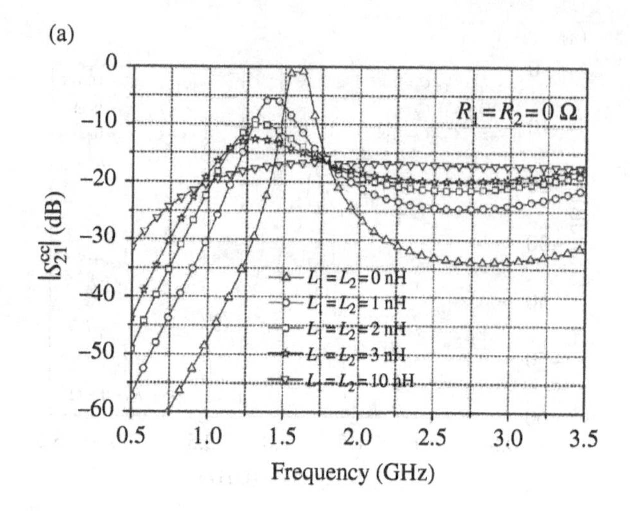

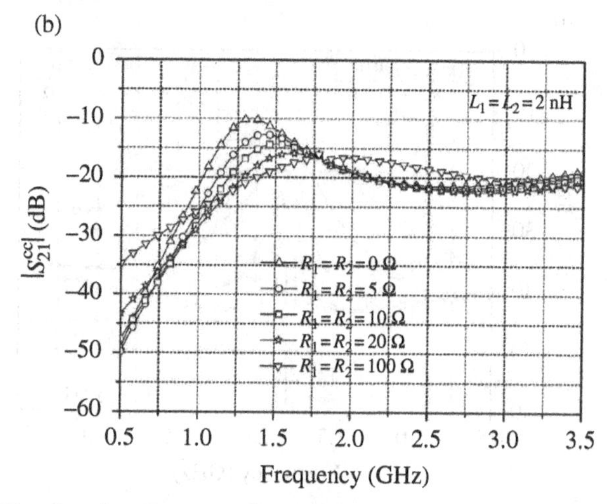

Figure 12.43 Simulated common-mode response for the RL loaded version of the filter in Figure 12.40(a) and Figure 12.41. (a) Response curves for several values of L with $R = 0\ \Omega$. (b) Response curves for several values of R with fixed $L = 2.0$nH. Reprinted with permission from Ref. [4]; copyright 2010 IEEE.

compared with the simulations for the case of the conventional filter without lumped elements. In addition, the DM has been barely altered by the presence of the capacitor. However, compared with the balanced filters with intrinsic CM rejection studied in Section 12.2, this balanced filter offers relatively poor CM reduction.

12.3.1.2 Folded SIR Loaded Filter As it has already been pointed out, the use of SIRs instead of UIRs gives an extra degree of freedom that allows for the improvement of the out-of-band behavior of coupled-resonator filters. This means that if the methodology explained in the previous subsection (introduction of lumped elements in the symmetry plane of open-loop resonator filters) is applied to filters based on SIRs, more compact filters with better out-of-band performance (for both DM and CM responses) could be expected [9, 10]. An example of this kind of filters will be discussed in the next paragraphs.

Capacitor–Resistor Loaded FSIR Figure 12.44 shows the basic structure of a loaded SIR and its equivalent circuits for DM and CM operation. When the FSIR is excited by a differential signal, the symmetry plane has a vanishing voltage (virtual ground). Therefore, the loading components placed in this plane should not affect the electrical response. However, for CM excitation, the symmetry plane is a virtual open circuit, and the loading components, which are connected in parallel with this open-circuit virtual condition, will strongly affect the CM transmission. It was previously discussed that, in order to reduce CM transmission, the best option is to introduce a combination of series-connected capacitors and resistors in the symmetry plane (inductances barely improve CM performance). It is then desirable to analyze the behavior of the resonator for the two excitation regimes (differential and common) when a capacitor or a resistor is located in the symmetry plane. For DM operation (odd excitation), the equivalent circuit of the SIR is the one depicted in Figure 12.44(b), which is identical to a

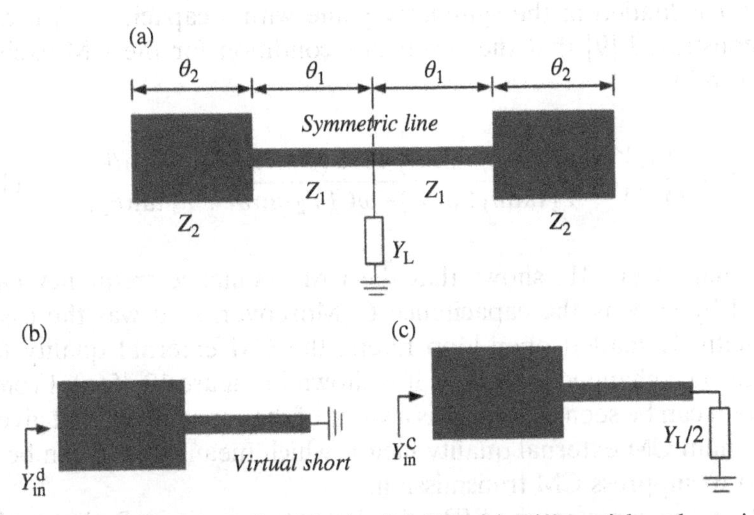

Figure 12.44 (a) Representation of a loaded SIR. (b) Differential-mode equivalent circuit. (c) Common-mode equivalent circuit. Reprinted with permission from Ref. [9]; copyright 2010 IEEE.

quarter-wavelength short-circuited SIR. The resonance condition for this resonator is well known and can be written as [23, 47]

$$\tan\theta_1 \tan\theta_2 = R_Z = \frac{Z_2}{Z_1} \tag{12.28}$$

For identical electrical lengths of the two involved transmission-line sections ($\theta_1 = \theta_2$), the following relationship is fulfilled:

$$\theta_1 = \theta_2 = \tan^{-1}\sqrt{R_Z} \tag{12.29}$$

and the ratio of the first spurious resonance frequency, f_s^d, to the fundamental mode resonance frequency, f_0^d, is given by

$$\frac{f_s^d}{f_0^d} = \frac{\pi}{\tan^{-1}\sqrt{R_Z}} - 1 \tag{12.30}$$

From (12.30) it is obvious that when $R_Z \leq 1$, then $f_s^d/f_0^d \geq 3$ and vice versa, that is, when $R_Z \geq 1$ then $f_s^d/f_0^d \leq 3$. With this information the dimensions of the SIRs can be chosen in such way that R has a suitable value to achieve a wide stopband. Note that this cannot be done if open-loop UIRs are used instead of SIRs.

When the resonator is excited with a CM signal (even symmetry), the equivalent circuit of the SIR is the one in Figure 12.44(c). If the resonator is loaded in the symmetry plane with a capacitor, C, it can be demonstrated [9] that the resonance condition for the CM excitation is given by

$$0 = \frac{2Y_1(Y_1\tan\theta_1 + Y_2\tan\theta_2) + \omega C(Y_1 - Y_2\tan\theta_1\tan\theta_2)}{2Y_1(Y_2 - Y_1\tan\theta_1\tan\theta_2) - \omega C(Y_2\tan\theta_1 + Y_1\tan\theta_2)} \tag{12.31}$$

Equation (12.31) shows that the CM resonance frequency can be tuned by varying the capacitance C. Moreover, as it was the case for capacitively loaded open-loop filters, the CM external quality factor can also be changed with C, as it is shown in Figure 12.45(a). From this figure it can be seen that there is a value of the capacitance that gives the maximum CM external quality factor, which means that C can be used again to suppress CM transmission.

If, on the contrary, the SIR is loaded with a resistor, R, the condition for CM resonance is given by the expression

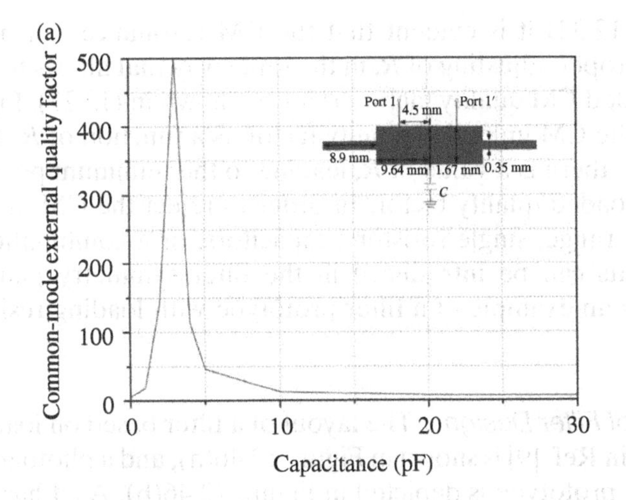

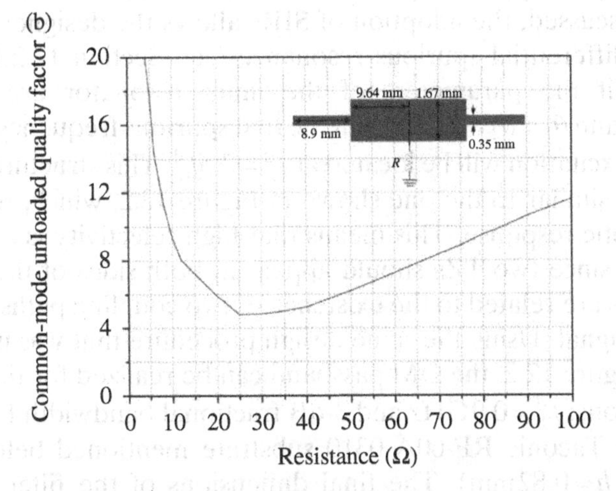

Figure 12.45 (a) Common-mode external quality factor as a function of C. The substrate (Taconic RF-60A-0310) parameters are $\varepsilon_r = 6.03$ and thickness $h = 0.82$mm. (b) Common-mode unloaded quality factor as a function of R for the same structure and substrate. Reprinted with permission from Ref. [9]; copyright 2010 IEEE.

$$0 = \frac{AR + j\left(BR^2 + E\right)}{4R^2\,Y_1^2(Y_2 - Y_1\tan\theta_1\tan\theta_2)^2 + (Y_2\tan\theta_1 + Y_1\tan\theta_2)^2} \tag{12.32}$$

where

$$A = 2Y_1^2Y_2^2\left(1 + \tan^2\theta_1 + \tan^2\theta_2 + \tan^2\theta_1\tan^2\theta_2\right)$$
$$B = 4Y_1^2Y_2(Y_1\tan\theta_1 + Y_2\tan\theta_2)(Y_2 - Y_1\tan\theta_1\tan\theta_2)$$
$$E = Y_2(Y_2\tan\theta_1\tan\theta_2 - Y_1)(Y_2\tan\theta_1 + Y_1\tan\theta_2)$$

From (12.32) it is evident that the CM resonance can be changed through proper adjusting of R, in the same way that allows for adjusting the unloaded CM quality factor, as it was shown in (12.27). Figure 12.45 (b) plots the CM unloaded quality factor as a function of R. This figure shows that there is a value of R, leading to the minimum possible value of the unloaded quality factor. In order to reject the CM in the widest frequency range, single resistors, capacitors, or a combination of such components can be introduced in the filter symmetry plane. Let us now study an example of a filter prototype with loading resistors.

Example of Filter Design The layout of a filter based on loaded FSIRs proposed in Ref. [9] is shown in Figure 12.46(a), and a photograph of the fabricated prototype is depicted in Figure 12.46(b). As it has been previously discussed, the adoption of SIRs allows the designer to push up the first differential spurious resonance (see Section 12.2.1). For this purpose, if the parameters of the inner resonator are chosen as $R_Z = 0.15$ and $\theta_1 = \theta_2 = 21.2°$, then the first spurious frequency for the differential excitation will be located at $f_s^d = 7.5 f_0^d$. This structure has a configuration similar to the one shown in Figure 12.2, which, recall, had a quasi-elliptic response. This means that high selectivity is expected for this filter, since two TZs should appear at both sides of the passband. These TZs are related to the existence of two coupling paths for the differential signal. Using the same design procedure that was used for the filter in Figure 12.2, the DM passband can be realized for the following specifications: $f_0^d = 0.9$GHz and 3-dB fractional bandwidth FBW = 13%. The same Taconic RF-60A-0310 substrate mentioned before is used ($\varepsilon_r = 6.03$, $h = 0.82$mm). The final dimensions of the filter layout are (all in mm) [9] $W = 1.2$, $W_1 = 6.6$, $W_2 = 7.3$, $W_3 = 0.3$, $W_4 = 0.25$, $l_1 = 5.5$, $l_2 = 8.15$, $l_3 = 15$, $l_4 = 8.95$, $l_t = 5.5$, $g_1 = 1.5$, $g_2 = 1.2$, and $g_3 = g_4 = 0.1$. The simulated and measured data for the designed and fabricated filter are plotted in Figure 12.47. The filter has been loaded with four identical resistors of value $R_1 = R_2 = R_3 = R_4 = 33\,\Omega$. The measured DM passband is centered at about 0.9 GHz with a 3-dB FBW of 12.2%. A very wide upper stopband can be observed. More specifically, the DM transmission is below −35 dB up to $7.5 f_0^d$. Concerning CM performance, the suppression of such signal is better than 35 dB up to about $7.06 f_0^d$. Moreover, the filter is quite compact ($0.125\lambda_g \times 0.186\lambda_g$). For comparison purposes, the simulated filter response without loading elements for CM operation is shown in Figure 12.48(a). This figure shows how the CM mode out-of-band response of the filter without loading

(a)

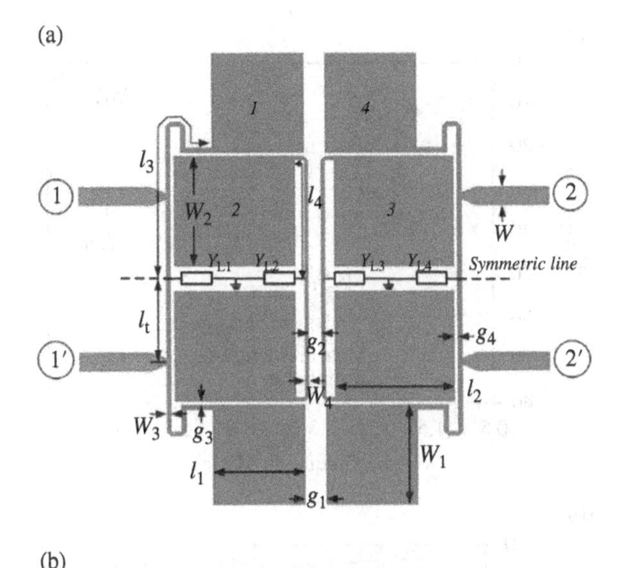

(b)

Figure 12.46 (a) Layout of the balanced bandpass filter with loaded folded SIRs proposed in Ref. [9] for wideband rejection. (b) A photograph of the fabricated prototype. Reprinted with permission from Ref. [9]; copyright 2010 IEEE.

elements has a spurious peak at about $4.54f_0^{\mathrm{d}}$, which is not present in the filter response when loaded with appropriate resistors.

If capacitor loading is employed instead of resistor loading, the same CM performance is obtained, as illustrated in Figure 12.48(b). This figure demonstrates that there is not too much difference between capacitor and resistor loading.

Although placing lumped components along the longitudinal symmetry plane of a filter is a good strategy to reject the CM signal in balanced

(a)

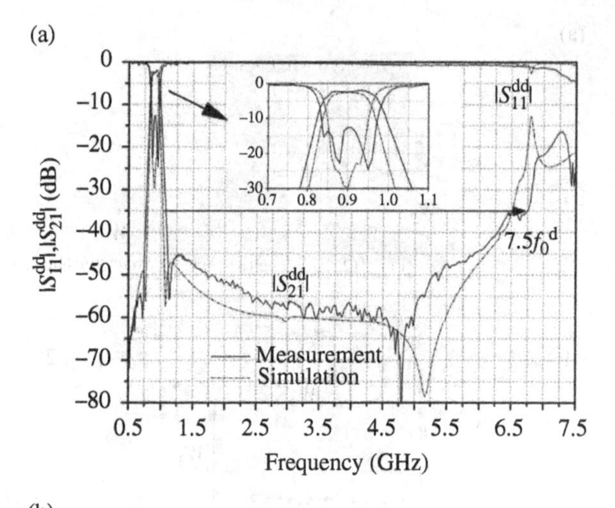

(b)

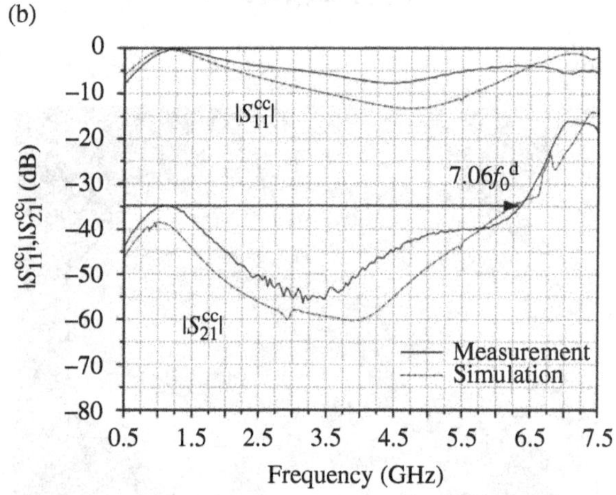

Figure 12.47 Measured and simulated data for balanced filter in Figure 12.46 (resistor-loaded FSIRs). A wide stopband is observed. (a) Differential-mode response. (b) Common-mode response. Reprinted with permission from Ref. [9]; copyright 2010 IEEE.

filters without meaningfully affecting the DM response, an alternative solution based on the etching of a printed distributed resonant element (for instance, an open stub) along the filter symmetry plane [22] is also possible. This approach is especially interesting from the point of view of easy fabrication, since fabrication problems related to the welding of lumped components are overcome. On the other hand, lumped capacitors, inductors, and resistors tend to exhibit parasitic effects at high frequencies whose accurate modeling is not a trivial task. The employment

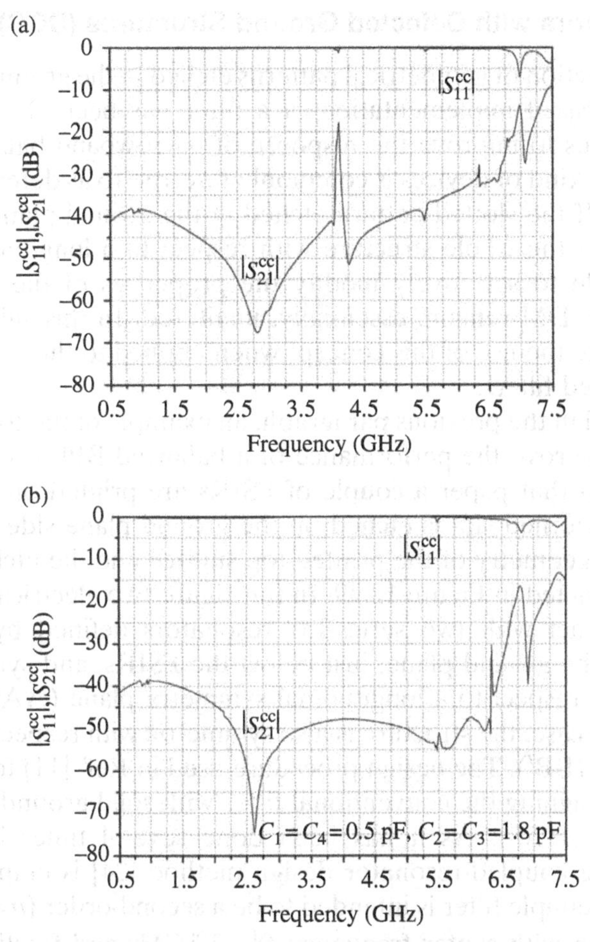

Figure 12.48 (a) Common-mode response without loading elements of the filter in Figure 12.46. (b) Common-mode response with loading capacitors instead of resistors. Reprinted with permission from Ref. [9]; copyright 2010 IEEE.

of strictly planar printed components makes it easier both fabrication and modeling and, at the same time, is cost effective. Since the configuration in Ref. [22] can be used to design wideband balanced BPFs, it has been described in detail in Chapter 5, which is focused on wideband differential filters. However, the reader should note that, when the structure proposed in Ref. [22] is used, the differential passband FBW can be increased at the expense of the CM rejection performance. Thus the layout proposed in Ref. [22] can be advantageously used to implement relatively narrowband responses.

12.3.2 Filters with Defected Ground Structures (DGS)

The introduction of geometrical patterns etched in the ground plane of a microstrip-based implementation of a filter has been shown to offer good chances to improve the response of narrowband balanced BPFs [11]. Such a kind of devices is commonly referred to as defected ground structures. If the slotted pattern etched in the ground plane keeps the symmetry of the whole structure with respect to a longitudinal plane, it is possible to selectively modify the properties of the CM signal, whereas the DM remain, essentially, unaltered. In this subsection we will illustrate the use of this concept when applied to the design of balanced printed filters.

As stated in the previous paragraph, an example of the use of a DGS aimed to improve the performance of a balanced BPF is explained in Ref. [11]. In that paper a couple of FSIRs are printed on a substrate where a patterned slot is etched on the ground plane side of the substrate. The geometry of the printed top surface and the etched ground plane is depicted in Figure 12.49. In this figure two electrically coupled FSIRs interact with two series-LC resonators defined by the DGS located at the ground plane, just below the FSIRs, and symmetrically placed with respect to a longitudinal symmetry plane (AA′ in the figure). In this case, the structure is also symmetric with respect to a transverse plane (BB′). The design procedure used in Ref. [11] for this filter starts by designing the conventional filter with solid ground plane (i.e., without the DGS). As it has been done several times in previous sections, the coupled-resonator design method [24] is employed. The proposed example filter is intended to be a second-order ($n = 2$) Butterworth design with center frequency $f_0^{\mathrm{d}} = 2.5\,\mathrm{GHz}$ and fractional bandwidth FBW = 10%. The substrate parameters are relative dielectric constant $\varepsilon_{\mathrm{r}} = 5.9$ and thickness $h = 0.508\,\mathrm{mm}$. The required external quality factor and coupling coefficient turn out to be $Q_{\mathrm{e}} = 14.142$ and $M_{1,2} = 0.071$. The physical dimensions leading to those values are (in mm) $l_1 = 4.75$, $l_2 = 4.71$, $l_3 = 2.81$, $l_4 = 3.83$, $w_0 = 0.75$, $w_1 = 0.6$, $s_1 = 0.2$, and $t = 2.7$. The simulated results for this prototype are shown in Figure 12.50(a). It is possible to find DM and CM lumped-element equivalent circuits for the structure in Figure 12.49(a). Such circuits are shown in Figure 12.50(b). In order to understand the influence of the DGS once introduced in the conventional layout, it will be very convenient to compare the equivalent circuits of both implementations (the conventional, with solid continuous ground plane, and the proposed one, with the modified ground plane). The numbered sections in Figure 12.49 and the lumped elements in Figure 12.50(b) are clearly

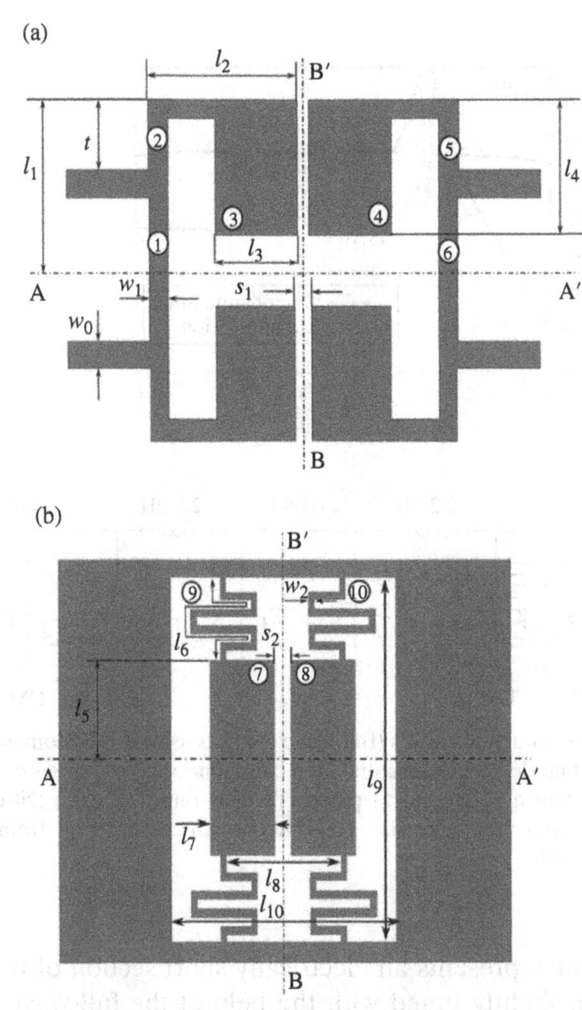

Figure 12.49 (a) Layout of the FSIRs balanced filter with modified (DGS) ground plane reported in Ref. [11]. (a) FSIRs in the top plane and (b) DGS in the ground plane. Reprinted with permission from Ref. [11]; copyright 2015 IEEE.

connected. Thus, L_1 is the inductance associated with Section 1 (and 6). This section behaves as a short-ended transmission-line length in the case of DM operation. Conversely, C_1 is the capacitance of those sections for CM excitation (note that AA′ plane is a virtual open circuit in that situation). The remaining parameters are obviously the same for DM and CM operation: L_2 is the inductance of sections 2 and 5, C_2 is the coupling capacitance between sections 3 and 4, and C_3 is the capacitance between sections 3 and 4 and the ground plane. The values of the elements have been analytically approximated (considering that

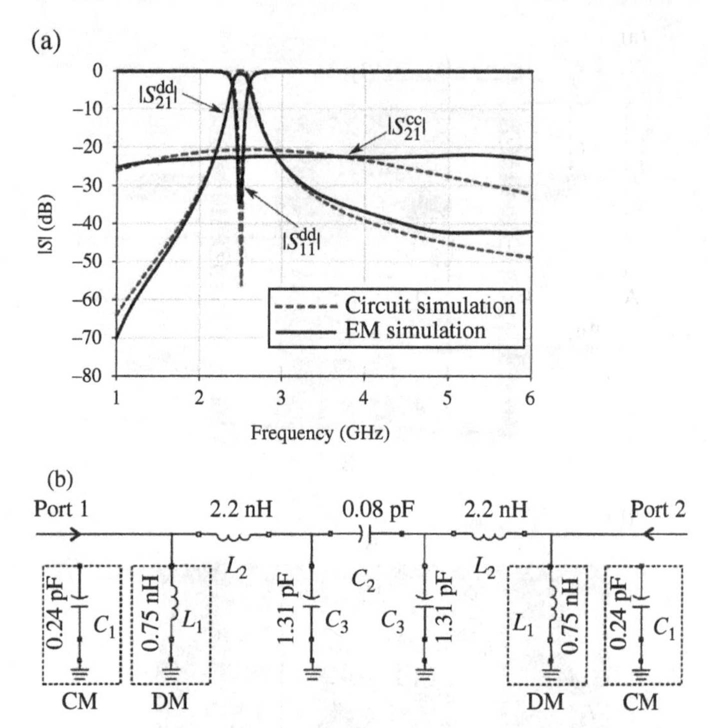

Figure 12.50 (a) Simulated results (full-wave analysis using ADS Momentum [25] and equivalent circuit model) for differential- and common-mode responses of the structure in Figure 12.49 with a solid ground plane (DGS is removed). (b) Differential- and common-mode equivalent circuits. Reprinted with permission from Ref. [11]; copyright 2015 IEEE.

each parameter represents an electrically short section of transmission line) and then slightly tuned with the help of the full-wave simulator, *ADS Momentum* [25]. Simulation results for this filter together with its CM and DM equivalent circuit responses are plotted in Figure 12.50(a). Note that a reasonably good agreement between circuit models and full-wave simulations is appreciated. The inspection of the filter response reveals a relatively poor CM rejection level (around 20 dB within the differential passband) when compared with other available balanced BPFs, as well as poor symmetry of the differential passband. Once the conventional filter has been designed, the next step consists in introducing the DGS shown in Figure 12.49(b), which must be symmetrically aligned with the FIRs. Since the ground plane has been modified, the DM and CM equivalent circuits must be modified too. The new equivalent circuits are depicted in Figure 12.51. Considering that the AA′ plane is a virtual open circuit for CM operation, the return

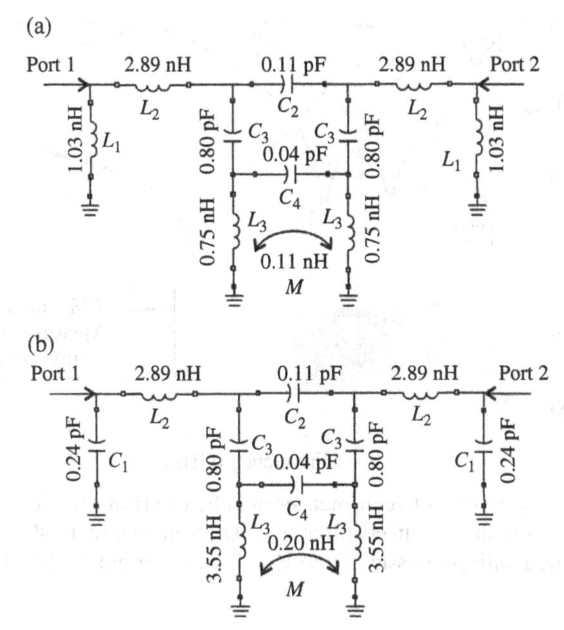

Figure 12.51 (a) DM equivalent circuit and (b) CM equivalent circuit for the filter with modified ground plane depicted in Figure 12.49. Reprinted with permission from Ref. [11]; copyright 2015 IEEE.

current is forced to flow toward the ground plane through the meandered lines (sections 9 and 10), whose inductance value is represented by L_3. However, AA′ behaves as a virtual short circuit for DM operation, in such a way that the displacement current flowing between sections 3 and 7 (or 4 and 8) has two different paths to flow toward the ground plane: a high-impedance path (sections 9 and 10) and a low-impedance path (sections 7 and 8). The current will then mainly flow through the low-impedance path, which is represented by a significantly smaller value of L_3. On the other hand, C_4 accounts for the electric coupling between sections 7 and 8, and M accounts for the magnetic coupling between sections 9 and 10 (CM operation) or 7 and 8 (DM operation). Moreover, it should be appreciated that, for the new filter geometry, C_3 represents the capacitance between sections 3 and 4 and the bottom plane patches (sections 7 and 8). The dimensions (in mm) for the new designed filter [11] are $l_1 = 5.05$, $l_2 = 4.21$, $l_3 = 2.31$, $l_4 = 3.83$, $l_5 = 4.15$, $l_6 = 7.9$, $l_7 = 2.11$, $l_8 = 2.82$, $l_9 = 12.1$, $l_{10} = 7.42$, $w_2 = 0.2$, $t = 2$, and $s_2 = 0.6$. The new set of dimensions for the FSIRs is slightly different from the one corresponding to the conventional FSIR filter design in order to compensate for the small perturbation introduced by the slotted ground plane in the DM response. The simulated

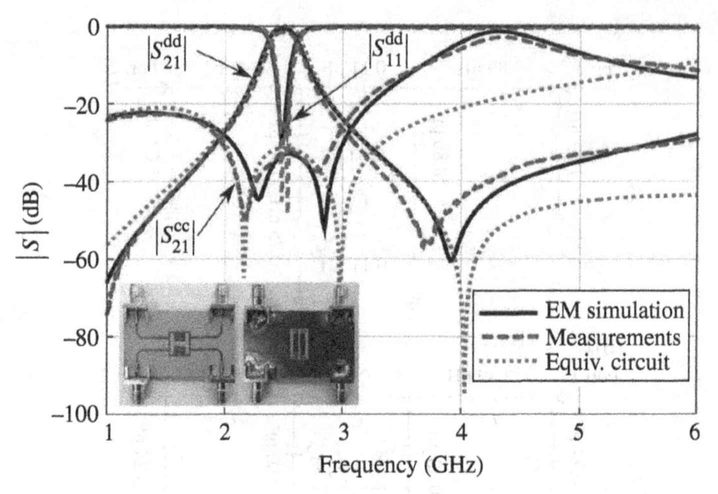

Figure 12.52 Comparison between measurements, electromagnetic simulations, and circuit simulations for the balanced filter with DGS in Figure 12.49 and reported in Ref. [11]. Reprinted with permission from Ref. [11]; copyright 2015 IEEE.

and measured results for the DGS filter are shown in Figure 12.52, where equivalent circuit simulations have been included for comparison purposes. Comparing the response of this filter with the one in Figure 12.50(a), several conclusions can be attained:

- An additional TZ located at $1.5f_0^d$ appears in the DM response of the DGS implementation. This TZ improves DM selectivity, also leading to a more symmetric passband. The presence of this TZ is clearly explained with the help of the equivalent circuit for DM operation.
- The introduction of the slotted ground plane does not meaningfully deteriorate the differential passband performance.
- The CMRR has been improved as much as about 15 dB within the differential passband. However, CM rejection has been degraded in the out-of-band region when compared with the conventional (solid ground plane) version of the same filter.

12.3.2.1 Control of the Transmission Zeros The position of both DM and CM TZs can be controlled by adjusting the values of the circuit elements accounting for the coupling between the top layer resonators and the bottom layer pattern (see the inset in Figure 12.53). The position of the TZs is given by [11]

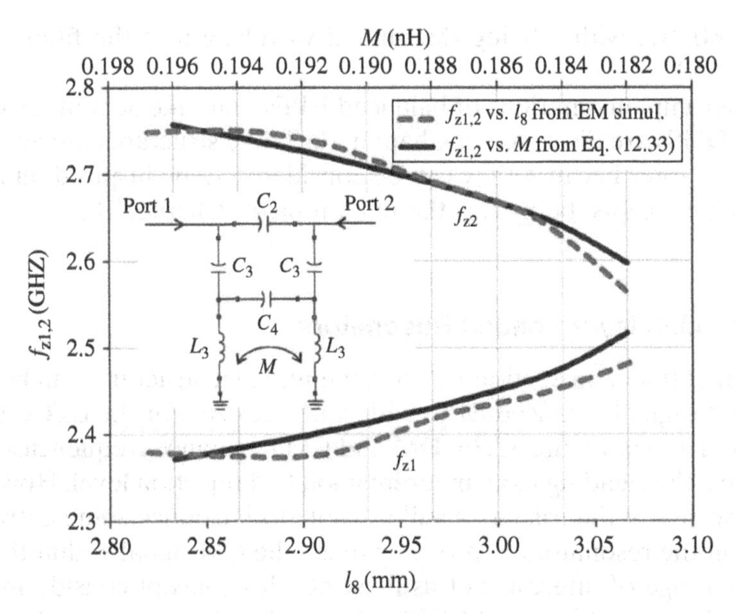

Figure 12.53 Location of the CM transmission zeros as a function of l_8 and M. Full-wave simulations and the predictions of (12.33). The inset shows the equivalent circuit section used for the calculation of the embedded TZs. Reprinted with permission from Ref. [11]; copyright 2015 IEEE.

$$Z_{C3}^2(2Z_{L4}+Z_{C4})+2Z_{C3}Z_{C4}Z_{L4}+Z_{L4}^2(2Z_{C3}+Z_{C2}+Z_{C4})$$
$$+Z_{C3}Z_M(4Z_{L4}+2Z_{C4})+Z_{C2}Z_M(Z_{C4}+Z_{L4})+2Z_{L4}Z_MZ_{C4}=0 \quad (12.33)$$

where $L_4=L_3-M$ and Z_{Xi} represent the impedance of the X_i component (X is C or L and i is the index number, i=2,3,4). The location of the TZs can easily be adjusted by proper selection of the values of the circuit components (which, in turn, are controlled by the structural dimensions). As an example, Figure 12.53 compares the CM TZ frequencies as a function of M using both Equation (12.33) and numerical simulations. Note that M is mainly dependent on the value of l_8, which is continuously varied, whereas the remaining parameters are kept constant. This figure clearly shows that there is an apparent correlation between the geometrical parameter l_8 and the circuit parameter M, as expected from the circuit interpretation of the physical structure. This figure also demonstrates the suitability of Equation (12.33) to control the position and bandwidth of the CM rejected band. As a final note, it is worth mentioning that the structure is quite compact

($0.15\lambda \times 0.21\lambda$, with λ being the guided wavelength at the filter center frequency).

Other implementations of balanced BPFs that take advantage of the use of DGSs are discussed in Chapter 10. These structures are analyzed in that chapter because they can be considered to be inspired on metamaterial concepts, being this the main topic of Chapter 10.

12.3.3 Multilayer Loaded Resonators

It was mentioned in Section 12.2.5 that multilayer structures can be used for the design of balanced BPFs with compact size. For the LTCC structure depicted in Figure 12.36, DM and CM resonance frequencies were far apart, thus leading to an inherent good CM rejection level. However, if this were not the case, it is still possible to introduce some extra elements in the resonators to properly reject the CM signal within the frequency range of interest. Let us illustrate this concept considering the layout shown in Figure 12.54 (this layout has been proposed in Ref. [18]). The structure is composed of two half-wavelength open-loop

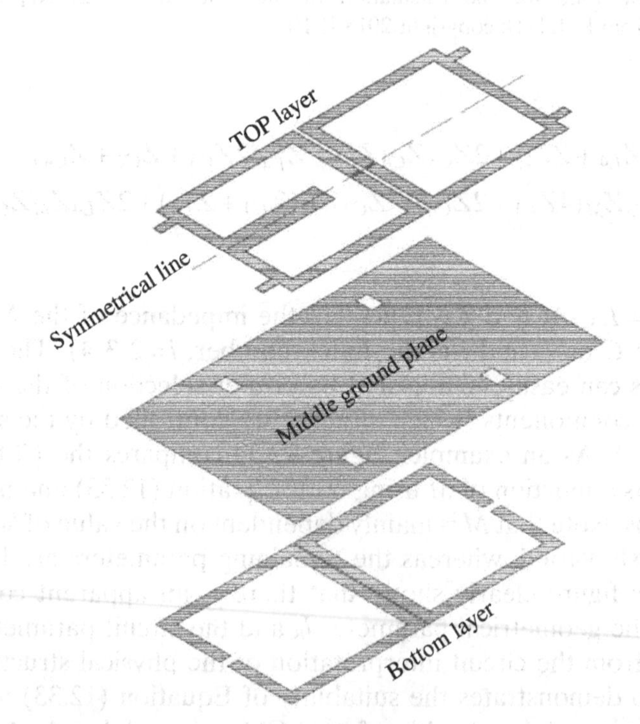

Figure 12.54 Multilayer balanced BPF with a stub-loaded open-loop resonator [18]. Reprinted with permission from Ref. [18]; copyright 2012 IEEE.

resonators printed on the top layer of a multilayer structure, which are coupled to a second pair of similar resonators located at the bottom layer. One of the top layer resonators is loaded with a centered open stub. The coupling is achieved by means of apertures made in the common ground plane, which is sandwiched between the two dielectric layers supporting the resonators. The open-loop resonators etched on the top layer are electrically coupled, while resonators in the bottom layer are magnetically coupled (see Figure 12.55). Under DM operation, the longitudinal symmetry plane behaves as a virtual short circuit, being the equivalent half circuit for the structure the one depicted in Figure 12.56(a). Since for CM operation the symmetry plane is a virtual open circuit, the corresponding equivalent half circuit will be the one drawn in Figure 12.56(b). The circuit in Figure 12.56(a) behaves as a differential BPF that can be designed using the standard coupled-resonator design procedure [24]. On the other hand, if we focus the attention on the circuit in Figure 12.56(b), we can verify that the introduction of the stub along the symmetry plane of resonator #1 (top layer) makes that, for CM operation, this resonator has a different resonance frequency of those resonators #2, #3, and #4. This forces a weak coupling for the CM signal and, consequently, the desired poor CM transmission.

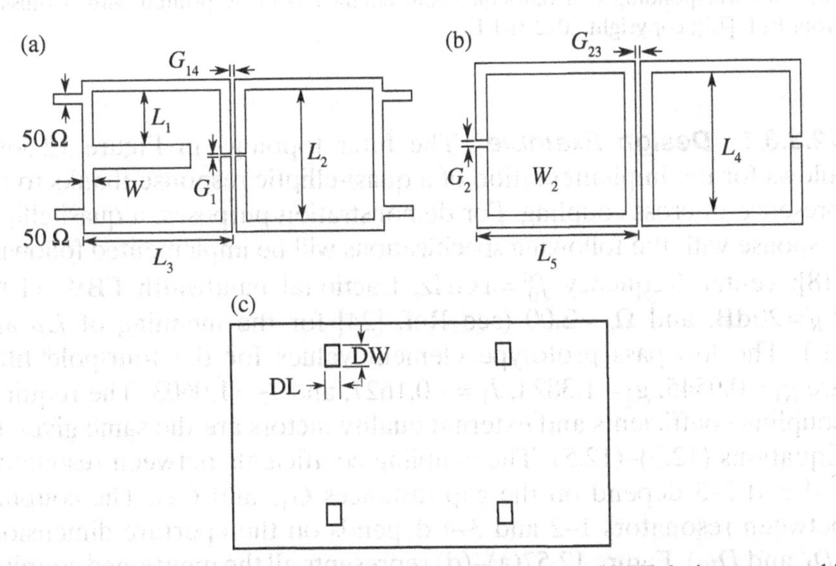

Figure 12.55 Layout of the multilayer structure for balanced BPF design proposed in Ref. [18]. (a) Top layer resonators (two electrically coupled open loops, one of them including a centered stub). (b) Bottom layer resonators. (c) Slotted middle ground plane (the slots provide a coupling mechanism between top and bottom resonators).

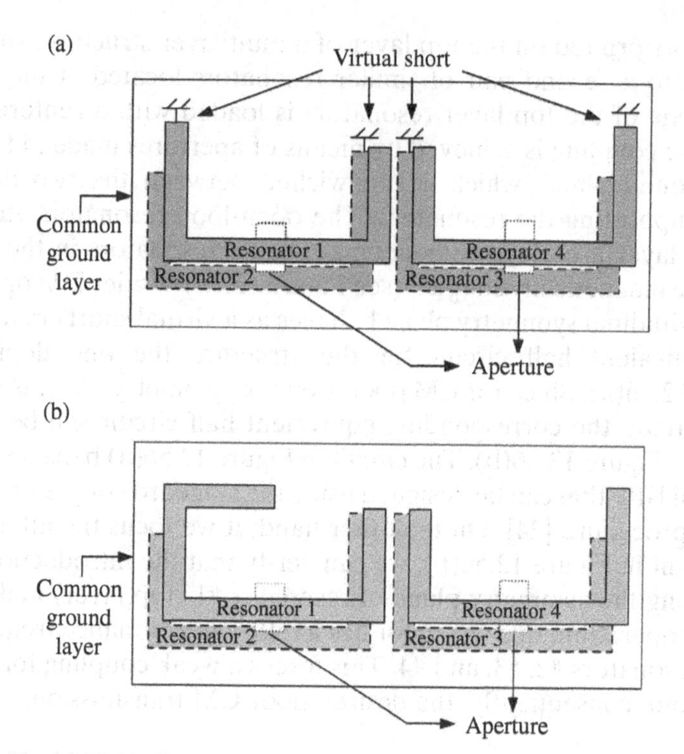

Figure 12.56 (a) DM half-equivalent circuit for the structure in Figures 12.54 and 12.55; (b) the corresponding CM half-equivalent circuit model. Reprinted with permission from Ref. [18]; copyright 2012 IEEE.

12.3.3.1 Design Example The filter topology in Figure 12.56(a) allows for the implementation of a quasi-elliptic response thanks to the presence of cross coupling. For demonstration purposes, a quasi-elliptic response with the following specifications will be implemented following [18]: center frequency $f_0^{\rm d} = 1\,\text{GHz}$, fractional bandwidth FBW = 10%, $L_R = 20\text{dB}$, and $\Omega_a = 2.00$ (see Ref. [24] for the meaning of L_R and Ω_a). The low-pass prototype element values for the four-pole filter are $g_1 = 0.9545$, $g_2 = 1.3824$, $J_1 = -0.1627$, and $J_2 = 1.0603$. The required coupling coefficients and external quality factors are the same given by Equations (12.3)–(12.5). The coupling coefficients between resonators 1–4 and 2–3 depend on the gap distances G_{14} and G_{23}. The coupling between resonators 1–2 and 3–4 depends on the aperture dimensions ($D_{\rm L}$ and $D_{\rm W}$). Figure 12.57(a)–(d) represents all the mentioned coupling coefficients as a function of their respective dimensional control variables. It can be seen that the greater the aperture dimensions, the greater the coupling between resonators 1–2 and 3–4. From these

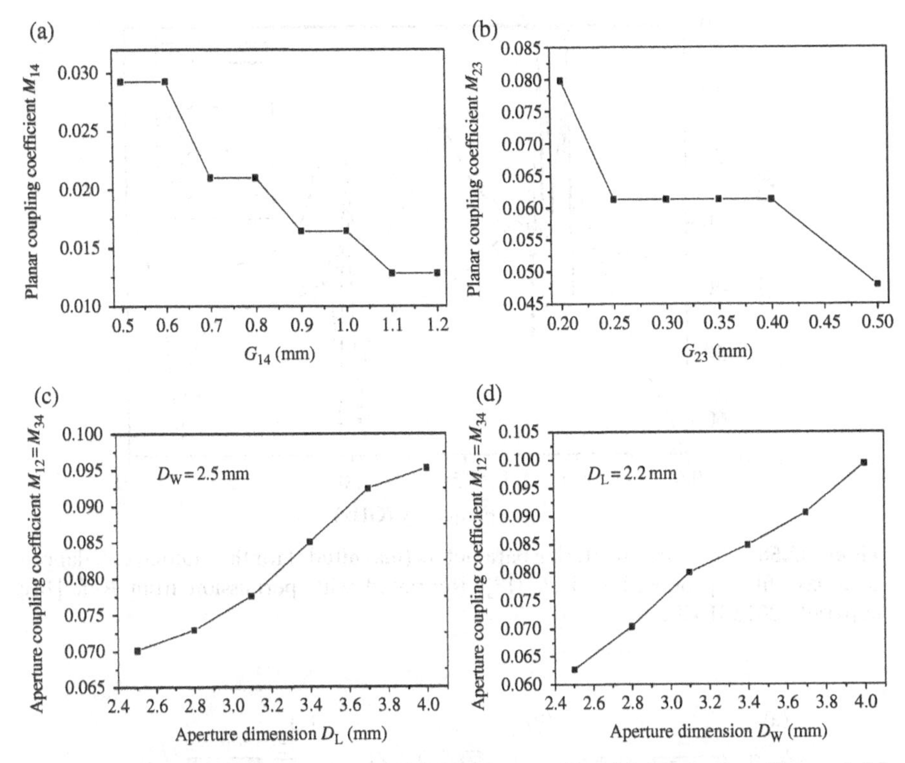

Figure 12.57 Coupling coefficients: (a) M_{14} versus G_{14}, (b) M_{23} versus G_{23}, and (c) $M_{12} = M_{34}$ versus D_L for a fixed value of $D_W = 2.5$ mm and (d) $M_{12} = M_{34}$ versus D_W for a fixed value of $D_L = 2.2$ mm. These parameters correspond to the quasi-elliptic filter in Ref. [18]. Reprinted with permission from Ref. [18]; copyright 2012 IEEE.

curves, the final filter dimensions (in mm) are determined to be the following ones: $W = 2$, $L_1 = 9$, $L_2 = 22$, $L_3 = 26$, $L_4 = 22$, $L_5 = 26$, $G_1 = 0.4$, $G_2 = 1.4$, $G_{14} = 0.8$, and $G_{23} = 0.2$. The simulated response for the proposed quasi-elliptic balanced BPF is shown in Figure 12.58. Good CM suppression is observed in the whole considered frequency range. Specifically, it is about −30 dB at the center frequency. Note the two TZs at both sides of the differential passband, which are typical from quasi-elliptic response.

12.4 COUPLED LINE BALANCED BANDPASS FILTER

In Chapter 5 it was demonstrated that parallel-coupled lines can be used to design wideband balanced BPFs. However, parallel-coupled lines are also suitable for the implementation of narrowband differential filters, as it will be illustrated in this section.

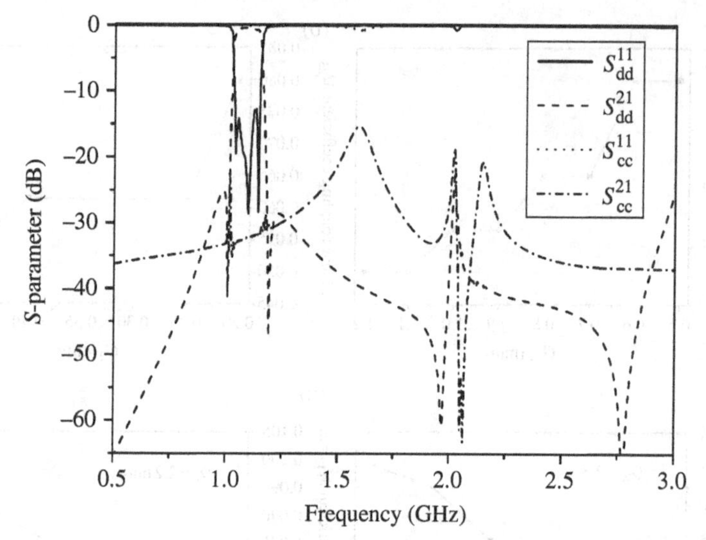

Figure 12.58 Simulated scattering parameters (magnitude) for the multilayer balanced bandpass filter proposed in Ref. [18]. Reprinted with permission from Ref. [18]; copyright 2012 IEEE.

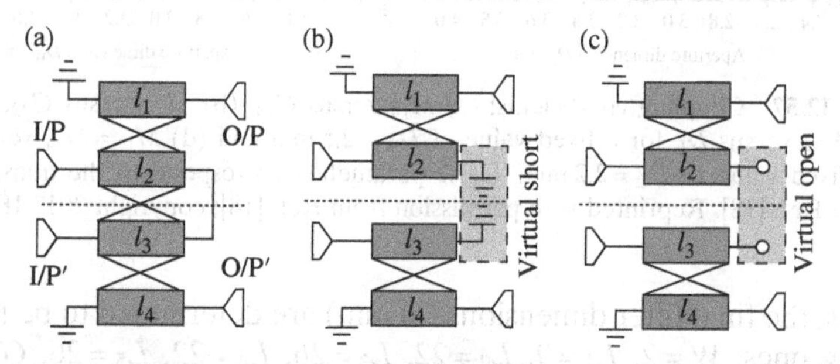

Figure 12.59 (a) Circuit structure for the balanced filter (type I) suggested in Ref. [21]. (b) Differential-mode equivalent circuit. (c) Common-mode equivalent circuit. Reprinted with permission from Ref. [21]; copyright 2007 IEEE.

Since the seminal paper by Seymour B. Cohn [48], parallel-coupled lines have been extensively used for the design of single-ended microwave filters. Recently [20, 21], it has been demonstrated that parallel-coupled lines make it also possible to design filters for balanced operation providing the desired DM bandpass response with good CM rejection. The layout of two possible configurations that can be adopted for designing coupled line balanced BPFs is shown in Figures 12.59 and 12.60. The authors of Ref. [21] refer to those configurations as type I and type II, respectively. The main difference between them is the

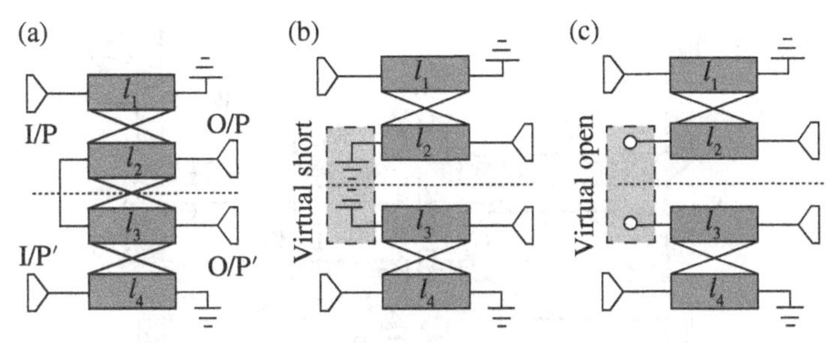

Figure 12.60 (a) Circuit structure for the balanced filter (type II) proposed in Ref. [21]. (b) Differential-mode equivalent circuit. (c) Common-mode equivalent circuit. Reprinted with permission from Ref. [21]; copyright 2007 IEEE.

position of the input/output lines and the short-circuited ends. Note that the half circuits for DM and CM operation are also shown in Figures 12.59 and 12.60. For DM excitation, the equivalent circuits in Figures 12.59(b) and 12.60(b) represent quarter-wavelength coupled line sections with their two ends being short-circuited. These circuits exhibit a bandpass frequency response with a center frequency located at f_0^{d} [26]. On the other hand, for CM operation, the equivalent circuits in Figures 12.59(c) and 12.60(c) exhibit an all-stop response [26] around f_0^{d} since they consists of two quarter-wavelength coupled line sections with one of its ends short-circuited and the other end terminated in an open circuit. Let us now discuss in the following paragraphs one of the two designs (for the interested reader, the first design, with poorer performance, can be found in Ref. [21]).

12.4.1 Type-II Design Example

A possible physical implementation of the circuit in Figure 12.60(a) is shown in Figure 12.61. For DM, this layout can be represented by the equivalent circuit in Figure 12.62(a). This circuit is composed of two bandpass-response coupled line sections with modal impedances $Z_{0\mathrm{e}}$ and $Z_{0\mathrm{o}}$, which are connected by means of a series capacitor (C_{12}) and are grounded at the two ends. The capacitor is realized in the physical structure (Figure 12.61(a)) by means of the gap G_1. The mutual inductance M_{cross} represents the coupling between the two inductances associated with the via holes. It provides a cross-coupled path that introduces two TZs at the lower and upper sides of the differential passband. The level of this inductive cross coupling (and, therefore, the position of the TZs) can be controlled by means of the gap distance between the via holes (G_2). Since the signal level through the cross-coupled path is

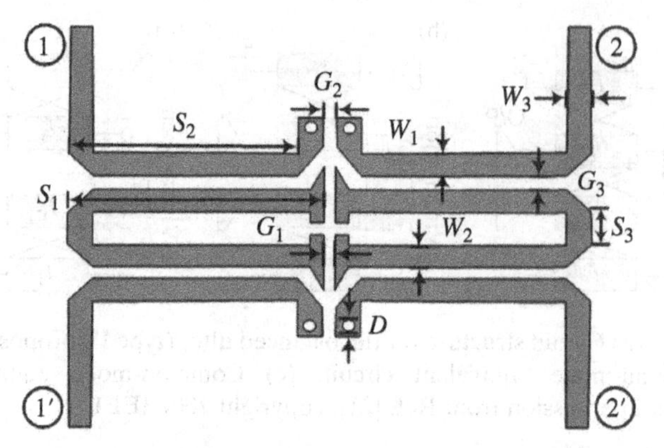

Figure 12.61 Physical layout (microstrip implementation) for the type II balanced filter proposed in Ref. [21]. Reprinted with permission from Ref. [21]; copyright 2007 IEEE.

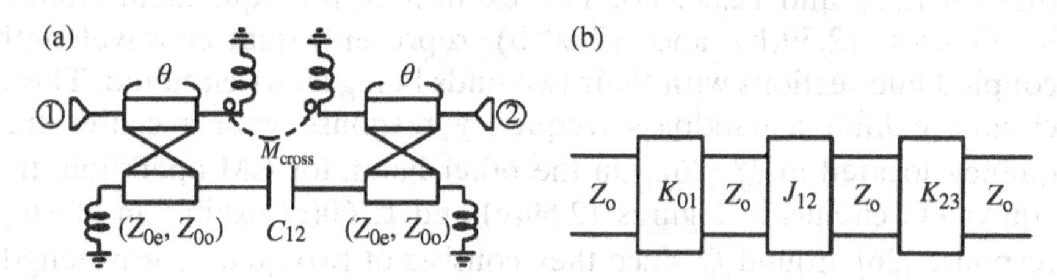

Figure 12.62 (a) DM equivalent half circuit for the second-order balanced filter in Figure 12.61 [21]. (a) $M_{cross} \neq 0$ and (b) $M_{cross} = 0$. Reprinted with permission from Ref. [21]; copyright 2007 IEEE.

significantly lower than the signal level through the main path, for design purposes, M_{cross} can be ignored. This results into the equivalent circuit in Figure 12.62(b). In this circuit the J-inverter, with two transmission lines of characteristic impedance Z_0 and length θ, models the quarter-wavelength open-ended coupled line sections of the same length [21]. On the other hand, the K-inverter replaces the ensemble formed by the inductor vias, L, and the two transmission-line sections of impedance Z_0 and length $\phi/2$ [49]. If in Figure 12.62(b) $\theta - \phi = \pi/2$, the resulting equivalent circuit corresponds to a second-order end-coupled quarter-wavelength resonator BPF. Based on the circuit in Figure 12.62(b), a microstrip BPF has been designed [21] with the following specifications: second-order ($n = 2$) Butterworth response, DM center frequency $f_0^d = 2\text{GHz}$, and 3-dB fractional bandwidth FBW = 10%. With these specifications, the obtained circuit parameters are

$$(Z_{0e}, Z_{0o}) = (64.56\Omega, 40.99\Omega) \qquad (12.34)$$

$$\theta = 86.9^\circ \text{ at } 2\,\text{GHz} \tag{12.35}$$

$$C_{12} = 0.088\,\text{pF} \tag{12.36}$$

A comprehensive description of the design methodology for this type of filters can be found in Refs. [50, 51]. The procedure can be summarized as follows:

1. According to the filter specifications (viz., FBW, center frequency, and reference impedance, usually $Z_0 = 50\,\Omega$), the required J- and K-inverter values (see Figure 12.62(b)) are obtained following the conventional filter synthesis technique for a quarter-wavelength resonator BPF [49].
2. From the J-inverter calculated in the first point, the modal impedances are obtained, whereas the K-inverters are used to calculate ϕ [49].
3. The required coupled line length, θ, is obtained from the condition $\theta - \phi = \pi/2$.
4. Finally, the cross coupling is introduced and its value is chosen to properly allocate the position of the desired TZs.

The particular filter used as example in this section was implemented [21] using an FR4 substrate ($\varepsilon_r = 4.3$, thickness $h = 1\,\text{mm}$, and loss tangent $\tan\delta = 0.02$). The final physical dimensions for the topology in Figure 12.61 result to be (in mm) $S_1 = 17.8$, $S_2 = 16.8$, $S_3 = 2.7$, $W_1 = 1.9$, $W_2 = 1.9$, $W_3 = 1.7$, $G_1 = 0.3$, $G_2 = 0.5$, $G_3 = 0.3$, and $D = 1$. The simulated and measured response for the DM is depicted in Figure 12.63(a). The presence of two TZs at the lower and upper sides of the passband is clearly observed. The zeros come from the fact that the signals through the two paths in Figure 12.62(a) have the same magnitude and are nearly 180° out of phase at those two frequency points. Thus they will cancel out each other at those two frequencies [52]. It is clear that the creation of the TZs would not be possible without the inclusion of cross coupling. The measured insertion loss level is about 2.3 dB, which can be mainly attributed to the high loss tangent typical of FR4 substrates. The use of substrates with lower losses would improve this figure of merit.

Let us now focus our attention on the CM response. The equivalent half circuit for CM operation is shown in Figure 12.64. If the inductances associated with the via holes and their cross coupling are ignored, the response of the studied structure would be the one corresponding to an all-stop coupled line section [26]. However, the existence of those

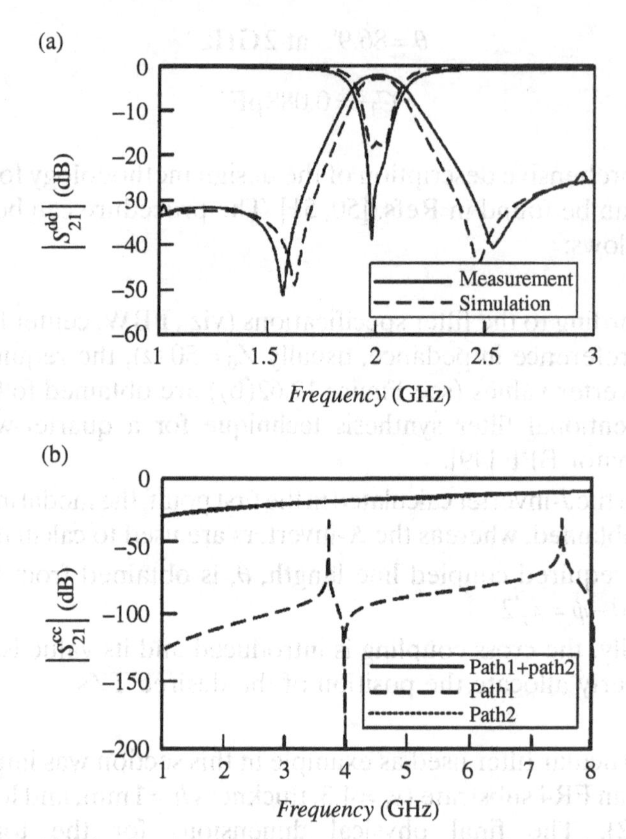

Figure 12.63 (a) Simulated and measured responses (differential mode) for the second-order balanced filter whose layout is depicted in Figure 12.61. (b) Common-mode simulated response of the signals transmitted along the two paths in Figure 12.64. Reprinted with permission from Ref. [21]; copyright 2007 IEEE.

parasitic inductances and the cross coupling between them precludes the all-stop behavior of the structure. This is illustrated in Figure 12.63(b), where the CM response is simulated along the two transmission paths. This figure shows that CM response is mainly determined by the signal path #2. Considering that, for CM operation, the resonator behaves as a $\lambda/2$ open-end resonator along the path #1, its resonance frequency should be located at $f_0^c = 2f_0^d = 4\,\text{GHz}$. This resonance is observed in the form of a very weak peak in Figure 12.63(b) along path #1. However, its contribution to the overall CM response can be totally neglected. The measured DM and CM insertion losses are plotted in Figure 12.65. CM rejection is better than 25 dB from 0.5 to 8.5 GHz. No spurious response is observed for the DM response either for the CM one.

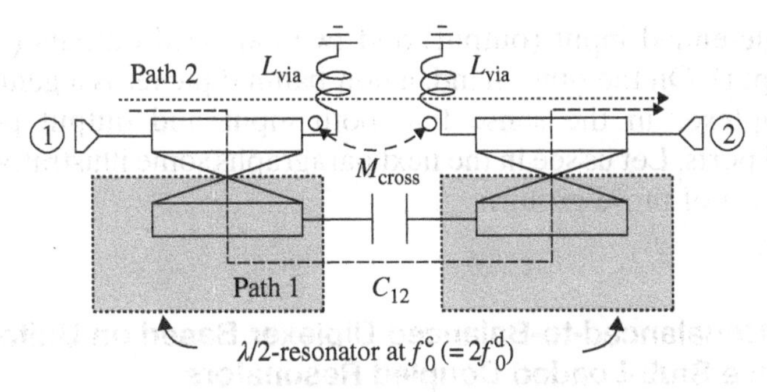

Figure 12.64 Common-mode equivalent half circuit of the second-order balanced filter in Figure 12.61. Reprinted with permission from Ref. [21]; copyright 2007 IEEE.

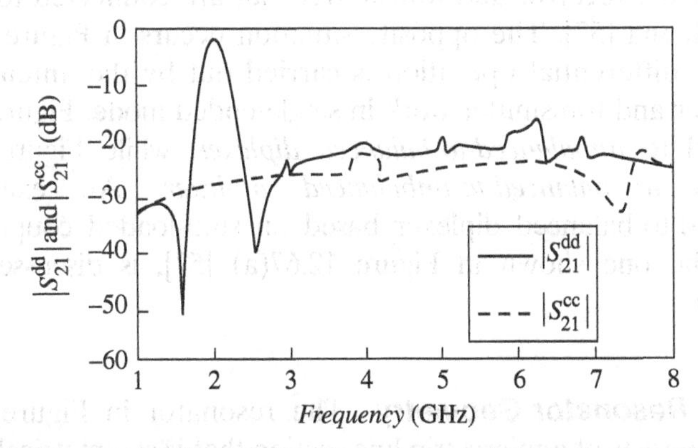

Figure 12.65 Measured differential- and common-mode insertion loss of the second-order balanced filter in Figure 12.61. Reprinted with permission from Ref. [21]; copyright 2007 IEEE.

The CMRR can be improved by increasing the filter order but at the cost of a more complex design process, higher losses, and larger size [21].

12.5 BALANCED DIPLEXERS

The last section of this chapter is devoted to make a brief review of the still scarce literature on balanced diplexers. Diplexers are passive devices whose purpose is to split signals in a common port into two different ports, according to their different frequencies. Diplexers would be the simplest version of multiplexers. The balanced versions of these devices are nowadays under development. Balanced diplexers can be roughly classified into two categories: (i) balun diplexers [53–57] and (ii) non-balun diplexers [53, 54, 58–62]. Balun diplexers are composed of a single-ended input (output) and two balanced outputs (or a balanced input). On the other hand, a non-balun diplexer is a genuine balanced diplexer in the sense that both input and output ports are balanced ports. Let us see in the next paragraphs some illustrative examples of each of those groups.

12.5.1 Unbalanced-to-Balanced Diplexer Based on Uniform Impedance Stub-Loaded Coupled Resonators

Figure 12.66 shows two common architectures of RF front ends using balun diplexers. In Figure 12.66(a) the differential operation is performed by the receiver and transmitter that are connected to a single-ended antenna [57]. The opposite situation occurs in Figure 12.66(b), where the differential operation is carried out by the antenna, while the receiver and transmitter work in single-ended mode. Figure 12.66(a) is referred to *unbalanced-to-balanced diplexer*, while Figure 12.66(b) represents a *balanced-to-unbalanced diplexer*. An example of unbalanced-to-balanced diplexer based on stub-loaded coupled UIRs, such as the one shown in Figure 12.67(a) [57], is discussed in this subsection.

12.5.1.1 Resonator Geometry The resonator in Figure 12.67(a) basically consists of a microstrip line section that is symmetrically loaded with two open stubs. Due to the symmetry with respect to the middle plane shown in that figure, even- and odd-mode analysis can be applied to the structure to study the conditions for DM and CM resonances. The odd- and even-mode equivalent circuits are shown in Figure 12.67(b) and (c), respectively. From those figures, the resonance conditions for

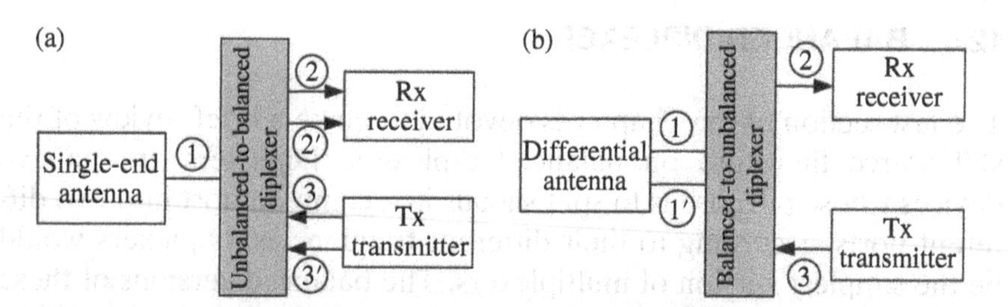

Figure 12.66 Two typical RF front ends based on balun diplexers. (a) Unbalanced-to-balanced diplexer and (b) balanced-to-unbalanced diplexer. Reprinted with permission from Ref. [57]; copyright 2011 IEEE.

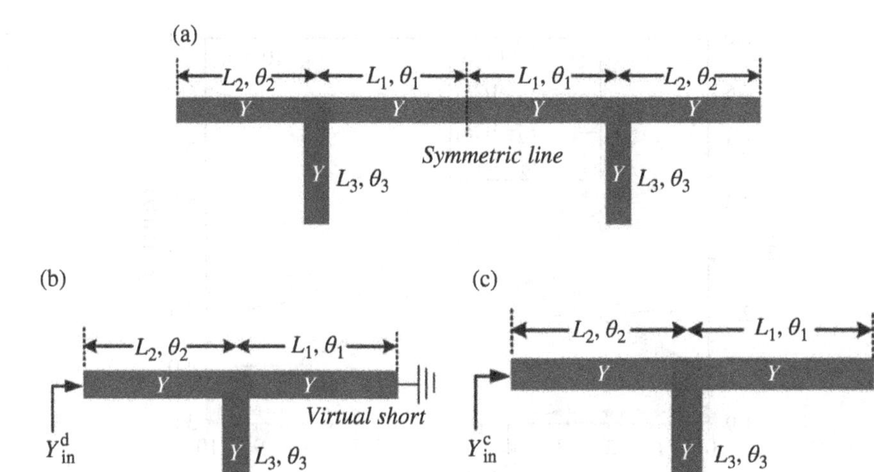

Figure 12.67 (a) Stub-loaded UIR. (b) DM equivalent circuit. (c) CM equivalent circuit for the stub-loaded resonator in (a). Reprinted with permission from Ref. [57]; copyright 2011 IEEE.

DM (superscript "d") and CM (superscript "c") excitations can be easily derived:

$$Y_{in}^{d} = 0 = \frac{\tan\theta_2 + \tan\theta_3 - \cot\theta_1}{1 - (\tan\theta_3 - \cot\theta_1)\tan\theta_2} \tag{12.37}$$

$$Y_{in}^{c} = 0 = \frac{\tan\theta_1 + \tan\theta_2 + \tan\theta_3}{1 - (\tan\theta_1 + \tan\theta_3)\tan\theta_2} \tag{12.38}$$

The usefulness of Equations (12.37) and (12.38) becomes clear by looking over Figure 12.68. This figure shows the fundamental CM resonance frequency (f_0^c) and the ratio of CM to DM fundamental resonance frequencies (f_0^c/f_0^d) as a function of L_3. Several values of L_2 are considered, whereas a fixed value of L_1 has been enforced ($L_1 = 10$mm). From the inspection of this figure, it is clear that it is possible to increase the separation between f_0^c and f_0^d by modifying the lengths of the three transmission-line sections involved in the resonators (see Figure 12.67). This feature allows the designer to extend the bandwidth for high CM rejection. This flexibility comes from the fact that, for a given fixed value of one of the transmission-line sections of the basic resonator, there are many choices of the other two electrical lengths compatible with the desired DM resonance frequency. This gives place to many different possible values of the CM resonance.

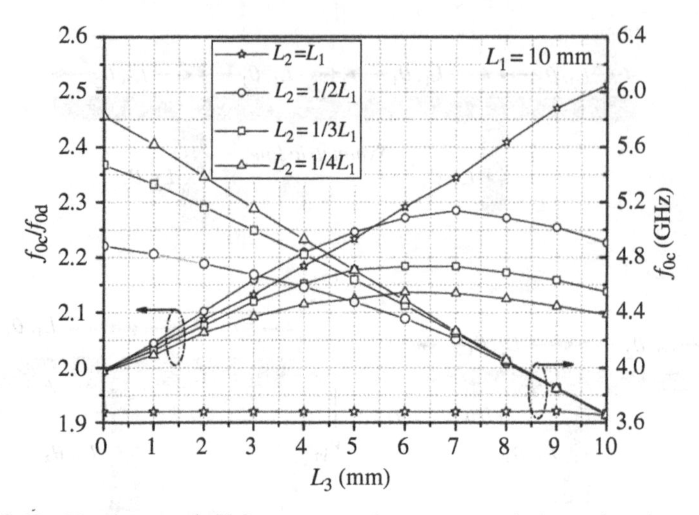

Figure 12.68 Fundamental CM resonance frequency and the ratio of common- to differential-mode fundamental resonance frequencies as a function of L_3 for several values of L_1 and L_2. Reprinted with permission from Ref. [57]; copyright 2011 IEEE.

12.5.1.2 Unbalanced-to-Balanced Diplexer Design The basic idea behind the balun diplexer design in Ref. [57] is very simple. It lies on the design of two different four-pole balun filters based on the resonators discussed earlier. Each balun filter generates a differential passband. Then both balun filters are joined by means of a T junction to create the balun diplexer. The design specifications for the specific implementation considered in Ref. [57] are:

1. First DM passband: $\mathrm{FBW}^1 = 10\%$, $f_{01}^{\mathrm{d}} = 1.84\,\mathrm{GHz}$
2. Second DM passband: $\mathrm{FBW}^2 = 10\%$, $f_{02}^{\mathrm{d}} = 2.45\,\mathrm{GHz}$

The required coupling coefficients and external quality factors can be, once again, obtained by means of the fundamental equations reported in Ref. [24] and handled several times along this chapter. The substrate used to implement the balun filters in Ref. [57] had a dielectric constant $\varepsilon_{\mathrm{r}} = 6.03$, a thickness $h = 0.82\mathrm{mm}$, and a loss tangent $\tan\delta = 0.0038$. The layouts of the two balun filters are depicted in Figure 12.69(a) and (b), respectively. The design procedure for this type of balun filters is very similar to the one explained for filters in Section 12.2 except for the fact that the filters in that section had balanced input and output ports. A balanced input port has been added (dotted line) to the filter drawing in Figure 12.69(a) in order to illustrate this possibility. The key

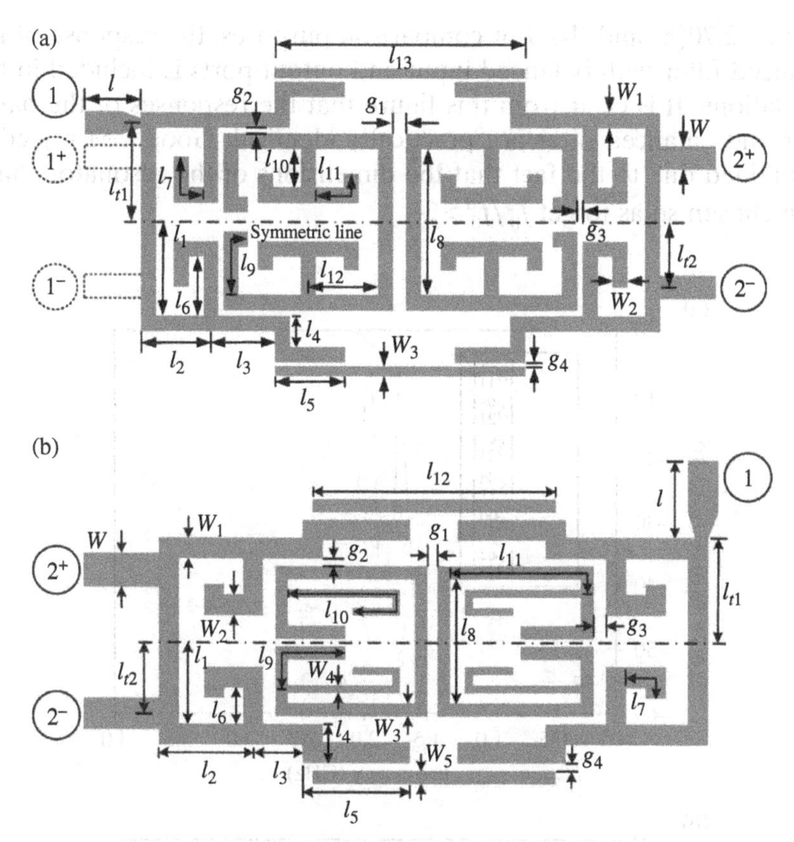

Figure 12.69 (a) Lower-band (Rx) balun filter. Dimensions are (in mm) $W=1.2$, $W_1=W_2=0.8$, $W_3=0.5$, $g_1=0.69$, $g_2=g_3=0.3$, $g_4=0.2$, $l=3$, $l_{t1}=5.15$, $l_{t2}=3.37$, $l_1=4.9$, $l_2=3.75$, $l_3=3.45$, $l_4=1.6$, $l_5=3.755$, $l_6=3.1$, $l_7=3.2$, $l_8=7.6$, $l_9=3.8$, $l_{10}=2$, $l_{11}=3$, $l_{12}=3.75$, $l_{13}=13.39$. (b) Upper-band (Tx) balun filter. Dimensions are (in mm) $W=1.2$, $W_1=W_2=0.8$, $W_3=W_5=0.5$, $W_4=0.3$, $g_1=0.4$, $g_2=g_4=0.3$, $g_3=0.515$, $l=3$, $l_{t1}=4$, $l_{t2}=2.76$, $l_1=3.2$, $l_2=3.75$, $l_3=1.95$, $l_4=1.5$, $l_5=4.37$, $l_6=1.4$, $l_7=2$, $l_8=4.8$, $l_9=4$, $l_{10}=6.8$, $l_{11}=6.15$, $l_{12}=9.7$. Reprinted with permission from Ref. [57]; copyright 2011 IEEE.

point here is that the external quality factor for the single port (Q_e^s) and the one of the differential port (Q_e^d) must be equal. This means that two different curves for Q_e^s (vs. l_{t1}) and Q_e^d (vs. l_{t2}) must be plotted to fulfill this condition. The coupling coefficients depend on the gap distances g_1, g_2, g_3, and g_4. By drawing curves similar to those in Figure 12.4(b), (c), and (d), the necessary values of g_i ($i=1 \ldots 4$) can be extracted. In both balun filters, θ_1 and θ_2 have been chosen to be equal such that CM resonance frequency is not affected by the length θ_3 of the loaded stubs (see Figure 12.68). The simulated responses of the balun filters are shown in

Figure 12.70(a) and (b). For comparison purposes, the response of the balanced filter with balanced input and output ports is included in the simulations. It is clear from this figure that the responses of the balun filter and balanced filter are practically identical. Good CM rejection is obtained due to the fact that the dimensions of the resonators have been chosen so as to get $f_0^c/f_0^d > 2$.

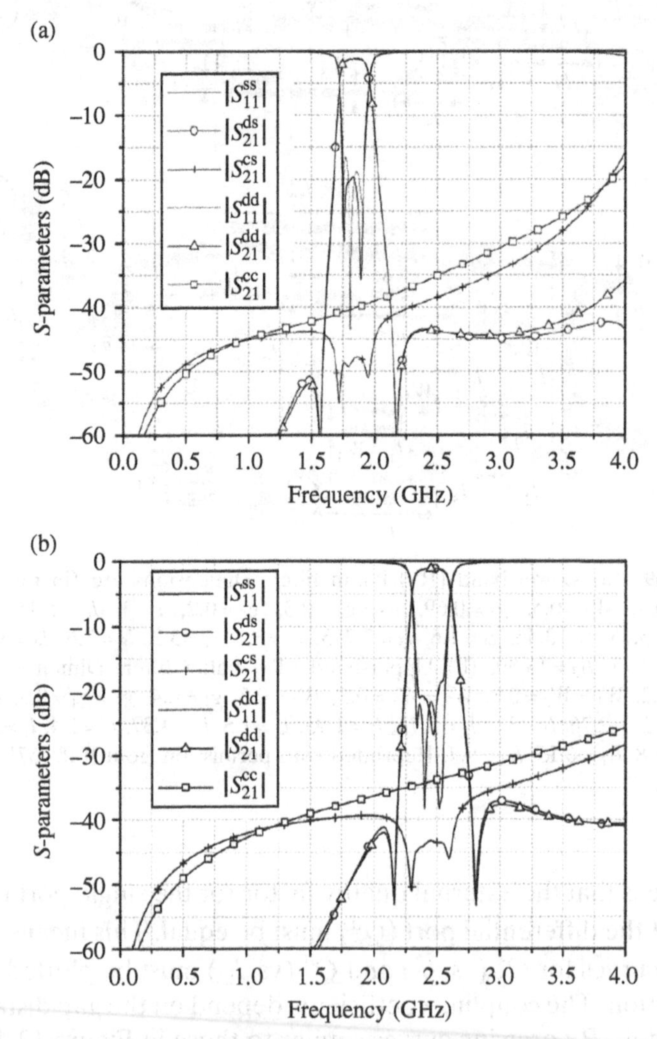

Figure 12.70 Scattering parameters (magnitude, simulations) for the structures shown in (a) Figure 12.69(a), and (b) Figure 12.69(b), under differential- and common-mode operation conditions. The response of the balanced filter (balanced input and output ports) is included for comparison. Reprinted with permission from Ref. [57]; copyright 2011 IEEE.

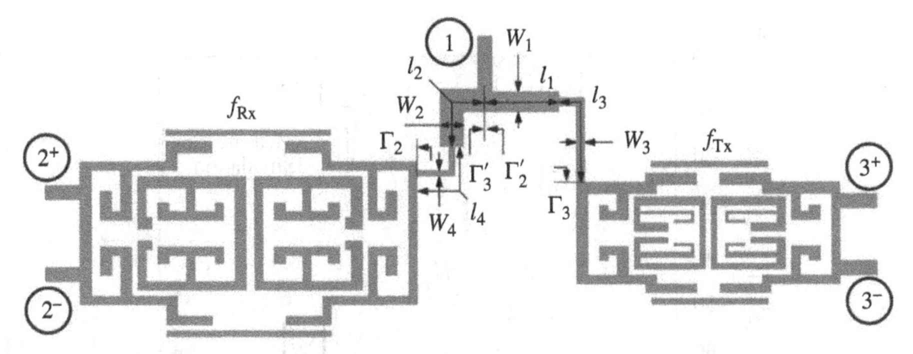

Figure 12.71 Final layout of the balun diplexer designed in Ref. [57]. f_{Rx} and f_{Tx} are the center frequencies of the Rx and Tx passbands, respectively. Dimensions (in mm) are $W_1 = 1.88$, $W_2 = 2$, $W_3 = 0.72$, $W_4 = 0.5$, $l_1 = 6$, $l_2 = 5.8$, $l_3 = 8.68$, and $l_4 = 5.15$. The remaining dimensions are those in Figure 12.69. Reprinted with permission from Ref. [57]; copyright 2011 IEEE.

Once each of the balun filters has separately been designed, a T junction is used to connect them to the same input port. This will finally provide the desired balun diplexing operation. The final layout of the balun diplexer is shown in Figure 12.71. In this figure, in order to achieve high isolation and low return loss, the reflection coefficient Γ'_2 must satisfy the conditions $|\Gamma'_2(f_{Tx})| \approx 1$ and $|\Gamma'_2(f_{Rx})| \approx 0$. Similar conditions are required for Γ'_3, that is, $|\Gamma'_3(f_{Tx})| \approx 0$ and $|\Gamma'_3(f_{Rx})| \approx 1$. These requirements can be easily fulfilled by adding transmission-line sections between the common port 1 and the two balun filters. SIRs have been used instead because its lengths are shorter than the corresponding ones of the equivalent UIRs.

The simulated and measured results for the structure in Figure 12.71 are shown in Figure 12.72(a) and (b) for both DM and CM operations, respectively. If the responses shown in Figure 12.70(a) and (b) are compared with the ones in Figure 12.72(a) and (b), it can be observed that not too much difference is appreciated for both DM and CM performance. Good isolation between channels is also observed in Figure 12.72 for the differential excitation, an aspect that is critical for the proper operation of the balun diplexer.

12.5.2 Example Two: Balanced-to-Balanced Diplexer Based on UIRs and Short-Ended Parallel-Coupled Lines

A balanced diplexer can be also designed by combining two different balanced BPFs with a proper balanced input. Figure 12.73(a) shows

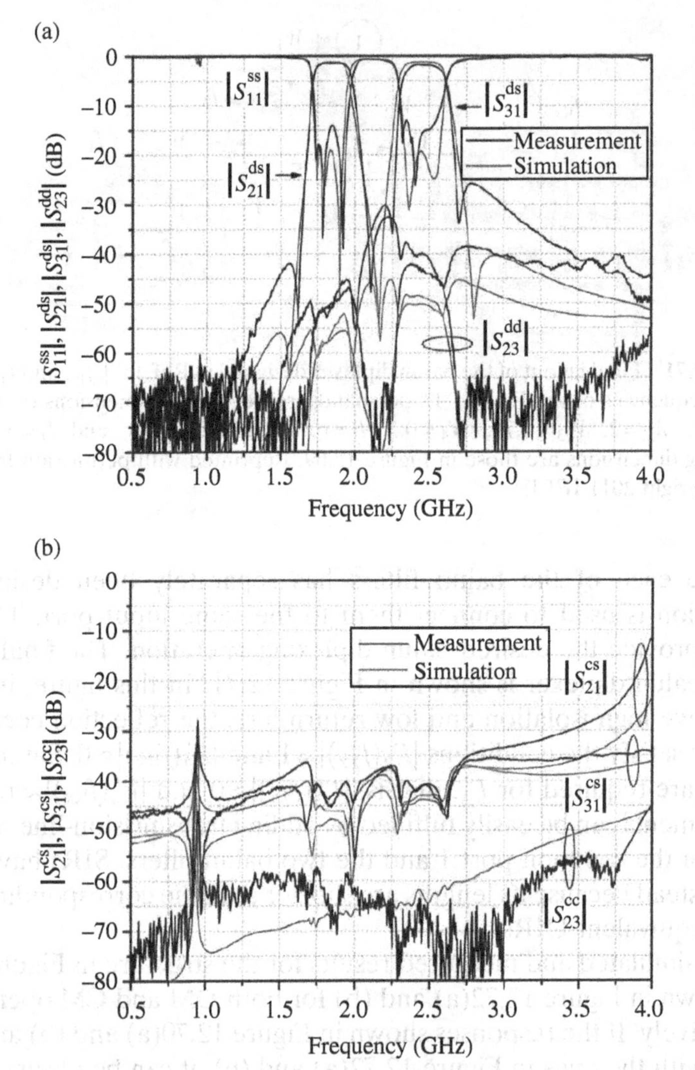

Figure 12.72 Simulated and measured responses of the designed [57] unbalanced-to-balanced diplexer. (a) Differential-mode response and (b) common-mode response. Reprinted with permission from Ref. [57]; copyright 2011 IEEE.

the layout of the balanced BPF used for the implementation of the balanced diplexer suggested in Ref. [58].

Under DM operation, the resonators can be treated as shorted quarter-wavelength resonators (see Figure 12.73b). On the other hand, for CM excitation the UIRs can be seen as open-ended half-wavelength resonators, as it is illustrated in Figure 12.73(c). For the configurations

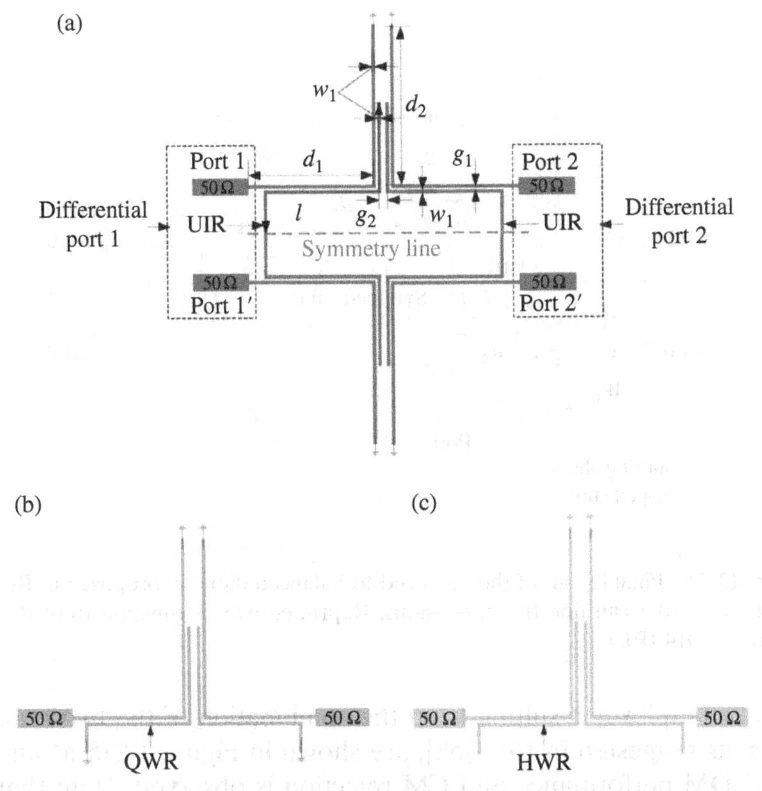

Figure 12.73 (a) Layout of the balanced BPF proposed in Ref. [58] to be used in the implementation of a balanced diplexer. (b) DM equivalent circuit and (c) CM equivalent circuit. Reprinted with permission from Ref. [58]; copyright 2014 IEEE.

in Figure 12.73(b) and (c), CM has no resonance within the DM passband. This ensures good CM suppression in the desired frequency range. As usual in this chapter, the filter can be designed using the coupled-resonator design procedure based on the use of the external quality factor, Q_e, and coupling coefficient, M. The external quality factor is determined by d_1, while the coupling coefficient is controlled by the gap distance g_2. The final layout of the diplexer is shown in Figure 12.74.

The lower-band balanced BPF has been designed to be centered at $f_{01}^{d} = 2.46\,\text{GHz}$ with a 3-dB fractional bandwidth FBW = 8.1%. For the upper passband, $f_{02}^{d} = 3.65\,\text{GHz}$ and FBW = 4.9%. Both filters are of Butterworth type and they have been implemented using a Duroid 5880 substrate. The simulated and measured results for the balanced-to-

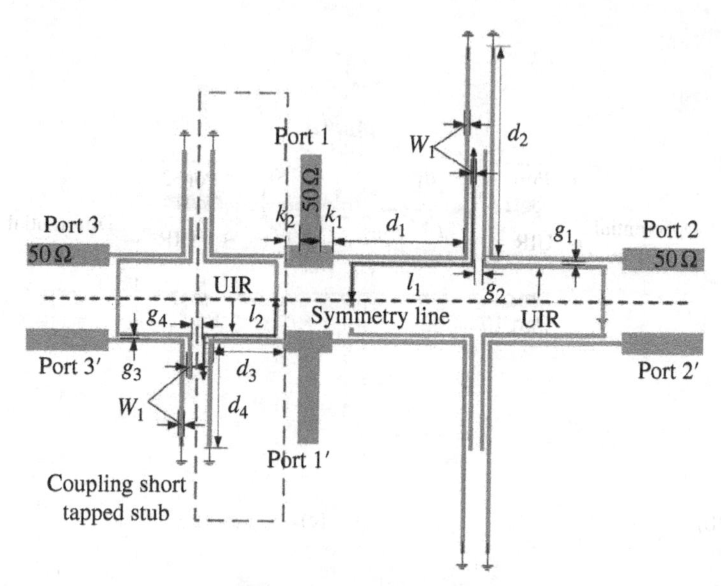

Figure 12.74 Final layout of the balanced-to-balanced diplexer proposed in Ref. [58], where the reader can find the dimensions. Reprinted with permission from Ref. [58]; copyright 2014 IEEE.

balanced diplexer resulting from the combination of the two balanced BPFs, as suggested in Ref. [58], are shown in Figure 12.75(a) and (b). Good DM performance and CM rejection is observed. Note that two TZs appear at both sides of the two differential passbands, which originated from the capacitive coupling between the input and output ports. This helps to improve the selectivity. Finally, the DM isolation between bands is better than 33 dB, as illustrated in Figure 12.76.

12.6 SUMMARY

In this chapter, the state of the art of planar narrowband balanced BPFs has been reviewed. Different approaches followed by a number of researchers have been explained and discussed. Many of these filters are implemented using coupled resonators, since this choice is simple and especially suitable for narrowband applications. Most of the proposed structures are symmetrical versions of their single-ended counterparts, that is, they are inspired in available single-ended designs. The DM response is then obtained using conventional methods and the geometry of the layout is chosen in such a way that strong CM rejection is achieved in the widest possible frequency band including, of course, the passband of the differential response. Typically, the CM response

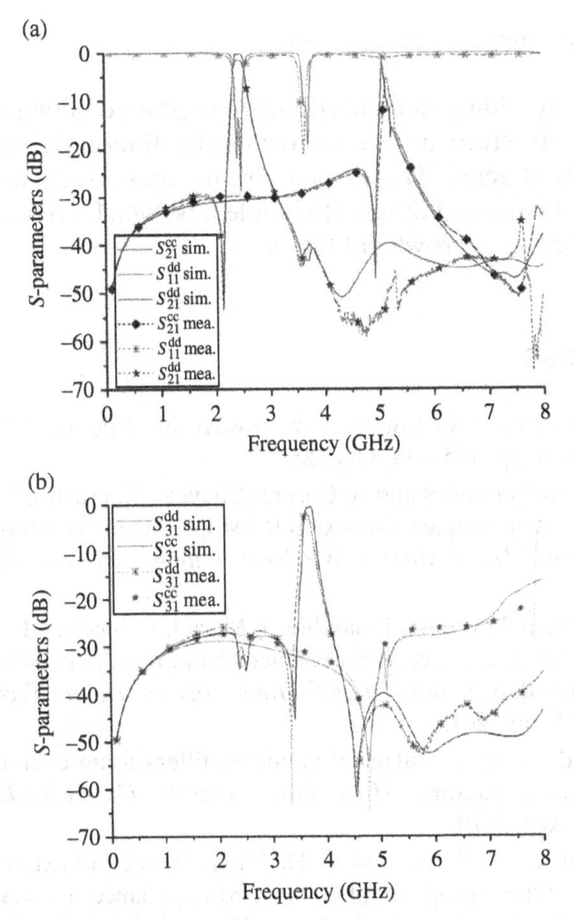

Figure 12.75 Simulated and measured DM and CM responses for the balanced-to-balanced diplexer proposed in Ref. [58]. (a) Lower passband. (b) Upper passband. Reprinted with permission from Ref. [58]; copyright 2014 IEEE.

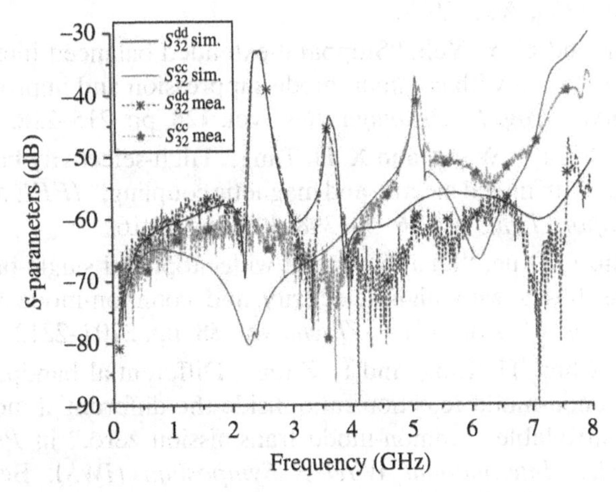

Figure 12.76 Simulated and measured DM and CM isolation between bands for the diplexer in Figure 12.75. Reprinted with permission from Ref. [58]; copyright 2014 IEEE.

is improved by adding elements (lumped or printed) along the symmetry plane of the structure, in such a way that the differential response is not meaningfully affected. The chapter also devotes a section to the implementation of balanced differential diplexers, which are passive devices closely related to narrowband filters.

REFERENCES

1. D. G. Swanson, "Narrow-band microwave filter design," *IEEE Microw. Mag.*, vol. **8**, pp. 105–114, Oct. 2007.
2. J. L. Olvera-Cervantes and A. Corona-Chavez, "Microstrip balanced bandpass filter with compact size, extended-stopband and common-mode noise suppression," *IEEE Microw. Wireless Compon. Lett.*, vol. **23**, pp. 530–532, Oct. 2013.
3. A. Fernández-Prieto, A. Lujambio, J. Martel, F. Medina, F. Mesa, and R. R. Boix, "Simple and compact balanced bandpass filters based on magnetically coupled resonators," *IEEE Trans. Microw. Theory Tech.*, vol. **63**, pp. 1843–1853, Jun. 2015.
4. J. Shi and Q. Xue, "Balanced bandpass filters using center-loaded half-wavelength resonators," *IEEE Trans. Microw. Theory Tech.*, vol. **58**, pp. 970–977, Apr. 2010.
5. C. H. Wu, C. H. Wang, and C. H. Chen, "Stopband-extended balanced bandpass filter using coupled stepped-impedance resonators," *IEEE Microw. Wireless Compon. Lett.*, vol. **17**, pp. 507–509, Jul. 2007.
6. C. H. Wu, C. H. Wang, and C. H. Chen, "Balanced coupled-resonator bandpass filters using multisection resonators for common-mode suppression and stopband extension," *IEEE Trans. Microw. Theory Tech.*, vol. **55**, pp. 1756–1763, Aug. 2007.
7. S.-C. Lin and C.-Y. Yeh, "Stopband-extended balanced filters using both $\lambda/4$ and $\lambda/2$ SIRS with common-mode suppression and improved passband selectivity," *Prog. Electromagn. Res.*, vol. **128**, pp. 215–228, May 2012.
8. T. Yan, D. Lu, J. Wang, and X. H. Tang, "High-selectivity balanced bandpass filter with mixed electric and magnetic coupling," *IEEE Microw. Wireless Compon. Lett.*, vol. **26**, pp. 398–400, Jun. 2016.
9. J. Shi and Q. Xue, "Dual-band and wide-stopband single-band balanced bandpass filters with high selectivity and common-mode suppression," *IEEE Trans. Microw. Theory Tech.*, vol. **58**, pp. 2204–2212, Aug. 2010.
10. J. Shi, J. Chen, H. Tang, and L. Zhou, "Differential bandpass filter with high common-mode rejection ratio inside the differential-mode passband using controllable common-mode transmission zero," in *Proceedings of 2013 IEEE International Wireless Symposium (IWS)*, Beijing, China, pp. 1–4, 14–18 Apr. 2013.

11. A. Fernández-Prieto, J. Martel, F. Medina, F. Mesa, and R. R. Boix, "Compact balanced FSIR bandpass filter modified for enhancing common-mode suppression," *IEEE Microw. Wireless Compon. Lett.*, vol. **25**, pp. 154–156, Mar. 2015.
12. A. Fernández-Prieto, J. Bhatker, A. Lujambio, J. Martel, F. Medina, and R. R. Boix, "Balanced bandpass filter based on magnetically coupled coplanar waveguide folded-stepped impedance resonators," *Electron. Lett.*, vol. **52**, pp. 1229–1231, Jul. 2016.
13. J. Z. Chen and J. Chen, "Compact balanced bandpass filter using interdigital line resonator with high common-mode noise suppression," *Microw. Opt. Technol. Lett.*, vol. **54**, pp. 918–920, Apr. 2012.
14. J. Shi, J.-X. Chen, and Q. Xue, "A novel differential bandpass filter based on double-sided parallel-strip line dual-mode resonator," *Microw. Opt. Technol. Lett.*, vol. **50**, pp. 1733–1735, Jul. 2008.
15. X. Gao, W. Feng, and W. Che, "High selectivity differential bandpass filter using dual-behavior resonators," *Prog. Electromagn. Res. Lett.*, vol. **53**, pp. 89–94, May 2015.
16. A. K. Horestani, M. Durán-Sindreu, J. Naqui, C. Fumeaux, and F. Martín, "S-shaped complementary split ring resonators and their application to compact differential bandpass filters with common-mode suppression," *IEEE Microw. Wireless Compon. Lett.*, vol. **24**, pp. 149–151, Mar. 2014.
17. J. X. Chen, C. Shao, Q. Y. Lu, H. Tang, and Z. H. Bao, "Compact LTCC balanced bandpass filter using distributed-element resonator," *Electron. Lett.*, vol. **49**, pp. 354–356, Feb. 2013.
18. M. Li, H. Lee, J. G. Park, and J. C. Lee, "A narrow bandpass balanced filter with back-to-back structure," in *Proceedings of the 5th Global Symposium on Millimeter Waves (GSMM)*, Harbin, China, pp. 295–298, 27–30 May 2012.
19. N. Janković and V. Crnojević-Bengin, "Balanced bandpass filter based on square patch resonators," in *Proceedings of the 12th International Conference on Telecommunication in Modern Satellite, Cable and Broadcasting Services (TELSIKS)*, Nis, Serbia, pp. 189–192, 14–17 Oct. 2015.
20. Y. X. Zhou, J. D. Ye, D. X. Qu, X. J. Zhong, and J. F. Zhang, "A balanced bandpass filter with open stubs centrally loaded," in *Proceedings of 2014 IEEE International Conference on Communication Problem-Solving (ICCP)*, Beijing, China, pp. 429–430, 5–7 Dec. 2014.
21. C. H. Wu, C. H. Wang, and C. H. Chen, "Novel balanced coupled-line bandpass filters with common-mode noise suppression," *IEEE Trans. Microw. Theory Tech.*, vol. **55**, pp. 287–295, Feb. 2007.
22. L. L. Qiu and Q. X. Chu, "Balanced bandpass filter using stub-loaded ring resonator and loaded coupled feed-line," *IEEE Microw. Wireless Compon. Lett.*, vol. **25**, pp. 654–656, Oct. 2015.
23. M. Makimoto and S. Yamashita, *Microwave Resonators and Filters for Wireless Communications*. New York: Springer, 2000.

24. J.-S. Hong, *Microstrip Filters for RF/Microwave Applications*, 2nd edition. Hoboken: John Wiley & Sons, Inc., 2011.
25. ADS-Momentum, Keysight Technologies, Santa Rosa, 2015 [online]. http://www.keysight.com (accessed on August 3, 2017).
26. D. M. Pozar, *Microwave Engineering*, 3rd edition. Hoboken: John Wiley & Sons, Inc., 2005.
27. L. Zhu, P. M. Wecowski, and K. Wu, "New planar dual-mode filter using cross-slotted patch resonator for simultaneous size and loss reduction," *IEEE Trans. Microw. Theory Tech.*, vol. **47**, pp. 650–654, May 1999.
28. J. X. Chen, C. H. K. Chin, and Q. Xue, "Double-sided parallel-strip line with an inserted conductor plane and its applications," *IEEE Trans. Microw. Theory Tech.*, vol. **55**, pp. 1899–1904, Sep. 2007.
29. C. Quendo, E. Rius, and C. Person, "Narrow bandpass filters using dual-behavior resonators," *IEEE Trans. Microw. Theory Tech.*, vol. **51**, pp. 734–743, Mar. 2003.
30. C. Quendo, E. Rius, and C. Person, "Narrow bandpass filters using dual-behavior resonators based on stepped-impedance stubs and different-length stubs," *IEEE Trans. Microw. Theory Tech.*, vol. **52**, pp. 1034–1044, Mar. 2004.
31. C. Quendo, E. Rius, A. Manchec, Y. Clavet, B. Potelon, J. F. Favennec, and C. Person, "Planar tri-band filter based on dual behavior resonator (dbr)," in *Proceedings of the 35th European Microwave Conference*, Paris, France, vol. **1**, pp. 1–4, 4–6 Oct. 2005.
32. E. Rius, C. Quendo, A. Manchec, Y. Clavet, C. Person, J. Favennec, G. Jarthon, O. Bosch, J. Cayrou, and P. Mo, "Design of microstrip dual behavior resonator filters: A practical guide," *Microw. J.*, vol. **49**, pp. 72–83, Dec. 2006.
33. P. H. Deng and J. T. Tsai, "Design of microstrip cross-coupled bandpass filter with multiple independent designable transmission zeros using branch-line resonators," *IEEE Microw. Wireless Compon. Lett.*, vol. **23**, pp. 249–251, May 2013.
34. K. Lim, S. Pinel, M. Davis, A. Sutono, C.-H. Lee, D. Heo, A. Obatoynbo, J. Laskar, M. Tantzeris, and R. Tummala, "RF-system-on-package (SOP) for wireless communications," *IEEE Microw. Mag.*, vol. **3**, pp. 88–99, Mar. 2002.
35. K. Brownlee, S. Bhattacharya, K. Shinotani, C. P. Wong, and R. Tummala, "Liquid crystal polymer for high performance SOP applications," in *Proceedings of the 8th IEEE International Symposium on Advanced Packaging Materials*, Stone Mountain, GA, USA, pp. 249–253, 3–6 Mar. 2002.
36. D. C. Thompson, O. Tantot, H. Jallageas, G. E. Ponchak, M. Tentzeris, and J. Papapolymerou, "Characterization of liquid crystal polymer (LCP) material and transmission lines on LCP substrate from 30-100 GHz," *IEEE Trans. Microw. Theory Tech.*, vol. **52**, pp. 1343–1352, Apr. 2004.

37. L. Devlin, G. Pearson, J. Pittock, and B. Hunt, "RF and microwave component development in LTCC," in *Proceedings of IMAPS Nordic 38th Annual Conference*, Oslo, Norway, 23–26 Sep. 2001.
38. Y. Imanaka, *Multilayered Low Temperature Cofired Ceramics LTCC Technology*. New York: Springer, 2005.
39. R. Zhang and R. R. Mansour, "Dual-band dielectric-resonator filters," *IEEE Trans. Microw. Theory Tech.*, vol. **57**, pp. 1760–1766, Jul. 2009.
40. M. Memarian and R. R. Mansour, "Quad-mode and dual-mode dielectric resonator filters," *IEEE Trans. Microw. Theory Tech.*, vol. **57**, pp. 3418–3426, Dec. 2009.
41. J. X. Chen, Y. Zhan, W. Qin, Z. H. Bao, and Q. Xue, "Novel narrow-band balanced bandpass filter using rectangular dielectric resonator," *IEEE Microw. Wireless Compon. Lett.*, vol. **25**, pp. 289–291, May 2015.
42. Y. Zhan, J. Li, W. Qin, and J. X. Chen, "Low-loss differential bandpass filter using $TE_{01\delta}$-mode dielectric resonators," *Electron. Lett.*, vol. **51**, no. 13, pp. 1001–1003, 2015.
43. J.-X. Chen, Y. Zhan, W. Qin, Z.-H. Bao, and Q. Xue, "Analysis and design of balanced dielectric resonator bandpass filters," *IEEE Trans. Microw. Theory Tech.*, vol. **64**, pp. 1476–1483, May 2016.
44. R. J. Cameron, C. M. Kudsia, and R. R. Mansour, *Microwave Filters for Communication Systems*. Hoboken: John Wiley & Sons, Inc., 2007.
45. X. Y. Zhang, J. X. Chen, Q. Xue, and S. M. Li, "Dual-band bandpass filters using stub-loaded resonators," *IEEE Microw. Wireless Compon. Lett.*, vol. **17**, pp. 583–585, Aug. 2007.
46. R. Ludwig and P. Bretchko, *RF Circuit Design: Theory and Applications*. Upper Saddle River: Prentice-Hall, 2000.
47. M. Sagawa, M. Makimoto, and S. Yamashita, "Geometrical structures and fundamental characteristics of microwave stepped-impedance resonators," *IEEE Trans. Microw. Theory Tech.*, vol. **45**, pp. 1078–1085, Jul. 1997.
48. S. B. Cohn, "Parallel-coupled transmission-line-resonator filters," *IRE Trans. Microw. Theory Tech.*, vol. **6**, pp. 223–231, Apr. 1958.
49. G. L. Matthaei, L. Young, and E. M. T. Jones, *Microwave Filters, Impedance-Matching Networks, and Coupling Structures*. Dedham: Artech House, 1980.
50. C.-C. Chen, Y.-R. Chen, and C.-Y. Chang, "Miniaturized microstrip cross-coupled filters using quarter-wave or quasi-quarter-wave resonators," *IEEE Trans. Microw. Theory Tech.*, vol. **51**, pp. 120–131, Jan. 2003.
51. Y.-S. Lin, C.-H. Wang, C.-H. Wu, and C. H. Chen, "Novel compact parallel-coupled microstrip bandpass filters with lumped-element k-inverters," *IEEE Trans. Microw. Theory Tech.*, vol. **53**, pp. 2324–2328, Jul. 2005.
52. C.-H. Wang, Y.-S. Lin, and C. H. Chen, "Novel inductance-incorporated microstrip coupled-line bandpass filters with two attenuation poles," in *Proceedings of 2004 IEEE MTT-S International Microwave Symposium*, Fort Worth, TX, USA, vol. **3**, pp. 1979–1982, 6–11 Jun. 2004.

53. C.-H. Lee, C.-I. G. Hsu, and P.-H. Wen, "Balanced and balun diplexers designed using center-grounded uniform-impedance resonators," *Microw. Opt. Technol. Lett.*, vol. **56**, no. 3, pp. 555–559, 2014.
54. P.-H. Wen, C.-I. G. Hsu, C.-H. Lee, and H.-H. Chen, "Design of balanced and balun diplexers using stepped-impedance slot-line resonators," *J. Electromagn. Waves Appl.*, vol. **28**, no. 6, pp. 700–715, 2014.
55. C. H. Wu, C. H. Wang, and C. H. Chen, "A novel balanced-to-unbalanced diplexer based on four-port balanced-to-balanced bandpass filter," in *Proceedings of the 38th European Microwave Conference*, Amsterdam, Netherlands, pp. 28–31, 27–31 Oct. 2008.
56. Z. H. Bao, J. X. Chen, E. H. Lim, and Q. Xue, "Compact microstrip diplexer with differential outputs," *Electron. Lett.*, vol. **46**, pp. 766–768, May 2010.
57. Q. Xue, J. Shi, and J. X. Chen, "Unbalanced-to-balanced and balanced-to-unbalanced diplexer with high selectivity and common-mode suppression," *IEEE Trans. Microw. Theory Tech.*, vol. **59**, pp. 2848–2855, Nov. 2011.
58. Y. Zhou, H. W. Deng, and Y. Zhao, "Compact balanced-to-balanced microstrip diplexer with high isolation and common-mode suppression," *IEEE Microw. Wireless Compon. Lett.*, vol. **24**, pp. 143–145, Mar. 2014.
59. H. L. Chan, C. H. Lee, and C. I. G. Hsu, "Balanced dual-band diplexer design using microstrip and slot-line resonators," in *Proceedings of 2015 Asia-Pacific Microwave Conference (APMC)*, Nanjing, China, vol. **3**, pp. 1–3, 6–9 Dec. 2015.
60. C. H. Lee, C. I. G. Hsu, S. X. Wu, and P. H. Wen, "Balanced quad-band diplexer with wide common-mode suppression and high differential-mode isolation," *IET Microw. Antennas Propagat.*, vol. **10**, no. 6, pp. 599–603, 2016.
61. H. Deng, Y. Zhao, Y. Fu, Y. He, and X. Zhao, "High selectivity and cm suppression microstrip balanced bpf and balanced-to-balanced diplexer," *J. Electromagn. Waves Appl.*, vol. **27**, no. 8, pp. 1047–1058, 2013.
62. W. Jiang, Y. Huang, T. Wang, Y. Peng, and G. Wang, "Microstrip balanced quad-channel diplexer using dual-open/short-stub-loaded resonator," in *Proceedings of the IEEE MTT-S International Microwave Symposium (IMS)*, San Francisco, CA, USA, pp. 1–3, 22–27 May 2016.

CHAPTER 13

DUAL-BAND BALANCED FILTERS BASED ON LOADED AND COUPLED RESONATORS

Jin Shi[1] and Quan Xue[2]

[1]School of Electronics and Information, Nantong University, Nantong, China
[2]School of Electronic and Information Engineering, South China University of Technology, Guangzhou, China

The emerging development of modern communication technology places critical requirements on highly integrated systems with multiple functions. Undoubtedly, multiband and dual-band microwave components are one possible solution to this need. In addition, balanced circuits can further enhance immunity to spurious interference or harmonics in the systems. Therefore, the design of dual-band balanced filters has become a hot topic in recent years, and two objectives must be carefully addressed in the design of such filters: the construction of two differential-mode passbands and the suppression of the common-mode signal inside and outside of the two differential-mode passbands. At least two approaches can be applied to achieve the first goal. In the first approach, each resonator produces two controllable resonant

Balanced Microwave Filters, First Edition. Edited by Ferran Martín, Lei Zhu, Jiasheng Hong, and Francisco Medina.

frequencies, and in the second approach, the resonator has only one resonant frequency but can produce two differential-mode passbands by introducing transmission zeros into the middle of the bands. Common-mode suppression of a dual-band balanced filter is notably different from that of a single-band one because in a single-band balanced filter, the common-mode operating frequency is generally located far away from the differential-mode operating frequency, whereas in the dual-band one, the common-mode frequency is located near or between the two differential-mode operating frequencies. Thus, common-mode suppression in a dual-band balanced filter is enabled by enhancing the common-mode absorption or by reducing the common-mode coupling, which can be achieved using a loaded or coupled resonator. Therefore, this chapter addresses the analysis and design of dual-band balanced filters using loaded and coupled resonators.

13.1 DUAL-BAND BALANCED FILTER WITH LOADED UNIFORM IMPEDANCE RESONATORS

13.1.1 Center-Loaded Uniform Impedance Resonator

The center-loaded uniform impedance resonator is a type of resonator used to create a dual-band balanced filter. Figure 13.1(a) shows a typical

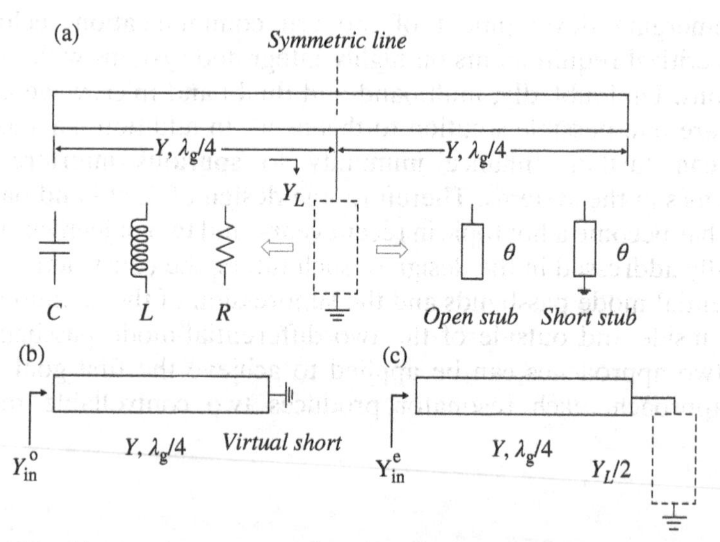

Figure 13.1 (a) Center-loaded uniform impedance resonator. (b) Odd-mode equivalent circuit. (c) Even-mode equivalent circuit. Reprinted with permission from Ref. [1]; copyright 2010 IEEE.

center-loaded uniform impedance resonator in which the center-loaded element can be lumped element (e.g., capacitor, inductor, or resistor) or stub (e.g., open or short stub). Under odd- and even-mode operation, the plane along the symmetric line is an electric and magnetic wall, respectively. Therefore, the odd- and even-mode equivalent circuits can be obtained as shown in Figure 13.1(b) and (c). Thus, the loaded elements only operate under even-mode operation.

Under odd-mode operation, the odd-mode resonant frequency f_o can be written as follows:

$$f_o = \frac{c}{4\lambda_g\sqrt{\varepsilon_{eff}}} \tag{13.1}$$

where c is the velocity of light in free space, ε_{eff} is the effective permittivity, and λ_g is the guided wavelength. The odd-mode resonant frequency does not change with the loaded element.

Under even-mode operation, the resonant performance varies with the loaded elements. For the capacitor-loaded uniform impedance resonator, the even-mode resonant condition can be written as follows:

$$\frac{\omega C + 2Y\tan\theta_0}{2Y - \omega C\tan\theta_0} = 0 \tag{13.2}$$

where C is the value of the capacitor, Y is the admittance of the microstrip line, θ_0 is the electric length at f_0, and ω is the angular frequency. The even-mode resonant condition for the inductor-loaded uniform impedance resonator can be described as follows:

$$\frac{2Y\omega L\tan\theta_0 - 1}{2Y\omega L + \tan\theta_0} = 0 \tag{13.3}$$

Figure 13.2 presents the calculated even-mode resonant frequencies with different values of C and L according to Equations (13.2) and (13.3). Figure 13.2 shows that the even-mode resonant frequencies (f_e) decrease as the values of C or L increase and the ranges extend from $2f_o$ to f_o and from f_o to 0.

When the loaded elements are stubs, we assume that θ is the electric length of the open or short stub at f_o. For the short-stub-loaded uniform impedance resonator, the even-mode resonant condition can be written as follows:

$$\frac{2\tan(\theta f_e/f_o)\tan\theta_0 - 1}{2\tan(\theta f_e/f_o) + \tan\theta_0} = 0 \tag{13.4}$$

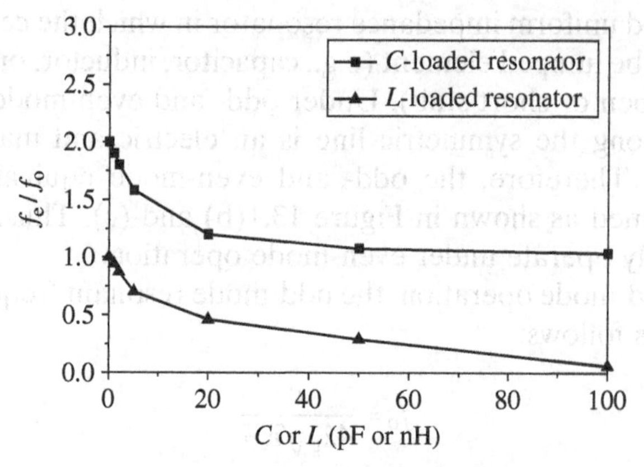

Figure 13.2 Even-mode resonant frequencies of the uniform impedance resonator with different values of C and L.

where $\theta_0 = (f_e/f_o)\cdot 90°$. For the open-stub-loaded uniform impedance resonator, the even-mode resonant condition can be described as follows:

$$\frac{\tan(\theta f_e/f_o) + 2\tan\theta_0}{2 - \tan(\theta f_e/f_o)\tan\theta_0} = 0 \tag{13.5}$$

Figure 13.3 shows the even-mode resonant frequencies of the resonator with a loaded open or short stub according to Equations (13.4) and (13.5). This figure shows that the even-mode resonant frequencies of the open-stub-loaded uniform impedance resonator are reduced from $2f_o$ to f_o when θ is increased from 0° to 90° and the even-mode resonant

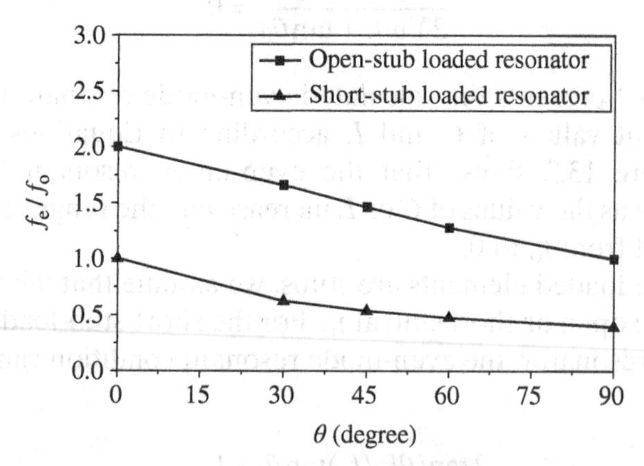

Figure 13.3 Even-mode resonant frequencies of the uniform impedance resonator with different electric length (θ) of the open stub and short stub.

frequencies of the short-stub-loaded uniform impedance resonator change from f_o to $f_o/2$.

For the resistor-loaded uniform impedance resonator, the even-mode input admittance can be written as follows:

$$Y_{\text{in}}^{\text{c}} = Y\frac{2RY + 2RY\tan^2\theta_0}{4R^2Y^2 + \tan^2\theta_0} + jY\frac{4R^2Y^2\tan\theta_0 - \tan\theta_0}{4R^2Y^2 + \tan^2\theta_0} \tag{13.6}$$

where R is the resistance of the loaded resistor. Thus, the resonant condition can be obtained as follows:

$$\frac{4R^2Y^2\tan\theta_0 - \tan\theta_0}{4R^2Y^2 + \tan^2\theta_0} = 0 \tag{13.7}$$

In addition, the unloaded quality factor is more important than the resonant frequency for the resistor-loaded resonator because low Q_u means high absorption for the even-mode signal. Figure 13.4 shows the simulated even-mode Q_u of a resistor-loaded uniform impedance resonator. This figure shows that the even-mode Q_u can be greatly reduced because of the effect of the loaded resistor and an optimum resistance occur under which the proposed resonator has the minimum even-mode Q_u.

Therefore, the loaded capacitor, inductor, and open/short stub are primarily used to control the even-mode resonant frequency, and the loaded resistor is used in even-mode absorption.

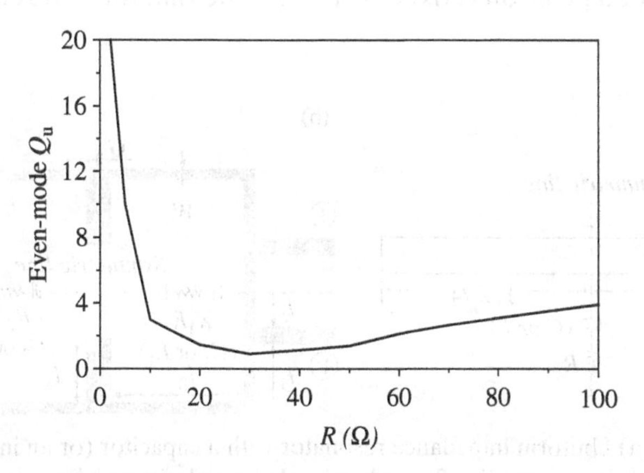

Figure 13.4 Even-mode unloaded quality factor versus resistance based on lossless microstrip line.

13.1.2 Dual-Band Balanced Filter Using the Uniform Impedance Resonator with Center-Loaded Lumped Elements

For a balanced filter in which the resonators are symmetrical along the center line of the filter, the differential- and common-mode performances are determined by the odd- and even-mode feature of the resonators, respectively. Thus, the loaded elements of the resonators control the common-mode suppression of a balanced filter. For a uniform impedance resonator with loaded lumped elements, the effect of the capacitor or inductor is different from that of the resistor. Thus, a uniform impedance resonator with a capacitor (or an inductor) and a resistor loaded in series can benefit from both the capacitor (or inductor) and resistor. Figure 13.5 exhibits the uniform impedance resonator with lumped elements loaded in series as well as an example of a balanced filter with two such resonators to analyze the effect of series loaded elements [1].

For the resonator with a capacitor and a resistor loaded in series, the even-mode resonant condition can be written as follows:

$$\frac{Y}{(2Y-\omega C\tan\theta_0)^2+(2YR\omega C)^2} \times \left[2Y\omega C(1-\tan^2\theta_0)+(4Y^2R^2\omega^2C^2+4Y^2-\omega^2C^2)\tan\theta_0\right]=0 \tag{13.8}$$

Equation (13.8) shows that the even-mode resonant frequency can be adjusted by changing the value of C. Thus, the common-mode external quality factor of the filter in Figure 13.5 can be changed by adjusting C even if the feed position is fixed. If we assume that the reference plane of

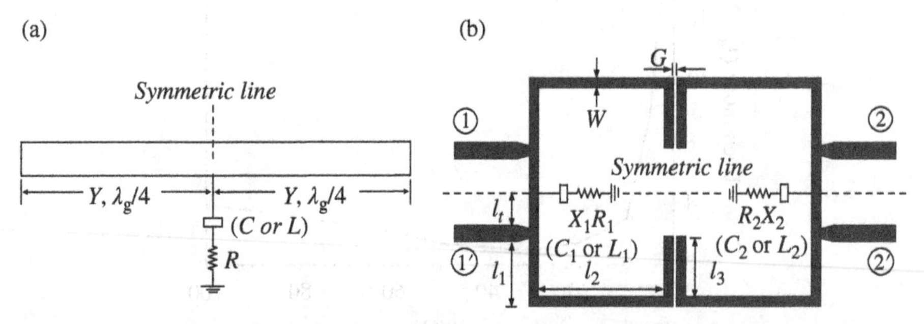

Figure 13.5 (a) Uniform impedance resonator with a capacitor (or an inductor) and a resistor loaded in series. (b) Second-order balanced filter using such resonators. Reprinted with permission from Ref. [1]; copyright 2010 IEEE.

the even-mode input impedance of the resonator is located at the open end, the even-mode quality factor can be obtained as follows:

$$Q_{\mathrm{C}}^{\mathrm{e}} = \left|\frac{1}{R}\left(\frac{\cos(2\theta_0)}{\omega C} + \frac{Y\sin(2\theta_0)}{\omega^2 C^2} - \frac{\sin(2\theta_0)}{4Y}\right) + RY\sin(2\theta_0)\right| \tag{13.9}$$

Equation (13.9) shows that a value of R occurs for the minimum $Q_{\mathrm{C}}^{\mathrm{e}}$.

When the loaded elements are an inductor and a resistor in series, the resonant condition can be described as follows:

$$\frac{Y}{(2Y\omega L + \tan\theta_0)^2 + (2YR)^2} \times \left[2Y\omega L\left(\tan^2\theta_0 - 1\right) + \left(4Y^2R^2 + 4Y^2\omega^2L^2 - 1\right)\tan\theta_0\right] = 0 \tag{13.10}$$

Thus, the value L can also control the even-mode resonant frequency and the common-mode external quality factor of the filter. According to the even-mode input impedance, the even-mode quality factor can be presented as follows:

$$Q_{L}^{\mathrm{e}} = \left|\frac{1}{R}\left(Y\omega^2L^2\sin(2\theta_0) - \omega L\cos(2\theta_0) - \frac{\sin(2\theta_0)}{4Y}\right) + RY\sin(2\theta_0)\right| \tag{13.11}$$

Similarly, a value of R occurs for the minimum Q_{L}^{e}.

To study the effect of the series lumped elements, the differential- and common-mode responses of the filter in Figure 13.5 without loaded elements are shown in Figure 13.6, and the common-mode responses with different values of lumped elements are shown in Figure 13.7. Figure 13.7(a) and (b) shows how $|S_{\mathrm{cc21}}|$ varies with the capacitance and the resistance when the capacitor and the resistor are loaded in series, respectively. It can be seen from Figure 13.7(a) that the center frequency of the common-mode response is primarily affected by the capacitor, which is equivalent to that described in Equation (13.8). Furthermore, the common-mode suppression reaches a maximum value when the capacitance is 11.6 pF. From Figure 13.7(b), we note that it is possible to identify an optimum resistance that delivers maximum common-mode suppression. The optimum resistance value is 0.5 Ω.

Figure 13.7(c) and (d) illustrates that $|S_{\mathrm{cc21}}|$ varies with the inductance and resistance when the inductor and resistor are loaded in series,

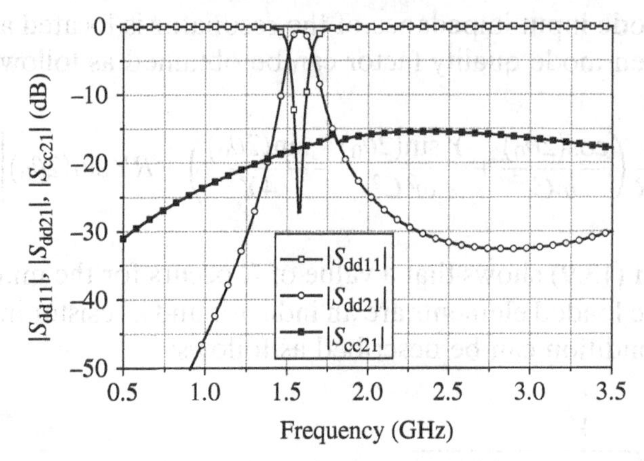

Figure 13.6 Simulated differential- and common-mode response without loaded elements. Reprinted with permission from Ref. [1]; copyright 2010 IEEE.

respectively. Figure 13.7(c) shows that the center frequency of the common-mode response is primarily affected by the inductor and is less than f_o, which is equivalent to that described in Equation (13.10). However, Figure 13.7(c) and (d) shows that the common-mode suppression does not demonstrate remarkable improvements compared with the filter that use the unloaded resonator because of the limited common-mode external quality factor of the center-loaded resonator. Therefore, according to the aforementioned analysis and simulation, the uniform impedance resonator with the capacitor and resistor loaded in series can greatly improve the common-mode suppression. Moreover, the differential-mode response does not change when adjusting the capacitance and resistance.

Based on the advantages of the proposed resonator, a dual-band balanced filter is designed. Figure 13.8 shows the structure of the proposed dual-band balanced filter, which is a fourth-order filter with a symmetrical structure consisting of four uniform impedance resonators loaded with a capacitor and resistor in series, where the loads are marked as Y_{Ln} ($n = 1, 2, 3,$ and 4) and the odd-mode resonant frequencies of the four resonators are nearly equivalent. The first step in the filter design realizes the differential-mode dual-band bandpass response. Figure 13.9 illustrates the differential-mode feeding and coupling scheme and indicates that two coupling paths occur. These two coupling paths result in two differential-mode transmission zeros for the filter. When the magnetic coupling between resonators 2 and 3 is weak, the two transmission zeros are located between the two passbands, thus demonstrating the dual-band differential-mode response.

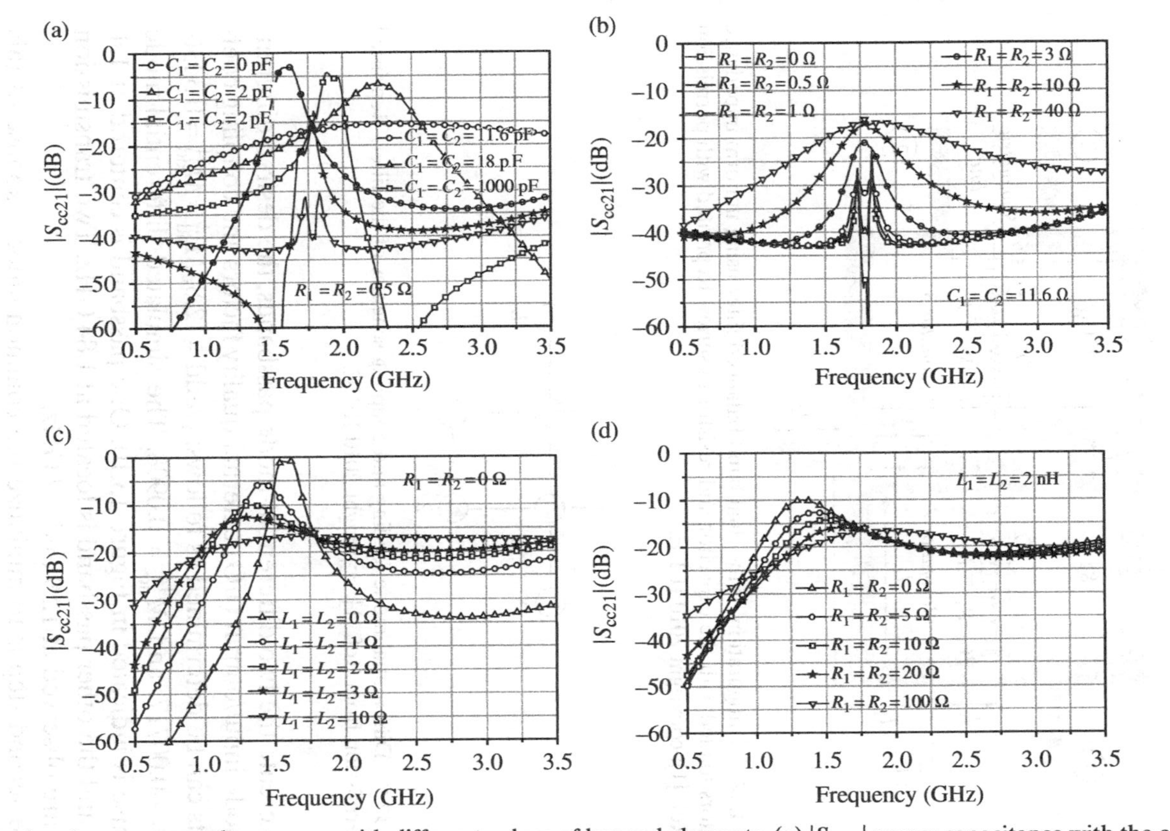

Figure 13.7 Simulated common-mode response with different values of lumped elements. (a) $|S_{cc21}|$ versus capacitance with the capacitor and the resistor loaded. (b) $|S_{cc21}|$ versus resistance with the capacitor and the resistor loaded. (c) $|S_{cc21}|$ versus inductance with the inductor and the resistor loaded. (d) $|S_{cc21}|$ versus resistance with the inductor and the resistor loaded. Reprinted with permission from Ref. [1]; copyright 2010 IEEE.

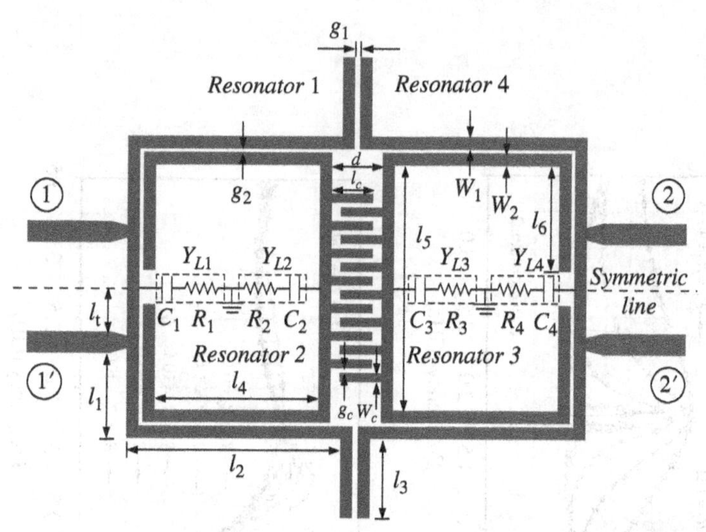

Figure 13.8 Configuration of the dual-band balanced filter using uniform impedance resonators loaded with the capacitor and resistor in series. Reprinted with permission from Ref. [1]; copyright 2010 IEEE.

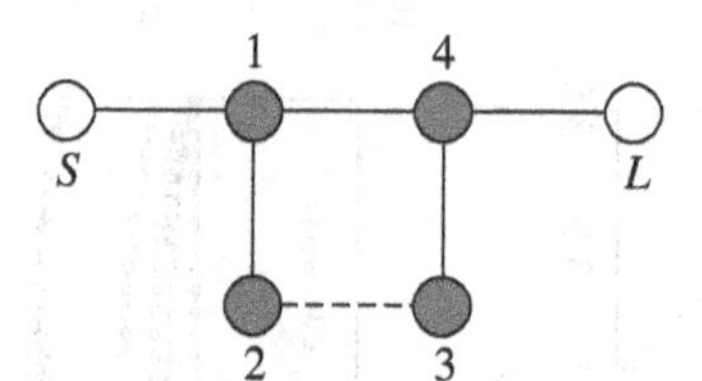

Figure 13.9 Differential-mode feeding and coupling schemes of the filter. Reprinted with permission from Ref. [1]; copyright 2010 IEEE.

To create the two differential-mode passbands, the desired differential-mode input/output (I/O) external quality factors and coupling coefficients can be determined as follows: $Qedd = 12.8$, $M_{\text{dd}}^{12} = M_{\text{dd}}^{34} = 0.3436$, $M_{\text{dd}}^{23} = -0.00717$, and $M_{\text{dd}}^{14} = 0.0908$. The simulated differential-mode response is indicated in Figure 13.10. One passband is centered at 1.4 GHz, and the other passband is located at 1.84 GHz. Two transmission zeros are observed at 1.59 and 1.7 GHz.

The second step is to minimize the common-mode response of this dual-band balanced filter by tuning the loaded elements. Figure 13.10 shows that the common-mode suppression in the entire frequency range from 0.5 to 3.5 GHz is low without the loaded lumped elements. After adding the loads Y_{L1} and Y_{L4}, the common-mode suppression near the differential-mode passband can be improved by 10 dB, whereas the

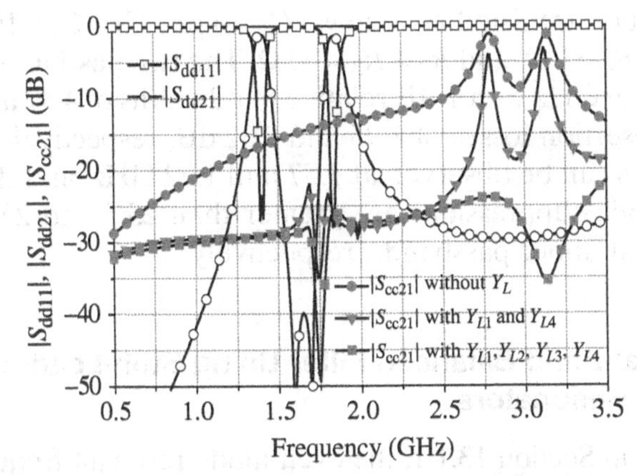

Figure 13.10 Simulated results of the dual-band balanced filter with different load combinations. Reprinted with permission from Ref. [1]; copyright 2010 IEEE.

common-mode suppression near the second harmonic of the differential-mode passband remains poor. To widen the common-mode rejection range, Y_{L1}, Y_{L2}, Y_{L3}, and Y_{L4} are all loaded at the center of the respective uniform impedance resonator. Figure 13.10 shows that the common-mode suppression near the second harmonic frequency is greatly improved. Furthermore, the common-mode suppression near the fundamental frequency is also slightly improved when four capacitors and four resistors are loaded.

One prototype was fabricated, and the measured differential- and common-mode responses are shown in Figure 13.11. The loaded

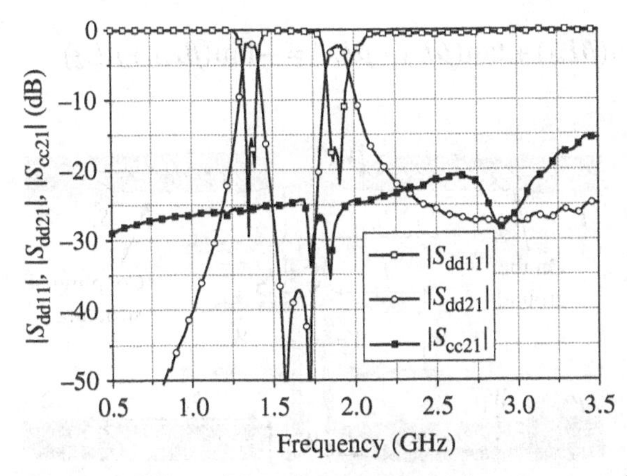

Figure 13.11 Measured differential- and common-mode responses of the dual-band balanced filter. Reprinted with permission from Ref. [1]; copyright 2010 IEEE.

element values are listed as follows: $C_1 = 4$ pF, $C_2 = C_3 = 1000$ pF, $C_4 = 3.6$ pF, $R_1 = R_4 = 0\ \Omega$, and $R_2 = R_3 = 33\ \Omega$. The two passbands are located at 1.37 and 1.9 GHz with 1-dB relative bandwidths of 4.37 and 4.2% and minimum insertion losses of 1.95 and 2.32 dB, respectively. Two transmission zeros can be observed at 1.57 and 1.72 GHz, and the measured common-mode suppressions are greater than 25.5 and 25.4 dB within the lower and upper passbands, respectively.

13.1.3 Dual-Band Balanced Filter Using Stub-Loaded Uniform Impedance Resonators

As analyzed in Section 13.1.1, the even-mode resonant frequency can be controlled through the uniform impedance resonator with a center-loaded stub. In addition, if the filter only contains two single-frequency resonators, it is impossible to produce a dual-band balanced filter. Thus, the uniform impedance resonator with three open stubs is proposed to construct a dual-band balanced filter with only two resonators. Figure 13.12 shows the configuration of the dual-band balanced bandpass filter [2].

The odd- and even-mode equivalent circuits of the stub-loaded uniform impedance resonator are shown in Figure 13.13. For odd-mode operation, the center point of the resonator is a virtual short circuit, and the resonant condition can be derived as follows:

$$\tan(\beta L_4) = \cot(\beta L_3) - \tan(\beta L_1 + \beta L_2) \tag{13.12}$$

For even-mode operation, the center stub still operates and the resonant condition can be calculated as follows:

$$\tan(\beta L_4) + \tan(\beta L_3 + \beta L_5) = -\tan(\beta L_1 + \beta L_2) \tag{13.13}$$

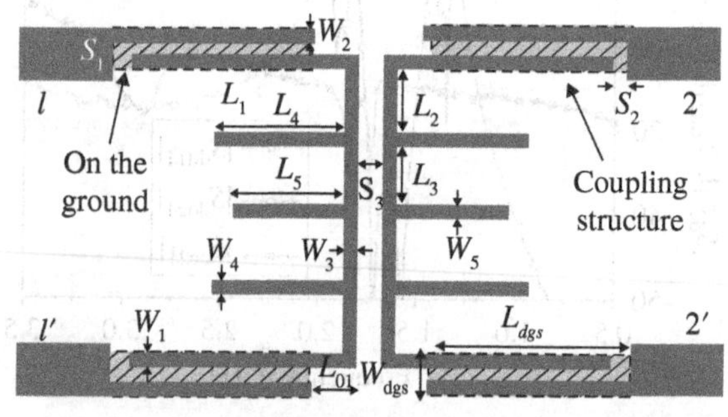

Figure 13.12 Configuration of the dual-band balanced filter. Reprinted with permission from Ref. [2]; copyright 2015 IEEE.

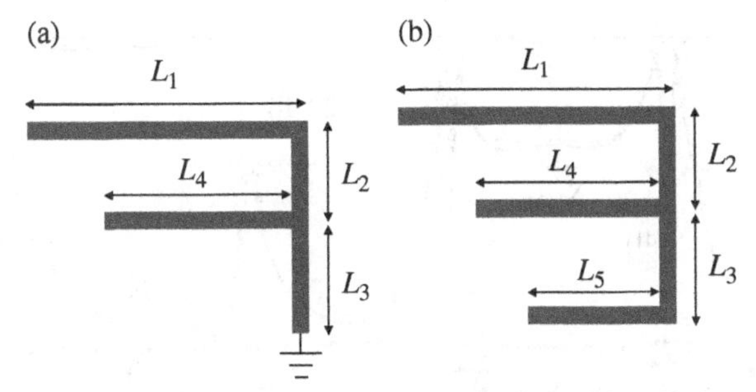

Figure 13.13 Equivalent circuit of the stub-loaded uniform impedance resonator. (a) Odd mode. (b) Even mode. Reprinted with permission from Ref. [2]; copyright 2015 IEEE.

According to Equations (13.12) and (13.13), two odd-mode and two even-mode resonant frequencies can be obtained. Figure 13.14 shows the four resonant frequencies with different lengths of center stub (L_5). Obviously, the even-mode frequencies can be tuned independently from the odd-mode frequencies by changing L_5. Therefore, the common-mode response can be shifted away from the differential-mode passbands.

For the filter design, the I/O coupling and the coupling between the two resonators should also be considered in addition to the resonant frequencies. The defected ground structures under the I/O coupling structures are used to increase the coupling between the I/O ports and resonators. The coupling coefficient between the two resonators can be tuned by changing the gap width S_3. Finally, a dual-band balanced

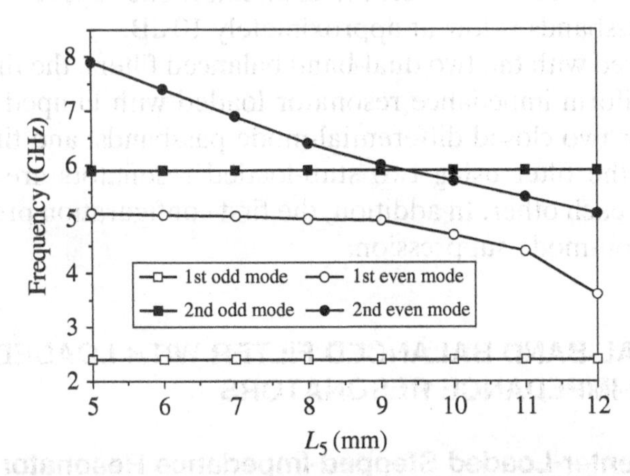

Figure 13.14 Effects of the center-loaded open stub with a length of L_5 on the resonant frequencies of the stub-loaded uniform impedance resonator. Reprinted with permission from Ref. [2]; copyright 2015 IEEE.

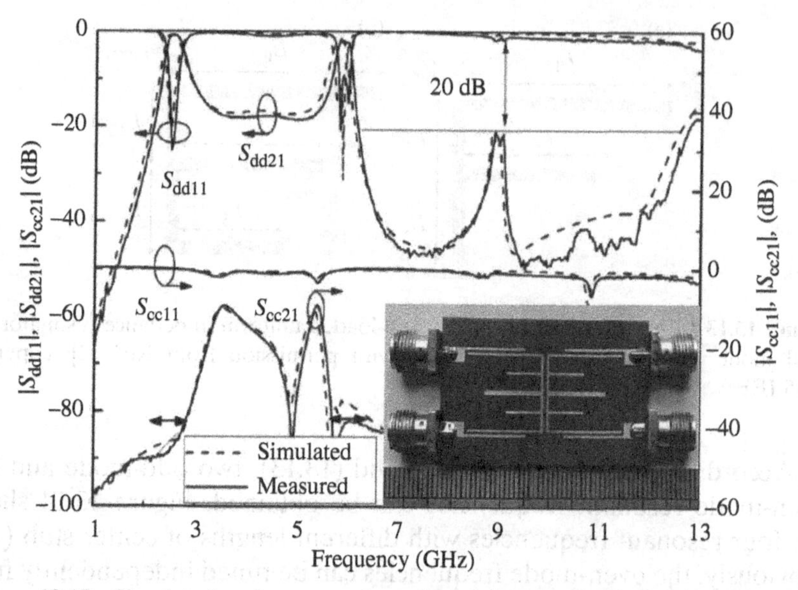

Figure 13.15 Simulated and measured results of the proposed dual-band bandpass filter. Reprinted with permission from Ref. [2]; copyright 2015 IEEE.

filter operating at 2.5 and 5.8 GHz can be obtained. Figure 13.15 shows the simulated and measured results of the dual-band balanced filter and the 3-dB fractional bandwidths of 12.9 and 4.5%, respectively. The minimum insertion losses are 0.77 and 1.56 dB for the two differential-mode passbands. For the common-mode response, the measured common-mode suppressions inside the two differential-mode passbands are greater than 38 dB. However, the common-mode suppression between the two passbands is low at approximately 10 dB.

Compared with the two dual-band balanced filters, the filter containing the uniform impedance resonator loaded with lumped elements is suitable for two closed differential-mode passbands, and the two passbands for the filter using two stub-loaded resonators are located far away from each other. In addition, the first configuration presents superior common-mode suppression.

13.2 DUAL-BAND BALANCED FILTER WITH LOADED STEPPED-IMPEDANCE RESONATORS

13.2.1 Center-Loaded Stepped-Impedance Resonator

The stepped-impedance resonator is commonly used to realize the dual-band filter. Figure 13.16(a) shows a center-loaded stepped-impedance

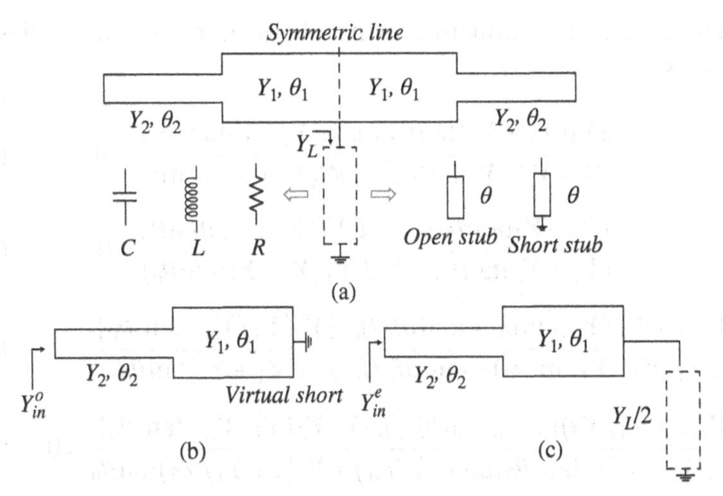

Figure 13.16 (a) Center-loaded stepped-impedance resonator with different elements. (b) Odd-mode equivalent circuit. (c) Even-mode equivalent circuit. Reprinted with permission from Ref. [3]; copyright 2010 IEEE.

resonator for which the loaded elements can consist of a capacitor, inductor, resistor, and open/short stub. The symmetrical planes are equivalent to an electric wall and a magnetic wall for odd- and even-mode operations, respectively. The odd- and even-mode equivalent circuits are shown in Figure 13.16(b) and (c), respectively. Similar to the loaded uniform impedance resonator, the loaded elements of the stepped-impedance resonator only operate under even-mode operations.

Assuming $\theta_1 = \theta_2$, according to the odd-mode equivalent circuits, the odd-mode resonant condition can be written as follows [4]:

$$\theta_1 = \theta_2 = \tan^{-1}\sqrt{\frac{Y_1}{Y_2}} \tag{13.14}$$

$$\frac{f_{o2}}{f_{o1}} = \frac{\pi}{\tan^{-1}\sqrt{\frac{Y_1}{Y_2}}} - 1 \tag{13.15}$$

where f_{o1} and f_{o2} are the first and second odd-mode resonant frequencies, respectively.

For even-mode operations, the resonant condition changes with the loaded elements and the resonant conditions for the resonator with the loaded capacitor, inductor, open stub, or short stub can be described as follows:

$$\frac{2Y_1(Y_1+Y_2)\tan\theta_0+\omega C(Y_1-Y_2\tan^2\theta_0)}{2Y_1(Y_2-Y_1\tan^2\theta_0)-\omega C(Y_1+Y_2)\tan\theta_0}=0 \tag{13.16}$$

$$\frac{(Y_1-Y_2\tan^2\theta_0)-2\omega LY_1(Y_1+Y_2)\tan\theta_0}{(Y_1+Y_2)\tan\theta_0+2\omega LY_1(Y_2-Y_1\tan^2\theta_0)}=0 \tag{13.17}$$

$$\frac{2Y_1(1+Y_1/Y_2)\tan\theta_0+\tan(\theta f_e/f_{o1})Y_3(Y_1/Y_2-\tan^2\theta_0)}{2Y_1(Y_2-Y_1\tan^2\theta_0)-\tan(\theta f_e/f_{o1})Y_3(Y_1+Y_2)\tan\theta_0}=0 \tag{13.18}$$

$$\frac{2Y_1(1+Y_1/Y_2)\tan\theta_0\tan(\theta f_e/f_{o1})-Y_3(Y_1/Y_2-\tan^2\theta_0)}{2Y_1(1-Y_1/Y_2\tan^2\theta_0)\tan(\theta f_e/f_{o1})+Y_3(1+Y_1/Y_2)\tan\theta_0}=0 \tag{13.19}$$

where $\theta_0=\theta_1 f_e/f_{o1}$, f_e is the even-mode resonant frequency, and θ is the electric length of the loaded open or short stub at f_{o1}. According to Equations (13.16)–(13.19), the calculated frequency ratios (f_{o1}/f_{o1}, f_{o2}/f_{o1}, f_{e1}/f_{o1}, f_{e2}/f_{o1}, where f_{e1} and f_{e2} are the first and second even-mode resonant frequencies, respectively) of the stepped-impedance resonator loaded with a capacitor, inductor, or open/short stub are illustrated in Figure 13.17.

Figure 13.17 shows that f_{o1} and f_{o2} do not change. For the capacitor- or inductor-loaded stepped-impedance resonator, f_{e1} changes from $1.6f_{o1}$ to f_{o1} or from f_{o1} to $0.3f_{o1}$, and f_{e2} varies from $3.2f_{o1}$ to $2.2f_{o1}$ or from $2.2f_{o1}$ to $1.6f_{o1}$ when C or L increases, respectively. For the

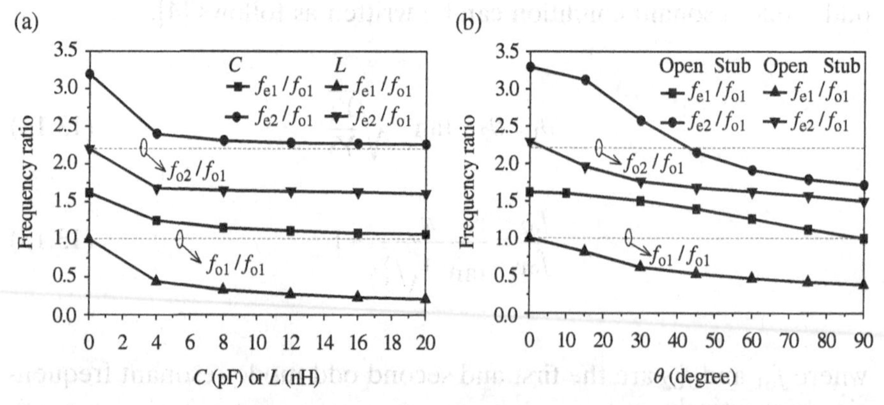

Figure 13.17 Calculated frequency ratios with (a) different values of C or L ($Y_1=0.024$ S, $Y_2=0.012$ S) and (b) different electric lengths (θ) of the loaded open stub or short stub.

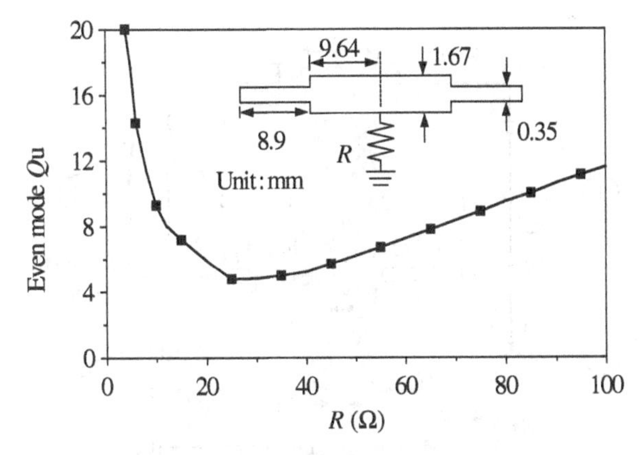

Figure 13.18 Even-mode Q_u with different R based on lossless microstrip line. Reprinted with permission from Ref. [3]; copyright 2010 IEEE.

stub-loaded stepped-impedance resonator, f_{e1} changes from $1.6f_{o1}$ to f_{o1} or from f_{o1} to $0.3\,f_{o1}$, and f_{e2} varies from $3.2f_{o1}$ to $1.6f_{o1}$ or from $2.2f_{o1}$ to $1.7f_{o1}$ when the electric length of the open or short stub increases, respectively.

For the resistor-loaded stepped-impedance resonator, the even-mode Q_u can also be greatly influenced by the resistance. Figure 13.18 shows the simulated even-mode Q_u of the resistor-loaded stepped-impedance resonator. At the optimum resistance, the minimum even-mode Q_u occurs for the resonator. Therefore, the resistor can be used to absorb the even-mode signal.

13.2.2 Dual-Band Balanced Filter Using Stepped-Impedance Resonators with Center-Loaded Lumped Elements

A dual-band balanced filter [3] using four stepped-impedance resonators with loads Y_{Ln} ($n = 1, 2, 3$, or 4) is shown in Figure 13.19. The differential-mode feeding and coupling schemes are shown in Figure 13.20, where the coupling between resonators 1 and 4 is realized using short microstrip lines. The first step for the filter design produces the differential-mode dual-band response. For the two differential-mode passbands located at 2.45 and 5.55 GHz, $\theta_1 = 56.4°$ at 2.45 GHz, and $Z_1 = 41.2\,\Omega$ and $Z_2 = 84.9\,\Omega$ according to Equations (13.14)–(13.15). The differential-mode external quality factor is determined by the tap position l_t. The differential-mode coupling coefficients between the resonators can be tuned by changing the gaps (g_1, g_2, g_3, g_4, and g_5). Because cross coupling occurs in the coupling scheme, differential-mode

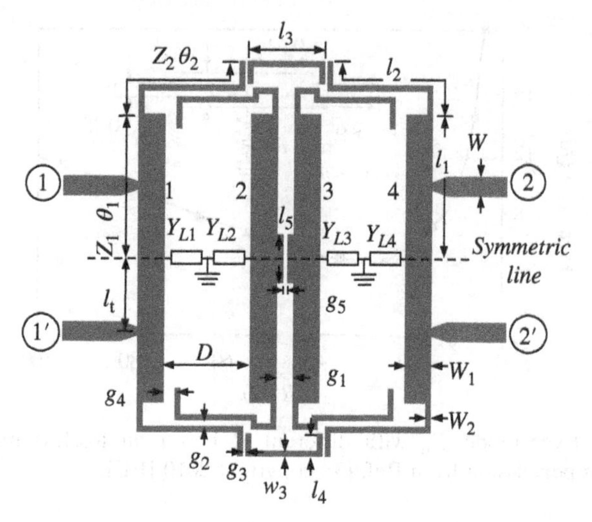

Figure 13.19 Configuration of the balanced dual-band bandpass filter designed using center-loaded stepped-impedance resonators. Reprinted with permission from Ref. [3]; copyright 2010 IEEE.

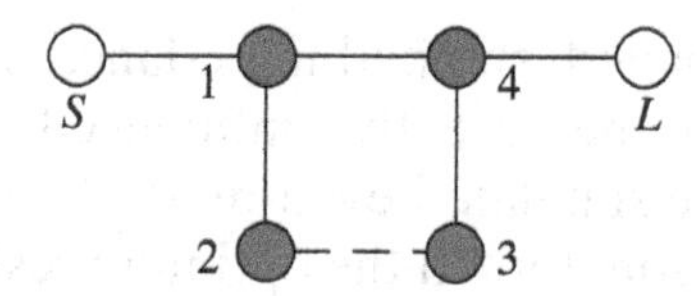

Figure 13.20 Differential-mode feeding and coupling scheme of the dual-band balanced filter. Reprinted with permission from Ref. [3]; copyright 2010 IEEE.

transmission zeros can be realized, and the positions of the transmission zeros are determined by the coupling coefficients.

Figure 13.21 presents the simulated differential- and common-mode responses of the filter without loaded elements. As shown in Figure 13.21, two differential-mode passbands with four transmission zeros can be observed. However, the common-mode suppression of the filter without the loaded lumped elements is not sufficient because of the common-mode passband near 4.2 GHz.

The second step is to improve the common-mode suppression level while holding the differential-mode response unchanged, which also verifies the theory in Section 13.2.1. Therefore, the common-mode performance of the filter should be studied in cases of different loaded lumped elements. According to the study, the inductor has a rather limited improving effect on the common-mode suppression. If the loads are all capacitors, then the minimum common-mode suppression between 1 and 7 GHz is approximately 15 dB as shown in Figure 13.22(a). However,

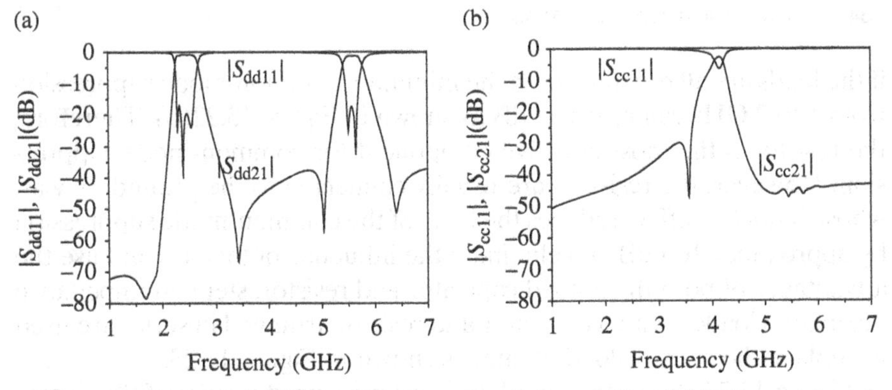

Figure 13.21 Simulated results of the dual-band balanced filter. (a) Differential-mode response. (b) Common-mode response without loaded lumped elements. Reprinted with permission from Ref. [3]; copyright 2010 IEEE.

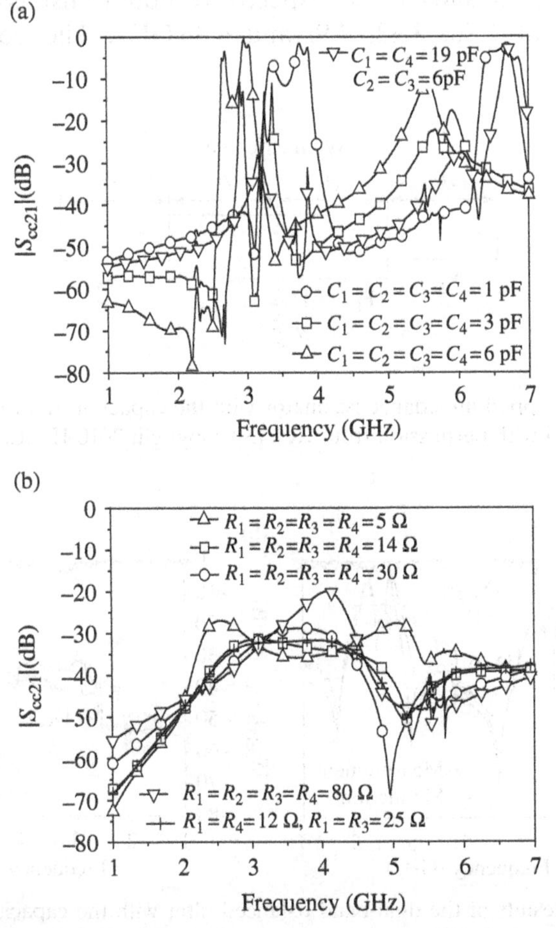

Figure 13.22 Common-mode response of the dual-band balanced bandpass filter. (a) Center-loaded resonators with capacitors. (b) Center-loaded resonators with resistors. Reprinted with permission from Ref. [3]; copyright 2010 IEEE.

if the loads are all resistors, then the minimum common-mode suppression from 1 to 7 GHz can reach 32 dB as shown in Figure 13.22(b). Therefore, the resistor is the most effective component for common-mode suppression. However, the resistors are usually connected to the ground by vias, whose inductive effect reduces the level of the common-mode suppression by approximately 7 dB. To eliminate the influence of the vias and use the advantages of both the loaded capacitor and resistor, stepped-impedance resonators loaded with a capacitor and resistor arranged in series are used to replace the resistor-loaded one, as shown in Figure 13.23.

Figure 13.24 shows the simulated and measured results of the prototype with a capacitor and resistor loaded in series, where $R_1 = R_4 = 10\,\Omega$, $R_2 = R_3 = 22\,\Omega$, $C_1 = C_4 = 1$ pF, and $C_2 = C_3 = 1.5$ pF. The two differential-mode passbands are located at 2.46 and 5.56 GHz with 1-dB relative bandwidth of 16.26 and 6.65%, respectively. Four transmission zeros can be observed at 1.76, 3.42, 4.9, and 6.36 GHz. The common-mode

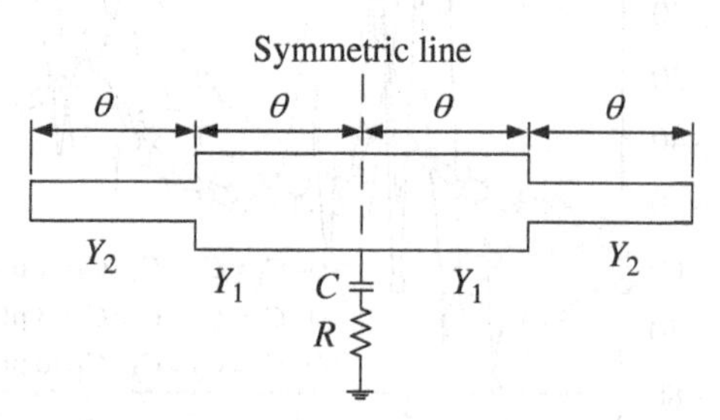

Figure 13.23 Stepped-impedance resonator with the capacitor and resistor loaded in series. Reprinted with permission from Ref. [3]; copyright 2010 IEEE.

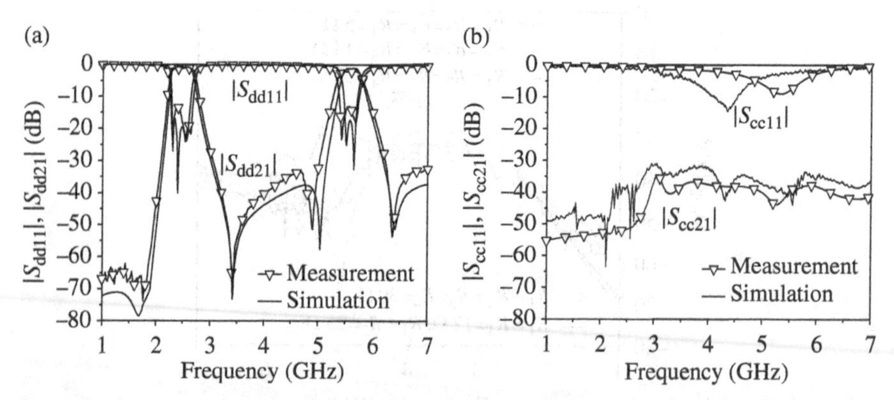

Figure 13.24 Results of the dual-band balanced filter with the capacitor and resistor loaded in series. (a) Differential-mode response. (b) Common-mode response. Reprinted with permission from Ref. [3]; copyright 2010 IEEE.

suppression level is greater than 31 dB from 1 to 7 GHz and originates from three sources: reflection caused by the poor common-mode external quality factor, low common-mode coupling coefficient caused by different resonant frequencies, and absorption of the resistors.

13.2.3 Dual-Band Balanced Filter Using Stub-Loaded Stepped-Impedance Resonators

The filter using stub-loaded stepped-impedance resonators [5] shown in Figure 13.25 can also deliver differential-mode dual-band response with high common-mode suppression. The design method for the two differential-mode passbands is similar to that of the filter with a capacitor and resistor loaded in series. The difference is that cross coupling does not occur to produce transmission zeros. To achieve differential-mode dual-band passbands at 2.45 GHz (2.4–2.484 GHz) and 5.25 GHz (5.15–5.35 GHz), Y_1/Y_2 should be 2.42, and θ_1 should equal to 57.27° at 2.45 GHz. Therefore, the physical structure of the stepped-impedance resonator can be achieved. The initial goal for the fractional bandwidths of the two passband is 7%; therefore, the differential-mode external quality factor $Qedd = 10.93$ and the differential-mode coupling coefficient $M_{\mathrm{dd}}^{\mathrm{AB}} = M_{\mathrm{dd}}^{\mathrm{CA}} = 0.059$ and $M_{\mathrm{dd}}^{\mathrm{BC}} = 0.038$. Thus, the widths and lengths of the coupling gaps of the coupled stepped-impedance resonators as well as the position of the feed point can be extracted.

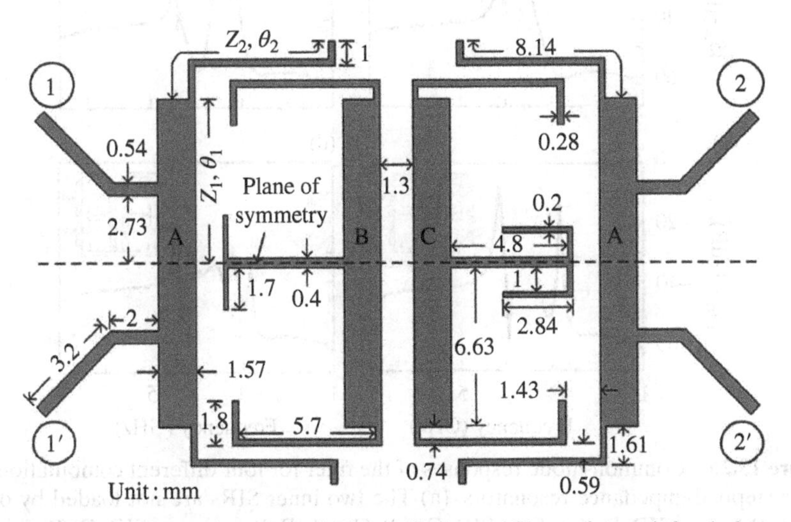

Figure 13.25 Configuration of the dual-band balanced filter using stub-loaded stepped-impedance resonators. Reprinted with permission from Ref. [5]; copyright 2010 IEEE.

According to the analysis in Figure 13.16, the loaded stub changes the even-mode resonant frequency, whereas the odd-mode resonant frequency is maintained. Therefore, the T-shaped open stubs do not change the differential-mode response, and the stepped-impedance resonators with different open stubs have different even-mode resonant frequencies. Figure 13.26 shows the common-mode responses of the filter for four different combinations of the inner stepped-impedance resonators. The maximum common-mode suppression can be achieved when different stubs are loaded with resonators because the even-mode resonant frequencies of the resonators A, B, and C are different from each other as shown in Figure 13.27. Therefore, the common-mode coupling between resonators can be minimized.

The results from a filter prototype using the stub-loaded stepped-impedance resonators are shown in Figure 13.28. The measured first and second differential-mode passbands are centered at 2.44 and 5.25 GHz with 3 dB frequency ranges of 2.34–2.55 and 5.14–5.38 GHz, respectively. The minimum insertion losses of the two passbands are 2.4 and 2.82 dB, respectively. The common-mode suppression is greater than 25 dB in the range of 1–7 GHz.

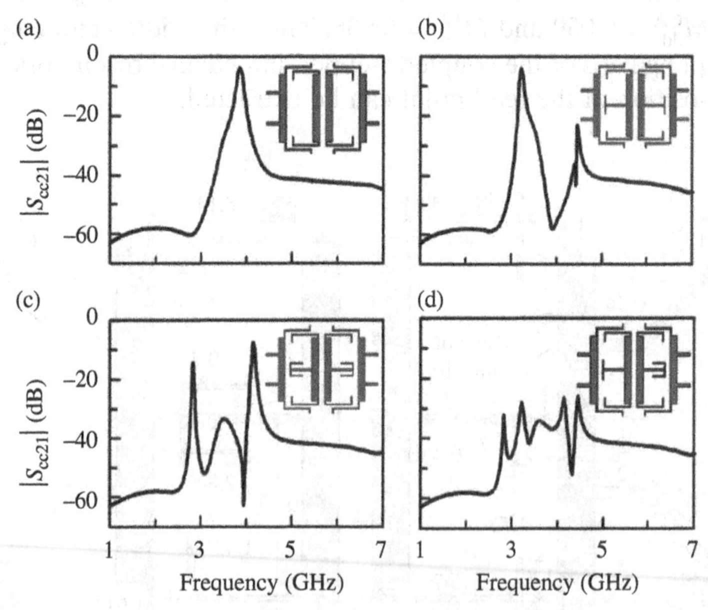

Figure 13.26 Common-mode responses of the filter for four different combinations of inner stepped-impedance resonators. (a) The two inner SIRs are not loaded by open stubs. (b) Two SIR B. (c) Two SIR C. (d) One SIR B and one SIR C. Reprinted with permission from Ref. [5]; copyright 2010 IEEE.

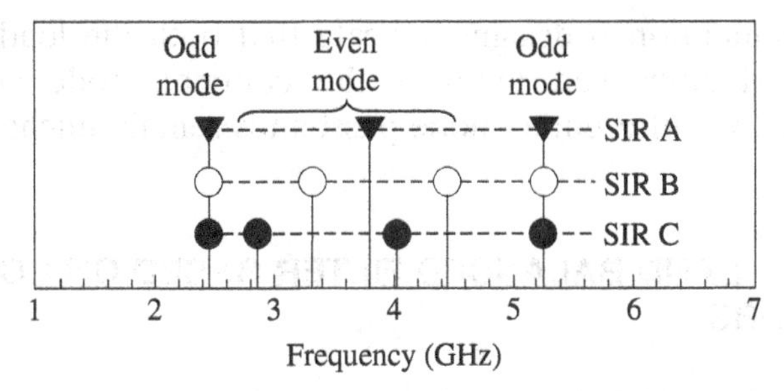

Figure 13.27 Resonant-frequency spectra for stepped-impedance resonators A, B, and C. Reprinted with permission from Ref. [5]; copyright 2010 IEEE.

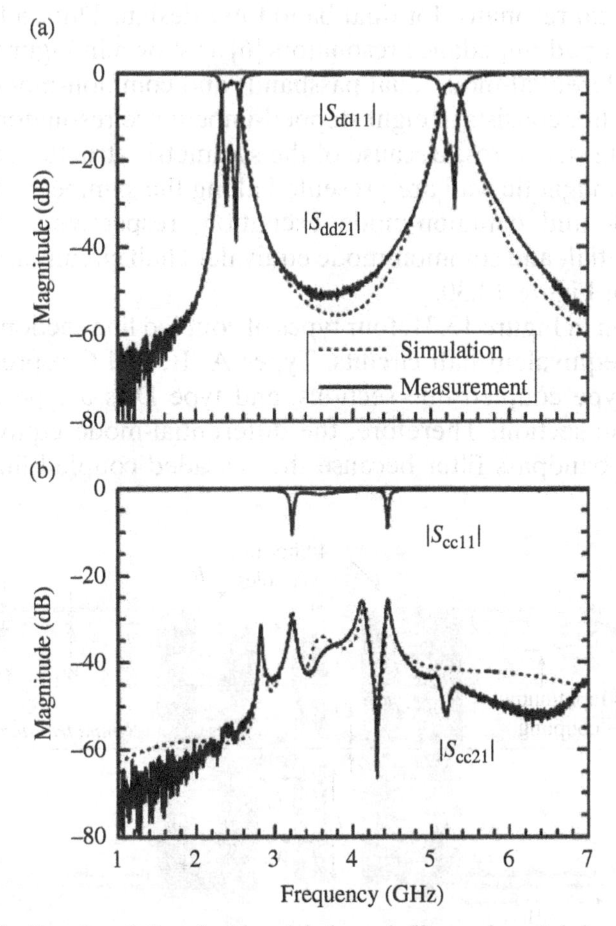

Figure 13.28 Results of the dual-band balanced filter using stub-loaded stepped-impedance resonators. (a) Differential-mode response. (b) Common-mode response. Reprinted with permission from Ref. [5]; copyright 2010 IEEE.

The aforementioned designs indicate that both the loaded lumped elements and stubs can improve the common-mode suppression, whereas the two differential-mode passbands remain unchanged.

13.3 DUAL-BAND BALANCED FILTER BASED ON COUPLED RESONATORS

13.3.1 Dual-Band Balanced Filter with Coupled Stepped-Impedance Resonators

For the design of the balanced filter, common-mode suppression can also be achieved by coupled resonators. The stepped-impedance resonator is a typical resonator for dual-band filter design. Thus, a filter using coupled stepped-impedance resonators [6] as shown in Figure 13.29 can achieve differential-mode dual passbands and common-mode suppression. This filter consists of eight stepped-impedance resonators and four I/O coupling structures. Because of the symmetric structure, an electric wall and a magnetic wall are presented along the symmetric line under differential- and common-mode excitation, respectively. Therefore, the differential- and common-mode equivalent half circuits are achieved as shown in Figure 13.30.

As shown in Figure 13.31, four types of coupled line sections are present in the equivalent half circuits. Types A, B, and C represent three bandpass-type coupled line sections, and type D is a type of all-stop coupled line section. Therefore, the differential-mode equivalent half circuit is a bandpass filter because the cascaded coupled line sections

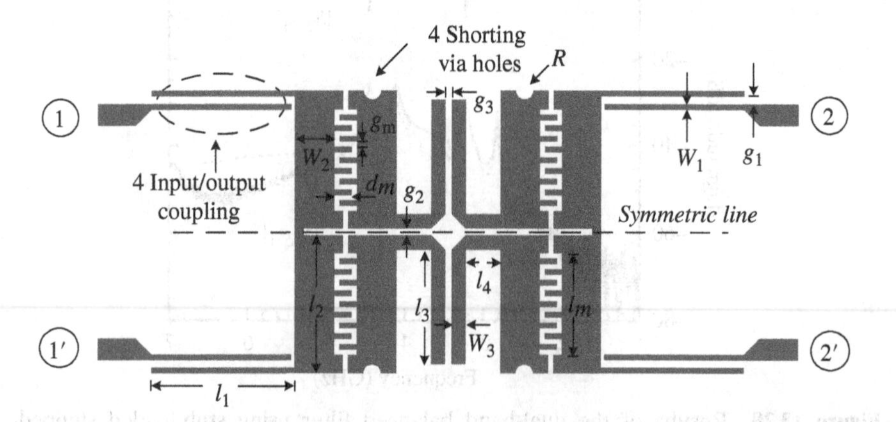

Figure 13.29 Configuration of the dual-band balanced filter using coupled stepped-impedance resonators. Reprinted with permission from Ref. [6]; copyright 2010 IEEE.

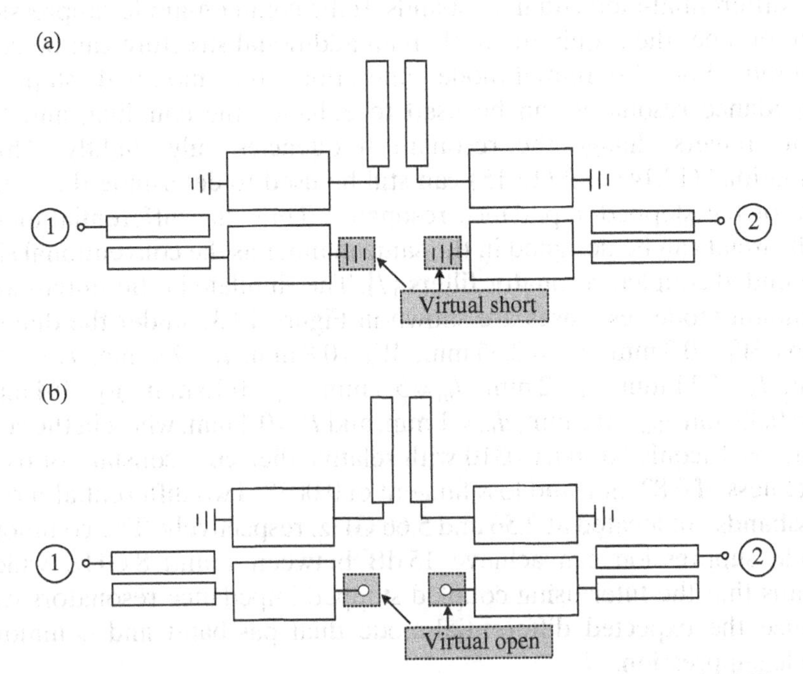

Figure 13.30 Equivalent half circuits of the dual-band balanced filter. (a) Differential mode. (b) Common mode. Reprinted with permission from Ref. [6]; copyright 2010 IEEE.

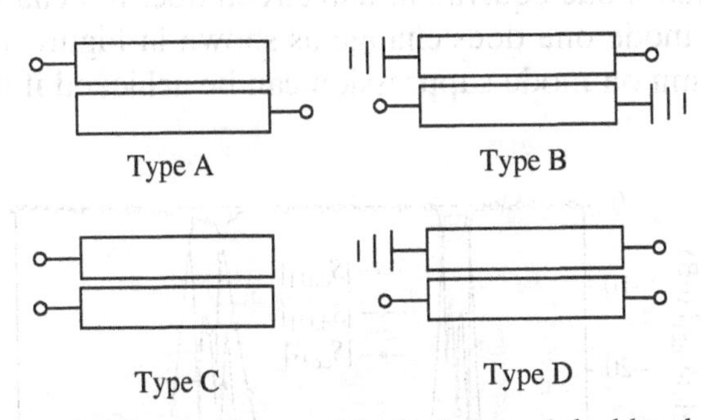

Figure 13.31 Coupled line sections used in the balanced dual-band filter. Reprinted with permission from Ref. [6]; copyright 2010 IEEE.

are all bandpass types. In contrast, the common-mode equivalent half circuit is bandstop at any frequency because of the type D coupled line sections.

The procedures for the filter design are described as follows. First, the filter structure is determined according to the desired specifications of the differential-mode dual passbands. If the common-mode suppression cannot meet the requirement, then an additional structure can be considered. For differential-mode operation, the modified stepped-impedance resonator can be used to enhance the coupling, and the short fingers change the resonant frequencies only slightly. Thus, Equations (13.14) and (13.15) can still be used to determine the initial size of the stepped-impedance resonator. Thus, the differential-mode half circuit can be designed in the same manner as the conventional single-ended coupled-resonator filters [7]. The simulated differential- and common-mode responses are shown in Figure 13.32 under the dimensions $W_1 = 0.3$ mm, $W_2 = 2.35$ mm, $W_3 = 0.8$ mm, $l_1 = 7.9$ mm, $l_2 = 7.43$ mm, $l_3 = 6.11$ mm, $l_4 = 2$ mm, $l_m = 5.4$ mm, $g_1 = 0.13$ mm, $g_2 = 0.4$ mm, $g_3 = 0.35$ mm, $g_m = 0.2$ mm, $d_m = 1$ mm, and $R = 0.4$ mm, wherein the substrate is Taconic RF-60A-0310 with relative dielectric constant of 6.03, thickness of 0.82 mm, and loss tangent of 0.0038. Two differential-mode passbands are located at 2.56 and 5.66 GHz, respectively. The common-mode suppression can achieve 15 dB between 1 and 8 GHz, which means that the filter using coupled stepped-impedance resonators can realize the expected differential-mode dual passband and common-mode suppression.

If high common-mode suppression is required, one long open stub (L_1) and one short open stub (L_2) can be added along the symmetric line to improve the common-mode suppression as shown in Figure 13.33(a). The differential-mode equivalent half circuit does not change, although the common-mode one does change as shown in Figure 13.33(b). The maximum common-mode suppression can be achieved if the minimum

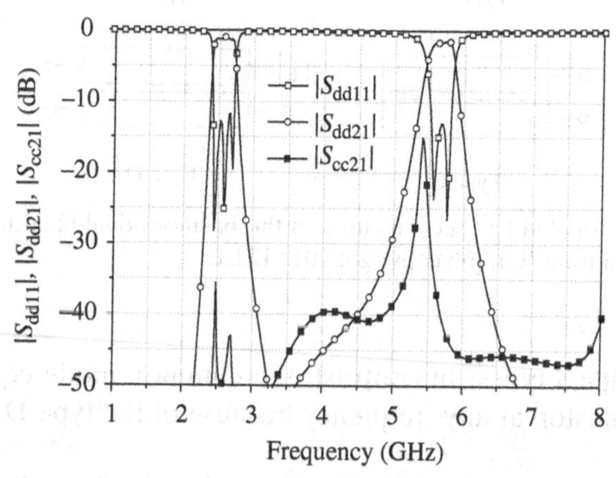

Figure 13.32 Simulated responses of the dual-band balanced filter shown in Figure 13.29. Reprinted with permission from Ref. [6]; copyright 2010 IEEE.

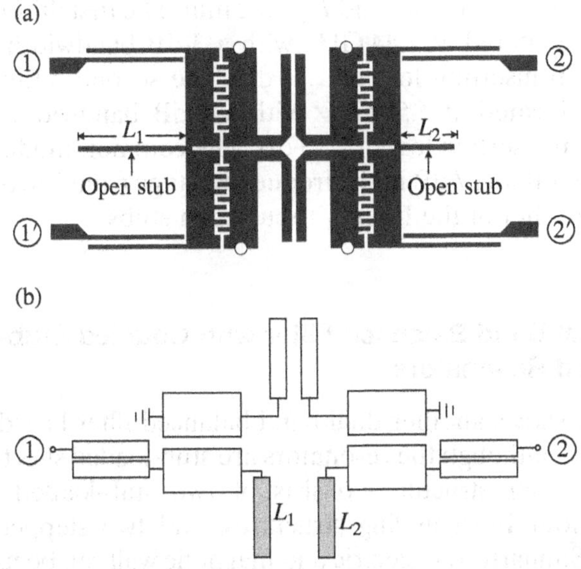

Figure 13.33 (a) Configuration of the proposed balanced dual-band bandpass filter with two open stubs. (b) Common-mode equivalent half circuit. Reprinted with permission from Ref. [6]; copyright 2010 IEEE.

value is realized for the common-mode coupling between the resonators by optimizing L_1 and L_2.

Figure 13.34 shows the simulated and measured results of the dual-band balanced filter using coupled stepped-impedance resonators and

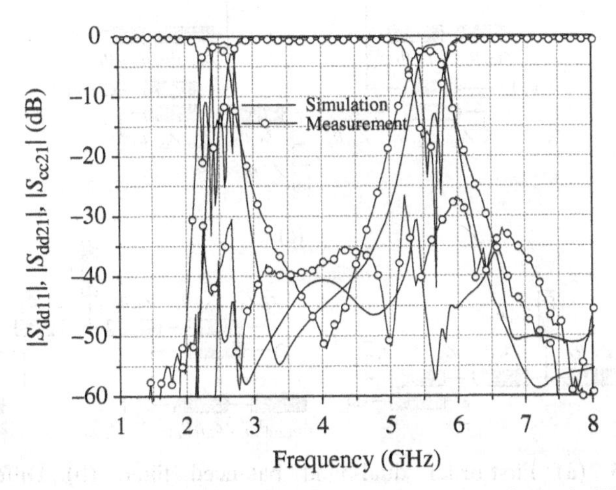

Figure 13.34 Simulated and measured responses of the filter shown in Figure 13.33(a). Reprinted with permission from Ref. [6]; copyright 2010 IEEE.

open stubs with $L_1 = 11$ mm and $L_2 = 5.2$ mm. The first differential-mode passband is centered at 2.44 GHz with a 3-dB bandwidth of 400 MHz and minimum insertion loss of 1.78 dB. The second differential-mode passband is located at 5.57 GHz with a 3-dB bandwidth of 480 MHz and minimum insertion loss of 2.53 dB. The common-mode suppression level is greater than 27 dB in the frequency range from 1 to 8 GHz, which is better than that of the filter without open stubs.

13.3.2 Dual-Band Balanced Filter with Coupled Stub-Loaded Short-Ended Resonators

Figure 13.35 shows another dual-band balanced filter based on coupled resonators [8], although the resonators are stub-loaded short-ended resonators. The filter structure consists of two stub-loaded short-ended resonators, four I/O coupling structures, and two stepped-impedance open stubs. Similarly, the electric and magnetic wall can be used in differential- and common-mode analyses because of the symmetrical structure. Thus, the differential- and common-mode half circuits are shown in Figure 13.35(b) and (c).

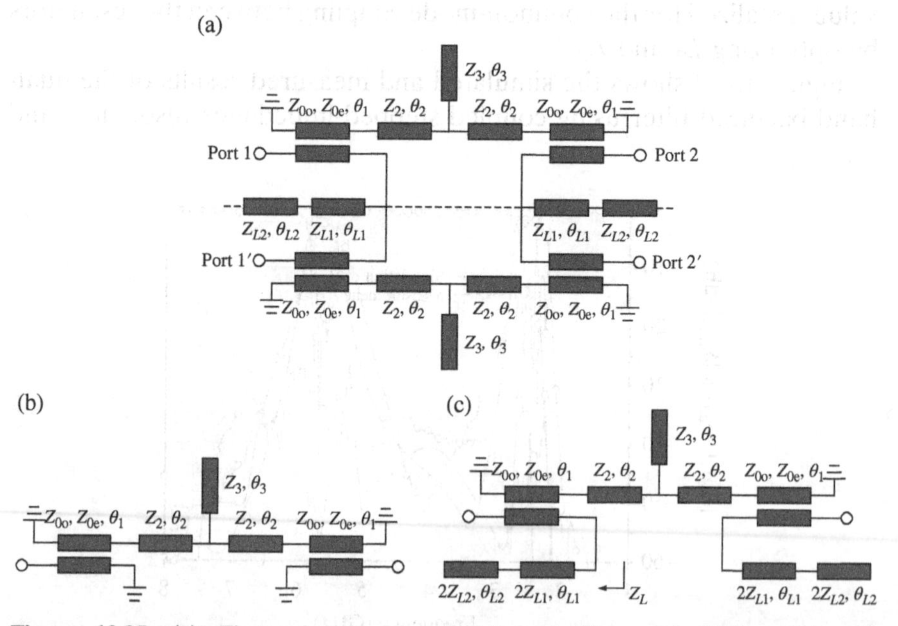

Figure 13.35 (a) First-order dual-band balanced filter. (b) Differential-mode equivalent half circuit. (c) Common-mode equivalent half circuit. Reprinted with permission from Ref. [8]; copyright 2015 IEEE.

For the differential-mode operation, the operating frequency is determined by the resonant frequency of the stub-loaded short-ended resonator as shown in Figure 13.36, and the even- and odd-mode resonant conditions can be formulated as follows:

$$2Z_2Z_3 = Z_2^2 \tan\theta_2 \tan\theta_3 + Z_1Z_2 \tan\theta_1 \tan\theta_3 + 2Z_1Z_3 \tan\theta_1 \tan\theta_2 \quad (13.20)$$

$$\frac{Z_1 \tan\theta_1 + Z_2 \tan\theta_2}{Z_1 - Z_2 \tan\theta_1 \tan\theta_2} = 0 \quad (13.21)$$

By adjusting the electrical length θ_2 in Figure 13.35, the spurious passbands caused by the odd-mode resonant frequencies can be removed [9]. Thus, the two differential-mode passbands are designed with the even-mode resonant frequencies of the resonator.

For the common-mode operation, the stepped-impedance open stubs can relocate the common-mode transmission zeros at the differential-mode resonant frequencies to improve the common-mode suppression. The common-mode transmission zeros can be determined as follows:

$$\frac{\sqrt{Z_{0e}^2 + X_L^2}\sin(\theta_e - \varphi_e) - \sqrt{Z_{0e}^2 + X_L^2}\sin(\theta_o - \varphi_o)}{\tan\theta_3 = 0} \quad (13.22)$$

where

$$X_L = \frac{2Z_{L1}(Z_{L2} - Z_{L1}\tan\theta_{L1}\tan\theta_{L2})}{Z_{L1}\tan\theta_{L2} + Z_{L2}\tan\theta_{L1}}, \quad \varphi_e = \tan^{-1}\left(\frac{X_L}{Z_{0e}}\right), \quad \varphi_o = \tan^{-1}\left(\frac{X_L}{Z_{0o}}\right)$$

Good common suppression with $Z_{L1} = 89.2\ \Omega$ and $Z_{L2} = 13.2\ \Omega$ can be realized by analyzing the common-mode equivalent circuit corresponding to different stepped-impedance open stubs. The simulated frequency ratios, which are defined as the frequencies of common-mode

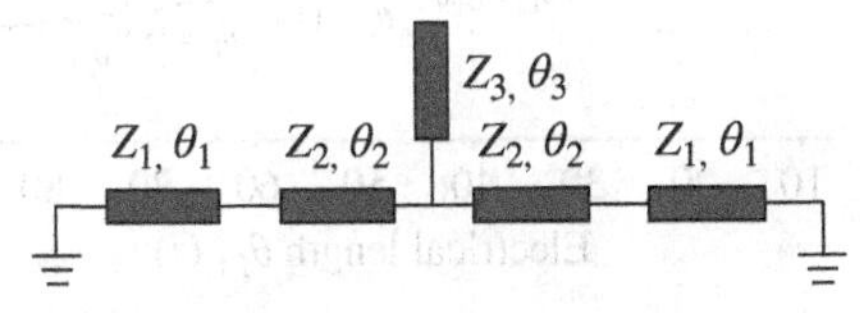

Figure 13.36 Stub-loaded short-ended resonator. Reprinted with permission from Ref. [8]; copyright 2015 IEEE.

transmission zeros versus the first differential-mode operating frequency f_0, are demonstrated in Figure 13.37 for variable θ_{L1} and θ_{L2} at f_0. The common-mode transmission zeros decrease with increasing θ_{L1} and θ_{L2}. When $\theta_{L1} = 91.8°$ and $\theta_{L2} = 24.5°$, five common-mode transmission zeros are located at $f_{CT1} \approx f_0$, $f_{CT2} \approx 2.11f_0$, $f_{CT3} \approx 2.77f_0$, $f_{CT4} \approx 4.19f_0$, and $f_{CT5} \approx 4.90f_0$. Thus, wideband common-mode suppression is available, and the first and third common-mode transmission zeros can be used to enhance the common-mode suppression inside the differential-mode passbands.

Based on the aforementioned analysis, a second-order dual-band balanced filter centered at 900 MHz and 2.45 GHz is designed, as shown in Figure 13.38. The design procedures can be described as follows. The first step confirms the differential-mode dual passbands. The resonator parameters can be obtained from Equation (13.20) and the parametric study of θ_3/θ_1 and Z_3/Z_1. For this filter, the desired frequency ratio of the two passbands is 2.72, and the resonator parameters are optimized at $W_1 = 1.5$ mm, $W_3 = 1.4$ mm, and $\theta_3 = 0.28\theta_1$. Dual passbands are expected with 3-dB fractional bandwidths of 5 and 3%. For the dimensions $L_1 = 44$ mm, $S_1 = 0.2$ mm, and $S_2 = 1.4$ mm, the external quality factors for the two passbands of the filter are $Q_{eI} = 33.8$ and $Q_{eII} = 85.3$, and the coupling coefficients of the two passbands $K_I = 0.055$ and $K_{II} = 0.01$ can satisfy the required bandwidths.

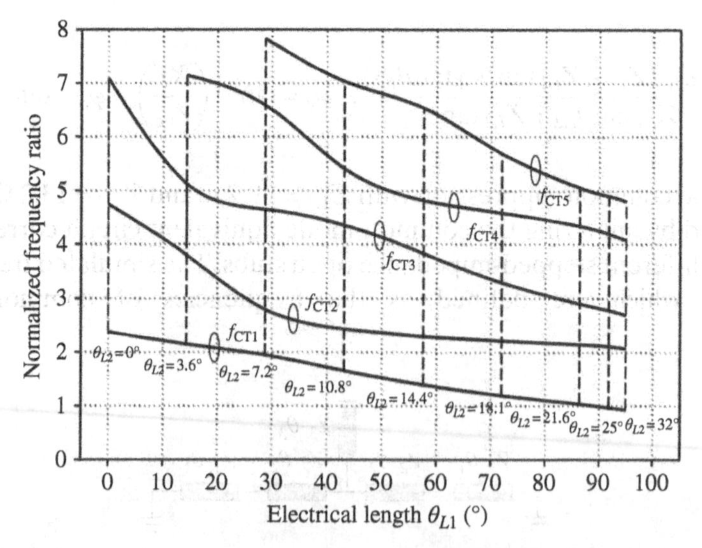

Figure 13.37 Frequency ratios of common-mode transmission zeros to f_0 with different θ_{L1} and θ_{L2} at f_0. Reprinted with permission from Ref. [8]; copyright 2015 IEEE.

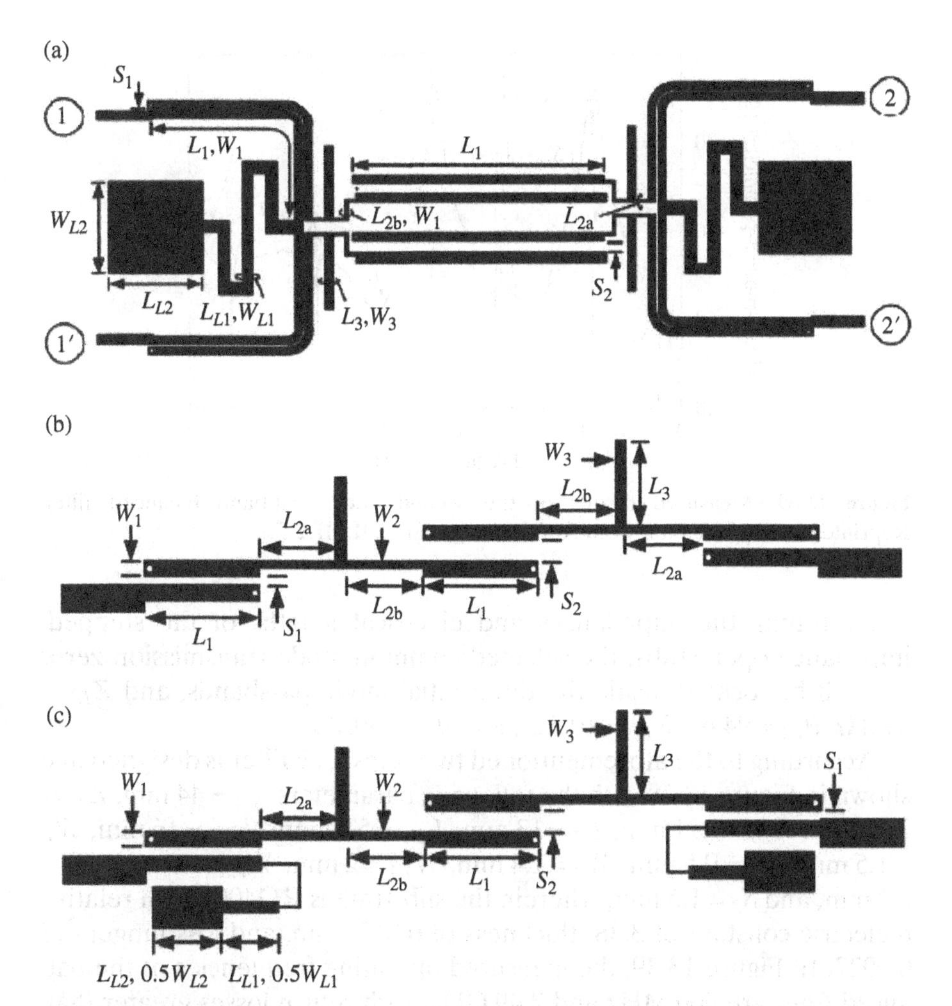

Figure 13.38 (a) Layout of the second-order dual-band balanced filter. (b) Differential-mode equivalent half circuit. (c) Common-mode equivalent half circuit. Reprinted with permission from Ref. [8]; copyright 2015 IEEE.

The second step realizes common-mode suppression. According to the common-mode equivalent half circuit, the common-mode transmission zeros can be determined as follows:

$$\left(\frac{Z_{0\mathrm{e}}\sin(\theta_{\mathrm{e}}-\varphi_{\mathrm{e}})}{|\cos\varphi_{\mathrm{e}}|}-\frac{Z_{0\mathrm{o}}\sin(\theta_{\mathrm{o}}-\varphi_{\mathrm{o}})}{|\cos\varphi_{\mathrm{o}}|}\right)\frac{A}{\tan\theta_3}=0 \tag{13.23}$$

where

$$A=Z_{0\mathrm{e}}\sin\theta_{\mathrm{e}}-Z_{0\mathrm{o}}\sin\theta_{\mathrm{o}}$$

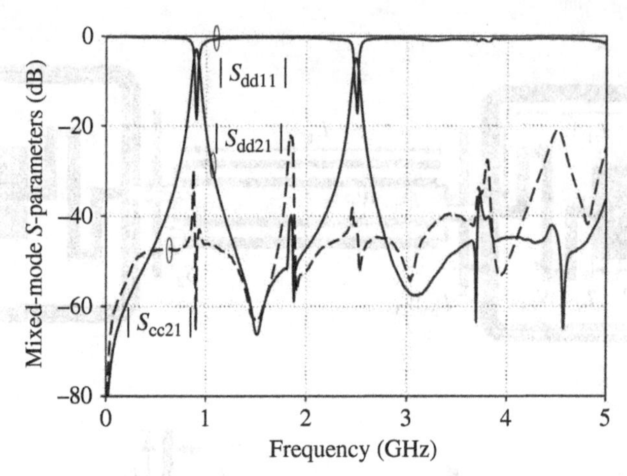

Figure 13.39 Measured results of the second-order dual-band balanced filter. Reprinted with permission from Ref. [8]; copyright 2015 IEEE.

By tuning the impedances and electrical lengths of the stepped-impedance open stubs, the selected common-mode transmission zeros can still be located inside the differential-mode passbands, and $Z_{L1} = 59.8\ \Omega$, $\theta_{L1} = 94.6°$, $Z_{L2} = 10.5\ \Omega$, and $\theta_{L2} = 30.7°$.

According to the aforementioned two steps, one filter is designed and shown in Figure 13.38 with the following parameters: $L_1 = 44$ mm, $L_{2a} = 2.71$ mm, $L_{2b} = 7.71$ mm, $L_3 = 13$ mm, $L_{L1} = 51.4$ mm, $L_{L2} = 16$ mm, $W_1 = 1.5$ mm, $W_2 = 0.5$ mm, $W_3 = 1.4$ mm, $W_{L1} = 2$ mm, $W_{L2} = 15$ mm, $S_1 = 0.2$ mm, and $S_2 = 1.5$ mm, wherein the substrate is RO4003 with relative dielectric constant of 3.38, thickness of 0.8128 mm, and loss tangent of 0.0027. In Figure 13.39, the measured operating frequencies of the balanced filter are 900 MHz and 2.49 GHz, with return losses greater than 18.5 and 16.1 dB and measured 3-dB fractional bandwidths of 3.6 and 2.1%, respectively. The in-band common-mode suppression reaches 30 and 40 dB for the two passbands, respectively. Six common-mode transmission zeros are obtained in the frequency range of interest from 0 to 5 GHz.

13.4 SUMMARY

This chapter focuses on dual-band balanced filters using different loaded and coupled resonators. Two types of resonators, that is, uniform impedance resonators and stepped-impedance resonators, are primarily used for the implementation of dual-band balanced filters. The even- and odd-mode equivalent circuit models of the filters are proposed and analyzed to determine the initial parameters of the filters. In addition, to enhance the common-mode suppression between the two passbands, selected loading techniques of the stub or lumped elements are proposed. The measured results of the prototypes are consistent with the theoretical expectations, thus verifying the validity of the proposed design strategies.

REFERENCES

1. J. Shi and Q. Xue, "Balanced bandpass filters using center-loaded half-wavelength resonators," *IEEE Trans. Microw. Theory Tech.*, vol. **58**, no. 4, pp. 970–977, Apr. 2010.
2. F. Wei, Y. J. Guo, P. Y. Qin, and X. W. Shi, "Compact balanced dual- and tri-band bandpass filters based on stub loaded resonators," *IEEE Microw. Wireless Compon. Lett.*, vol. **25**, no. 2, pp. 76–78, Feb. 2015.
3. J. Shi and Q. Xue, "Dual-band and wide-stopband single-band balanced bandpass filters with high selectivity and Common-Mode Suppression," *IEEE Trans. Microw. Theory Tech.*, vol. **58**, no. 8, pp. 2204–2212, Aug. 2010.
4. M. Sagawa, M. Makimoto, and S. Yamashita, "Geometrical structures and fundamental characteristics of microwave stepped-impedance resonators," *IEEE Trans. Microw. Theory Tech.*, vol. **45**, no. 7, pp. 1078–1085, Jul. 1997.
5. C. H. Lee, C. I. G. Hsu, and C. C. Hsu, "Balanced dual-band BPF with stub-loaded SIRs for common-mode suppression," *IEEE Microw. Wireless Compon. Lett.*, vol. **25**, no. 2, pp. 70–72, Feb. 2010.
6. J. Shi and Q. Xue, "Novel balanced dual-band bandpass filter using coupled stepped-impedance resonators," *IEEE Microw. Wireless Compon. Lett.*, vol. **20**, no. 1, pp. 19–21, Jan. 2010.
7. M. Makimoto and S. Yamashita, "Bandpass filters using parallel coupled stripline stepped impedance resonators," *IEEE Trans. Microw. Theory Tech.*, vol. **28**, no. 12, pp. 1413–1417, Dec. 1980.
8. L. Yang, W. W. Choi, K. W. Tam, and L. Zhu, "Balanced dual-band bandpass filter with multiple transmission zeros using doubly short-ended resonator coupled line," *IEEE Trans. Microw. Theory Tech.*, vol. **63**, no. 7, pp. 2225–2232, Jul. 2015.
9. A. Riddle, "High performance parallel coupled microstrip filters," in *IEEE MTT-S International Microwave Symposium Digest*, New York, pp. 427–430, May 1988.

CHAPTER 14

DUAL-BAND BALANCED FILTERS IMPLEMENTED IN SUBSTRATE INTEGRATED WAVEGUIDE (SIW) TECHNOLOGY

Wen Wu, Jianpeng Wang, and Chunxia Zhou

Ministerial Key Laboratory, JGMT, Nanjing University of Science and Technology, Nanjing, China

Balanced bandpass filters (BPFs), due to their high immunity to the environmental noise and low electromagnetic interference (EMI), are important as a critical circuit block in building a modern wireless communication system. In particular, the dual-band balanced BPFs play a significant role in many applications since the requirement for multiband operation is highly demanded. In the past few years, the dual-band balanced circuits were constructed mostly by line-based resonators. In this context, dual-band balanced BPFs have been proposed with lumped-element loaded resonators [1, 2], stub-loaded resonators [3, 4], coupled complementary split-ring resonators [5], and double-layer microstrip-to-slotline transitions [6]. However, when the operating frequency goes into millimeter-wave band, they are hardly employable

Balanced Microwave Filters, First Edition. Edited by Ferran Martín, Lei Zhu, Jiasheng Hong, and Francisco Medina.

because of their drawbacks of high radiation loss, low power handling capability, and low Q-factor. In recent years, the substrate integrated waveguide (SIW) has been receiving tremendously increasing attention as it successfully overcomes the drawbacks of classical line-based structures, while maintaining low-cost, compact size, and good integration [7]. In this chapter, we will introduce two types of balanced filters with dual operation bands utilizing this potential SIW technology [8, 9].

14.1 SUBSTRATE INTEGRATED WAVEGUIDE (SIW) CAVITY

Figure 14.1 shows a typical SIW cavity configuration. The cavity is constructed entirely on a dielectric substrate with a thickness of h, which is usually much smaller than the width w and length l of the cavity. The cavity's sidewalls are formed by metallic via fences, while the top and bottom surfaces are formed by metallic planes. The posts have a diameter of d and the pitch length between two adjacent posts is denoted by p. These two important parameters should be properly chosen in order to minimize the radiation loss. According to Ref. [10], the radiation loss can be negligible, for an electrically small post, namely, $d < 0.2\lambda$, where λ is the wavelength in the dielectric material, under the ratio d/p of 0.5. Furthermore, the radiation loss tends to decrease as the post gets smaller with a constant ratio of d/p, which is conditioned by the fabrication process.

Following the work in Ref. [11], all the resonant frequencies of TE_{m0n} modes in an SIW cavity can be determined by formula

$$f_{\mathrm{TE}_{m0n}} = \frac{c}{2\sqrt{\varepsilon_r}}\sqrt{\frac{m^2}{w_e^2} + \frac{n^2}{l_e^2}} \tag{14.1a}$$

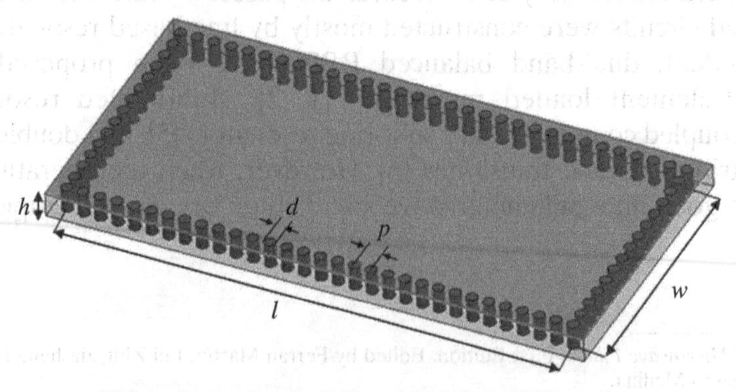

Figure 14.1 Configuration of a single SIW cavity.

w_e and l_e, the equivalent width and length of the SIW cavity, are given as

$$w_e = w - \frac{d^2}{0.95p} \quad l_e = l - \frac{d^2}{0.95p} \tag{14.1b}$$

where w and l are the real width and length of the SIW cavity, c is the velocity of light in the vacuum, and ε_r is the relative permittivity of the substrate.

14.2 CLOSELY PROXIMATE DUAL-BAND BALANCED FILTER DESIGN

Figure 14.2(a) illustrates the field pattern of TE_{102} mode in a single SIW cavity. As can be observed from the figure, the E-field distribution inside the SIW cavity is odd-symmetric with respect to the symmetrical plane A–A′, which indicates the opposite direction and same intensity of the E-field distributions in two symmetrical sides. Figure 14.2(b) depicts the calculated E-field distribution from full-wave simulation software HFSS, which verifies the E-field distribution properties of the TE_{102} mode. Thus, we can conclude that the E-field distribution of the TE_{102} mode exhibits two distinctive properties, that is, identical magnitude and out of phase, as highly demanded in the design of a balanced circuit.

Based on the analysis introduced earlier, the E-field distribution properties of TE_{102} mode can be applied for the demonstration and design of balanced circuits. Figure 14.3 depicts the physical configuration of the designed dual-band balanced BPF. Herein, cavities 1 and 2 are readily occupied by TE_{102} mode, while others are arranged with TE_{101} mode. Two pairs of differential termination ports are both directly connected with 50 Ω microstrip lines via slot-coupling structures. In

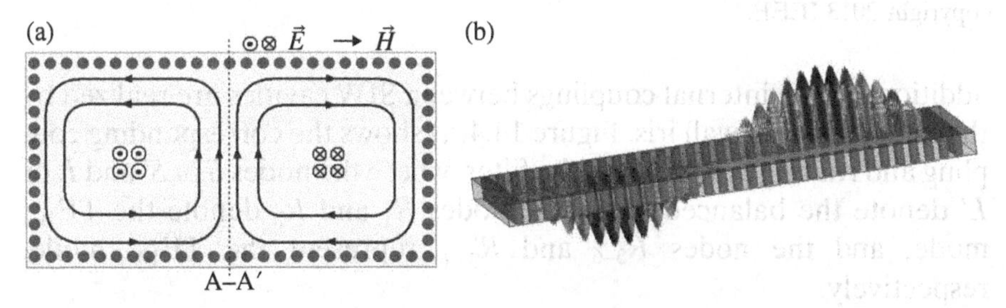

Figure 14.2 (a) Field pattern of TE_{102} mode in a SIW cavity. (b) E-field distribution of TE_{102} mode. Reprinted with permission from Ref. [8]; copyright 2013 IEEE.

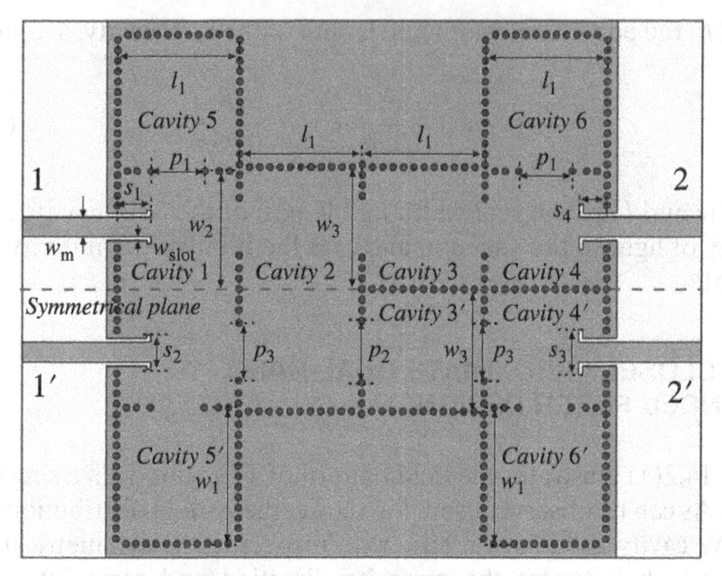

Figure 14.3 Layout of a dual-band balanced BPF. Reprinted with permission from Ref. [8]; copyright 2013 IEEE.

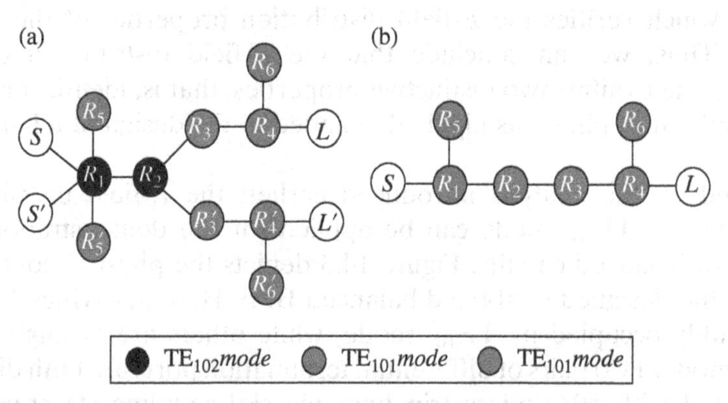

Figure 14.4 Coupling and routing scheme of the filter. (a) Original 4-port circuit. (b) Equivalent 2-port circuit under DM operation. Reprinted with permission from Ref. [8]; copyright 2013 IEEE.

addition, all the internal couplings between SIW cavities are realized by the magnetic post-wall iris. Figure 14.4(a) shows the corresponding coupling and routing schematic of the filter, where the nodes S or S' and L or L' denote the balanced ports, the nodes R_1 and R_2 denote the TE_{102} mode, and the nodes R_{3-6} and R'_{3-6} represent the TE_{101} mode, respectively.

The common-mode (CM) signal is firstly launched by port 1 and 1′, and then translated into two out-of-phase signals which will be canceled out with each other in cavity 1, thus resulting in the desired CM suppression. On the contrary, when passing through cavities 1 and 2, the DM signal can be converted to two in-phase signals and the desired response can proceed. In addition, due to the symmetric property of the filter, a virtual-short or perfect electric conductor would appear along its symmetrical plane under the DM operation. Consequently, as illustrated in Figure 14.4(b), the equivalent 2-port circuit schematic topology could be achieved, and then utilized to analyze the specified DM response of the filter. Note that similar to the node R_{3-6}, R_1 and R_2 now stand for the TE_{101} mode due to a virtual-short conductor appearing along the symmetrical plane. Meanwhile, the principal mechanism of generating the desired dual-band response is similar to Ref. [1], that is, creating a stopband inside a wide virtual passband by inserting transmission zeros (TZs). Specifically, cavities 5 and 6 are introduced to bring in a TZ for band-to-band rejection. In general, such split-type topology is highly suitable for the filter design with closely proximate dual passbands.

To verify the design concept, the central frequencies of two passbands in the DM response are chosen at $f_1 = 9.41$ GHz and $f_2 = 9.96$ GHz with three transmission poles in each passband. Besides, the return loss in each band is better than 20 dB, while the normalized TZ is set at $\Omega = 0$ with the insertion loss higher than 40 dB. In order to achieve the previously specified dual-band frequency response, following the work in Ref. [12], synthesized coupling matrix is expressed as

$$\begin{bmatrix} & S & R_1 & R_2 & R_3 & R_4 & R_5 & R_6 & L \\ S & 0 & 0.7176 & 0 & 0 & 0 & 0 & 0 & 0 \\ R_1 & 0.7176 & 0 & 0.451 & 0 & 0 & 0.73 & 0 & 0 \\ R_2 & 0 & 0.451 & 0 & 0.811 & 0 & 0 & 0 & 0 \\ R_3 & 0 & 0 & 0.811 & 0 & 0.451 & 0 & 0 & 0 \\ R_4 & 0 & 0 & 0 & 0.451 & 0 & 0 & 0.73 & 0.7176 \\ R_5 & 0 & 0.73 & 0 & 0 & 0 & 0 & 0 & 0 \\ R_6 & 0 & 0 & 0 & 0 & 0.73 & 0 & 0 & 0 \\ L & 0 & 0 & 0 & 0 & 0.7176 & 0 & 0 & 0 \end{bmatrix}. \quad (14.2)$$

Additionally, the central frequency of the virtual wide passband can be given as $f_{TE_{101}} = \sqrt{f_1 f_2} = 9.68$ GHz, and the fractional bandwidth of FBW = 10%. Here, the virtual resonant frequency of TE_{101} mode is already confirmed; thus the initial physical size of the SIW cavity can be determined easily by using Equation (14.1). It should be noted that the internal coupling coefficients between SIW cavities and external quality factor are the key parameters to obtain desired specifications. Referring to Ref. [13], the required internal coupling coefficients can be obtained by means of adjusting the width of post-wall iris. As for the external quality factor, this configuration also serves as some tuning elements, such as the length and widths of coupling slot, and the position of the feed line, to satisfy design requirements. With resorting to the commercial full-wave electromagnetic simulator HFSS, all the geometric parameters illustrated in Figure 14.3 have been optimized as follows: $w_m = 2.3$, $w_{\text{slot}} = 0.45$, $w_1 = 15.75$, $w_2 = 13.65$, $w_3 = 14.1$, $l_1 = 14.4$, $p_1 = 6.3$, $p_2 = 7.12$, $p_3 = 6.7$, $s_1 = 4.05$, $s_2 = 3.9$, $s_3 = 5.2$, and $s_4 = 3.25$. (Unit : mm).

The filter is fabricated on a 0.79 mm-thick RT/Duroid 5880 substrate with $\varepsilon_r = 2.2$, $\tan\delta = 0.0009$. The diameter of the metallic via-holes is 0.8 mm and the spacing between two adjacent via-holes is around 1.2 mm. The photograph of the fabricated circuit is depicted in Figure 14.5. The designed circuit without feeding lines occupies an area of 2.87 $\lambda_g \times 2.95$ λ_g, where λ_g is the guided wavelength at central frequency of the virtual wide passband.

The simulated and measured responses are recorded in Figure 14.6 with good agreement. In the DM response, the measured passbands are centered at 9.47 and 9.96 GHz, with 3 dB bandwidths of 270 and 310 MHz, respectively. The minimum in-band insertion losses are measured to be 1.89 and 1.73 dB, respectively. The measured insertion losses are mainly attributed to a pair of SMA connectors and feeding lines (about 0.7 dB by measurement), dielectric substrate and conductor. In particular, the TZ which contributes to the middle stopband can be observed at 9.69 GHz with an insertion loss of 38 dB. As expected, in

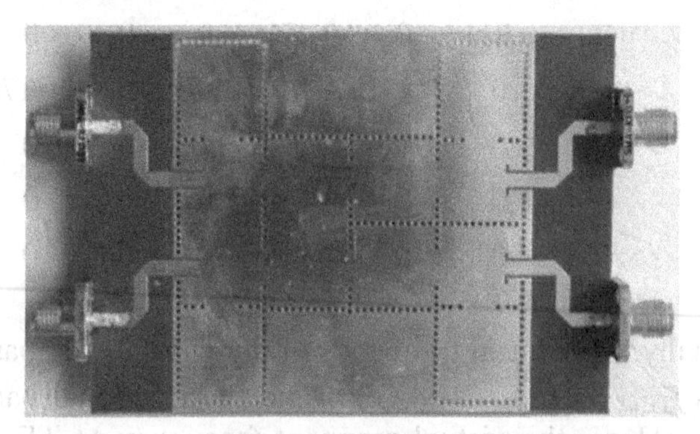

Figure 14.5 Photograph of the fabricated dual-band balanced BPF. Reprinted with permission from Ref. [8]; copyright 2013 IEEE.

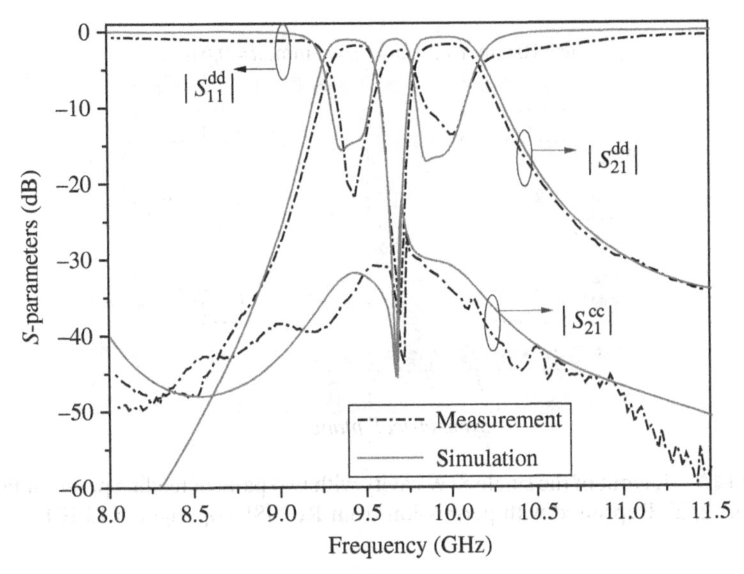

Figure 14.6 Simulated and measured performances of the dual-band balanced BPF. Reprinted with permission from Ref. [8]; copyright 2013 IEEE.

the CM response, the measured signal is suppressed with rejection greater than 30 dB in DM passbands.

14.3 DUAL-BAND BALANCED FILTER DESIGN UTILIZING HIGH-ORDER MODES IN SIW CAVITIES

Following the closely adjacent dual-band balanced BPF introduced in the previous section, this section will deal with a new dual-band balanced filter using high-order resonant modes in SIW cavities. Specifically, in the DM response, the first passband is derived by the dominant modes, while the second passband is obtained by employing the higher-order modes in the SIW cavities.

Firstly, let's pay attention to the resonant modes in a single SIW cavity. Figure 14.7 depicts a rectangular SIW cavity in a dielectric substrate with $\varepsilon_r = 2.2$, $\tan\delta = 0.0009$, and $h = 0.79$ mm. Its two pairs of external feeding points are denoted as port 1, 1′ and port 2, 2′, respectively. Using the Equation (14.1), all the resonant frequencies of TE_{m0n} modes can be calculated. When n is an even integer, *E*-field distribution inside the SIW cavity is odd-symmetric with respect to the symmetrical plane as shown in Figure 14.7, thus *E*-field distribution of resonant modes in the cavity could be in the statuses of out-of-phase under CM operation and in-phase under DM operation. When n is an odd integer, *E*-field

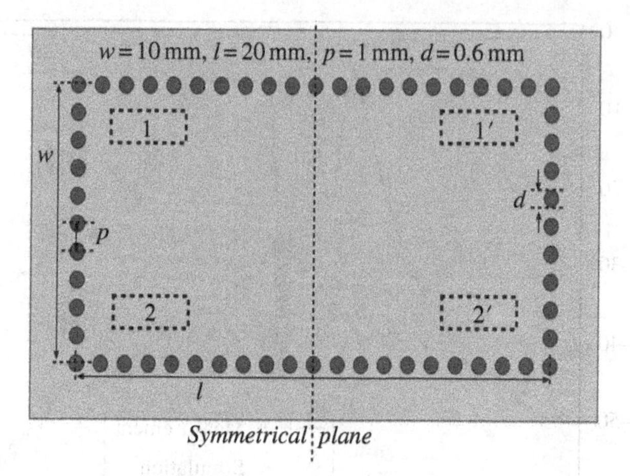

Figure 14.7 Layout of the single SIW cavity with two pairs of feeding points at Port 1, 1′ and Port 2, 2′. Reprinted with permission from Ref. [8]; copyright 2013 IEEE.

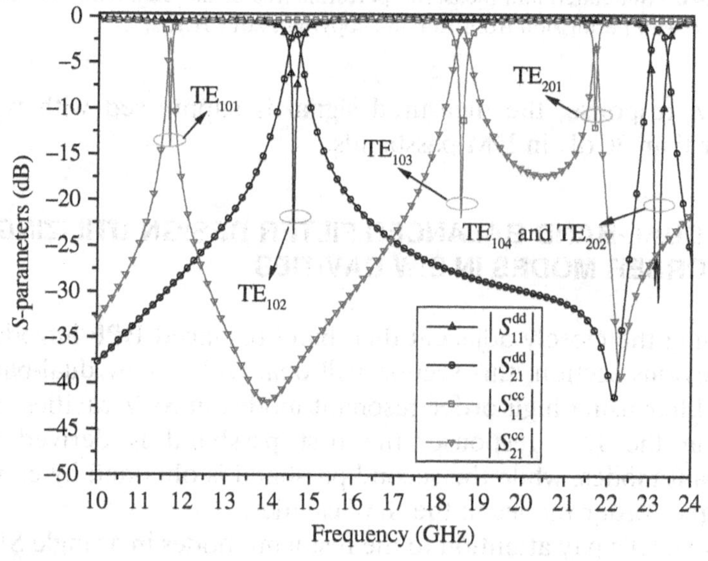

Figure 14.8 Frequency responses of DM and CM S-parameters.

distribution is even-symmetric with respect to the symmetric plane, so E-field distribution of resonant modes in the cavity could be in the statuses of out-of-phase under DM operation and in-phase under CM operation. For verification, frequency responses under DM and CM excitations are provided in Figure 14.8 to support the aforementioned descriptions. The first six resonant modes in the cavity are visibly excited in DM and CM responses, respectively. As illustrated in the figure, the signal can pass only at the TE_{101}, TE_{103}, and TE_{201} modes under CM excitation, while under DM excitation the signal can propagate only at the TE_{102}, TE_{104}, and TE_{202} modes. These properties imply that the cavity can be used to construct a balanced section to design a DM filter with good CM suppression if the resonant modes are properly selected. In the following discussion, the dominant TE_{102} mode and the higher-order TE_{202} mode are applied to achieve dual-band response in the DM response.

As depicted in Figure 14.9, a practical dual-band differential BPF is presented with six SIW cavities. Similarly, this SIW filter is also externally fed by planar 50 Ω microstrip lines via coupling slots, and the internal coupling among all the SIW cavities is realized via magnetic post-wall iris. The balanced section in the filter is constructed between cavities 2 and 3, aiming to selectively transmit DM signals while highly suppressing CM signals. It should be mentioned that, in the first DM passband, the six cavities are occupied by either TE_{101} or TE_{102} mode, while the second DM passband is constructed by higher-order modes with either TE_{201} or TE_{202} mode. Figure 14.10 shows the coupling schematic topologies of these two DM passbands.

As for the implementation of this balanced dual-band filter, the key step is to realize DM dual-band response. Under DM excitation, TE_{101} and TE_{201} modes of the cavities 1, 1′, 4, and 4′ are excited, while TE_{102} and TE_{202} of cavities 2 and 3 are excited. The first DM passband frequency f_{DM1} is constructed by TE_{101} mode of the cavities 1, 1′, 4, and 4′ and TE_{102} mode of cavities 2 and 3. The second DM passband

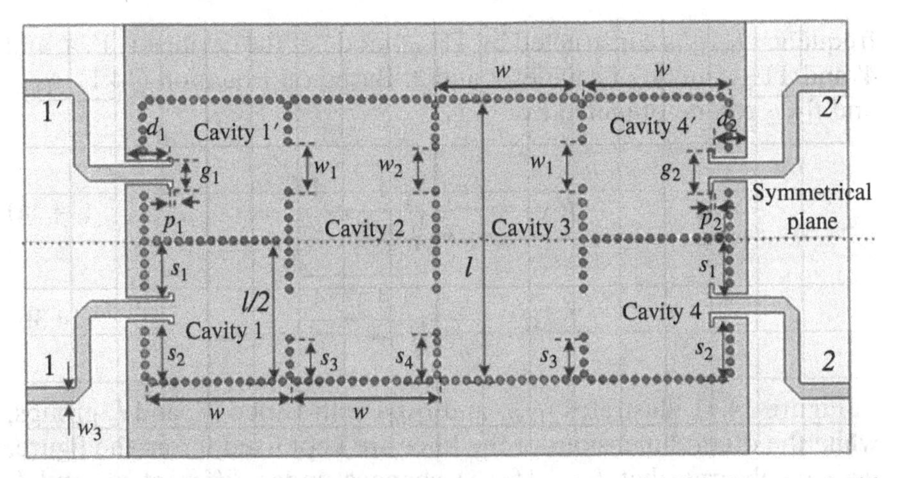

Figure 14.9 Configuration of the designed dual-band balanced BPF. Reprinted with permission from Ref. [9]; copyright 2015 IEEE.

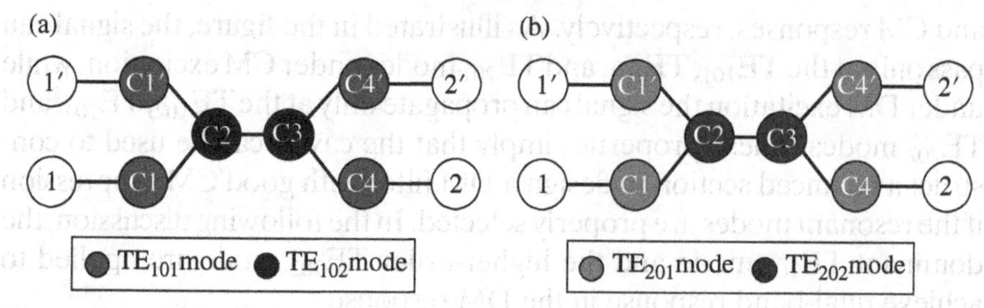

Figure 14.10 Schematic topologies of (a) the first DM passband and (b) the second DM passband. Reprinted with permission from Ref. [9]; copyright 2015 IEEE.

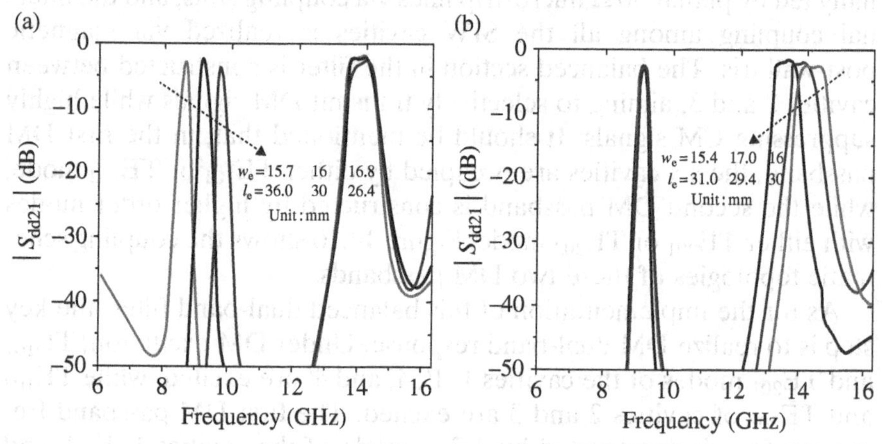

Figure 14.11 DM frequency responses of the filter with varied w_e and l_e groups. (a) Variation of the location for the first passband. (b) Variation of the location for the second passband. Reprinted with permission from Ref. [9]; copyright 2015 IEEE.

frequency f_{DM2} is constructed by TE_{201} mode of the cavities 1, 1′, 4, and 4′ and TE_{202} mode of cavities 2 and 3. Based on Equation (14.1), f_{DM1} and f_{DM2} can be obtained by

$$f_{DM1} = \frac{c}{2\sqrt{\varepsilon_r}}\sqrt{\frac{1}{w_e^2} + \frac{4}{l_e^2}} \tag{14.3a}$$

$$f_{DM2} = \frac{c}{2\sqrt{\varepsilon_r}}\sqrt{\frac{4}{w_e^2} + \frac{4}{l_e^2}} \tag{14.3b}$$

Figure 14.11 illustrates f_{DM1} and f_{DM2} with varied w_e and l_e groups, while the other dimensions of the filter are kept fixed. From the figure, we can observe that f_{DM1} (f_{DM2}) changes under different w_e and l_e groups, while f_{DM2} (f_{DM1}) mostly keeps constant. Thus, the two passbands can be controlled individually by choosing w_e and l_e properly. Moreover, the bandwidths of two passbands can be tuned by adjusting the couplings among cavities (i.e., w_1 and w_2). When couplings get stronger (w_1 or w_2 increases), the bandwidths of both passbands become wider simultaneously. Thus, it is hard to individually control the bandwidths of the two passbands.

Based on the aforementioned analysis, the presented filter shows controllable DM center frequencies and good CM suppression. After fine optimization through the full-wave electromagnetic simulator HFSS, final dimensions displayed in Figure 14.9 are derived as follows: $w = 16$, $l = 30$, $d_1 = 4.7$, $d_2 = 3.7$, $g_1 = 3.2$, $g_2 = 4.5$, $p_1 = 0.45$, $p_2 = 0.45$, $s_1 = 6$, $s_2 = 6.55$, $s_3 = 4.5$, $s_4 = 5.3$, $w_1 = 5.3$, $w_2 = 4.7$, and $w_3 = 1.56$ (all in mm). The circuit is fabricated on an RO5880 substrate with a dielectric constant of 2.2 and a thickness of 0.508 mm, and the photograph of the fabricated circuit is shown in Figure 14.12. Simulated and experimental results provided in Figure 14.13 are in good agreement. It can be seen that the central frequencies of two DM passbands are located at 9.15 and 14.08 GHz, and the CM rejections in DM passbands are 49 and 31 dB, respectively. However, the resonant frequencies of TE_{103}, TE_{201}, and TE_{203} modes ($f_{TE_{103}}$, $f_{TE_{201}}$, $f_{TE_{203}}$) are inherently emerged in the DM stopband, and the CM suppression at $f_{TE_{201}}$ is only 9.5 dB, which will greatly affect the ability to suppress CM rejection over the entire bandwidth.

To tackle this problem, one simple and effective approach is introduced herein. Four slotlines are etched on the top metal plane in cavities 2 and 3 as shown in Figure 14.14 with other dimensions of the filter unchanged. Note that these four slotlines should be placed at the direction perpendicular to the maximum surface current of the TE_{201} mode in cavities 2 and 3. It will significantly interrupt the surface current flow and

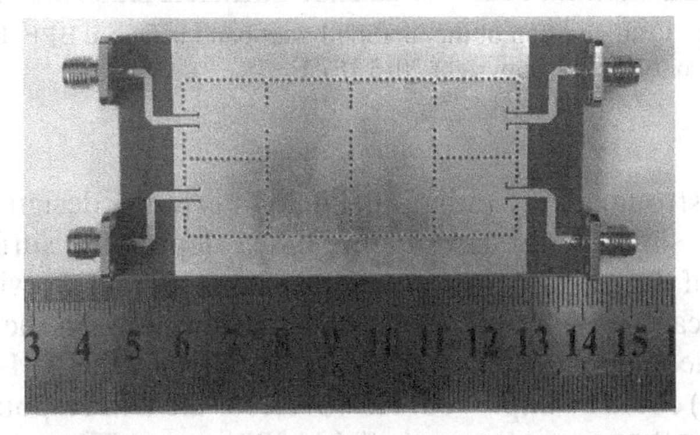

Figure 14.12 Photograph of the fabricated dual-band balanced BPF. Reprinted with permission from Ref. [9]; copyright 2015 IEEE.

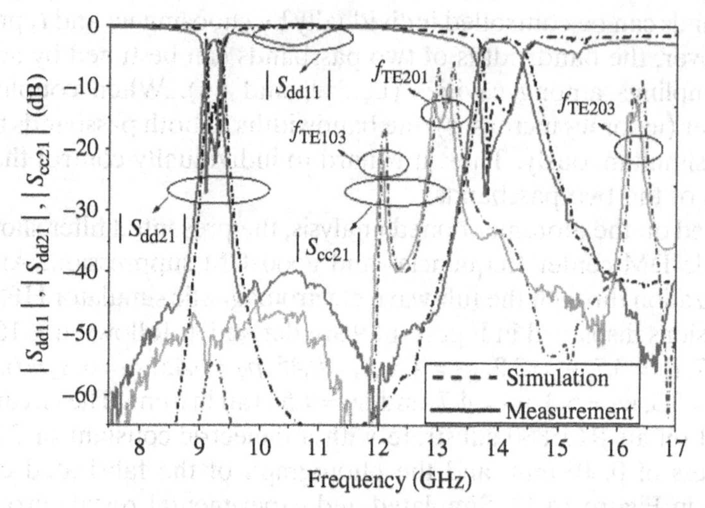

Figure 14.13 Simulated and measured performances of the dual-band balanced BPF. Reprinted with permission from Ref. [9]; copyright 2015 IEEE.

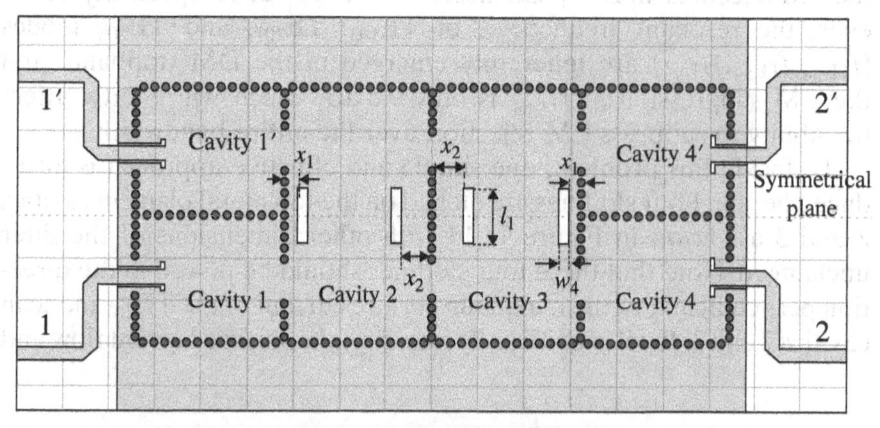

Figure 14.14 Configuration of the modified dual-band balanced BPF. Reprinted with permission from Ref. [9]; copyright 2015 IEEE.

introduce strong energy radiation, analogous to the design principle of waveguide slot antenna [14]. Figure 14.15 depicts the surface current densities of TE_{201} mode on the metal plane without or with four slotlines. As can be observed, the surface current flow on the slot etched metal plane is effectively interrupted. In this way, the CM rejection at $f_{TE_{201}}$ (RJ1) could be improved. However, for the DM response, slotlines have little influence on magnetic field of TE_{102} and TE_{202} modes in cavities 2 and 3. Thus, the two DM passbands are almost unaffected by

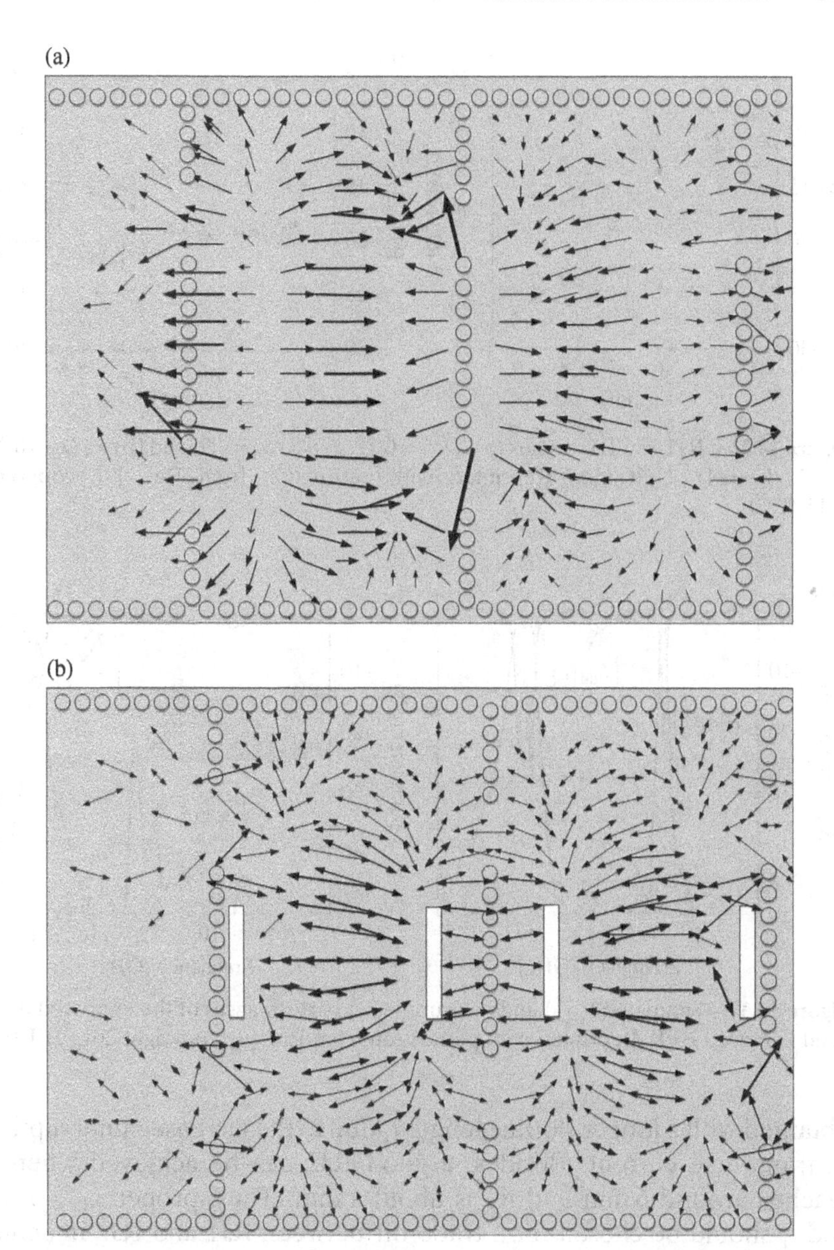

Figure 14.15 Surface current densities of TE_{201} mode on the metal plane (a) without and (b) with slotlines. Reprinted with permission from Ref. [9]; copyright 2015 IEEE.

slotlines. Meanwhile, the CM rejection in the second DM passband (RJ2) is affected by the physical size of the slotlines to a certain degree.

Figure 14.16 shows RJ1 and RJ2 with varied slotline lengths l_1 and widths w_4, respectively. As shown in the figure, a better RJ1 can be

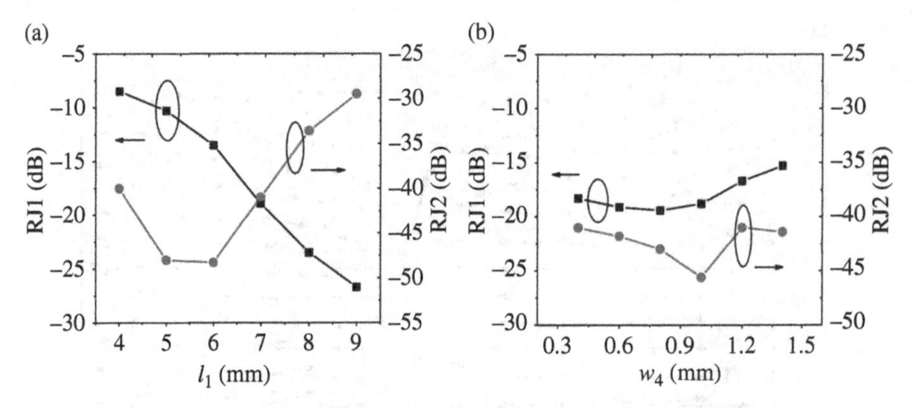

Figure 14.16 RJ1 and RJ1 versus (a) l_1 ($x_1 = 0.75, x_2 = 3, w_4 = 0.9$) and (b) w_4 ($x_1 = 0.75, x_2 = 3, l_1 = 6.4$). Unit : mm. Reprinted with permission from Ref. [9]; copyright 2015 IEEE.

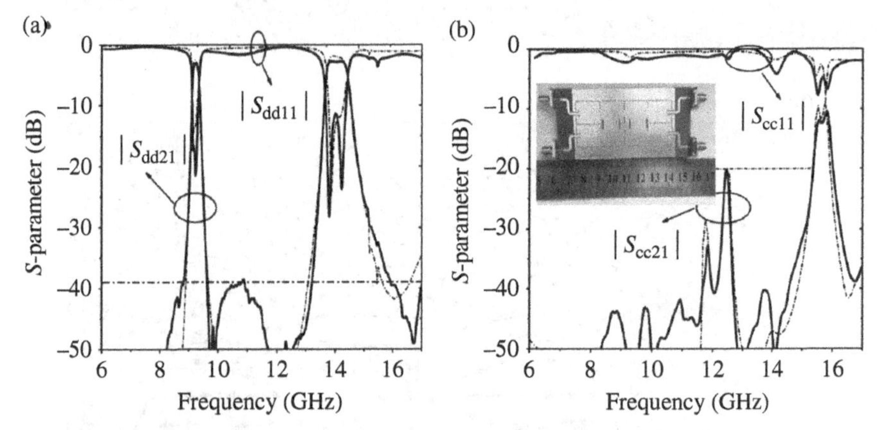

Figure 14.17 Simulated (- - -) and measured (—) performances of the improved dual-band balanced BPF. Reprinted with permission from Ref. [9]; copyright 2015 IEEE.

obtained with a longer slotline length l_1 due to the increased interruption of transverse current. Besides, a good RJ2 can be achieved when l_1 reaches around 6 mm and w_4 is about 1 mm. Thus, proper x_1, x_2, w_4, and l_1 should be chosen for a trade-off between RJ1 and RJ2 in actual applications. After a fine-tuning process is conducted, final dimensions are determined and provided as $x_1 = 0.75$, $x_2 = 2.95$, $w_4 = 0.9$, and $l_1 = 6.4$ (all in mm).

The fabricated circuit of the improved balanced filter is shown in the inset of Figure 14.17(b). Figure 14.17 shows the final performance of both simulated and measured results. From the figure, we can see that two DM passbands are centered at 9.23 and 14.05 GHz with 3 dB bandwidth of 260 and 780 MHz, respectively. In each passband, the measured minimum insertion losses are 2.9 and 2.7 dB, which are mainly attributed to a pair of SMA connectors and meandered feed lines. Besides, for the CM response, the measured CM rejections are above 48 and 40 dB for the lower and higher passbands, respectively. In addition, the average CM suppression of the improved filter is better than 20 dB in the frequency range from 0 to 15.20 GHz. Obviously, the attenuation of CM response at $f_{\mathrm{TE}_{201}}$ has been improved by 10.5 dB, which further indicates the simplicity and effectiveness of the improve method in CM suppression.

14.4 SUMMARY

In this chapter, the design of dual-band balanced BPFs based on the SIW technology is presented. By utilizing the inherent properties of the electric field distributions under different resonant modes in the SIW cavities, two types of dual-band balanced BPFs are successfully designed and proposed. For the first type of dual-band balanced BPF, it is constructed by utilizing the TE_{101} and TE_{102} modes, where the CM signals are effectively suppressed by the balanced sections resonating at the TE_{102} mode. As for the second type of dual-band balanced BPF, it is designed by the adoption of TE_{101}, TE_{102}, TE_{201}, and TE_{202} modes, where the CM signals are effectively suppressed by the balanced sections resonating at the TE_{102} and TE_{202} modes. Both the operation principles and design procedures for the two filters are clearly clarified. Simulation and measured results are provided with good agreement, indicating that the proposed dual-band filters have the properties of high DM selectivity and good CM suppression. With all these good properties, it is obvious that the methods introduced in this chapter are applicable for the design of dual-band balanced BPF in the SIW technology.

REFERENCES

1. J. Shi and Q. Xue, "Balanced bandpass filters using center-loaded half-wavelength resonators," *IEEE Trans. Microw. Theory Tech.*, vol. **58**, no. 4, pp. 970–977, Apr. 2010.
2. J. Shi and Q. Xue, "Dual-band and wide-stopband single-band balanced bandpass filters with high selectivity and common-mode suppression," *IEEE Trans. Microw. Theory Tech.*, vol. **58**, no. 8, pp. 2204–2212, Aug. 2010.

3. X. H. Wu, F. Y. Wan, and J. X. Ge, "Stub-loaded theory and its application to balanced dual-band bandpass filter design," *IEEE Microw. Wireless Compon. Lett.*, vol. **26**, no. 4, pp. 231–233, Apr. 2016.
4. L. Yang, W.-W Choi, K.-W. Tam, and L. Zhu, "Balanced dual-band bandpass filter with multiple transmission zeros using doubly short-ended resonator coupled line," *IEEE Trans. Microw. Theory Tech.*, vol. **63**, no. 7, pp. 2225–2232, Jul. 2015.
5. F. Wei, P.-Y. Qin, Y. J. Guo, C. Ding, and X. W. Shi, "Compact balanced dual- and tri-band BPFs based on coupled complementary split-ring resonators (C-CSRR)," *IEEE Microw. Wireless Compon. Lett.*, vol. **26**, no. 2, pp. 107–109, Feb. 2016.
6. K. Wang, L. Zhu, S.-W. Wong, D. Chen, and Z.-C. Guo, "Balanced dual-band BPF with intrinsic common-mode suppression on double-layer substrate," *Electron. Lett.*, vol. **51**, no. 9, pp. 705–707, Apr. 2015.
7. X.-P. Chen and K. Wu, "Substrate integrated waveguide cross-coupled filter with negative coupling structure," *IEEE Trans. Microw. Theory Tech.*, vol. **56**, no. 1, pp. 142–149, Jan. 2008.
8. X. Xu, J. P. Wang, G. Zhang, and J. X. Chen, "Design of balanced dual-band bandpass filter based on substrate integrated waveguide," *Electron. Lett.*, vol. **49**, no. 20, pp. 1278–1280, Sep. 2013.
9. Y. J. Shen, H. Wang, W. Kang, and W. Wu, "Dual-band SIW differential bandpass filter with improved common-mode suppression," *IEEE Microw. Wireless Compon. Lett.*, vol. **25**, no. 2, pp. 100–102, Feb. 2015.
10. D. Deslandes and K. Wu, "Single-substrate integration technique of planar circuits and waveguide filters," *IEEE Trans. Microw. Theory Tech.*, vol. **51**, no. 2, pp. 593–596, Feb. 2003.
11. Y. Cassivi, L. Perregrini, P. Arcioni, M. Bressan, K. Wu, and G. Conciauro, "Dispersion characteristics of substrate integrated rectangular waveguide," *IEEE Microw. Wireless Compon. Lett.*, vol. **12**, no. 2, pp. 333–335, Feb. 2002.
12. S. Amari, "Synthesis of cross-coupled resonator filters using an analytical gradient-based optimization technique," *IEEE Trans. Microw. Theory Tech.*, vol. **48**, no. 9, pp. 1559–1564, Sep. 2000.
13. J.-S. Hong and M. J. Lancaster, *Microstrip Filters for RF/Microwave Applications.* New York: John Wiley & Sons, Inc., 2001.
14. R. S. Elliott, *Antenna Theory and Design*. New York: John Wiley & Sons, Inc., 2003, ch. 3.

PART 5

Other Balanced Circuits

CHAPTER 15

BALANCED POWER DIVIDERS/ COMBINERS

Lin-Sheng Wu, Bin Xia, and Jun-Fa Mao

Key Laboratory of Ministry of Education of Design and Electromagnetic Compatibility of High-Speed Electronic Systems, Shanghai Jiao Tong University, Shanghai, PR China

15.1 INTRODUCTION

As indicated in the previous chapters of this book, balanced RF circuits have more merits for modern communication systems than their single-ended counterparts. For example, balanced filters and diplexers show good common-mode suppression and high immunity to noise [1–4]; balanced and differentially driven antennas are not sensitive to the perturbations of ground plane and have wide impedance bandwidth and weak cross polarization [5–7]; balanced amplifiers have low noise, good input and output return losses, and good linearity and stability [8–10]; balanced mixers provide good port isolation and conversion efficiency [11, 12]; balanced oscillators can directly provide exact antiphase signals from an oscillator without the need for external baluns or resonators [13, 14].

Balanced Microwave Filters, First Edition. Edited by Ferran Martín, Lei Zhu, Jiasheng Hong, and Francisco Medina.

Based on the balanced passive components and active circuits, a fully balanced transceiver architecture can be constructed with higher immunity to the environmental noise compared with the single-ended signaling [15]. Figure 15.1(a) shows a fully balanced RF front end, where the output power of the balanced power amplifier (PA) is delivered to the array with two balanced antenna elements and the received signal from the balanced antenna is transmitted to the input port of the balanced low-noise amplifier (LNA). It is easy to understand that a power divider/combiner is required in this front end whose input and output ports are both in the balanced form. The configuration shown in Figure 15.1(b) can be used for this purpose, which is directly built up by three baluns and a single-ended power divider/combiner. However, the circuit size and in-band insertion loss may be relatively large.

Therefore, it is valuable to develop a balanced-to-balanced or differential-mode power divider/combiner, with its diagram shown in Figure 15.1(c). The balanced ports A, B, and C are composed by the three pairs of single-ended ports 1 and 4, ports 2 and 3, and ports 6 and 5, respectively. To the best of our knowledge, previous research on power divider/combiner has been mainly focused on single-ended components [16–18], and only few balanced-to-balanced power dividers have been reported. In Ref. [19], a differential power divider is proposed by using shielded broadside-coupled striplines. However, the isolation between its differential output ports and the suppression of common

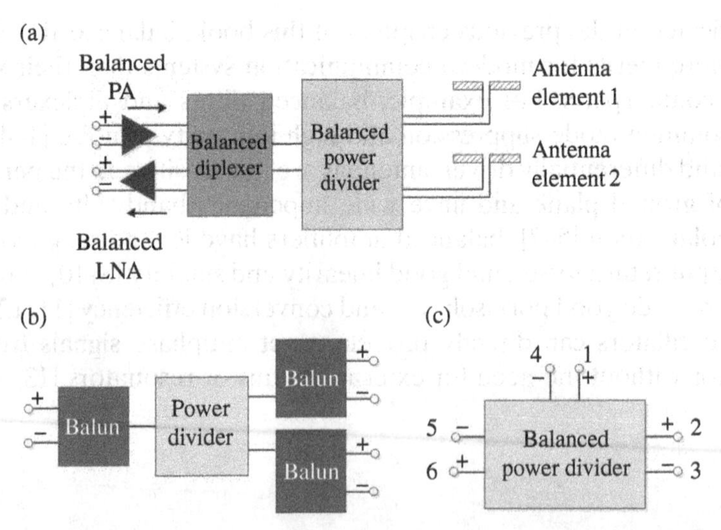

Figure 15.1 (a) A fully balanced RF front end; (b) the diagram of a power divider/combiner composed of three baluns and a single-ended power divider/combiner; and (c) a balanced-to-balanced power divider/combiner. Reprinted with permission from Ref. [20]; copyright 2012 IEEE.

mode are not considered in the design. Thus, this component cannot be used as a balanced-to-balanced power combiner.

In this chapter, several kinds of balanced-to-balanced power divider/combiner have been discussed. Microstrip and substrate integrated waveguides (SIWs) have been used to implement the balanced-to-balanced power divider/combiner. The balanced-to-balanced power divider/combiner with arbitrary power division ratio, filtering response, or dual-band performance is also presented.

15.2 BALANCED-TO-BALANCED WILKINSON POWER DIVIDER WITH MICROSTRIP LINE

At first, the balanced-to-balanced Wilkinson power dividers implemented using microstrip technology will be introduced and discussed in this section [20].

15.2.1 Mixed-Mode Analysis

15.2.1.1 Mixed-Mode Scattering Matrix of a Balanced-to-Balanced Power Divider As shown in Figure 15.1(c), the proposed balanced-to-balanced power divider/combiner is a six-port component. Let $[\mathbf{V}^+]$ and $[\mathbf{V}^-]$ represent the normalized incident and reflected wave vectors, respectively, and $[\mathbf{S}^{\text{std}}]$ represent the 6×6 scattering matrix of the proposed balanced-to-balanced power divider/combiner. The relationship between them can be written by

$$[\mathbf{V}^-] = [\mathbf{S}^{\text{std}}][\mathbf{V}^+] \tag{15.1}$$

Of course, the balanced-to-balanced power divider/combiner is a reciprocal six-port network. Then we have $S_{mn} = S_{nm}$, where S_{mn} and S_{nm} are the elements of $[\mathbf{S}^{\text{std}}]$, and $m, n = 1, \ldots, 6$. In our design, the power divider/combiner is a symmetric structure, which means ports 1, 2, and 3 are symmetric to ports 4, 5, and 6, that is, $[\mathbf{S}^{\text{std}}]$ can be written by

$$[\mathbf{S}^{\text{std}}] = \begin{bmatrix} [S_\text{U}] & [S_\text{D}] \\ [S_\text{D}] & [S_\text{U}] \end{bmatrix} = \begin{bmatrix} a & b & c & d & e & f \\ b & g & h & e & i & j \\ c & h & k & f & j & l \\ d & e & f & a & b & c \\ e & i & j & b & g & h \\ f & j & l & c & h & k \end{bmatrix} \tag{15.2}$$

where $[S_\text{U}]$ and $[S_\text{D}]$ are both 3×3 submatrices.

When the balanced ports are defined, as shown in Figure 15.1(c), the sequence of ports in $[\mathbf{S}^{\mathrm{std}}]$ should be reordered to 1, 2, 6, 4, 3, and 5 by using the matrix transformation [21], according to the port assignment of A^+, B^+, C^+, A^-, B^-, and C^-. Then, the mixed-mode scattering matrix $[\mathbf{S}^{\mathrm{mm}}]$ can be obtained by the definitions of differential and common modes as

$$[\mathbf{S}^{\mathrm{mm}}] = \begin{bmatrix} [S^{\mathrm{dd}}] & [S^{\mathrm{dc}}] \\ [S^{\mathrm{cd}}] & [S^{\mathrm{cc}}] \end{bmatrix} \tag{15.3a}$$

$$[S^{\mathrm{dd}}_{3\times3}] = \frac{1}{2}\begin{bmatrix} 2(a-d) & b-c-e+f & b-c-e+f \\ b-c-e+f & g-2h+k & -i+2j-l \\ b-c-e+f & -i+2j-l & g-2h+k \end{bmatrix} \tag{15.3b}$$

$$[S^{\mathrm{dc}}_{3\times3}] = \frac{1}{2}\begin{bmatrix} 0 & b+c-e-f & -b-c+e+f \\ b-c+e-f & g-k & i-l \\ -b+c-e+f & -i+l & -g+k \end{bmatrix} \tag{15.3c}$$

$$[S^{\mathrm{cd}}_{3\times3}] = \frac{1}{2}\begin{bmatrix} 0 & b-c+e-f & -b+c-e+f \\ b+c-e-f & g-k & -i+l \\ -b-c+e+f & i-l & -g+k \end{bmatrix} \tag{15.3d}$$

$$[S^{\mathrm{cc}}_{3\times3}] = \frac{1}{2}\begin{bmatrix} 2(a+d) & b+c+e+f & b+c+e+f \\ b+c+e+f & g+2h+k & i+2j+l \\ b+c+e+f & i+2j+l & g+2h+k \end{bmatrix} \tag{15.3e}$$

where the 3 × 3 submatrices $[S^{\mathrm{dd}}]$ and $[S^{\mathrm{cc}}]$ indicate the differential- and common-mode scattering matrices, respectively, $[S^{\mathrm{dc}}]$ indicates the conversion from common to differential mode, and $[S^{\mathrm{cd}}]$ indicates the conversion from differential to common mode. Due to the reciprocity, we have

$$[S^{\mathrm{dd}}_{3\times3}] = [S^{\mathrm{dd}}_{3\times3}]^{\mathrm{T}},\quad [S^{\mathrm{cc}}_{3\times3}] = [S^{\mathrm{cc}}_{3\times3}]^{\mathrm{T}},\quad [S^{\mathrm{cd}}_{3\times3}] = [S^{\mathrm{dc}}_{3\times3}]^{\mathrm{T}} \tag{15.4}$$

where the superscript "T" means the transpose of a matrix.

15.2.1.2 Constraint Rules of Balanced-to-Balanced Power Divider

In order to provide ideal performances of balanced power dividing and combining, the proposed balanced-to-balanced power divider should satisfy the following constraint rules:

1. When a differential-mode signal is fed into the balanced port A, no differential-mode power should be reflected, that is, $a-d=0$, no power should be converted to the common-mode output at the balanced ports B and C, that is, $b+c-e-f=0$, and the differential-mode output at the balanced ports B and C should satisfy $|b-c-e+f|=\sqrt{2}/2$. From (15.3d), we can also find that no power is converted to the common-mode reflection of the balanced port A in this case, due to the symmetry of the structure.
2. When a common-mode noise goes into the balanced port A, no noise in the form of differential or common mode should be outputted at the balanced ports B and C, that is, $b-c+e-f=0$ and $b+c+e+f=0$. From (15.3c), it can be found that no power is converted to the differential-mode reflection of the balanced port A in this case.
3. When the differential-mode signal is fed into the balanced port B (C), no power should be converted into the differential-mode and common-mode reflection, that is, $g-2h+k=0$ and $g-k=0$, and no power should be transmitted to the other balanced port C (B) in the form of differential or common modes, that is, $-i+2j-l=0$ and $i-l=0$.
4. When a common-mode noise goes into the balanced port B (C), no power is reflected in the form of differential mode, that is, $g-k=0$, and no noise in the form of differential mode or common mode should be outputted at the balanced port A and C (B), that is, $b+c-e-f=0$, $-i+l=0$, $b+c+e+f=0$ and $i+2j+l=0$.

Based on the above constraint rules, the following equations are derived:

$$a=d \tag{15.5a}$$

$$b=-c=-e=f \tag{15.5b}$$

$$|b|=\frac{\sqrt{2}}{4} \tag{15.5c}$$

$$g=h=k \tag{15.5d}$$

$$i=j=l=0 \tag{15.5e}$$

15.2.1.3 Odd- and Even-Mode Scattering Matrices of Balanced-to-Balanced Power Divider Since the six-port balanced-to-balanced power divider/combiner is a symmetric circuit, the odd-/even-mode method is applied for analysis. For even-mode excitation, we have $v_1^+ = v_4^+$, $v_2^+ = v_5^+$, $v_3^+ = v_6^+$, where v_n^+ is the normalized incident wave of the nth single-ended port. For odd-mode excitation, we have $v_1^+ = -v_4^+$, $v_2^+ = -v_5^+$, $v_3^+ = -v_6^+$. Then, the even-mode and odd-mode scattering matrices are calculated by

$$[S_e] = [S_U] + [S_D], \quad [S_o] = [S_U] - [S_D] \tag{15.6}$$

Substituting the constraint rules of (15.5) into (15.6), $[S_e]$ and $[S_o]$ can be obtained by

$$[S_e] = \begin{bmatrix} 2a & 0 & 0 \\ 0 & g & g \\ 0 & g & g \end{bmatrix}, \quad [S_o] = \begin{bmatrix} 0 & 2b & -2b \\ 2b & g & g \\ -2b & g & g \end{bmatrix} \tag{15.7}$$

If one can find a symmetric six-port circuit whose even- and odd-mode scattering matrices have the same forms as those in (15.7), a balanced-to-balanced power divider/combiner will be achieved.

15.2.2 A Transmission-Line Balanced-to-Balanced Power Divider

In order to realize a balanced-to-balanced power divider/combiner based on the aforementioned theory, the circuit shown in Figure 15.2 (a) is utilized [20]. It consists of seven sections of transmission lines and four resistances, where $\theta_{14} = \theta_{25} = \theta_{36} = \pi$ at the central frequency f_0.

15.2.2.1 Even-Mode Circuit Model As shown in Figure 15.2(b), it is easy to obtain the even-mode circuit model from Figure 15.2(a), which is a three-port network with each port matched to Z_0. Obviously, we have $S_e^{11} = -1$ and $S_e^{12} = S_e^{13} = S_e^{21} = S_e^{31} = 0$ at f_0. The two-port **ABCD** matrix between ports 2 and 3 for the even-mode circuit model can be calculated by

$$\begin{bmatrix} A_e^{23} & B_e^{23} \\ C_e^{23} & D_e^{23} \end{bmatrix} = \begin{bmatrix} 1 & 0 \\ -\dfrac{j\cot\theta_{12}}{Z_{12}} & 1 \end{bmatrix} \begin{bmatrix} 1 & 0 \\ \dfrac{1}{R_2} & 1 \end{bmatrix} \begin{bmatrix} \cos\theta_{23} & jZ_{23}\sin\theta_{23} \\ \dfrac{j\sin\theta_{23}}{Z_{23}} & \cos\theta_{23} \end{bmatrix} \begin{bmatrix} 1 & 0 \\ \dfrac{1}{R_3} & 1 \end{bmatrix} \tag{15.8}$$

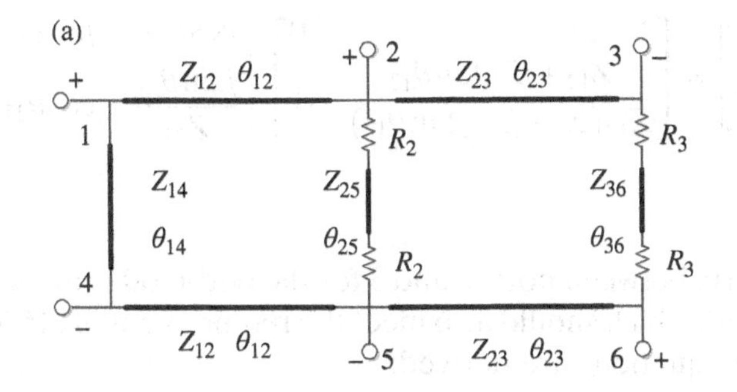

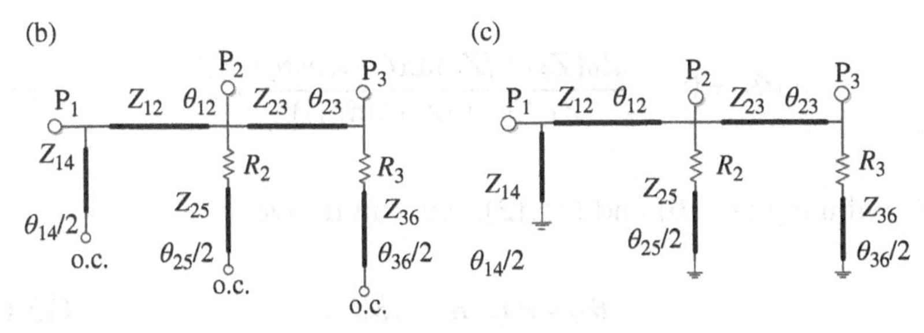

Figure 15.2 (a) The proposed balanced-to-balanced power divider/combiner; (b) the even-mode equivalent circuit model; and (c) the odd-mode equivalent circuit model. Reprinted with permission from Ref. [20]; copyright 2012 IEEE.

The S-matrix between ports 2 and 3 for the even-mode model can be calculated from the corresponding **ABCD** matrix [22], and the derived even-mode S-parameters must meet the requirement of (15.7). Therefore, the following equations should be satisfied:

$$A_e^{23} = D_e^{23}, \quad \frac{B_e^{23}}{Z_0} - C_e^{23} Z_0 = 2 \tag{15.9}$$

Applying (15.5e) into (15.9), we will have

$$\begin{aligned} &\cot\theta_{12}\sin\theta_{23} = 0, \quad \left(\frac{1}{R_2} - \frac{1}{R_3}\right)\sin\theta_{23} = 0, \\ &\cos\theta_{23} = -\frac{2}{Z_0}\frac{R_2R_3}{R_2+R_3}, \quad \left(\frac{Z_{23}}{Z_0} - \frac{Z_0}{Z_{23}} - \frac{Z_0Z_{23}}{R_2R_3}\right)\sin\theta_{23} = 0 \end{aligned} \tag{15.10}$$

15.2.2.2 *Odd-Mode Circuit Model*

From Figure 15.2(a), we can also obtain the odd-mode circuit model of the balanced-to-balanced power divider/combiner, as shown in Figure 15.2(c). When the port 1 is terminated by Z_0, the two-port **ABCD** matrix between ports 2 and 3 at f_0 is given by

$$\begin{bmatrix} A_o^{23} & B_o^{23} \\ C_o^{23} & D_o^{23} \end{bmatrix} = \begin{bmatrix} 1 & 0 \\ \dfrac{Z_{12}+jZ_0\tan\theta_{12}}{Z_{12}(Z_0+jZ_{12}\tan\theta_{12})} & 1 \end{bmatrix} \begin{bmatrix} \cos\theta_{23} & jZ_{23}\sin\theta_{23} \\ \dfrac{j\sin\theta_{23}}{Z_{23}} & \cos\theta_{23} \end{bmatrix} \tag{15.11}$$

The S-matrix between ports 2 and 3 for the odd-mode circuit model is then calculated, which should also meet the requirement of (15.7). Then, the following equations are derived:

$$\sin\theta_{23}=0, \quad -\frac{Z_0(Z_{12}+jZ_0\tan\theta_{12})\cos\theta_{23}}{Z_{12}(Z_0+jZ_{12}\tan\theta_{12})}=2 \tag{15.12}$$

Combining (15.10) and (15.12), one can derive

$$\theta_{23}=n\pi, \quad n=1,3,5\ldots \tag{15.13a}$$

$$\frac{1}{R_2}+\frac{1}{R_3}=\frac{2}{Z_0} \tag{15.13b}$$

$$\theta_{12}=m\frac{\pi}{2}, \quad m=1,3,5\ldots \tag{15.13c}$$

$$Z_{12}=\frac{\sqrt{2}Z_0}{2} \tag{15.13d}$$

It should be indicated that the aforementioned equations are valid only at the central frequency.

According to (15.13), the three-port scattering matrices of the even- and odd-mode circuit models at f_0 are derived as

$$[S_e]=\begin{bmatrix} -1 & 0 & 0 \\ 0 & -1/2 & -1/2 \\ 0 & -1/2 & -1/2 \end{bmatrix}, \quad [S_o]=\begin{bmatrix} 0 & -j\sqrt{2}/2 & j\sqrt{2}/2 \\ -j\sqrt{2}/2 & -1/2 & -1/2 \\ j\sqrt{2}/2 & -1/2 & -1/2 \end{bmatrix} \tag{15.14}$$

Obviously, (15.14) is one of the special cases of (15.7), where $b=-j\sqrt{2}/4$ and $a=g=-1/2$. Note that the value of b also satisfies (15.5b).

15.2.2.3 Scattering Matrix of the Balanced-to-Balanced Power Divider According to (15.6) and (15.14), the *S*-matrix of balanced-to-balanced power divider/combiner can be deduced at the central frequency f_0 by

$$[S_U] = \begin{bmatrix} -1/2 & -j\sqrt{2}/4 & j\sqrt{2}/4 \\ -j\sqrt{2}/4 & -1/2 & -1/2 \\ j\sqrt{2}/4 & -1/2 & -1/2 \end{bmatrix}, \quad [S_D] = \begin{bmatrix} -1/2 & j\sqrt{2}/4 & -j\sqrt{2}/4 \\ j\sqrt{2}/4 & 0 & 0 \\ -j\sqrt{2}/4 & 0 & 0 \end{bmatrix} \tag{15.15}$$

The mixed-mode *S*-parameters are then calculated by

$$[\mathbf{S}^{\mathrm{mm}}] = \begin{bmatrix} 0 & j\sqrt{2}/2 & j\sqrt{2}/2 & 0 & 0 & 0 \\ j\sqrt{2}/2 & 0 & 0 & 0 & 0 & 0 \\ j\sqrt{2}/2 & 0 & 0 & 0 & 0 & 0 \\ 0 & 0 & 0 & -1 & 0 & 0 \\ 0 & 0 & 0 & 0 & -1 & 0 \\ 0 & 0 & 0 & 0 & 0 & -1 \end{bmatrix} \tag{15.16}$$

At f_0, it is seen from (15.11) that the proposed component provides equal power division without loss, perfect matching for each balanced port, and perfect isolation between the balanced ports B and C when differential-mode excitation is applied; it also provides total reflection and perfect isolation between balanced ports when common-mode excitation is applied, and it also guarantees that no conversion between differential and common modes will take place.

15.2.3 Theoretical Result

From (15.13), it can be found that:

1. The characteristic impedances Z_{14}, Z_{25}, Z_{36}, and Z_{23} of the half-wavelength transmission lines can be selected arbitrarily, which will not affect the performance of the proposed power divider at the central frequency.
2. The loaded resistances R_2 and R_3 only need to satisfy (15.13), whose detailed values will not affect the performance at the central frequency.

For miniaturization and simple design, the following values are selected:

$$\theta_{14} = \theta_{23} = \theta_{25} = \theta_{36} = \pi \text{ at } f_0 \tag{15.17a}$$

$$\theta_{12} = \frac{\pi}{2} \text{ at } f_0 \tag{15.17b}$$

$$Z_{14} = Z_{23} = Z_{25} = Z_{36} = Z_x \tag{15.17c}$$

$$R_2 = R_3 = Z_0 = 50\ \Omega \tag{15.17d}$$

Based on (15.17), a prototype of the balanced-to-balanced power divider/combiner is designed with the central frequency of $f_0 = 2.0$ GHz. We set $Z_x = 50\ \Omega$. The theoretical results are shown in Figure 15.3, where lossless transmission lines and ideal lumped resistors are used. Since the balanced ports B and C are symmetric, some mixed-mode S-parameters are omitted here. Note that the S-parameter identified as S_{ddAB} refers to the differential- to common-mode transmission from port B to port A in Figure 15.1 and the other S-parameters use similar naming rules. The value of $S_{\text{cdAA}} = S_{\text{dcAA}}$ is always equal to zero due to the symmetry between ports 1 and 4, and the corresponding theoretical curves are not plotted in Figure 15.3.

As shown in Figure 15.3(b), the differential-mode transmission coefficient and the common-mode reflection coefficients reach their maximum of $|S_{\text{ddAB}}| = -3.01$ dB and $|S_{\text{ccAA}}| = |S_{\text{ccBB}}| = 0$ dB at $f_0 = 2.0$ GHz, respectively. All the other mixed-mode S-parameters approach zero at f_0. Comparing the magnitudes of S_{cd}- and S_{dc}-parameters in Figure 15.3(b), we can find that the two groups of curves have almost the same values around the central frequency, which conforms to (15.4). It can be concluded that all the requirements of the balanced-to-balanced power divider/combiner have been satisfied at the central frequency.

If a value of $|S_{\text{ddAB}}|$ better than −4 dB and values of $|S_{\text{ddAA}}|$, $|S_{\text{ddBB}}|$, $|S_{\text{ddBC}}|$, $|S_{\text{ccAB}}|$, $|S_{\text{ccBC}}|$, $|S_{\text{cdBB}}|$, $|S_{\text{cdAB}}|$, $|S_{\text{cdBC}}|$, $|S_{\text{dcBB}}|$, $|S_{\text{dcAB}}|$, $|S_{\text{dcBC}}|$ all better than −15 dB should be satisfied simultaneously, an operating band can be achieved from 1.72 to 2.28 GHz, that is, the achieved fractional bandwidth (FBW) of about 28%.

15.2.4 Simulated and Measured Results

As shown in Figure 15.4, a prototype is realized with microstrip lines and surface-mounted resistors and fabricated on a F4B substrate, with a relative permittivity of $\varepsilon_r = 2.65$, a loss tangent of $\tan\delta = 0.003$, and a

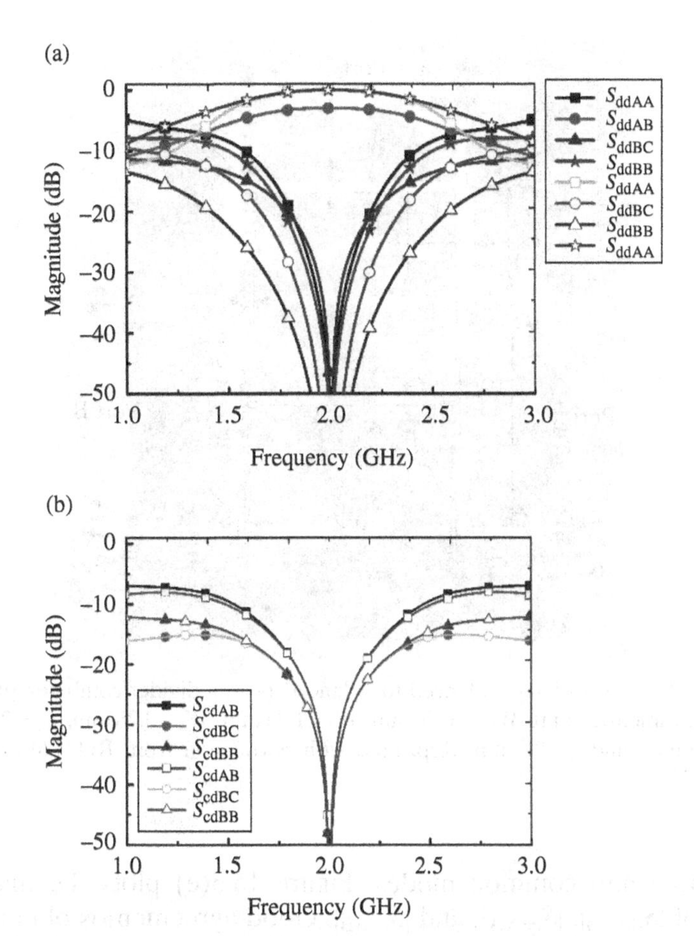

Figure 15.3 The theoretical mixed-mode S-parameters of the balanced-to-balanced power divider/combiner prototype: (a) S_{dd} and S_{cc}; (b) S_{cd} and S_{dc}. Reprinted with permission from Ref. [20]; copyright 2012 IEEE.

thickness of $h = 0.73$ mm. The total size is about $0.5 \times 0.75\ \lambda_g$, where λ_g is the guided wavelength of a 50-Ω microstrip line on the substrate at 2.0 GHz.

The six-port S-parameters are simulated by the commercial software ANSYS HFSS and measured with the four-port vector network analyzer (VNA) Agilent E5071C. Then, the mixed-mode S-parameters are extracted by using (15.2) and (15.3). Figure 15.5(a) and (b) shows the comparison between simulated and measured transmission coefficients, reflection coefficients, and isolation between the balanced ports B and C for differential- and common-mode operation, respectively. Figure 15.5(c) and (d) provides the mode conversions between

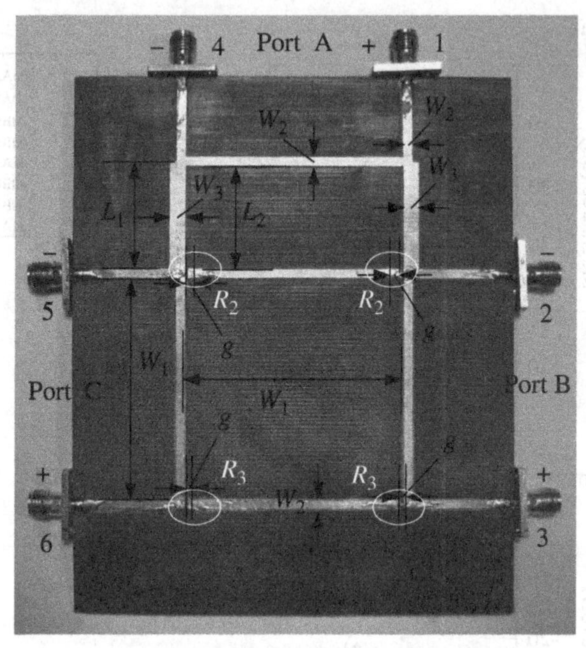

Figure 15.4 Photo of the balanced-to-balanced power divider/combiner prototype. Its critical dimensions are $W_1 = 50.76$ mm, $W_2 = 1.95$ mm, $W_3 = 3.28$ mm, $L_1 = 24.88$ mm, $L_2 = 23.91$ mm, and $g = 0.5$ mm. Reprinted with permission from Ref. [20]; copyright 2012 IEEE.

differential and common modes. Figure 15.5(e) plots the magnified curves of $|S_{ddAB}|$, $|S_{ccAA}|$, and $|S_{ccBB}|$. Good agreement is obtained.

The measured and simulated zeros of $|S_{ddAA}|$, $|S_{ddBB}|$, $|S_{ddBC}|$, $|S_{cdAB}|$, $|S_{cdBB}|$, $|S_{cdBB}|$, $|S_{ccBC}|$, and $|S_{ccAB}|$ are a little deviated from the designed central frequency of $f_0 = 2$ GHz. The deviation is mainly caused by the discontinuities of microstrip lines and the parasitic effect of the surface-mounted resistors. There is a little discrepancy between the curves in Figure 15.5(c) and (d) due to the fabrication tolerance. The simulated $|S_{cdAA}|$ and $|S_{dcAA}|$ are always better than −50 dB, while the measured ones are better than −35 dB. The difference between them is mainly caused by the imperfect symmetry of the fabricated prototype, which is also due to the tolerance of fabrication and the surface-mounted resistors.

In the measurements, the maximum differential-mode transmission coefficient of the prototype is $|S_{ddAB}| = -3.2$ dB at 2.04 GHz, the best differential-mode isolation is 47.2 dB at 2.13 GHz, and the common-mode reflection coefficients of $|S_{ccAA}|$ and $|S_{ccBB}|$ reach their maximum of −0.16 and −0.2 dB at 2.11 and 2.06 GHz, respectively.

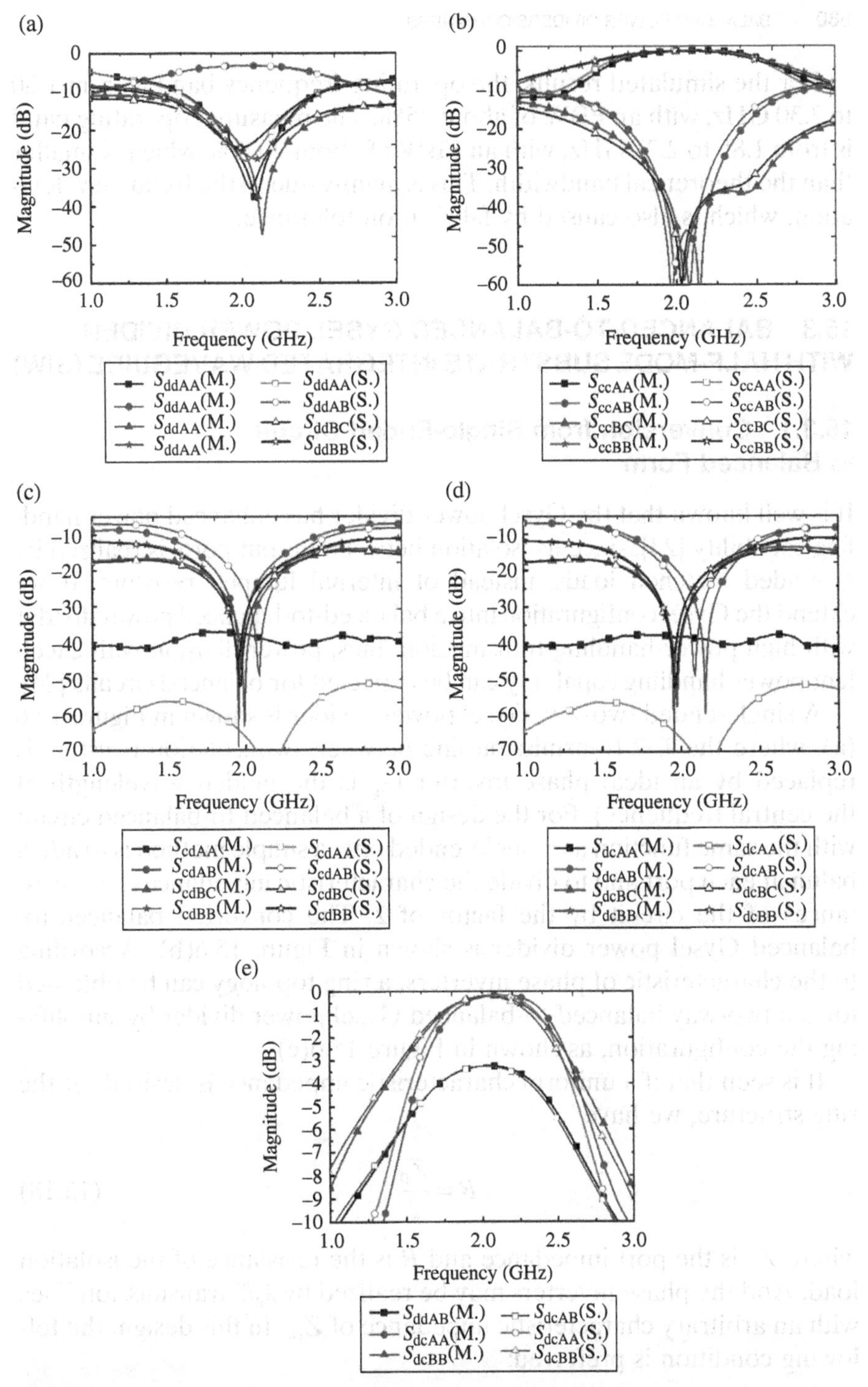

Figure 15.5 Simulated and measured mixed-mode *S*-parameters of the prototype: (a) $|S_{dd}|$; (b) $|S_{cc}|$; (c) $|S_{cd}|$; (d) $|S_{dc}|$; and (e) the magnified curves of $|S_{ddAB}|$, $|S_{ccAA}|$, and $|S_{ccBB}|$. M., measured results; S., simulated results. Reprinted with permission from Ref. [20]; copyright 2012 IEEE.

For the simulated results, the operating frequency band is from 1.80 to 2.30 GHz, with an FBW of about 25%. The measured operating band is from 1.85 to 2.27 GHz, with an FBW of about 20.8%, which is smaller than the theoretical bandwidth. This is mainly due to the frequency deviation, which is also caused by fabrication tolerance.

15.3 BALANCED-TO-BALANCED GYSEL POWER DIVIDER WITH HALF-MODE SUBSTRATE INTEGRATED WAVEGUIDE (SIW)

15.3.1 Conversion from Single-Ended Circuit to Balanced Form

It is well known that the Gysel power divider has enhanced power handling capability [23], since its isolation between output ports is realized by grounded matched loads, instead of internal lumped resistors. If we extend the Gysel configuration into a balanced-to-balanced power divider with high power handling transmission lines, power division with excellent power handling capability can be expected for balanced circuits [24].

A single-ended two-way Gysel power divider is shown in Figure 15.6 (a), where the $\lambda_g/2$ transmission line between two isolation resistors is replaced by an ideal phase inverter (λ_g is the guided wavelength at the central frequency). For the design of a balanced-to-balanced circuit with the same function as a single-ended one, a simple method is to add a balun at each port and to divide the characteristic impedances and resistances of the circuit by the factor of 2. The converted balanced-to-balanced Gysel power divider is shown in Figure 15.6(b). According to the characteristic of phase inverters, a ring topology can be obtained for the two-way balanced-to-balanced Gysel power divider by simplifying the configuration, as shown in Figure 15.6(c).

It is seen that if a uniform characteristic impedance is desired for the ring structure, we have

$$R = \frac{Z_p}{2} \tag{15.18}$$

where Z_p is the port impedance and R is the resistance of the isolation load. And the phase inverters may be realized by $\lambda_g/2$ transmission lines with an arbitrary characteristic impedance of Z_{pi}. In this design, the following condition is preferred:

$$Z_{pi} = \frac{\sqrt{2}Z_p}{2} \tag{15.19}$$

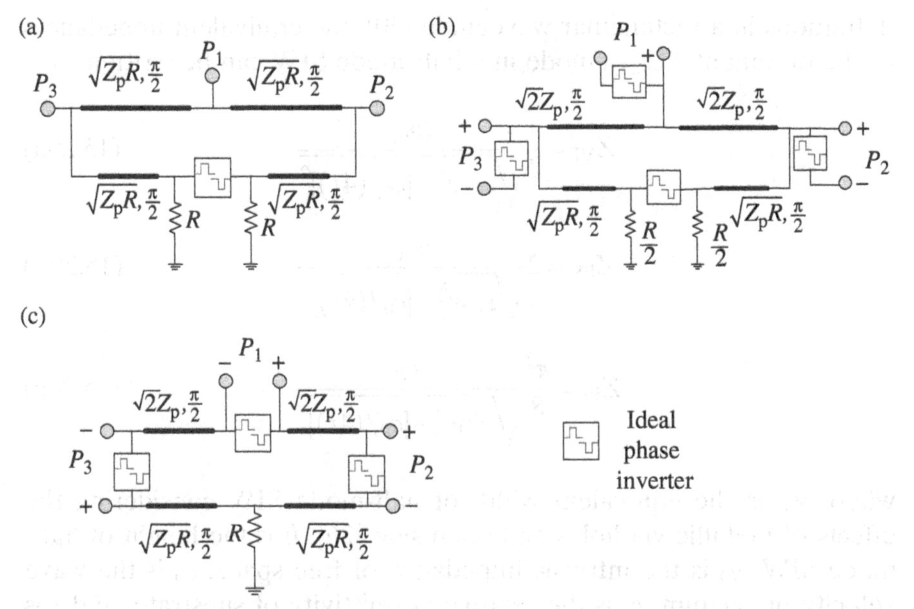

Figure 15.6 Configurations of Gysel power dividers: (a) a conventional single-ended one; (b) a balanced-to-balanced one directly converted from its single-ended counterpart; and (c) a simplified balanced-to-balanced one. Reprinted with permission from Ref. [24]; copyright 2012 IEEE.

Then, the balanced-to-balanced Gysel power divider can be built up with a ring structure whose circumference is 2.5 λ_g.

15.3.2 Half-Mode SIW Ring Structure

SIWs have drawn much attention recently [25], due to their low loss, low cost, easy integration, etc. Half-mode SIW is a type of miniaturized SIW [26]. Power dividers have been realized with SIW and half-mode SIW [27–29]. One of their obvious merits is good power handling capability.

In order to implement the balanced-to-balanced Gysel power divider in Figure 15.6(c) with excellent power handling capability, a 2.5-λ_g half-mode SIW ring is utilized. The same as conventional metal waveguides, there are non-unique definitions of voltage, current, and characteristic impedance for a half-mode SIW. The so-called equivalent characteristic impedances Z_{VI}, Z_{PV}, and Z_{PI} are defined by voltage–current, power–voltage, and power–current, respectively. Based on the general definitions in a rectangular waveguide [30], the equivalent impedances of the dominant $TE_{0.5,0}$ mode in a half-mode SIW can be written as

$$Z_{VI} = \frac{\pi}{2}\frac{h\eta_0}{\sqrt{\varepsilon_r w_e^2 - [c_0/(4f)]^2}} \tag{15.20a}$$

$$Z_{PV} = 2\frac{h\eta_0}{\sqrt{\varepsilon_r w_e^2 - [c_0/(4f)]^2}} \tag{15.20b}$$

$$Z_{PI} = \frac{\pi^2}{8}\frac{h\eta_0}{\sqrt{\varepsilon_r w_e^2 - [c_0/(4f)]^2}} \tag{15.20c}$$

where w_e is the equivalent width of half-mode SIW considering the effects of metallic via holes and open side [26], h is the height of half-mode SIW, η_0 is the intrinsic impedance of free space, c_0 is the wave velocity in vacuum, ε_r is the relative permittivity of substrate, and f is the operating frequency. When the substrate is preselected, all the equivalent impedances at a certain operating frequency f increase with the increasing of h and decrease with the increasing of w_e.

As shown in Figure 15.7, the half-mode SIW ring is connected with six 50-Ω microstrip ports for convenient measurement. A 50-Ω load is used for isolation, and a quarter-wavelength microstrip line with the

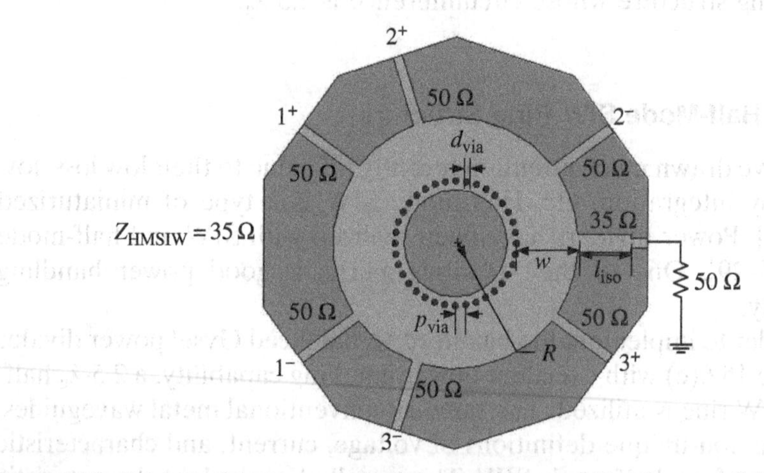

Figure 15.7 Layout of our proposed balanced-to-balanced Gysel power divider based on a half-mode SIW ring structure. Reprinted with permission from Ref. [24]; copyright 2012 IEEE.

characteristic impedance of $\sqrt{2}Z_p/2$ =35 Ω is inserted to make the input impedance of the isolation load looked from the half-mode SIW ring be $R = Z_p/2 = 25$ Ω. The half-mode SIW ring should have an equivalent impedance of $Z_{\mathrm{HMSIW}} = 35$ Ω. Since the TE-mode half-mode SIW is integrated with quasi-TEM-mode microstrip lines, it is very important to decide which equivalent impedance definition should be applied for the half-mode SIW in this case.

It is noted that the appropriate equivalent impedance of half-mode SIW looked from microstrip is related to their connection form or the transition between half-mode SIW and microstrip. Because the microstrip lines are directly connected to the open side of the half-mode SIW, where the electric field of the dominant TE mode in half-mode SIW reaches its maximum, it is reasonable to assume that the equivalent voltage of half-mode SIW has the same value as that of microstrip lines. Further, around the open side, the transversal surface current on the top metal plane of the half-mode SIW is close to zero, while the longitudinal surface current reaches its maximum. Note that the equivalent current of waveguide is just defined by the longitudinal surface current. Therefore, the definition of Z_{VI} is preferred in this configuration.

A prototype is designed with the operating frequency of $f_0 = 5$ GHz. The Taconic RF-35A2 substrate is selected, whose relative permittivity and height are $\varepsilon_r = 3.5$ and $h = 0.762$ mm, respectively. The diameter and spacing of via holes are $d_{\mathrm{via}} = 1.0$ mm and $p_{\mathrm{via}} = 1.8$ mm, respectively. It is obtained that $w = 10.2$ mm by using (15.15), together with the relationship between the equivalent width w_e and the physical width w of the half-mode SIW [26]. The guided wavelength in the half-mode SIW at f_0 is $\lambda_g = 49.0$ mm. Then, the radius of the ring structure is approximated to be $R = (5\lambda_g)/(4\pi) = 19.5$ mm.

15.3.3 Results and Discussion

The measured mixed-mode S-parameters of the prototype are plotted in Figure 15.8 and the simulated ones are also obtained with ANSYS HFSS. All the measured results agree well with the simulated ones, except that the measured levels of S_{dc11} and S_{dc22} are about 15 dB higher than the simulated ones. The values of S_{dc11} are very sensitive to the structure symmetry, which has been slightly disturbed by the fabrication tolerance. The discrepancy of S_{dc22} may be caused by the impedance tolerance of microstrip lines at ports 2^+ and 2^-. However, the measured S_{dc11} is still below −30 dB, still good enough to suppress the conversion between differential and common modes at the balanced input pair.

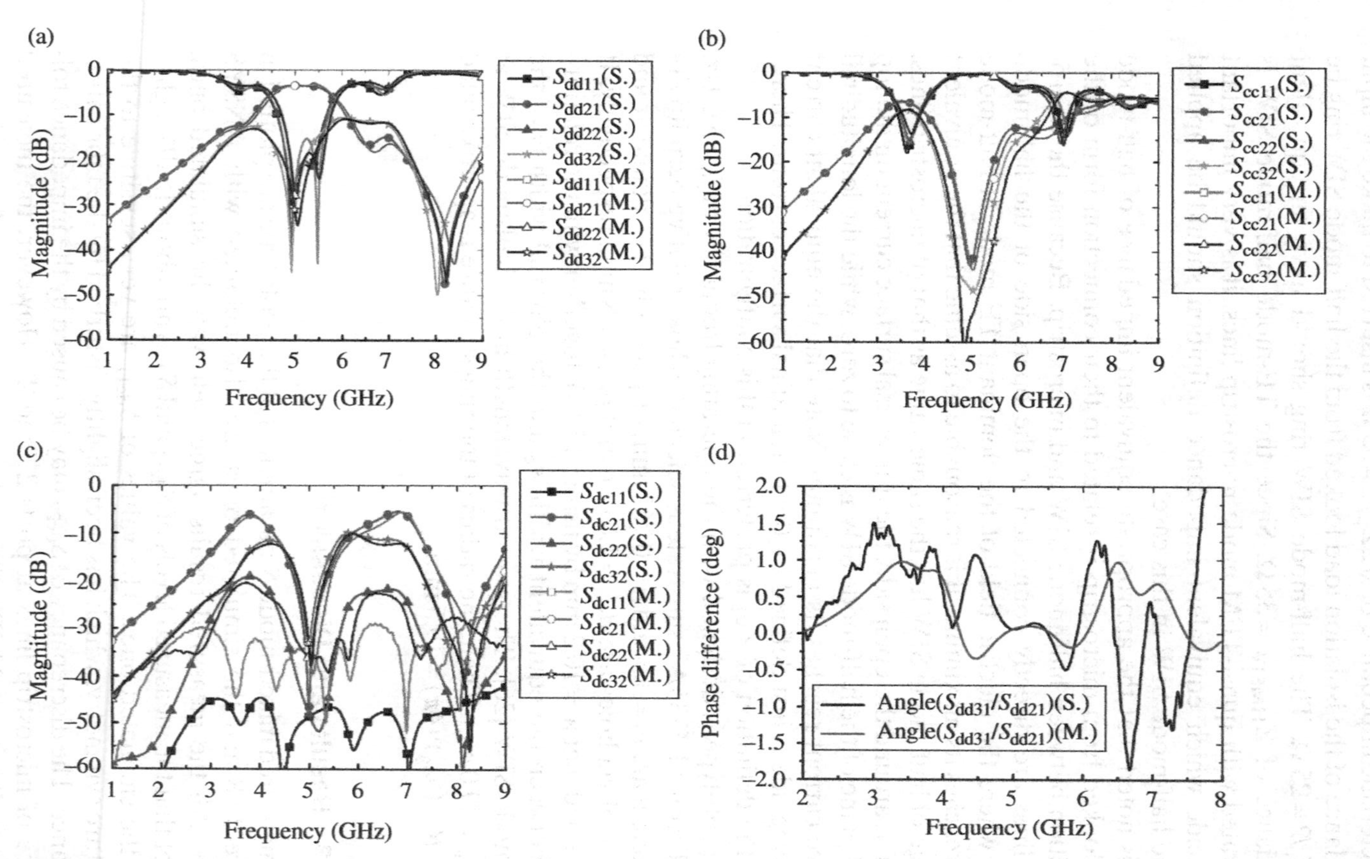

Figure 15.8 Measured and simulated mixed-mode S-parameters of the balanced-to- balanced Gysel power divider prototype: (a) S_{dd}; (b) S_{cc}; (c) S_{dc}; and (d) phase difference between S_{dd31} and S_{dd21}. M., measured results; S., simulated results. Reprinted with permission from Ref. [24]; copyright 2012 IEEE.

And the measured S_{dc22} is still lower than –25 dB within the operating band.

The simulated and measured central frequency is 5.12 and 5.20 GHz, respectively. The little frequency shift is mainly due to the tolerance of relative permittivity. Both the simulated and measured central insertion losses of S_{dd21} are 3.4 (0.4 + 3.0) dB. The measured operating bandwidth is from 4.82 to 5.49 GHz, that is, 13%, with S_{dd11}, S_{dd22}, S_{dd32}, S_{cc21}, S_{cc32} and all the S_{dc}- and S_{cd}-parameters below –15 dB. The measured in-band phase difference between S_{dd31} and S_{dd21} is within ±0.2°. The desired performance of the balanced-to-balanced Gysel power divider has been implemented with the half-mode SIW ring structure, and a high power handling capability can be expected.

15.4 BALANCED-TO-BALANCED GYSEL POWER DIVIDER WITH ARBITRARY POWER DIVISION

15.4.1 Analysis and Design

In many cases, power dividers are designed with equal power division to different outputs. However, in some cases, power dividers/combiners with unequal power division ratios are also desired, especially for feeding networks of antenna arrays. Single-ended power dividers with arbitrary power division have attracted much attention, due to their design flexibility for RF systems. Here, we will discuss the power divider with arbitrary power division ratio in the balanced form [31].

The diagram of the proposed arbitrary balanced-to-balanced power divider is shown in Figure 15.9(a). $[\mathbf{S}^{std}]$ represents its standard 6×6 scattering matrix. The differential-mode power should be divided with arbitrary ratio. If the ratio of the power that are divided to balanced ports B and C is $1:k^2$ and the ratio of $|S_{ddAB}|$ to $|S_{ddAC}|$ is $1:k$, the

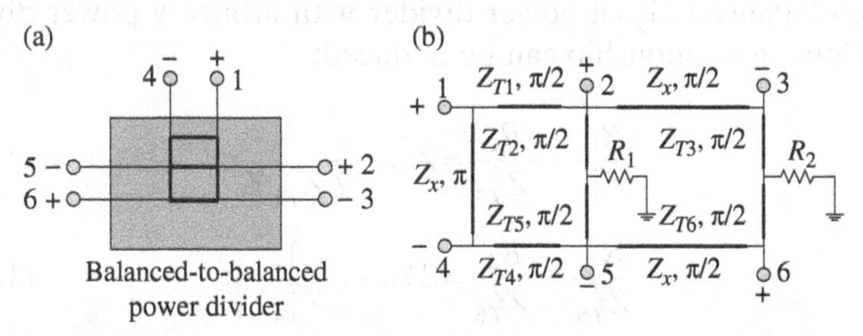

Figure 15.9 Proposed balanced-to-balanced power divider with arbitrary power division: (a) its diagram; and (b) the structure. Reprinted with permission from Ref. [31]; copyright 2013 IEEE.

proposed component with arbitrary division should satisfy the following constraint rules at the central frequency as before:

1. $|S_{ddAA}| = |S_{ddBB}| = |S_{ddCC}| = |S_{ddBC}| = 0$;
   we set $|S_{ddAB}| = 1/\sqrt{1+k^2}$, $|S_{ddAC}| = k/\sqrt{1+k^2}, k > 0$.
2. $|S_{ccAA}| = |S_{ccBB}| = |S_{ccCC}| = 1$; $|S_{ccAB}| = |S_{ccAC}| = |S_{ccBC}| = 0$.
3. $|S_{dcAA}| = |S_{dcBB}| = |S_{dcCC}| = |S_{dcAB}| = |S_{dcAC}| = |S_{dcBC}| = 0$.
4. $|S_{cdAA}| = |S_{cdBB}| = |S_{cdCC}| = |S_{cdAB}| = |S_{cdAC}| = |S_{cdBC}| = 0$.

So, the following equations can be derived:

$$S_{11} = S_{14} = S_{41} = S_{44} \tag{15.21a}$$

$$S_{12} = -S_{13} = -S_{42} = S_{43} = S_{21} = -S_{24} = -S_{31} = S_{34} \tag{15.21b}$$

$$|S_{12}| = \frac{1}{2\sqrt{1+k^2}} \tag{15.21c}$$

$$S_{16} = -S_{15} = -S_{46} = S_{45} = S_{61} = -S_{64} = -S_{51} = S_{54} \tag{15.21d}$$

$$|S_{16}| = \frac{k}{2\sqrt{1+k^2}} \tag{15.21e}$$

$$S_{22} = S_{23} = S_{32} = S_{33} = S_{66} = S_{65} = S_{56} = S_{55} \tag{15.21f}$$

$$|S_{11}| = |S_{55}| = \frac{1}{2} \tag{15.21g}$$

$$S_{25} = S_{26} = S_{35} = S_{36} = S_{52} = S_{53} = S_{62} = S_{63} = 0 \tag{15.21h}$$

The balanced-to-balanced power divider shown in Figure 15.9(b) is realized with transmission lines and resistors. It can be regarded as a balanced-to-balanced Gysel power divider with arbitrary power division. The following relationship can be deduced:

$$\frac{R_1}{Z_{T2}^2} + \frac{R_2}{Z_{T3}^2} = 2Y_0 - \frac{1}{2\,Z_{T1}^2\,Y_0} \tag{15.22a}$$

$$\frac{R_1}{Z_{T5}^2} + \frac{R_2}{Z_{T6}^2} = 2Y_0 - \frac{1}{2\,Z_{T4}^2\,Y_0} \tag{15.22b}$$

$$\frac{R_1}{Z_{T2}Z_{T5}} + \frac{R_2}{Z_{T3}Z_{T6}} = \frac{1}{2Z_{T1}Z_{T4}Y_0} \tag{15.22c}$$

According to (15.21), S_{11} and S_{22} are set to $-1/2$ here, while S_{12} and S_{21} are set to $-j/2\sqrt{1+k^2}$. Then, the following relationship can be presented:

$$Z_{T1} = \frac{1}{2Y_0}\sqrt{1+k^2}, \quad Z_{T4} = \frac{1}{2Y_0}\sqrt{1+\frac{1}{k^2}} \tag{15.23a}$$

$$Z_{T2} = \sqrt{R_{1c}(1+1/k^2)/Y_0} = \frac{Z_{T5}}{k}, \quad Z_{T3} = \sqrt{R_{2c}(1+1/k^2)/Y_0} = \frac{Z_{T6}}{k} \tag{15.23b}$$

$$R_1 = (2-m)R_{1c}, \quad R_2 = mR_{2c} \tag{15.23c}$$

where the positive values of R_{1c}, R_{2c}, and m can be selected arbitrarily.

15.4.2 Results and Discussion

Based on the aforementioned analysis, a balanced-to-balanced power divider prototype is developed with the power division ratio between balanced B and C of $1:3^2$. The design parameters are chosen as $m=1$, $R_{1c}=R_{2c}=25\ \Omega$, and $Z_x=50\ \Omega$. The other critical characteristic impedances and loaded resistances are calculated by (15.23), which are $Z_{T1}=79.06\ \Omega$, $Z_{T2}=Z_{T3}=37.27\ \Omega$, $Z_{T4}=26.35\ \Omega$, $Z_{T5}=Z_{T6}=111.8\ \Omega$, and $R_1=R_2=25\ \Omega$. As shown in Figure 15.10, the prototype is fabricated on an F4B substrate with relative permittivity of $\varepsilon_r=2.65$, loss tangent of $\tan\delta=0.003$, and thickness of $h=0.73$ mm. The balanced-to-balanced power divider occupies an area of about $0.5\times0.75\ \lambda_g^2$, where λ_g is the guided wavelength at the central frequency.

The six-port network is simulated with ANSYS HFSS and measured by the VNA Agilent E5071C. In Figure 15.11, the values of $[S^{dd}]$, $[S^{dc}]$, $[S^{cc}]$ are obtained from the simulated and measured S-parameters of the six-port single-ended network. Figure 15.11(a) and (b) shows a comparison between the simulated and measured transmission coefficients, reflection coefficients, and isolation between the balanced ports B and C for differential- and common-mode operations, respectively. Figure 15.11(c) and (d) provides a comparison between the simulated and measured differential-to-common mode conversions. The curves of $|S_{ddAB}|$, $|S_{ddAC}|$, $|S_{ccAA}|$, $|S_{ccBB}|$, and $|S_{ccCC}|$ are magnified in Figure 15.11(e).

The theoretical maximum differential-mode transmission coefficients of the prototype are $|S_{ddAB}|=-10$ dB and $|S_{ddAC}|=-0.46$ dB at the

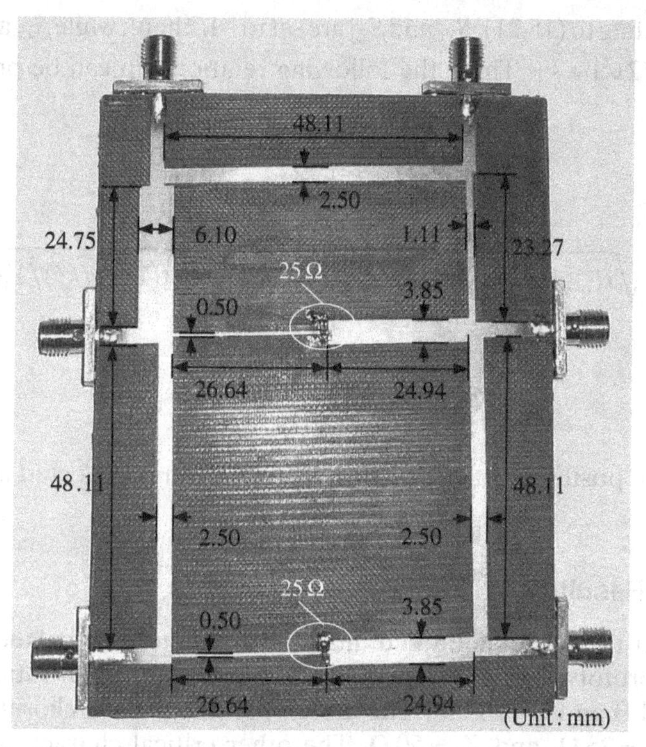

Figure 15.10 Photo of the fabricated prototype of the balanced-to-balanced power divider with arbitrary power division. Reprinted with permission from Ref. [31]; copyright 2013 IEEE.

central frequency of 2.0 GHz. In the measurements, the maximum differential-mode transmission coefficient of the prototype is $|S_{ddAB}| = -10.32$ dB at 1.96 GHz, $|S_{ddAC}| = -0.94$ dB at 1.95 GHz, the best differential-mode isolation is 61.72 dB at 1.88 GHz, and the common-mode reflection coefficients of $|S_{ccAA}|$ and $|S_{ccBB}|$ reach their maximum of −0.16 and −0.2 dB at 2.11 and 2.06 GHz, respectively. The conversions between differential- and common-mode signals are successfully suppressed around the central frequency.

The same rules as before are used to determine the bandwidths for measured and simulated results. An operating band from 1.71 to 2.13 GHz is achieved in simulations, with an FBW of about 21%. The measured operating band is from 1.74 to 2.10 GHz with an FBW of about 18%. Both of them are narrower than the theoretical value of 23%, due to the tolerances of practical distributed circuit and fabrication.

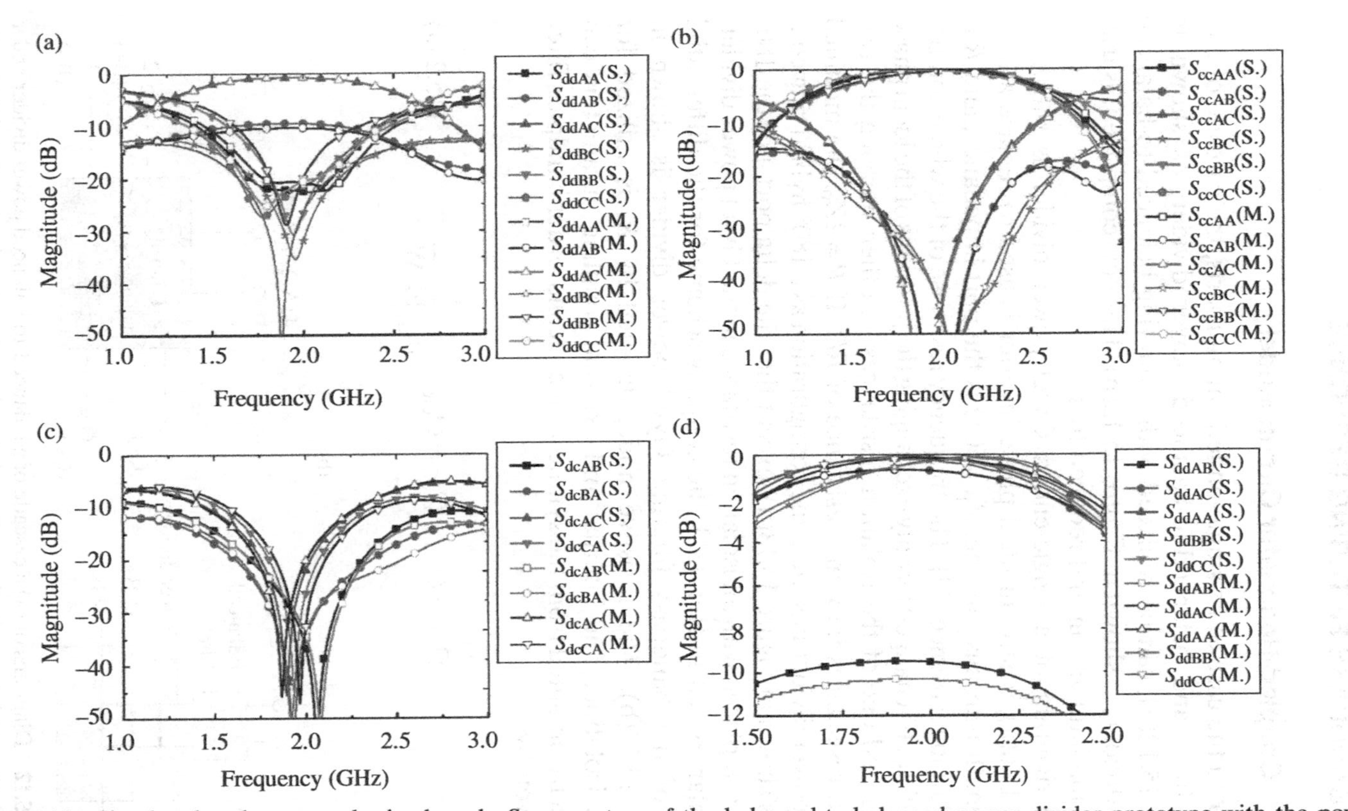

Figure 15.11 Simulated and measured mixed-mode S-parameters of the balanced-to-balanced power divider prototype with the power division ratio of $1:3^2$: (a) S_{dd}; (b) S_{cc}; (c) S_{dc}; and (d) the magnified curves of $|S_{ddAB}|$, $|S_{ddAC}|$, $|S_{ccAA}|$, $|S_{ccBB}|$, and $|S_{ccCC}|$. M. and S. represent the measured and simulated results, respectively. Reprinted with permission from Ref. [31]; copyright 2013 IEEE.

15.5 BALANCED-TO-BALANCED GYSEL POWER DIVIDER WITH BANDPASS FILTERING RESPONSE

15.5.1 Coupled-Resonator Circuit Model

Similar to the case of its single-ended counterparts, the cascade of balanced filter and balanced-to-balanced power divider, as shown in Figure 15.12(a), can be replaced with a filtering balanced-to-balanced power divider, as shown in Figure 15.12(b). This new component is useful for miniaturization and reduced loss.

The diagram of a single-ended Gysel power divider is shown in Figure 15.6(a), where an ideal phase inverter is used to replace a half-wavelength transmission line, $Z_\text{p} = 50\,\Omega$ is the port impedance, and R is the isolation resistance [22]. The frequency response of the whole component will not be changed by only exchanging the positions of the ideal phase inverter and one of the isolation resistances and then merging the two shunted resistances into one resistance of $R/2$. If $R = 2Z_\text{p}$, the modified Gysel power divider has the same configuration as a 180° hybrid ring coupler, as shown in Figure 15.13(a), except that the delta port 4 of the 180° hybrid is replaced with an isolation resistance in the Gysel power divider.

Similar to that in Ref. [32], the coupling scheme for the differential mode of the bandpass filtering Gysel power divider is shown in Figure 15.13(b). The resistive node, denoted by R, is utilized for perfect isolation. The internal and external coupling parameters can be directly obtained from the synthesis of a second-order bandpass filter and given by

$$Q_{\text{eA}} = Q_{\text{eB}} = Q_{\text{eC}} = Q_{\text{e}R} = \frac{1}{m_{S1}^2 \text{FBW}} \tag{15.24a}$$

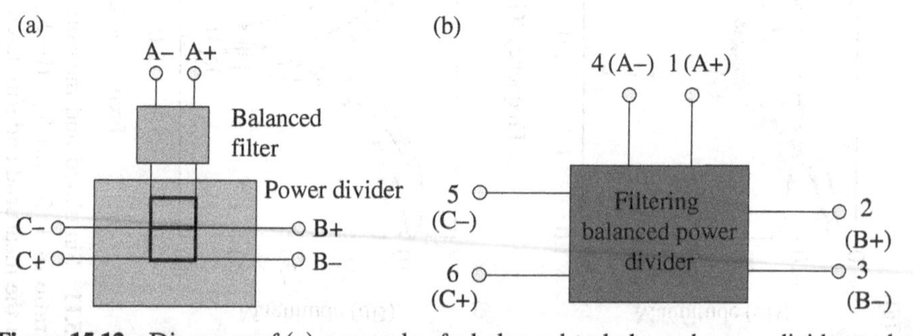

Figure 15.12 Diagrams of (a) a cascade of a balanced-to-balanced power divider and a balanced filter and (b) a filtering balanced-to-balanced power divider. Reprinted with permission from Ref. [33]; copyright 2013 IEEE.

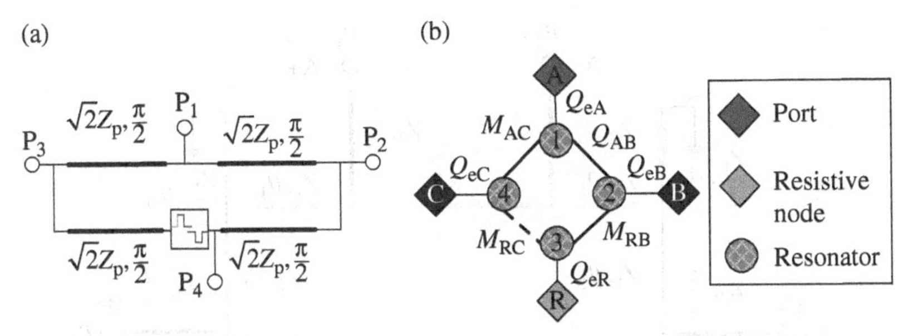

Figure 15.13 Schematics of (a) a 180° hybrid ring coupler and (b) the bandpass filtering balanced-to-balanced Gysel power divider. Reprinted with permission from Ref. [33]; copyright 2013 IEEE.

$$M_{\mathrm{AB}} = M_{\mathrm{AC}} = M_{RB} = -M_{RC} = \frac{m_{12}\mathrm{FBW}}{\sqrt{2}} \tag{15.24b}$$

where FBW is the fractional bandwidth and m_{S1} and m_{12} are the normalized external and internal coupling coefficients of the two-port bandpass filter prototype, respectively, which are synthesized according to the specifications of bandpass filtering response. The positive and negative terminals of the balanced port C are defined reversely to those of port B, which is equivalent to introducing an ideal phase inverter near port C, just leading to the negative sign of M_{RC}. Therefore, the coupled-resonator circuit model for the design of bandpass 180° hybrid coupler [32] can be extended for bandpass filtering balanced-to-balanced Gysel power dividers.

15.5.2 Realization in Transmission Lines

In order to realize the coupling scheme in Figure 15.13(b), the ring structure loaded with shorted stubs is built up [33], as shown in Figure 15.14. This topology is also very similar to the power divider in Ref. [22]. The ring can be regarded as four microstrip resonators coupled through shorted stubs. Each of the balanced port nodes A, B, and C consists of two terminals, marked with positive and negative signs. The half-wavelength section between the positive and negative terminals of each balanced port is utilized to reject common-mode signal and mode conversion. Since there is only one resistance of value R_i used to implement the resistive node, no half-wavelength section is required for this node. Then, three resonators are full-wavelength and the other one is half-wavelength, which are connected with balanced ports and isolation resistance, respectively. This means the dominant passband is

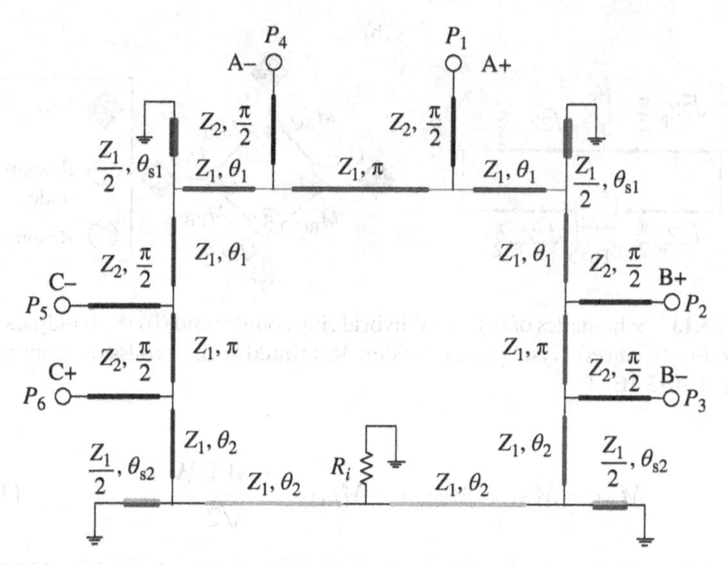

Figure 15.14 Schematic of the balanced-to-balanced Gysel power divider with bandpass filtering response, where the upper part corresponds to resonator 1, the left and right parts correspond to resonators 2 and 3, respectively, and the lower part corresponds to resonator 4. Reprinted with permission from Ref. [33]; copyright 2013 IEEE.

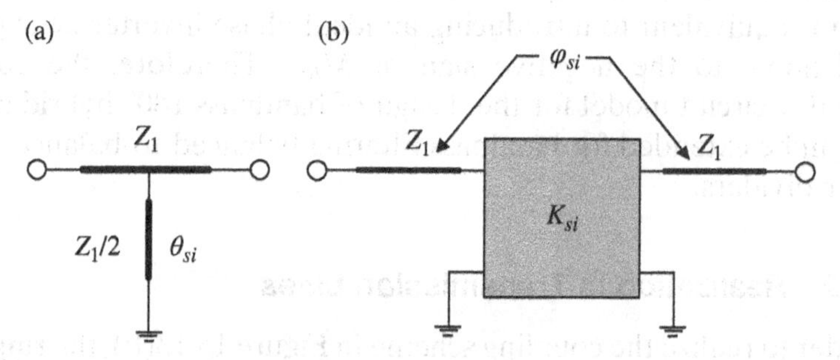

Figure 15.15 Equivalent circuit model of the shorted stub, where $i = 1$ and 2. Reprinted with permission from Ref. [33]; copyright 2013 IEEE.

contributed to the second resonance of resonators 1–3 and the dominant resonance of resonator 4.

15.5.2.1 Internal Coupling Coefficient When the characteristic impedance of the shorted stub is half the characteristic impedance of the ring, the stub is equivalent to a K-inverter and two resident transmission lines [34, 35], as shown in Figure 15.15. The inverter impedance and the electrical length of the transmission lines are given by [34]

$$K_{Si} = Z_1 \tan\frac{\theta_{Si}}{2} \approx \frac{Z_1\theta_{Si}}{2}, \quad \varphi_{Si} = \frac{\theta_{Si}}{2} \tag{15.25}$$

where $i = 1$ and 2. Then, the electrical lengths θ_1 and θ_2 are determined by

$$\theta_i = \frac{\pi - \theta_{Si}}{2} \tag{15.26}$$

The input impedance of a short-ended microstrip resonator looked at from one end, with its characteristic impedance of Z_1 and the total electrical length of θ_{total}, is given by

$$Z_{\text{in}} = jZ_1 \tan\theta_{\text{total}} \tag{15.27}$$

The reactance slope parameter is derived as

$$X = \frac{Z_1\theta_{\text{total}}}{2\cos^2\theta_{\text{total}}} \tag{15.28}$$

At the central frequency, the reactance slope parameter of resonators 1–3 with $\theta_{\text{total}} = 2\pi$ is

$$X_{\text{full}} = Z_1\pi \tag{15.29}$$

and the reactance slope parameter of resonator 4 with $\theta_{\text{total}} = \pi$ is

$$X_{\text{half}} = \frac{Z_1\pi}{2} \tag{15.30}$$

Then, the two internal coupling coefficients M_{AB} and M_{AC} between the input resonator and the two output resonators are calculated by

$$M_{\text{AB}} = M_{\text{AC}} = \frac{K_{S1}}{\sqrt{X_{\text{full}}X_{\text{full}}}} \approx \frac{\theta_{S1}}{2\pi} \tag{15.31}$$

while the other two internal coupling coefficients, $M_{R\text{B}}$ and $M_{R\text{C}}$, between resonator 4 and the two output resonators are calculated by

$$M_{R\text{B}} = M_{R\text{C}} = \frac{K_{S2}}{\sqrt{X_{\text{full}}X_{\text{half}}}} \approx \frac{\theta_{S2}}{\sqrt{2}\pi} \tag{15.32}$$

By substituting (15.22c) and (15.23a) into (15.21e), we have

$$\theta_{S1} \approx \sqrt{2} m_{12} \pi \mathrm{FBW}, \quad \theta_{S2} \approx m_{12} \pi \mathrm{FBW} \tag{15.33}$$

15.5.2.2 External Q Factor The external Q factor for the differential mode of each port can be calculated according to the stored energy and the externally coupled power. When differentially driven at the central frequency, the voltage distribution of resonators 1–3 is shown in Figure 15.16, where V_0 is the peak voltage in the resonator.

The stored energy is calculated by

$$W_s = \int_0^l \frac{1}{2} C_0 (V_0 \sin\beta x)^2 dx = \frac{C_0 V_0^2}{4}\left(l - \frac{\sin 2\beta l}{2\beta}\right) \tag{15.34}$$

where C_0, l, and β are the distributed capacitance, length, and phase constant of the resonator, respectively. Note that $\beta l = 2\pi$ and $\beta = \omega_0 Z_1 C_0$, where ω_0 is the central angular frequency. Then we have

$$W_s = \frac{C_0 V_0^2 l}{4} = \frac{\pi V_0^2}{2\omega_0 Z_1} \tag{15.35}$$

The externally coupled power of the differential pair is given by

$$P_e = 2\frac{V_0^2}{2(Z_2^2/Z_p)} = \frac{V_0^2 Z_p}{Z_2^2} \tag{15.36}$$

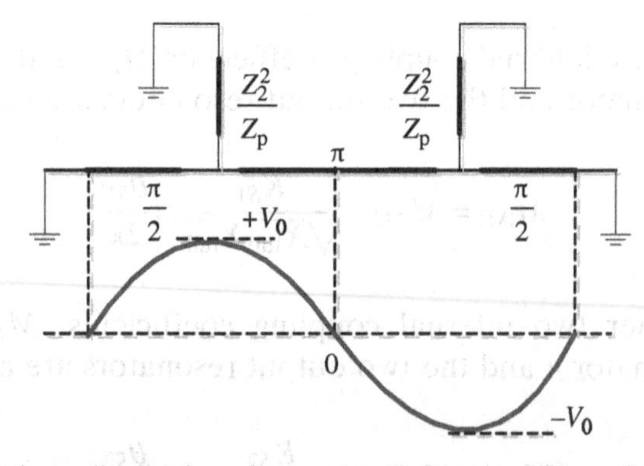

Figure 15.16 Voltage distribution of resonators 1–3 under differential excitation at the central frequency. Reprinted with permission from Ref. [33]; copyright 2013 IEEE.

Therefore, the external Q factor is obtained by

$$Q_{\mathrm{eA}} = Q_{\mathrm{eB}} = Q_{\mathrm{eC}} = \frac{\omega_0 W_{\mathrm{s}}}{P_{\mathrm{e}}} = \frac{\pi Z_2^2}{2Z_1 Z_{\mathrm{p}}} \tag{15.37}$$

The following design equation for the characteristic impedance Z_2 can be derived by substituting (15.24b) into (15.21d):

$$Z_2 = \frac{1}{m_{S1}} \sqrt{\frac{2Z_1 Z_{\mathrm{p}}}{\pi \mathrm{FBW}}} \tag{15.38}$$

For resonator 4 and the resistive node, a similar analysis can be carried out. The value Q_{eR} of the corresponding external Q factor is given by

$$Q_{eR} = \frac{\pi R_{\mathrm{i}}}{2Z_1} \tag{15.39}$$

According to (15.21d) and (15.26), the isolation resistance R_{i} is determined by

$$R_{\mathrm{i}} = \frac{2Z_1}{m_{S1}^2 \pi \mathrm{FBW}} \tag{15.40}$$

According to the aforementioned analysis and derivation, the proposed bandpass filtering balanced-to-balanced Gysel power divider can be designed with (15.25), (15.21f), (15.33), (15.25), and (15.27) when Z_1 is preselected. The physical dimensions of the component are directly determined by the characteristic impedances and electrical lengths of the transmission line sections. Then, the design will be successfully completed.

15.5.3 Results and Discussion

As shown in Figure 15.17, a prototype of the bandpass filtering balanced-to-balanced Gysel power divider is designed with the stub-loaded sections. In order to miniaturize the component, the technique of stub loading is utilized [33]. The stepped-impedance stub-loaded transmission line is used. By selecting proper parameters for the loaded stubs, the slow-wave effect can be achieved and tuned. The equivalent value of Z_1 is selected to be 36.5 Ω. Then, we have $Z_2 = 98.4$ Ω and

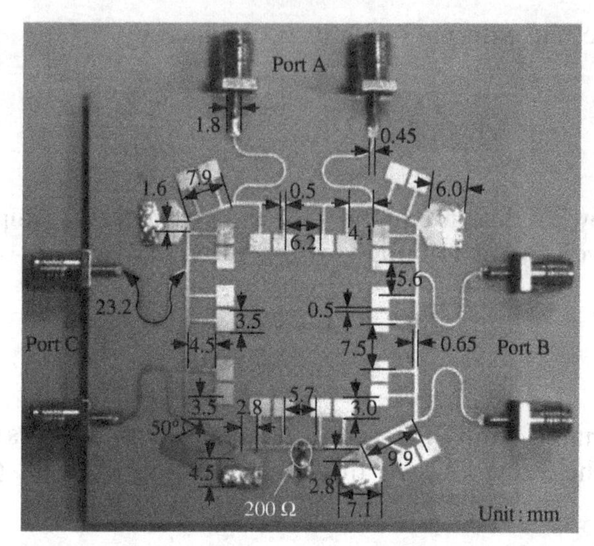

Figure 15.17 Photo of the fabricated prototype of the stub-loaded bandpass filtering balanced-to-balanced Gysel power divider. Reprinted with permission from Ref. [33]; copyright 2013 IEEE.

$R_i = 200\ \Omega$. The occupied area of the miniaturized prototype is about $0.67 \times 0.63\ \lambda_g^2$, reduced by about 50% when compared with the filtering power divider without stub loading.

The simulated and measured results are compared in Figure 15.18. Some curves are not plotted for brevity. In Figure 15.18(a), the measured differential-mode passband is centered at 2.01 GHz with the bandwidth of 7.7% and the central insertion loss of 0.86 + 3 dB, while the simulated passband is centered at 1.99 GHz with the bandwidth of 7.4% and the central insertion loss of 0.71 + 3 dB. The measured in-band differential-mode return loss is better than 15.8 dB, very close to the simulated value of 16.8 dB. The measured in-band isolation is better than 27.4 dB. The significant spurious in the upper stopband is located around 3.01 GHz, due to the frequency-dependent shrinking factor of the stub-loaded lines. It is seen from Figure 15.18(b) that the in-band magnitude difference between two outputs is within ±0.3 dB while the phase difference is within ±4°.

As shown in Figure 15.18(c), the measured common-mode transmissions between the different balanced ports are always lower than −20 dB from 1.39 to 2.45 GHz. For the measured mode conversions in Figure 15.18(d) and (e), their magnitudes are lower than −20 dB from 1.35 to 2.47 GHz. In Figure 15.18(d), good agreement is observed between the simulated and measured mode-conversion responses,

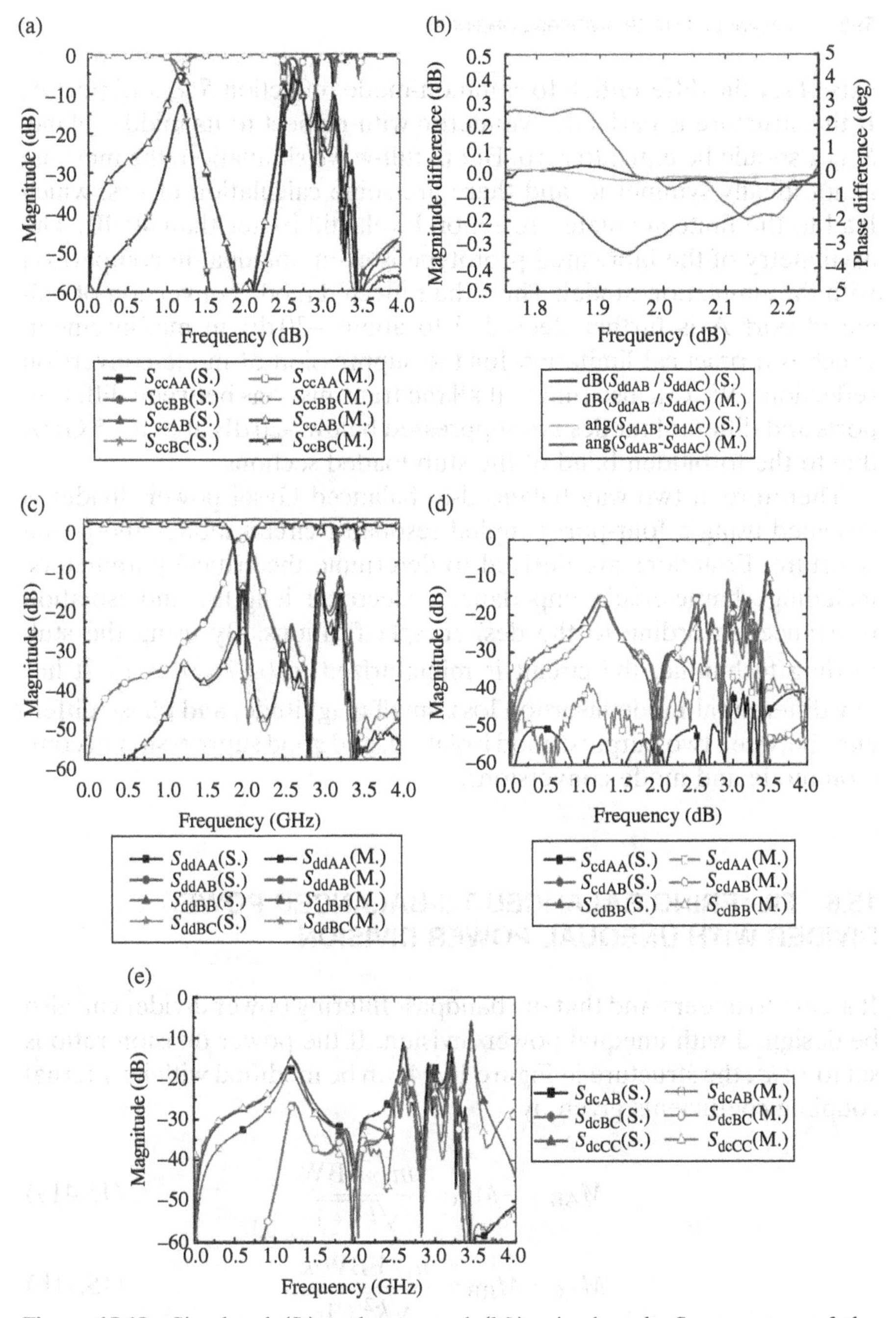

Figure 15.18 Simulated (S.) and measured (M.) mixed-mode *S*-parameters of the miniaturized bandpass filtering balanced-to-balanced Gysel power divider prototype loaded with stubs: (a) S_{ddAA}, S_{ddBB}, S_{ddAB}, and S_{ddBC}; (b) magnitude and phase differences of differential-mode transmissions; (c) S_{ccAA}, S_{ccBB}, S_{ccAB}, and S_{ccBC}; (d) S_{cdAA}, S_{cdAB}, and S_{cdBB}; and (e) S_{dcAB}, S_{dcBC}, and S_{dcCC}. M. and S. represent the measured and simulated results, respectively. Reprinted with permission from Ref. [33]; copyright 2013 IEEE.

except for the differential- to common-mode reflection S_{cdAA} of port A. If the structure is perfectly symmetric with respect to its middle plane, S_{cdAA} should be equal to zero. But in full-wave simulation, the meshing is not ideally symmetric, and there are some calculation errors, which lead to the finite simulated rejection level, still better than 40 dB. The asymmetry of the fabricated prototype is even nonideal in comparison with the simulation model. Thus, the reflected mode conversion of balanced port A is further degraded to about –30 dB in measurement, which is a practical limitation for the suppression of mode-conversion reflections. We can also find that all the transmissions between different ports and different modes are suppressed below –20 dB above 3.5 GHz, due to the forbidden band of the stub-loaded sections.

Therefore, a two-way balanced-to-balanced Gysel power divider is designed using a four-port coupled-resonator circuit model and a ring structure. Equations are derived to determine the critical parameters, including characteristic impedances, electrical lengths, and isolation resistance, according to the desired specifications. By using the stub loading technique, the circuit is miniaturized to $0.67 \times 0.63\ \lambda_g^2$. It has low differential-mode insertion loss, small magnitude, and phase difference between two outputs, good isolation, and good suppression of common mode and mode conversions.

15.6 FILTERING BALANCED-TO-BALANCED POWER DIVIDER WITH UNEQUAL POWER DIVISION

It is easy to understand that the bandpass filtering power divider can also be designed with unequal power division. If the power division ratio is set to $1 : k^2$, the structure in Figure 15.14 can be modified with its internal coupling coefficients given by

$$M_{AB} = -M_{RC} = \frac{m_{12} \cdot \mathrm{FBW}}{\sqrt{k^2+1}} \tag{15.41a}$$

$$M_{AC} = M_{RB} = \frac{m_{12} \cdot \mathrm{FBW} \cdot k}{\sqrt{k^2+1}} \tag{15.41b}$$

The critical characteristic impedances and electrical lengths of the component are then calculated by using the equations in Section 15.5.2, such as (15.33).The measured results are shown in Figure 15.19. Both the bandpass filtering response and unequal power division are clearly observed.

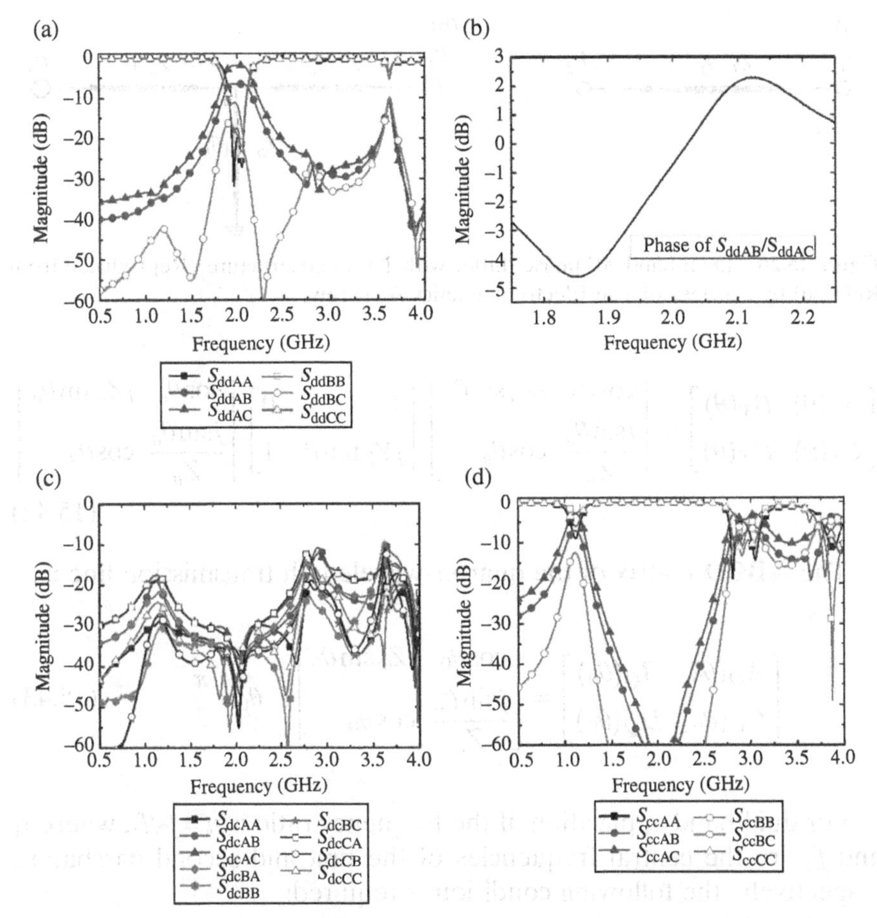

Figure 15.19 Measured mixed-mode S-parameters of the balanced-to-balanced power divider with bandpass filtering response and arbitrary power division ratio: (a) S_{dd}; (b) the phase difference between S_{ddAB} and S_{ddAB}; (c) S_{dc}; and (d) S_{cc}.

15.7 DUAL-BAND BALANCED-TO-BALANCED POWER DIVIDER

15.7.1 Analysis and Design

To further extend the applications of balanced-to-balanced power dividers, a dual-band balanced-to-balanced power divider is designed here [36]. As shown in Figure 15.20, the T-shaped unit is used to replace the quarter-wavelength transmission line [37]. By using matrix transformation, the dual frequency 90° phase shifter can be deduced for the balanced-to-balanced power divider. The S-parameter of the T-shaped unit is calculated by cascading the $ABCD$ matrices:

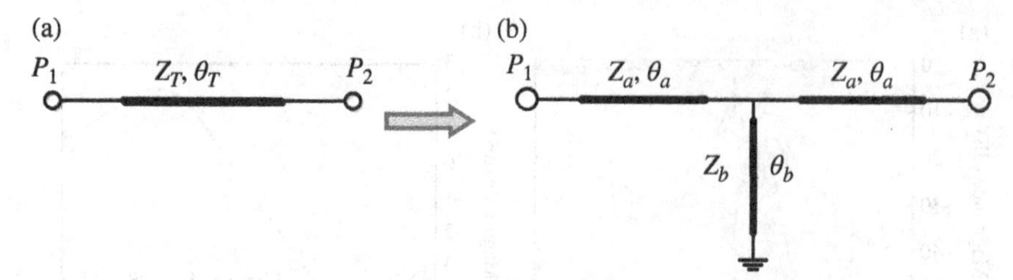

Figure 15.20 Dual-band 90° phase shifter with T-shaped structure. Reproduced from Ref. [36] by courtesy of The Electromagnetics Academy.

$$\begin{bmatrix} A_T(\theta) & B_T(\theta) \\ C_T(\theta) & D_T(\theta) \end{bmatrix} = \begin{bmatrix} \cos\theta_a & jZ_a\sin\theta_a \\ \dfrac{j\sin\theta_a}{Z_a} & \cos\theta_a \end{bmatrix} \begin{bmatrix} 1 & 0 \\ jY_b\tan\theta_b & 1 \end{bmatrix} \begin{bmatrix} \cos\theta_a & jZ_a\sin\theta_a \\ \dfrac{j\sin\theta_a}{Z_a} & \cos\theta_a \end{bmatrix} \tag{15.42}$$

The **ABCD** matrix of the quarter-wavelength transmission line is

$$\begin{bmatrix} A_Q(\theta_0) & B_Q(\theta_0) \\ C_Q(\theta_0) & D_Q(\theta_0) \end{bmatrix} = \begin{bmatrix} \cos\theta_0 & jZ_c\sin\theta_0 \\ \dfrac{j\sin\theta_0}{Z_c} & \cos\theta_0 \end{bmatrix}, \quad \theta_0 = \frac{\pi}{2} \tag{15.43}$$

For dual-band application, if the frequency ratio is $n = f_2/f_1$, where f_1 and f_2 are the central frequencies of the first and second passbands, respectively, the following condition is required:

$$\tan\theta_a = \pm\frac{Z_c}{Z_a}, \quad \tan n\theta_a = \pm\frac{Z_c}{Z_a} \tag{15.44a}$$

$$\tan\theta_b = \frac{2Z_b}{Z_a\tan 2\theta_a}, \quad \tan n\theta_b = \frac{2Z_b}{Z_a\tan 2n\theta_a} \tag{15.44b}$$

The parameters can be chosen as

$$\theta_a = \frac{\pi}{n+1}, \quad \theta_b = \frac{2\pi}{n+1}, \quad Z_a = \frac{Z_c}{\tan\theta_a}, \quad Z_b = \frac{Z_c\tan^2 2\theta_a}{2\tan\theta_a}, \quad \text{for } 1 < n < 3 \tag{15.45a}$$

$$\theta_a = \frac{\pi}{n+1}, \quad \theta_b = \frac{\pi}{n+1}, \quad Z_a = \frac{Z_c}{\tan\theta_a}, \quad Z_b = \frac{Z_c}{2}\tan 2\theta_a, \quad \text{for } n > 3 \tag{15.45b}$$

15.7.2 Results and Discussion

The T-shaped unit is used to replace the quarter-wavelength transmission lines of the balanced-to-balanced Wilkinson power divider in Ref. [20], and the dual operating frequencies are designed for WLAN applications, that is, 2.45 and 5.2 GHz, with $n = 2.12$. Then, we use (15.44a) to determine the parameters as $\theta_{a1} = \theta_{a2} = 57.65°$, $\theta_{b1} = \theta_{b2} = 115.30°$, $Z_{a1} = 22.40\ \Omega$, $Z_{b1} = 50.14\ \Omega$, $Z_{a2} = 32.41\ \Omega$, and $Z_{b2} = 62.09\ \Omega$, where "1" corresponds to the T-shaped unit with the equivalent impedance of $Z_y = 35.35\ \Omega$, while "2" corresponds to that with the equivalent impedance of $Z_x = 48\ \Omega$. As shown in Figure 15.21, the prototype is realized with microstrip lines and surface-mounted resistors. It is fabricated on an F4B substrate with a relative permittivity of $\varepsilon_r = 2.65$, a loss tangent of $\tan\delta = 0.003$, and a thickness of $h = 0.73$ mm.

The six-port S-parameters are simulated with ANSYS HFSS and measured with the VNA Agilent E5071. Figure 15.22(a) and (b) shows a comparison between the simulated and measured transmission coefficients, reflection coefficients, and isolation between the balanced ports

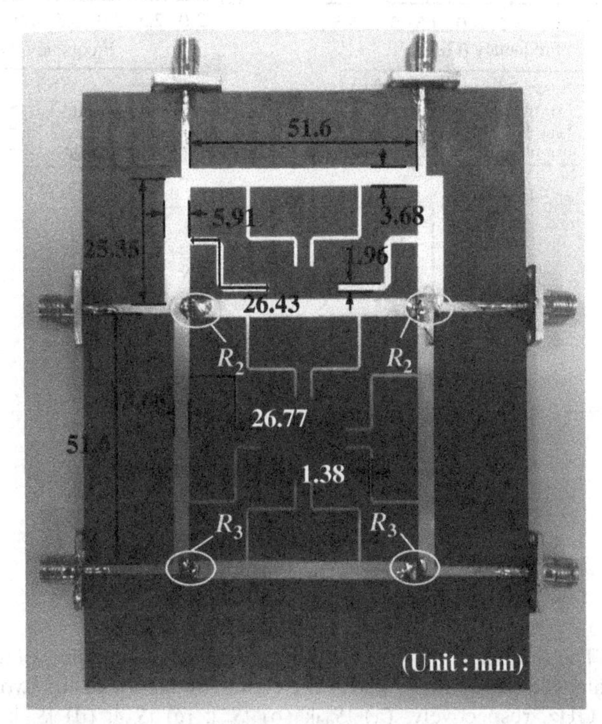

Figure 15.21 Photo of the balanced-to-balanced power divider/combiner prototype. Reproduced from Ref. [36] by courtesy of The Electromagnetics Academy.

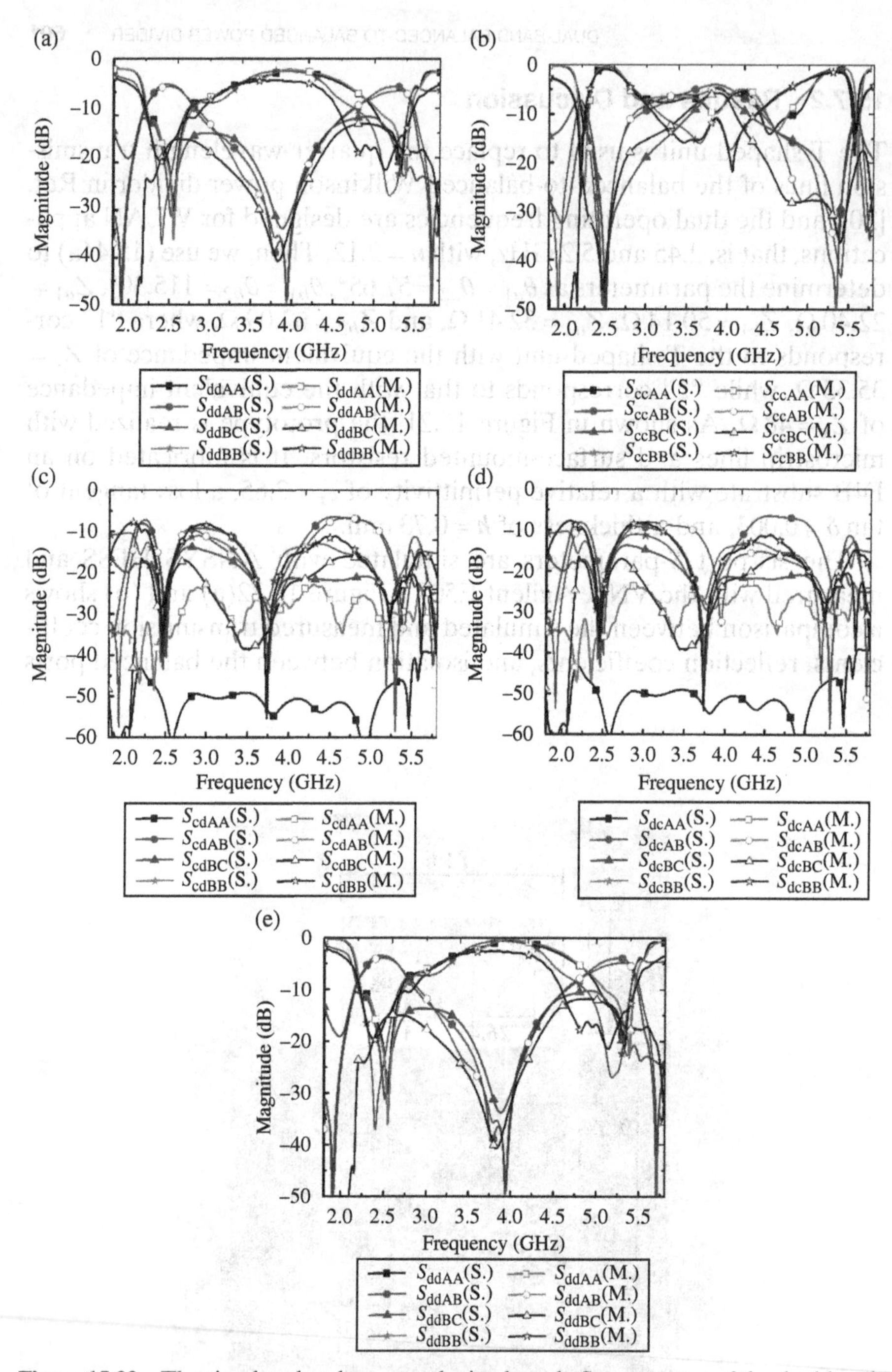

Figure 15.22 The simulated and measured mixed-mode *S*-parameters of the dual-band balanced-to-balanced power divider with the center frequency of the two passband at 2.45 and 5.25 GHz, respectively, (a) $|S_{dd}|$; (b) $|S_{cc}|$; (c) $|S_{cd}|$; (d) $|S_{dc}|$; and (e) the magnified curves of $|S_{ddAB}|$, $|S_{ccAA}|$, and $|S_{ccBB}|$. M. and S. represent the measured and simulated results, respectively. Reproduced from Ref. [36] by courtesy of The Electromagnetics Academy.

B and C for differential- and common-mode operations, respectively. Figure 15.22(c) and (d) provides a comparison between the simulated and measured differential-to-common and common-to-differential mode conversions. Figure 15.22(e) shows the magnified curves of $|S_{ddAB}|$, $|S_{ccAA}|$, and $|S_{ccBB}|$.

In the measurements, the maximum differential-mode transmission coefficient of the prototype is $|S_{ddAB}| = -3.92$ dB at 2.47 GHz, $|S_{ddAB}| = -4.76$ dB at 5.3 GHz, and the common-mode reflection coefficients of $|S_{ccAA}|$ and $|S_{ccBB}|$ reach their maximum of −0.29 and −0.84 dB at 2.51 and 2.50 GHz and reach −0.54 and −1.43 dB at 5.38 and 5.36 GHz, respectively. The first operating band is from 2.40 to 2.55 GHz, and the second band is from 5.25 to 5.30 GHz. The maximum isolation for the first and second bands are 17.8 and 17.6 dB, respectively. The difference between the simulated and measured results is mainly caused by the nonideal symmetry of the fabricated prototype, which is also due to the tolerance of fabrication and the mounting of resistors.

15.8 SUMMARY

Differential transmission lines and circuits have been utilized more and more widely, due to their significant advantages in comparison with the single-ended counterparts. As the key component in a balanced RF transceiver, the concept of balanced-to-balanced power divider/combiner is proposed, which divides the input differential-mode power into several ways with good port isolation and common-mode suppression. The balanced-to-balanced power dividers are designed in Wilkinson and Gysel topologies and realized in microstrip line and half-mode SIW. They are also designed with arbitrary power division ratio, bandpass filtering response, and dual-band operation. By using the proposed balanced-to-balanced power dividers, a power divider and three baluns, or further together with a bandpass filter, are designed collaboratively to construct only one component. Then, the system will be significantly simplified, with reduced overall circuit size and losses.

REFERENCES

1. C.-H. Wu, C.-H. Wang, and C.-H. Chen, "Novel balanced coupled-line bandpass filters with common mode noise suppression," *IEEE Trans. Microw. Theory Tech.*, vol. **55**, no. 2, pp. 287–295, Feb. 2007.

2. T.-B. Lim, and L. Zhu, "A differential mode wideband bandpass filter on microstrip line for UWB application," *IEEE Microw. Wireless Compon. Lett.*, vol. **19**, no. 10, pp. 632–634, Oct. 2009.
3. J. Shi, and Q. Xue, "Novel balanced dual-band bandpass filter using coupled stepped-impedance resonators," *IEEE Microw. Wireless Compon. Lett.*, vol. **20**, no. 1, pp. 19–21, Jan. 2010.
4. Q. Xue, J. Shi, and J.-X. Chen, "Unbalanced-to-balanced and balanced-to-unbalanced diplexer with high selectivity and common-mode suppression," *IEEE Trans. Microw. Theory Tech.*, vol. **59**, no. 11, pp. 2848–2855, Jan. 2011.
5. R. Meys, and F. Janssens, "Measuring the impedance of balanced antennas by an S-parameter method," *IEEE Antennas Propag. Mag.*, vol. **40**, no. 6, pp. 65–68, Dec. 1998.
6. Y.-P. Zhang, "Design and experiment on differentially-driven microstrip antennas," *IEEE Trans. Antennas Propag.*, vol. **55**, no. 10, pp. 2701–2708, Oct. 2007.
7. E. B. Kaldjob, B. Geck, and H. Eul, "Impedance measurement of properly excited small balanced antennas," *IEEE Antennas Wireless Propag. Lett.*, vol. **8**, no. 6, pp. 65–68, Dec. 2009.
8. R. S. Engelbrecht, and K. Kurokawa, "A wideband low noise L-band balanced transistor amplifier," *Proc. IEEE*, vol. **53**, no. 3, pp. 237–247, Mar. 1965.
9. K. Kurokawa, "Design theory of balanced transistor amplifiers," *Bell Syst. Tech. J.*, vol. **44**, pp. 1675–1698, Oct. 1965.
10. J.-D. Jin, and S. S. H. Hsu, "A 0.18 μm CMOS balanced amplifier for 24-GHz applications," *IEEE J. Solid-State Circuits*, vol. **43**, no. 2, pp. 440–445, Feb. 2008.
11. S. A. Maas, "Novel single device balanced resistive HEMT mixers," *IEEE Trans. Microw. Theory Tech.*, vol. **43**, no. 12, pp. 2863–2867, Dec. 1995.
12. P.-Y. Chiang, C.-W Su, S.-Y. Luo, R. Hu, and C.-F. Jou, "Wide-IF-band CMOS mixer design," *IEEE Trans. Microw. Theory Tech.*, vol. **58**, no. 4, pp. 831–840, Apr. 2010.
13. K. W. Kobayashi, A. K. Oki, L. T. Tran, J. C. Cowles, A. Gutierrez-Aitken, F. Yamada, T. R. Block, and D. C. Streit, "A 108-GHz InP-HBT monolithic push-push VCO with low phase noise and wide tuning bandwidth," *IEEE J. Solid-State Circuits*, vol. **34**, no. 9, pp. 1225–1232, Sep. 1999.
14. D. Baek, S. Ko, J.-G. Kim, D.-W. Kim, and S. Hong, "Ku-band InGaP-GaAs HBT MMIC VCOs with balanced and differential topologies," *IEEE Trans. Microw. Theory Tech.*, vol. **52**, no. 4, pp. 1353–1359, Apr. 2004.
15. C.-H. Wang, Y.-H. Cho, C.-S. Lin, H. Wang, C.-H. Chen, D.-C. Niu, J. Yeh, C.-Y. Lee, and J. Chern, "A 60 GHz transmitter with integrated antenna in 0.18 um SiGe BiCMOS technology," in *IEEE International Solid-State Circuit Conference Technical Digest*, San Francisco, CA, Feb. 2006, pp. 186–187.

16. Y. Wu, Y. Liu, Q. Xue, S. Li, and C. Yu, "Analytical design method of multiway dual-band planar power dividers with arbitrary power division," *IEEE Trans. Microw. Theory Tech.*, vol. **58**, no. 12, pp. 3832–3841, Dec. 2010.
17. Y. Wu, Y. Liu, and Q. Xue, "An analytical Approach for a novel coupled-line dual-band Wilkinson power divider," *IEEE Trans. Microw. Theory Tech.*, vol. **59**, no. 2, pp. 286–294, Feb. 2011.
18. A. Genc, and R. Baktur, "Dual- and triple-band Wilkinson power dividers based on composite right- and left-handed transmission lines," *IEEE Trans. Microw. Theory Tech.*, vol. **1**, no. 3, pp. 327–334, Mar. 2011.
19. J. W. May, and G. M. Rebeiz, "A 40–50-GHz SiGe 1:8 differential power divider using shielded broadside-coupled striplines," *IEEE Trans. Microw. Theory Tech.*, vol. **56**, no. 7, pp. 1575–1581, Jul. 2008.
20. B. Xia, L.-S. Wu, and J.-F. Mao, "A new balanced-to-balanced power divider/combiner," *IEEE Trans. Microw. Theory Tech.*, vol. **60**, no. 9, pp. 287–295, Sep. 2012.
21. D. E. Bockelman, and W. R. Eisenstadt, "Combined differential and common mode scattering parameters: theory and simulation," *IEEE Trans. Microw. Theory Tech.*, vol. **43**, no. 7, pp. 1530–1539, Jul. 1995.
22. J.-S. Hong, and M. J. Lancaster, *Microstrip Bandpass Filters for RF/Microwave Applications*. New York: John Wiley & Sons, Inc., 2001, ch.1.
23. U. H. Gysel, "A new *N*-way power divider/combiner suitable for high-power applications," in *IEEE MTT-S International Microwave Symposium Digest*, Pola Alto, CA, May 1975, pp. 116–118.
24. L.-S. Wu, B. Xia, and J.-F. Mao, "A half-mode substrate integrated waveguide ring for two-way power division of balanced circuit," *IEEE Microw. Wireless Compon. Lett.*, vol. **22**, no. 7, pp. 333–335, Jul. 2012.
25. D. Deslandes and K. Wu, "Integrated microstrip and rectangular waveguide in planar form," *IEEE Microw. Wireless Compon. Lett.*, vol. **11**, no. 2, pp. 68–70, Feb. 2001.
26. Q.-H. Lai, C. Fumeaux, W. Hong, and R. Vahldieck, "Characterization of the propagation properties of the half-mode substrate integrated waveguide," *IEEE Trans. Microw. Theory Tech.*, vol. **57**, no. 8, pp. 1996–2004, Aug. 2009.
27. K.-J. Song, Y. Fan, and Y.-H. Zhang, "Eight-way substrate integrated waveguide power divider with low insertion loss," *IEEE Trans. Microw. Theory Tech.*, vol. **56**, no. 6, pp. 1473–1477, Jun. 2008.
28. H.-Y. Jin and G.-J. Wen, "A novel four-way Ka-band spatial power combiner based on HMSIW," *IEEE Microw. Wireless Compon. Lett.*, vol. **18**, no. 8, pp. 515–517, Aug. 2008.
29. D.-S. Eom, J.-D. Byun, and H.-Y. Lee, "Multilayer substrate integrated waveguide four-way out-of-phase power divider," *IEEE Trans. Microw. Theory Tech.*, vol. **57**, no. 12, pp. 3469–3476, Dec. 2009.

30. S. A. Schelkunoff, "Impedance concept in wave guides," *Quar. Appl. Math.*, vol. **2**, no. 1, pp. 1–15, Apr. 1944.
31. B. Xia, L.-S. Wu, S.-W. Ren, and J.-F. Mao, "A balanced-to-balanced power divider with arbitrary power division," *IEEE Trans. Microw. Theory Tech.*, vol. **61**, no. 8, pp. 2831–2840, Aug. 2013.
32. L.-S. Wu, B. Xia, W.-Y. Yin, and J.-F. Mao, "Collaborative design of a new dual-bandpass 180° hybrid coupler," *IEEE Trans. Microw. Theory Tech.*, vol. **61**, no. 3, pp. 1053–1066, Mar. 2013.
33. L.-S. Wu, Y.-X. Guo, and J.-F. Mao, "Balanced-to-balanced Gysel power divider with bandpass filtering response," *IEEE Trans. Microw. Theory Tech.*, vol. **61**, no. 12, pp. 4052–4062, Dec. 2013.
34. G. L. Matthaei, L. Young, and E. M. T. Jones, *Microwave Filters, Impedance-Matching Networks and Coupling Structures*. Norwood, MA: Artech House, 1980.
35. H. Uchida, N. Yoneda, Y. Konishi, and S. Makino, "Bandpass directional couplers with electromagnetically-coupled resonators," in *IEEE MTT-S International Microwave Symposium Digest*, San Francisco, CA, Jun. 2006, pp. 1563–1566.
36. B. Xia and J.-F. Mao, "A new dual band balanced-to-balanced power divider," *Prog. Electromagn. Res. C*, vol. **37**, pp. 53–66, 2013.
37. Y.-L. Wu, Y.-A. Liu, Y. Zhang, J. Gao, H. Zhou and Q. Xue, "A dual band unequal Wilkinson power divider without reactive components," *IEEE Trans. Microw. Theory Tech.*, vol. **57**, no. 1, pp. 216–222, Jan. 2009.

CHAPTER 16

DIFFERENTIAL-MODE EQUALIZERS WITH COMMON-MODE FILTERING

Tzong-Lin Wu and Chiu-Chih Chou

Graduate Institute of Communication Engineering, National Taiwan University, Taipei, Taiwan

16.1 INTRODUCTION

The balanced filters described in the previous chapters are mostly for analog applications. In this chapter, we present two balanced filters for digital applications. The main difference is that the signals running on the differential pair are square waves, instead of sinusoidal waves. It is known that the power spectral density of a random digital signal with peak-to-peak V_0, unit interval T_b, system impedance Z_0, and at frequency f, is given by [1]

$$S(f) = \frac{V_0^2}{4Z_0} T_b \left(\frac{\sin \pi f T_b}{\pi f T_b} \right)^2 \tag{16.1}$$

which gives the average power per unit bandwidth (BW), with the unit (W/Hz). According to this formula, a random digital signal has

Balanced Microwave Filters, First Edition. Edited by Ferran Martín, Lei Zhu, Jiasheng Hong, and Francisco Medina.

significant spectral power from dc to several times the bit rate ($1/T_b$), namely, the signal has a very broad spectrum. Therefore, when transmitting digital signals from one point to other, we would like the channel to be ideally all-pass, in order to ensure the best quality of the digital signal. This consideration is different from the analog filters presented earlier.

The transfer function (S_{dd21}) of a digital channel is a frequency-domain description of the channel response. However, due to the broadband nature of the digital signal, from the curve of the transfer function, the designer cannot tell if this channel is good enough to meet the communication specification or not. Consequently, when evaluating the quality of a digital channel, we have to look at the "eye diagram," in addition to the transfer function [2]. A large eye opening of the eye diagram means that the difference of the voltage level between logic 1 and logic 0 is large, and thus the digital signals can be determined at the receiver with little bit error. This is another different consideration for digital application.

As an intrinsic nature of a transmission line, the frequency-dependent loss of interconnects causes higher attenuation for high-frequency components. The direct consequence of this low-pass channel is that the rising and falling time of a step input will be longer, that is, the signal takes longer time to change from one state to the other. For the eye diagram, both the eye height and eye width will become smaller after the signal passes through such channel. If the transmission line is very long, the eye might be completely closed, resulting in significant bit error rate. In such situation, the channel needs to be "equalized." This means that we put an (differential-mode) equalizer either at the transmitter side or at the receiver side to compensate this low-pass response. The concept of equalizer will be described in more detail later.

In practice, the transmission line need not be very long. For example, the CPU chip and the memory chip may be very close to each other. However, as the clock rate of digital signal soars to several GHz, the channel loss of an interconnect of several to tens of centimeters might be enough to make the eye completely closed. Moreover, the current high-speed serial links such as USB, SATA, and PCI Express are running toward tens of Gb/s [3]. In such situations, an equalizer for the channel is usually a necessity.

For digital applications, the common-mode response of a differential channel is also important. Specifically, studies [4] have shown that the common-mode current can dominate the radiated emission. The generation of common-mode current, however, is sometimes unavoidable because there are various sources such as transmitter imbalance, return-path imbalance, transmission-line length mismatch, transmission-line bend, and crosstalk. Hence, to ensure that the equipment conforms to EMC regulations, using a common-mode filter (CMF) in a differential channel to suppress (filter out) the common-mode current that causes radiated emission is a straightforward solution.

Several types of equalizers and CMFs have been discussed by many researchers [5–14]. Among various kinds of equalizers, passive equalizers have the advantages of low power consumption and wide BW up to several GHz. For example, using the concept of reflection gain [5], the signal loss can be compensated by inserting inductance or a high-impedance line between the signal trace and matched termination. Parasitic inductance from PCB may be used to replace the inductor or the high-impedance line to realize the high-pass filter [6]. In addition, the preemphasis of high-frequency components and de-emphasis of low-frequency components may be, respectively, effected by exploiting the near-end crosstalk (NEXT) and reflection from the parallel-connected branches, which can be implemented with a defected ground structure (DGS) [7, 8]. Reference [9] has studied the optimization of passive equalizers.

For the suppression of multi-GHz common-mode noise, several CMFs in DGS type with compact and low-cost structures have been proposed [10, 11] that can be integrated into a PCB. In terms of cost, the DGS type is better than other types of CMF such as ferrite-material-based common-mode chokes or CMFs implemented on a low-temperature co-fired ceramic (LTCC) process [12–14].

Although many equalizers and CMFs have been proposed, up to now it has not drawn much attention to combine equalizers with CMFs into one single circuit in order to further reduce size and cost. In this chapter, we present two innovative circuits [15, 16] with dual function of the differential-mode equalizer and common-mode filter (DME–CMF). Both designs are analyzed in even mode and odd mode, while emphases are placed on different parts. The first design is a completely new circuit topology, and we put more effort in the analysis and optimization of the odd mode, namely, the equalizer part. The common-mode response, although analytically derived, is not studied in great detail due to some physical limitations, as will be explained in due course. For the second design, its common-mode rejection is achieved by using a previously studied method. Hence little detail will be given in this part. The novelty of the second design is that, by introducing an additional shunt stub, we achieve differential-mode equalization and widen the common-mode suppression BW at the same time. The function and effect of this additional shunt stub is thus the main focus of the second design.

In the ensuing parts of this chapter, we will briefly introduce the concepts of equalizer and CMF. Then the two designs will be covered subsequently. Both designs are verified with frequency-domain and time-domain measurements, with circuit photos provided.

16.2 DESIGN CONSIDERATIONS

16.2.1 Equalizer Design

As described in the previous section, the frequency-dependent loss of interconnects causes higher attenuation for high-frequency components, which is equivalently a low-pass filtering effect. Consequently, a typical approach to achieve the equalization is to find the inverse response of the interconnect for compensation, that is, a high pass response. Figure 16.1 illustrates the basic concept of equalization using a passive equalizer. Ideally, both the magnitude and propagation velocity of the differential insertion loss (S_{dd21}) are flattened within the BW of the equalizer.

Figure 16.2 shows three typical passive equalizer topologies [2, 9]: *RC* type, *RL* type, and T junction. All these topologies offer a high-pass response that is required for an equalizer. However, the properties of these three topologies are quite different. First of all, the number of components used is two for *RL* and *RC* types, while the T junction needs five components. If the reactive components, namely, the *L* and *C*, are realized by PCB layout such as transmission lines, the required number of components would only be the number of resistors. In this case, however, the effective BW, that is, the BW in which the transmission line behaves inductively or capacitively as we designed, has to be carefully controlled. Second, although more costly, the T junction can achieve

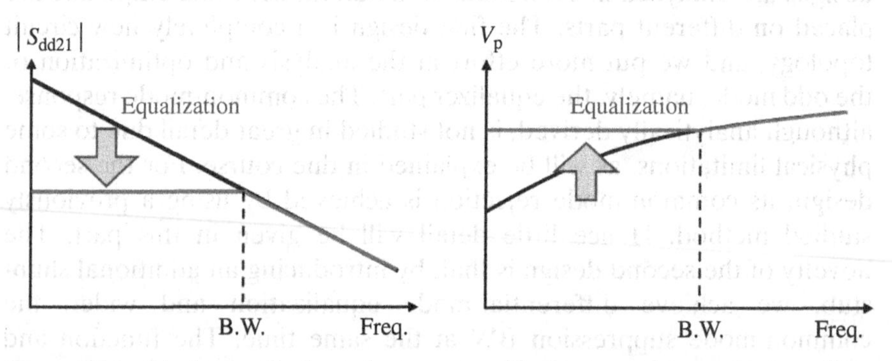

Figure 16.1 Ideal overall channel response with the use of a passive equalizer. Reprinted with permission from Ref. [15]; copyright 2013 IEEE.

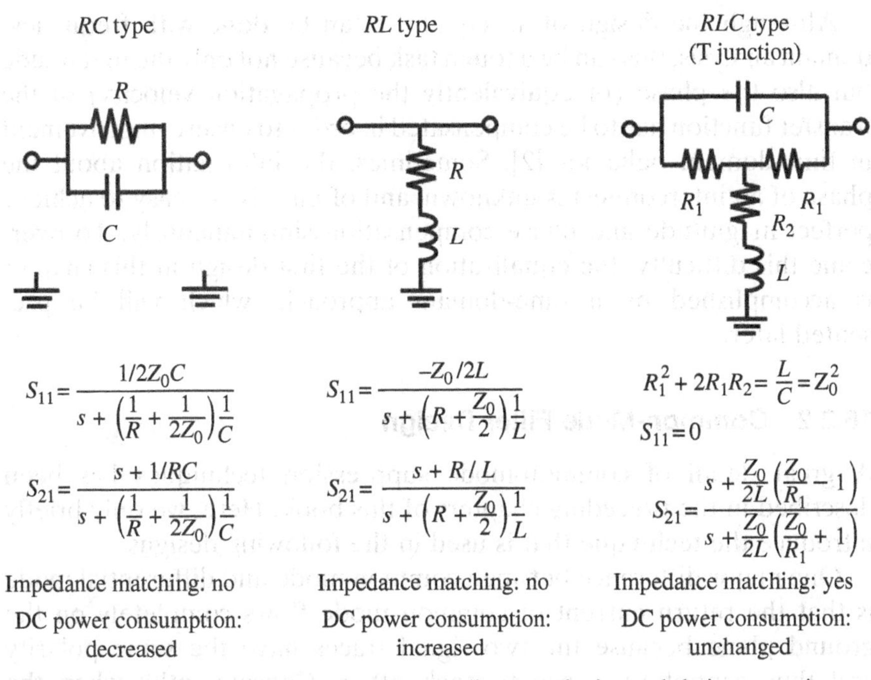

Figure 16.2 Three common passive equalizer topologies and their properties.

impedance matching by properly choosing the *RLC* values, while the *RL* and *RC* types cannot. This means that a T-junction equalizer can produce zero reflection, while an *RL*- or *RC*-type equalizer will inevitably generate reflected waves in the channel. Nevertheless, if the source and receiver ends are properly matched, this reflection will not affect the receiver eye diagram. Finally, the dc power consumptions of these three types of equalizer are different. To understand this phenomenon, we note that at dc, all inductors can be viewed as short circuit, while all capacitors can be viewed as open circuit. For *RL* type, because a resistor is connected with the port impedance of port 2 in parallel, the total impedance seen from port 1 is reduced. Therefore, more current will be drawn from the driver of port 1, thus increasing the total dc power consumption. For *RC* type, on the contrary, because a resistor is connected with the port impedance of port 2 in series, the total impedance seen from port 1 is increased. Hence, less current will be drawn from the driver of port 1, and the dc power consumption is decreased. Finally, for the T junction, if the impedance is designed to be exactly matched, then since the impedance seen from port 1 is the same, as if there were nothing added in the channel, the amount of current drawn from the port-1 driver will be the same, resulting in the same dc power consumption.

Although the design of an equalizer can be done with frequency-domain analysis, this can be a tough task because not only the magnitude but also the phase (or equivalently the propagation velocity) of the transfer function has to be compensated in order to ensure improvement in time-domain behavior [2]. Sometimes, the information about the phase of an interconnect is unknown, and often it is not easy to achieve perfect magnitude and phase compensation simultaneously. To overcome this difficulty, the equalization of the first design in this chapter is accomplished by a time-domain approach, which will be presented later.

16.2.2 Common-Mode Filter Design

A great detail of common-mode suppression techniques has been described in the preceding chapters of this book. Here, we only briefly introduce the technique that is used in the following designs.

One major difference between common mode and differential mode is that the return current of common mode flows completely on the ground plane because the two signal traces have the same polarity and thus cannot reference to each other. Consequently, when the ground plane encounters discontinuities, the common mode will be severely disrupted, while the differential mode is less affected. Thus, by properly design (etching) the ground plane below the differential traces, we can achieve various common-mode suppression performance with minimal effect on differential mode. This kind of common-mode filtering is called the "defected ground structure" or "patterned ground structure."

The simplest DGS is to etch a slotline perpendicular to the differential trace as in Ref. [10]. This slotline can be, to first order, modeled as a parallel LC resonator inserted into the return current path for common mode. At the resonance frequency of the LC resonator, the common mode sees an open in its return path, thus forming the stop band of the common mode. This is the basic principle of DGS-type CMF. In actual designs, the ground patterns are carefully arranged, for example, Ref. [11], so as to produce the desired common-mode response.

Another way of producing common-mode rejection is to add a "mushroom" structure below the differential pair as in Ref. [14]. In this method, a three-layer PCB is usually required. The differential pair runs on the first layer; a large patch is placed at the second layer below the differential pair serving as the reference plane of the transmission line between the first and second layers. The patch is connected from its center to the third layer, the system ground, by a via, thus forming a mushroomlike structure. The mushroom can be modeled simply as a parallel *LC* resonator connecting the second-layer patch and the third-layer ground as in Ref. [14], or more accurately by several other transmission lines as in Ref. [17].

The detailed design procedure for DGS and mushroom CMFs can be found in the reference papers. Here, we briefly point out some advantages and disadvantages of the DGS-type and mushroom-type CMFs. First, the DGS type only needs a two-layer PCB, while the mushroom needs three. If the whole PCB only contains two layers, then obviously the DGS type is preferred because it reduces cost. However, if the PCB has more than two layers, then the mushroom type is more feasible because the third layer will significantly affect the performance of the DGS type but not the mushroom type (it is in the design of mushroom type). Second, the analysis of DGS type is usually simpler than the mushroom type. Physical explanations for the common-mode suppression are then easier to identify for DGS type. On the other hand, due to the complexity of mushroom type, a variety of possible common-mode responses may be realized by mushroom type within the same PCB area. The analysis of mushroom type, however, is more complicated, and sometimes a computer program is needed to find out the proper design values. Third, a potential problem for the DGS type is that it may cause unwanted common-mode emission due to the ground pattern. It is well known that a half-wavelength slotline is an efficient radiator. Since the digital signal is broadband and contains almost all frequency components, care should be taken for this problem. The mushroom type, however, does not have this issue.

In the following designs, the first one is more like the DGS type, while the second one is the mushroom type. For the first one, a slight difference from the usual DGS type is that the differential pair is connected to the ground by a via. This is due to the DME design. Another difference is that the added resistor, originally for the de-emphasis of differential mode, also contributes to the common-mode suppression. Specifically, part of the common-mode noise will be reflected to the source, while part of it will be absorbed by the resistor. This part will be clearer when we see the measurement results later.

16.3 FIRST DESIGN

16.3.1 Proposed Topology

The proposed DME–CMF and its equivalent circuit model are shown in Figure 16.3(a) and (b), respectively. The differential line, of length l_1,

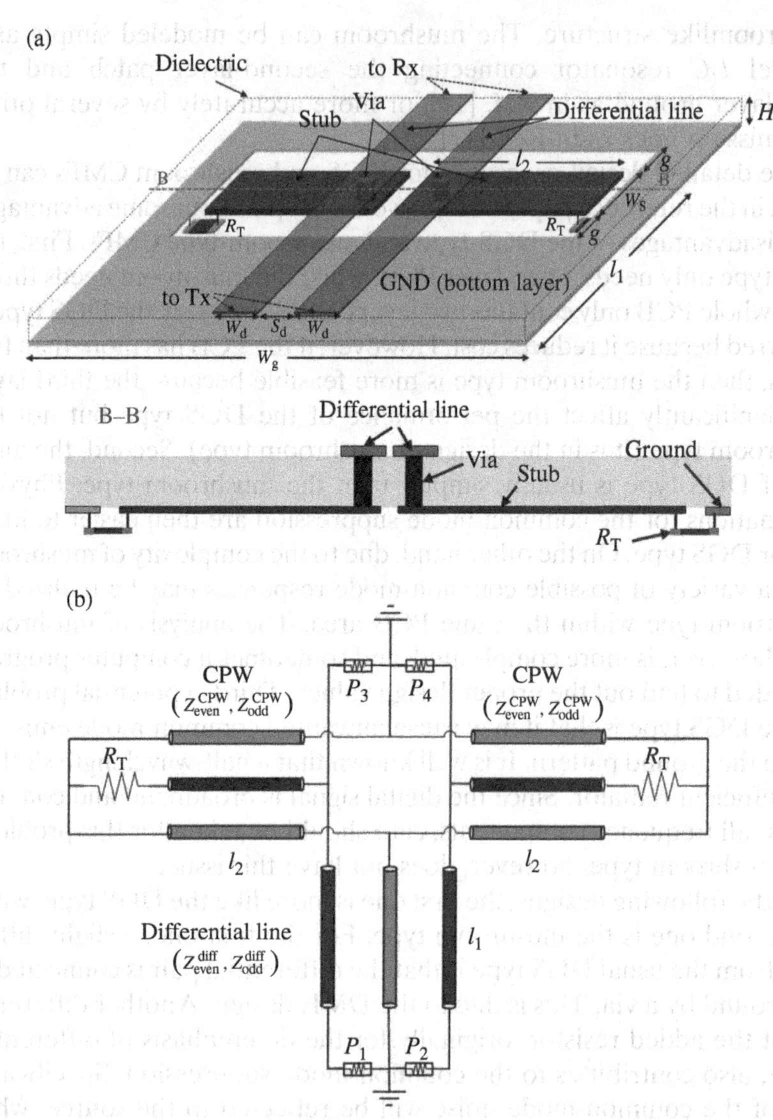

Figure 16.3 (a) Physical configuration of a two-layer structure realizing the DME–CMF and (b) the equivalent circuit model of the proposed DME–CMF. Reprinted with permission from Ref. [15]; copyright 2013 IEEE.

has odd- and even-mode characteristic impedances Z_{odd}^{diff} and Z_{even}^{diff}, respectively, while the CPW, of length l_2 on each side of the differential pair, has odd- and even-mode characteristic impedances Z_{odd}^{CPW} and Z_{even}^{CPW}, respectively. The CPW is the primary part of the proposed DME–CMF. The main advantage of using CPW is that it can provide a large value of $Z_{\mathrm{even}}^{\mathrm{CPW}}$, which will be proved essential later in time-domain analysis.

In this section, we will frequently refer to the "even" and "odd" modes of CPW structure. For clarifying these terms, the field distributions of these two modes are depicted in Figure 16.4(a) and (b) [18], respectively. The even mode is the usual quasi-TEM mode of CPW transmission line, in which the voltages of the two side conductors are equal. On the other hand, the odd mode can be thought of as a "coupled-slotline" mode. If we denote the characteristic impedance of the quasi-TEM CPW transmission line as Z_0, then $Z_{\mathrm{even}}^{\mathrm{CPW}}$ equals $2Z_0$, because by "even" and "odd" we refer to the half circuit. For even mode, the current is halved when considering half circuit, while the voltage remains the same. This results in the previous relation.

In the following, we will begin our analysis based on the equivalent model instead of the physical design. Later in Section 16.3.4, we will go back to see the performance of the actual circuit.

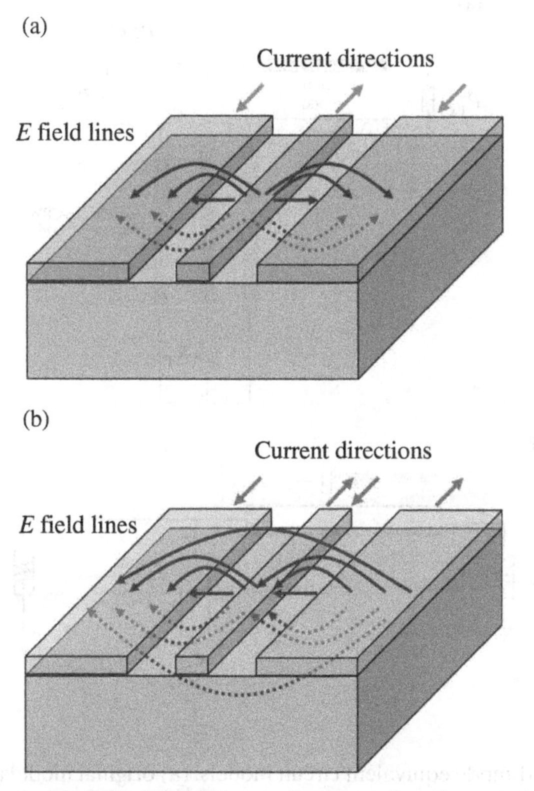

Figure 16.4 The electric field distribution and current directions of the (a) even mode and (b) odd mode of CPW structure.

16.3.2 Odd-Mode Analysis

With a perfect electric conductor (PEC) plane inserted at the symmetric plane of the differential line in Figure 16.3(b), shorting the ground associated with the differential pair and the ground associated with the CPW, we can obtain the odd-mode equivalent circuit model shown in Figure 16.5(a), referred to as the original model here, and the simplified model shown in Figure 16.5(b) replacing the CPW in the original model with a single-ended transmission line with characteristic impedance equal to $Z_{\mathrm{even}}^{\mathrm{CPW}}/2$. From the simplified model, it is apparent that the CPW shows inductive behavior and thus the CPW with the resistor R_{T} serves as an RL-type high-pass filter within the frequency range f satisfying

$$0 < f < \frac{c}{4l_2\sqrt{\varepsilon_{\mathrm{even}}^{\mathrm{CPW}}}} \tag{16.2}$$

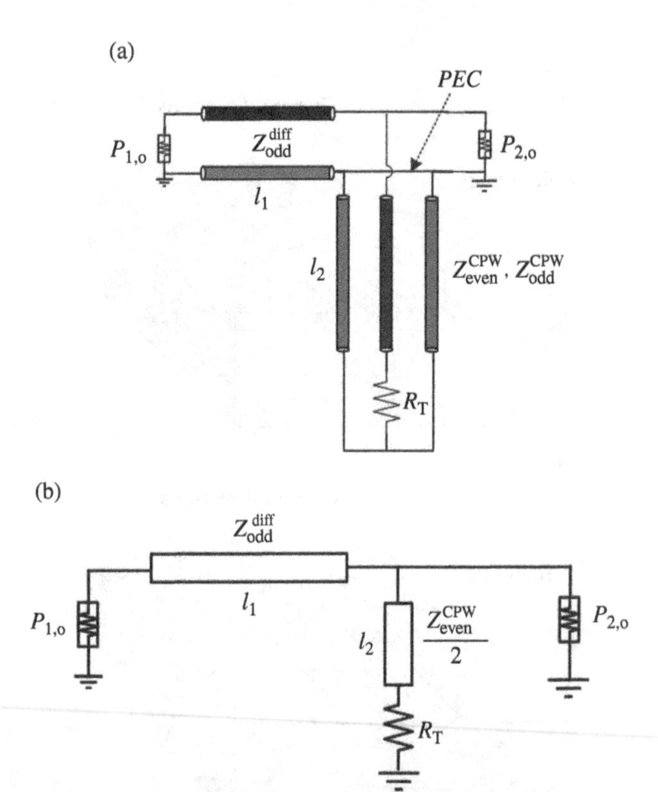

Figure 16.5 Odd-mode equivalent circuit models: (a) original model and (b) simplified model with the CPW modeled by a single-ended transmission line and the ground lines omitted. Reprinted with permission from Ref. [15]; copyright 2013 IEEE.

where $\varepsilon_{\text{even}}^{\text{CPW}}$ denotes the even-mode effective dielectric constant of the CPW. This frequency range is also the BW of the proposed equalizer.

16.3.2.1 *Equalizer Optimization in Time Domain* To analyze the proposed equalizer, we assume that the DME–CMF is placed just before the receiver, after a long differential pair of package and PCB traces. In view of the odd-mode equivalent circuit model shown in Figure 16.5(b), it is clear that given the geometry of the lossy differential pair preceding the DME–CMF, the DME can be optimized by proper selections of l_2, $Z_{\text{even}}^{\text{CPW}}$, and R_{T}. This part will be done with time-domain analysis examining the step response of the overall system including the differential pair and the DME–CMF. The common-mode behavior is ignored for the moment.

The odd-mode equivalent circuit model including both the lossy differential pair and the proposed equalizer is shown in Figure 16.6, with the length, characteristic impedance, and attenuation constant of the lossy differential line denoted as l_1, $Z_{\text{odd}}^{\text{diff}}$, and α, respectively, and matched source/load impedance assumed. The lossy effect of the differential line is considered the major cause of signal attenuation because its length l_1 is in general much greater than l_2, the length of the CPW; consequently the signal distortion due to the CPW is neglected. In addition, only the dielectric loss α_{d} is considered, and the conductor loss α_{c} is ignored for simplicity, that is, $\alpha = \alpha_{\text{d}}$.

Next, we are going to formulate the step response $v_{\text{R}}(t)$, which is the voltage at node A where the superposition of multiple reflections from the CPW occurs. We use the idealized trapezoidal shape to express the input step function $v_{\text{s}}(t)$, with rise time t_{r} and amplitude V_{h}, as

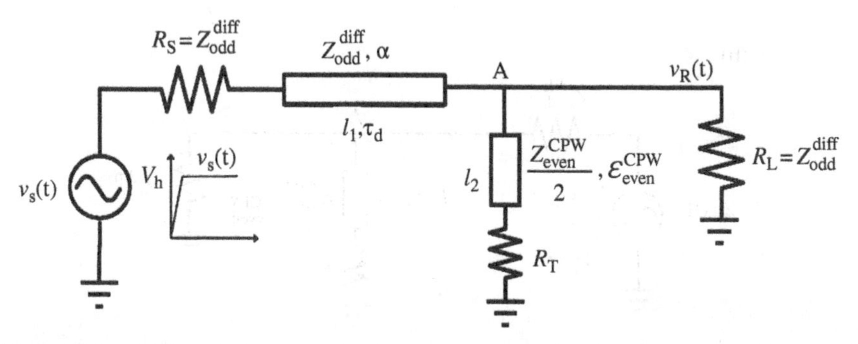

Figure 16.6 Odd-mode half circuit of overall transmission-line system for differential-mode optimization. Reprinted with permission from Ref. [15]; copyright 2013 IEEE.

$$v_s(t) = V_h\left[\frac{t}{t_r}u(t) - \frac{t-t_r}{t_r}u(t-t_r)\right] \tag{16.3}$$

where $u(t)$ is the unit step function. The input step $v_s(t)$, after being launched, experiences distortion when traveling down the transmission line of length l_1 and time delay τ_d and becomes a deformed step function $v_A(t)$ immediately before it arrives at node A such that the reflected and the transmitted wave have not yet been set up and the DME–CMF and R_L have not yet been seen. The physical meaning of $v_A(t)$, as illustrated in Figure 16.7(a), is that $v_A(t)$ is equivalent to the voltage at node A when the DME–CMF and R_L are replaced with an infinitely long transmission line matched to the existing differential line. To obtain the expression for the deformed step function, we need to compute the convolution between $v_s(t)$ and the impulse response of the l_1 long lossy line, which is given by the Equation (19) of [19]:

$$h(t) = \frac{B}{\pi(B^2+t^2)}u(t) \tag{16.4a}$$

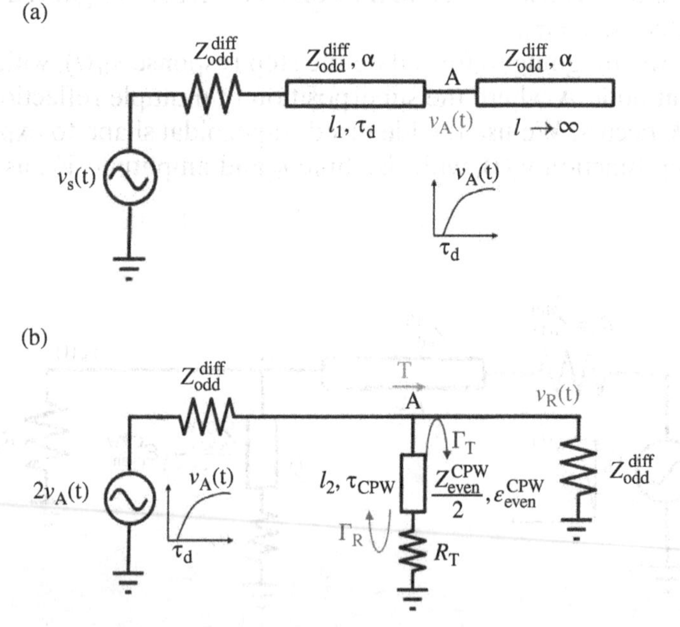

Figure 16.7 (a) Physical meaning of $v_A(t)$. (b) Remove the lossy transmission line in (a) and revise the input. Reprinted with permission from Ref. [15]; copyright 2013 IEEE.

where B was defined in Ref. [19] by

$$B = G_d Z_0 \frac{l_1}{2} \tag{16.4b}$$

Note that B is roughly of the order 10^{-9}, as given in Ref. [19]. The attenuation constant is related to G_d by

$$\alpha_d = \frac{G_d Z_0}{2} \omega \tag{16.5}$$

Thus, we could also express B as

$$B = \frac{\alpha_d l_1}{2\pi f} \tag{16.6}$$

Convolution of the first part of $v_s(t)$ in (16.3) with the $h(t)$ earlier gives

$$V_h \frac{t}{t_r} u(t) \otimes h(t) = \frac{V_h}{t_r} \int_0^{\infty} (t-\tau) u(t-\tau) h(\tau) d\tau = \frac{V_h}{t_r} \int_0^{t} (t-\tau) \frac{B}{\pi(B^2+\tau^2)} d\tau$$

$$= \frac{V_h B}{\pi t_r} \int_0^{t} \frac{t}{B^2+\tau^2} - \frac{\tau}{B^2+\tau^2} d\tau = \frac{V_h B}{\pi t_r} \left[\frac{t}{B} \tan^{-1} \frac{t}{B} - \frac{1}{2} \ln\left(\frac{B^2+t^2}{B^2} \right) \right] \tag{16.7}$$

The second part of $v_s(t)$ is different by only a time shift. Thus the convolution is similar. In (16.4a), we neglected the channel time delay τ_d. To include the delay, all the t in (16.7) should be changed to $(t-\tau_d)$. Together, we obtain

$$v_A(t) = \frac{V_h B}{\pi t_r} [X(t) u(t-\tau_d) - X(t-t_r) u(t-\tau_d-t_r)] \tag{16.8a}$$

where

$$X(t) = \frac{t-\tau_d}{B} \tan^{-1}\left(\frac{t-\tau_d}{B} \right) - \frac{1}{2} \ln\left(\frac{(t-\tau_d)^2 + B^2}{B^2} \right) \tag{16.8b}$$

To analyze the response at node A, denoted as $v_R(t)$, we can simplify the circuit by eliminating the lossy differential line, while keeping the same reflection mechanism at the junction, and changing the input step $v_s(t)$ to a new voltage source equal to $2v_A(t)$ activated at $t = \tau_d$, as shown in the resulting model (Figure 16.7b). Three coefficients are defined at the junction: the transmission coefficient seen from the source impedance toward node A, the reflection coefficient looking into R_T, and the reflection coefficient seen from the CPW toward node A, denoted as T, Γ_R, and Γ_T, respectively, as given by

$$T = \frac{2\left(Z_{\text{odd}}^{\text{diff}} \| \left(Z_{\text{even}}^{\text{CPW}}/2\right)\right)}{\left(Z_{\text{odd}}^{\text{diff}} \| \left(Z_{\text{even}}^{\text{CPW}}/2\right)\right) + Z_{\text{odd}}^{\text{diff}}} = \frac{Z_{\text{even}}^{\text{CPW}}}{Z_{\text{even}}^{\text{CPW}} + Z_{\text{odd}}^{\text{diff}}} \tag{16.9a}$$

$$\Gamma_R = \frac{R_T - \left(Z_{\text{even}}^{\text{CPW}}/2\right)}{R_T + \left(Z_{\text{even}}^{\text{CPW}}/2\right)} = \frac{2R_T - Z_{\text{even}}^{\text{CPW}}}{2R_T + Z_{\text{even}}^{\text{CPW}}} \tag{16.9b}$$

$$\Gamma_T = \frac{\left(Z_{\text{odd}}^{\text{diff}}/2\right) - \left(Z_{\text{even}}^{\text{CPW}}/2\right)}{\left(Z_{\text{odd}}^{\text{diff}}/2\right) + \left(Z_{\text{even}}^{\text{CPW}}/2\right)} = \frac{Z_{\text{odd}}^{\text{diff}} - Z_{\text{even}}^{\text{CPW}}}{Z_{\text{odd}}^{\text{diff}} + Z_{\text{even}}^{\text{CPW}}} \tag{16.9c}$$

The multiple reflections on the CPW, superposed at node A, can be fully determined by these three factors. Here, it is noted that for the equalizer to achieve de-emphasis for low-frequency components, both Γ_R and Γ_T should be less than 0, which can be fulfilled when R_T and $Z_{\text{odd}}^{\text{diff}}$ are both smaller than $Z_{\text{even}}^{\text{CPW}}/2$, according to (16.9b) and (16.9c).

The multiple reflection details are illustrated in Figure 16.8. First, the voltage wave $v_A(t)$ arrives at node A at the time $t = \tau_d^-$, and then two identical transmitted waves $v_1(t)$ are built up: one becomes a part of

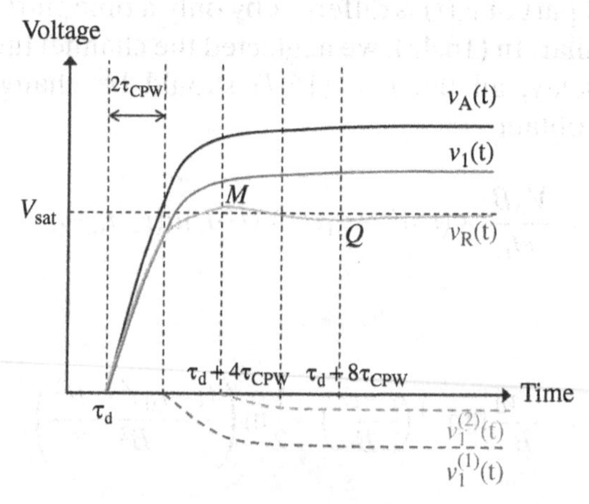

Figure 16.8 Waveform and superposition mechanism at node A. Reprinted with permission from Ref. [15]; copyright 2013 IEEE.

the step response $v_R(t)$ and the other travels down the CPW, with $v_1(t)$ given by

$$v_1(t) = T \cdot v_A(t). \tag{16.10}$$

The voltage wave that goes down the CPW produces a reflected wave at the CPW-R_T junction, which then flies back to node A and contributes to another part of $v_R(t)$ with waveform $v_1^{(1)}(t)$, as given by

$$v_1^{(1)}(t) = \Gamma_R(1+\Gamma_T)\cdot v_1(t-2\tau_{CPW}) = T\Gamma_R(1+\Gamma_T)\cdot v_A(t-2\tau_{CPW}) \tag{16.11}$$

where τ_{CPW} denotes the one-way time delay of the CPW. The voltage wave arriving at node A from the CPW will undergo reflection again due to impedance mismatch. The next waveform $v_1^{(2)}(t)$ that constitutes the total step response $v_R(t)$ is given by

$$v_1^{(2)}(t) = \Gamma_R^2\Gamma_T(1+\Gamma_T)\cdot v_1(t-4\tau_{CPW}) \tag{16.12}$$

Note that since $\Gamma_R < 0$ and $\Gamma_T < 0$, we have $v_1^{(1)}(t) < 0$ and $v_1^{(2)}(t) < 0$ after they take on values. Following this analysis, we can express the subsequent reflected waves that constitute parts of $v_R(t)$ as, for $n \in \mathbb{N}$,

$$v_1^{(n)}(t) = \Gamma_R^n\Gamma_T^{n-1}(1+\Gamma_T)\cdot v_1(t-2n\tau_{CPW}) \tag{16.13}$$

Finally, the step response $v_R(t)$ can be obtained as

$$v_R(t) = v_1(t) + \sum_{n=1}^{\infty} v_1^{(n)}(t) \tag{16.14}$$

The steady-state value V_{sat} for the step response is given by

$$V_{sat} = V_h \frac{Z_{odd}^{diff} || R_T}{Z_{odd}^{diff} || R_T + Z_{odd}^{diff}} = \frac{V_h R_T}{2R_T + Z_{odd}^{diff}} \tag{16.15}$$

which is obtained by simply replacing the CPW transmission lines with shorting wires.

Having the expression for the step response, we can then optimize the design parameters l_2, Z_{even}^{CPW}, and R_T, of the DME. First, according to Ref. [8], the optimal CPW length l_2 should be selected such that $2\tau_{\mathrm{CPW}}$ equals 0.5 UI (UI is the unit interval) in order to provide a BW up to the data rate of the digital signals. Specifically, setting $f = 1/\mathrm{UI}$ in (16.2), we obtain the formula for l_2:

$$l_2 = \frac{c \cdot \mathrm{UI}}{4\sqrt{\varepsilon_{\mathrm{even}}^{\mathrm{CPW}}}} \tag{16.16}$$

Next, the two remaining parameters, $Z_{\mathrm{even}}^{\mathrm{CPW}}$ and R_{T}, can be determined using the eye height estimation method of Ref. [20] based on the step response $v_{\mathrm{R}}(t)$.

Consider a typical "tritonic" [20] step response $v_{\mathrm{R}}(t)$ as shown in Figure 16.8. It possesses a local maximum above V_{sat} at point M when $t = \tau_{\mathrm{d}} + 4\tau_{\mathrm{CPW}}$, followed by a local minimum of V_{Q} below V_{sat} at point Q when $t = \tau_{\mathrm{d}} + 8\tau_{\mathrm{CPW}}$. From (16.14), $v_{\mathrm{R}}(t)$ will continuously increase from $t = \tau_{\mathrm{d}}$ to $t = \tau_{\mathrm{d}} + 4\tau_{\mathrm{CPW}}$ if $v_1^{(1)}(t)$ decreases more slowly than $v_1(t)$ increases from $t = \tau_{\mathrm{d}} + 2\tau_{\mathrm{CPW}}$ to $t = \tau_{\mathrm{d}} + 4\tau_{\mathrm{CPW}}$. After that, $v_{\mathrm{R}}(t)$ continuously decreases until $t = \tau_{\mathrm{d}} + 8\tau_{\mathrm{CPW}}$ if $v_1^{(1)}(t) + v_1^{(2)}(t)$ decreases faster than $v_1(t)$ increases from $t = \tau_{\mathrm{d}} + 4\tau_{\mathrm{CPW}}$ to $t = \tau_{\mathrm{d}} + 8\tau_{\mathrm{CPW}}$. Finally, $v_{\mathrm{R}}(t)$ slowly saturates to V_{sat}. For a linear channel with this step response, the eye height can be estimated by [20]

$$V_{\mathrm{eye}} = 2V_{\mathrm{Q}} - V_{\mathrm{sat}} \tag{16.17}$$

Fixing R_{T} and thus V_{sat} for the moment, we see that the eye height (16.17) can be maximized if V_{Q} is maximized, which, as an intuitive way, can be achieved by choosing a larger $Z_{\mathrm{even}}^{\mathrm{CPW}}$ to increase the transmission coefficient T according to (16.9a), (16.10), and (16.14). As a result, we suggest to choose the parameter $Z_{\mathrm{even}}^{\mathrm{CPW}}$ as large as possible under the limit of the manufacturing process. The last parameter R_{T} is optimized by, with the predetermined l_2 and $Z_{\mathrm{even}}^{\mathrm{CPW}}$, sweeping R_{T} through a feasible interval and choosing the value that gives the largest eye height using (16.14) and (16.17).

Consider an example as listed in Table 16.1, where $l_1 = 800$ mm and data rate = 8 Gb/s. Before optimization (l_2, $Z_{\mathrm{even}}^{\mathrm{CPW}}$, R_{T}) = (8 mm, 150 Ω,

Table 16.1 Parameters for Odd-Mode *S*-Parameter Simulation

Parameters	$Z_{\mathrm{odd}}^{\mathrm{diff}}$	$Z_{\mathrm{even}}^{\mathrm{diff}}$	$Z_{\mathrm{odd}}^{\mathrm{CPW}}$	$Z_{\mathrm{even}}^{\mathrm{CPW}}$	l_1	l_2	R_{T}
Value	50 Ω	100 Ω	70 Ω	150 Ω → 200 Ω	800 mm	8 mm → 6.3 mm	25 Ω → 38 Ω

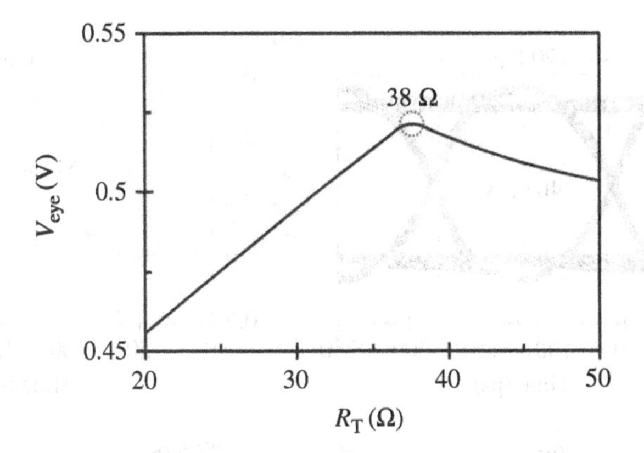

Figure 16.9 Design curve of R_T based on predetermined l_2 and Z_{even}^{CPW}. Reprinted with permission from Ref. [15]; copyright 2013 IEEE.

25 Ω). According to the design procedure, l_2 is 6.3 mm from (16.16), Z_{even}^{CPW} is chosen as 200 Ω, and R_T is determined by parameter sweep as demonstrated in Figure 16.9, where a maximum eye height V_{max} = 517 mV is obtained when R_T = 38 Ω. Consequently, after optimization, (l_2, Z_{even}^{CPW}, R_T) = (6.3 mm, 200 Ω, 38 Ω).

The eye diagrams before optimization, after optimization, and without the DME–CMF, as shown in Figure 16.10 obtained from ADS time-domain simulation using an 8 Gb/s PRBS source with amplitude 2 V, demonstrate an eye height improvement from 465 to 514 mV and an eye width improvement from 120.8 to 122.0 ps and that both cases are much better than the case without DME–CMF as in Figure 16.10(c), which verifies the effectiveness of the proposed optimization procedure.

16.3.3 Even-Mode Analysis

Two even-mode equivalent circuit models are shown in Figure 16.11, where the complete model (a) is quite similar to the odd-mode equivalent circuit model as in Figure 16.5(a) except that the PEC at the symmetric plane of the differential pair is replaced by a perfect magnetic conductor (PMC) and thus the two grounds associated with the CPW are not connected at the junction. The simplified model (b) is obtained by omitting the grounds of the transmission lines in Figure 16.11(a). The model indicates that the DME–CMF is connected in series between the lossy transmission line and the load (represented by a port) under common-mode transmission.

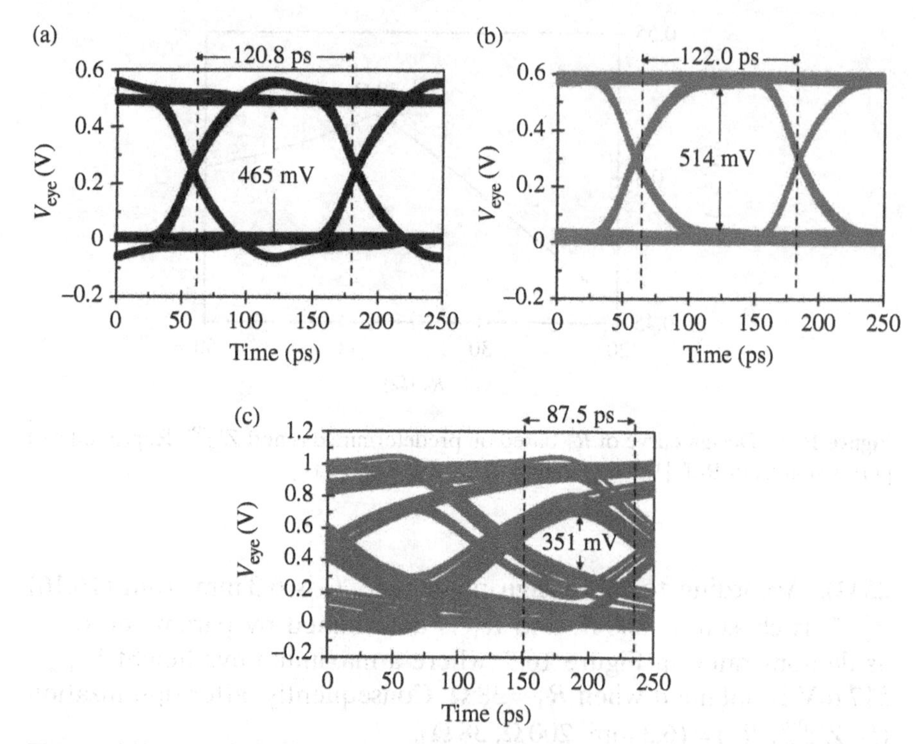

Figure 16.10 Eye diagrams from ADS circuit simulation. (a) Non-optimized case. (b) Optimized case. (c) Without the DME–CMF. Reprinted with permission from Ref. [15]; copyright 2013 IEEE.

Based on the even-mode equivalent circuit and the parameters in Table 16.1, we perform S-parameter simulation in ADS and observe the even-mode response. The results are shown in Figure 16.12. From the curve of S_{cc21}, it is seen that the proposed DME–CMF provides broadband filtering effect on common-mode noise.

Next, we will analyze the common-mode response based on the equivalent circuit model. Formula for S_{cc21} will be derived, and two important points—the dc loss and minimum of $|S_{cc21}|$—are highlighted. However, as will be discussed later, the analysis presented here does not provide a simple way for the design of the common-mode response.

For simplicity, all the transmission-line losses of the model are neglected. To begin with, the CPW with R_T is regarded as a two-port network, and the S-parameter of this network is derived. Exploiting the symmetry of the CPW, the coupled line used to model the CPW is decomposed into the corresponding even mode and odd mode, as illustrated in Figure 16.13. The input impedances for the even mode and the odd mode can be expressed as

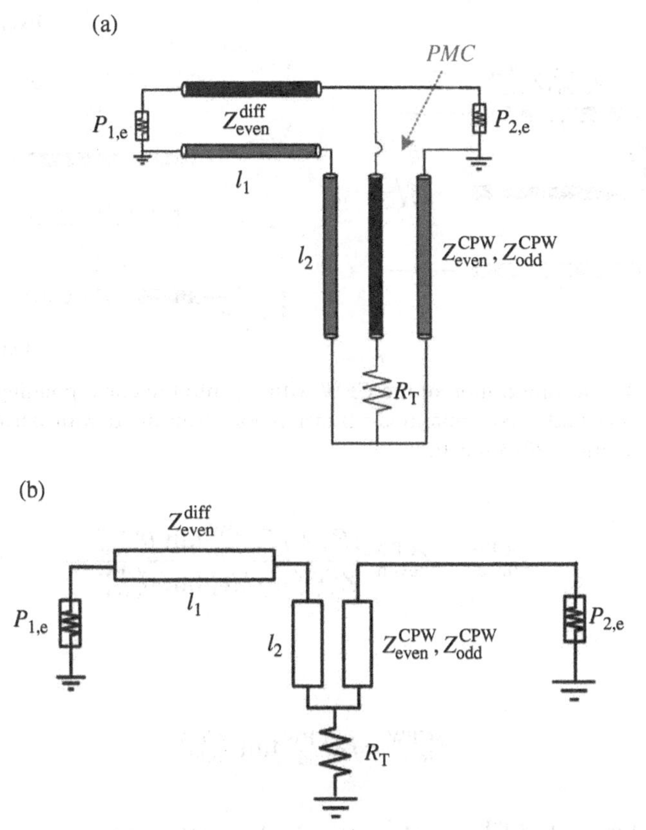

Figure 16.11 Even-mode equivalent circuit models: (a) original model and (b) simplified model with the grounds of the transmission lines omitted. Reprinted with permission from Ref. [15]; copyright 2013 IEEE.

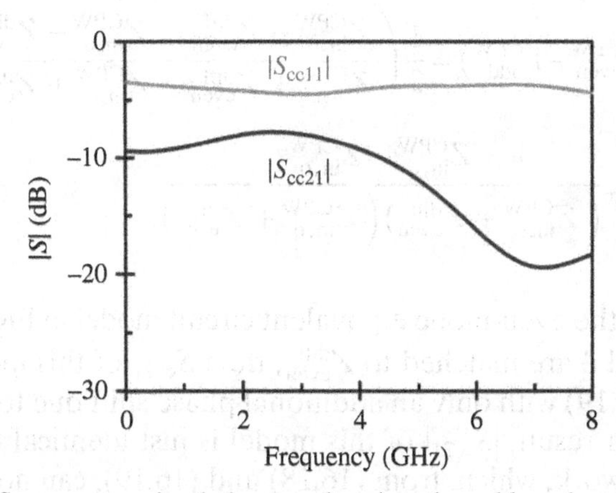

Figure 16.12 *S*-parameter simulation result using the odd-mode equivalent circuit model. Reprinted with permission from Ref. [15]; copyright 2013 IEEE.

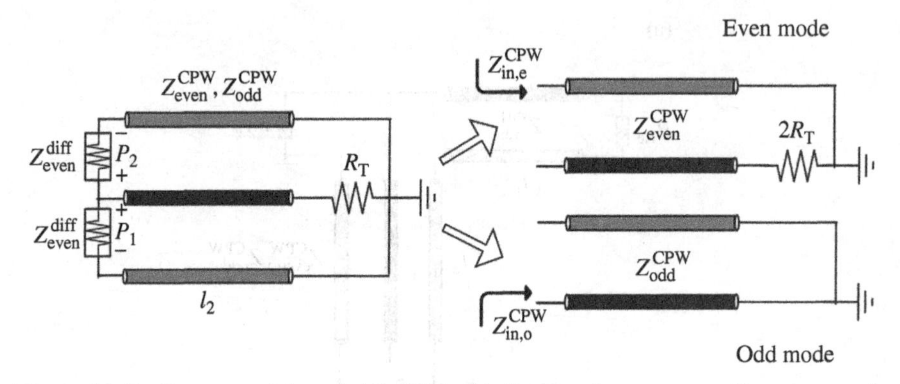

Figure 16.13 Decomposition of the CPW with R_{T} into the corresponding even mode and odd mode under common-mode transmission. Reprinted with permission from Ref. [15]; copyright 2013 IEEE.

$$Z_{\text{in,e}}^{\text{CPW}} = Z_{\text{even}}^{\text{CPW}} \frac{2R_{\text{T}} + j\, Z_{\text{even}}^{\text{CPW}} \tan \theta_{\text{even}}^{\text{CPW}}}{Z_{\text{even}}^{\text{CPW}} + j2R_{\text{T}} \tan \theta_{\text{even}}^{\text{CPW}}} \tag{16.18a}$$

and

$$Z_{\text{in,o}}^{\text{CPW}} = jZ_{\text{odd}}^{\text{CPW}} \tan \theta_{\text{odd}}^{\text{CPW}} \tag{16.18b}$$

where $\theta_{\text{even}}^{\text{CPW}}$ and $\theta_{\text{odd}}^{\text{CPW}}$ are the electrical lengths of the even mode and odd mode of the CPW, respectively. Then, the S_{cc21} of this two-port network is given by

$$\begin{aligned} S_{\text{cc21}} &= \frac{1}{2}\left(\Gamma_{\text{even}}^{\text{CPW}} - \Gamma_{\text{odd}}^{\text{CPW}}\right) = \frac{1}{2}\left(\frac{Z_{\text{in,e}}^{\text{CPW}} - Z_{\text{even}}^{\text{diff}}}{Z_{\text{in,e}}^{\text{CPW}} + Z_{\text{even}}^{\text{diff}}} - \frac{Z_{\text{in,o}}^{\text{CPW}} - Z_{\text{even}}^{\text{diff}}}{Z_{\text{in,o}}^{\text{CPW}} + Z_{\text{even}}^{\text{diff}}}\right) \\ &= Z_{\text{even}}^{\text{diff}} \frac{Z_{\text{in,e}}^{\text{CPW}} - Z_{\text{in,o}}^{\text{CPW}}}{\left(Z_{\text{in,e}}^{\text{CPW}} + Z_{\text{even}}^{\text{diff}}\right)\left(Z_{\text{in,o}}^{\text{CPW}} + Z_{\text{even}}^{\text{diff}}\right)} \end{aligned} \tag{16.19}$$

Return to the even-mode equivalent circuit model in Figure 16.11(a). If ports 1 and 2 are matched to $Z_{\text{even}}^{\text{diff}}$, then S_{cc21} of this model is determined by (16.19) with only an additional phase shift due to the differential pair. As a result, $|S_{\text{cc21}}|$ of this model is just identical to that of the two-port network, which, from (16.18) and (16.19), can now be directly related to $Z_{\text{even}}^{\text{CPW}}$, $Z_{\text{odd}}^{\text{CPW}}$, l_2, and R_{T}.

At dc, $|S_{cc21}|$ can be simplified to

$$|S_{cc21}(f=0)| = \left|\frac{2R_T}{2R_T + Z_{even}^{diff}}\right| \tag{16.20}$$

because $Z_{in,e}^{CPW} = 2R_T$ and $Z_{in,o}^{CPW} = 0$. At the frequency such that $\theta_{odd}^{CPW} = \pi/2$ (l_2 equals to quarter wavelength for the odd mode of the CPW), $|S_{cc21}|$ can be approximated as

$$\left|S_{cc21}\left(\theta_{odd}^{CPW} = \frac{\pi}{2}\right)\right| = \left|\frac{-Z_{even}^{diff}}{Z_{in,e}^{CPW} + Z_{even}^{diff}}\right| \tag{16.21}$$

If the phase velocities of the even and odd mode of CPW are not significantly different, such that θ_{even}^{CPW} is nearly $\pi/2$, then the input impedance of the even mode can be estimated by $Z_{in,e}^{CPW} \approx \left(Z_{even}^{CPW}\right)^2/2R_T$, and the minimum of $|S_{cc21}|$ will roughly occur at this frequency. Take the case using the design parameters listed in Table 16.1, for example. Equation (16.20) indicates that $|S_{cc21}|$ at dc is about –9.4 dB and the minimum of $|S_{cc21}|$ occurs around 7.4 GHz, which is consistent with the simulation in Figure 16.12.

In the previous sections, we proposed a design procedure for the DME, in which the length l_2, the even-mode impedance Z_{even}^{CPW}, and the loading resistor, R_T, are well determined. Consequently, the only remaining parameter is the odd-mode impedance Z_{odd}^{CPW}. Conceptually, we can vary the value of Z_{odd}^{CPW} in a feasible range, plot S_{cc21} according to Equation (16.19), and see if the resulting S_{cc21} meets the common-mode suppression specification. However, the problem is that the odd-mode impedance of CPW, Z_{odd}^{CPW}, is rather frequency dependent, compared with Z_{even}^{CPW}. For demonstration, Figure 16.14 shows the characteristic impedances of the two modes of a typical CPW structure [18]. As can be seen, although Z_{even}^{CPW} is almost constant from dc to 10 GHz, Z_{odd}^{CPW} varies from 60 Ω at low frequency to about 74 Ω at 10 GHz. This means that we cannot simply substitute a particular value for Z_{odd}^{CPW} into Equation (16.19) and plot the S_{cc21} for the whole frequency range. Instead, we should use a frequency-dependent impedance formula if we know. This, however, is usually not easy to achieve. As a result, to accurately obtain the common-mode response, we need full-wave simulation.

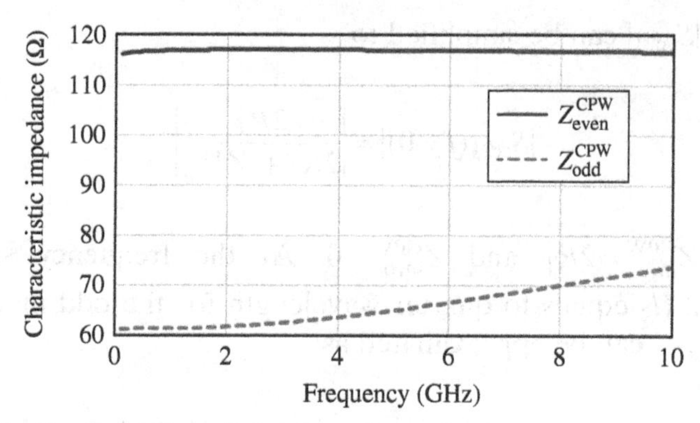

Figure 16.14 Simulated characteristic impedances for the even mode and odd mode of a CPW structure.

Finally, we point out that if the common-mode response does not meet the specification, trade-off can be made between the performance of DME and common-mode suppression. For example, varying $Z_{\text{even}}^{\text{CPW}}$ can change the minimum value of $|S_{\text{cc21}}|$ according to (16.21), and different values of R_{T} give different dc losses.

16.3.4 Measurement Validation

In Figure 16.3(a), the two-layer PCB, with thickness $H = 1$ mm, has on the top layer a pair of differential signal line with length l_1 (from Tx to the vias), width W_{d}, and spacing S_{d}, and on the bottom layer a pair of signal stub with length l_2 and width W_{s} embedded on a defected ground with spacing to the ground plane g, forming a CPW structure. Each of the stubs is individually connected to the middle position of the closer signal lines through a via and to the ground at the end away from the differential pairs by a resistor R_{T}. This pair of R_{T} may be SMD resistors and can be mounted on the backside of the ground plane or on the top layer with additional pads and vias for connection.

W_{d} and S_{d} are chosen such that $Z_{\text{even}}^{\text{diff}}$ and $Z_{\text{odd}}^{\text{diff}}$ are 100 and 50 Ω, respectively. According to the design procedure, l_2, $Z_{\text{even}}^{\text{CPW}}$, $Z_{\text{odd}}^{\text{CPW}}$, and R_{T} are determined as 7 mm, 128, 70, and 20 Ω, respectively. Table 16.2 lists the values of the corresponding geometric parameters along with H, W_{d}, and S_{d}.

Three test boards as shown in Figure 16.15 are fabricated and measured for verification and comparison: Board SD that contains the proposed DME–CMF with a short differential pair of length 20 mm,

Table 16.2 Geometric Parameters Used in Test Board Fabrication

Paramerters	H	W_d	S_d	W_g	l_1
Value	1 mm	0.85 mm	0.32 mm	50 mm	20 mm
Parameters	l_2	W_s	g	R_T	l_d
Value	7 mm	2 mm	0.5 mm	22 Ω	800 mm

Reprinted with permission from Ref. [15]; copyright 2013 IEEE.

(a)

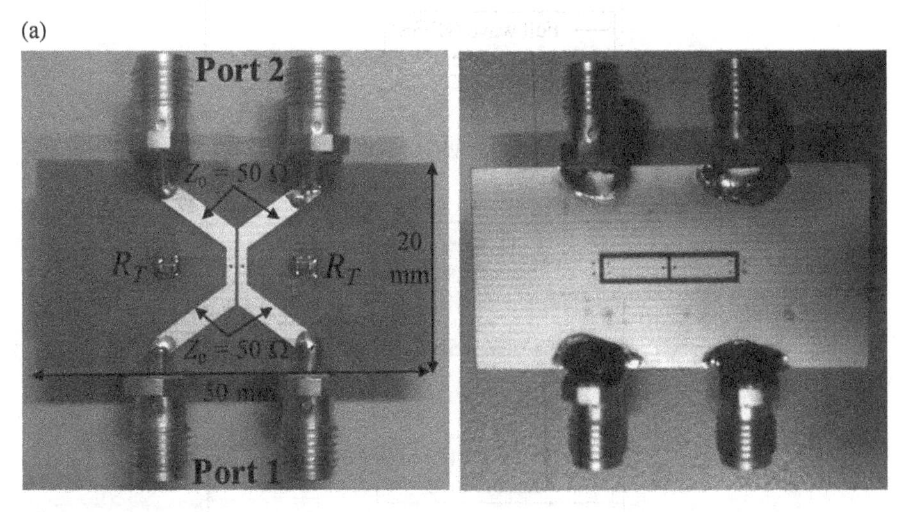

(b)

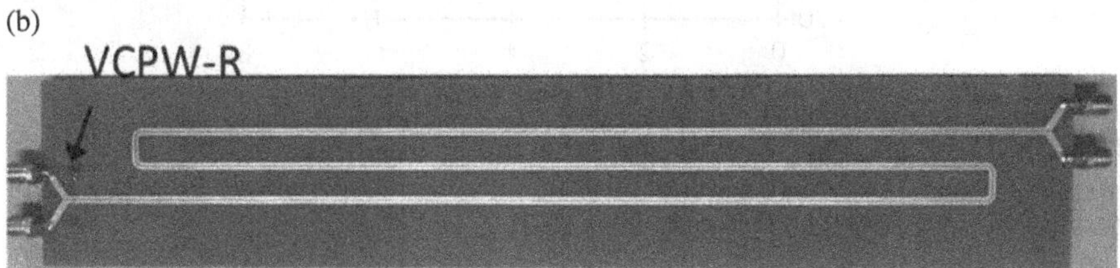

Figure 16.15 Photographs of the fabricated equalizer. (a) Top view and bottom view of Board SD. (b) Top view of Board LD. Reprinted with permission from Ref. [15]; copyright 2013 IEEE.

Board LD that contains a long differential pair of length 800 mm and the DME–CMF at its end, and Board REF that contains an 800-mm differential pair without the DME–CMF. Two 22-Ω SMD 0603 resistors serving as R_T are mounted on the top layer with additional vias and pads. The ends of the differential pair are separated into 50-Ω single-ended transmission lines with SMA connectors attached.

First, frequency-domain measurement by a network analyzer (Agilent N5230A PNA-L) is carried out on Board SD as shown in Figure 16.16, including the differential-mode $|S_{dd11}|$ and $|S_{dd21}|$ in (a)

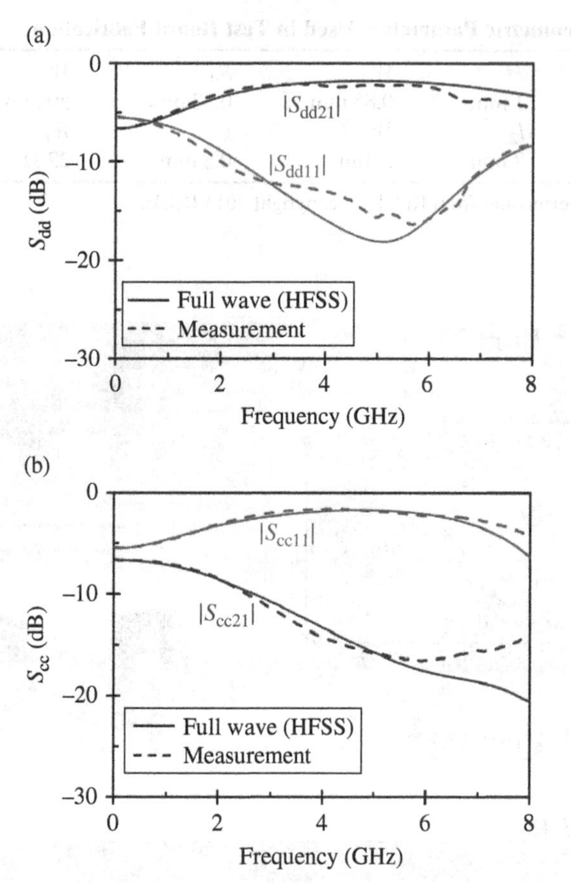

Figure 16.16 Comparison between measured and simulated S-parameter of Board SD: (a) differential-mode S-parameter and (b) common-mode S-parameter. Reprinted with permission from Ref. [15]; copyright 2013 IEEE.

and the common-mode $|S_{cc11}|$ and $|S_{cc21}|$ in (b), and the results are compared with full-wave simulation by HFSS, where good agreement can be seen in general. The dc attenuation of $|S_{dd21}|$ due to the resistance R_T is about 6.7 dB, and the high-pass region of the DME is from dc to about 4 GHz. The common-mode responses in (b) demonstrate the broadband filtering capability of the DME–CMF, though some discrepancy exists between the measurement and simulation results above 6 GHz, which may be due to the parasitics of the SMD resistors and connecting vias. It is noted that $|S_{cc21}|$ displays the same dc attenuation as $|S_{dd21}|$, since at dc all transmission lines behave simply as shorting wires and the common mode sees exactly what the differential mode sees. As the frequency increases, the common-mode suppression is significantly enhanced due to the interaction between the resistance R_T and the effect of the CPW.

Figure 16.17 provides a deeper look at the common-mode suppression of the proposed structure, in which the measurement result shown in Figure 16.16(b) is demonstrated in terms of the absorbed power and reflected power of the common mode defined by

$$\text{Absorbed power } (\%) = 1 - |S_{cc11}|^2 - |S_{cc21}|^2 \tag{16.22a}$$

$$\text{Reflected power } (\%) = |S_{cc11}|^2. \tag{16.22b}$$

Note that, after full-wave simulation, the total radiated power from the circuit was found to be negligible; thus in calculating the absorbed power by (16.22a), the radiated power was ignored. As can be seen, the absorption and reflection show an inverse trend relative to each other with extreme values both around 4.5 GHz, which together suppress the common mode greatly from dc to 8 GHz.

Next, Board LD and Board REF are measured in frequency domain as shown in Figure 16.18, where for Board REF significant loss is found and for Board LD the equalization ability that flattens $|S_{dd21}|$ from dc to about 4 GHz with a dc attenuation of 6.7 dB due to R_T is observed.

Time-domain eye diagram measurement results by using Agilent 54855A digital oscilloscope at the receiving end and Anritsu MP1763C pulse pattern generator at input launching an 8 Gb/s 2^{10} PRBS source with amplitude 2 V are shown in Figure 16.19, where it can be seen that Board REF corresponds to a totally closed eye, in sharp contrast to the

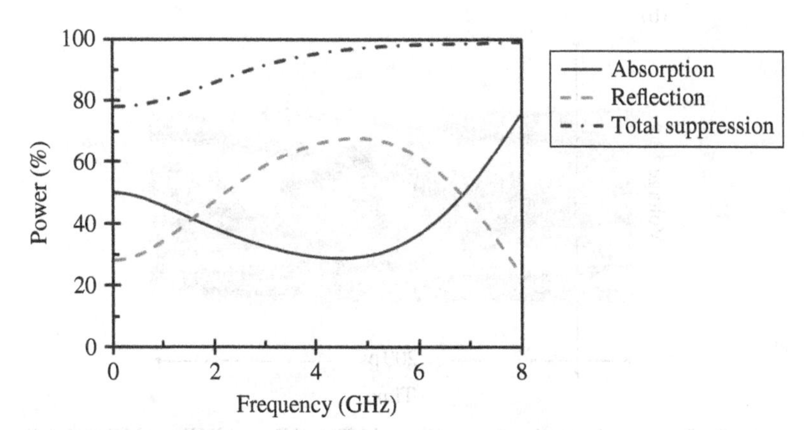

Figure 16.17 Mechanism of the common-mode suppression of Board SD. Reprinted with permission from Ref. [15]; copyright 2013 IEEE.

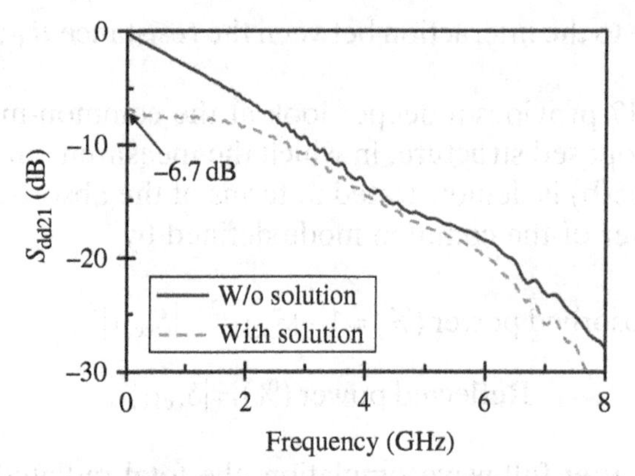

Figure 16.18 Measured differential insertion loss of Board LD and Board REF. Reprinted with permission from Ref. [15]; copyright 2013 IEEE.

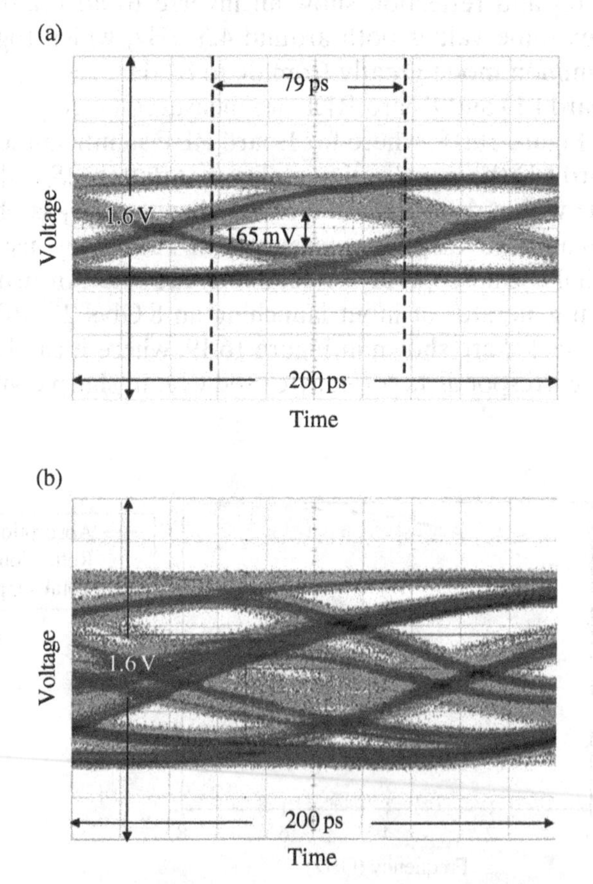

Figure 16.19 Measured eye diagrams of Board LD and Board REF at 8 Gb/s. (a) Board LD. (b) Board REF. Reprinted with permission from Ref. [15]; copyright 2013 IEEE.

Table 16.3 Measured Data of Eye Diagrams at Different Data Rates

	Board LD		Board REF	
	Eye height (mV)	Eye width (%)	Eye height (mV)	Eye width (%)
5 Gb/s	344	74.5	362	67.5
6 Gb/s	279	63.6	247	58.8
7 Gb/s	215	53.2	141	39.2
8 Gb/s	165	63.2	~0	~0

case of Board LD. The effectiveness of Board LD can also be seen at other data rates as listed in Table 16.3, despite the fact that the design parameters have not been optimized. On the whole, the higher the data rate is, the more improvement the proposed solution provides.

16.4 SECOND DESIGN

In the first design, we used two resistors to achieve differential-mode equalization. In this section, we develop a DME–CMF that only uses one resistor, which reduces cost. The key idea is that, since both traces of the differential transmission line have to see the resistor in order to have equalization effect, we can connect the two traces directly by a resistor R. Then, under differential-mode excitation in which a PEC is placed at the symmetry plane, each trace will see a resistor $R/2$ shunt to ground. Next by properly introducing inductors into the design, an RL-type equalizer is realized. This shunt RL branch, in this design, is called the differential shunt stub (DSS).

In this design, the common-mode suppression is achieved by the mushroom structure, which has previously been studied [14, 17]. The specialty of this design is that the added DSS (the resistive branch) not only equalizes the differential-mode signal but also largely increases the BW of common-mode suppression by introducing an additional transmission zero. Because of this, the circuit size can be significantly reduced. In the following, we mainly focus on the effect of adding the DSS to the differential mode and common mode. The design of the mushroom CMF without the DSS is not presented here.

16.4.1 Proposed Circuit and Analysis

The proposed circuit as shown in Figure 16.20 is a four-port network with differential inputs at ports 1 and 3 and differential outputs at ports

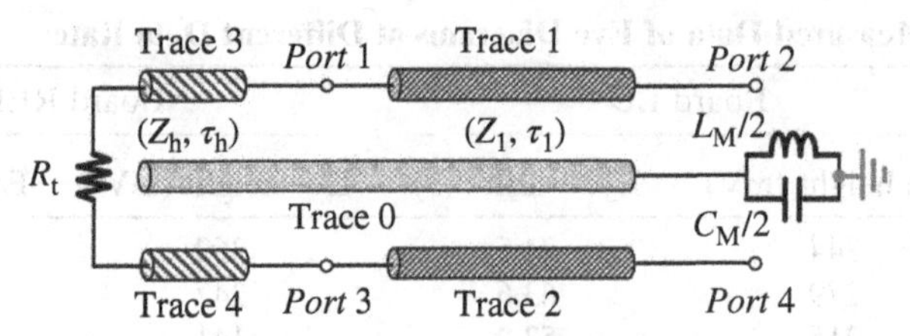

Figure 16.20 Proposed circuit for realizing differential-mode equalizer and common-mode filter. Reprinted with permission from Ref. [16]; copyright 2014 IEEE.

2 and 4 and consists of two identical transmission lines (traces 1 and 2) with characteristic impedance Z_1 and delay time τ_1 and another two identical transmission lines (traces 3 and 4) with propagation characteristics (Z_h, τ_h), where typically Z_h is larger than Z_1. These four transmission lines all return from the common reference (trace 0), which is connected to a LC resonator. A resistor R_t is connected between traces 3 and 4. Ports 1 and 3 are set at the connection points of traces 1 and 3 and traces 2 and 4, respectively. It should be noted that the reference points of ports 1–4 are the ground, that is, the point with zero voltage, instead of the common reference, that is, trace 0. This will be more clear later when the circuit layout is presented.

Due to the symmetry of the circuit, even-/odd-mode analysis can be applied to obtain the half-circuit models. The odd-mode half circuit as shown in Figure 16.21(a), obtained by setting a short-circuit condition (or PEC) on the symmetry plane (trace 0), will increase the insertion loss at low frequency, or equalize, for the differential digital signals through the design of the DSS (trace 3) and the terminated resistor R_t. The insertion loss of high-frequency components will become small when the Z_1 of trace 1 is matched to the load impedance of differential signals.

Under even-mode excitation, an open-circuit condition (or PMC) is set on the symmetry plane, and the equivalent circuit is as shown in Figure 16.21(b). Viewed between ports 1 and 2, the transmission line of trace 1 is shunt connected with the open-ended DSS with the common reference (trace 0), which is connected to ground through a parallel LC resonator, L_M, and C_M. As shown in Figure 16.21(b), the transfer impedance Z_{up} between ports 1 and 2 without including the LC resonator can be derived as

$$Z_{up} = \frac{-jZ_1 Z_h}{Z_1 \tan(\omega\tau_h)\cos(\omega\tau_1) + Z_h \sin(\omega\tau_1)} \tag{16.23}$$

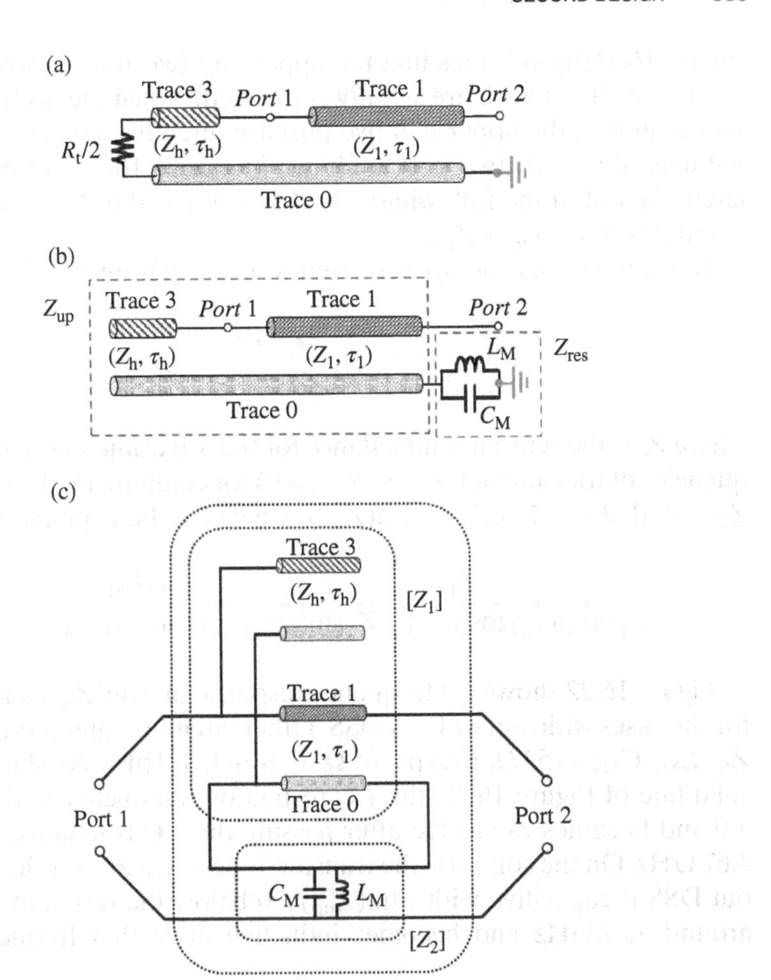

Figure 16.21 Half-circuit models: (a) odd mode, (b) even mode, and (c) reorganization of the even mode. Reprinted with permission from Ref. [16]; copyright 2014 IEEE.

This is the Z_{21} of the upper-half two-port network where trace 0 is connected directly to ground. The transfer impedance Z_{res} of the LC resonator is

$$Z_{res} = (j\omega L_M)||\left(\frac{1}{j\omega C_M}\right) = \frac{j\omega L_M}{1-\omega^2 L_M C_M} \tag{16.24}$$

This is the Z_{21} of the LC resonator when viewed as a two-port network where port 1 is directly connected to port 2 (note: these ports 1 and 2 are not the ports 1 and 2 in Figure 16.21) with a parallel LC resonator shunted to ground. It is understood that the circuit shown in

Figure 16.21(b) indicates that the upper-half two-port network and the LC two-port network are *serially* connected, which means that the reference nets of the upper-half two port are directly connected to the signal nets of the LC two port as shown in Figure 16.21(c) (also, see the circuit layout in the following). By this manner, the Z_{21} between ports 1 and 2 is $Z_{21} = Z_{up} + Z_{res}$.

It is known that the S_{21} is related to the Z-parameters by [21]

$$S_{21} = \frac{2Z_{21}Z_0}{(Z_{11}+Z_0)(Z_{22}+Z_0)-Z_{12}Z_{21}} \tag{16.25}$$

where Z_0 is the reference impedance for the S-parameters. Thus, the frequencies of transmission zeros ($S_{cc21} = 0$) for common mode will occur at $Z_{21} = 0$, that is, $-\text{Im}(Z_{up}) = \text{Im}(Z_{res})$, which can be expressed as

$$\frac{Z_1 Z_h}{Z_1 \tan(\omega\tau_h)\cos(\omega\tau_1) + Z_h \sin(\omega\tau_1)} = \frac{\omega L_M}{1-\omega^2 L_M C_M} \tag{16.26}$$

Figure 16.22 shows the frequency response of $-\text{Im}(Z_{up})$ and $\text{Im}(Z_{res})$ for the cases with and without DSS. Other circuit parameters are (Z_1, τ_1, Z_h, L_M, C_M) = (52 Ω, 67.8 ps, 65 Ω, 0.61 nH, 1.9 pF). As shown by the solid line of Figure 16.22, the LC resonator is inductive with $\text{Im}(Z_{res}) > 0$ and becomes capacitive after passing the LC resonance frequency 4.67 GHz. On the contrary, the transfer impedance Z_{up} for the case without DSS is capacitive with $-\text{Im}(Z_{up}) > 0$ below the resonant frequency around 7.37 GHz and becomes inductive after that frequency. This

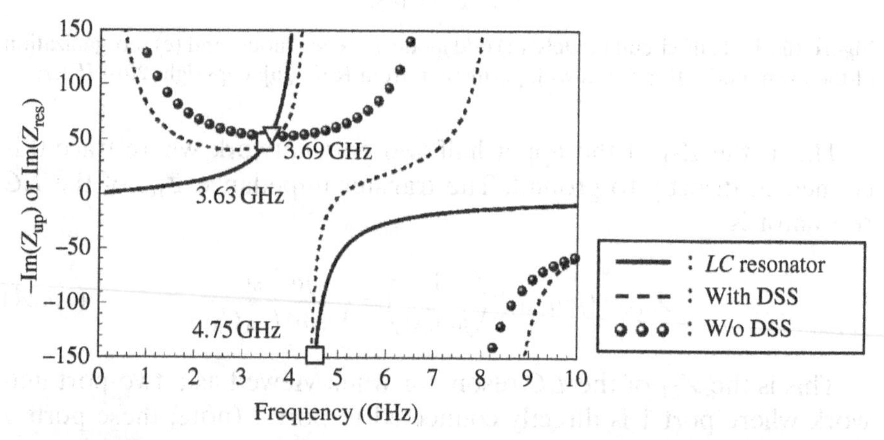

Figure 16.22 The transmission zeros for the cases with and without the DSS. Reprinted with permission from Ref. [16]; copyright 2014 IEEE.

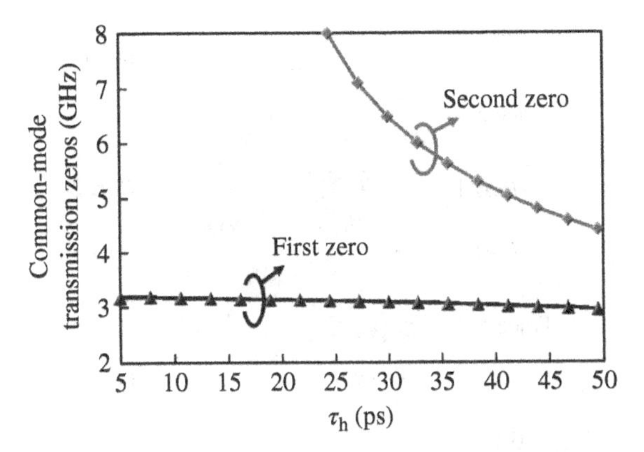

Figure 16.23 Positions of CM transmission zeros obtained by sweeping the length of DSS. Reprinted with permission from Ref. [16]; copyright 2014 IEEE.

resonant frequency occurs at half-wavelength of trace 1. It is seen that there is only one crossing point at around 3.69 GHz for the case without DSS, whereas there is an additional zero around 4.75 GHz when the DSS with $\tau_h = 45$ ps is added. The reason is that the resonant frequency of trace 1 is moved from 7.37 to 4.32 GHz after adding this DSS. The additional crossing point occurs as Z_{up} becomes inductive and the LC resonator becomes capacitive. In Figure 16.23, the length (or delay time) of DSS, that is, τ_h, is swept to observe the frequency change of the zeros with all other parameters fixed. It is observed that the second zero is shifted to a lower frequency when τ_h becomes larger, while the first zero is almost unchanged. This behavior implies we could design the frequency of the second zero by suitably choosing the length of the DSS. In addition, the equalization characteristic can then be designed by changing the terminated resistance R_t to flat the differential-mode insertion loss, as discussed in the first design of this chapter.

16.4.2 Realization and Measurement

16.4.2.1 Realization The top and side views of the circuit layout in three-layer PCB with the embedded substrate thicknesses, h_1 and h_2, are as shown in Figure 16.24(a) and (b), where the symmetrical differential lines (traces 1 and 2) are located at the top layer, meandered for reducing the circuit size, with length and width denoted as l_1 and w_1, respectively. A square patch of side length p at the second layer is used as the trace 0 for the return reference of traces 1 and 2, while the solid ground plane is located at the third layer. Together with a via of diameter d

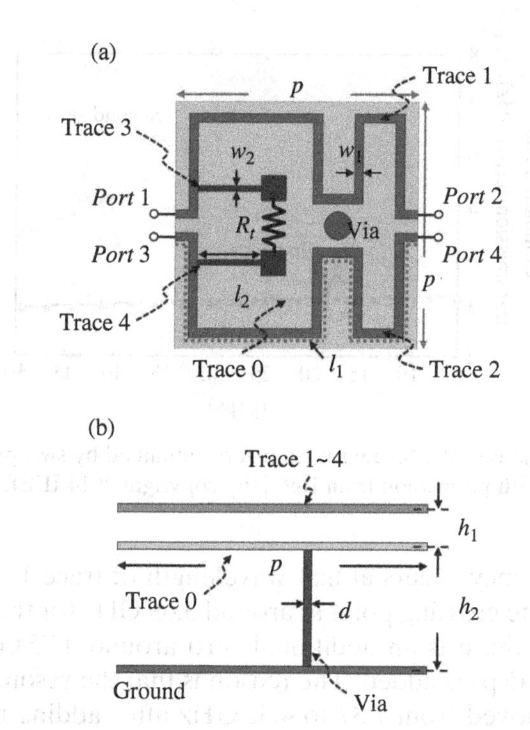

Figure 16.24 Circuit layout in 3-layer PCB: (a) top view and (b) side view. Reprinted with permission from Ref. [16]; copyright 2014 IEEE.

connecting the square patch and the bottom ground plane, a mushroom structure *LC* resonator is formed, where the via contributes to the *L* and the patch contributes to the *C*. A pair of symmetrical DSS (traces 3 and 4) of length l_2 and width w_2 is routed on the top layer, with ends connected by a resistor with resistance R_t.

16.4.2.2 Common-Mode Noise Suppression

The FR4 substrate of $\varepsilon_r = 4.2$ and $\tan\delta = 0.025$ is used to fabricate this circuit. Physical dimensions are summarized as $(h_1, h_2, l_1, w_1, l_2, w_2, p, d) = (0.1, 0.9, 12.1, 0.16, 8.5, 0.1, 5, 0.3)$ in mm unit, which has the same propagation characteristics as discussed in Figure 16.22. The photos of the fabricated circuits and *S*-parameter results are shown in Figure 16.25. For the circuit without DSS, the $|S_{cc21}|$ and $|S_{dd21}|$ between full-wave simulation, measurement, and equivalent models as shown in Figure 16.25(a) demonstrate reasonably good agreement. The $|S_{dd21}|$ is less than 1 dB from dc to 8 GHz, whereas the common-mode transmission zero is at 3.7 GHz and the stopband BW ($|S_{cc21}| < -20$ dB) is about 0.5 GHz with fractional

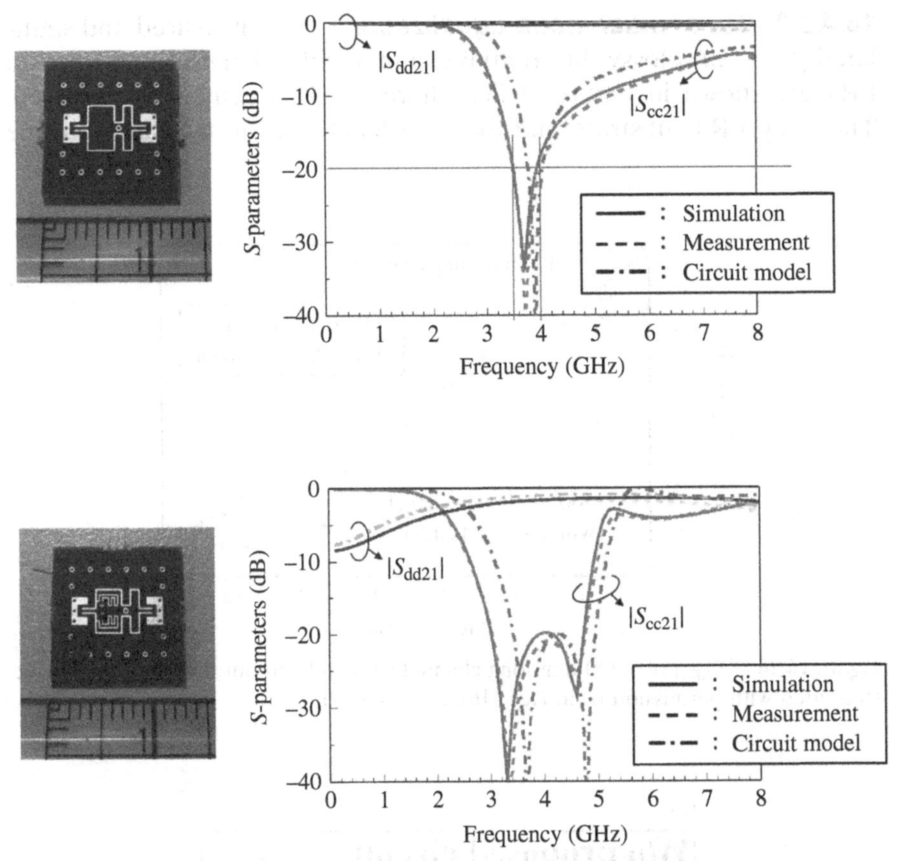

Figure 16.25 Comparison among full-wave simulation, measurement, and equivalent circuit models: (a) without equalizer and (b) with equalizer. Reprinted with permission from Ref. [16]; copyright 2014 IEEE.

bandwidth (FBW) of 13%. *S*-parameter results for the circuit with DSS and $R_t = 30\,\Omega$ from full-wave simulation, measurement, and equivalent circuit model are also in good agreement as shown in Figure 16.25(b). As expected, an additional zero for $|S_{cc21}|$ is observed around 4.5 GHz due to the DSS, close to the prediction in Figure 16.22, which thus enhances the common-mode stopband BW to 1.7 GHz with FBW of 46%. Another function for this DSS combined with the R_t is to equalize differential signals by making the $|S_{dd21}|$ a high-pass filter. In the fabricated circuit, an 8-dB attenuation at dc and a high-pass response up to about 4 GHz are realized, which helps to improve the eye diagram opening as shown in the following text.

16.4.2.3 Differential-Mode Equalization The measured and simulated $|S_{dd21}|$ of a lossy differential channel with a length of 400 mm on FR4 are shown in Figure 16.26, where good agreement is observed. The lossy FR4 substrate and the long length of the transmission line

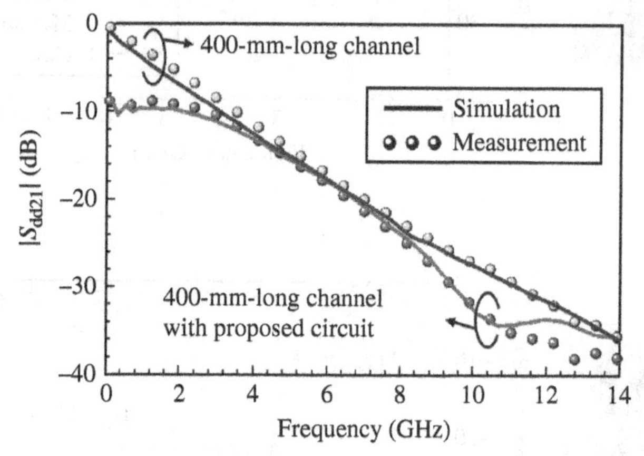

Figure 16.26 $|S_{dd21}|$ of a 400-mm-long channel with and without the proposed circuit. Reprinted with permission from Ref. [16]; copyright 2014 IEEE.

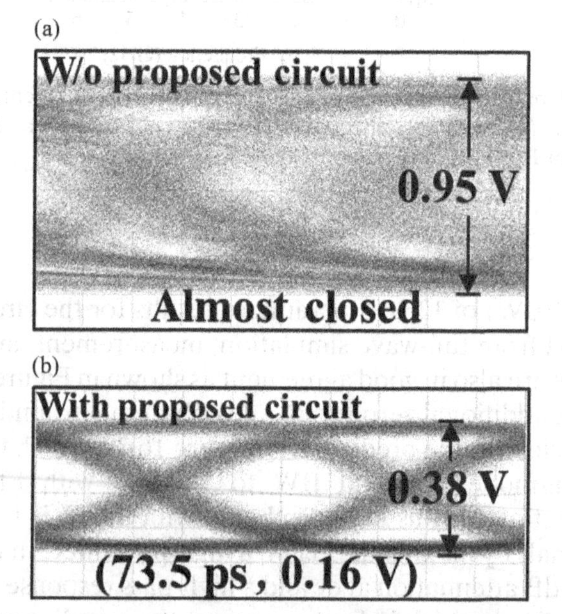

Figure 16.27 Time-domain eye diagram measurements: (a) reference board and (b) solution board. Reprinted with permission from Ref. [16]; copyright 2014 IEEE.

cause the insertion loss to increase significantly for high frequencies, for example, –8 dB at 3 GHz and –36 dB at 14 GHz. When cascaded with the designed circuit at the end of this channel, the frequency response becomes flat (or equalized) below 3 GHz, where the $|S_{dd21}|$ variation is smaller than 2 dB.

The time-domain eye diagram responses for the long channel without and with the proposed circuit are shown in Figure 16.27(a) and (b), respectively, where a differential pseudorandom binary sequence (PRBS) signal with 8-Gb/s data rate and 2-V peak-to-peak voltage is fed to the input. It is observed that the eye pattern at output for the lossy channel alone is nearly closed, while the eye pattern is clearly opened as the proposed circuit is cascaded at the end of the channel, with measured eye width and height being 73.5 ps and 0.16 V, respectively. Though the voltage swing is reduced from 0.95 to 0.38 V after cascading this circuit, the clearly opened eye is more important for the receivers in high-speed circuits to distinguish the 0/1 bit streams.

16.5 SUMMARY

In this chapter, we presented two designs of differential-mode equalizer/common-mode filter. The equalizers of both designs were in the form of *RL* type, while CMFs were DGS type and mushroom type. In the first circuit, a time-domain step-by-step design procedure for the DME was proposed, while for the second circuit, we analyzed the effect of the DSS added to the mushroom CMF for widening the common-mode suppression BW. Experimental results were presented for verifying the effectiveness of the proposed circuits. Both circuits can be used in high-speed digital channel for equalizing the channel loss in order to have a better signal integrity, as well as eliminating common-mode noise for reducing electromagnetic interference.

REFERENCES

1. C. R. Paul, *Introduction to Electromagnetic Compatibility*, 2nd ed., Hoboken, NJ: John Wiley & Sons, Inc., 2006, pp. 151–154.
2. S. H. Hall and H. L. Heck, *Advanced Signal Integrity for High Speed Digital System Design*, Hoboken, NJ: John Wiley & Sons, Inc., 2009.
3. S. Rylov, S. Reynolds, D. Storaska, B. Floyd, M. Kapur, T. Zwick, S. Gowda, and M. Sorna, "10+ Gb/s 90 nm CMOS serial link demo in CBGA package," *IEEE J. Solid-State Circuits*, vol. **40**, no. 9, pp. 1987–1991, Sep. 2005.

4. C. R. Paul, “A comparison of the contributions of common-mode and differential-mode currents in radiated emissions,” *IEEE Trans. Electromagn. Compat.* vol. **31**, no. 2, pp. 189–193, May 1989.
5. W.-D. Guo, F.-N. Tsai, G.-H. Shiue, and R.-B. Wu, “Reflection enhanced compensation of lossy traces for best eye-diagram improvement using high impedance mismatch,” *IEEE Trans. Adv. Packag.*, vol. **31**, no. 3, pp. 619–626, Aug. 2008.
6. E. Song, J. Kim, J. Kim, and J. Cho, “A compact, low-cost, and wide-band passive equalizer design using multi-layer PCB parasitics,” *Proceedings of the IEEE Electrical Performance of Electronic Packaging Conference*, Austin, TX, 2010, pp. 165–168.
7. E. Song, J. Cho, J. Kim, Y. Shim, G. Kim, and J. Kim, “Modeling and design optimization of a wideband passive equalizer on PCB based on near-end crosstalk and reflections for high-speed serial data transmission,” *IEEE Trans. Electromagn. Compat.*, vol. **52**, no. 2, pp. 410–420, May 2010.
8. Y. Shim, W. Lee, E. Song, J. Cho, and J. Kim, “A compact and wide-band passive equalizer design using a stub with defected ground structure for high speed data transmission,” *IEEE Microw. Wireless Compon. Lett.*, vol. **20**, no. 5, pp. 256–258, May 2010.
9. L. Zhang, W. Yu, Y. Zhang, R. Wang, A. Deutsch, G. A. Katopis, D. M. Dreps, J. Buckwalter E. S. Kuh, and C.-K. Cheng, “Analysis and optimization of low-power passive equalizers for CPU-Memory links,” *IEEE Trans. Compon. Packag. Manuf. Technol.*, vol. **1**, no. 9, pp. 1406–1418, Sep. 2011.
10. W.-T. Liu, C.-H. Tsai, T.-W. Han, and T.-L. Wu, “An embedded common-mode suppression filter for GHz differential signals using periodic defected ground plane,” *IEEE Microw. Wireless Compon. Lett.*, vol. **18**, no. 4, pp. 248–250, Apr. 2008.
11. S.-J. Wu, C.-H. Tsai, T.-L. Wu, and T. Itoh, “A novel wideband common-mode suppression filter for GHz differential signals using coupled patterned ground structure,” *IEEE Trans. Microw. Theory Tech.*, vol. **57**, no. 4, pp. 848–855, Apr. 2009.
12. K. Yanagisawa, F. Zhang, T. Sato, K. Yanagisawa, and Y. Miura, “A new wideband common-mode noise filter consisting of Mn-Zn ferrite core and copper/polyimide tape wound coil,” *IEEE Trans. Magn.*, vol. **41**, no. 10, pp. 3571–3573, Oct. 2005.
13. B.-C. Tseng and L.-K. Wu, “Design of miniaturized common-mode filter by multilayer low-temperature co-fired ceramic,” *IEEE Trans. Electromagn. Compat.*, vol. **46**, no. 4, pp. 571–579, Nov. 2004.
14. C.-H. Tsai and T.-L. Wu, “A broadband and miniaturized common-mode filter for gigahertz differential signals based on negative-permittivity metamaterials,” *IEEE Trans. Microw. Theory Tech.*, vol. **58**, no. 1, pp. 195–202, Jan. 2010.

15. Y.-J. Cheng, H.-H. Chuang, C.-K. Cheng, and T.-L. Wu, "Novel differential-mode equalizer with broadband common-mode filtering for Gb/s differential-signal transmission," *IEEE Trans. Compon. Packag. Manuf. Technol.*, vol. **3**, no. 9, pp. 1578–1587, Sep. 2013.
16. C.-Y. Hsiao and T.-L. Wu, "A novel dual-function circuit combining high-speed differential equalizer and common-mode filter with an additional zero," *IEEE Microw. Wireless Compon. Lett.*, vol. **24**, no. 9, pp. 617–619, Sep. 2014.
17. C.-Y. Hsiao, C.-H. Tsai, C.-N. Chiu, and T.-L. Wu, "Radiation suppression for cable-attached packages utilizing a compact embedded common-mode filter," *IEEE Trans. Compon. Packag. Manuf. Technol.*, vol. **2**, no. 10, pp. 1696–1703, Oct. 2012.
18. Y.-J. Cheng, "A novel differential-mode equalizer with broadband common-mode filtering for Gbps differential signaling," M.S. thesis, Graduate Institute of Communication Engineering, National Taiwan University, Taipei, Taiwan, ROC, 2012.
19. W.-D. Guo, J.-H. Lin, C.-M. Lin, T.-W. Huang, and R.-B. Wu, "Fast methodologies for determining eye diagram characteristics of lossy transmission lines," *IEEE Trans. Adv. Packag.*, vol. **32**, no. 1, pp. 175–183, Feb. 2009.
20. L. Zhang, W. Yu, H. Zhu, A. Deutsch, G. A. Katopis, D. M. Dreps, E. Kuh, and C.-K. Cheng, "Low power passive equalizer optimization using tritonic step response," in *Proceedings of the IEEE/ACM Design Automation Conference*, Anaheim, CA, pp. 570–573, 8–13 Jun. 2008.
21. D. M. Pozar, *Microwave Engineering*, 3rd ed, Hoboken, NJ: John Wiley & Sons, Inc., 2005, p. 187.

INDEX

Balanced Microwave Filters, First Edition. Edited by Ferran Martín, Lei Zhu, Jiasheng Hong, and Francisco Medina.

WILEY SERIES IN MICROWAVE AND OPTICAL ENGINEERING

Professor Kai Chang, Editor
Texas A&M University

The Wiley Series in *Microwave and Optical Engineering* publishes authoritative treatments of foundational areas central to *Microwave and Optical Engineering* as well as research monographs in hot-topic emerging technology areas. The series was founded in 1988 and to date includes over 80 titles.

BALANCED MICROWAVE FILTERS
by Ferran Martín, Lei Zhu, Jiasheng Hong, Francisco Medina

ENABLING TECHNOLOGIES FOR HIGH SPECTRAL-EFFICIENCY COHERENT OPTICAL COMMUNICATION NETWORKS
by Xiang Zhou

PASSIVE MACROMODELING: THEORY AND APPLICATIONS
by Stefano Grivet-Talocia, Bjorn Gustavsen

ARTIFICIAL TRANSMISSION LINES FOR RF AND MICROWAVE APPLICATIONS
by Ferran Martín

RADIO-FREQUENCY INTEGRATED-CIRCUIT ENGINEERING
by Cam Nguyen

FUNDAMENTALS OF MICROWAVE PHOTONICS
by V. J. Urick, Keith J. Williams, Jason D. McKinney

RADIO PROPAGATION AND ADAPTIVE ANTENNAS FOR WIRELESS COMMUNICATION NETWORKS, 2nd Edition
by Nathan Blaunstein, Christos G. Christodoulou

MICROWAVE NONCONTACT MOTION SENSING AND ANALYSIS
by Changzhi Li, Jenshan Lin

PHOTONIC SENSING: PRINCIPLES AND APPLICATIONS FOR SAFETY AND SECURITY MONITORING
by Gaozhi Xiao (Editor), Wojtek J. Bock (Editor)

FUNDAMENTALS OF OPTICAL FIBER SENSORS
by Zujie Fang, Ken Chin, Ronghui Qu, Haiwen Cai, Kai Chang (Series Editor)

COMPACT MULTIFUNCTIONAL ANTENNAS FOR WIRELESS SYSTEMS
by Eng Hock Lim, Kwok Wa Leung

DIODE LASERS AND PHOTONIC INTEGRATED CIRCUITS, 2nd Edition
by Larry A. Coldren, Scott W. Corzine, Milan L. Mashanovitch

INVERSE SYNTHETIC APERTURE RADAR IMAGING WITH MATLAB ALGORITHMS
by Caner Ozdemir

SILICA OPTICAL FIBER TECHNOLOGY FOR DEVICES AND COMPONENTS: DESIGN, FABRICATION, AND INTERNATIONAL STANDARDS
by Kyunghwan Oh, Un-Chul Paek

MICROWAVE BANDPASS FILTERS FOR WIDEBAND COMMUNICATIONS
by Lei Zhu, Sheng Sun, Rui Li

SUBSURFACE SENSING
by Ahmet S. Turk, Koksal A. Hocaoglu, Alexey A. Vertiy

RF AND MICROWAVE TRANSMITTER DESIGN
by Andrei Grebennikov

MICROSTRIP FILTERS FOR RF/MICROWAVE APPLICATIONS, 2nd Edition
by Jia-Sheng Hong

FUNDAMENTALS OF WAVELETS: THEORY, ALGORITHMS, AND APPLICATIONS, 2nd Edition
by Jaideva C. Goswami, Andrew K. Chan

Radio Frequency Circuit Design, 2nd Edition
by W. Alan Davis

TIME AND FREQUENCY DOMAIN SOLUTIONS OF EM PROBLEMS USING INTEGRAL EQUATIONS AND A HYBRID METHODOLOGY
by B. H. Jung, T. K. Sarkar, Y. Zhang, Z. Ji, M. Yuan, M. Salazar-Palma, S. M. Rao, S. W. Ting, Z. Mei, A. De

LASER DIODES AND THEIR APPLICATIONS TO COMMUNICATIONS AND INFORMATION PROCESSING
by Takahiro Numai

FIBER-OPTIC COMMUNICATION SYSTEMS, 4th Edition
by Govind P. Agrawal

SOLAR CELLS AND THEIR APPLICATIONS, 2nd Edition
by Lewis M. Fraas, Larry D. Partain

MICROWAVE IMAGING
by Matteo Pastorino

EM DETECTION OF CONCEALED TARGETS
by David J. Daniels

PHASED ARRAY ANTENNAS, 2nd Edition
by Robert C. Hansen

ELECTROMAGNETIC SIMULATION TECHNIQUES BASED ON THE FDTD METHOD
by W. Yu

PARALLEL SOLUTION OF INTEGRAL EQUATION-BASED EM PROBLEMS IN THE FREQUENCY DOMAIN
by Y. Zhang, T. K. Sarkar

ADVANCED INTEGRATED COMMUNICATION MICROSYSTEMS
by Joy Laskar, Sudipto Chakraborty, Anh-Vu Pham, Manos M. Tantzeris

ANALYSIS AND DESIGN OF AUTONOMOUS MICROWAVE CIRCUITS
by Almudena Suarez

PHYSICS OF MULTIANTENNA SYSTEMS AND BROADBAND PROCESSING
by T. K. Sarkar, M. Salazar-Palma, Eric L. Mokole

THE STRIPLINE CIRCULATORS: THEORY AND PRACTICE
by J. Helszajn

ELECTROMAGNETIC SHIELDING
by Salvatore Celozzi, Rodolfo Araneo, Giampiero Lovat

LOCALIZED WAVES
by Hugo E. Hernandez-Figueroa (Editor), Michel Zamboni-Rached (Editor), Erasmo Recami (Editor)

METAMATERIALS WITH NEGATIVE PARAMETERS: THEORY, DESIGN AND MICROWAVE APPLICATIONS
by Ricardo Marqués, Ferran Martín, Mario Sorolla

HIGH-SPEED VLSI INTERCONNECTIONS, 2nd Edition
by Ashok K. Goel

OPTICAL SWITCHING
by Georgios I. Papadimitriou, Chrisoula Papazoglou, Andreas S. Pomportsis

ELECTRON BEAMS AND MICROWAVE VACUUM ELECTRONICS
by Shulim E. Tsimring

ASYMMETRIC PASSIVE COMPONENTS IN MICROWAVE INTEGRATED CIRCUITS
by Hee-Ran Ahn

ADAPTIVE OPTICS FOR VISION SCIENCE: PRINCIPLES, PRACTICES, DESIGN AND APPLICATIONS
by Jason Porter, Hope Queener, Julianna Lin, Karen Thorn, Abdul A. S. Awwal

PHASED ARRAY ANTENNAS: FLOQUET ANALYSIS, SYNTHESIS, BFNS AND ACTIVE ARRAY SYSTEMS
by Arun K. Bhattacharyya

RF/MICROWAVE INTERACTION WITH BIOLOGICAL TISSUES
by Andre Vander Vorst, Arye Rosen, Youji Kotsuka

HISTORY OF WIRELESS
by T. K. Sarkar, Robert Mailloux, Arthur A. Oliner, M. Salazar-Palma, Dipak L. Sengupta

INTRODUCTION TO ELECTROMAGNETIC COMPATIBILITY, 2nd Edition
by Clayton R. Paul

APPLIED ELECTROMAGNETICS AND ELECTROMAGNETIC COMPATIBILITY
by Dipak L. Sengupta, Valdis V. Liepa

MULTIRESOLUTION TIME DOMAIN SCHEME FOR ELECTROMAGNETIC ENGINEERING
by Yinchao Chen, Qunsheng Cao, Raj Mittra

FUNDAMENTALS OF GLOBAL POSITIONING SYSTEM RECEIVERS: A SOFTWARE APPROACH, 2nd Edition
by James Bao-Yen Tsui

MICROWAVE RING CIRCUITS AND RELATED STRUCTURES, 2nd Edition
by Kai Chang, Lung-Hwa Hsieh

FUNDAMENTALS OF GLOBAL POSITIONING SYSTEM RECEIVERS: A SOFTWARE APPROACH
by James Bao-Yen Tsui

MICROSTRIP FILTERS FOR RF/MICROWAVE APPLICATIONS
by Jia-Shen G. Hong, M. J. Lancaster

ARITHMETIC AND LOGIC IN COMPUTER SYSTEMS
by Mi Lu

RADIO FREQUENCY CIRCUIT DESIGN
by W. Alan Davis, Krishna Agarwal

SMART ANTENNAS
by T. K. Sarkar, Michael C. Wicks, M. Salazar-Palma, Robert J. Bonneau

WAVELETS IN ELECTROMAGNETICS AND DEVICE MODELING
by George W. Pan

PLANAR ANTENNAS FOR WIRELESS COMMUNICATIONS
by Kin-Lu Wong

RF AND MICROWAVE CIRCUIT AND COMPONENT DESIGN FOR WIRELESS SYSTEMS
by Kai Chang, Inder Bahl, Vijay Nair

COMPACT AND BROADBAND MICROSTRIP ANTENNAS
by Kin-Lu Wong

SPHEROIDAL WAVE FUNCTIONS IN ELECTROMAGNETIC THEORY
by Le-Wei Li, Xiao-Kang Kang, Mook-Seng Leong

THEORY AND PRACTICE OF INFRARED TECHNOLOGY FOR NONDESTRUCTIVE TESTING
by Xavier P. Maldague

COPLANAR WAVEGUIDE CIRCUITS, COMPONENTS, AND SYSTEMS
by Rainee N. Simons

ELECTROMAGNETIC FIELDS IN UNCONVENTIONAL MATERIALS AND STRUCTURES
by Onkar N. Singh (Editor), Akhlesh Lakhtakia (Editor)

INFRARED TECHNOLOGY: APPLICATIONS TO ELECTRO-OPTICS, PHOTONIC DEVICES AND SENSORS
by Animesh R. Jha

ANALYSIS METHODS FOR RF, MICROWAVE, AND MILLIMETER-WAVE PLANAR TRANSMISSION LINE STRUCTURES
by Cam Nguyen

RF AND MICROWAVE WIRELESS SYSTEMS
by Kai Chang

ANALYSIS AND DESIGN OF INTEGRATED CIRCUIT-ANTENNA MODULES
by K. C. Gupta (Editor), Peter S. Hall (Editor)

ELECTROMAGNETIC OPTIMIZATION BY GENETIC ALGORITHMS
by Yahya Rahmat-Samii (Editor), Eric Michielssen (Editor)

ELECTRODYNAMICS OF SOLIDS AND MICROWAVE SUPERCONDUCTIVITY
by Shu-Ang Zhou

OPTICAL FILTER DESIGN AND ANALYSIS: A SIGNAL PROCESSING APPROACH
by Christi K. Madsen, Jian H. Zhao

OPTICAL CHARACTER RECOGNITION
by Shunji Mori, Hirobumi Nishida, Hiromitsu Yamada

SIGE, GAAS, AND INP HETEROJUNCTION BIPOLAR TRANSISTORS
by Jiann S. Yuan

DESIGN OF NONPLANAR MICROSTRIP ANTENNAS AND TRANSMISSION LINES
by Kin-Lu Wong

ELECTROMAGNETIC PROPAGATION IN MULTI-MODE RANDOM MEDIA
by Harrison E. Rowe

OPTICAL SEMICONDUCTOR DEVICES
by Mitsuo Fukuda

SUPERCONDUCTOR TECHNOLOGY: APPLICATIONS TO MICROWAVE, ELECTRO-OPTICS, ELECTRICAL MACHINES, AND PROPULSION SYSTEMS
by Animesh R. Jha

NONLINEAR OPTICAL COMMUNICATION NETWORKS
by Eugenio Iannone, Francesco Matera, Antonio Mecozzi, Marina Settembre

INTRODUCTION TO ELECTROMAGNETIC AND MICROWAVE ENGINEERING
by Paul R. Karmel, Gabriel D. Colef, Raymond L. Camisa

ADVANCES IN MICROSTRIP AND PRINTED ANTENNAS
by Kai Fong Lee (Editor), Wei Chen (Editor)

OPTOELECTRONIC PACKAGING
by Alan R. Mickelson (Editor), Nagesh R. Basavanhally (Editor), Yung-Cheng Lee (Editor)

ACTIVE AND QUASI-OPTICAL ARRAYS FOR SOLID-STATE POWER COMBINING
by Robert A. York (Editor), Zoya B. Popovic (Editor)

PHASED ARRAY-BASED SYSTEMS AND APPLICATIONS
by Nicholas Fourikis

NONLINEAR OPTICS
by E. G. Sauter

FINITE ELEMENT SOFTWARE FOR MICROWAVE ENGINEERING
by Tatsuo Itoh (Editor), Giuseppe Pelosi (Editor), Peter P. Silvester (Editor)

INTEGRATED ACTIVE ANTENNAS AND SPATIAL POWER COMBINING
by Julio A. Navarro, Kai Chang

FUNDAMENTALS OF MICROWAVE TRANSMISSION LINES
by Jon C. Freeman

HIGH-FREQUENCY ELECTROMAGNETIC TECHNIQUES: RECENT ADVANCES AND APPLICATIONS
by Asoke K. Bhattacharyya

HIGH-FREQUENCY ANALOG INTEGRATED CIRCUIT DESIGN
by Ravender Goyal

COHERENT OPTICAL COMMUNICATIONS SYSTEMS
by Silvello Betti, Giancarlo De Marchis, Eugenio Iannone

MICROWAVE SOLID-STATE CIRCUITS AND APPLICATIONS
by Kai Chang

MICROWAVE DEVICES, CIRCUITS AND THEIR INTERACTION
by Charles A. Lee, G. Conrad Dalman

MICROSTRIP CIRCUITS
by Fred Gardiol

COMPUTATIONAL METHODS FOR ELECTROMAGNETICS AND MICROWAVES
by Richard C. Booton